The Lanthanides and Actinides

Synthesis, Reactivity, Properties and Applications

The Lanthanides and Actinides

Synthesis, Reactivity, Properties and Applications

Editors

Stephen T. Liddle
David P. Mills
Louise S. Natrajan

The University of Manchester, UK

World Scientific

NEW JERSEY • LONDON • SINGAPORE • BEIJING • SHANGHAI • HONG KONG • TAIPEI • CHENNAI • TOKYO

Published by

World Scientific Publishing Europe Ltd.

57 Shelton Street, Covent Garden, London WC2H 9HE

Head office: 5 Toh Tuck Link, Singapore 596224

USA office: 27 Warren Street, Suite 401-402, Hackensack, NJ 07601

Library of Congress Cataloging-in-Publication Data
Names: Liddle, Stephen T. (Stephen Taylor), editor. | Mills, David P., editor. |
Natrajan, Louise S., editor.
Title: The lanthanides and actinides : synthesis, reactivity, properties and applications / editors,
Stephen T. Liddle, David P. Mills, Louise S. Natrajan, The University of Manchester, UK.
Description: New Jersey : World Scientific, 2022. | Includes bibliographical references.
Identifiers: LCCN 2021009805 | ISBN 9781800610156 (hardcover) |
ISBN 9781800610163 (ebook) | ISBN 9781800610170 (ebook other)
Subjects: LCSH: Rare earth metals. | Actinide elements.
Classification: LCC QD172.R2 L28 2022 | DDC 546/.41--dc23
LC record available at https://lccn.loc.gov/2021009805

British Library Cataloguing-in-Publication Data
A catalogue record for this book is available from the British Library.

For any available supplementary material, please visit
https://www.worldscientific.com/worldscibooks/10.1142/Q0298#t=suppl

Desk Editors: Aanand Jayaraman/Michael Beale/Shi Ying Koe

Typeset by Stallion Press
Email: enquiries@stallionpress.com

Printed in Singapore

To our co-workers, past and present, for making f-element chemistry so much fun to research.

https://doi.org/10.1142/9781800610163_fmatter

Preface

f-block chemistry is a vibrant field spanning many areas of endeavour. Although it is a decades-old discipline, every year sees new and exciting advances that genuinely redefine our understanding in still quite transformative — rather than incremental — ways. Looking back at the original book proposal paperwork, it was stated that this book could legitimately cover *periodicity, natural occurrence and extraction, separations, electronic structure, coordination chemistry, organometallic chemistry, small molecule activation, catalysis, organic synthesis applications, magnetism, spectroscopy, computation, materials, photonics, solar cell technology, biological imaging, and technological applications*, and, gratifyingly, that has indeed proven to essentially be the case.

Much has gone on since this book was first proposed. We were originally approached by Imperial College Press, who over time have become part of the World Scientific Publishing Company. Consequently, with the organisational changes that resulted we have enjoyed working with several editors that have come and gone. One of us moved institutions during this time, and to cap it all off the coronavirus global pandemic arrived. So, it has at times felt like events have conspired to stop this project ever seeing the light of day, but we persevered and got there eventually. The result is 18 chapters covering essentially all the aforementioned areas.

With some irony given technological advances and the modern day speed of publishing research, it seems to have become harder to assemble a book nowadays, and certainly to do so and keep it all still relevant and up-to-date at the point of publication is now a formidable task. To spread the load we enlisted the help of experts from around the world in the field spanning a healthy range of career stages and diverse backgrounds. While we have made some efforts to harmonise the structure of individual chapters we have done so lightly to let the style of individual authors come through and make each chapter a fresh style to readers. We are immensely grateful to the chapter authors, because the result is, we hope, a wide-ranging volume of knowledge that will be useful to seasoned operators and neophytes alike and provide a resource from which to inspire future advances in the field.

We would like to take this opportunity to thank authors for their patience and efforts, and the editors that we have worked with for their support.

https://doi.org/10.1142/9781800610163_fmatter

About the Editors

Stephen Liddle is Professor and Head of Inorganic Chemistry at The University of Manchester. He received his BSc (Hons) and PhD degrees from Newcastle University. After postdoctoral positions at Edinburgh, Newcastle, and Nottingham universities, he started his independent career with a Royal Society University Research Fellowship in 2007. He moved to Manchester in 2015 to his current position which also includes co-directorship of the Centre for Radiochemistry Research. His research interests span metal–ligand multiple bonding, metal–metal bonding, small molecule activation and single-molecule magnetism, with a particular focus on the early actinides.

David Mills obtained his MChem (Hons) in 2004 and PhD degree in 2008, both at Cardiff University, and postdoctoral work at the University of Nottingham. He started his independent academic career as a Lecturer in Chemistry at The University of Manchester in 2012, where he is now a Professor. His research interests are in non-aqueous synthetic chemistry, predominantly with the f-elements.

Louise Natrajan is a Reader in Chemistry at The University of Manchester. She received her MChem degree from the University of York (UK) and PhD from the University of Nottingham (UK). She took up postdoctoral positions at the CEA, Grenoble, France and at The University of Manchester. During 2009–2018 she was an EPSRC then a Leverhulme Trust funded fellow. Her current research focuses on the coordination chemistry and optical properties of the f-elements, with a particular focus on the early actinides.

https://doi.org/10.1142/9781800610163_fmatter

Acknowledgements

We would like to thank the chapter authors, who have stayed with this book through thick and thin, the editorial staff at Imperial College Press and now World Scientific Publishing Company for their guidance throughout this project, and the funding bodies that made all the work described in this book possible in the first place.

https://doi.org/10.1142/9781800610163_fmatter

Contents

https://doi.org/10.1142/9781800610163_fmatter

List of Abbreviations

An	Actinide ion
Bipy	2,2′-bipyridine
CCI	Cation–cation interaction
Cp	Cyclopentadienyl and cyclopentadienyl derivatives
Cp*	1,2,3,4,5-pentamethyl cyclopentadienyl
Cp^{Me}	1-methyl cyclopentadienyl
Cy	Cyclohexyl
Dipp	2,6-diisopropyl phenyl
DME	1,2-dimethoxyethane
DMP	2,6-bis(Mes) phenyl
DMAP	4-dimethylaminopyridine
EXAFS	Extended X-ray absorption fine structure
^{i}Bu	Isobutyl
^{i}Pr	Isopropyl
ITI	Inverse trans influence
LMCT	Ligand to metal charge transfer
Me	Methyl
Mes	2,4,6-trimethyl phenyl
Mes*	2,4,6-tritertbutyl phenyl
MO	Molecular orbital
Ph	Phenyl
Phen	1,10-phenanthroline
^{t}Bu	Tertbutyl
THF	Tetrahydrofuran
Trip	2,4,6-triisopropyl phenyl
XANES	X-ray absorption near edge structure
Xyl	2,6-dimethyl phenyl

https://doi.org/10.1142/9781800610163_0001

Chapter 1

The Chemistry of Rare-Earth Metals

Thomas Behrsing,‡, Glen B. Deacon*,§, and Peter Junk†,¶*

*School of Chemistry, Monash University, Clayton, Vic. 3800, Australia

†College of Science & Engineering, James Cook University, Townsville, Qld., 4811, Australia

‡thomas.behrsing@monash.edu

§glen.deacon@monash.edu

¶peter.junk@jcu.edu.au

1.1 Introduction: The Rare-Earth Elements

The rare-earth elements comprise scandium and yttrium from Group 3 of the Periodic Table (Figure 1.1) plus the lanthanoids (IUPAC)/lanthanides (widely used), comprising the 15 elements from lanthanum to lutetium.[1] Until recently the lanthanoids were defined as the 14 lanthanum-like elements from Ce to Lu. This distinction, which coupled lanthanum with scandium and yttrium, makes sense in terms of electronic configurations of the trivalent ions, but was abandoned by IUPAC reluctantly under pressure of common usage. However, it is unlikely that a change will occur from lanthanoid to lanthanide as the "ide" suffix carries anionic implications.

Despite enormous recent advances in molecular organolanthanoid and lanthanoid metal-organic chemistry, and earlier advances in their solid-state chemistry, rare-earth elements rarely have a prominent place in undergraduate courses at universities. This may be a

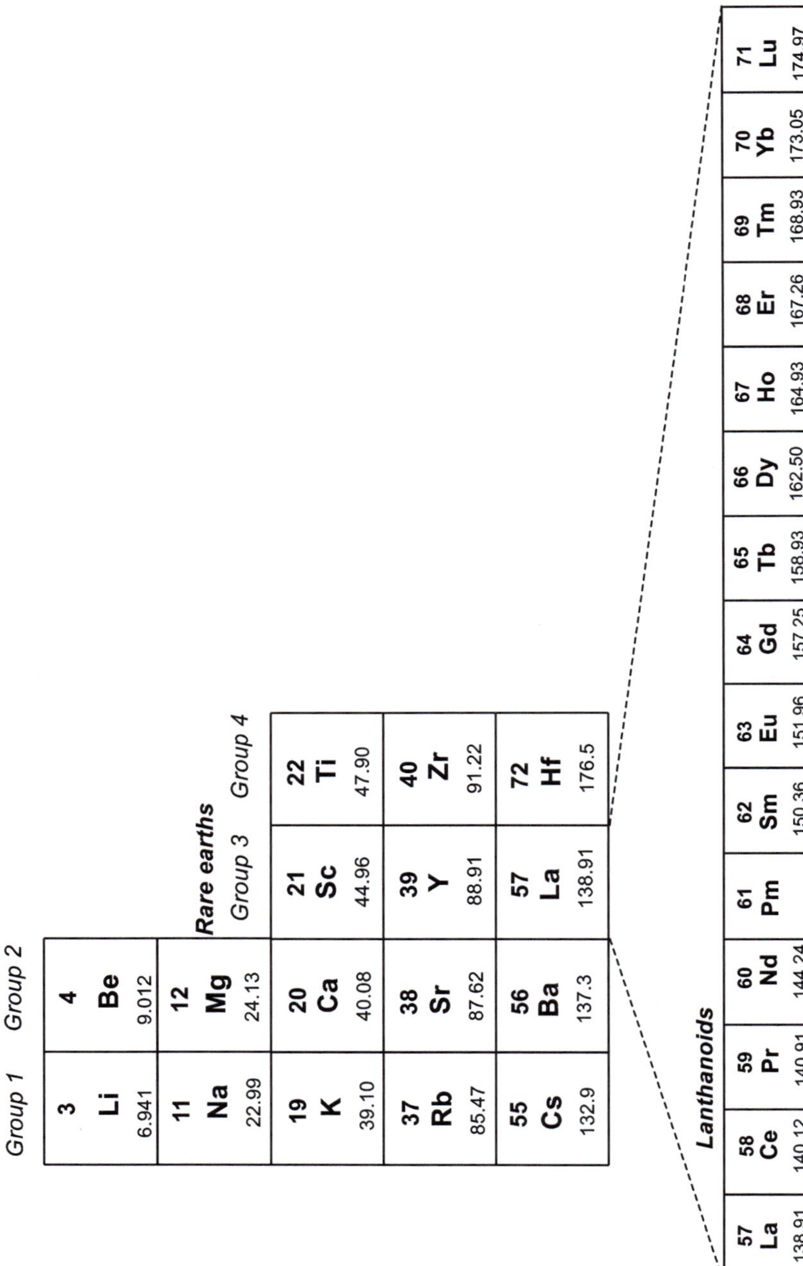

Figure 1.1. Rare-earth elements in the periodic table.

historic carry-over of the long time it took to separate the elements and establish their identity, as well as incorrect perceptions from their chemical similarities. However, the elements are far from rare and have considerable differences in their chemistry. They need a mnemonic as an aid to memory. One such published in "Chemistry in Australia" reads "**L**ittle **C**ute **P**eople **N**eed **P**lenty **S**ex **E**very **G**iven **T**ime **D**espite **H**aving **E**nough **T**hrough **Y**oung **L**ove" and serves the purpose for the lanthanoids.[2]

1.2 General Properties

1.2.1 *Electronic configurations of atoms and trivalent ions*

The electronic configurations are summarised in Table 1.1. The most striking feature is the progression of the electronic configurations for Ln^{3+} ions, from $4f^0$ for La^{3+} to $4f^{14}$ for Lu^{3+}. It can also be seen why IUPAC for years classified La with Group 3 and began the lanthanoids with Ce, as Sc^{3+}, Y^{3+} and La^{3+} have in common the electronic configuration of the preceding inert gas, whereas the filling of the $4f$ level begins with cerium. A case based on chemical similarity can be made that Group 3 should be Sc, Y and Lu, and the ionic radius of Y^{3+} is closer to that of Lu^{3+} than La^{3+}. Scandium, yttrium and lanthanum elements have [inert gas]$\mathrm{n}d(\mathrm{n}+1)s^2$ configurations. Cerium to lutetium have mainly a $4f^{\mathrm{n}}6s^2$ configuration where n is one greater than the f electrons in Ln^{3+}, but Ce, Gd and Lu have $4f^{\mathrm{n}-1}5d6s^2$ configurations where n−1 is the same number of f electrons as in the trivalent ion.

1.2.2 *Atomic and ionic radii and the lanthanoid contraction*

The notable feature of the ionic (Table 1.1) radii is the contraction from La^{3+} to Lu^{3+}. For the Ln^{3+} ions, the decrease is ca 20% and far exceeds what is expected across a period. The f electrons inadequately screen each other from the nuclear charge leading to an increase in effective nuclear charge across the series and hence a decrease in radii. A consequence is that stabilities of complexes with a specific ligand increase from La^{3+} to Lu^{3+}. This change in

Table 1.1. Rare-earth elements.

Atomic no./name/symbol	Electronic configuration Atom	M^{3+}	(IR Å)[a]	Oxidation state common; unusual[b]	Magnetic moment for $Ln^{3+}(\mu B)$[c]
21 Scandium (Sc)	[Ar] $3d4s^2$	[Ar]	0.87	**III**; 0, I, II	—
39 Yttrium (Y)	[Kr] $4d5s^2$	[Kr]	1.01	**III**; 0, II	—
57 Lanthanum (La)	[Xe] $5d^1 6s^2$	[Xe]	1.16	**III**; 0, II	—
58 Cerium (Ce)	$4f^1 5d^1 6s^2$	$4f^1$	1.14	**III, IV**; II	2.4
59 Praseodymium (Pr)	$4f^3 6s^2$	$4f^2$	1.13	**III, IV**; 0, II	3.5
60 Neodymium (Nd)	$4f^4 6s^2$	$4f^3$	1.11	**III**; 0, II	3.6
61 Promethium (Pm)	$4f^5 6s^2$	$4f^4$	1.09	**III**	2.8
62 Samarium (Sm)	$4f^6 6s^2$	$4f^5$	1.08	**II, III**; 0	1.5
63 Europium (Eu)	$4f^7 6s^2$	$4f^6$	1.07	**II, III**	3.6
64 Gadolinium (Gd)	$4f^7 5d^1 6s^2$	$4f^7$	1.05	**III**; 0, II	8.0
65 Terbium (Tb)	$4f^9 6s^2$	$4f^8$	1.04	**III, IV**; 0, II	9.6
66 Dysprosium (Dy)	$4f^{10} 6s^2$	$4f^9$	1.03	**III**; 0, II	10.5
67 Holmium (Ho)	$4f^{11} 6s^2$	$4f^{10}$	1.02	**III**; 0, II	10.5
68 Erbium (Er)	$4f^{12} 6s^2$	$4f^{11}$	1.00	**III**; 0, II	—
69 Thulium (Tm)	$4f^{13} 6s^2$	$4f^{12}$	0.99	**III**; II	7.5
70 Ytterbium (Yb)	$4f^{14} 6s^2$	$4f^{13}$	0.98	**II, III**	4.4
71 Lutetium (Lu)	$4f^{14} 5d^1 6s^2$	$4f^{14}$	0.98	**III**; 0, II	—

Notes: [a]Ionic radius for eight coordination;[3] [b]Refs. [4–7]. [c]Ref. [8].

stability finds practical application in the separation of rare earths by solvent extraction or ion exchange, where the extraction/elution order is $Lu^{3+} \ldots La^{3+}$. Because of the contraction, the radius of Lu^{3+} is smaller than that of Y^{3+} in the previous period, with Er^{3+} and Ho^{3+} near the size of Y^{3+} (Table 1.1). Consequently, the chemistry and geochemistry of Y^{3+} resembles that of the later lanthanoids and it can be classified as a "heavy" rare earth despite its atomic number.

A further consequence of the lanthanoid contraction is the similarity in size of the third transition series to those of their second transition series counterparts. A fascinating area of research is the effect of the lanthanoid contraction on the coordination numbers and structures of complexes. One might expect the large decrease in ionic radius would lead to a decrease in coordination number in interactions with the same ligand. While there are cases that illustrate this, e.g. $[LaCp_3]_n$ is an 11 coordinate polymer, $[YbCp_3]$ is a 9 coordinate monomer, and $[LuCp_3]_n$ is an 8 coordinate polymer.[9,10] However, all

[$LnCp_3$(thf)] (thf = tetrahydrofuran) complexes are 10 coordinate and pseudo-tetrahedral.[9,10] As a further example, [$LnCl_3(thf)_n$] (n = 2, 3, 3.5, 4) complexes diminish in coordination number from La (8) to Lu (6).[11] There is also an apparently anomalous seven coordinate [$YCl_3(thf)_2]_n$ which does not fit the ionic radius correlation.[12] Furthermore, structural breaks can occur at any point in the lanthanoid series. In the [$LnCl_3(thf)_n$] series, there is a change from La (CN = 8) to Ce (CN = 7)[11] even though the ionic radius difference is only 0.02 Å.[3] For a specific metal, the same ligand set can provide complexes with different structures and/or coordination numbers. Thus, holmium salicylate (salH) crystallises as monomeric nine coordinate [$Ho(salH)_3(H_2O)_3$].$3H_2O$, dimeric nine coordinate [$Ho_2(salH)_6(H_2O)_4$].$4H_2O$ and polymeric eight coordinate [$Ho(salH)_3(H_2O)_2]_n$.$2H_2O$ under virtually the same conditions.[13a] Similarly, anhydrous dysprosium cinnamate has been obtained in both seven and nine coordinate forms.[13b] It is this glorious uncertainty that makes lanthanoid coordination chemistry exciting.

1.2.3 *Oxidation states*

The oxidation states of rare-earth elements have been transformed in the last 30 years.[4–7,14–19] Prior to that, it was known that Ln^{3+} was the stable oxidation state for all elements. In addition Sm^{II}, Eu^{II} and Yb^{II} could be attained under normal laboratory conditions employing inert atmospheres for Sm^{III} and Yb^{II}.[8]

$$\begin{aligned} Ln^{3+} + e^- \rightarrow Ln^{2+} \quad E^0 &= -1.5\ \mathrm{V(Sm)} \\ &= -1.05\ \mathrm{V(Yb)} \\ &= -0.34\ \mathrm{V(Eu)\ vs.\ SHE} \end{aligned}$$

On the basis of these potentials, only Eu^{2+} is stable in water. Associated ligands can have a major effect on the potentials and the tris(cyclopentadienyls) in thf are much harder to reduce.[14c]

$$\begin{aligned} LnCp_3 + e^- \rightarrow LnCp_3^- \quad E^0 &= -2.5\ \mathrm{V(Sm)} \\ &= -1.5\ \mathrm{V(Yb)} \\ &= -1.0\ \mathrm{V(Eu)\ vs.\ Fc^+/Fc} \end{aligned}$$

In the case of Ln^{IV}, only Ce^{IV} is stable under normal laboratory conditions.[8]

$$Ce^{4+} + e^- \rightarrow Ce^{3+} \quad E^0 = 1.3 - 1.7 \text{ V vs. SHE}$$
$$\text{(depending on the condition)}$$

Despite the huge thermodynamic factor in favour of reduction, $Ce(SO_4)_2$ and $(NH_4)_2[Ce(NO_3)_6]$ are kinetically inert in aqueous solution and thus find wide use in analytical chemistry and organic synthesis. Again ligands have a major influence. Cerium(III) acetylacetonate ($Ce(acac)_3(H_2O)_n$) suspended in benzene is reported to be oxidised to $[Ce(acac)_4]$ by air,[20a] and the anhydrous compounds are quite stable. However, the hydrate $[Ce(acac)_4].10H_2O$ rapidly undergoes reduction in air or when exposed to X-rays.[20b] The E^0 value for

$$[Ce(acac)_4] + e^- \rightarrow [Ce(acac)_4]^- \quad E^- = +0.2 \text{ V vs. SHE}$$

shows why it can be prepared by atmospheric oxidation.[20] A recent study has challenged the oxidation state assignment on the basis of magnetic data[20c] Pr^{IV} and Tb^{IV} have limited existence,[8] the former as fluoride derivatives and both in the commercially available mixed valence oxide compounds Pr_6O_{11} and Tb_4O_7. Very recently molecular Tb^{IV} and Pr^{IV} complexes have been reported.[19]

In molecular rare-earth complexes, the first major oxidation state breakthrough was the totally unexpected (except to FGN Cloke[4]) discovery of zero-valent compounds by Cloke, e.g. $[Gd(C_6H_3(t\text{Bu})_3\text{-}1,3,5)_2]$.[4] Not only were complexes of bulky arenes obtained but also of bulky heteroarenes, e.g. pyridines. Such compounds were obtained for elements having a $4f^{n-1}5d^16s^2$ configuration (Table 1.1) or easy promotion to it, e.g. Y, Nd, Gd → Er, Lu.[5] In addition, they are not formed for the elements normally giving stable Ln^{II} complexes, *viz* Sm, Eu and Yb. The synthesis involved condensation of metal atoms with arenes/heteroarenes at low temperatures.[4,5]

Solid-state divalent and subvalent halides have been known for many years.[21–23] Complexes LnX_2 (Ln = Eu, Sm, Yb, Nd, Dy, Tm) are ionic and divalent, but other elements give solids best formulated as $[Ln^{III}(X^-)_2(e^-)]$, and have pseudo-metallic behaviour. In addition, phases with other stoichiometries, e.g. GdCl; TbCl are known.[23] On the other hand, "molecular" divalent compounds are a more recent development,[7,14–18] notably by W.J. Evans at U.C. Irvine.[7a,b] Now divalent organometallics are known for all rare earths except the radioactive Pm.[7] A large number are of the form [K(18-crown-6) or (2,2,2-cryptand)][$LnCp'_3$] (Cp′ = bulky cyclopentadienyl, e.g. $C_5H_4SiMe_3$[7b]). Recently, Ln^{II} derivatives of very bulky silylamides have been prepared.[17]

1.3 Discovery, Location and Abundance of Rare Earths

1.3.1 *Discovery of the elements*

The lanthanoid elements were first discovered in 1787 when Carl Axel Arrhenius identified the mineral "ytterbite" near Ytterby, Sweden. Over the next 150 years or so, the lanthanoids were progressively discovered and isolated (see Figure 1.2). Significant chemists involved in the discoveries were Johann Gadolin (yttrium), Ekeberg (isolated beryllium from ytterbite), Berzelius and Hisinger (cerium), Klaproth (cerium), Mosander (didymia, lanthanum, terbium and erbium), Berlin (erbium oppositely named terbium by Mosander), Delafontaine (terbium oppositely named erbium by Mosander), Boisbaudran (samarium, gadolinium and europium), Marignac (gadolinium), Crookes (europium), Demarçay (europium), Moseley (identified element 61, promethium had yet to be discovered), Spedding (separated and purified the lanthanoids and actinoids in the Manhattan project). Worthy of mention here again is the town of Ytterby in Sweden, of significance for having the elements yttrium, ytterbium, terbium and erbium named after it.

Even though the rare-earth minerals are relatively abundant, they are widely spread and since the properties of the rare-earth elements are exceedingly similar in the Ln^{III} state, their isolation into pure elements was problematic. Separation techniques such as liquid–liquid extraction and ion exchange were developed in the mid-1900s and aided in their isolation.

1.3.2 *Ores, locations, miners and producers of rare earths and geopolitical considerations*

The term "rare earths" is a misnomer, as the elements are far from rare in crustal abundance,[24] though it may well reflect public ignorance of them. The most abundant is cerium, which is present

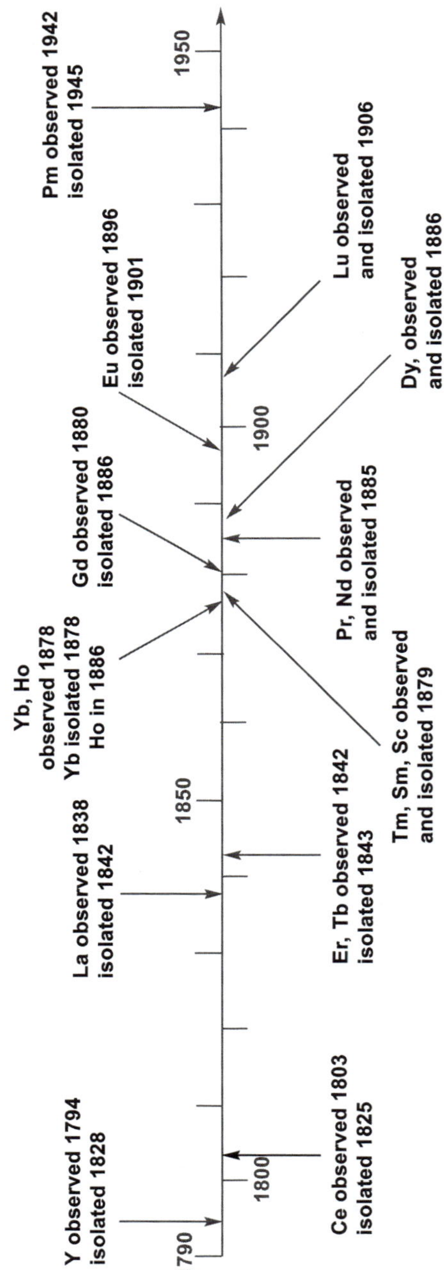

Figure 1.2. Timeline of discovery of the rare-earth elements.

in similar amounts to zinc and copper, and is more prevalent than lead or tin, as indeed are lanthanum, neodymium, yttrium and (surprisingly) the very expensive scandium. Even the least abundant, thulium, has a greater prevalence than mercury, bismuth and cadmium, none of which are considered rare.

The principal rare-earth ores are monazite, bastnaesite and xenotime. Monazite is mainly a light rare earth (La, Ce, Pr and Nd) phosphate (93–94%). It is often a component of mineral sands, which are primarily a source of ilmenite, rutile and zircon. Bastnaesite is even more dominantly a light rare-earth source (ca 99%) and is often termed a "fluorocarbonate" but is a fluoride/carbonate mixture, the FCO_3^- ion being unknown. Xenotime is substantially (60%) YPO_4 with approximately 30% of other heavy rare earths.[25] Lopanite contains Ce, Ti and Nb, and lateritic ion adsorption clays contain rare earths. Swedish rare-earth minerals such as gadolinite no longer play a significant role, despite their historic importance. As heavy rare-earth metals and oxides generally command higher prices, bastnaesite is the least attractive ore for development. Scandium is not normally a component of rare-earth ores, but is found in some uranium ores and some Co/Ni ores, e.g. Australian mines near Greenvale, North Queensland.[25] Tailings from uranium mines may become a source if applications of Sc increase.

Mirroring its dominance (at one stage ca 95%, but now estimated as nearer 80%) in the supply of separated rare earths, China has the largest proven rare earth reserves by a large margin, followed by Brazil, Australia, India and USA.[26] Actual reserves are probably much larger, as low world prices over much of the last two decades have dampened exploration. In addition, rare earths in tailings are unlikely to be included.

A brief dot point history of rare-earth production and suppliers has been provided in a recent report.[27] For a few decades, US (Molycorp) dominated the supply of separated rare earths. Australian monazite from WA, NSW and Qld mineral sands was exported to France for several years, but was discontinued in the late 1980s owing to issues over disposal of radioactive residues. However, during that period, 1980–2000, China, taking advantage of abundant resources and low labour costs, was gradually driving competitors out of business, culminating with the mothballing of Molycorp's Mountain Pass plant in 2002. In this period, socio-economic issues associated with the world supply of rare earths relying upon a single country were largely ignored. However, in 2010, China banned rare earth exports to Japan in response to a territorial dispute and subsequently (2011) imposed general export restrictions leading to a huge spike in rare-earth prices in June–August 2011. World leaders became suddenly aware of the strategic importance of rare earths. Thus, President Obama stressed the importance of neodymium (RE magnets, catalysts for artificial rubber, military uses), Chancellor Merkel undertook a visit to Mongolia, a potential alternative source of rare earths, and Germany set up a scientific committee to examine the situation. The United States Department of Energy in its 2010 Critical Materials Strategy report identified dysprosium as the element that was most critical in terms of import reliance.[28] Molycorp reopened the Mountain Pass plant and floated shares at $14 in 2010, and the price peaked at $74 (all monetary values quoted are in US Dollars).[29] Using some of these funds, Molycorp acquired the metal and alloy manufacturer Santoku America (formerly Research Chemicals, built by the Nuclear Corporation of America (NUCOR), then Rhône-Poulenc then Rhodia Rare Earths, then Santoku America, and now post Molycorp collapse resurrected by Eutektix), AS Silmet Estonia, one of two European rare-earth producers (the other is Treibacher, a private Austrian company) and the Canadian Neo Materials Technologies. Around this time, the share prices of Australian rare-earth companies such as Lynas, (now Lynas Rare Earths), and Arafura resources also peaked.

However, rare-earth metal prices declined initially rapidly in 2011, then progressively through 2012 and 2013, as Chinese exports, boosted by smuggled supplies, came into balance with demand. In addition, in 2014, the world trade organisation ordered Beijing to end curbs on rare-earth exports.[30] This was accompanied by a rapid decline in share prices of rare-earth companies. Molycorp filed for bankruptcy with a closing 2015 share price of $0.36.[31] However, there are signs and predictions that rare-earth prices are trending upwards again. The Chinese government is now attempting to curb rare earth smuggling and is crucially introducing new environmental controls to limit supply again.[27]

With the collapse of Molycorp, the way has opened up for other potential producers. In particular, Australian companies, operating in a politically stable environment, are becoming attractive. Foremost of these Lynas Rave Earths, which has the low radioactivity Mt Weld deposit, is operating a processing plant in Malaysia and is already producing rare-earth oxides. It has also announced a joint rare-earth magnet venture with Siemens.[32] Nevertheless, the company has warned of survival prospects,[33] though a long-term debt restructure was later announced.[34] Operations have been delayed by environmental protests, as the plant has been located in an area with past (non-rare-earth operations) pollution. Arafura Resources are developing the Nolans rare earth/phosphate/uranium deposit and the company aims to be a major supplier of rare earths. However, the reduced rare-earth prices have taken their toll, and the original plan to build a processing plant at Whyalla, SA, has been abandoned (2013)[35] in favour of a plant nearer the deposit. The main shareholder is Chinese (9.37%) in contrast to the Federal Government blocking a Chinese takeover of Lynas. Despite a share price in the range 5c–7c in 2015, currently 16c it is still operational,[36] and mining should begin soon.[37] Alkane Resources is a multi-commodity miner with a zirconia, rare earth prospect at Dubbo, NSW. A pilot plant has been built by ANSTO and a mining lease for the project was recently granted.[38] This project has now been demerged from Alkane Resources and listed as Australian Strategic Minerals.

Astron Limited is of particular interest as it owns the Donald zirconia/titania/monazite, etc., resource, which covers a large area of the Wimmera, and was originally owned by WIM Minerals, a Rio Tinto subsidiary, which abandoned the deposit owing to processing problems. Although the rare-earth component is subsidiary to the zirconia, etc., it is still a vast resource with some Xenotime. The holding company is based in Hong Kong and there is considerable Chinese involvement.[39] The company also has a major deposit in Senegal. There are at least a dozen other Australian companies with rare-earth interests, while the tailings dam of BHP's Roxby Downs project is a potential source. Two other companies warrant mention. Hastings Technology Metals Ltd., which has substantial Singapore involvement, has the Yangibama and Brockman prospects, both of which are rich in heavy rare earths,[40] and thus are of potential long-term value. Australian Mines has, as its flagship project, the SCONI nickel–cobalt–scandium prospect which is of interest in this chapter because it is a rare scandium source.

Postscript: Since this account was originally written, there have been notable developments in the geo-politics of the rare earths. In 2019, as part of the US–China trade war, China has threatened to cut off rare-earth exports to the US.[41] This feature together with the fact that Lynas Rare Earths is now fully producing separated rare earths in their Malaysian plant, thereby providing an alternative to Chinese suppliers, has seen the Lynas share price rise to above $5, though now nearer $1.55. It is estimated that Lynas Rare Earths is now supplying about 8% of world demand and has the potential to see this rise.[42] Because of Malaysian environmental concerns over radioactive residues (common in rare-earth processing), Lynas is establishing a cracking and leaching plant in the Goldfields area of WA to do the pre-processing in Australia to remove the radioactive U and Th before shipping the ore to Malaysia.[43] This should not be over-burdensome, as the Mt Weld ore body has relatively low radioactive content. The company has now a new operating licence in Malaysia till early March 2023. The firm is looking to set up a processing plant in Texas in association with Blue Line,[42] and has received Department of Defence funding for the first stage of the project.[44] One socio-economic factor to consider is the potential for China to control the world rare-earth market from a near monopoly position; in such a scenario, commercial opportunities in other countries could be quashed, as exemplified by the collapse of Molycorp. The Mountain Pass mine of Molycorp has now been bought from bankruptcy and operates as MP Materials. The rare-earth concentrate is processed in China by Shenghi Resources (a shareholder) and currently provide 15% of RE oxides. The Chinese government has also made a determined effort to crack down on illegal exports which depressed prices. In a recent announcement, Apple has said that it will in future only use recycled rare earths in its products.[42]

1.4 Uses of the Rare Earths

1.4.1 *General uses (see also Chapter 17)*

Unseparated rare-earth elements, typically in the form of Mischmetal, which is an alloy of rare-earth elements in various naturally occurring proportions, can be used for a variety of applications. Mischmetal typically includes approximately 50% cerium and 25% lanthanum, with smaller amounts of neodymium and praseodymium. It is used for the deoxygenation and desulfurisation of steel and in the flint ignition device of many lighters and torches. Unseparated rare earths are also used in unusual areas such as agriculture where several reports from China have laid claims that crop growth is enhanced,[45a,b] while Swiss industry with Ludwigs Maximillian University, Germany, have demonstrated with Lancer®, Sanocer® (main component, cerium citrate) clear benefits and no toxicity or environmental drawbacks in uses as animal feed supplements.[45b,c] Approval for use in the EU is being sought.[45d] Rare earths are widely used as petroleum cracking catalysts, generally as mixed oxides.

1.4.2 *Specific uses of rare earths*

Scandium, being rather different from the other rare earths, has some standalone properties among these elements. Scandium/aluminium alloy (0.1–0.5% Sc) provides considerably more strength to areas affected by heat, e.g. welded joints, and is also used in baseball bats and bicycle frames. However, its main use is in some aerospace parts, particularly in Russian MiG 21 and MiG 29 aircraft. Scandium compounds are used in high-intensity discharge lamps and light bulbs, and scandium triflate is used in organic synthesis as a strong Lewis acid catalyst.

(a) *Luminescent properties*: Yttrium has its primary application in TV screens where Y_2O_3 is combined with europium and this combination is also used in LEDs. Cerium and terbium are also used in LED television screens as phosphor components. The green terbium phosphor is used in combination with blue and red europium phosphors, creating the technology used in trichromatic lighting. Erbium is also combined with europium isotopes to provide specific fluorescent properties. Neodymium also has application in fluorescent technology.

(b) *Laser technology*: Neodymium, samarium, dysprosium, holmium, erbium, thulium and ytterbium have applications in lasers in both commercial and military uses, optical cables, medical and dental lasers and in solid-state lasers where Nd-doped yttrium aluminium garnet (Nd:YAG) is in use.

(c) *Alloys*: As for scandium (aforementioned), rare earths have found application in alloying with other metals, where small amounts add strength to many metals. Thus, Y is used in Al and Mg alloys, Ce in aluminium and Fe alloys, and La and Yb in steel alloys. Pr is used as an alloying agent for metals in aircraft engines, Gd is present in metal alloys to resist high temperature and Tb in several alloys, such as the magnetostrictive alloy Terfenol-d ($Tb_xDy_{1-x}Fe_2$, x $\sim$0.3).
(d) *Magnets*: $Nd_2Fe_{14}B$ is the strongest permanent magnetic material currently known, and is used in electric motors in hybrid vehicles, microphones, loudspeakers, headphones and computer hard disks and in wind turbines for power generation. Dy has been added to $Nd_2Fe_{14}B$ to improve its magnetic properties and Pr for anticorrosion properties. $SmCo_5$ has application in some magnets and was the strongest magnet known prior to $Nd_2Fe_{14}B$. Ho has the highest magnetic strength of all elements and is used as a component in powerful magnets.
(e) *Superconductors*: Both yttrium and thulium have found application in high-temperature superconductors, e.g. $YBa_2Cu^{II}_{3-\delta}Cu^{III}_{\delta}O_{6.5+0.5\delta}$. Despite the excitement caused by their discovery, it seems that a rise in the critical temperature is needed before there are wide applications.
(f) *Batteries*: A major use of lanthanum is in nickel metal hydride batteries in hybrid vehicles where approximately 8–10 kg of the metal is used. While Pm has little use due to its radioactivity, it is used in nuclear batteries and has potential to be used as a power source in space vehicles and satellites.
(g) *Medical*: The properties of many rare earths lend themselves to a variety of medical uses. Yttrium radioisotopes have been used in the treatment of cancers such as lymphoma, leukaemia, ovarian, pancreatic and bone cancers. A radioactive isotope of Sm can be used to treat severe pain associated with bone cancers. Europium is used in screening for genetic diseases, such as Down's Syndrome. Probably most importantly, gadolinium, due to its strongly paramagnetic properties (seven unpaired electrons), is extensively used in MRI imaging. Although highly stable complexes, generally macrocylic, are used, evidence of Gd

toxicity leading to nephrogenic systemic fibrosis, particularly in patients with impaired renal function, has emerged.[46]

(h) *Ceramics*: A very large amount of cerium is used as a ceramic component in the catalytic converter of automobiles, and is the largest use of rare earths in the US. Yttria-stabilised zirconia is a thermally stable refractory, a solid electrolyte in fuel cells and is used for oxygen sensing. The introduction of yttria stabilises the high temperature cubic form of zirconia.

(i) *Mischmetal uses*: As noted, mischmetal (comprising ca 50% cerium and 25% lanthanum with smaller amounts of praseodymium and neodymium) has found widespread usage in the flints of cigarette lighters and other fire starters. More important is its metallurgical use (aforementioned).

(j) *Glass colouring and decolouring*: The pale green colour of praseodymium has been used to colour cubic zirconias and glasses. Er_2O_3 in sunglasses increases colour perception. By contrast, CeO_2 is used as a decolourant in glass.

(k) *Catalysts*: Ln^{3+} ions are strong Lewis acids and have found many applications in organic synthesis particularly where Lewis acids are used as catalysts. Neodymium (as neodymium carboxylates) has recently found application with aluminium alkyls to catalyse the synthesis of artificial rubber.[47] A wide variety of compositions has been used for these pseudo-Ziegler catalyst mixtures, with other added components such as transition metal salts. While lutetium is very expensive and therefore has few commercial uses, it has found application in petroleum cracking.

(l) *Uses in nuclear reactors*: Samarium, gadolinium, dysprosium and holmium have all found application in nuclear reactors due to their ability to absorb neutrons arising from nuclear fission.

(m) *Critical defence uses of rare-earth elements*: Military applications have been found for many of the rare-earth metals and while some substitutes have been found for these critical metals, they are generally not as effective. Military uses include: Lanthanum in night vision goggles, neodymium in laser range-finders, guidance systems, communications and magnets, europium in fluorescent and phosphorescent materials, erbium in fibre-optic data transmission and samarium in permanent magnets, guided weapons and in stealth technology.

1.4.3 *Rare-earth outlook*

Uses of rare earths (see also Chapter 17) as pollution control ceramics in automobiles, as permanent magnets and rechargeable batteries are certain to increase as demand for electric and hybrid automobiles, computers, electronics and portable equipment expands.[48] For example, the market for cerium and neodymium in automobile catalytic converters and petroleum refining catalysts is expected to grow by 6–8% per year. The demand will be not only for high purity separated materials but also high purity mixed products.

The use of rare earths in magnets is expected to increase by 10–16% per year in the immediate future.[49] The growth of rare earths in rechargeable NiMH batteries, mostly for hybrid vehicles, will expand similarly. Increased use in portable equipment, such as camcorders, cellular telephones, compact disc players, digital cameras, digital video disc players, laptop computers and MPEG audio-layer-3 players will also mean a greater requirement of NiMH batteries. (An article on recycling has prognoses for various uses, as part of projected streams for recycling.[48b])

Further developments that will ensure increased rare earth use are applications such as fibre optics, medical applications including dental and surgical lasers, magnetic resonance imaging, medical contrast agents, medical isotopes and positron emission tomography scintillation detectors. The use of rare-earth alloys in magnetic refrigeration is also expected to contribute to greater rare-earth requirements.[50] New bulk uses, particularly in corrosion inhibitors, and also in animal feed supplements, would greatly increase demand, and increase the viability of the many startup mining projects in Australia (Section 1.3.2). Equally relevant is what premium end users are prepared to pay for security of supply as offered by Australia.

Rare-earth carboxylates have emerged as an attractive alternative to chromates to inhibit the corrosion of mild steel, a major construction material.[51–53] Although chromates are cheap and effective, the adverse health effects have forced the need for replacements.[54] Following separate observation of inhibitor properties of rare-earth ions,[55] and of carboxylate ions,[56] rare-earth carboxylates have made considerable progress as dual purpose inhibitors[51–53,57] with the rare earth acting as a cathodic inhibitor and the carboxylate anions as anodic inhibitors.[57–59] In particular, cerium salicylate,[58] lanthanum 4-hydroxy-cinnamate[59] and yttrium 3-(4′-methylbenzoyl)propanoate[60] show excellent inhibitor properties and are being further developed.

1.5 General Chemistry Properties of Rare-Earth Elements and Compounds

Sections 1.2.1–1.2.3, namely electronic configurations, ion and atomic sizes and oxidation states, outline the main determinants in rare-earth chemistry.

1.5.1 *Separation of the lanthanoids*

Similar properties in the usual +III oxidation state have made separation of the elements challenging, as they are found together in ores. Early separations were based on fractional crystallisation of salts, were time-consuming and tedious, and led to erroneous claims. Current separation methods are based on (i) less usual oxidation states (see Section 1.2.3) in the case of cerium and europium and (ii) variations in complex stability resulting from ionic radii differences.

Prior to the separation of the lanthanoids, other metals, e.g. iron, thorium and uranium, have to be removed.[61] The last two can lead to environment problems. Ores with low radioactivity like Mt Weld in Western Australia are therefore very attractive. An accessible tailings dam of a uranium mine is a viable solution for storage of U/Th waste. For example, the ill-fated SX Holdings separation plant planned to use the Roxby Downs Tailings Dam.

From an acidic solution of mixed Ln^{3+} ions, cerium is first removed based on the existence of the readily available Ce^{IV} oxidation state and the lower basicity of CeO_2 than $Ln(OH)_3$.

$$Ln^{3+} \xrightarrow{OH^-} Ln(OH)_3 \xrightarrow[100-500^\circ C]{O_2 \text{ from air}} Ln(OH)_3 + CeO_2 \xrightarrow[-CeO_2]{H^+} Ln^{3+}$$

The whole process can be performed as a single step with careful control of pH. High purity CeO_2 is separated and can be used for ceramics. Alternatively, bastnaesite may be roasted to give CeO_2 and mixed Ln^{III} oxides which can be separated by pH control as mentioned earlier. Europium is removed utilising the existence of a stable divalent state and the low solubility of europium(II) sulphate. However, this separation is not quantitative, but the losses and the cost of zinc are offset by the value of Eu. Also, residual Eu^{II} oxidised to Eu^{III}, can be recovered by solvent extraction.

$$Ln^{3+} \xrightarrow[H_2SO_4]{Zn} EuSO_4 \downarrow + Ln^{3+}$$

The principal method for the remaining rare earths is currently solvent extraction, mainly in China (though Lynas Rare Earths in Malaysia makes a contribution). Cation exchange has also been used but is more expensive. With cation exchange, Ln^{3+} ions are eluted by citrate, EDTA or 2-hydroxy*iso*butyrate ions, for example. The elution order is $Lu^{3+} \rightarrow La^{3+}$ corresponding to the stability order for complexes (aforementioned) and Y^{3+} between Er^{3+} and Ho^{3+} is according to ion size. In the case of solvent extraction, tributyl phosphate/HNO_3 can be used, but more commonly di(2-ethylhexyl) phosphate (phosphoric acid), $(Me(CH_2)_3CH(Et)CH_2O)P(O)OH$/HCl or H_2SO_4 in kerosene is the eluent. A range of alternative but related extractants is available. The extraction order is $Lu^{3+} \rightarrow La^{3+}$, corresponding to decreasing complex stability. Solvent extraction is usually operated under counter-current conditions.[62]

1.5.2 *Rare-earth recycling*

China currently dominates the production of separated rare-earth element output with a greater than 60% share, although it has less than one half of global rare-earth resources.[48b] Surprisingly, perhaps, a lack of global recycling incentives and poor collection methods contribute to a low rare-earth recycling rate estimated in 2013 to be less than 1%. Typically, the potential recyclable sources of rare-earth elements consist of pre-consumer scrap/residue, industrial residues, end-of-life products and tailings dams or landfill (waste residue) sites.

The cost of collection and removal of unwanted materials prevents the waste being a feedstock for existing separation plants.

This low recycling rate contrasts with recycling efforts in place for the base and precious metals such as iron, copper, gold and silver. Recently, however, work by Binnemans[63] has led to the development of extraction of rare earths from recyclates using ionic liquids in a simpler, more economic and more environmentally friendly process in which the ionic liquid can be regenerated. In addition, a ligand-based method for the rapid separation of Nd and Dy from magnets has newly been reported by Schelter[64] and the ligands can also be reused. The development and commercialisation of these novel techniques have the potential to significantly amplify current rare-earth recycling rates. Recycling operations can also lead to intrinsically more simple separation processes where the rare earth of interest is present in greater relative amounts than in ores or mischmetal alloys. As an example, a simple classical method has been designed for recovery of Eu and Y from fluorescent tubes.[65]

Another advantage of improving the rate of rare-earth recycling is a reduction in "waste rare earths". Recycling can regenerate the specific rare earth(s) required, whereas primary extraction generates the metals in their natural distribution. The recent decision by Apple to use only recycled rare earths in their products should give a major boost to recycling.

1.5.3 *Metal oxide formation*

Rare-earth oxides are the primary industrial product. After solvent extraction separation, the Ln^{3+} ions are back extracted from the organic solvent by acid and then precipitation with OH^- and thermal dehydration of the hydroxides/hydrated oxides gives the oxide.

$$Ln^{3+}(org) \xrightarrow{H_2O/H^+} Ln^{3+}(aq) \xrightarrow{OH^-} Ln(OH)_3 \xrightarrow{heat} Ln_2O_3$$

Because of the high basicity of Ln_2O_3, the last step requires protection from atmospheric moisture and carbon dioxide. If ion exchange is used, the aqueous eluate is treated with oxalic acid to give $Ln_2(C_2O_4)_3$ which is decomposed into the oxide.

$$Ln^{3+} \xrightarrow{C_2O_4^{2-}} Ln_2(C_2O_4)_3 \xrightarrow{heat} \text{basic carbonate} \xrightarrow{heat} Ln_2O_3$$

Again, the final step requires exclusion of CO_2 and water. It should be noted that Pr and Tb form mixed oxidation state oxides, Pr_6O_{11} and Tb_4O_7, and cerium is obtained as CeO_2 after earlier isolation (aforementioned).

1.5.4 *Metal halide formation*

Unseparated rare-earth chlorides are often an initial product of a mineral separation process, e.g. as planned by Arafura Resources for the Nolans project. Hydrated rare-earth halides (Cl, Br and I) can be prepared by dissolution of the oxides (Section 1.5.3) in the appropriate aqueous acid, evaporation and crystallisation.

$$\mathrm{Ln_2O_3 + 6HX \xrightarrow{H_2O} 2LnX_3 \cdot nH_2O + 3H_2O (X = Cl, Br, I)}$$

However, care must be exercised to avoid any oxy-halide contamination. In the case of commercial Tb_4O_7, reduction occurs during dissolution in HCl.

The rare-earth trifluorides and CeF_4 are insoluble and can be precipitated from aqueous solution.

$$\mathrm{Ln^{3+} + 3F^- \xrightarrow{H_2O} LnF_3 \downarrow \quad Ce^{4+} + 4F^- \xrightarrow{H_2O} CeF_4 \downarrow}$$

Anhydrous rare-earth halides, LnX_3 (X = Cl, Br, I), can be prepared by heating Ln_2O_3 with ammonium halides.[66] As water is also formed, there is the potential for some hydrolysis, and the pure trichlorides are purified by sublimation.

$$\mathrm{Ln_2O_3 + 6NH_4X \xrightarrow{heat} 2LnX_3 + 3H_2O + 6NH_3}$$

Alternatively, if $(NH_4)_{3-n}[LnCl_{6-n}]$ (n = 0 or 1) complexes are prepared from the oxide, their pyrolysis gives the anhydrous trichlorides.[66]

$$\mathrm{Ln_2O_3 + 6HCl + 2(3-n)NH_4Cl \longrightarrow 2(NH_4)_{3-n}[LnCl_{6-n}] + 3H_2O}$$

$$\mathrm{(NH_4)_{3-n}[LnCl_{6-n}] \xrightarrow{heat} LnCl_3 + (3-n)NH_4Cl}$$

The pure sublimed trichlorides have very low solubility in organic solvents, constituting a problem in their use in metathesis reactions in non-aqueous solvents. $[LnCl_3(solv)_n]$ (solv = tetrahydrofuran (thf) or 1,2-dimethoxyethane (dme)) complexes are more suitable and can be prepared by prolonged refluxing of the anhydrous chlorides with the donor solvent in an inert atmosphere. More conveniently, $[LnCl_3(solv)_n]$ are prepared by direct oxidation of the metals with an excess of C_2Cl_6.[67,68] Working up with pentane removes the chlorinated organic compounds, leaving the chlorides in high yield and purity.

$$2Ln + 3C_2Cl_6 \xrightarrow{solv} 2[LnCl_3(solv)_n] + 3C_2Cl_4$$

Divalent halides are prepared by reduction of LnX_3 with Ln metal or hydrogen. SmI_2, widely used in organic synthesis and YbI_2 are obtained by oxidation of Sm or Yb metal in an inert atmosphere.

$$\text{e.g. } Sm + 2(CH_2I)_2 \xrightarrow{thf} [SmI_2(thf)_5] + C_2H_4$$

Alternatively, CH_2I_2, I_2 or HgI_2 can be used as the oxidant. Although crystals of $[SmI_3(thf)_5]$ can be isolated,[69] they readily lose thf giving $[SmI_2(thf)_2]$.[70] The less usual diiodides, LnI_2 (Ln = Nd, Dy, Tm) are obtained by the direct reaction of Ln metal and iodine[6]. Evans has provided a detailed preparation of DyI_2 in quite a large scale,[71] and the method can plausibly be adopted for Nd and Tm.

1.5.5 *Rare-earth carboxylates*

These compounds form the basis of innumerable coordination polymers because of the large Ln^{3+} ion size (Section 1.2.2) and the versatility of carboxylate binding modes (Figure 1.3).[72]

The carboxylates are readily prepared by a variety of ways. Commonly, metathesis reactions between rare-earth salts and alkali metal carboxylates in aqueous alcoholic media are used.[72]

$$LnX_3 + 3RCO_2M \longrightarrow Ln(O_2CR)_3 + 3MX;\ M = Na, K, NH_4;$$
$$X = Cl, NO_3,\ \text{etc.}$$

A variation of the aforementioned equation is the reaction between the Ln metal salt and carboxylic acid followed by addition of a metal hydroxide until precipitation of the rare-earth

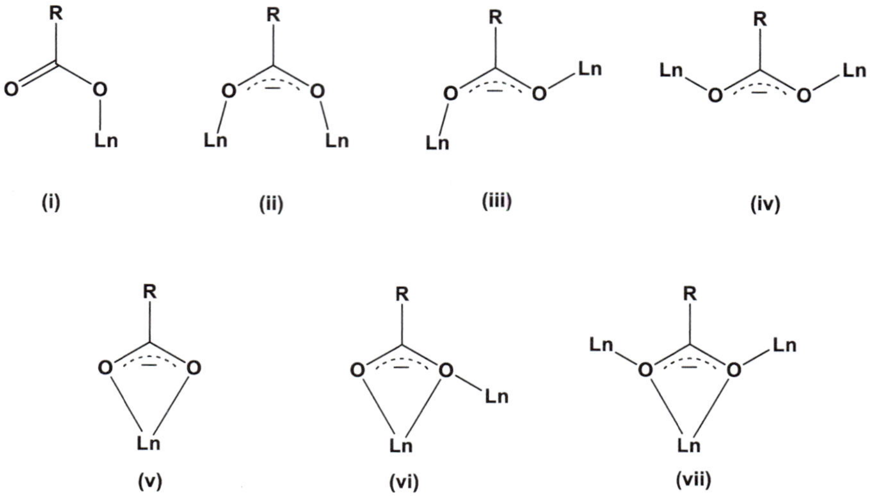

Figure 1.3. Some coordination modes of the carboxylate ion with rare-earth metals: (i) $\kappa(O)$; (ii) Z,Z-μ-*1*κ*(O):2*$\kappa(O')$; (iii) Z,E-μ-*1*κ*(O):2*$\kappa(O')$; (iv) E,E-μ-*1*κ*(O):2*$\kappa(O')$; (v) $\kappa(O,O')$; (vi) Z,Z-μ-*1*κ*(O)-2*$\kappa(O,O')$; (vii) E,E-μ_3-*1*κ*(O):2*κ*(O,O'):3*$\kappa(O')$.

carboxylate occurs.

$$LnX_3 + 3RCO_2H \xrightarrow{MOH} Ln(O_2CR)_3 + 3MX + 3H_2O$$
$$\text{(M, X as aforementioned)}$$

Alternatively, reaction of the acid or the anhydride with rare-earth carbonates, hydroxides or oxides is possible, or in this case of aromatic carboxylic acids, reactions with the corresponding acetates are viable, e.g.

$$3(RCO)_2O + Ln_2O_3 \longrightarrow 2Ln(O_2CR)_3$$
$$6\ RCO_2H + Ln_2O_3 \longrightarrow 2Ln(O_2CR)_3 + 3H_2O$$
$$Ln(OAc)_3 + 3ArCO_2H \longrightarrow Ln(O_2CAr)_3 + 3HOAc$$

Direct reactions of the highly electropositive metals with carboxylic acids are also possible, especially if elevated temperatures are used.[72e]

1.5.6 *Rare-earth salts*

A large range of hydrated rare-earth(III) salts have been prepared by reaction of rare-earth oxides (or hydroxides or carbonates) with the appropriate acid.

$$Ln_2O_3 + 6HX \longrightarrow 2LnX_3 + 3H_2O$$

This method provides hydrated chlorides, nitrates, triflates, acetates, sulphates, bromates, bromides, perchlorates, etc. The chlorides, nitrates, triflates and acetates are commercially available, the first two as hexa- or hepta-hydrates which are highly hygroscopic. Most oxyanions coordinate to some extent owing to the oxophilic nature and high Lewis acidity of Ln^{3+}. The oxides are strongly basic and absorb CO_2 from the air giving carbonates, which are also commercially available. The triflates which are important Lewis acid catalysts, and which form hydrates, $[Ln(H_2O)_9](O_3SCF_3)_3$ from aqueous solution, are readily dehydrated at 180–200°C without hydrolysis. This enables them to be recovered after catalytic use and aqueous workup, a property which is a major advantage over the chlorides which partly hydrolyse into oxychlorides on attempted dehydration.

Rare-earth salts undergo hydrolysis in solution giving $[Ln(H_2O)_8OH]^{2+}$ or more complex species, and the solutions are acidic. Hydrated hydroxides are precipitated on addition of base. Single crystals of the hydroxides have been grown by hydro-thermal methods[73] and the coordination numbers decrease from nine (La-Er) to eight (Lu, Sc). Novel hydroxolanthanoid cages have been crystallised utilising amino acid or β-diketonate coligands.[74]

The solubility properties are analogous to those of alkaline earth salts. Thus, the fluorides, oxalates, phosphates, carbonate, hydroxides and chromates are insoluble. In addition, the sulfates have low solubility. The double sulfates $Ln_2(SO_4)_3.Na_2SO_4.xH_2O$ have been used to provide qualitative separation between light (La-Eu) and heavy (Gd-Lu, Y) rare earths, in that the former are insoluble in aqueous sodium sulfate where the latter have limited solubility. Double nitrates were used in separation of rare earths by fractional crystallisation.

A range of cerium(IV) compounds are commercially available, reflecting their importance as one electron oxidants in organic synthesis and in analytical chemistry. Ammonium cerium(IV) nitrate and sulfate as well as ceric sulfate are the most commonly used and the double salts are coordination complexes, e.g. $[NH_4]_2[Ce(NO_3)_6]$ which has the 12-coordinate hexanitrocerate(IV) anion (isocoordinate with the Ce^{III} analogue). Cerium(IV) fluoride is insoluble in water like the Ce(III) analogue. In aqueous solution, Ce^{4+} hydrolyses extensively and polymerisation of the hydrolysis products occurs.

1.5.7 *Preparation of the metals*

Rare-earth metals are highly electropositive and oxidise readily in air. E^0 values for the reduction of the trivalent ions are in the range −1.8 to −2.4 V; hence, powerful reductants are needed to form the metals from metal fluorides, or oxides or chlorides. The oxides can be converted to the trifluorides by treatment with HF under argon, and then reduction by calcium can be employed.

$$2LnF_3 + 3Ca \longrightarrow 3CaF_2 + 2Ln$$

Alternatively, metal oxides can be directly used. After heating in dry air to decompose hydroxide and carbonate impurities, the oxides are reduced by the most electropositive rare-earth metal, La, at high temperatures (ca 1500°C).

$$Ln_2O_3 + La \longrightarrow La_2O_3 + 2Ln$$

Industrially, electrochemical smelting is commonly used. Rare-earth chlorides in metal chloride fluxes, e.g. NaCl, KCl and/or $CaCl_2$, are reduced at a Mo, W or Fe cathode. The anode is usually carbon.

$$LnCl_6{}^{3-} + 3e^- \longrightarrow Ln + 6Cl^-$$

The "light" rare earths, La, Ce, Pr, Nd and mischmetal, an alloy of these metals, are usually made electrochemically.

The alloy magnets $Nd_2Fe_{14}B$ and $SmCo_5$ are prepared by heating the elements at very high temperatures. These commercial magnets contain considerable Dy to enhance magnetic properties and Pr to reduce corrosion.

1.5.8 *Coordination chemistry*

Only a brief account is given as the topic is covered in Chapter 4. Earlier comprehensive accounts are given in Refs. [75–77]. The rare earths prefer coordination numbers 8–10, whereas the smaller Sc^{3+} prefers 6–8. With application of steric constraints, coordination numbers of 2–7 can be achieved.[78] For example, three coordination is effected with bis(trimethylsilyl)amide, bulky aryloxide and bis(trimethylsilyl)methanide ligands[9,78,79] and the ultimate two-coordination for Ln^{II} with tris(trimethylsilyl)methanide ligands,[80] but in several cases, there are additional agostic CH–Ln interactions. With low coordination number complexes, steric saturation prevents coordination saturation. Steric repulsion (mainly) and electronic repulsion between ligands decide the degree to which coordination saturation can be achieved. Small bite nitrate and nitrosodicyanomethanide ligands give 12 coordinate $[Ln^{III}(\kappa^2\text{-}O_2NO)_6]^{3-}$, $[Ce^{IV}(\kappa^2\text{-}O_2NO)_6]^{2-}$,[76,77] and $[Ln(\eta^2\text{-}ONC(CN)_2)_6]^{3-}$[81,82] complexes. As the f-orbitals are inner orbitals, the f-electron configuration exerts no control over coordination number and stereochemistry. Although steric factors are mainly determinative, electronic *trans*-influence has been observed.[83]

Ln^{3+} ions in aqueous solution, in the presence of weakly coordinating anions such as triflate, are considered to be either nine coordinate tricapped trigonal prismatic $[Ln(H_2O)_9]^{3+}$ ($La^{3+} \rightarrow Eu^{3+}$), or eight coordinate square antiprismatic $[Ln(H_2O)_8]^{3+}$ ($Dy^{3+} \rightarrow Lu^{3+}$), with intermediate elements having a mixture of nine- and eight-coordinate species.[76] In contrast, isolated hydrated triflates, bromates and ethyl sulphates all contain $[Ln(H_2O)_9]^{3+}$ tricapped trigonal prismatic complex ions, despite the lanthanoid contraction, which is associated with a decrease in Ln–O bond lengths of 0.22Å across the series.[76] On the other hand, water-complexed lanthanoid perchlorates contain six-coordinate $[Ln(H_2O)_6]^{3+}$ and are coordination and sterically unsaturated.[76] The sum of the steric coordination numbers of the ligands is (5.04) as against a sterically normal value of 7.56 for the nona-aqua complexes.[84]

Coordinating anions compete with water as ligands. Hydrated lanthanoid nitrates crystallise as $[Ln(\kappa^2\text{-}O_2NO)_3(H_2O)_5].H_2O$ (Ln=La, Ce), $[Ln(\kappa^2\text{-}O_2NO)_3(H_2O)_4].2H_2O$ (Ln=Pr-Dy, Y) or $[Lu(\kappa^2\text{-}O_2NO)_3(H_2O)_3]$, Ln being 11, 10 and 9-coordinate, respectively. Chloro-lanthanoid hydrates vary from dimeric $[Ln(H_2O)_7(\mu\text{-}Cl)_2\ Ln(H_2O)_7]$ complex should have a 4+ charge (Ln=La, Ce are nine-coordinate whereas $[LnCl_2(H_2O)_6]^+$ ions (Ln=Pr-Lu) are eight-coordinate and monomeric.[76]

Because rare-earth complexes are generally labile, isolation of isomers (geometric, coordination and optical) is relatively uncommon. However, some examples are known, e.g. Ref. [85].

1.5.9 *Spectral and magnetic properties*

The rare-earth ions apart from Sc^{3+}, Y^{3+}, La^{3+}, Lu^{3+} and Yb^{2+} show distinctive f←f absorptions in their electronic spectra.[8,75] These may be found from the near UV through the visible region into the near IR region. The bands are sharp but of low intensity (ε 1–10). Because of their distinctive nature they can be used analytically to identify lanthanoid ions and for quantitative analysis.[86] The spectra are generally independent of ligands because the compounds have essentially ionic bonding and the f orbitals are inner orbitals, hence, are comparatively unaffected by the ligand field (splitting ~100 cm^{-1}). For example, there are close similarities between the visible spectra of aqueous Nd^{3+} and of neodymium 2,6-diphenylphenolate in thf,[86,87a] despite the differences in ligands and solvents. In some cases, e.g. $[Yb(Cp)_2(dme)]$,[87b] charge transfer absorptions (ε 200–10000) are observed and swamp any f←f absorptions in the region. In addition, for Eu^{II}, Yb^{II}, Sm^{II}, d←f bands are observed and have significant intensities (ε 100–1000), e.g. Sm^{2+} has a blood red colour.[87c] A surprising feature of the spectra of [K(2,2,2-cryptand)][$Ln^{II}(Cp')_3$] ($Cp' = C_5H_4SiMe_3$; Ln=Pr, Gd, Tb, Ho, Er, Lu, Y) is that they have a common visible absorption at ca 500 nm ($\varepsilon \approx$ 4000 $M^{-1}cm^{-1}$) with a weaker maximum near 750 cm^{-1}. The spectra are attributed to a $4f^n5d^1$ configuration for all Ln^{II} complexes,[7b] with the more intense band showing mainly M-L d→π^* charge transfer and the weaker is d→d and d→π^* in character.[7b] However, Nd^{II} and Dy^{II} show variable character[16] with, for example, the visible-near IR spectrum of [K(18-crown-6)INdCp'_2] ($Cp'=C_5H_2tBu_3$-1,2,4) having similarities to the spectrum of Pm^{3+} (isoelectronic with Nd^{2+}).[14b]

Many lanthanoid ions have luminescent properties, which are employed in numerous applications as sensors and in bioimaging, and as phosphors.[88] The most common lanthanoid ions used are Eu^{3+} (red), Tb^{3+} (green) and Eu^{2+} (blue-red depending on the environment). Holmium salts spectacularly change colour from pale yellow to red with appropriate irradiation. Due to the f-f

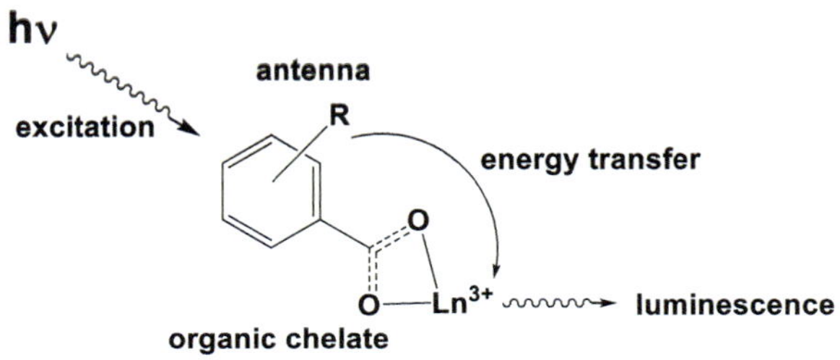

Figure 1.4. Illustration of the antenna effect, in this case an aromatic carboxylate onto the lanthanoid ions aids in luminescence.

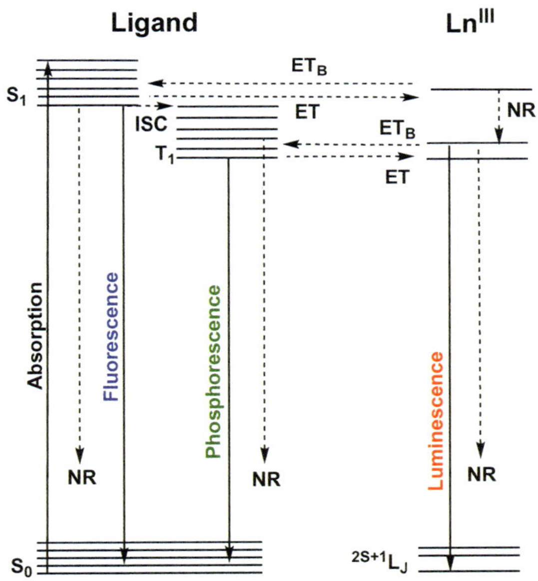

Figure 1.5. Jablonski diagram showing excitation of the organic antenna (an organic chromophore) followed by energy transfer to the lanthanoid *via* either the singlet or triplet state of the antennae. ET = energy transfer, ET_B = back energy transfer, NR = non-radiative decay (usually by energy matched vibrations in the complex or solvent).

transitions being Laporte forbidden, excitation of lanthanoid ions can be indirectly approached by using the so-called antennae effect (see Figure 1.4) where a ligand is capable of absorbing and being excited by UV light. The energy is subsequently transferred onto the lanthanoid ion by inter-system crossing and then relaxing from an excited lanthanoid state. Judicious choice of ligand is required to ensure the singlet excited state of the organic ligand is higher in energy than the excited state orbitals of the lanthanoid ion (Figure 1.5).

The magnetic moments (see Table 1.1) of lanthanoid ions are also independent of ligands for the same reasons as the electronic spectra. Approximate values for the ions are given in Table 1.1. La^{3+} and Lu^{3+} (Yb^{2+}) are diamagnetic. Only for Gd^{3+} and to some extent Nd^{3+} is the value near the spin-only value. In most cases, values agree with values derived from J states from the Russell–Saunders coupling scheme. Of increasing interest is the formation of complexes with single molecule magnetic moment behaviour, particularly for Dy^{III} species.[89]

1.5.10 *Uses in organic synthesis*

This topic has been discussed in detail elsewhere, but some lead references are given.[90] Samarium iodide has become an exceptional one-electron reductant, trivalent lanthanoid triflates are Friedel–Crafts acylation catalysts, and there are a range of other Lewis acid roles. The trivalent halides are used in Luche reductions and facilitate acetal formation. Exciting recent developments include unusual reactivity from reactions of SmI_2 by addition of traces of water.[90c]

1.5.11 *Organometallics, organoamides, organooxides, organophosphides, etc. — Highly reactive compounds*

These topics are considered in detail in Chapters 6 and 8. This area has developed rapidly and spectacularly following the discovery of the $LnCp_3$ species by Wilkinson,[91] the failure of transmetallation by Gilman[92] and pioneering alkoxide studies by Bradley.[93] A range of relevant reviews and books are listed[4,6,76–79] and a range of representative syntheses are described in detail.[94]

The main synthetic route to these classes of compounds is metathesis between lanthanoid trichlorides (often solvated by thf, dme, etc.) for Ln^{III} compounds or lanthanoid diiodides for Ln^{II} compounds and appropriate alkali metal reagents.

$$LnX_n + nML \longrightarrow Ln(L)_n + nMX \quad (n = 2, 3)$$

These reactions normally take place in polar organic solvents, e.g. thf, dme, Et_2O, py, which are usually coordinated to the target product. Because of the usual choice of reaction solvents, more soluble $[LnX_3(solv)_n]$ or $[LnI_2(thf)_2]$ reagents are helpful. Another important route is protolysis where usually a stronger acid displaces a weaker acid.

$$Ln(L)_n + nL'H \longrightarrow Ln(L')_n + nLH\ (n = 2, 3)$$

There are a very few exceptions to the acidity rule, a spectacular example being displacement of C_5Me_3H (pKa ~30) by benzene (pKa 43) and toluene.[95]

$$[Y(C_5Me_5)_3] + PhH \longrightarrow [Y(C_5Me_5)_2Ph] + C_5Me_5H$$

The bis(trimethylsilyl)amides, $Ln^{III}(N(SiMe_3)_2)_3$ and Ln^{II} $(N(SiMe_3)_2)_2$ (and $N(SiHMe_2)_2$ analogues) are commonly used in such reactions,[79a–d] but stabilised benzyls $[(Ln(o\text{-}Me_2NC_6H_4CH_2)_3]$[96] and $Ln(AlMe_4)_3$[79c] are also popular in this regard.

An alternative strategy is to use free Ln metals as reagents,[97] in redox transmetallation reactions (see Ref. [97a] for specific examples), e.g.

$$Ln + 3TlCp \longrightarrow 3Tl + [LnCp_3]$$

redox transmetallation/protolysis,[94]

$$Ln + n/2HgAr_2 + nLH \longrightarrow n/2Hg + nArH + Ln(L)_n\ (n = 2 \text{ or } 3)$$

or direct reaction

$$Ln + nLH \longrightarrow LnLn + n/2H_2$$

either at elevated temperature, usually activated by Hg metal,[98] or with activation of Ln metal by a catalytic amount of iodine in a donor solvent.[99]

Metathesis reactions can have complications through retention of the alkali metal co-product either physically or by complexation. The protolysis syntheses are simple in work-up, but the reagents, e.g. bis(trimethylsilyl)amides, are extremely expensive to purchase and can be challenging to prepare. Metal-based reactions are generally easy to perform and simple to work up, but the use of mercury, thallium and lead reagents can be a deterrent but see Ag and Bi alternatives.[97c]

These compounds have a rich structural and reactivity history including in stabilisation of low coordination numbers (Chapters 4, 6, 8).[78,79]

References

1. IUPAC Compendium of Chemical Terminology, Gold Book Version 2.3.3, 2014-02-24; goldbook.iupac.org/pdf/goldbook.pdf.
2. G. B. Deacon, *Chem. Aust.*, 1991, *58*, 162.
3. R. D. Shannon, *Acta. Cryst.*, 1976, *A32*, 751.
4. F. G. N. Cloke, *Comprehensive Organometallic Chemistry II* (Ed. E. W. Able, F. G. A. Stone, G. Wilkinson), Vol. 4, Ch. 1, p. 1, Pergamon, Oxford, 1995.
5. F. G. N. Cloke, *Chem. Soc. Rev.*, 1993, *22*, 17.
6. M. W. Bochkarev, *Coord. Chem. Rev.*, 2004, *248*, 835.
7. (a) W. J. Evans, *Inorg. Chem.*, 2007, *46*, 3435; (b) M. R. MacDonald, J. E. Bates, J. W. Ziller, F. Furche, W. J. Evans, *J. Am. Chem. Soc.*, 2013, *135*, 9857.
8. F. A. Cotton, G. Wilkinson, *Advanced Inorganic Chemistry*, 5th Edn., Ch. 20, Wiley, New York, 1988.
9. F. T. Edelmann, Scandium, Yttrium and the Lanthanide and Actinide Elements, Excluding their Zero Oxidation State Complexes, in *Comprehensive Organometallic Chemistry II* (Ed. E. W. Abel, F. G. A. Stone, G. Wilkinson), Vol. 4, pp. 11–212, Elsevier Science, Amsterdam, 1995.
10. G. B. Deacon, P. MacKinnon, R. S. Dickson, G. N. Pain, B. O. West, *Appl. Organomet. Chem.*, 1990, *4*, 439.
11. G. B. Deacon, T. Feng, P. C. Junk, B. W. Skelton, A. H. White, A. N. Sobolev, *Aust. J. Chem.*, 1998, *51*, 75, and references therein.
12. P. Sobota, J. Utko, S. Szafert, *Inorg. Chem.*, 1994, *33*, 5203.
13. (a) T. Behrsing, G. B. Deacon, J. Luu, P. C. Junk, B. W. Skelton, A. H. White, *Polyhedron*, 2016, *120*, 69; (b) G. B. Deacon, C. M. Forsyth, M. Hilder, S. G. Leary, C. Bromant, I. Pantenberg, G. Meyer, B. W. Skelton, A. H. White, *Z. Anorg. Allg. Chem.*, 2008, *634*, 91.
14. (a) P. B. Hitchcock, M. F. Lappert, L. Maron, A. V. Protchenko, *Angew. Chem. Int. Ed.*, 2008, *43*, 1488; (b) F. Jaroschik, A. Momin, F. Nief, X. F. Le Goff, G. B. Deacon, P. C. Junk, *Angew. Chem. Int. Ed.*, 2009, *48*, 1117; (c) A. M. Bond, G. B. Deacon, R. H. Newnham, *Organometallics*, 1986, *5*, 2312.
15. P. L. Arnold, F. G. N. Cloke, J. F. Nixon, *Chem. Commun.*, 1998, 797.
16. M. E. Fieser, M. R. MacDonald, B. T. Krull, J. E. Bates, J. W. Ziller, F. Furche, W. J. Evans, *J. Am. Chem. Soc.*, 2015, *137*, 369.
17. C. A. P. Goodwin, N. F. Chilton, G. F. Vettese, E. M. Pineda, I. F. Crowe, J. W. Ziller, R. E. P. Winpenny, W. J. Evans, D. P. Mills, *Inorg. Chem.*, 2016, *55*, 10057.

18. F. Jaroschik, F. Nief, X. F. Le Goff, L. Ricard, *Organometallics*, 2007, *26*, 1123.
19. (a) N. T. Rice, J. A. Popov, D. R. Russo, J. Bacsa, E. R. Batista, P. Yang, J. Telsev, H. S. LaPievve, *J. AM. Chem. Soc.*, 2019, *141*, 13222; (b) C. T. Palumbo, I. Ziwkovic, R. Scopelliti, M. Mazzanti, *J. Am. Chem. Soc.*, 2019, *141*, 9827; (c) A. R. Willaver, C. T. Palumbo, I. Zivlcovic, F. F. Tirani, I. Dovair, L. Marov, M. Mazanti, *J. Am. Chem. Soc.* DOI: 10/1021/jacs.0c01204.
20. (a) T. Pinnavaid, R. Fay, *Inorg. Synth.*, 1970, *12*, 77; (b) T. Behrsing, A. M. Bond, G. B. Deacon, C. M. Forsyth, M. Forsyth, K. J. Kamble, B. W. Skelton, A. H. White, *Inorg. Chem. Acta*, 2003, *352*, 229; (c) R. L. Halbach, G. Nocton, C. H. Booth, C. Marou, R. A. Andersen, *Inorg. Chem.*, 2018, *57*, 7290.
21. G. Meyer, *Chem. Rev.*, 1988, *88*, 93.
22. G. Meyer, *Z. Anorg. Allg. Chem.*, 2007, *633*, 2537.
23. A. Simon, H. Mattausch, N. Holzer, *Angew. Chem. Int. Ed. Engl.*, 1976, *15*, 624.
24. N. N. Greenwood, H. G. Earnshaw, *Chemistry of the Elements*, 2nd Edn., Elsevier, Amsterdam, 1997, Appendix 4.
25. Rare Earths — AIMR 2013 — Australia Mines Atlas, http://www.australianminesatlas.gov.au/aimr/commodity/rare_earths.html.
26. The Statistics Portal: Rare Earth Reserves Worldwide as of 2014, http://www.statista.com/statistics/277268/rare-earth-reserves-by-country.
27. Rare Earth Prices have Turned, June 17, 2014, http://www.mining.com/rare-earth-prices-at-turning-point-33322/.
28. M. P. Mills, "Tech's Mineral Infrastructure — Time to Emulate China's Rare Earth Policies." *Forbes*, January 1, 2011, https://www.forbes.com/sites/markpmills/2011/01/01/techs-mineral-infrastructure-time-to-emulate-chinas-rare-earth-policies/#6d56db983b47.
29. https://www.businessinsider.com.au/molycorp-decline-in-2014-2014-9.
30. *Financial Times*, March 28, 2014.
31. https://www.bloomberg.com/news/articles/2015-06-25/molycorp-files-for-bankruptcy-proposes-debt-restructuring-plan.
32. Siemens Global, July 7, 2011.
33. http://www.smh.com.au/business/lynas-warns-of-uncertain-survival-prospects-20150313-1434by.html.
34. Lynas, shareholder announcement, August 17, 2015.
35. https://thewest.com.au/news/wa/arafura-abandons-whyalla-plan-ng-ya-352500.
36. http://investorintel.com/technology-metals-intel/arafura-resources.

37. https://www.abc.net.au/news/rural/2014-09-03/rare-earth-mine-on-track-for-central-australia/5715100.
38. http://www.alkane.com.au/.
39. http://www.astronlimited.com/.
40. http://hastingstechmetals.com/.
41. https://roskill.com/market-report/rare-earths/.
42. https://www.forbes.com/sites/greatspeculations/2019/11/06/australia-may-be-the-saving-grace-for-the-rare-earth-metals-market/#45edd09941cd.
43. https://www.australianmining.com.au/news/lynas-mounds-government-backing-for-plant-construction/.
44. https://www.wsj.com/amp/articles/pentagon-invests-in-strategic-metals-mine-seeking-to-blunt-chinese-dominance-11587924001.
45. (a) Z. Hu, H. Richter, G. Sparovek, E. Schnug, *J. Plant Nutrition*, 2004, *27*, 183; (b) K. Redling, *Doctoral Thesis*, LMU, 2006; (c) M. I. He, U. Wehr, W. A. Rambeck, *J. An. Physio. & An. Nut.*, 2010, *94*, 86; (d) S. Gruhne, *Zehentmeyer AG*, personal communication. W. A. Rambeck, LMU, personal communication.
46. M. Rogosnitzky, S. Branch, *Biometals*, 2016, *29*, 365.
47. (a) A. Fischback, R. Anwander, *Adv. Polym. Sci.*, 2006, *204*, 155; (b) M. Zimmermann, K.W. Törnroos, R. Anwander, *Angew. Chem. Int. Ed.*, 2008, *47*, 775; (c) L. Friebe, O. Nuyken, W. Obrecht, *Adv. Polym. Sci.*, 2006, *204*, 1.
48. (a) References cited in the US Geological Survey 2008 Minerals Yearbook, October 2010, see: http://minerals.usgs.gov/minerals/pubs/commodity/rare_earths/myb1-2008-raree.pdf; (b) K. Binnemans, P. T. Jones, B. Blanpain, T. Van Gerven, Y. Yang, Allan Walton, M. Buchert, *J. Clean. Prod.*, 2013, *51*, 1.
49. D. Kingsnorth, 2008, Rare Earths Supply — Alternatives to China, Littleton, CO, Society for Mining, Metallurgy and Exploration Inc. meeting and exhibit, Salt Lake City, UT, February 26, 2009, Presentation, 25 pages.
50. K. Gschneidner, Jr., V. Pecharsky, 2008, Magnetic Refrigeration/Heat Engines, Littleton, CO, Society for Mining, Metallurgy and Exploration Inc. meeting and exhibit, Salt Lake City, UT, February 26, 2009, Presentation, 20 pages.
51. M. Forsyth, M. Seter, B. Hinton, G. B. Deacon, P. Junk, *Aust. J. Chem.*, 2011, *64*, 812.
52. A. E. Somers, G. B. Deacon, B. R. W. Hinton, D. R. MacFarlane, P. C. Junk, M. Forsyth, *J. Ind. Inst. Sci.*, 2016, *96*, 285.
53. (a) G. B. Deacon, C. M. Forsyth, M. Forsyth, P. C. Junk, W. W. Lee, *Chem. Aust.*, 2008, *75*, 18; (b) *Rare Earth-based Corrosion Inhibitors*

(Ed. M. Forsyth, B. Hinton), Woodhead Publishing, Cambridge, 2014.
54. (a) D. M. L. Goodgame, A. M. Joy, *J. Inorg. Biochem.*, 1986, *26*, 219; (b) M. Costa, *Crit. Rev. Toxicol.*, 1997, *27*, 431; (c) R. Codd, C. T. Dillon, A. Levina, P. A. Lay, *Coord. Chem. Rev.*, 2001, *216–217*, 537.
55. (a) B. P. F. Goldie, J. J. McCarroll, *Inhibiting Corrosion Using Cations*, Eur. Pat. App. EP 89810, 1983; (b) B. P. F. Goldie, J. J. McCarroll, *Inhibiting Metal Corrosion in Aqueous Systems*, Eur. Pat. Appl. EU 136860, 1985.
56. (a) V. S. Muralidharan, R. Sethuraman, S. Krishanamoory, *Corros. Sci. Eng.*, 1988, *4*, 705; (b) M. Frey, S. G. Harris, J. M. Holmes, D. A. Nation, S. Parsons, P. A. Tasker, S. J Teat, R. E. P. Winpenny, *Angew. Chem. Int. Ed.*, 1998, *37*, 3245; (c) M. Frey, S. G. Harris, J. M. Holmes, D. A. Nation, S. Parsons, P. A. Tasker, R. E. P. Winpenny, *Chem. Eur. J.*, 2000, *6*, 1407.
57. M. Forsyth, K. Wilson, T. Behrsing, C. M. Forsyth, G. B. Deacon, A. Phanasgoanker, *Corrosion*, 2002, *58*, 953.
58. M. Forsyth, C. M. Forsyth, K. Wilson, T. Behrsing, G. B. Deacon, *Corros. Sci.*, 2002, *44*, 2651.
59. F. Blin, S. G. Leary, G. B. Deacon, P. C. Junk, M. Forsyth, *Corros. Sci.*, 2006, *48*, 404.
60. (a) A. E. Somers, G. Talbi, C. N. de Bruin-Dickason, S. Hanf, M. Forsyth, G. B. Deacon, P. C. Junk, B. Hinton, Corrosion and Prevention 2016, AC Conference, Auckland, NZ, November 13–16, 2016 (The Australian Corrosion Association: Kerrimuir, Vic,); (b) A. E. Somers, G. Talbi, C. N. de Bruin-Dickason, M. Forsyth, G. B. Deacon, B. Hinton, in NACE Corrosion 2016, Section 3, Vancouver, Canada, March 6–10, 2016 (NACE International: Houston, TX).
61. Z. Zhu, Y. Pranolo, C. Y. Cheng, *Min. Eng.*, 2015, *77*, 185.
62. F. Xie, T.A. Zhang, D. Dreisinger, F. Doyle., *Min. Eng.*, 2014, *56*, 10.
63. (a) D. Dupont, K. Binnemans, *Green Chem.*, 2015, *17*, 856; (b) D. Dupont, K. Binnemans, *Green Chem.*, 2015, *17*, 2150.
64. J. A. Bogart, C. A. Lippincott, P. J. Carroll, E. J. Schelter, *Angew. Chem. Int. Ed.*, 2015, *54*, 8222.
65. T. Lorenz, P. Fröhlich, M. Bertau, *Nachr. Chem.*, 2015, *63*, 984.
66. G. Meyer, Binary Lanthanide(III) Halides, *Synthesis of Lanthanide and Actinide Compounds* (Ed. G. Meyer, L. R. Morss), Kluwer, Dordrecht, 1991.
67. G. B. Deacon, T. Feng, P. C. Junk, B. W. Skelton, A. H. White, A. N. Sobolev, *Aust. J. Chem.*, 1998, *51*, 15.
68. G. B. Deacon, T. Feng, P. C. Junk, G. Meyer, N. M. Scott, B. W. Skelton, A. H. White, *Aust. J. Chem.*, 2000, *53*, 853.

69. W. J. Evans, T. S. Gummersheimer, J. W. Ziller, *J. Am. Chem. Soc.*, 1995, *117*, 8999.
70. J. L. Namy, P. Giraud, H. B. Kagan, P. E. Caro, *Lanthanides and Actinides, Synthetic Methods of Organometallic and Inorganic Chemistry* (Hersmann/Brauer), in P. T. Edelmun (Ed. W.A. Herrmann, G. Thieme), Vol. 6, p. 26, Stuttgart, 1997.
71. W. J. Evans, N. T. Allen, P. S. Workman, J. C. Meyer, *Inorg. Chem.*, 2003, *42*, 3097.
72. (a) R. Wang, Z. Zheng, *Rare Earth Coordination Chemistry: Fundamentals and Applications* (Ed. C. Huang), Ch. 3, John Wiley and Sons, 2010; (b) S. R. Batten, S. M. Neville, D. R. Turner, *Coordination Polymers — Design, Analysis and Application*, Ch. 6 Royal Society of Chemistry, Cambridge, 2009; (c) A. Ouchi, Y. Suzuki, Y. Ohki, Y. Koizumi, *Coord. Chem. Rev.*, 1988, *92*, 29; (d) G. B. Deacon, S. Hein, P. C. Junk, T. Jüstel, W. Lee, D. R. Turner, *CrystEngComm.*, 2007, *9*, 1110; (e) G. Meyer, *J. Alloys Compd.*, 2000, *113*, 300.
73. (a) S. Mroczkowski, J. Eckert, H. Meisner, J. C. Duran, *J. Cryst. Growth*, 1970, *7*, 333; (b) D. F. Mullika, W. O. Milligan, *J. Inorg. Nucl. Chem.*, 1980, *43*, 223.
74. (a) Z. Zheng, *Chem. Commun.*, 2001, 2521; (b) D. T. Thielemann, I. Fernández, P. Roesky, *Dalton Trans.*, 2010, *39*, 6661, and references therein.
75. (a) S. Cotton, *Lanthanide and Actinide Chemistry*, Wiley, UK, 2006; (b) H. C. Aspinall, *Element Chemistry*, Oxford University Press, Oxford, 2020.
76. S. Cotton, Scandium, Yttrium and Lanthanides, in *Comprehensive Coordination Chemistry II*, Vol. 3, Ch. 32, Pergamon, Oxford, 2004.
77. F. A. Hart, Scandium, Yttrium and the Lanthanides, in *Comprehensive Coordination Chemistry I*, Vol. 3, Ch. 39, Pergamon, Oxford, 1987.
78. F. Ortu, D. P. Mills, Chapter 306 — Low Coordinate Rare Earth and Actinide Complexes, in *Handbook on the Physics and Chemistry of Rare Earths* (Eds. J.-C. G. Bünzli, V. K. Pecharsky), Elsevier B. V., Amsterdam, 2019, *55*, 1.
79. (a) *Rare Earth Coordination Chemistry, Fundamentals and Applications*, C. Huang Ed., Chichester, 2010; (b) *Comp. Organomet. Chem. III*, Vol. 4, Ch. 4.01, M. Bockmann (Ed.), Elsevier, 2007; (c) M. Zimmermann, R. Anwander, *Chem. Rev.*, 2010, *110*, 6194; (d) R. Kempe, *Angew. Chem. Int Ed.*, 2000, *39*, 469; (e) M. F. Lappert, A. Protchenko, P. P. Power, A. Seeber, *Metal Amide Chemistry*, Wiley, UK, 2009; (f) D. C. Bradley, R. C. Mehrotra, I. P. Rothwell, A. Singh, *Alkoxo and Aryloxo Derivatives of Metals*, Academic Press, London, 2001; (g) T. J. Boyle, L. A. M. Otley, *Chem. Rev.*, 2008, *108*, 1896.

80. C. Eaborn, P. B. Hitchcock, K. Izod, J. D. Smith, *J. Am. Chem. Soc.*, 1994, 116, 12071.
81. A. S. R. Chesman, D. R. Turner, E. Izgorodina, S. R. Batten, G. B. Deacon, *Dalton Trans.*, 2007, 1371.
82. A. S. R. Chesman, D. R. Turner, G. B. Deacon, S. R. Batten, *Eur. J. Inorg. Chem.*, 2010, 2798.
83. K. Krogh-Jespersen, M. D. Raonanelli, J. H. McIman, T. J. Einge, J. G. Brennan, *Inorg. Chem.*, 2010, *49*, 552.
84. J. Marçalo, A. Pires de Matos, *Polyhedron*, 1989, *8*, 2431.
85. (a) L. N. Punta, K. A. Lyssenko, I. S. Pekareva, *Mat. Phys.*, 2010, *108*, 557; (b) M. Hokansson, M. Vestergren, B. Gustafsson, G. Hilmersson, *Angew. Chem. Int. Ed.*, 1999, *38*, 2119; (c) J. M. Bakker, G. B. Deacon, C. M. Forsyth, P. C. Junk, M. Wiecko, *Eur. J. Inorg. Chem.*, 2010, *18*, 2813; (d) H. Bauer, J. Blanc, D. L. Ross, *J. Am. Chem. Soc.*, 1964, *86*, 5125; (e) R. C. Holz, L. C. Thompson, *Inorg. Chem.*, 1993, *32*, 5251; (f) M. Woods, Z. Kovacs, S. Zhang, D. Sherry, *Angew. Chem. Int. Ed.*, 2003, *42*, 5889; (g) H. G. Brittain, J. F. Desreux, *Inorg. Chem.*, 1984, *23*, 4459.
86. D. C. Stewart, D. Kato, *Anal. Chem.*, 1958, *30*, 164.
87. (a) G. B. Deacon, S. Nickel, P. MacKinnon, E. R. T. Tiekink, *Aust. J. Chem.*, 1990, *43*, 1245; (b) Herrmann-Brauer, *Synthetic Methods of Inorganic and Organimetallic Chemistry*, Ed. W. A. Hermann (Ed. F. T. Edelmann), Vol. 6, p. 55, Thieine, Verlag, Stuttgart, 1997; (c) F. D. S. Butement, *Trans. Faraday Soc.*, 1948, *44*, 617.
88. (a) J.-C. G. Bünzli, C. Piguet, *Chem. Soc. Rev.*, 2005, *34*, 1048; (b) S. V. Eliseeva, J.-C. G. Bünzli, *Chem. Soc. Rev.*, 2010, *39*, 189; (c) J.-C. G. Bünzli, *Chem. Rev.*, 2010, *110*, 2729.
89. D. N. Woodruff, R. E. P. Winpenny, R. A. Layfield, *Chem. Rev.*, 2013, *113*, 5110.
90. (a) K. Gopalaiah, H. B. Kagan, *Chem. Rec.*, 2013, *13*, 187; (b) K. Gopalaiah, H. B. Kagan, *New J. Chem.*, 2008, *32*, 607, and references therein; (c) M. Szostak, N. J. Fazakerley, D. Parmar, D. J. Procter, *Chem. Rev.*, 2014, *114*, 5959.
91. J. M. Birmingham, G. W. Wilkinson, *J. Am. Chem. Soc.*, 1956, *78*, 42.
92. H. Gilman, R. G. Jones, *J. Organic Chem.*, 1945, *10*, 505.
93. D. C. Bradley, R. C. Mehrotra, D. P. Gaur, *Metal Alkoxides*, Academic Press, London, 1978.
94. (a) F. T. Edelmann, Lanthanides and Actinides, in *Synthetic Methods in Organometallic and Inorganic Chemistry*, Vol. 6, pp. 26–143, Herrmann-Brauer, Stuttgart, 1997; (b) M. F. Lappert, A. Singh, R. G. Smith, *Inorg. Synth.*, 1990, *27*, 164.
95. W. J. Evans, B. L. Davis, T. M. Champagne, J. W. Ziller, *Proc. Nat. Acad. Sci.*, 2006, *34*, 12678.

96. S. Harder, C. Ruspic, N. N. Bhriain, F. Berkermann, M. Schürmann, *Z. Naturforsch.*, 2008, *63b*, 267.
97. (a) G. B. Deacon, C. M. Forsyth, *Inorganic Chemistry Highlights* (Ed. G. Meyer, D. Naumann, L. Wesemann), Ch. 7, Wiley-VCH, Weinheim, 2002; (b) G. B. Deacon, M. E. Hossain, P. C. Junk, M. Salehisaki, *Coord. Chem. Rev.*, 2017, *340*, 247; (c) Z. Guo, R. Huo, Y. Q. Tan, V. Blair, G. B. Deacon, P. C. Junk, *Coord. Chem. Rev.*, 2020, 213232.
98. G. B. Deacon, C. M. Forsyth, S. Nickel, *J. Organomet. Chem.*, 2002, *647*, 50.
99. S. Hamidi, G. B. Deacon, P. C. Junk, P. Neumann, *Dalton Trans.*, 2012, *41*, 3541.

https://doi.org/10.1142/9781800610163_0002

Chapter 2

The Chemistry of the Actinides

Robert J. Baker
School of Chemistry, University of Dublin Trinity College,
Dublin 2, Ireland
bakerrj@tcd.ie

2.1 Discovery and Synthesis of the Actinide Elements

2.1.1 *Position in the periodic table*

The actinide (common usage) or actinoid (IUPAC) elements comprise the elements from actinium (Ac) to lawrencium (Lr). All are radioactive, and for many, handling requires specialist facilities to protect workers from the radiation and incorporation of these into the body. For example, the alpha emissions from ^{239}Pu are about 10^8 per mg per minute, which can be readily shielded. However, if this is incorporated into the body via inhalation or ingestion, significant long-term health effects are to be expected. The use of gloveboxes and other methods of completely separating the experiment from the person are the only way trans-uranium chemistry can be conducted on more than tracer scales. Therefore, it is not a surprise that broadly speaking the chemistry of the actinides, especially the trans-uranium elements, is much less well developed than that of the lanthanides.

Unlike the lanthanoid elements, which are characterised by a similarity in properties, the actinide elements are perhaps best characterised by their trends in reactivity. Moreover, only actinium, thorium, protactinium and uranium occur in nature, while the remainder have been synthesised. Some of the later elements were first made on atom-at-a-time scales, unquestionably a phenomenal feat in synthesis and characterisation.

The actinide series was first postulated by Seaborg in 1944[1] to account for the fact that the attempts to synthesise Am and Cm failed. This suggested they did not fit the periodic table as written at that time. In this older version, uranium was considered as a group 5 element, Np under Re and Pu under Os. The actinide concept moved Ac and all the heavier elements into a period of its own: the actinides. When the experiments were designed according to this concept, Am and Cm were successfully separated and characterised, leading to the recognition of the actinide series as an analogue to the lanthanide series and enabling the chemistry of the later actinides to be predicted and rationalised.

2.1.2 *Discovery and synthesis*[2]

2.1.2.1 *Actinium* ($_{89}Ac$)

The discovery of actinium has been credited to two people who almost simultaneously reported a new element from residues of pitchblende processing. In 1899, André-Louis Debierne, working in the laboratory of Marie and Pierre Curie, discovered a new radioactive substance whose chemistry followed titanium, but after six months the activity had decreased significantly. He named this substance actinium. However, in light of current knowledge it is doubtful, though not certain, that the initial preparation contained any actinium. These results were challenged in 1904, when Friedrich Giesel isolated a substance he called emanium that had the chemical properties expected of actinium. Debierne, supported by Rutherford and the Curies, argued that this was the same substance he isolated and half-life determinations reinforced his claim for the discovery.[3] Nevertheless, the discovery of actinium allowed the decay chain of ^{235}U to be understood and completed.

2.1.2.2 *Thorium* ($_{90}Th$)

In 1815, the Swedish chemist Jacob Berzelius analysed a rare mineral from a copper mine in Falun, Sweden. Assuming it contained a new element he called it thorium, after the Norse god of thunder. Subsequent analysis showed this to be just yttrium orthophosphate.

A further mineral was supplied to Berzelius in 1825 from Norway, and this time it did contain a new element that was given the same name. The half-life of ^{232}Th is over 14.05 billion years, so is a primordial nuclide. The amount of heat generated by radioactive decay of ^{232}Th, along with uranium isotopes and ^{40}K is thought to account for 80% of the earth's internal heat.[4] Indeed it is noteworthy that the heat generated by the earth was central to Lord Kelvin's arguments in 1862 where he limited the age of the Earth, provided no new source [of heat] was discovered; in 1904, Rutherford postulated that the new source of heat was radioactive decay (albeit of radium).

2.1.2.3 *Protactinium* ($_{91}$*Pa*)

In constructing his periodic table, Mendeleev left gaps where he predicted as yet undiscovered elements. One of these gaps was between Th and U and he anticipated that the highest oxide would be X_2O_5, akin to niobium and tantalum. Sir William Crookes noticed a new radioactive substance in uranium ores in 1900 and, although he could not isolate it, named it uranium-X. This turned out to be two unknown compounds UX_1 (^{234}Th) and UX_2 (^{234m}Pa). The first people to isolate ^{234m}Pa were Fajans and Göhring in 1913, and they named it Brevium due to the short half-life of 1.15 minutes. ^{231}Pa, with a half-life of 32,000 years, was discovered independently by Otto Hahn and Lise Meitner in Germany and Frederick Soddy and John Cranston in Great Britain. The name brevium was therefore no longer accurate and Meitner proposed protoactinium (as this is the progenitor of actinium), and later shortened to protactinium. ^{231}Pa was first isolated in 1927 as 2 mg of Pa_2O_5, and elemental Pa prepared in 1934. In the same year, Graue and Kading treated 5.5 tonnes of pitchblende residues and isolated 0.5 g of pure K_2PaF_7.

While Pa is one of the most rare elements (crustal abundance = 1.4×10^{-6} ppm), it has found uses as a radiometric dating of sediments which are up to 175,000 years old, notably to model the movements of North Atlantic water during glacial melting during the last ice age.[5]

2.1.2.4 *Uranium* ($_{92}$*U*)

The uses of uranium predate its discovery. It was used as a glaze in Roman villas in at least 79 AD,[6] and mined for this purpose in the Middle Ages, most notably at the Hapsburg Silver mines in Joachimsthal, Bohemia (now Jáchymov in the Czech Republic). The discovery of uranium in 1789 is credited to Martin Klaproth who dissolved pitchblende in nitric acid and precipitated a yellow solid by the addition of sodium hydroxide. He then treated this with charcoal and the black powder that was isolated was assumed to be the metal. He named this after the planet Uranus that had been discovered a few years earlier. This black powder was likely uranium oxide, and the metal was finally isolated in 1841. Becquerel discovered radioactivity in 1896 as uranium containing salts were found to fog a photographic plate. Marie and Pierre Curie discovered radium in pitchblende and this created a market for uranium ores that far outstripped colorants for glass. The discovery of nuclear fission propelled uranium to a central position in science in the 1940s onward.

2.1.2.5 *Neptunium* ($_{93}$*Np*)

Neptunium is the first man-made element and is found in nature only in very minute amounts. The discovery of neptunium is interlinked with the discovery of nuclear fission. Frédéric Joliot and Irène Joliot-Curie showed in 1934 that bombarding elements with alpha particles could induce radioactivity, and this won them the Nobel Prize in Chemistry in 1935.[7] Using the newly discovered neutron, Enrico Fermi in Rome bombarded naturally occurring uranium with slow neutrons (made from a radon/beryllium generator) and claimed the synthesis of new trans-uranic elements as their products were heavier than lead. This finding was questioned by Ida Noddack who suggested that all elements past lead should also have been discounted. Noddack suggested that nuclear fission was occurring, although the scientific community did not take this seriously. In 1938, Fermi received the Nobel Prize in Physics for his discovery of "the existence of new radioactive elements produced by neutron irradiation, and for his related discovery of nuclear reactions brought about by slow neutrons".[8] The Joliot-Curies also discovered lanthanum in this mixture, but incorrectly suggested that element 93 showed properties similar to lanthanum. Otto Hahn, working in Berlin, positively identified barium as one of these products and Lise Meitner, then in Stockholm, correctly interpreted that upon bombardment with neutrons the uranium atom splits into roughly half, which resulted in the 1944 Nobel Prize in Chemistry awarded to Hahn.[9] So, despite the Nobel Prize being awarded for the "discovery" of Np in 1935, the first *synthesis* was conducted in 1939 in the University of California, Berkley by McMillan and he and Seaborg received the Nobel Prize in Chemistry in 1951 "for their discoveries in the *chemistry* of the trans-uranium elements".[10] In these experiments, thin layers of uranium oxide were bombarded with neutrons from a cyclotron and two radioactive decays were measured. One of these corresponded to the known ^{239}U and one with a half-life of 2.3 days that was unknown. Initial chemical separation of this was thwarted by the fact that at this time element 93 was predicted to be in group 7 and have similar properties to rhenium. However, in 1940 McMillian and Abelson's experiments showed that this species with a 2.3 day half-life was similar to uranium in its chemistry, and not a fission product. The name neptunium was suggested, following the ordering of the planets. A much longer lived isotope, ^{237}Np ($t_{1/2} = 2.1 \times 10^6$ years), was identified in 1942 and the first weighable quantity (10 μg) isolated from uranium irradiated in a nuclear reactor in 1944. This was formed by the $^{238}U(n,2n)^{237}U$ reaction followed by a beta decay of ^{237}U to form ^{237}Np. In 1945, the first sample of metallic Np was isolated.

$$^{238}_{92}U + {}^{1}n \longrightarrow {}^{239}_{92}U \xrightarrow[23\,min]{\beta^-} {}^{239}_{93}Np$$

Neptunium has been found in nature, in some uranium ores. It is thought that it is formed by ^{238}U absorbing neutrons arising from ^{235}U fission.[11] The typical ratio of ^{237}Np:U in uranium bearing ores are of the order 10^{-12}.

2.1.2.6 *Plutonium* ($_{94}Pu$)

Following the discovery of Np, Seaborg postulated that the β-decay of ^{239}Np should form element 94. In order to synthesise enough to characterise the element, uranium was bombarded with deuterons to form trace quantities of ^{238}Pu in 1940. The most important isotope, ^{239}Pu, was prepared by neutron bombardment of ^{238}U and found to be fissionable in 1941. The first macroscopic amount of ^{239}Pu, 1 μg, was prepared in 1942 by neutron bombardment of 90 kg of uranyl nitrate (Figure 2.1). Heavier isotopes, ^{240}Pu to ^{244}Pu, can be formed by successive neutron capture reactions, and ^{244}Pu, ^{245}Pu and ^{246}Pu first

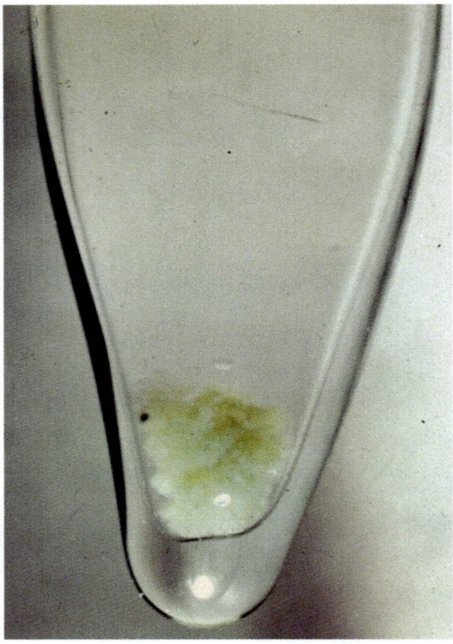

Figure 2.1. Twenty micrograms of pure plutonium hydroxide in capillary tube, September 1942. Copyright (2010) The Regents of the University of California, through the Lawrence Berkeley National Laboratory.

identified in debris from a thermonuclear explosion. Separation of Pu and other actinides (e.g. Np) formed in these neutron bombardment experiments is now typically by ion exchange. The name plutonium was proposed to follow the planetary series established for uranium and neptunium. One of the features of ^{238}Pu is the fact that its decay gives significant amount of heat (its specific power is $0.57\,\mathrm{W\,g^{-1}}$), and the oxide, PuO_2, has been used for power generation in space applications.

Plutonium is found in microscopic amounts in natural uranium ores; the nature of the ore dictates the amount but is of the order of micrograms for Pu per 100 metric ton of uranium ore. Given the half-life of ^{239}Pu ($t_{1/2} = 24,100$ years) it is not primordial in origin but rather formed via neutron capture reactions of ^{238}U to form ^{239}Np and then ^{239}Pu via beta decay. ^{244}Pu has also been noted in uranium ores, and this has a long half-life (8.00×10^7 years) that has been shown to be the remnants of stellar debris that coalesced during the formation of the solar system.

Since these early efforts to prepare plutonium, it has now been synthesised on much larger scales as part of defence activities and generation of electricity by nuclear power. The initial scale up of synthesis and isolation from micrograms to an industrial scale is an impressive feat of chemistry and chemical engineering.

$$^{238}_{92}\mathrm{U} + ^{2}_{1}\mathrm{H} \longrightarrow ^{238}_{93}\mathrm{Np} \xrightarrow[\text{2,1 days}]{\beta^-} ^{238}_{94}\mathrm{U}$$

$$^{238}_{92}\mathrm{U} + ^{1}_{0}\mathrm{n} \longrightarrow ^{239}_{92}\mathrm{U} \xrightarrow[\text{23.5 min}]{\beta^-} ^{239}_{93}\mathrm{Np} \xrightarrow[\text{2.36 days}]{\beta^-} ^{239}_{94}\mathrm{Pu}$$

2.1.2.7 *Americium* ($_{95}Am$)

Americium was discovered in 1944–1945, just after the discovery of Cm. For both these elements the recognition of the actinide series was important as early attempts assumed that these would behave somewhat like Pu with a readily accessible +4 oxidation state. The synthesis involved multiple neutron capture reactions of ^{239}Pu to form ^{241}Pu ($t_{1/2}$ = 14.3 years) which then decays via a beta emission to form ^{241}Am. However, ^{241}Am can also undergo a neutron capture reaction to form ^{242}Am which decays via a beta emission ($t_{1/2}$ = 16.02 h) to form ^{242}Cm.

$$^{239}\mathrm{Pu} \xrightarrow{+2\,^{1}\mathrm{n}} ^{241}\mathrm{Pu} \xrightarrow{\beta^-} ^{241}\mathrm{Am} \xrightarrow{\text{N.C.}} ^{242}\mathrm{Am} \xrightarrow{\beta^-} ^{242}\mathrm{Cm}$$

Separation of Am from Cm and the other fission products formed proved to be a challenge and ion-exchange was used over the purely chemical separations initially used for Np and Pu. The first compound of ^{241}Am, as the hydroxide, was isolated in late 1945 (Figure 2.2). It was not until 1951 that the first sample of Am metal was prepared. To do this, AmO_2 was first fluorinated using gaseous HF then reduced with barium metal at 1100°C.[12] The name americium was proposed by analogy with its lanthanide homologue europium.

2.1.2.8 *Curium* ($_{96}Cm$)

Curium was first prepared in the summer of 1944 by bombardment of ^{239}Pu with helium ions at the Berkeley 60 inch cyclotron to form ^{242}Cm (Figure 2.3). This isotope was firstly characterised by its alpha

Figure 2.2. The brown material in the glass tube contains the first sample of americium, produced in 1944. Copyright (2010) The Regents of the University of California, through the Lawrence Berkeley National Laboratory.

decay to form the known ^{238}Pu. ^{242}Cm can also be prepared by neutron bombardment of ^{239}Pu (i.e. a beta decay of ^{242}Am, as aforementioned). In 1947, the first pure compound was isolated as the hydroxide, although the first weighable amount was an impure oxide (40 μg) isolated in 1952. In 1960, the US Atomic Energy Commission initiated a programme of irradiating 20.5 kilograms of ^{239}Pu in a nuclear reactor in Savannah River to form 630g of ^{243}Am and ^{244}Cm combined. The isotope ^{248}Cm can be prepared in >99% isotopic purity by the decay of ^{252}Cf (see Section 2.1.2.10). The name curium was proposed in analogue with its lanthanide homologue; gadolinium was named to honour the chemist Gadolin.

$$^{239}\text{Pu} + {}^{4}\text{He} \longrightarrow {}^{242}\text{Cm} + {}^{1}\text{n}$$

2.1.2.9 *Berkelium* ($_{97}$*Bk*)

The synthesis of berkelium was achieved in 1949 by bombardment of 7 mg ^{241}Am with helium ions. Using the actinide concept it was predicted that Bk would elute off an ion exchange column before Cm

Figure 2.3. A sample of a Cm containing compound in a glass tube. Copyright (2010) The Regents of the University of California, through the Lawrence Berkeley National Laboratory.

and Am, and indeed this occurred. The name was proposed as the laboratory in University of California, Berkeley, had been the first to synthesise this element; the analogous lanthanide, terbium, was derived from its place of discovery, Ytterby. Macroscopic amounts (0.6 μg) of ^{249}Bk were isolated in 1958 via a long-term irradiation of 8 g of ^{239}Pu, which began in 1952. The first Bk complex that was structurally characterised was BkO_2 prepared in 1962, where 2 ng were used for X-ray diffraction. Perhaps most impressive has been the synthesis of 22 mg of ^{249}Bk in 2009, which was used as a target for the synthesis of $_{117}$Ts. The berkelium sample (Figure 2.4) was prepared by a 250 day neutron bombardment of an Am and Cm target followed by effective purification to yield the sample contaminated with only 1.7 ng of ^{252}Cf, which arises from an additional product of the neutron bombardment.[13]

$$^{241}\mathrm{Am} + {}^{4}\mathrm{He} \longrightarrow {}^{243}\mathrm{Bk} + 2\,{}^{1}\mathrm{n}$$

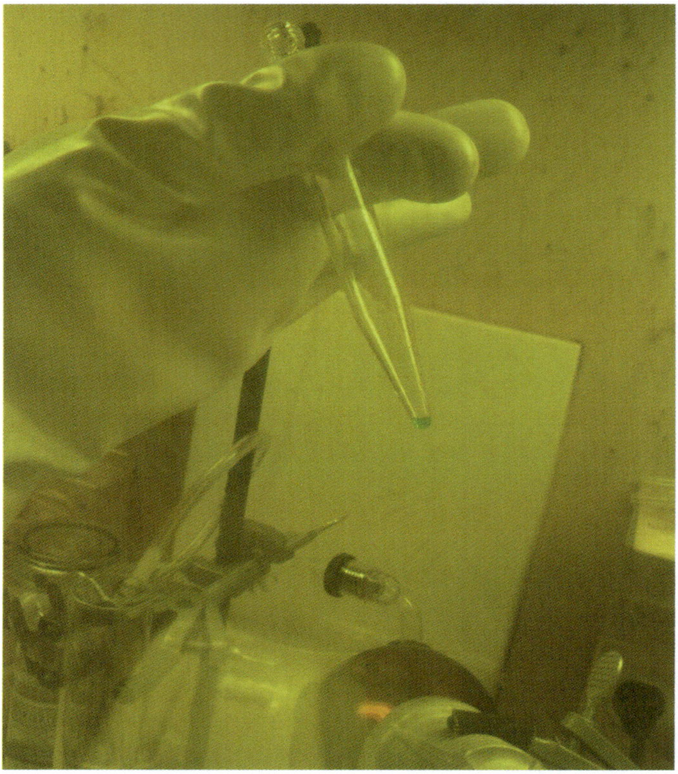

Figure 2.4. The blue-green solution at the tip of the vial is 22.2 mg of ^{249}Bk used as a target for the synthesis of ^{293}Ts and ^{294}Ts. Image courtesy of U.S. Department of Energy, Oak Ridge National Laboratory.

2.1.2.10 *Californium* (${}_{98}$*Cf*)

The synthesis of californium was achieved in 1950, by bombardment of micrograms of ^{242}Cm with helium ions accelerated in a cyclotron to form ^{245}Cf. Impressively, the amount of californium produced in this first experiment was $\sim$ 5000 atoms. Ion exchange techniques allowed the isolation of this and a significant amount of fundamental data has been obtained using tracer studies with these small amounts. The first macroscopic amount of californium was isolated in 1963, by long-term irradiation of ^{239}Pu, as 0.3 μg of ^{249}CfOCl. The amount of californium produced for further research is in the microgram range and is essentially the last element that chemists can study (albeit requiring laboratories equipped with the specialised facilities that can handle the intense radiation and neutron flux that is associated with this element).

$$^{242}Cm + {}^{4}He \longrightarrow {}^{245}Cf + {}^{1}n$$

2.1.2.11 *Einsteinium ($_{99}Es$) and fermium ($_{100}Fm$)*

Elements 99 and 100 were first produced, not by meticulous planning, but in the debris of the first thermonuclear explosion on November 1, 1952. During this explosion, a few *moles* of neutrons were produced on a nanosecond timescale which led to unprecedented neutron capture reactions of ^{238}U. Some of the isotopes captured by filter papers flown on aeroplanes close to the detonation site were the unknown ^{244}Pu and ^{246}Pu, which led to the realisation that heavier elements could be present. Ion exchange chromatography of coral debris allowed the positive identification of einsteinium, albeit with just 200 atoms. A more rational synthesis for both elements, via the intense neutron bombardment of plutonium for 2–3 years, or by bombarding uranium atoms with ^{16}O ions were subsequently developed. The first macroscopic einsteinium compound was prepared in 1961 where about 10 ng of ^{253}Es ($t_{1/2}$ = 20.47 days) was isolated. The names were given to honour these scientists. Coincidentally, Fm is the last element that can be synthesised by neutron capture reactions; the first experiments using this technique were conducted by Fermi. While einsteinium can be prepared on scales of up to 2 mg by neutron irradiation, it can also be isolated from the beta decay of ^{253}Cf. However, fermium is only produced on a picogram scale and metallic Fm has not been produced. Indeed all the known chemistry of fermium has been elucidated from tracer studies.

$$^{238}U \xrightarrow[6\beta^-]{+15\,n} {}^{253}Cf \xrightarrow{\beta^-} {}^{253}Es \xrightarrow{\beta^-} {}^{254}Es \xrightarrow{\beta^-} {}^{254}Fm$$

$$^{238}U + {}^{16}O \longrightarrow {}^{250}Fm$$

2.1.2.12 *Mendelevium ($_{101}Md$)*

The synthesis of mendelevium is a heroic tour de force in both synthesis and analysis. The first step was to produce enough ^{253}Es through neutron bombardment of Pu over one year — this produced a sample of 10^9 atoms of ^{253}Es, which was electroplated onto a gold target. The alpha beam was directed at the rear of the target so that any atoms of the new element formed would recoil onto a second "catcher" foil. This greatly simplified purification and allowed the target to be reused (remembering its half-life is only 20 days). This purification problem was especially important as calculations suggested that from the 10^9 atoms of Es *one atom* of ^{256}Md would be produced per experiment. After irradiation with alpha particles the gold foil was dissolved in aqua regia, and an ion exchange column used to remove the gold, einsteinium and other elements. The solution was then dried, re-dissolved and passed through a second ion exchange column and then dried. Element 101 was characterised by the decay of the daughter ^{256}Fm, formed by an electron capture reaction of ^{256}Md; the detectors were wired to the fire alarm so every decay event was signalled by the ringing of bells. In total 17 atoms of mendelevium were identified during these experiments, but the reactions have been repeated on a much larger scale and more than one million atoms per hour were subsequently prepared. The chemistry of Md has all been based on tracer scale work. The name mendelevium was suggested to recognise Mendeleev who first used the concept of the periodic table to predict properties for unknown elements — a key concept in trans-uranium chemistry. However, as a US team was proposing to honour a Russian scientist at the height of the Cold War, special permission from the US government was required.

$$^{253}\mathrm{Es} + {}^{4}\mathrm{He} \longrightarrow {}^{256}\mathrm{Md} + {}^{1}\mathrm{n}$$

$$^{256}\mathrm{Md} \xrightarrow[\text{capture}]{\text{electron}} {}^{256}\mathrm{Fm} \xrightarrow{} \begin{smallmatrix}\text{spontaneous}\\ \text{fission}\end{smallmatrix}$$

2.1.2.13 *Nobelium ($_{102}$No) and lawrencium ($_{103}$Lr)*

Nobelium can be thought of as the first controversial element. The first report of element 102 came from a multinational group in the Nobel Institute in Stockholm in 1957 whereby ^{244}Cm was irradiated with ^{13}C ions and, on the basis of ion-exchange chromatography and an 8.5 MeV alpha emission with a half-life of 10 minutes, this group suggested the name nobelium. However, this result could not be replicated by the US or Russian groups. In 1958, the US group announced the positive identification of element 102 by the reaction of a target of ^{244}Cm and ^{246}Cm with ^{12}C and ^{13}C ions. Nobelium decays with a 2.3 second half-life to form fermium, and 11 atoms were identified. Further investigations revealed that the actual sequence was as shown below. The Nobel group misidentified the Cf alpha decay as the decay of element 102. A Russian group based in Dubna correctly identified the half-life of ^{254}No and chemical identification of ^{252}Fm. However, as the symbol No and name nobelium were in common use, this was suggested for the name. Metallic nobelium has not been produced.

$$^{244}\text{Cm} + {}^{12}\text{C} \longrightarrow {}^{252}\text{No} + 4\,{}^{1}\text{n}$$

$$^{252}\text{No} \xrightarrow[t_{1/2}=2.3\,\text{sec}]{\alpha} {}^{248}\text{Fm} \xrightarrow[t_{1/2}=36\,\text{sec}]{\alpha} {}^{244}\text{Cf} \xrightarrow[t_{1/2}=19\,\text{min}]{\alpha}$$

Nobelium is known to exist in the +2 oxidation state due to the favourable f^{14} electronic configuration so characteristic position in the elution curves does not hold for No. In over 600 experiments the chemical behaviour of nobelium was elucidated and found to be significantly different from trivalent actinides, but bears a resemblance to the divalent group 2 elements. Notably these experiments were conducted on about 50,000 atoms of ^{255}No.

Lawrencium was similarly synthesised in 1961. 3 μg of Californium isotopes 249–252 were bombarded with ^{10}B and ^{11}B to form a new element with a measured half-life of ~8 seconds. Initially assigned to ^{257}Lr it was later determined to be ^{258}Lr. The longer lived ^{256}Lr, with a half-life of 30 seconds was prepared and identified by its decay sequence, as shown below. Tracer experiments, on 1,500 atoms of ^{256}Lr enabled the identification of the most stable +3 oxidation state, in line with the actinide concept. The name lawrencium was given after E.O. Lawrence who invented the cyclotron in 1939, which proved instrumental in the synthesis of the trans-uranium elements.

$$^{243}_{95}\text{Am} + {}^{18}_{8}\text{O} \longrightarrow {}^{256}_{103}\text{Lr} \xrightarrow{\alpha} {}^{101}_{252}\text{Md} \xrightarrow{\text{E.C.}} {}^{252}_{100}\text{Fm} \xrightarrow{\alpha} {}^{252}_{98}\text{Cf}$$

2.1.2.14 *Isotopes*

All elements of the actinide series have more than one isotope. The half-lives vary from the very long (primordial) to extremely short (microseconds). Table 2.1 lists the most important and long-lived isotopes.

Table 2.1. Longest lived or most important isotopes of the actinide elements and their major decay pathways.[14]

Actinide	Mass	Half-life	Decay	Product
Ac	227	21.772 y	β^-	^{227}Th
Th	232	1.405×10^{10} y	α	^{228}Ra
Pa	231	3.276×10^4 y	α	^{227}Ac
U	234	2.455×10^5 y	α	^{230}Th
U	235	7.04×10^8 y	α	^{231}Th
U	238	4.468×10^9 y	α	^{234}Th
Np	237	2.144×10^6 y	α	^{233}Pa
Pu	239	24 100 y	α	^{235}U
Pu	240	6561 y	α	^{236}U
Pu	242	3.73×10^5 y	α	^{238}U
Am	241	432.6 y	α	^{237}Np
Am	243	7370 y	α	^{239}Np
Cm	244	18.11 y	α	^{240}Pu
Cm	245	8500 y	α	^{241}Pu
Cm	246	4760 y	α	^{242}Pu
Cm	247	1.56×10^7 y	α	^{243}Pu
Cm	248	3.48×10^5 y	α	^{244}Pu
Cm	250	8.3×10^3 y	S.F.	various
Bk	247	1380 y	α	^{243}Am
Bk	249	330 d	α	^{245}Am
Cf	249	351 y	α	^{245}Cm
Cf	250	13.08 y	α	^{246}Cm
Cf	251	898 y	α	^{247}Cm
Cf	252	2.6470 y	α	^{248}Cm
Es	252	471.7 d	α	^{248}Bk
Es	253	20.47 d	α	^{249}Bk
Es	254	275.7 d	α	^{250}Bk
Es	255	39.8 d	β^-	^{255}Fm
Fm	257	100.5 d	α	^{253}Cf
Md	258	51.5 d	α	^{254}Es
No	259	58 min	α,	^{255}Fm
			EC	^{259}Md
Lr	260	180 s	α	^{256}Md

Notes: S.F. = spontaneous fusion; E.C. = electron capture.

2.2 Occurrence and Extraction

As already stated Ac, Th, Pa and U are naturally occurring (the concentrations of Np and Pu in ores are very small). Given the very low

Figure 2.5. Sample of Monazite, (Ce, La, Nd, Th) PO_4 from Lemmenjoki, Finland. From the author's collection.

abundance of Ac (ca. 0.2 mg per tonne of uranium ore) and Pa (0.3–3 ppm in uranium ore), coupled with the high radioactivity there is little commercial interest in mining these elements; for applications surrounding medical uses (e.g. ^{225}Ac) it is prepared using cyclotron methods. Thorium is relatively abundant in the earth's crust (ca. 9 ppm) and is comparable to boron or lead. Thorium is commercially extracted from the mineral monazite, a lanthanide phosphate (Figure 2.5), with thorium concentrations of 10^5–10^6 ppm. Monazite is quite inert and two common methods are used, namely acid or alkaline digestion. Under acidic conditions Monazite is dissolved in sulfuric acid and the pH adjusted using NH_4OH to ~1.0 whereupon hydrated $Th(PO_4)_2$ precipitates. This is subsequently treated with alkali to form thorium hydroxide and removal of phosphate or sulfate ions. Digestion of monazite in hot concentrated NaOH followed by acidification of the solid allows for the precipitation of $Th(OH)_4$. The thorium hydroxide is then dissolved in acid, usually nitric acid, and further purified via solvent extraction, commonly kerosene containing Bu_3PO.

(a) (b)

Figure 2.6. (a) Green crystals of meta-torbernite, $Cu(UO_2)_2(PO_4)_2.8H_2O$ and (b) uraninite (black deposits) along with another uranyl mineral (yellow). Both from the author's collection.

In 1961, the United Kingdom's Atomic Energy Authority produced 127 grams of 99.9% pure ^{231}Pa by processing 60 tonnes of waste material from uranium ore processing in a 12-stage process. For many years this was the world's only significant supply and much of the Pa chemistry has been elucidated from this stock.

The major commercial ores of uranium are uraninite (UO_2) and carnotite ($K_2[(UO_2)_2(VO_4)_2] \cdot 3H_2O$), although there are about 200 uranium containing minerals that have been identified (Figure 2.6). In 2013, 59,531 tonnes of uranium was mined, with Kazakhstan, Australia and Canada accounting for 63% of this.[15] Extraction of the uranium from the ore is generally via crushing to a fine powder then leaching of the uranium. The ores are either leached out in the form of uranyl sulfate or uranyl carbonate. Purification and precipitation with ammonia forms $(NH_4)_2U_2O_7$, which is then heated to form U_3O_8.

Uranium is also found in low levels in seawater (ca. $14 \times 10^{-9}\,\text{mol}\,\text{dm}^{-3}$),[16] but given the large ocean volumes, this has the potential to be a rich source of uranium. Extensive research has found that amidoxime-based adsorbents (Figure 2.7) are useful in the selective sorption of uranium and a proof of concept test in seawater showed that a maximum uptake of 3.3 mg of U per gram of adsorbent after 8 weeks is possible.[17] While there are still problems with the competitive removal of vanadium and costs, the commercial removal of uranium from seawater may be a possibility.

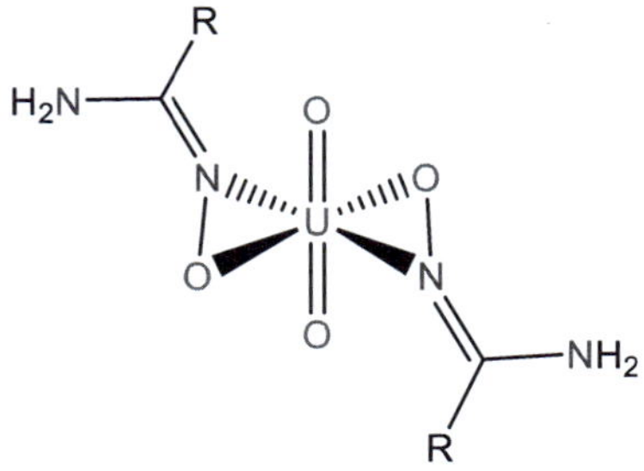

Figure 2.7. Coordination of amidoximes to the uranyl ion (R = alkyl group or polymer chain).

2.3 Periodicity, Electronic Configuration and Oxidation States

2.3.1 *Relativistic effects on 5f- and 6d-orbitals*

The actinide ions principally fill their 5f orbitals. While it is readily apparent how d-orbitals can be represented, for the f-orbitals a complication comes from the general and cubic set. The former is most often used and the latter is exclusively for cubic symmetry. The most important difference to the 4f orbitals of the lanthanides is that the 5f orbitals have an additional radial node. However, in order to fully understand periodicity, one also has to include relativistic effects. The special theory of relativity states that objects moving towards the speed of light gain mass due to the equivalence of kinetic energy and mass ($E = mc^2$). As electrons in heavier atoms, such as the actinides, are travelling close to the speed of light, the relativistic mass of these electrons increases. The consequence of this is that all s and p orbitals contract, termed direct relativistic orbital contraction. In turn, this increases the shielding on the d and f orbitals and leads to a radial expansion. In terms of the effects on the chemistry of the actinides this means that the 5f orbitals now extend further from the nucleus and can potentially engage in metal-ligand overlap and covalent character to the bonding (Figure 2.8). However, as the 5f orbitals do not effectively shield each other from Z_{eff}, the atomic number increases, the 5f-orbitals drop rapidly in energy and become more core-like. As the 5f, 6d, 7s and 7p orbitals are close to each other in energy for the early actinides then this suggests that multiple oxidation states are possible. As the f-orbitals contract, the

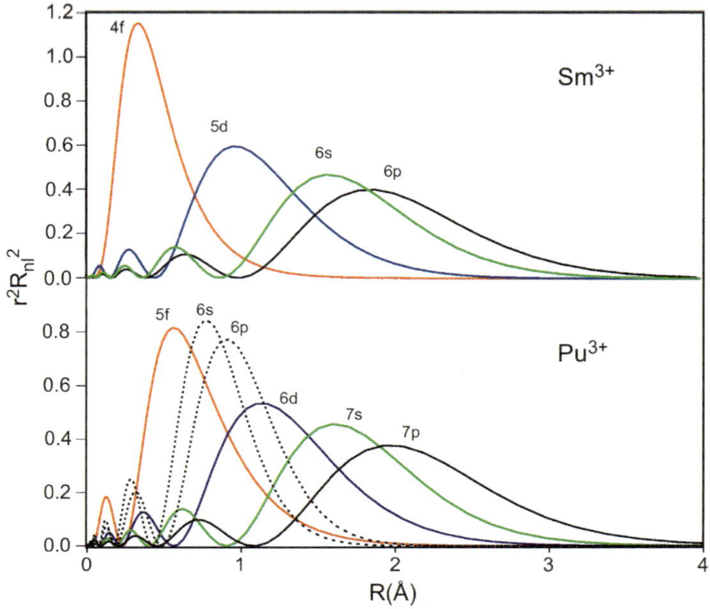

Figure 2.8. Relativistic Radial Distribution Functions for 4f (top) and 5f (bottom) orbitals. Reprinted from Ref. [20]. Copyright (2013) with permission from Elsevier.

later actinides have only one or two stable oxidation states. On the basis of this, one can already see that the chemistry of the early actinides will be transition metal-like while the late actinides more lanthanide-like. A stark example of the effect of relativistic effects is a comparison of the ground-state electronic configurations of No and Lr, which is predicted to be $[Rn]5f^{14}7s^2$ and $[Rn]5f^{14}7s^27p^1$. Contrast this to their lanthanide homologues Yb and Lu, $[Xe]4f^{14}6s^2$ and $[Xe]4f^{14}6s^25d^1$, as the 7p orbital is stabilized below the 6d orbital by relativistic effects. The measurement of the ionisation potential of ^{254}No ($t_{1/2}$ = 51.2 seconds)[18] and ^{256}Lr ($t_{1/2}$ = 27 seconds)[19] confirms this.

2.3.2 *Electronic configuration*

The electronic configurations of the actinide atoms are shown graphically in Figure 2.9. This shows that the early actinides have both the f and d orbitals involved due to the lower energy gap between

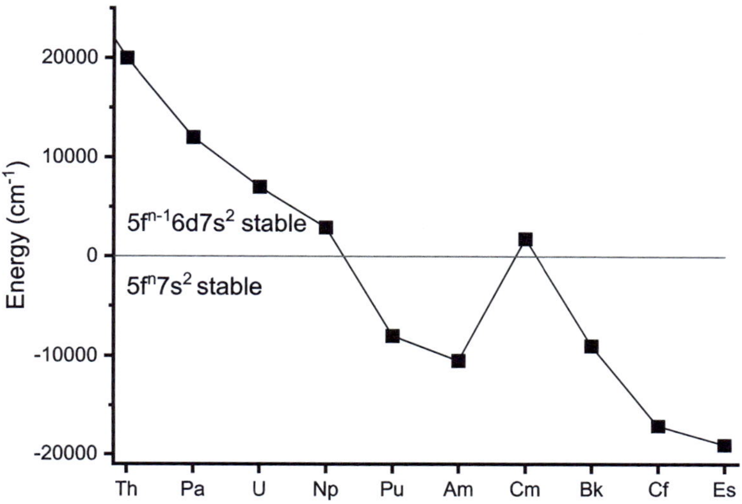

Figure 2.9. Relative energies of the $5f^n7s^2$ and $5f^{n-1}6d7s^2$ electron configurations in the gaseous actinide atoms.

these orbitals, while from Pu onwards only the f-orbitals are filled as the d orbitals are now higher in energy. This is with the exception of curium, which, due to the presence of a half filled shell, adds the electron to the d-orbital to obtain the electronic configuration $7s^25f^76d^1$. This is due to the fact that for the early actinides the 6d-orbitals are lower in energy than the 5f-orbitals, while as one traverses the series the f orbitals drop in energy and become lower in energy. This has been demonstrated theoretically for the compounds Cp_3An (An = Th–Pu) whereby the energy of the 5f orbitals drop across the series, while the energy of the d orbitals increase (Figure 2.10). At uranium the 5f and 6d orbitals are at very similar energy so that f-d mixing and hybridisation can occur. Recently experimental studies on the technologically important δ-phase of plutonium metal suggest that the electronic states are fluctuating on a sub-picosecond timeframe.[21]

2.3.3 *The actinide contraction*

The lanthanide contraction is a well-understood phenomena due to the core-like nature of the 4f orbitals (Chapter 1). It is no surprise that an actinide contraction also exists. Data for the ionic and

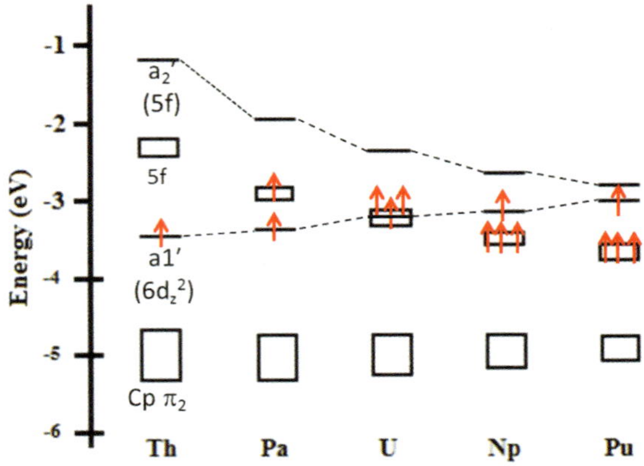

Figure 2.10. Metal-based orbital energies for the $5f^{n-1}6d^1$ electron configurations of Cp_3An (Cp = cyclopentadienyl). Adapted with permission from Ref. [70]. Copyright (1989) American Chemical Society.

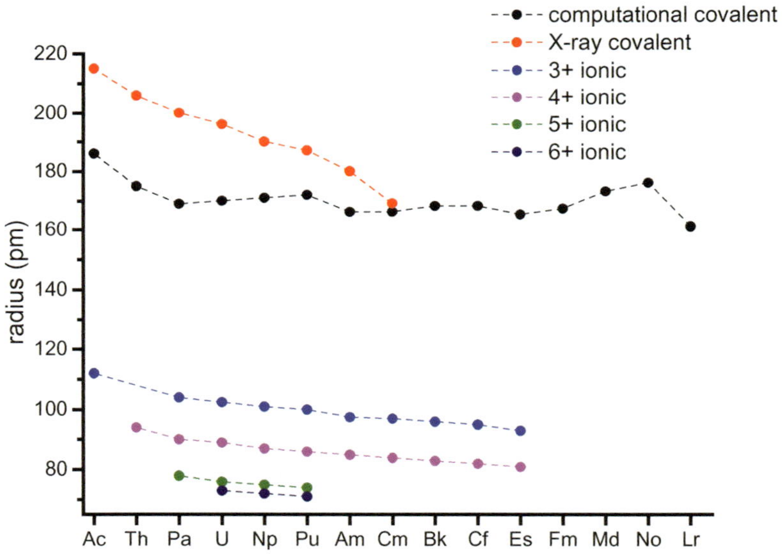

Figure 2.11. Ionic and covalent radii for the actinide ions.

covalent radii are shown in Figure 2.11. The actinide contraction is clearly present in all experimental data. The differences in the ionic radii of the +3, +4, +5 and +6 ions of a certain actinide are also apparent and as might be expected given the higher charge.

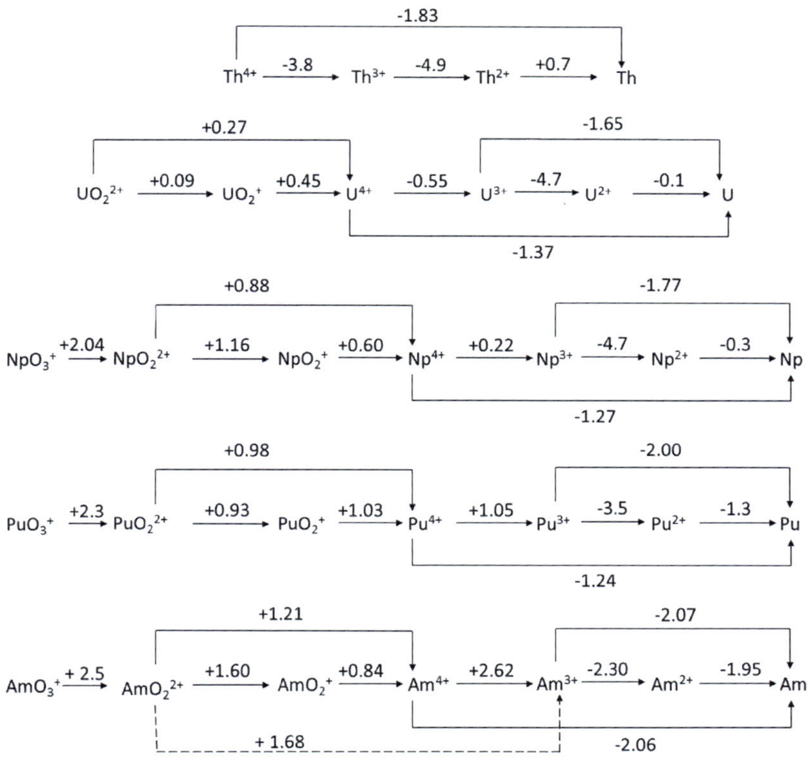

Figure 2.12. Selected standard reduction potentials of actinide ions in acidic aqueous solutions (vs. the Standard Hydrogen Electrode).[24]

Some of the implications of the actinide contraction are that the trans-actinide elements (6d) are smaller than the 5d elements, in a similar way to the 4d/5d elements due to the lanthanide contraction. The covalent radii are given from two sources, a computational route excluding M–X bonds[22] and those deduced from X-ray crystal structures.[23]

2.3.4 *Electrode potentials*

While for the lanthanides the ionisation energies could be used to rationalise the preference of oxidation states, for the actinides some of these values have not been measured. However, the redox potentials can give an insight into the stability of actinide oxidation states. Figure 2.12 shows the standard reduction potentials (vs. the standard hydrogen electrode) for selected actinide elements. As can be seen from this figure there is a variety in oxidation states for the

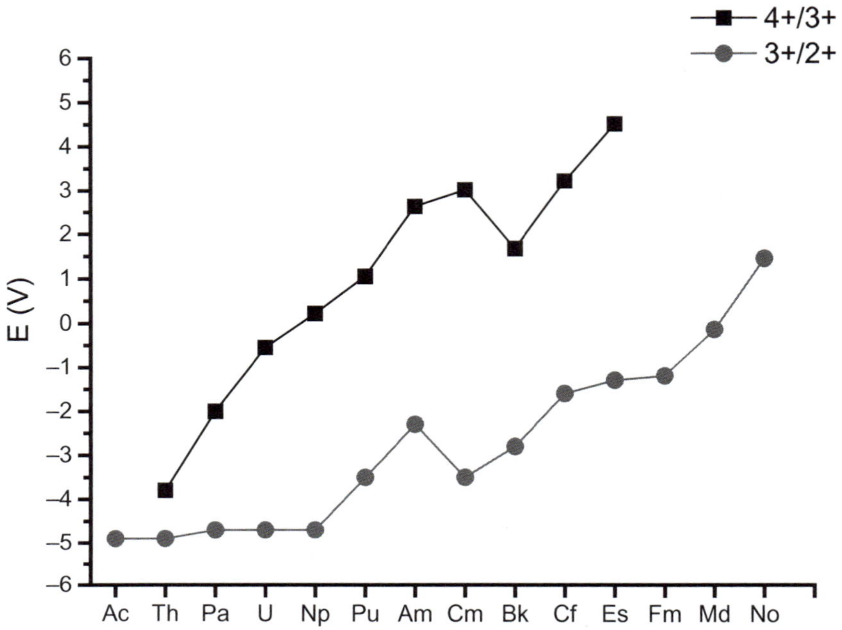

Figure 2.13. Plot of the redox potentials for the +4/ + 3 and +3/ + 2 actinide ions in acidic solution.

early actinides, due to the low energy gaps between the 5f and 6d orbitals. Thorium clearly has a much more stable +4 oxidation state which will be difficult to reduce to Th(III). According to these redox potentials Th(II) should not be stable in aqueous media. For uranium to americium the situation is more complex with a number of oxidation states that could be achievable. Figure 2.13 shows the redox potentials for the $An^{4+} \rightarrow An^{3+}$ and $An^{3+} \rightarrow An^{2+}$ for the series. This gives the relative stability of the 4+ and 3+ ions as the actinide series is transversed. The 4+ ion becomes destabilised with respect to the 3+ ion, with the exception of Bk(IV) ($5f^7$) and towards the later actinides the +2 oxidation state becomes more stable.

2.3.5 *Oxidation states*

We will conclude this section with the discussion of the oxidation states of the actinides. There is a significant difference compared to the lanthanides, where the +3 oxidation state is prevalent. For

Table 2.2. Known oxidation states of the actinide ions. Bold are the most common in aqueous solutions.

Ac	Th	Pa	U	Np	Pu	Am	Cm	Bk	Cf	Es	Fm	Md	No	Lr
	2		2			2			2	2	2	2	**2**	
3	3	3	3	3	3	**3**	**3**	**3**	**3**	**3**	**3**	**3**	3	**3**
	4	4	4	4	**4**	4	4	4	4					
		5	5	**5**	5	5								
			6	6	6	6								
				7	7									

the actinides there is a distinction where the early actinides have variable oxidation state but the later ones have the +3 oxidations states commonly. Table 2.2 shows the known oxidation states of the actinides; those in bold are the most common in aqueous environment.

2.3.5.1 +1 *oxidation state*

The only actinide with a +1 oxidation state is mendelevium, although this is somewhat controversial and not confirmed.

2.3.5.2 +2 *oxidation state*

The +2 oxidation state begins at americium and becomes increasingly more stable from Cf. While for Am these only exist in a few solids, and not in aqueous solution, $CfCl_2$ can be prepared at modest redox potentials. Recent reports on the coordination chemistry of ^{249}Cf suggest that a metastable Cf(II) is present,[25] while further evidence from gas phase reactions have also shown Cf(II) to be accessible.[26] For Md and No the +2 oxidation state is possible in water and, due to the enhanced stability from a full shell, No(II) is the common oxidation state ($5f^{14}$). In non-aqueous solutions, more oxidation states are possible, and some notable recent results have been the synthesis of Th(II)[27] and U(II)[28] despite redox potentials that would suggest these are inaccessible. $[Cp_3An]^-$ ions[28a] have an electronic configuration U(II) = $5f^36d^1$, Th(III) = $6d^1$ and Th(II) = $6d^2$ while a $5f^4$ U(II) has also been prepared[28b] (Figure 2.14).

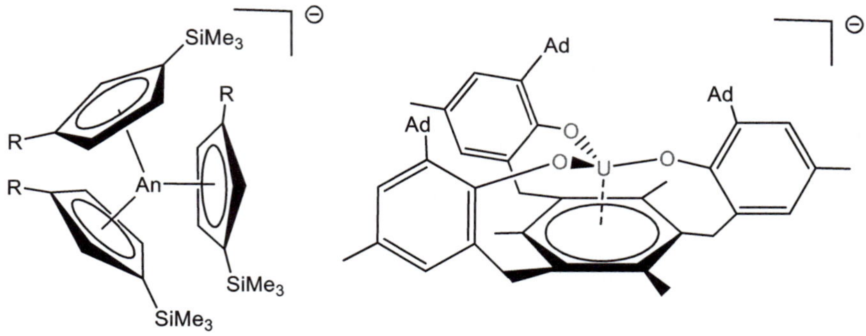

An = U, R = H, $SiMe_3$
An = Th, R = $SiMe_3$

Figure 2.14. Recent examples of the synthesis of Th(II) and U(II) compounds.

2.3.5.3 +3 *oxidation state*

With the exception of Th and Pa, all actinides have an accessible +3 oxidation state in aqueous solutions. This is the only oxidation state for Ac, but the high radioactivity, small sample sizes due to low abundance and closed shell — i.e. spectroscopically silent — electronic configuration has meant that Ac chemistry is very underdeveloped. Recent advances in X-ray spectroscopy have allowed EXAFS and XANES spectroscopy to be utilised on 28 μg sample sizes.[29] The +3 oxidation state becomes favoured from Am onwards, with a break at Pu where the localisation of f-electrons becomes apparent, i.e. the trend towards lanthanide-like behaviour. Given the short half-life of ^{249}Bk (330 days), difficulty in producing more than milligram quantities, and its high radioactivity it is not surprising that Bk chemistry is underdeveloped. Only recently has the first crystal structure of a Bk compound been reported, namely a $Bk^{III}L_3$ (L = dipicolinate) compound (Figure 2.15).[30] A number of Th(III) compounds have been prepared in non-aqueous media, and these have a $6d^1$ ground-state. Uranium(III) compounds in non-aqueous solutions are prevalent with a large amount of coordination and organometallic chemistry reported.[31]

2.3.5.4 +4 *oxidation state*

This is the most stable oxidation state for Th and Pu. Pa(IV) can be stabilised[33] and U(IV) is also stable in the absence of oxygen. Am(IV)

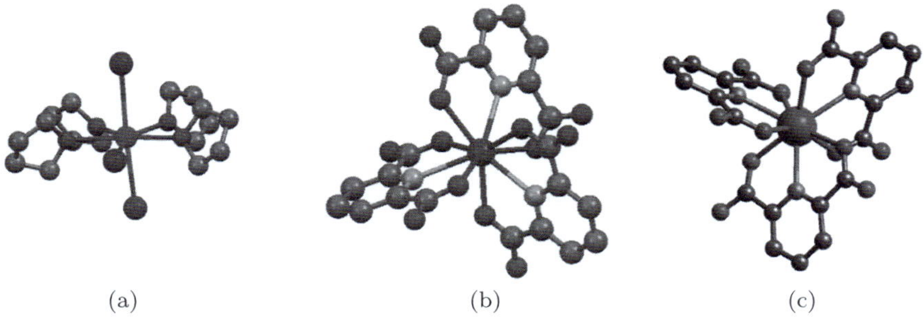

Figure 2.15. Crystal structure of $[UI_3(THF)_4]$ (a),[32] $[AmL]^3$ (b),[25] and $[BkL]^3$ (c),[30] where L = 2, 6-pyridinedicarboxylate.

is very unstable and disproportionates rapidly. The +4 state for Am and Cm are only accessible in solids. As these have high activities, the reduction by radiolysis also becomes apparent. For example, $^{244}CmF_4$ is reduced by 1% per minute by alpha radiolysis. Bk(IV) is known and stabilised due to its electronic configuration $5f^7$ although there are few compounds known. In non-aqueous conditions Th–Pu(IV) are readily accessible, and the coordination and organometallic chemistry of these, with the exception of Pa, is preponderant (see for example Figure 2.16).[31]

2.3.5.5 +5 *oxidation state*

The +5 oxidation state is the most common for Pa and Np. In aqueous media, U to Am exist as the actinyl moiety, $[AnO_2]^+$. Uranyl(V) is unstable to disproportionation, while neptunyl(V) will disproportionate only under acidic conditions or high concentrations; it is the most stable oxidation state. Strong oxidising agents are required to form americyl(V) ions and only under neutral or alkaline media. In non-aqueous media, a number of uranyl(V) compounds have been prepared, usually via reduction of a stable $[UO_2]^{2+}$ complex, which relies on steric effect to prevent disproportionation (Figure 2.17).[35]

2.3.5.6 +6 *oxidation state*

This is the most common oxidation state for uranium and accessible for Np, Pu and Am. Again these exist as actinyl ions $[AnO_2]^{2+}$. The neptunyl(VI) ion is commonly encountered in the pH range of 3–4 and is not as stable as the uranyl(VI) or plutonyl(VI) ion. Strong oxidising agents in acidic media give Am(VI) over Am(V) as the

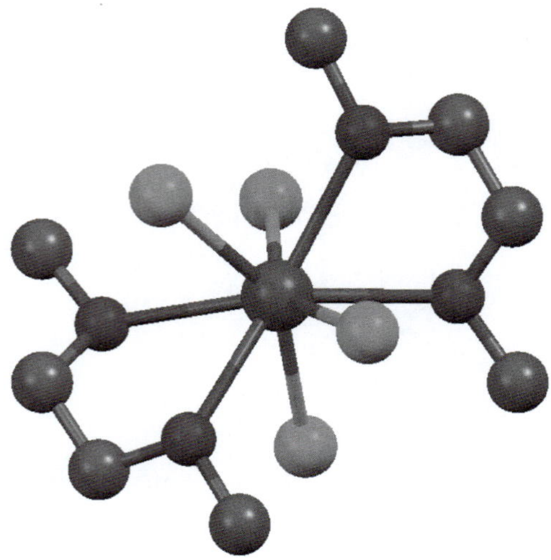

Figure 2.16. Structure of $[NpCl_4(DME)_2]$.[34]

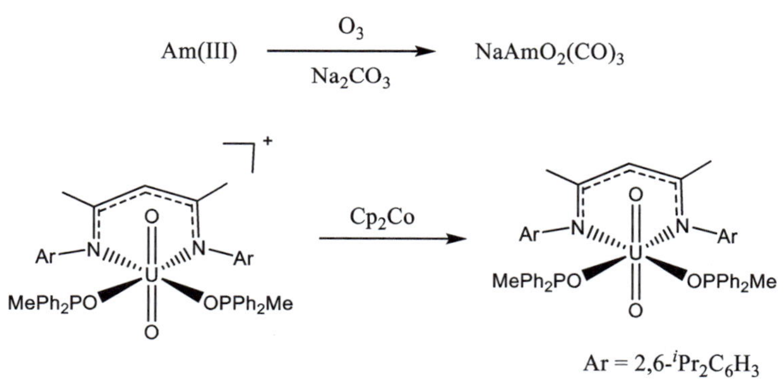

Figure 2.17. Preparation of americyl(V) and uranyl(V) complexes from Am(III) and uranyl(VI) precursors, respectively.

oxidation of +5 to +6 is easier than +3 to +5.

$$3/2S_2O_8^{2-} + Am(III) \longrightarrow 3SO_4^{2-} + AmO_2^{2+} + 4H$$

2.3.5.7 +7 *and* +8 *oxidation states*

The +7 oxidation state only exists for Np and Pu and in strongly basic solutions, prepared generally via reaction with O_3. The structures for both Np and Pu (Figure 2.18) are based around an

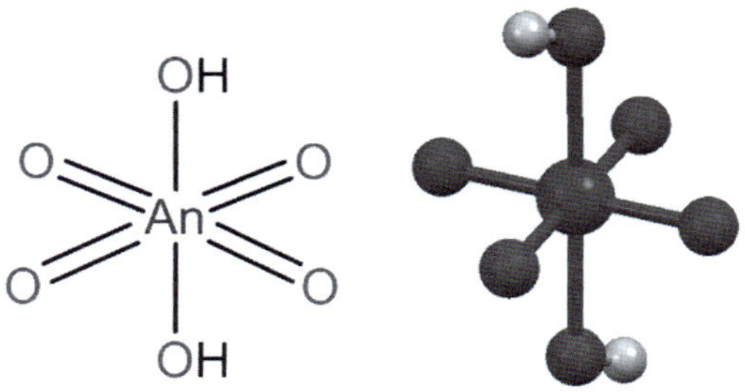

Figure 2.18. Solid-state structure of the core of $Cs_3[PuO_4(OH)_2].3H_2O$.[39]

$[AnO_4(OH)_2]^{3-}$ core; interestingly, this features two long OH axial bonds and 4 short planar oxo ligands (1.86–1.91 Å). These species are rather unstable in solution and the spontaneous reduction of Pu(VII) to Pu(VI) is rapid (ca. 3 hours).[36] Higher oxidation states such as Pu(VIII) are not confirmed,[37] and computational studies have suggested that PuO_4 is actually $[PuO_2(O_2)]$, i.e. Pu(V).[38]

2.3.5.8 *Plutonium*

The chemistry of plutonium is extremely complex and deserves to be treated on its own. Part of the complexity arises from the redox potential, all of which are very similar. Thus, in aqueous solution Pu can exist in +3 to +6 oxidation states; +3 and +4 are more stable in acid and +5 and +6 in basic solutions. The kinetics of the redox processes are such that more than one oxidation state can simultaneously exist in solution. Disproportion reactions are also well studied. Under acidic conditions with no complexing agents, Pu(IV) will disproportionate to Pu(III) and $[PuO_2]^{2+}$ via a transient $[PuO_2]^+$ intermediate.

$$3\,Pu^{4+} + 2\,H_2O \leftrightharpoons 2\,Pu^{3+} + [PuO_2]^{2+} + 4H^+$$

Pu(V) also undergoes disproportionation reactions depending on the conditions of the experiment. The two limiting reactions are:

$$3[PuO_2]^{4+} + 2\,H^+ \leftrightharpoons Pu^{3+} + [PuO_2]^{2+} + 2\,H_2O$$

and

$$2[PuO_2]^+ + 4\,H^+ \leftrightharpoons Pu^{4+} + [PuO_2]^{2+} + 2\,H_2O$$

What this means is that with a starting oxidation state of +4 at a pH of 0–2 significant concentrations of Pu(III) and Pu(VI) will exist at equilibrium in 1 M acid. At pH 1 it is mainly Pu(III) and at pH 2 it is a mixture of Pu(III) and Pu(V). The times required for these equilibria are relatively fast so that in a 2×10^{-3} M solution of Pu(IV) in 1 M $HClO_4$ half the Pu(IV) will have changed oxidation state in 3 hours.

2.3.6 *Summary*

To summarise the trends in oxidation states of the actinides, there are more observed than as for the lanthanides due to the close energy of 7s, 6d and 5f orbitals. Early actinides resemble d-block metals (variable oxidation state, covalency), while late actinides resemble lanthanides (preference for +3 oxidation state). Plutonium is a complex element that has no parallels in the periodic table.

2.4 Structure and Bonding

The structures of the cations commonly encountered are dependent on the oxidation state. They are either spherical (+2 to +4) as for lanthanides and transition metals, or exist as the actinyl unit for +5 and +6 ions. In most cases the bonding is predominantly ionic, but as the 5f orbitals extend further from the nucleus than the 4f orbitals, there is a possibility for some covalency in the bonding. To what extent this covalency occurs is a hot topic of research in this field. For the spherical cations the structures are mainly dependent upon the steric demand of the ligand and high coordination numbers are commonly observed.

2.4.1 *Actinyl structures and bonding*

2.4.1.1 *Bonding in the –yl unit*

Due to the propensity of the early actinides to high oxidation states it is only U-Am that features the –yl geometry. One of the main geometric features is the linear O=An=O fragment which arises from the interaction of the 5f orbitals and the p-orbitals on the oxygen. In transition metal chemistry, the cis-dioxo species is much more commonly encountered. A qualitative MO scheme for uranyl is shown in Figure 2.19.[40] The 5f and 6d orbitals can participate in bonding

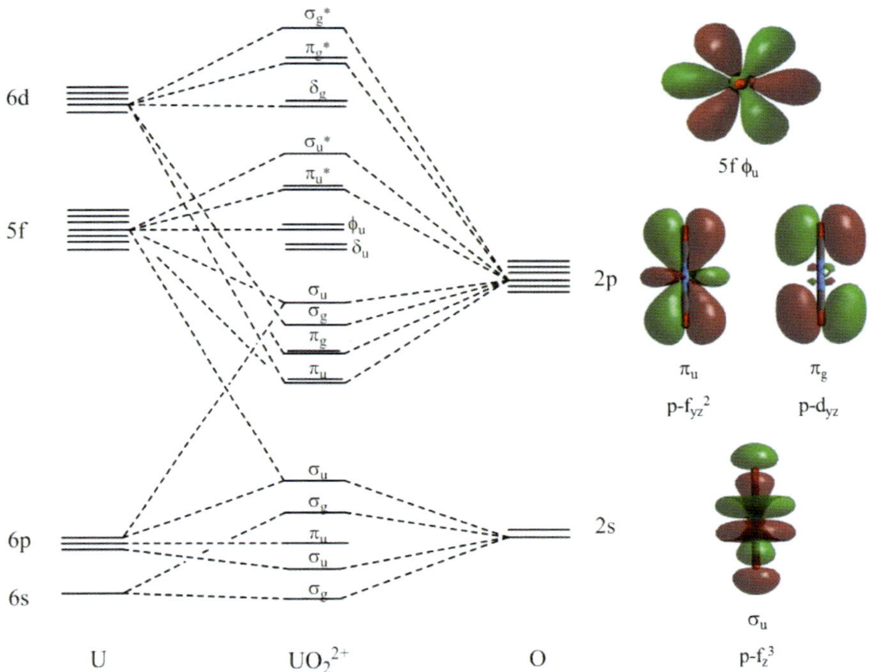

Figure 2.19. Qualitative MO diagram for the bonding in uranyl and graphical representations of selected molecular orbitals.

to the oxygen atoms; thus, U–O σ bonds are formed by the interaction of the O 2p σ orbitals with a uranium $6d_z{}^2$ orbital (σ_g) and a hybrid orbital formed by mixing $5f_z{}^3$ with the $6p_z\,(\sigma_u)$. U–O π bonds are formed by an interaction of O 2pπ orbitals with uranium 6dπ (π_g) and 5fπ (π_u) orbitals. Therefore, formally the uranyl has a bond order of 3. Interestingly, the π-orbitals are lower in energy than the σ-orbitals, due to a small destructive overlap of the 2p and σ_u orbitals. The LUMO consists of non-bonding 5f orbitals of δ_u and φ_u symmetry as they have no match with the ligand orbitals.

This bonding picture also holds for the Np-Am –yl ions, in both the +5 and +6 oxidation states. For $[AnO_2]^{2+}$ ions, Np has one, Pu two and Am three electrons in the formally non-bonding f orbitals; however, the stability decreases across the series as the f orbitals become more core-like. Similar arguments can be made for the +5 oxidation states.

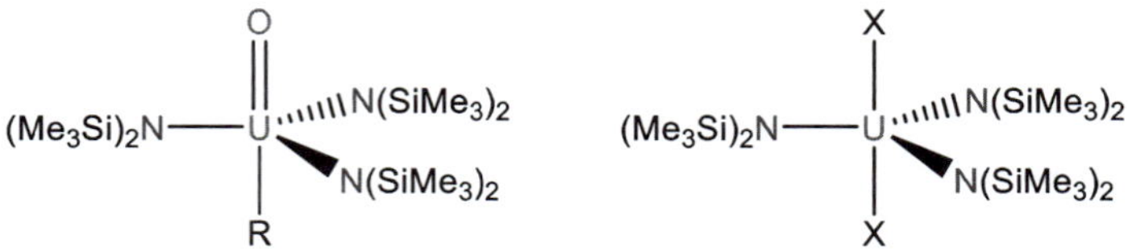

Figure 2.20. Examples of mono oxo uranyl(VI) and uranium(V) displaying an ITI (R = Me, C ≡ CPh; X = F, Cl, Br, N_3, NCS).

The bonding scheme also explains a number of interesting facets of the chemistry of the actinyl ions. The electronic absorption spectrum of the uranyl ion has been extensively studied and the typical bands are LMCT from the oxygen to the uranium orbitals. Similarly the photoluminescence spectra can be understood.[41] For the uranyl(V) ion the f-f transitions are also visible.[42] The optical spectroscopy of the other actinides is difficult to fully assign. For example, the emission in a neptunyl(VI) complex [$NpO_2(TPIP)_2(Ph_3PO)$], (TPIP = tetraphenylimidodiphosphinate) is a combination of LMCT from both the O_{yl} and ligand.[43]

An interesting observation was noted in the structures of $[PaOCl_5]^-$ and $[UOCl_5]^-$, whereby the trans An–Cl bonds are shorter than the cis An–Cl bond lengths, and this was termed the inverse-trans influence (ITI). The reason postulated for this has been that the hybridisation of the 6p and 5f orbitals can enhance the σ-bonding to the strongly bound trans directing ligand, leading to a partial hole in the 6p shell towards the trans ligand. This hole enhances 5f overlap in the trans position resulting in a shortening of the trans bond. There are now a number of examples where it has been observed, although it is still a subtle effect.[44] More recent examples featuring a mono-oxo uranyl(VI)[45] and uranyl(V)[46] have been reported (Figure 2.20).

2.4.1.2 *Structures of the –yl unit*

2.4.1.2.1 Generalities

A consequence of the linear arrangement of the O=An=O fragment is that the coordination of other ligands occurs in the equatorial plane, with the most common geometry being pentagonal bipyramidal for uranyl (Figure 2.21). There are less data available for Np and

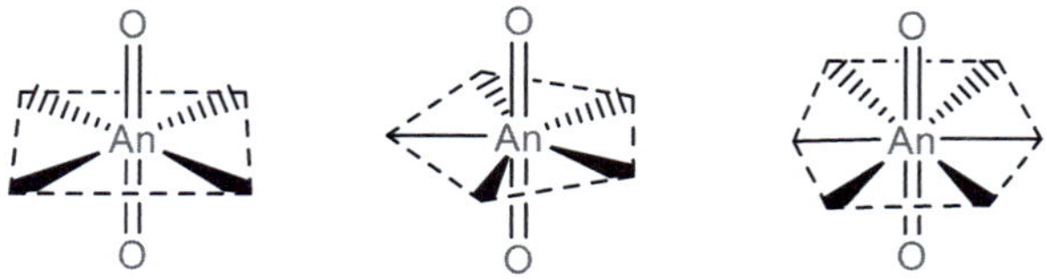

Figure 2.21. Common geometries of the actinyl ion.

Table 2.3. Bond lengths (Å) of the aqueous $[AnO_2(H_2O)_5]^{n+}$ ion from EXAFS data.[50]

	+5		+6	
Actinide ion	**An = O**	**An–O**	**An = O**	**An–O**
U	—	—	1.76	2.41
Np	1.83	2.51	1.75	2.41
Pu	1.81	2.47	1.74	2.42
Am	1.77	2.48	1.69	2.44

Pu, but the common coordination numbers are 4 and 5. Interestingly there are only few examples where either the O=U=O angle is not linear due to electronic (e.g. in uranyl formohydroxamates)[47] or steric (e.g. in $[UO_2(O\text{-}2,6\text{-}^tBu_2C_6H_3)_2(THF)_2])$[48] effects. There are examples of out-of-plane coordination, particularly with bulky or chelating N-donor ligands and one of the most distorted is in the complex $[UO_2(phen)_3][OTf]_2$.[49]

The bond lengths of the corresponding $[AnO_2]^{2+}$ and $[AnO_2]^{+}$ have been determined in acidic solution and are shown in Table 2.3. As can be seen the actinide contraction is evident in both the oxidation states and, because of the difference in charges, the An=O bond lengths in the +5 oxidation state are longer than in the +6 oxidation state.

2.4.1.2.2 Cation–cation interactions

These interactions are a unique type of interaction for the –yl fragments of the actinides. They were first noted in the 1960s where shorter than expected bond distances between a $[UO_2]^{2+}$ and $[NpO_2]^{+}$ cation in aqueous solutions,[51] and have been noted for $[UO_2]^{2+} \ldots [UO_2]^{2+}$, $[UO_2]^{2+} \ldots [UO_2]^{+}$, $[NpO_2]^{+} \ldots [NpO_2]^{+}$ and

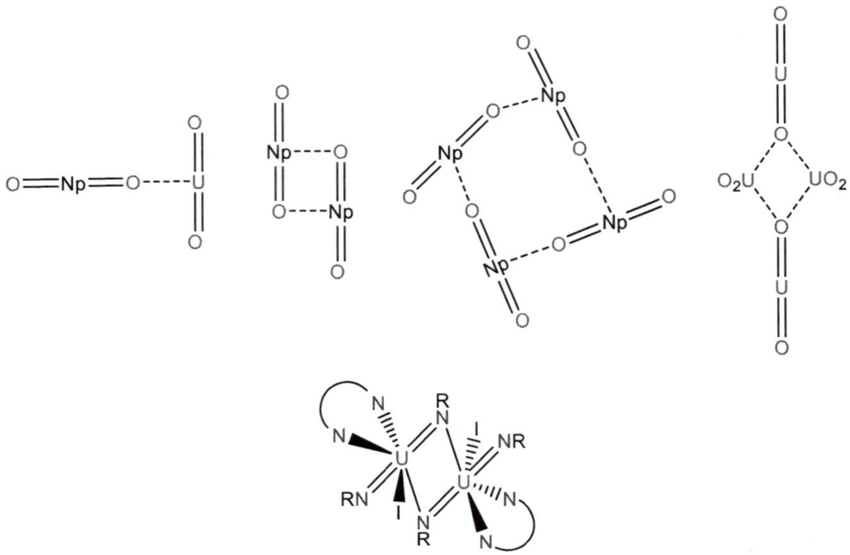

Figure 2.22. Typical CCI geometries.

mixed $[UO_2]^{2+} \ldots [AnO_2]^+$ (An = Np, Pu, Am). As the –yl oxygens have lone pairs, they can also act as ligands to other –yl fragments in a Lewis Acid-Base bonding model.[52] The presence of CCIs in uranyl(V) compounds has been postulated to be a mechanism for the inner sphere disproportionation reactions[53] and if steric bulk is incorporated into uranyl(V) compounds this does stabilise this unusual oxidation state of uranium. CCIs are much more prevalent for $[NpO_2]^+$ ions[54] compared to $[UO_2]^{2+}$ and $[NpO_2]^{2+}$,[55] ions, and this can be explained as the differing electronic configurations which leads to differing Lewis basicity of the oxygen atoms. There are very few reported examples of Pu featuring CCIs, although this may be simply a consequence of the practicalities of undertaking Pu chemistry.[56] There are a number of different arrangements that the CCIs can take, with the more common ones being linear, T-shaped or diamond (Figure 2.22). The major structural influence of these CCIs is that extended structures are also possible, and while this is more prevalent for the U(VI) compounds, some examples of Np have been reported.[57] The other structural feature of CCIs is the elongation of the An=O bond length and a corresponding shift in the Raman spectrum. Interestingly the presence of CCIs significantly

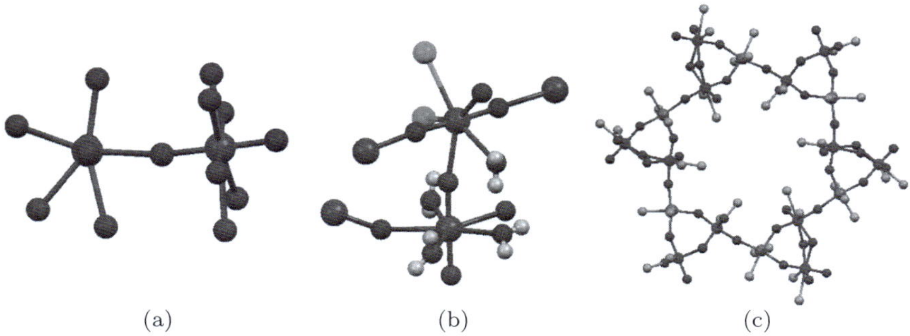

Figure 2.23. Structures of CCIs: (a) Np(VI) involved in a T-shaped interaction;[61] (b) Np(V) CCIS that builds up into a framework;[57] (c) uranyl(V)-Mn CCIs that generate unusual magnetic behaviour.[60a]

changes the magnetic properties of the paramagnetic compounds, notably neptunyl, due to the closer proximity of the metal centres than would normally be the case.[58] The isoelectronic imido fragments $[U(=NR)_2]^{n+}$ have been synthesised and some of these also feature CCIs.[59]

More recently, the term CCI has been extended to the interaction of any –yl unit with another cation (Figure 2.23). This has primarily been transition metals and uranyl(V) or uranyl(VI) and in the former case, unusual magnetic properties (e.g. single molecule magnets) have resulted.[60] The presence of CCIs also gives different topologies to the structure of the formed compounds.

2.4.2 *Spherical An^{n+} ions*

For those actinides in the +4 and +3 oxidation states, along with Pa(V), in aqueous solutions the ions are hydrated similarly to those of the lanthanides.[62] However, the number of coordinated water molecules continues to be contentions, possibly due to the differing measurement techniques (EXAFS, emission spectroscopy, X-ray diffraction) used in these studies. For the +4 oxidation state Th could be 9 or 10 and U–Pu in solution is likely to be 9 while for berkelium, the coordination number drops to 8.[33] In the solid state, there are fewer examples reported. Fully aquated thorium is the only member of the series to be structurally characterised, as the $[Th(H_2O)_{10}]Br_4$ species (Figure 2.24).[63] For the +3 oxidation states in solution, it

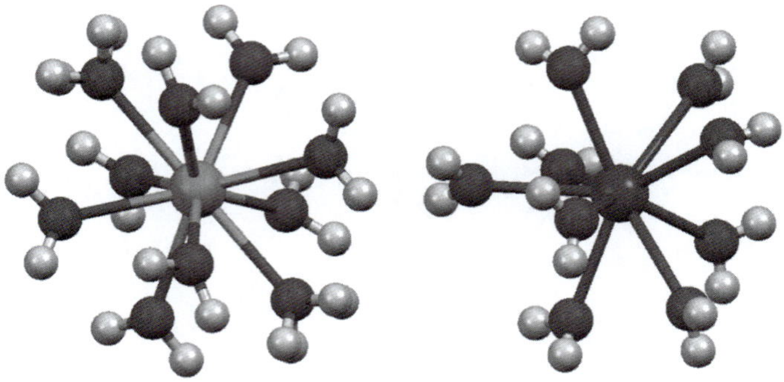

Figure 2.24. Solid-state structures of the cationic component in $[Th(H_2O)_{10}]Br_4$ (left)[63] and $[Cm(H_2O)_9][CF_3SO_3)]_3$ (right).[66]

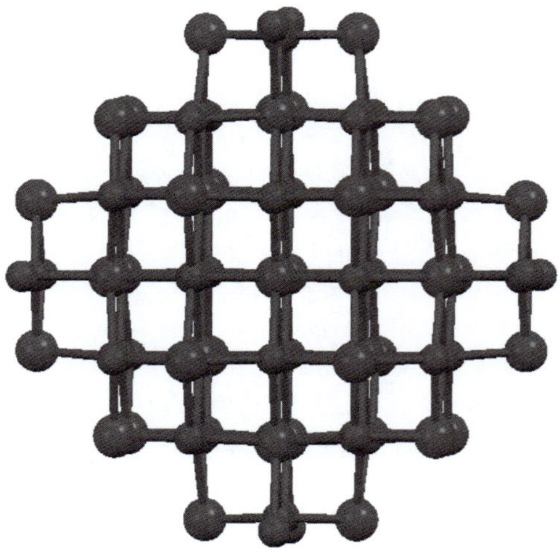

Figure 2.25. $[Pu_{38}O_{56}]^{40+}$ core in $Li_{14}(H_2O)_{20}[Pu_{38}O_{56}Cl_{54}(H_2O)_8]$.[67]

appears that 9 coordination number exists throughout the stable species (U to Cf),[64] in contrast to the lanthanides where a decrease in coordination number from 9 at Sm to 8 at Dy occurs. In the solid state, a series of structurally characterised examples of U-Cm are all 9-coordinate.[65] In all examples, the An–O bond decreases along the series and this is ascribed to the actinide contraction.

Given that the actinide ions in both +3 and +4 oxidation states are highly charged and electropositive, it is not surprising that hydrolysis of a coordinated water molecule is facile. Once a metal hydroxide is formed this can condense further into oligomeric hydroxo-bridged species or oxo-bridged species. These higher nuclear clusters have importance in environmental speciation of actinides and eventually lead to colloids. One of the more impressive examples has been the structural characterisation of $[Pu_{38}O_{56}]^{40+}$ cores (Figure 2.25) that has the same topology as bulk PuO_2.[67]

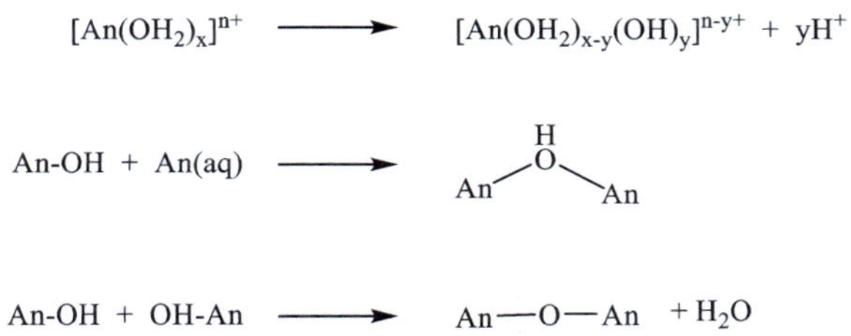

Figure 2.26. Hydrolysis reactions of aquated actinide cations.

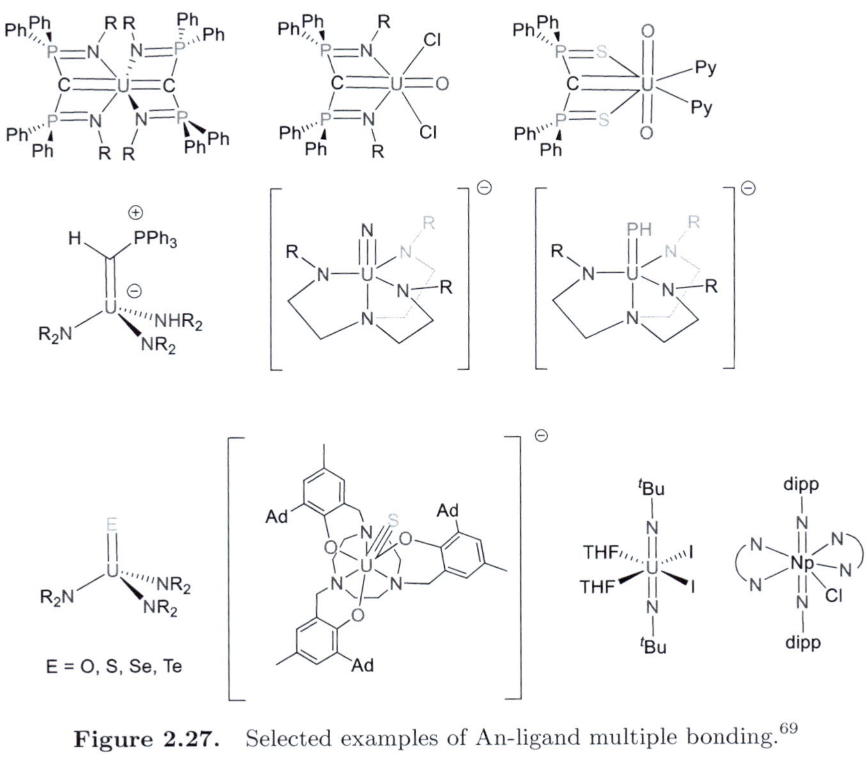

Figure 2.27. Selected examples of An-ligand multiple bonding.[69]

The –yl compounds of the actinides also undergo hydrolysis reactions and there are many examples of $[UO_2]^{2+}$ as this hydrolyses at pH > 3 and the nature of the hydrolysis products dependent upon pH, temperature, concentrations etc (Figure 2.26). Dimers, through octamers and chains and extended structures are all well known.

2.4.3 *Multiple bonding*

Covalency, or the propensity for the 5f/6d orbitals to overlap with ligand orbitals is a current hot topic in actinide chemistry. As described earlier, this is well known in the –yl compounds. It has been argued that in trivalent actinides this is important, but from plutonium this becomes less important and ionicity is the dominant bonding factor.[20] However, recent results on actinide borates suggest that at Cf covalency does return. Due to the difficulties in chemistry from Np onwards, the majority of examples featuring covalent bonding feature uranium in varying oxidation states. The field has burgeoned over the past decade and many compounds featuring U=L multiple bonding are known. Some recent examples are shown in Figure 2.27. There is only one example of a transuranic compound that features multiple bonding, namely $[Np^{V}(NDipp)_2(^tBu_2bipy)_2Cl]$ (Dipp = $2,6^iPr_2C_6H_3$).[68] Calculations on this and the U^V analogue show the participation of both 5f and 6d Np orbitals in forming the multiple bonds to nitrogen.

References

1. G. T. Seaborg, J. J. Katz and W. M. Manning (eds.), The Transuranium Elements: Research Papers, *Natl. Nucl. En. Ser., Div. IV*, 14B, McGraw-Hill, New York, 1949.
2. This section is based upon G. T. Seaborg and W. D. Loveland, *The Elements Beyond Uranium*, Wiley, New York, 1990; G. T. Seaborg, *Adventures in the Atomic Age: From Watts to Washington*, Farrar, Straus and Giroux, New York, 2001; L. R. Morss, N. M. Edelstein and J. Fuger, *The Chemistry of the Actinide and Transactinide Elements*, 4th edn., Springer, Dordrecht, The Netherlands, 2010.
3. J. P. Adloff, *Radiochim. Acta*, 2000, *88*, 123.
4. A. Gando, Y. Gando, K. Ichimura, H. Ikeda, K. Inoue, Y. Kibe, Y. Kishimoto, M. Koga, Y. Minekawa, T. Mitsui, T. Morikawa,

N. Nagai, K. Nakajima, K. Nakamura, K. Narita, I. Shimizu, Y. Shimizu, J. Shirai, F. Suekane, A. Suzuki, H. Takahashi, N. Takahashi, Y. Takemoto, K. Tamae, H. Watanabe, B. D. Xu, H. Yabumoto, H. Yoshida, S. Yoshida and S. Enomoto, *Nat. Geosci.*, 2011, *4*, 647.

5. J. F. McManus, R. Francois, J.-M. Gherardi, L. D. Keigwin and S. Brown-Leger, *Nature*, 2004, *428*, 834.
6. R. T. Günther and J. J. Manley, *Archaeologia*, 1912, *63*, 99.
7. The Nobel Prize in Chemistry 1935. http://www.nobelprize.org/nobel_prizes/chemistry/laureates/1935/. Accessed October 12, 2015.
8. The Nobel Prize in Physics 1938. http://www.nobelprize.org/nobel_prizes/physics/laureates/1938/. Accessed October 12, 2015.
9. The Nobel Prize in Chemistry 1944. http://www.nobelprize.org/nobel_prizes/chemistry/laureates/1944/. Accessed October 12, 2015.
10. The Nobel Prize in Chemistry 1951. http://www.nobelprize.org/nobel_prizes/chemistry/laureates/1951/. Accessed October 12, 2015.
11. D. F. Peppard, G. W. Mason, P. R. Gray and J. F. Mech, *J. Am. Chem. Soc.*, 1952, *74*, 6081.
12. E. F. Westrum and L. Eyring, *J. Am. Chem. Soc.*, 1951, *73*, 3396.
13. Yu. Ts. Oganessian, F. Sh. Abdullin, P. D. Bailey, D. E. Benker, M. E. Bennett, S. N. Dmitriev, J. G. Ezold, J. H. Hamilton, R. A. Henderson, M. G. Itkis, Yu. V. Lobanov, A. N. Mezentsev, K. J. Moody, S. L. Nelson, A. N. Polyakov, C. E. Porter, A. V. Ramayya, F. D. Riley, J. B. Roberto, M. A. Ryabinin, K. P. Rykaczewski, R. N. Sagaidak, D. A. Shaughnessy, I. V. Shirokovsky, M. A. Stoyer, V. G. Subbotin, R. Sudowe, A. M. Sukhov, Yu. S. Tsyganov, V. K. Utyonkov, A. A. Voinov, G. K. Vostokin and P. A. Wilk, *Phys. Rev. Lett.*, 2010, *104*, 142502.
14. All data from Tables of Physical & Chemical Constants. 4.6 Radioactive Elements. Kaye & Laby Online. Version 2.0, 2010. www.kayelaby.npl.co.uk.
15. Uranium 2014: Resources, Production and Demand, A Joint Report by the OECD Nuclear Energy Agency and the International Atomic Energy Agency. http://www.oecd-nea.org/ndd/pubs/2014/7209-uranium-2014.pdf. Accessed November 1, 2016.
16. J. Kim, C. Tsouris, R. T. Mayes, Y. Oyola, T. Saito, C. J. Janke, S. Dai, E. Schneider and D. Sachde, *Sep. Sci. Technol.*, 2013, *48*, 367.
17. J. Kim, C. Tsouris, Y. Oyola, C. J. Janke, R. T. Mayes, S. Dai, G. Gill, L.-J. Kuo, J. Wood, K.-Y. Choe, E. Schneider and H. Lindner, *Ind. Eng. Chem. Res.*, 2014, *53*, 6076.
18. M. Laatiaoui, W. Lauth, H. Backe, M. Block, D. Ackermann, B. Cheal, P. Chhetri, C. E. Düllmann, P. van Duppen, J. Even, R. Ferrer,

F. Giacoppo, S. Götz, F. P. Heßberger, M. Huyse, O. Kaleja, J. Khuyagbaatar, P. Kunz, F. Lautenschläger, A. K. Mistry, S. Raeder, E. M. Ramirez, T. Walther, C. Wraith and A. Yakushev, *Nature*, 2016, *538*, 495.

19. T. K. Sato, M. Asai, A. Borschevsky, T. Stora, N. Sato, Y. Kaneya, K. Tsukada, Ch. E. Düllmann, K. Eberhardt, E. Eliav, S. Ichikawa, U. Kaldor, J. V. Kratz, S. Miyashita, Y. Nagame, K. Ooe, A. Osa, D. Renisch, J. Runke, M. Schädel, P. Thörle-Pospiech, A. Toyoshima and N. Trautmann, *Nature*, 2015, *520*, 209.
20. M. L. Neidig, D. L. Clark and R. L. Martin, *Coord. Chem. Rev.*, 2013, *257*, 394.
21. M. Janoschek, P. Das, B. Chakrabarti, D. L. Abernathy, M. D. Lumsden, J. M. Lawrence, J. D. Thompson, G. H. Lander, J. N. Mitchell, S. Richmond, M. Ramos, F. Trouw, J.-X. Zhu, K. Haule, G. Kotliar and E. D. Bauer, *Sci. Adv.*, 2015, *1*, e1500188.
22. P. Pyykkö, M. Atsumi, *Chem. Eur. J.*, 2009, *15*, 186.
23. B. Cordero, V. Gómez, A. E. Platero-Prats, M. Revés, J. Echeverría, E. Cremades, F. Barragan and S. Alvarez, *Dalton Trans.*, 2008, 2832.
24. R. J. M. Konings, L. R. Morss and J. Fuger, *The Chemistry of the Actinide and Transactinde Elements* (Eds. L. R. Morss, N. M. Edelstein and J. Fuger), 4th edn., Ch. 19, Springer, Dordrecht, The Netherlands, 2010.
25. S. K. Cary, M. Vasiliu, R. E. Baumbach, J. T. Stritzinger, T. D. Green, K. Diefenbach, J. N. Cross, K. L. Knappenberger, G. Liu, M. A. Silver, A. E. DePrince, M. J. Polinski, S. M. Van Cleve, J. H. House, N. Kikugawa, A. Gallagher, A. A. Arico, D. A. Dixon and T. E. Albrecht-Schmitt, *Nat. Commun.*, 2015, *6*, 6827.
26. P. D. Dau, D. K. Shuh, M. Sturzbecher-Hoehne, R. J. Abergel and J. K. Gibson, *Dalton Trans.*, 2016, *45*, 12338.
27. R. R. Langeslay, M.E. Fieser, J.W. Ziller, F. Furche and W. J. Evans, *Chem. Sci.*, 2015, *6*, 517.
28. (a) M. R. MacDonald, M. E. Fieser, J. E. Bates, J.W. Ziller, F. Furche and W. J. Evans, *J. Am. Chem. Soc.*, 2013, *135*, 13310; (b) H. S. La Pierre, A. Scheurer, F. W. Heinemann, W. Hieringer and K. Meyer, *Angew. Chem. Int. Ed.*, 2014, 53, 7158.
29. M. G. Ferrier, E. R. Batista, J. M. Berg, E. R. Birnbaum, J. N. Cross, J. W. Engle, H. S. La Pierre, S. A. Kozimor, J. S. L. Pacheco, B. W. Stein, S. C. E. Stieber and J. J. Wilson, *Nat. Commun.*, 2016, *7*, 12312.
30. M. A. Silver, S. K. Cary, J. A. Johnson, R. E. Baumbach, A. A. Arico, M. Luckey, M. Urban, J. C. Wang, M. J. Polinski, A. Chemey, G. Liu, K.-W. Chen, S. M. Van Cleve, M. L. Marsh, T. M. Eaton, L. J. van de

Burgt, A. L. Gray, D. E. Hobart, K. Hanson, L. Maron, F. Gendron, J. Autschbach, M. Speldrich, P. Kögerler, P. Yang, J. Braley and T. E. Albrecht-Schmitt, *Science*, 2016, *353*, 888.

31. (a) C. J. Burns, M. P. Neu, H. Boukhalfa, K. E. Gutowski, N. J. Bridges and R. D. Rogers, *Comprehensive Coordination Chemistry II: From Biology to Nanotechnology* (Eds. J. A. McCleverty, T. J. Meyer), Ch. 3.3, Elsevier, Amsterdam, 2003; (b) F. T. Edelmann, *Comprehensive Organometallic Chemistry III From Fundamentals to Applications*, (Eds. R. H. Crabtree and D. M. P. Mingos), Ch. 4.02, Elsevier, Amsterdam, 2007.
32. D. L. Clark, A. P. Sattelberger, S. G. Bott and R. N. Vrtis, *Inorg. Chem.*, 1989, *28*, 1771.
33. N. lal Banik, V. Vallet, F. Réal, R. M. Belmecheri, B. Schimmelpfennig, J. Rothe, R. Marsac, P. Lindqvist-Reis, C. Walther, M. A. Denecke and C. M. Marquardt, *Dalton Trans.*, 2016, *45*, 453.
34. S. D. Reilly, J. L. Brown, B. L. Scott and A. J. Gaunt, *Dalton Trans.*, 2014, *43*, 1498.
35. (a) C. R. Graves and J. L. Kiplinger, *Chem. Commun.*, 2009, 3831; (b) P. L. Arnold, J. B. Love and D. Patel, *Coord. Chem. Rev.*, 2009, *253*, 1973.
36. M. R. Antonio, C. W. Williams, J. A. Sullivan, S. Skanthakumar, Y.-J. Hu and L. Soderholm, *Inorg. Chem.*, 2012, *51*, 5274.
37. W. Huang, P. Pyykkö and J. Li, *Inorg. Chem.*, 2015, *54*, 8825.
38. W. Huang, W.-H. Xu, J. Su, W. H. E. Schwarz and J. Li, *Inorg. Chem.*, 2013, *52*, 14237.
39. M. S. Grigoriev and N. N. Krot, *Acta Cryst.*, 2007, *E63*, i108.
40. R. G. Denning, *J. Phys. Chem. A*, 2007, *111*, 4125.
41. L. S. Natrajan, *Coord. Chem. Rev.*, 2012, *256*, 1583.
42. K. Mizuoka, S. Tsushima, M. Hasegawa, T. Hoshi and Y. Ikeda, *Inorg. Chem.*, 2005, *44*, 6211.
43. S. D. Woodall, A. N. Swinburne, N. lal Banik, A. Kerridge, P. Di Pietro, C. Adam, P. Kaden and L. S. Natrajan, *Chem. Commun.*, 2015, *51*, 5402.
44. (a) R. Denning, *Struct. Bond.*, 1992, *79*, 215; (b) H. S. La Pierre and K. Meyer, *Inorg. Chem.*, 2013, *52*, 529.
45. A. J. Lewis, P. J. Carroll and E. J. Schelter, *J. Am. Chem. Soc.*, 2013, *135*, 13185.
46. A. J. Lewis, K. C. Mullane, E. Nakamaru-Ogiso, P. J. Carroll and E. J. Schelter, *Inorg. Chem.*, 2014, *53*, 6944.
47. M. A. Silver, W. L. Dorfner, S. K. Cary, J. N. Cross, J. Lin, E. J. Schelter and T. E. Albrecht-Schmitt, *Inorg. Chem.*, 2015, *54*, 5280.

48. M. P. Wilkerson, C. J. Burns, D. E. Morris, R. T. Paine and B. L. Scott, *Inorg. Chem.*, 2002, *41*, 3110.
49. J. C. Berthet, M. Nierlich and M. Ephritikhine, *Dalton Trans.*, 2004, 2814.
50. C. Riddle, K. Czerwinski, E. Kim, P. Paviet, P. Weck, F. Poineau and S. Conradson, *J. Radioanal. Nucl. Chem.*, 2016, *309*, 1087.
51. J. C. Sullivan, A. J. Zielen and J. C. Hindman, *J. Am. Chem. Soc.*, 1961, *83*, 3373.
52. (a) B. Vlaisavljevich, P. Miro, D. Ma, G. E. Sigmon, P. C. Burns, C. J. Cramer and L. Gagliardi, *Chem. Eur. J.*, 2013, *19*, 2937; (b) M. L. McKee and M. Swart, *Inorg. Chem.*, 2005, *44*, 6975.
53. H. Steele and R. J. Taylor, *Inorg. Chem.*, 2007, *46*, 6311.
54. T. Z. Forbes, C. Wallace and P. C. Burns, *Can. Mineral.*, 2008, *46*, 1623.
55. S. Wang, J. Diwu, E. V. Alekseev, L. J. Jouffret, W. Depmeier and T. E. Albrecht-Schmitt, *Inorg. Chem.*, 2012, *51*, 7016.
56. I. A. Charushnikova, N. N. Krot and Z. A. Starikova, *Radiochim. Acta.*, 2007, *95*, 495.
57. G. B. Jin, *Inorg. Chem.*, 2016, *55*, 2612.
58. See for example: (a) G. B. Jin, S. Skanthakumar and L. Soderholm, *Inorg. Chem.*, 2011, *50*, 5203; (b) P. M. Almond, S. Skanthakumar, L. Soderholm and P. C. Burns, *Chem. Mater.*, 2007, *19*, 280; (c) T. Z. Forbes, P. C. Burns, L. Soderholm and S. Skanthakumar, *Chem. Mater.*, 2006, *18*, 1643.
59. (a) C. Camp, J. Pecaut and M. Mazzanti, *J. Am. Chem. Soc.*, 2013, *135*, 12101; (b) L. P. Spencer, E. J. Schelter, P. Yang, R. L. Gdula, B. L. Scott, J. D. Thompson, J. L. Kiplinger, E. R. Batista and J. M. Boncella, *Angew. Chem. Int. Ed.*, 2009, 48, 3795.
60. (a) V. Mougel, L. Chatelain, J. Pécaut, R. Caciuffo, E. Colineau, J.-C. Griveau and M. Mazzanti, *Nature Chem.*, 2012, *4*, 1011; (b) V. Mougel, L. Chatelain, J. Hermle, R. Caciuffo, E. Colineau, F. Tuna, N. Magnani, A. de Geyer, J. Pécaut and M. Mazzanti, *Angew. Chem. Int. Ed.*, 2014, *53*, 819.
61. S. Wang, J. Diwu, E. V. Alekseev, L. J. Jouffret, W. Depmeier and T. E. Albrecht-Schmitt, *Inorg. Chem.*, 2012, *51*, 7016.
62. K. E. Knope and L. Soderholm, *Chem. Rev.*, 2013, *113*, 944.
63. R. E. Wilson, S. Skanthakumar, P. C. Burns and L. Soderholm, *Angew. Chem. Int. Ed.*, 2007, *46*, 8043.
64. E. Galbis, J. Hernández-Cobos, C. den Auwer, C. Le Naour, D. Guillaumont, E. Simoni, R. R. Pappalardo and E. S. Marcos, *Angew. Chem. Int. Ed.*, 2010, *49*, 3811.

65. P. Lindqvist-Reis, C. Apostolidis, J. Rebizant, A. Morgenstern, R. Klenze, O. Walter, T. Fanghänel and R. G. Haire, *Angew. Chem. Int. Ed.*, 2007, *46*, 919.
66. S. Skanthakumar, M. R. Antonio, R. E. Wilson and L. Soderholm, *Inorg. Chem.*, 2007, *46*, 3485.
67. L. Soderholm, P. M. Almond, S. Skanthakumar, R. E. Wilson and P. C. Burns, *Angew. Chem. Int. Ed.*, 2008, *47*, 298.
68. J. L. Brown, E. R. Batista, J. M. Boncella, A. J. Gaunt, S. D. Reilly, B. L. Scott and N. C. Tomson, *J. Am. Chem. Soc.*, 2015, *137*, 9583.
69. (a) B. M. Gardner and S. T. Liddle, *Chem. Commun.*, 2015, *51*, 10589; (b) M. W. Rosenzweig, A. Scheurer, C. A. Lamsfus, F. W. Heinemann, L. Maron, J. Andrez, M. Mazzanti and K. Meyer, *Chem. Sci.*, 2016, *7*, 5857; (c) T. W. Hayton, *Chem. Commun.*, 2013, *49*, 2956; (d) T. W. Hayton, *Dalton Trans.*, 2010, *39*, 1145.
70. Bursten *et al.*, *J. Am. Chem. Soc.*, 1989, *111*, 2756.

https://doi.org/10.1142/9781800610163_0003

Chapter 3

Solid-State Chemistry: Synthesis and Structural Diversity in Lanthanide and Actinide Complexes

*Matthew L. Marsh, Frankie D. White, Wesley M. Potter, and Thomas E. Albrecht-Schoenzart**

Department of Chemistry & Biochemistry, Florida State University, 95 Chieftan Way, Tallahassee, FL, USA

* *talbrechtschoenzart@gmail.com*

3.1 Introduction

The solid-state chemistry of lanthanide and actinides has been of acute interest due to some of the novel properties seen in these series. The need to characterise these heavy and sometimes radioactive metals in the environment is also subtly woven throughout the narrative that follows. Solid-state systems can furthermore include a wide canvas of elementary anions that can be grouped in several ways including by size, charge, geometry, acid/base character, the ability to hydrogen bond, covalence, etc. For example, the nephelauxetic series (which can ascribe covalent properties) for a few of the systems seen throughout this chapter is given as follows:

$$F^- < H_2O < SO_4^{2-} < CO_3^{2-} < PO_4^{3-} < BO_3^{3-} < SiO_4^{4-} < I^- < O_2^{2-}$$

Studying each of these systems routinely helps to predict the structures, and ultimately, the environment the f-elements can be placed in. The ability to further mix and match these systems together can create very complicated systems and structures that will not be the aim of this work.

Depending upon the system, a variety of synthetic approaches have been utilised. Among these, the hydrothermal route has become a popular way to reach extremely pure crystals that are sufficient for studying physicochemical properties.[1] The use of an autoclave is designed for temperatures greater than 100 °C pressures greater than 1 bar. Solvents such as water lose viscosity with supercritical conditions, and other ions increase their mobilities in solution. Furthermore, redox changes can be attained, and a few Eu(II) examples will be seen throughout this chapter.

One disadvantage of hydrothermal routes is that there are often a range of hydrolysis products which can be difficult to predict depending on the variation of conditions. As an alternative, a range of solid-state techniques, flux routes and slow evaporation represents a few of the alternatives given. Other newer routes that use supramolecular synthons will not be detailed, as they would require an analysis too disparate from the general topics outlined here.[2]

Intermetallics have most notably provided some of the strongest known magnets such as $SmCo_5$ and $Nd_2Fe_{14}B$ in the world while hydroxides, oxides and oxyhalides all present cheap materials with a variety of desirable optical properties. Actinide oxides are commonly used as a fuel source in nuclear reactors. Halides, on the other hand, represent the easiest routes for synthetic and crystallisation strategies. Structural variety and routes to anhydrous and other redox states are a few of the examples that will be explored.

Oxyanions also represent a major class of materials. Nitrates, sulphates and perchlorates have been selected as prototype examples of soluble systems that can be used in synthesis while also demonstrating structural analogues. Most oxyanionic systems, however, are rather insoluble. Examples discussed will include the carbonates, silicates, phosphates, borates, vanadates, molybdates and iodates. Quite a structural diversity can be found among these different systems with large consequences for the fine-tuning of properties as well as nuclear waste storage and environmental chemistry.

3.2 Intermetallics, Oxides and Hydroxides

The f-element intermetallics are important in various fields due to their magnetic, electronic, optical and superconducting properties. Additionally, many of these materials have been shown to catalyse ammonia synthesis, olefin hydrogenation and methanation reactions. Effective models for understanding self-organised nanoscale architectures have been derived from these efforts, and due to their unusual type of H_2 sorption, they have been used to study the key relationships between sorptional properties and catalytic activity.[3]

The defined intermetallics generally consist of an extended network with lanthanide or actinide metal centres coordinated by transition metals (e.g. Fe, Ni and Co) and/or main group elements (e.g. B, Sn and Si). Such structures can be described by different polyhedra packing in complex patterns.[3] Due in part to the radioactive nature of (and relative difficulty in) handling actinide materials, the majority of f-element intermetallics found in the literature contain networks of lanthanide-transition metals coordinated by main group elements.

Common crystal-growth techniques used in synthesis include sintering, melt spinning, the flux-growth technique, hydrothermal synthesis, the Czochralski method, the traditional solid-state methods and arc melting together stoichiometric ratios of a desired product. Ln and An intermetallics crystallise in a variety of structure-types. Some of the more common types include the cubic $NaZn_{13}$-type (Fm-$3c$, No. 226), the tetragonal $ThMn_{12}$-type ($I4/mmm$, No. 139), the hexagonal Th_2Ni_{17}-type ($P6_3/mmc$, No. 194), the rhombohedral Th_2Zn_{17}-type ($R{-}3m$, No. 166) and the monoclinic $RE_3(Fe,Ti)_{29}$-type.[3] Also worth noting is the hexagonal $CaCu_5$-type ($P6/mmm$, No. 191) which is a fundamental structure-type for RE-TM intermetallics and the prototype for one of the world's strongest permanent magnets, $SmCo_3$.[4]

First discovered in the 1970s, polycrystalline samples of the similar stoichiometric intermetallic $SmCo_5$ are commonly synthesised by arc melting followed by sintering.[5] For industrial synthesis, the elements are melted in an induction furnace either under vacuum or an inert gas, pressed into pellets or ingots, sintered and then magnetised.

$SmCo_5$ crystallises in the tetragonal $P4_2/mnm$ (No. 136) space group and is isostructural with the $RENi_5$ series.

In the 1980s, $Nd_2Fe_{14}B$ was discovered to have an even greater coercive field than $SmCo_5$. Single crystals of $Nd_2Fe_{14}B$ have been grown using various solid-state methods, including sintering, melt spinning and the Czochralski method.[6] The industrial synthesis of this outstanding permanent magnet is the same as that for $SmCo_5$. Also like $SmCo_5$, $Nd_2Fe_{14}B$ crystallises in the $P4_2/mnm$ space group.[4] At room temperature the spins of $Nd_2Fe_{14}B$ order ferromagnetically, which follows the general observation that spins of lighter lanthanides and transition-metal intermetallics align ferromagnetically while the heavier ones tend towards antiferromagnetic ordering.[7] Such trends observed in lanthanide-containing compounds are not necessarily observed in actinide-containing compounds. For example, U_2Co_2Sn — a light actinide, and expected to order antiferromagnetically if it were in the lanthanide series — is a non-magnetic compound that orders ferromagnetically upon annealing under a stream of hydrogen.[7,8] This compound has the Mo_2FeB_2 structure-type (space group $P4/mbm$, No. 127) of which uranium is in a seven coordinated by six cobalt atoms and four tin atoms. U_2Co_2Sn has been synthesised via arc melting and the flux growth technique. Interestingly, upon additional hydrogenation of U_2Co_2Sn the magnetic ordering changes from ferromagnetic to antiferromagnetic. The absorption of hydrogen by U_2Co_2Sn does not result in a structure-type change; however, there is an increase in unit cell volume (1.4% for the α-hydrogenation and 8.1% for the β-hydrogenation).

The second group of materials detailed in this section, are hydroxides. Hydroxides have received much attention in recent years in nanochemistry due to their fluorescence and electron transport properties.[9] For example lanthanide nanorods can be synthesised via hydrothermal reactions using triethylamine as both a complexing and alkaline agent. Hydroxides can also be used as precursors for oxides by calcining them at 600 °C for 2 hours.

Lighter lanthanide hydroxides can be prepared by ageing amorphous hydrous lanthanides under water at 25 °C for one week to several months, or they can be synthesised in a Teflon-lined autoclave at 160 °C. They crystallise in the hexagonal $P6_3/m$ space group (No. 176) with the UCl_3 structure-type.[10] Heavier lanthanides yield crystalline products different from $P6_3/m$. However, $P6_3/m$ can also be obtained with the heavier lanthanides by ageing them in a strong sodium hydroxide solution with applied heat.

Lighter actinide hydroxides are prepared by other methods. For example, uranyl hydroxide (known to exist in many crystallographic forms) can be prepared by heating $UO_3 \cdot 2H_2O$ to 80 °C in air.[11] A second form can be isolated by heating anhydrous UO_3 in water at 180 °C. The hydrated plutonyl hydroxide, $PuO_2(OH)_2 \cdot H_2O$, can be isolated from precipitation of aqueous Pu(VI) in ammonia. The hydrated neptunium hydroxide, $NpO_2(OH)_3 \cdot nH_2O$, can be prepared by oxidising Np(VI) to Np(VII) with ozone, then neutralising the product. Farther along the series, Am(III)-Cf(III) all precipitate an amorphous gel when reacted with ammonia. If Am(III) is allowed to age, its gel becomes crystalline $Am(OH)_3$. This is isostructural with $Nd(OH)_3$ (CN = 9) and has the hexagonal $P6_3/m$ space group.

Finally, oxides represent one of the largest and most important groups of materials. While lanthanides are of importance due in part to their potential in ferromagnetics, ferroelectrics, semiconductors, phosphors and lasers, the actinides are important due to their refractory properties that make them ideal for the preparation of ceramic-based fuels to be used in nuclear reactors.[11,12] The Ln oxides crystallise in various space groups while a few of the more common ones observed are *Fm-3m* (No. 225), $C2/m$(No. 12) and *Ia*-3 (No. 206).[13] At room temperature, for example, Dy_2O_3, Ho_2O_3, Er_2O_3, Tm_2O_3, Yb_2O_3 and Lu_2O_3 all crystallise in the *Ia*-3 space group and have the rare-earth cubic C-type structure (Mn_2O_3). Like the Ln oxides, the An oxides adopt various structure-types and crystallise in different space groups. A common structure-type observed in the An oxides is the fluorite structure (space group *Fm*-3*m*, Figure 3.1(a)).[13] For example, ThO_2 crystallises in this space group and has been of particular interest as an alternative nuclear fuel (due in part to claims that its fuel cycle is more resistant to nuclear weapons proliferation). This structure is energetically favourable as it forms cuboids in which all atoms bonded to the central metal typically have the same bond lengths (Figure 3.1(b)). This allows for easy packing of polyhedra throughout the crystal structure.

However, other actinide oxide systems can become quite complicated. UO_2, for example, can either lose or gain oxygen interstitially into its crystal lattice, and this leads to different stoichiometric phases (i.e. UO_{2-x} and UO_{2+x}). Several anhydrous complex

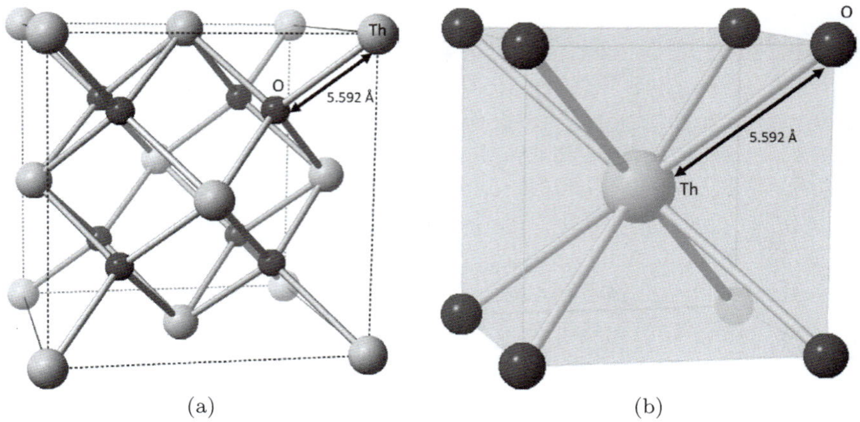

Figure 3.1. (a) A representation of the fluorite structure for ThO_2. (b) A signature cuboid with all Th–O bond lengths equal to 5.592 Å.

oxide phases are known for the actinides: monoxides (AnO), dioxides (AnO_2), trioxides (AnO_3), sesquioxides (An_2O_3) and pentoxides (An_2O_5) (just to name a few). The monoxide phases, known for Th-Am (inclusive) and Bk, have face-centred cubic symmetry with the NaCl structure-type. The dioxide phases are known for Th-Cf (inclusive) and adopt the face-centred fluorite structure-type. The only stable anhydrous actinide trioxide known is the uranium compound UO_3 which exists in several crystal modifications. The most common form, amorphous UO_3, is prepared by hydrating uranium peroxide and then heating it to 450 °C.[13] The sesquioxide phases, known for Pu–Cf (inclusive), possess the body-centred cubic lanthanide oxide C-type structure. In this cubic structure, the metal atom is in the centre of the unit cell and is surrounded by six oxygen atoms and two vacant sites. The pentoxide phases are known for protactinium, uranium and neptunium, and these are a bit more complicated than the other phases. U_2O_5, for example, has hexagonal symmetry, while Np_2O_5 has monoclinic symmetry.

Many synthesis methods have been used to grow Ln and An oxides: vapour condensation, skull-melting methods, flux methods, the Czochralski method, the sol-gel process and hydrothermal synthesis.[14] Each method may have limitations on the size of the crystals they can grow or other drawbacks. Flux methods, for example, often result in flux-inclusion into the crystal lattice; for a compound such as UO_2, such defects could have potentially dangerous side reactions in a nuclear chamber. Due to the high melting point of compounds such as UO_2 (2868 °C) or ThO_2 (3377 °C), the Czochralski method is not always a viable option. Hydrothermal growth conditions, on the other hand, offer an effective environment to grow complex and simple oxides of single crystal purity for both the lanthanides and actinides.[15]

In 2009, Kolis *et al.* used hydrothermal reactions to grow single crystals of ThO_2 over 8 mm long (Figure 3.2).[16] Using cesium fluoride as a mineraliser, the reaction was heated to 750 °C for up to 40 days at a time-generating pressures of 10–30 kpsi. A thermal gradient was used to promote crystal growth in which the nucleation site was at 750 °C and the growth region was 40–60 °C cooler. The growth of such large single crystals has enabled researchers to measure the thermal conductivity and thermal expansion of ThO_2 for the first time. ThO_2 was found to have a higher thermal conductivity and lower thermal expansion than UO_2. Therefore, in addition to minimising the amount of ^{235}U produced, ThO_2 results in less highly radioactive waste, has a higher melting point, higher corrosion resistivity, higher thermal conductivity and lower thermal expansion than UO_2. As a result, thorium-based fuel rods could be safer for the environment and are able to cool quicker than the uranium-based fuel rods.

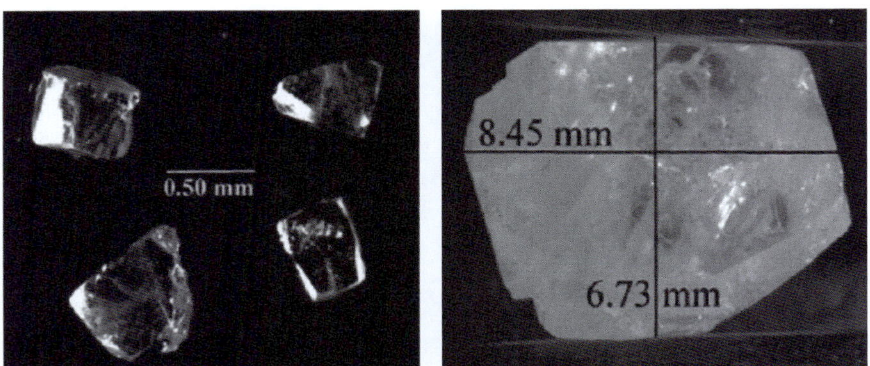

Figure 3.2. ThO_2 crystal (right) was grown using a smaller seed crystal (left) in 6 M CsF at 750 °C at 60 °C temperature gradient.[16]

Substitution of ions within the crystal lattice of Ln oxides has also gained attention due to the resulting ability to fine-tune properties.[17] Sn^{4+} and Zn^{2+} are typically chosen as substitution ions in Ln oxides due in part to the wide application capabilities of their metal oxides (SnO_2 and ZnO) and because the ionic radii of Sn^{4+} and Zn^{2+} are similar to those of the lanthanides. Sn^{4+} (0.81 Å), for example, is close in size to Tb^{4+} (0.87 Å) in an eight-fold coordination. It has been shown that when Sn^{4+} substitutes for Ce^{4+} in the fluorite-structured CeO_2, a higher oxygen storage capacity is achieved due to high coordination numbers and the availability of the Sn^{2+}/Sn^{4+} redox couple.

Recently, Nagarajan *et al.* examined the extent and effect of substitution of Th^{4+} with Sn^{4+} and Zn^{2+} cations in ThO_2 (Figure 3.3).[17] An epoxide-based sol-gel synthesis method was used to produce crystallite samples of ThO_2 in high purity. For this synthesis, aqueous thorium nitrate was dissolved in ethanol, propylene oxide was added while stirring, and the reaction was sonicated for 15 minutes. An example of the stoichiometric substitution of Sn^{4+} is given by Eq. (3.1):

$$\begin{aligned} &Th(NO_3)_4 \cdot 5H_2O + xSnCl_2 \cdot 2H_2O + C_3H_6O \\ &\xrightarrow{C_2H_6O} Th(Sn_x^{4+})O_{4x} + 3H_2O + C_3H_3O + 4HNO_3 + 2HCl \quad (3.1) \end{aligned}$$

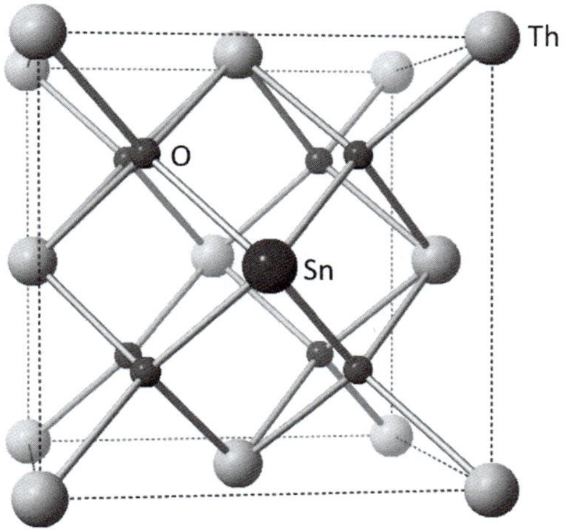

Figure 3.3. Representation of a Sn^{4+} cation substitution into the ThO_2 lattice.

Note that the propylene oxide serves to both dehydrate as well as to oxidise Sn^{2+}. It was found that the extent of substitution of Sn^{4+} was 50 mol% while the extent of substitution of Zn^{2+} was determined to be 15 mol%.

3.3 Halides

Lanthanide and actinide halides are one of the most well-studied class of materials due to both their fundamental and practical applications. For example, lanthanide fluorides can have outstanding optical properties while actinide fluorides are used as fuel material for enrichment as well as in molten salt reactor (MSR) processes. Additionally, the halides represent a convenient route for synthesis of more complex materials.

The trifluorides are easily precipitated from HF solution for trivalent cations such as the lanthanides and heavier actinides.[18] For Pu, a combination of a fluorinating gas such as HF or F_2 with a reductant such as H_2 is sufficient. A less hazardous route, however, was recently investigated with the use of NH_4HF_2 which dissociates into HF, H_2 and N_2 at 240 °C.[19] It is important to use a closed environment in order to contain the reducing agents and prevent oxidation to PuF_4. For very ionic systems like these, structure-type is heavily influenced by ionic radius and experiences a divide in both series according to this parameter. For Ln-Pm, the structure-type was first identified in and carries the name LaF_3. The LaF_3 structure-type is trigonal with the space group *P*-3*c*1 (No. 165). Oftentimes, the trifluorides can be difficult to crystallise and are notorious for exhibiting twinning.[20] In light of this, the proper refinement was only settled after neutron diffraction data was collected in 1985.[21] CmF_3, which should also be *P*-3*c*1, had also been misidentified and provoked some debate even after this event. More recently, calculations have suggested that the trigonal environment has an energetic advantage of $\Delta H = 9.1\,kJ/mol$ over its previously assigned structure.[22] As such, the CN drops from 11 to 9 and the Cm–F bonds become shorter which is more plausible. For Sm-Lu, the so-named YF_3 structure-type is adopted. This structure is orthorhombic with the space group *Pnma* (No. 62). BkF_3 is known to adopt this structure at low temperatures, yet when heated between 300 °C and 650 °C, it is able to transform into the *P*-3*c*1 structure.[23] Spectroscopic experiments have been able to conclude that the dimorphic Bk^{3+} sits in a less centrosymmetric position in the lower-temperature form.[24] The *Pnma* structure also should extend at least out to EsF_3, but this has only been inferred to date from decay-chain products.[25]

Unlike the lanthanide trihalides, the lanthanide tetrahalides represent a challenging class of complexes to make. The Ce^{4+}/Ce^{3+} redox couple of +1.61 V sets a benchmark for the entire series. Fluorine gas, however, is a strong oxidiser and can achieve the oxidation of both CeF_3 and TbF_3 to CeF_4 and TbF_4, respectively.[26] The synthesis of PrF_4 with this method is difficult, however, due to its decomposition at 90 °C, well below the temperatures required for the reaction to proceed. Instead, a room temperature synthesis involving photolytic cleavage of F_2 works very well for both the mixed-valent oxides of Tb and Pr (Eq. (3.2)):[27]

$$Pr_6O_{11} + nF_2 \xrightarrow{\text{aHF,UV}} PrF_4 \tag{3.2}$$

It should be noted that direct use of F_2 without photolysis causes a violent reaction that does not succeed in oxidising the metal. The black oxide loses colour after three days, which provides a good indicator for the reaction's progress.

Both UF_4 and its ammonium complexes can be used as efficient reagents for producing UN fuel material.[28] Nitrides carry particular advantages of fuel density and thermal conductivity, which can lower nuclear reactor temperature and thus save costs. However, there are many structures and phases of these materials that are sensitive to impurities depending upon the synthetic conditions.[29] UF_4, as the first example, has the ZrF_4 structure-type and crystallises in the monoclinic space group $C2/c$ (No. 15).[30] This is also true for Th-Cf. Since the metal's charge increases from +3 to +4 when comparing with the trifluorides, there is a greater tendency to adopt smaller coordination numbers — this is true here, as this series adopts a CN = 8. The U adopts a square antiprismatic polyhedron, and the structure extends in three-dimensions due to corner-sharing interactions. For the heavier actinides synthesis of the tetrafluorides can only be achieved by oxidation with fluorinating agents similarly to that described for the lanthanides.[31] For Np and Pu, mild reducing agents can suffice. The neptunium analogue, for example, can be precipitated from HF with the additive hydrazine hydrochloride.[32]

Complexation with ammonium provides a simple example of how to view these systems with additional cations. Sticking with the example of U, the addition of a cation typically limits itself to charge-balance the stable UF_8 polyhedron (F/U limit = 8).[33] Thus, the reaction of UF_4 and saturated NH_4F goes as follows (Eq. (3.3)):[34]

$$UF_4 + 4NH_4F \xrightarrow{80-130\ ^{\circ}C,\,6\,hr.} (NH_4)_4UF_8 \tag{3.3}$$

For PaF_4, the reaction was found to proceed in an autoclave.[35] Alternatively, the solid-state synthesis starting with the dioxide and NH_4HF_2 also works well (Th-Np).[36] Of course, a range of lower stoichiometric complexes are also known. An example TGA analysis of $(NH_4)_4PuF_8$ shows almost all of them (Eq. (3.4)):[34]

$$\underset{\text{pink}}{(NH_4)_4PuF_8} \xrightarrow{160\ ^{\circ}C} \underset{\text{green}}{(NH_4)_2PuF_6} \xrightarrow{210\ ^{\circ}C} \underset{\text{orange}}{7NH_4F \cdot 6PuF_4}$$

$$\xrightarrow{220\ ^{\circ}C} \underset{\text{light brown}}{\gamma(NH_4PuF_5)} \xrightarrow{290\ ^{\circ}C} \underset{\text{pink}}{NH_4F \cdot 3PuF_4} \xrightarrow{300\ ^{\circ}C} \underset{\text{pink}}{PuF_4} \tag{3.4}$$

Returning back to $(NH_4)_4UF_8$, it should be noted that the observed crystallographic behaviour is isostructural to UF_4.[37] However, there are a few important differences to note. First, the coordination polyhedron changes. Initially, it was noted that the $[UF_8]$ groups could be configured as an intermediate to both the distorted square antiprism polyhedron as well as the distorted trigonal dodecahedron. Penneman and Ryan gave this problem further treatment and noted that the value for *rms* deviations was less in the distorted trigonal dodecahedron.[38] The origin of this problem generally arises due to the very small energetic differences in both eight-coordination polyhedra. Therefore, differences in packing forces lead the complexed uranium fluoride to deviate (Figure 3.4, left). Second, the overall topography changes due to the hydrogen bonds that exist between the N-H $\cdots$ F atoms. The UF_8 polyhedra arrange into layers along [001], and the cations connect these layers via the H-bonding.

More recently, XAFS spectroscopy measurements on $(NH_4)_4NpF_8$ have confirmed many of the findings aforementioned.[36] Studies on $(NH_4)_4ThF_8$, however, find a transition to CN = 9 and crystallisation within the triclinic space group *P*-1 (No. 2, Figure 3.4, right).[39] What is strange about this structure is that while the Th adopts an

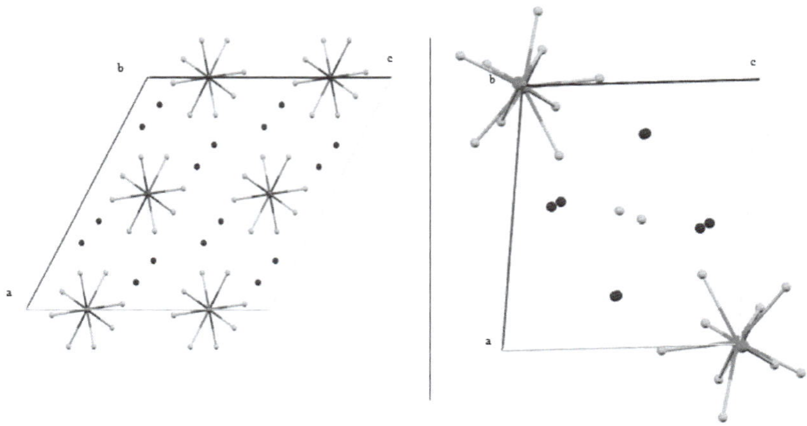

Figure 3.4. (Left) $(NH_4)_4UF_8$, like UF_4, crystallises in the monoclinic space group $C2/c$. Packing forces, however, cause its geometry to differ slightly. (Right) $(NH_4)_4ThF_8$ differs by crystallising in P-1, adopting CN = 9, and not coordinating all available F^- ions.[37,39]

increased tricapped trigonal prismatic geometry, it also fails to coordinate all of the F^- groups. The $[ThF_7]$ groups extend into chains along the b-axis, yet this happens to exclude one F^- that is surrounded by five NH_4^+ groups instead. The authors of the study point out that this complex is more stable than $(NH_4)_4UF_8$, and this should also have to rule out the possibility of an adduct.

Generally, there are a considerably smaller number of pentafluorides that exist and Pa and U have the largest number of stable pentafluorides.[40] For Np and Pu, the cations Cs^+, Rb^+ and NO^+ are all capable of stabilising the $[AnF_7]$ polyhedron.[41] For the +6 oxidation state, the most well-known compounds are UF_6 and PuF_6. Oxyfluorides of the type AnO_2F_2 have also been studied in detail, though the structural details of these will not be analysed here.

Chlorides and bromides share many similarities in synthesis as intermediate members of the halogen group. In many instances, the deliquescent properties of these compounds are a nuisance and several methods for dehydration are known. The use of $SOCl_2$ and CCl_4 have been common, for example. Another anhydrous synthesis method, the ammonium chloride/bromide route is shown and takes place between temperatures of 280 °C and 400 °C (Eq. (3.5)):[42]

$$Ln_2O_3 + 6NH_4Br \rightarrow 2LnBr_3 + 6NH_3 + 3H_2O \qquad (3.5)$$

This method also works when starting with a hydrated chloride or bromide. Depending upon whether using a lighter or heavier rare-earth element, the intermediate complex will be different which can affect the rate of the reaction ($(NH_4)_2LaBr_5$ and $(NH_4)_3YBr_6$, respectively). Furthermore, heat can be a problem due to another competing reaction to form oxychlorides and oxybromides. For example, in the series of lanthanides, the thermodynamic transition temperature for thermal decomposition of the tribromides to the oxybromides is given as follows (Eq. (3.6)):[43]

$$\underset{413\ °C}{\text{Pr}} > \underset{385\ °C}{\text{Nd}} > \underset{334\ °C}{\text{Pm}} > \underset{269\ °C}{\text{Sm}} > \underset{252\ °C}{\text{Gd}} \qquad (3.6)$$

Oxychloride temperatures tend to be higher, but the negative trend across the series has inspired a number of alternative low-temperature, anhydrous synthetic routes. One of these, a transmetallation reaction with the use of $HgCl_2$ proceeds as follows (Eq. (3.7)):[44]

$$3HgCl_2 + 2Ln \xrightarrow{50\ °C,\ THF} 2LnCl_3(THF)_3 + 3Hg \qquad (3.7)$$

The disadvantage to this procedure is that it requires Ln metal which can be an expensive investment for the goal of making an anhydrous halide. Instead, another easier method has been introduced by generating HCl as a reagent *in situ* (Eq. (3.8)):[45]

$$Ln_2O_3 + 3SOCl_2 + 2nDME \xrightarrow{RT} 2LnCl_3(DME)_n + 3SO_2 \qquad (3.8)$$

This reaction requires a catalytic amount of water to proceed while the DME stymies any attempt by the $SOCl_2$ to coordinate. If anhydrous HCl is used instead, this reaction produces the undesired $LnCl_3(H_2O)(DME)$. Clearly, this reaction is advantageous in a number of ways and provides an easier path to anhydrous chloride starting materials. Among actinides, this method might sometimes be inappropriate due to the concerns of combining radioactive material with peroxide-forming ethers. In place of that as well as the ammonium chloride route, a solid-state reaction has been recently developed and used successfully with Np and Am (Eq. (3.9)):[46]

$$2AmN + 3CdCl_2 \xrightarrow{450\ °C} 2AmCl_3 + 3Cd + N_2 \qquad (3.9)$$

In Eq. (3.9), a water-cooled cold finger traps the cadmium vapour that is released during the reaction. Furthermore, due to a lack of oxygenic impurities, there is no need for a sublimation step.

Structurally, the trichlorides and tribromides show more diversity in the phases adopted. The beginning of the series starts with the UCl_3 structure-type ($P6_3/m$, No. 176) followed by $PuBr_3$ ($Cmcm$, No. 63), $AlCl_3$ ($C2/m$, No. 12) and BiI_3 structure-types (R-$\mathit{3}$, No. 148).[47] UCl_3 is the ambient structure-type in the chlorides for La-Tb as well as U-Cf while $PuBr_3$ exists in Gd-Yb and Bk-Cf. In the lanthanides, for example, La-Gd possess CN = 9, Tb possesses CN = 8 and Dy-Lu a CN = 6. Among the four structures, the $PuBr_3$ type is the most efficiently packed and therefore resists changes under pressure despite its need to conform to the $AlCl_3$-structure at an ionic radius limit of 0.91 Å.[48] This has been demonstrated in both $BkCl_3$ and $CfCl_3$ up to 35 GPa.[49]

Compared with the fluorides, a larger anion such as bromide can decrease the coordination number. For example, $CmBr_3$ has a CN = 8 and has a layered structure which possesses relatively weak interactions holding the layers together.[22] CmI_3 exaggerates this trend even more with a CN = 6 and an even more unstable, layered structure. This, in terms of structure, is why I^- is considered the best leaving group in solid-state synthesis and synthesis in general. This trend can most readily be exploited in the thermal decomposition of halides to reach lower-valent complexes (Eqs. (3.10) and (3.11)):[44,50]

$$2SmI_3(THF)_3 \xrightarrow{220\ ^{\circ}C} 2SmI_2(THF)_{1.5} + I_2 + THF \quad (3.10)$$

$$2CfBr_3 \xrightarrow{580\text{–}650\ ^{\circ}C,\, H_2} 2CfBr_2 + Br_2 \quad (3.11)$$

Note that in the first reaction, Yb, with a less negative redox couple, thermally decomposes at 120 °. Also note that H_2 is important in the second reaction to provide a reducing environment and prevent oxybromide competition. In fact, the second reaction is more tricky with I^- because temperature fluctuations outside of 550 °C–570 °C have been reported to produce the oxyiodide.[51] Overall, it can be seen that the thermal decomposition temperature used is a function of both the metal's redox potential and halide choice.

Several lanthanide halide solvates have interesting structural features to highlight. Returning to the THF adducts, a variety of stoichiometry is possible. One example already seen is $LnCl_3(THF)_3$ which is referred to as the classical system.[52] This is mononuclear with CN = 6. Willey *et al.* point out that there are two other main classes of these, $LnCl_3(THF)_2$ and $LnCl_3(THF)_{3.5}$. The first is generally polymeric due to bridging (μ-Cl)$_2$ groups. This system generally includes lighter lanthanides with CN = 7. The second represents somewhat of a transition between CN and features pairs of cations (CN = 7) and anions (CN = 6) with the formula $[LnCl_2(THF)_5][LnCl_4(THF)_2]$ after autoionisation. Another adduct that shows a similar breadth of variety is $LnCl_3(DME)_n$. Polymeric structures are generally observed when $n = 1$ whereas monomeric structures dominate when $n = 2$.[45] Unlike the lanthanide chlorides of type $LnCl_3$, which exhibit a range of coordination numbers (6–9), DME's stabilising effect narrows this. The lighter lanthanides adopt a CN = 8 while the heavier lanthanides have a CN = 7. Nd represents a clear transition point in this series, as it can adopt both. For example, when $n = 2$ the lighter lanthanides experience a doubling of the unit due to two bridging μ_2-Cl atoms. This dinuclear structural unit is represented best by the formula $Ln_2Cl_2(\mu_2\text{-}Cl)_2(DME)_4$. It crystallises in the monoclinic space group $P2_1/n$ (No. 11) while the coordination polyhedron is intermediate between a dodecahedron and a bicapped trigonal prism. The mononuclear, seven-coordinate heavier lanthanides also crystallise in this space group although the coordination polyhedron is noted to be a distorted pentagonal bipyramid.

As was pointed out earlier, the synthesis of chlorides/bromides using the ammonium chloride/bromide route competes with oxyhalide formation. When oxyhalides are the desired product, raising the temperatures are necessary (Eq. (3.12)):[53]

$$Eu_2O_3 + 2NH_4Cl \xrightarrow{400\ ^\circ C} 2EuOCl + 2NH_3 + H_2O \tag{3.12}$$

Similar hydroxide-based chlorides can also be synthesised via solvothermal methods.[54] These materials are low-cost and can have highly desirable optical properties which can be exploited commercially. The lighter lanthanides, as shown in this example, crystallise in the tetragonal space group $P4/nmm$ (No. 129) and have a CN = 8. This difference in CN from the UCl_3 structure-type (CN = 9) is presumably due to the geometrical bonding differences seen in the new lattice environment. Two pairs of Cl^- ions bond from opposite sides, which effectively creates layers of positively charged LnO^+ and negatively charged Cl^-. The Ln has a distorted monocapped tetragonal antiprism coordination geometry.

Building from what has been presented so far, the next example features a structure that was made by mistake. Intending to reduce $YbCl_3$ to divalent $YbCl_2$, the researchers accidentally started with some YbOCl due to a mistake in the original ammonium chloride synthesis for starting material (Eq. (3.13)):[55]

$$YbOCl + 3YbCl_3 + 4Li \xrightarrow{750\text{–}950\ ^\circ\mathrm{C},\, 7\, days} Yb_4OCl_6 + 4LiCl \quad (3.13)$$

Interestingly, one Yb centre has a CN = 9 while another has a CN = 7. Furthermore, the Yb–O bond lengths are much shorter in this structure (2.243 and 2.287 Å) than in the comparable YbO (2.437 Å). A much different example of a recent structure examined is $[N(CH_3)_2][PuCl_6]$.[56] This chloride is an alkylated ammonium salt which is similar to the ammonium fluoride salts viewed prior. However, alkylation of an ammonium affects any structure quite seriously. Sometimes, various degrees of disorder are introduced and the number of packing arrangements can grow. This example crystallises in the cubic space group *Fd-3c* (No. 228) while Pu sits on a general position coordinated to six Cl^- ions (Figure 3.5, right). However, at 77 °C the structure undergoes a rotative-displacement transition, which moves it into the space group *Fm-3m* (No. 225) with the Pu on

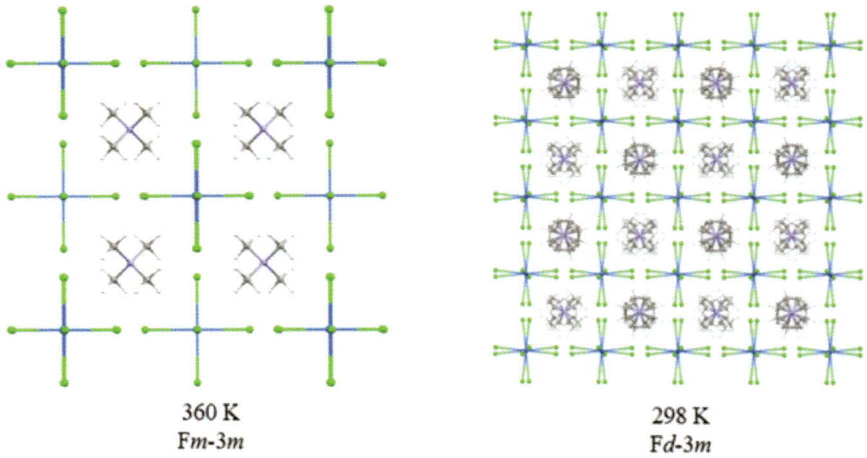

Figure 3.5. (Left) $[N(CH_3)_2][PuCl_6]$ at 77 °C. The Pu sits on an O_h site. (Right) Cooling to ambient temperatures causes a rotative displacement in the structure, which doubles the unit cell.[56]

an octahedral site (Figure 3.5, left). It was found that donor–acceptor interactions between methyl protons and Cl^- could likely be behind the phase transitions as tetramethylammonium cations rotate out of position and halve the unit cell at the higher temperature.

Such examples should indicate that there can be much more complexity to lanthanide and actinide halide structures depending upon substitutions in coordination, counterion choice and the synthetic conditions chosen. Furthermore, covalent contributions to the lattice energy increase in the order $F^- < Cl^- < Br^- < I^-$. One such calculation gives the covalent contribution to lattice energy as 10–14% in LnI_3 and 13–15% in AnI_3 vs. $(Ln/An)F_3$ which only have 3–7%.[57] There are many different variables to consider and with the ease of use, halides will continue to represent one of the most frequent structural motifs visited in lanthanide and actinide chemistry into the future.

3.4 Oxyanions

Lanthanide and actinide oxyanionic frameworks are of great significance in f-element chemistry, as they constitute a substantial number of the compounds utilised in the environment, industry and synthesis. In accordance with Pearson's Hard-Soft Acid-Base (HSAB) Theory, oxyanions often form very stable compounds with f-elements. By application, the f-elements are considered to behave, as hard Lewis acids while oxygens will act as hard Lewis bases. A large portion of the rare earth oxyanion structures consists of carbonates, phosphates, silicates, borates, nitrates and sulphates. Other important, but not as commonly studied systems include vanadates, molybdates, perchlorates and iodates. The synthesis, structure and bonding of the oxyanions that give rise to their properties are discussed in further detail in the following sections.

3.4.1 *Soluble oxyanion systems*

Rare-earth nitrates serve as excellent compounds for synthesis in f-element chemistry because of their high solubility in many solvents and ease of preparation. A general technique for making rare earth nitrates can be accomplished by dissolving the metal oxide in strong nitric acid (6M or higher) with gentle heat as shown by the following reaction (Eq. (3.14)):

$$x\mathrm{Ln_2O_{3(s)}} + (6x + 2n)\mathrm{HNO_{3(aq)}} \rightarrow 2\mathrm{Ln(NO_3)_3} \cdot n\mathrm{H_2O(s)} + \mathrm{H_2O} \quad (3.14)$$

The aforementioned equation is not completely balanced because of the many hydrated forms that can be obtained with different lanthanides. Across the $4f$-series, nitrates have been synthesised with degrees of hydration spanning from the monohydrate to hexahydrate forms although the latter type is the most common.[1] Anhydrous nitrates can be obtained several ways. The first and most preferable method is by heating the penta- or hexahydrate under vacuum at 100 °C.[58] Another reliable method, includes heating the appropriate sesquioxide or metal with N_2O_4 at 150 °C in an inert atmosphere.[59] Similarly to the halides, most oxyanions suffer competing reactions at higher temperatures. For example, heating lanthanide nitrates above 260 °C can result in their decomposition to oxynitrates, $LnO(NO_3)$.[60]

There are other interesting routes for preparing anhydrous lanthanide nitrates. However, these routes are not considered ideal, as they require crystallisation and purification from nitric acid. This route has been shown to actually incorporate nitric acid within the structure.[58,61] It should also be noted that there are lanthanides that can be more difficult in nitrate synthesis. These lanthanides include cerium, praseodymium and terbium, as their tetravalent oxidation states are accessible. For these elements, beginning with the metal may be more suitable.

In the heavier actinides (Am–Cf), the nitrate can be prepared in a similar manner by gently heating the metal oxide with strong nitric acid. This can be attributed to the similarities in oxidation states of the heavier actinides with the lanthanides. For the lighter actinides (Ac–Pu), obtaining the nitrate is more challenging because of the array of oxidation states available. Only the tetravalent oxidation state is obtainable for thorium in which thorium oxalate is treated with boiling, concentrated nitric acid.[62] Recently, there has also been another method developed for synthesising ultra-high purity thorium nitrate.[63] Uranium almost exclusively exists as the uranyl nitrate salt, $UO_2(NO_3)_2$. Similarly, pentavalent plutonium and neptunium can also exist as their actinyl nitrate salts, $PuO_2(NO_3)$ and $NpO_2(NO_3)$, respectively.[64]

Like nitrate, sulphate is an oxyanion which complexes strongly to lanthanides and actinides in acidic solutions. Sulphates tend to be very soluble and resist extensive synthetic and crystallographic manipulations. However, slow evaporation is a very common procedure towards getting crystals. Complex mixtures and robust synthetic approaches generally do not work well with sulphate because of its mild coordination ability, and both carbonates and phosphates outcompete it in the environment.

The bonding of nitrates to the metal centre in f-elements is very similar among all the lanthanides and actinides. Nitrates are almost solely inner-sphere coordinating and bidentate through two of the oxygen atoms. In the lighter lanthanides (La–Gd), a triclinic structure is observed for the hexahydrates, but in the heavier lanthanides and actinides many crystallographic forms are observed.[65] Therefore, an example of detailed structural characteristics are investigated with one unique, well-characterised compound: $Th(NO_3)_4 \cdot 4H_2O$.

Thorium(IV) nitrate tetrahydrate is unique as it is one of the few f-element binary nitrate salts obtainable in the tetravalent oxidation state. Charpin *et al.*,[66] published the first detailed structure of this compound in 1987 and showed that it crystallises in the monoclinic space group $P2_1/n$ (No. 11, Figure 3.6). All nitrates are coordinated in a bidentate manner, and all four waters are also coordinated in the equilateral plane making the thorium site a 12-coordinate icosahedron. This coordination number is larger than any examples shown for oxides, halides and other Ln nitrates. Furthermore, water–water hydrogen bonds (2.72 Å) and water–nitrate hydrogen bonds (2.92 Å) dominate the structure.

The hydrated sulphate $Th(SO_4)_2 \cdot 8H_2O$ shares a number of similarities with $Th(NO_3)_4 \cdot 4H_2O$ (Figure 3.7).[67] Both crystallise in monoclinic $P2_1/n$ (No. 11) and have high coordination numbers. NO_3^-, with a short N–O bond distance and a bite angle of 120 °C, can pack more efficiently and thus helps to give a CN = 12. SO_4^{2-}, while it has tetrahedral geometry, still has a longer idealised O–O contact distance of 2.43 Å (vs. 2.15 Å for nitrate) and therefore cannot pack as efficiently. The Th in this case has a CN = 10. Both systems bond in a bidentate fashion. The bonding was composed of four nitrates and four waters in the first example while there are only two sulphates with a complement of six waters in the second. Moreover, the Th–O_N bond distance is 2.59 Å while the bond distance for Th–O_S is

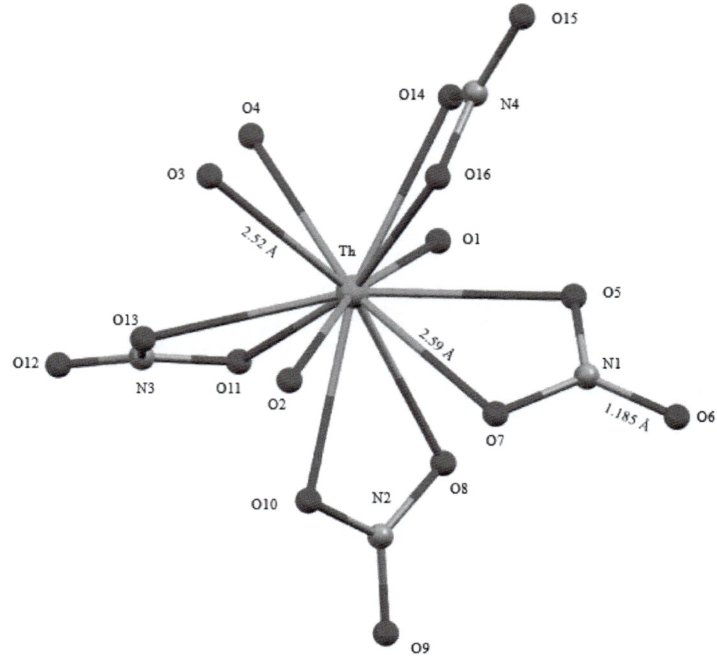

Figure 3.6. Representation of the 12-coordinate $Th(NO_3)_4 \cdot 4H_2O$.[66]

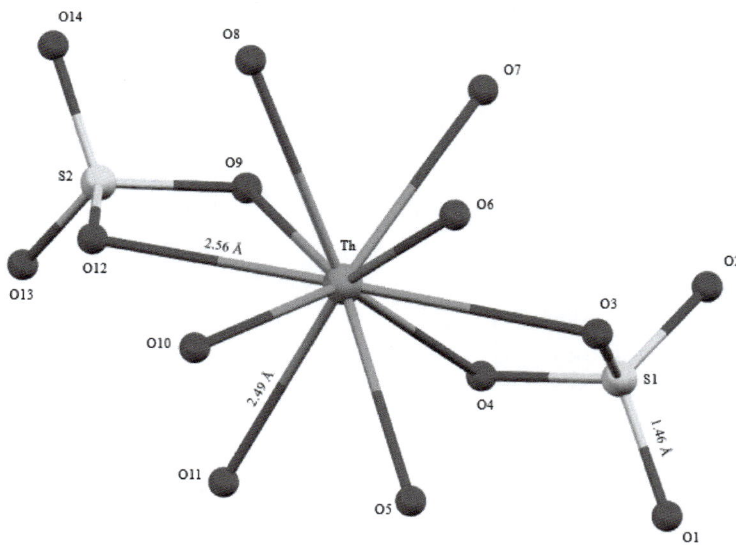

Figure 3.7. Representation of the 10-coordinate $Th(SO_4)_2 \cdot 8H_2O$.[67]

2.56 Å. Both of these bond distances are longer than for the average water bond distances (Th–O_W = 2.52 Å in nitrate and Th–O_W = 2.49 Å in sulphate). This is an interesting aspect considering nitrates and sulphates quite often out-compete water in coordinating to f-elements.

One important key difference between nitrates and sulphates in general is that sulphates can frequently bridge cation centres. When nitrates are the predominantly coordinating anion, they strongly favour bidentate coordination modes. This leaves a terminal N–O group that can participate in hydrogen bonding but does not bridge the structure. In contrast, sulphates can have much greater variety in coordination modes, and it can be very difficult predicting whether a sulphate will coordinate in a monodentate, bidentate or tridentate fashion. It is not uncommon to also have a mixture of binding modes with bridging in the structure. Additionally, the degree of hydrogen bonding in sulphates compared to nitrates can reach a frenetic count.

For example, advancing one position to the right of the series can produce $U(SO_4)_2 \cdot 4H_2O$.[68] The differing degree of hydration, sometimes tuneable to the synthetic conditions, has a large impact on the structure seen. In this case, the U(IV) takes a CN = 8 (distorted square antiprism) and crystallises in the orthorhombic space group *Pnma* (No. 62). This structure exclusively features monodentate coordination modes from sulphate, and the structure becomes more layered due to bridging parallel to the *bc*-plane while two more sulphate terminal oxygens participate in H-bonding to hold these layers together. More recently, a thorough description of the isostructural $Ce(SO_4)_2 \cdot H_2O$ has been given which details a second polymorph, the α-form, which adopts the space group *Fddd* (No. 70).[69] Unlike the Th example cited earlier, the Th–O bond distances are longer for water.

Sulphate and nitrate varieties can vary significantly depending upon oxidation states. While U(IV) sulphates can be made rather easily, nitrates are only known to complex the UO_2^+ bicameral unit. This could be due to the well-known oxidising characteristics of NO_3^- as opposed to the slightly reducing qualities of SO_4^{2-}. A very good example demonstrating this for nitrates is the very common $UO_2(NO_3)_2 \cdot 3H_2O$, which became very notable due to the rare triboluminescence property it exhibits.[70] A more recent, refined study of this crystal structure has shown that there are two symmetrically distinct uranium sites that each form hexagonal bipyramids.[71] Depending on the type of hydration, the dihydrate, trihydrate and hexahydrate forms will crystallise in the space groups $Cmc2_1$ (No. 36), $P1$ (No. 1) and $P2_1/c$ (No. 14), respectively.

Lower oxidation states can accommodate a wider span of hydration, and Ce(III) sulphate can have differing degrees of hydration from 4 to 16.[72] Greater degrees of hydration will generally accompany a decrease in bidentate binding modes.[73] A special case of higher coordination comes from the double salt $La_2[SO_4]_3 \cdot 9H_2O$.[74] This structure crystallises in the hexagonal space group $P6_3/m$ (No. 176) and features one La with CN = 12 and one with CN = 9. The 12-site La also notably only has bonds to six bidentate sulphate groups. It is thought that one U(III) in the same complex also has a CN = 12, while the similar $[NH_4]U[SO_4]_2 \cdot 4H_2O$ represents the first system where a U(III)–O bond length was determined.

In solution, it is known that bidentate sulphate linkages increase across the actinide series. This has been shown for $Th^{4+} < U^{4+} < Np^{4+} < Pu^{4+}$.[75] While the number of monodentate linkages increase when moving to the solid-state, it is unclear whether this can also be related to the ionic radius since many structural motifs can change so easily when moving across the series. However, there has been some evidence for this in $Cs_aAn(SO_4)_n \cdot xH_2O$ complexes which relate the increasing Lewis acidity to a greater degree of bidentate coordination.[76]

Recently, the synthesis of two new plutonium(IV) nitrates have been obtained that show minute differences in structure.[77] The two complexes, $[Pu_2(OH)_2(NO_3)_6(H_2O_4]_2 \cdot 11H_2O$ and $Pu_2(OH)_2(NO_3)_6(H_2O_4)_4 \cdot 2H_2O$, both contain Pu(IV) centres but crystallise in different space groups even though the basic moiety of the two compounds are the same. Once again, differences in hydration have a large impact on the resulting structure. The first compound crystallises in the orthorhombic space group $Fddd$ (No. 70), while the latter crystallises in the monoclinic space group $C2/c$ (No. 15). Both structures are based on hydroxo-bridged Pu^{IV} dimers that are 10-coordinate from three bidentate nitrate anions, two waters, and bridging hydroxides. The $Pu \cdots Pu$ dimer lengths of each compound are approximately 3.8 Å.

Although f-element perchlorates are not as abundant as nitrates or sulphates, they are worthy of mention as a recent study has shown the occurrence of perchlorates with nitrates and iodates in deserts on Earth.[78] Other than occurrence in nature, perchlorates also serve well in synthesis and solution studies as they generally are outer-sphere non-coordinating and have no interference with the bare ion chemistry of f-elements.[79] However, due to the danger of perchlorates in organics, this oxyanionic system has been on the decline in synthesis.

The anhydrous form of perchlorates can be prepared by adding chlorine hexoxide (Cl_2O_6) to lanthanide chlorides, perchlorates, or nitrates at 180 °C on a vacuum line.[80] They crystallise in the hexagonal space group $P6_3/m$ (No. 176), which is related to the UCl_3 crystal system mentioned in the previous section. The lanthanide perchlorates have several unique features. The lutetium structure, for example, does not crystallise in the $P6_3/m$ space group but rather the trigonal R3*c* (No. 161) unit cell. Additionally, depending on the temperature, the thulium and ytterbium structures can exist in either space group.[81] On the contrary, fewer actinide perchlorate studies have been pursued. One example of a mixed anionic structure with neptunium was recently synthesised. The $[Np(DMSO)_7(NO_3)](ClO_4)_3$ (DMSO = dimethyl sulfoxide) was prepared in a slow evaporation by dissolving a DMSO solution of neptunium perchlorate in 1M HNO_3 and further shows the outer-sphere coordination of perchlorate anions.[82]

3.4.2 *Insoluble oxyanion systems*

Carbonates are often given considerable attention in the environment due to the low solubility forms in which congregation of f-elements can occur. This happens to be even more important when referring to actinides because any build-up of radioactive materials in the environment would be a cause for serious concern. Regarding synthesis, carbonates are not an ideal starting material for many reactions because of their low solubility and mild reactivity. While it is possible to prepare carbonates of the type $Ln_x(CO_3)_y$ or $An_x(CO_3)_y$, it is more common to obtain them in the form of double metal alkali salts such as $ALn(CO_3)_2$ (where A^+ = alkali metal and Ln^{3+} = lanthanide).[83] The difficulty of making these carbonates is itself demonstrated by the observation that three different synthetic methods were employed. The binary form $Ln_2(CO_3)_3 \cdot xH_2O$ is made using the following reaction scheme that employs CO_2 under pressure (Eq. (3.15)):

$$2LnCl_{3(s)} + 3CO_{2(g)} + 3H_2O + 6C_6H_5NH_2 \rightarrow Ln_2(CO_3)_3 + 6C_6H_5NH_3Cl \quad (3.15)$$

The work done by Brauer and Wetzel goes into more detail on this synthesis.[84] An interesting aspect to note is the ability to obtain the Eu(II) carbonate. Since europium is the easiest of the lanthanides to reduce to the divalent oxidation state, it is possible to synthesise air stable forms of these carbonates by way of the europium(II) sulphate.[84] After reduction from $Eu^{3+} \rightarrow Eu^{2+}$ with zinc metal in acidic solution, the reaction scheme below allows the divalent carbonate to be obtained as shown by Eq. (3.16):

$$EuSO_4 + Na_2CO_3 \rightarrow EuCO_3 + Na_2SO_4 \quad (3.16)$$

This product can be easily filtered off due to its low solubility in water. The compound crystallises in the orthorhombic space group *Pmna* (No. 53, Figure 3.8).

Although carbonates are very stable under a number of conditions, they can be exceptionally reactive in the presence of acids. However, without the presence of an acid, these compounds are relatively heat resistant while the $Ln_2(CO_3)_3 \cdot xH_2O$ system can generally withstand temperatures up to 400 °C before decomposition sets in (Eq. (3.17)).[85] Dehydration of water is seen between 100 °C and 150 °C, and further heating to higher temperatures (~900 °C) eventually converts the carbonate to an oxide as shown by Eq. (3.18).

$$\sim 400\ ^{\circ}C\ Ln_2(CO_3)_2 \rightarrow Ln_2O_3 \cdot CO_2 \quad (3.17)$$

$$\sim 900\ ^{\circ}C\ Ln_2O_3 \cdot CO_2 \rightarrow Ln_2O_3 + CO_2 \quad (3.18)$$

The production of the oxide through this route is ideal if preparation of other lanthanide starting materials is needed. The loss of CO_2 as a gas also allows for the pure product of lanthanide nitrate, halide or sulphate salts to be obtained.

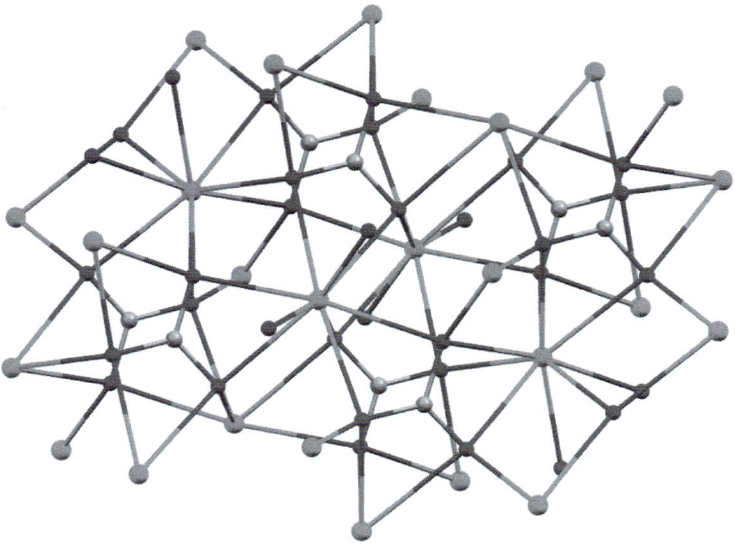

Figure 3.8. Packing structure of $EuCO_3$ that takes on the $CaCO_3$ structure-type. The Eu–O bond distances in Eu^{2+} are between 2.60 Å and 2.75 Å. This bond distance is longer than that of the Ca–O bond distances, which are between 2.45 Å and 2.65 Å.[84]

Compared to the lanthanides, very few studies on trivalent actinide carbonates have been conducted. This is most likely due to the lack of stability in the trivalent oxidation state for many of the lighter actinides. Among the tetravalent systems, plutonium(IV) and thorium(IV) carbonates have been the most well studied. Examples of these include $Pu(CO_3)_2$ and $Th(CO_3)_2 \cdot 0.5H_2O$. Plutonium(IV) carbonate is formed through the thermal decomposition of Pu(IV) oxalate.[86] On the other hand, thorium(IV) carbonate is prepared by heating $Th(OH)_4$ between 100 °C and 300 °C with CO_2 under pressure. In this example, care must be taken to avoid producing the mixed anionic system of hydroxide and carbonate.[87]

Pentavalent actinide carbonate chemistry has been largely dominated by neptunium. These systems exist as double metal or double cation carbonate systems of the formula type $A(NpO_2)CO_3 \cdot xH_2O$ ($A =$ alkali metal or NH_4^+).[88] An example of the ammonium carbonate synthesis is given by Eq. (3.19):

$$(NH_4)_2CO_3 + NpO_2^+ \rightarrow NH_4NpO_2CO_3 + NH_4^+ \qquad (3.19)$$

The Np(V) (and Am(V)) starting materials for these reactions can be obtained in solution by dissolving the oxide in 1*M* nitric acid followed by reduction with hydroxylamine chloride.[88]

The most dominant class of actinide carbonates are the hexavalent uranium carbonates. Rutherfordine, UO_2CO_3, exists naturally in nature but can be prepared in many ways.[89,90] A few methods include, but are not limited to: reacting CO_2 under pressure with UO_3, reacting alcoholic uranyl nitrate with CO_2 under pressure, or dry air oxidation of hygroscopic uranium(IV) oxycarbonate.[90] Other hexavalent carbonate systems that have been prepared include the neptunyl carbonate which is prepared at a pH = 5 in Eq. (3.20).

$$NpO_2(NO_3)_2 + Li_2CO_3 \rightarrow NpO_2CO_3 + 2LiNO_3 \tag{3.20}$$

The plutonyl carbonate (PuO_2CO_3) has also been prepared by the thermal decomposition of ammonium plutonyl carbonate or by the bubbling of CO_2 into a Pu(VI) solution at pH = 7.[91–93]

What gives rise to many of the properties of f-element carbonates is their chemical structure. In a study of aminocarbyne lanthanide carbonate salts, where the general anionic complexes consisted of $[Ln(CO_3)_4(H_2O)]^{5-}$ or $[Ln(CO_3)_4]^{5-}$, the carbonates all coordinate in a bidentate manner to the Ln^{3+} site.[37] In these structures, it is unclear whether the oxygens of the carbonate anions help form a distorted monocapped square antiprism or a distorted dodecahedron while the Ln^{3+} — O bond lengths range between 2.56 Å and 2.29 Å.[94] These bond lengths become shorter with the heavier lanthanides and calculations indicate that the covalency of the bond increases slightly. The strong coordination of carbonate anions to the lanthanide site and their ability to form extended structures most likely gives rise to their low solubility.

As mentioned previously, since there is not an excellent trend that exists among the actinide carbonates, a look at the structure and bonding of a few selected systems will be undertaken. Plutonium(VI) carbonate is particularly important because of the mobility in the environment with this oxidation state.[93] The mineral rutherfordine is also found in the environment. These two carbonate systems, along with the neptunyl carbonate, are isostructural as has been revealed by their powder XRD patterns.[90,92] The symmetry of these eight coordinate carbonates are orthorhombic and crystallise in the space group *Imm*2 (No. 44). The structures consist of hexagonal bipyramids and [CO_3] triangles and the hexagonal bipyramids form extended chains along the [001] plane.[90] The An–O bond distances are approximately 1.75 Å, which is an indication for a double or triple bond. The C atom of the carbonates are more distorted than usual, and this may be caused by the local connectivity of the polyhedra in the structure.[90,92]

Silicates are another large class of studied f-element materials. Lanthanide silicates are of importance due to their potential in electronics, semiconductors, sorption materials and lasers while actinide silicates have been of interest in the nuclear energy sector because they form refractory, stable compounds that can last for geological time scales.[95,97] Since most of the *f*-block silicates found in the literature are Ln silicates, they are given a more detailed analysis here.

Numerous solid-state techniques are used to synthesise Ln silicates. Other routes include hydrothermal syntheses and flux methods. Rocha *et al.*, for example, produced various layered silicates with Eu^{3+}, Er^{3+}, Tb^{3+} and Gd^{3+} using a hydrothermal approach.[95] An interesting advantage to layered silicates like these is that their structure can allow inclusion of a second or third type of Ln^{3+} ion into the crystal lattice, which enables the fine-tuning of properties.

The crystal structures of Ln silicates typically consist of either isolated SiO_4 tetrahedra or polysilicates of the forms Si_2O_7 and Si_3O_{10}.[96] The structures containing the isolated SiO_4 tetrahedra can be further divided into two different groups: one with isolated SiO_4 tetrahedra only, and one that contains SiO_4^{4-} and extra oxygen ions not bonded to silicon. Those with isolated SiO_4 tetrahedra only exist for divalent lanthanides, such as Eu^{2+}. For the ones containing two different types of anions, the Ln cations will typically exhibit 6 to 10-fold oxygen coordination.

The Ln_2O_3–SiO_2 phase diagram displays Ln_2O_3:SiO_2 compositions of 1:1, 1:2 and 7:9.[96] The 1:1 composition shows two different structure-types. The first structure-type is stable for the lighter lanthanides (La–Tb) and crystallises in the $P2_1/c$ (No. 14) space group; the second structure-type is stable for the heavier lanthanides (Dy–Lu) and crystallises in the $C2/c$ (No. 15) space group. On the other hand, the 1:2 composition displays several different polymorphs, mainly of the $Ln_2(Si_2O_7)$ structure-type. Finally, the 7:9 composition has been described as a cation-deficient variation of the oxyapatite structure and crystallises in the $P6_3/m$ (No. 176) space group. The 7:9 group is the most unique notably because it is the only isostructural group to include all the rare-earth elements.

Another class of materials that enables the fine-tuning of properties by interstitial intercalation are salt-inclusion solids.[97] Salt-inclusion phases of Ln and An silicates are a relatively new field of salt-inclusion chemistry and have gained increasing attention in academia and industry due to their luminescence and potential as waste-storage media and second-harmonic generators. Many of these salt-inclusion phases have been synthesised using mixed alkali halide eutectic fluxes where the salt ions have been shown to intercalate into the crystal framework.

In 2008, various new salt-inclusion samarium, europium and gadolinium silicates were synthesised using a KF-MoO_3 flux.[98] All three structures were found to be isostructural with one another, consisting of open-branched silicate single layers with 6-, 8- and 12-membered rings. These rings are further connected by LnO_6 octahedra to form a 3D framework. Finally, the F^- and K^+ are located in channels of the network and form F_2K_7 dimers structurally similar to those of Cl_2O_7.

Uranyl silicates are great candidates for nuclear waste management since they are able to hold radioactive elements within their crystal lattice without allowing them to escape easily into the environment.[97] Actinide(IV) silicates can be synthesised, for example, by heating SiO_2 with the appropriate dioxide.[96] An alternative method such as heating AnO_2F_2 with SiO_2 to 300 °C under pressure also works well.

In 2016, zur Loye *et al.*, used a CsF-CsCl flux to synthesise four new salt-inclusion uranyl silicates: $[Cs_3F][(UO_2)(Si_4O_{10})]$ (*Imma*, No. 74), $[Cs_2Cs_5F][(UO_2)_3(Si_2O_7)_2]$ ($P2_1/n$, No. 11), $[Cs_2Cs_5F][(UO_2)_2(Si_6O_{17})]$ ($P2_12_12$, No. 18) and $[Cs_9Cs_6Cl][(UO_2)_7(Si_6O_{17})_2(Si_4O_{12})]$ ($P1$, No. 2).[97] A simpler example of a similar stoichiometric reaction is given as follows (Eq. (3.21)):

$$2UO_2 + 3SiO_2 + 4CsF + O_2 \xrightarrow{900\ ^\circ C} 2Cs_2USiO_6 + 2SiF_2 \quad (3.21)$$

This would represent a 1:1 mole reaction between U:Si (and SiO_2 would become the limiting reagent). Additionally, a redox reaction takes place as U is oxidised to the hexavalent state.

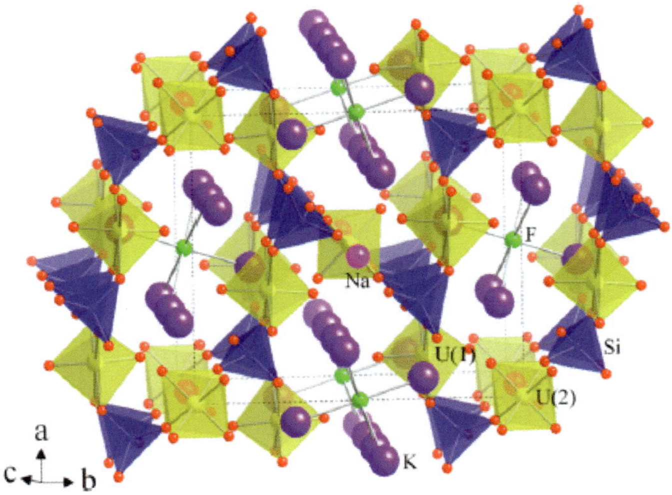

Figure 3.9. Crystal structure of $[NaK_6F][(UO_2)_3(Si_2O_7)_2]$, representative of all $[AB_6X][(UO_2)_3(Si_2O_7)_2]$ compounds.[99]

In these types of structures, isolated uranium polyhedra corner-share with Si_xO_y units to form 8-, 9-, 10- and 12-member uranyl silicate rings (Figure 3.9). The equatorial UO_6 oxygen atoms form these rings, where the axial uranyl oxygens point into the salt-inclusion channels. This structural property makes uranyl silicates ideal subjects for salt-inclusion phases. An additional interesting find in this study was that the surface-area-to-volume ratio of the flux reactions determined whether oxides or silicates were formed. A smaller surface-area-to-volume ratio was reported to favour the formation of silicates, while a larger ratio favoured the formation of oxide phases.[97,99]

Phosphates represent another tetrahedral class of oxyanions. Compared to sulphate, for example, the P–O bond is longer (1.53 Å) and can combine in a variety of linkages. Phosphates, however, carry an extra charge, which combine favourably with f-elements. It becomes difficult to study the structures of phosphates because precipitation and insoluble phases dominate the chemistry. Nevertheless, an interest in wasteform materials such as the monazite structure has fuelled the characterisation of these complexes.

The monazite structure (No. 11) is a desirable structure-type because it can accommodate both trivalent and tetravalent oxidation states. $PuPO_4$ has this structure as do other orthophosphates such as $PrPO_4$.[100] Pu has a CN = 9 and is edge-shared by the four oxygens of phosphate to extend in chains. From a greater perspective, the structure is arranged into tightly packed layers. Nevertheless, the flexibility of this structure allows for substitutions of the tetravalent state that minimise loss towards the integrity of the lattice. A common example is given by the cheralitic substitution as follows (Eq. (3.22)):

$$\begin{aligned}(1-2x)\mathrm{Ln}^{3+}\mathrm{PO}_4 + x\mathrm{An}^{4+}x\mathrm{M}^{2+} + 2x\mathrm{PO}_4^{3-} \\ \leftrightarrow \mathrm{Ln}^{3+}_{1-2x}\mathrm{M}^{2+}_x\mathrm{An}^{4+}_x(\mathrm{PO}_4)\end{aligned} \tag{3.22}$$

As seen from the equation, it is necessary to maintain charge balance by incorporating a divalent ion such as calcium along with the tetravalent actinide.[101]

Another type of insoluble oxyanion is borate. This anion has a much more polarisable B–O bond and is often able to create a variety of structural motifs inside the structures by itself. These elements are typically BO_3^- triangles and BO_4^- tetrahedra, but polyborate groups are also known to form. Furthermore, such complex structures can come about through a variety of synthetic methods, which include high-temperature solid-state melts, slow evaporation and hydrothermal boric acid flux reactions. Small changes in concentration, temperature, pH, pressure, cation size, stoichiometry or counterions can serve to alter the structural topology.

For example, use of either HI or Zn/Hg amalgam in a hydrothermal reaction with boric acid produces the complex europium borate $Eu[B_8O_{11}(OH)_4]$.[102] This is both an air and water-stable Eu(II) compound which demonstrates that borate can stabilise lower oxidation states. Another such example of this was shown from the stabilisation of divalent Cf in a SrB_4O_7 lattice.[103] In the europium borate example given, the Eu has a CN = 9 and displays a hula-hoop geometry (Figure 3.10). There are two BO_3 triangles and one BO_4 tetrahedra, and these arrange in such a way to give the structure a layered, 2D environment (Figure 3.10). Among borate structures many different layered topographies are known and obey a classification system, while the example given is classified as M-type.

A few additional oxyanions that are worth discussing include molybdates, vanadates and iodates. These oxyanions are not as

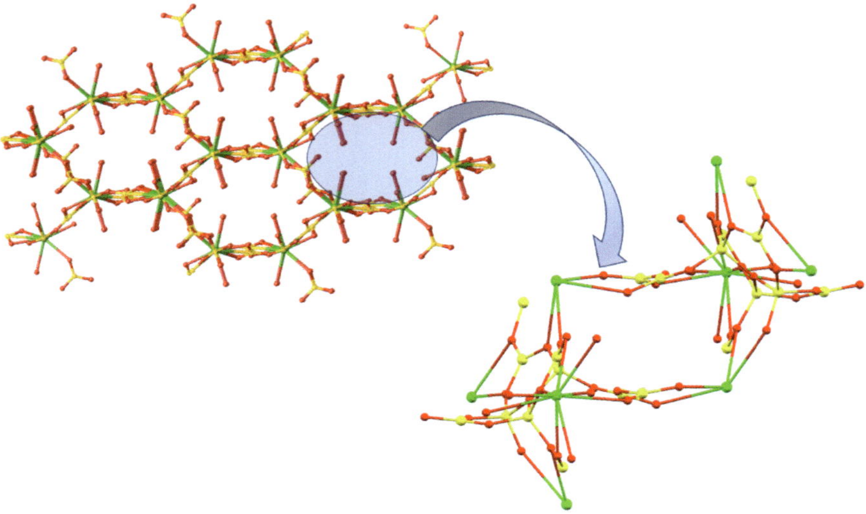

Figure 3.10. A representation of the M-type layered topography in borates. (Inset) Bridging units of Eu[$B_8O_{11}(OH)_4$] that adopt a hula-hoop coordination geometry. B: yellow spheres, O: red spheres, Eu: green spheres.[102]

common in the environment but have been known to exhibit unique properties that have potential applications for the design of phosphors and stimulated Raman scattering (SRS) crystals for lasers.[104,105]

Vanadates represent another tetrahedral oxyanion with V–O bond lengths usually greater than those for the earlier oxyanions presented. Among vanadates, for example, there has recently been the emergence of a new orange-emitting phosphor.[104] There has also been a series of tetragonal SRS crystals synthesised with the stoichiometric formula $LnVO_4$.[105] Kaminskii *et al.* have shown that this structure of lanthanum vanadate can be prepared hydrothermally by heating between 220 °C and 280 °C for 24–48 hours.[106] However, this route can also lead to by-products such as the LaV_3O_9 structure.[107]

$LnVO_4$ has a zircon-type structure that consists of isolated VO_4 tetrahedra surrounding the lanthanide site to form a dodecahedron (Figure 3.11). However, it is noteworthy that the cerium structure is dependent on pressure and is capable of taking three different types of structures: zircon, monazite and scheelite. As expected, the unit cell decreases in volume with these phase transitions.[107]

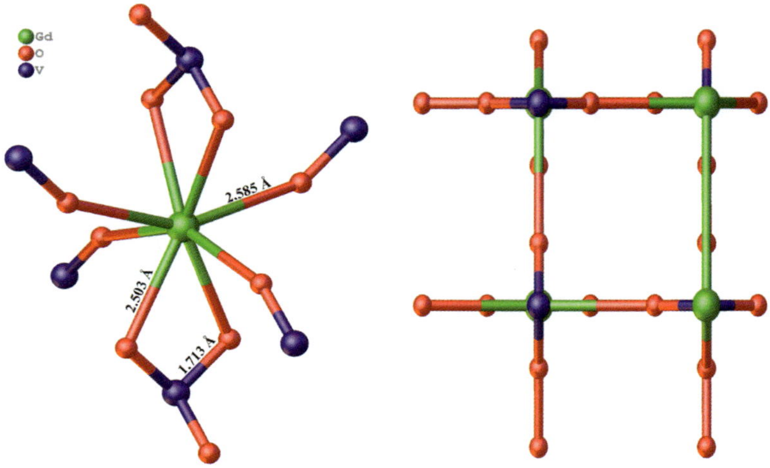

Figure 3.11. Structure of tetragonal $GdVO_4$. The packing view (right) shows the symmetry of the $LnVO_4$ compounds. It should be observed that the Ln–O bond distances in the equatorial plane differ from those in the axial plane with distances of 2.585 Å and 2.503 Å, respectively.[107]

In the zircon structure, the Ln–O bond distances of the compounds range from 2.44 Å to 2.96 Å. The VO_4^{3-} groups are bidentate, and there is also bridging between the metal centres. While vanadates have been well studied in the lanthanide series, this is not as true for the actinides. One notable study published in 2015 has shown the insertion of lanthanides into uranyl vanadate frameworks.[108] The $(UO_2)_3(VO_4)_2(H_2O)_3$ compounds consist of $UO_2{}^{2+}$ pentagonal bipyramids that coordinate to the VO_4 tetrahedra and has demonstrated a new path for synthesising 4*f*-5*f* compounds.

Molybdates are another system that has been studied within the f-elements that portray interesting properties. The silver-lanthanide system ($AgLnMo_2O_8$), for example, has been shown to exhibit luminescence and magnetic properties.[109] Molybdates have even been characterised with radioactive promethium.[110] As such, these compounds can form structures capable of withstanding high decomposition temperatures like carbonates. A study of $Eu_2Mo_5O_{18}$ and $Gd_2Mo_5O_{18}$ showed that these novel compounds could be obtained by the thermal decomposition of their polyoxomolybdate precursors at 750 °C.[111] Other lanthanide molybdates such as europium(II)molybdate (Eu_xMoO_4) have also been prepared and studied.[112]

Actinide molybdates such as uranium molybdate ($UMoO_6$) can be prepared by heating U_3O_8 with MoO_3 in air at 725 °C but can also be prepared from U_3O_8 with heat and anhydrous MoO_3.[113] The reaction is given as follows in Eq. (3.23):

$$1/3U_3O_8(s) + MoO_3(g) + 1/6O_2(g) \rightarrow UMoO_6(s) \qquad (3.23)$$

Lanthanide molybdates generally take the scheelite structure-type ($CaWO_4$). As expected, the structure of these compounds show a decrease in lattice parameters and unit cell volume from yttrium to lutetium, and the metal centres are eight coordinate with $Mo^VO_4^{3-}$ tetrahedra.[112] The thorium molybdate structure ($Th(MoO_4)_2$) has also been published and crystallises in the orthorhombic space group *Pbca* (No. 61).[114] However, this structure differs from the plutonium and uranium molybdate structures because it contains bridging molybdate tetrahedra in a square antiprismatic environment.

Finally, the iodates have been a nice system for studying gradual trends in $4f$ and $5f$ element structure and bonding. The f-element iodates are synthesised under simple hydrothermal conditions by heating the metal or metal nitrate with periodic acid at temperatures between 170 °C and 180 °C for 72 hours.[115–117] This preparation results in the formation of large single crystals that are often suitable for single crystal X-ray diffraction.

Iodate structures have been collected for every lanthanide except radioactive promethium. Additionally, structures for uranium through californium have also been obtained.[115–117] These compounds take on the type I $Ln(IO_3)_3$ structure except for the smaller, heavier ions of ytterbium, lutetium and californium which take on the monoclinic $Bi(IO_3)_3$ ($P2_1/n$ or No. 11) structure-type (Figure 3.12).[115] For the f-elements of the $Bi(IO_3)_3$ type structure, the environment can be described as a tricapped trigonal prism while the seven-coordinate structures have An/Ln–O bond distances ranging between 2.35 Å and 2.92 Å.[116] In the actinyl iodates of U, Np, and Pu, the $NpO_2(IO_3)_2 \cdot H_2O$ and $PuO_2(IO_3)_2 \cdot H_2O$ compounds have been determined to be isostructural. In these compounds, there is a pentagonal bipyramidal coordination environment around the metal centres. The axial An–O bond distances range from 1.73 to 1.78 Å whereas the equatorial bond distances range longer between 2.33 Å and 2.34 Å.[117]

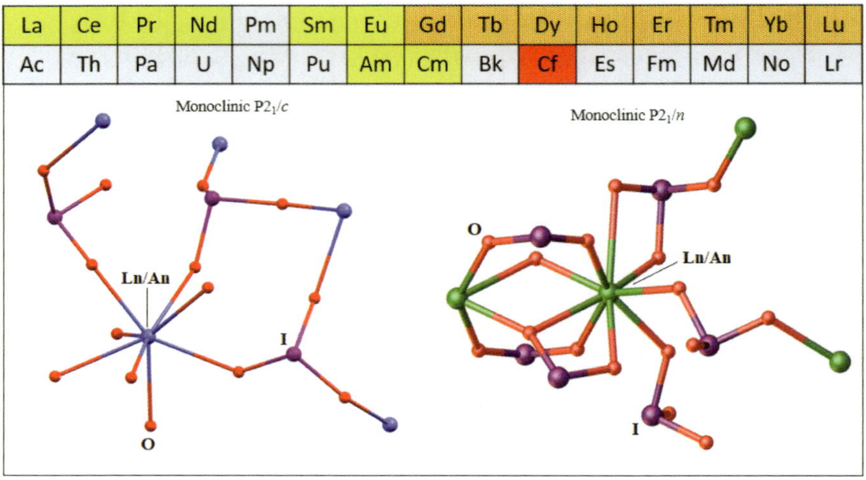

Figure 3.12. Lanthanide and actinide iodate structures. The two major types of f-element iodate compounds consist of the monoclinic $P2_1/c$ (yellow) on the left and monoclinic $P2_1/n$ (red) structures shown on the right. Elements highlighted in orange can take on both structure-types.[115]

3.5 Summary

As one can see from the aforementioned examples, there is an astounding amount of structural diversity in solid-state f-element chemistry. While fundamental aspects of this chemistry have been demonstrated, other recent, cutting-edge examples have shown that there is little to expect other than more intensive growth with this discipline in the future. In recent years, many more synthetic and analytical methods have been developed for this science, and the authors foresee that many new and exciting discoveries still remain undiscovered and lie ahead in the future.

References

1. A. Rabenau, *Angew. Chem.*, 1985, *24*, 1026–1040.
2. (a) N. Deifel and C. Cahill, *CrystEngComm*, 2009, *11*, 2739–2744; (b) N. Deifel and C. Cahill, *Compt. Rend. Chim.*, 2010, *13*, 747–754.
3. (a) M. K. Han, *Rare-Earth Transition-Metal Intermetallics: Structure-Bonding-Property Relationships*, 2006, Ph.D. Thesis, Iowa State University, Ames, IA; (b) I. R. Konenko, N. M. Parfenova, E. I.

Klabunovskii, E. M. Savitskii *et al.*, *Bull. Acad. Sci. USSR, Div. Chem. Sci.*, 1981, *30*, 749–753; (c) P. Chai, M. Abramchuk and M. Shatruk, *Crystals*, 2016, *6*, 165.
4. D. Givord, H. S. Li and J. M. Moreau, *Solid State Comm.*, 1984, *50*, 497–499.
5. (a) J. W. Walkiewicz, J., J. S. Winston and M. M. Wong, *Bureau of Mines Report of Investigations 7784*, 1973, U.S. Dept. of the Interior, Washington, D.C.; (b) A. Verma, P. Verma and R. K. Sidhu, *Bull. Mater. Sci.*, 1996, *19*, 539–548; (c) M. T. Mikhov, W. Gong and G. C. Hadjipanayis, *J. Appl. Phys.*, 1987, *61*, 3460.
6. D. E. Swets, *J. Cryst. Growth.*, 1986, *75*, 277–280.
7. (a) J. Herbst, J. Croat, F. Pinkerton and W. Yelon, *Phys. Rev. B*, 1984, *29*, 4176–4178; (b) K. Miliyanchuk, L. Havela, A. V. Kolomiets, S. Danis, L. C. J. Pereira and A. P. Goncalves, *Phys. B Cond. Mat.*, 2006, *378–380*, 983–984.
8. (a) R. P. Pinto, M. M. Amado, M. A. Salgueiro, M. E. Braga *et al.*, *J. Mag. Mat.*, 1995, *140–144*, 1371–1372; (b) F. R. Boer, K. Kindo, H. Nakotte, K. Prokes and V. Sechovsky, *Phys. B Cond. Mat.*, 1998, *246–247*, 129–134.
9. (a) N. Zhang, R. Yi, L. Zhou, G. Gao *et al.*, *Mat. Chem. Phys.*, 2009, *114*, 160–167; (b) X. Wang and Y. Li, *Angew. Chem.*, 2002, *41*, 4790–4793.
10. D. R. Dillin, *The Lanthanide-Hydroxides and Hydrous-Oxides*, 1972, Ph.D. Thesis, Baylor University, Waco, TX.
11. K. W. Bagnall, *The Actinide Elements*, Elsevier, Amsterdam, 1972.
12. C. Marin, *Synthesis and Applications of Lanthanide Sulfides and Oxides*, 2016, Ph.D. Thesis, University of Nebraska, Lincoln, NE.
13. R. J. Konings, O. Beneš, A. Kovács, D. Manara *et al.*, *J. Phys. Chem. Ref.*, 2014, *43*, 13101.
14. R. J. Bard and D. F. Bowersox, *LA Rep. No. 2076*, Los Alamos Scientific Laboratory, Los Alamos, New Mexico, 1957.
15. M. Bharathy, A. H. Fox, S. J. Mugavero and H. C. zur Loye, *Solid State Sci.*, 2009, *11*, 651–654.
16. M. Mann, D. Thompson, K. Serivalsatit, T. M. Tritt *et al.*, *Cryst. Growth Des.*, 2010, *10*, 2146–2151.
17. V. K. Tripathi and R. Nagarajan, *Inorg. Chem.*, 2016, *55*, 12798–12806.
18. L. B. Asprey, T. K. Keenan and F. H. Kruse, *Inorg. Chem.*, 1965, *4*, 985–986.
19. B. Claux, O. Beneš, E. Capelli, P. Souček and R. Meier, *J. Fluor. Chem.*, 2016, *183*, 10–13.
20. B. Maximov, *Acta Cryst.*, 1985, *B41*, 88–91.

21. A. Zalkin and D. H. Templeton, *Acta Cryst.*, 1985, *B41*, 91–93.
22. V. Milman, B. Winkler and C. J. Pickard, *J. Nucl. Mat.*, 2003, *322*, 165–179.
23. J. R. Peterson and B. B. Cunningham, *J. Inorg. Nucl. Chem.*, 1968, *30*, 1775–1784.
24. D. D. Ensor, J. R. Peterson, R. G. Haire and J. P. Young, *J. Inorg. Nucl. Chem.*, 1981, *43*, 1001–1003.
25. D. D. Ensor, J. R. Peterson, R. G. Haire and J. P. Young, *J. Inorg. Nucl. Chem.*, 1981, *43*, 2425–2427.
26. J. K. Gibson and R. G. Haire, *J. Less-Common Met.*, 1988, *144*, 123–131.
27. Z. Mazej, *J. Fluor. Chem.*, 2002, *118*, 127–129.
28. C. B. Yeamans, G. W. Chinthaka Silva, G. S. Cerefice, K. R. Czerwinski *et al.*, *J. Nucl. Mat.*, 2008, *374*, 75–78.
29. G. W. Chinthaka Silva, C. Yeamans, L. Ma, G. Cerefice *et al.*, *Chem. Mater.*, 2008, *20*, 3076–3084.
30. A. C. Larsen, R. B. Roof, Jr. and D. T. Cromer, *Acta Cryst.*, 1964, *17*, 555–558.
31. L. B. Asprey, F. H. Ellinger, S. Fried and W. H. Zachariasen, *J. Am. Chem. Soc.*, 1957, *79*, 5825.
32. T. K. Keenan and L. B. Asprey, *Inorg. Chem.*, 1969, *8*, 235–238.
33. R. A. Penneman, G. D. Sturgeon and L. B. Asprey, *Inorg. Chem.*, 1964, *3*, 126–129.
34. R. Benz, R. M. Douglass, F. H. Kruse and R. A. Penneman, *Inorg. Chem.*, 1963, *2*, 799–803.
35. L. B. Asprey, F. H. Kruse and R. A. Penneman, *Inorg. Chem.*, 1967, *6*, 544–548.
36. F. Poineau, C. M. Silva, C. B. Yeamans, G. S. Cerefice *et al.*, *Inorg. Chim. Acta*, 2016, *448*, 93–96.
37. A. Rosenzweig and D. T. Cromer, *Acta Cryst.*, 1970, *B26*, 38–44.
38. R. A. Penneman, R. R. Ryan and A. Rosenzweig, Structural Systematics in Actinide Fluoride Complexes, in *Rare Earths: Structure and Bonding*, 1973, Berlin, Springer, 1957, pp. 1–52.
39. R. R. Ryan, R. A. Penneman and A. Rosenzweig, *Acta Cryst.*, 1969, *B25*, 1958–1962.
40. (a) M. P. Eastman, P. G. Eller and G. W. Halstead, *J. Inorg. Nucl Chem.*, 1981, *43*, 2839–2842; (b) D. Brown and B. Whittaker, *J. Less-Common Met.*, 1982, *86*, 75–84.
41. P. G. Eller, J. G. Malm, B. I. Swanson and L. R. Morss, *J. Alloys & Compounds*, 1998, *269*, 50–56.
42. (a) G. Meyer, S. Dötsch and T. Staffel, *J. Less-Common Met.*, 1987, *127*, 155–160; (b) G. Meyer, P. Ax, *Mat. Res. Bull.*, 1982, *17*,

1447–1455; (c) M. D. Taylor and C. P. Carter, *J. Inorg. Nucl. Chem.*, 1962, *24*, 387–391.
43. (a) F. Weigel, V. Wishnevsky and R. Güldner, *J. Less-Common Met.*, 1980, *75*, 89–98; (b) F. Weigel, V. Wishnevsky and R. Güldner, *J. Less-Common Met.*, 1982, *85*, 137–143; (c) V. Wishnevsky, F. Weigel, M. Eiswirth and R. Schwankner, *J. Less-Common Met.*, 1983, *91*, 309–315; (d) V. Wishnevsky, W. Theissig and F. Weigel, *J. Less-Common Met.*, 1984, *99*, 321–329.
44. (a) G. B. Deacon and A. J. Koplick, *Inorg. Nucl. Chem. Lett.*, 1979, *5–6*, 263–265; (b) F. L. Carter and J. F. Murray, *Mat. Res. Bull.*, 1972, *7*, 519–523.
45. (a) D. B. Dell'Amico, F. Calderazzo, C. della Porta, A. Merigo *et al.*, *Inorg. Chim. Acta*, 1995, *240*, 1–3; (b) U. Baisch, D. B. Dell'Amico, F. Calderazzo, R. Conti *et al.*, *Inorg. Chim. Acta*, 2004, *357*, 1538–1548; (c) S. Petriček, A. Demšar, L. Golič and J. Košmrlj, *Polyhedron*, 2000, *19*, 199–204.
46. H. Hayashi, M. Takano, M. Kurata and K. Minato, *J. Nucl. Mat.*, 2013, *440*, 477–479.
47. H. Hayashi, M. Takano, M. Akabori and K. Minato, *J. Alloys & Compounds*, 2008, *456*, 243–246.
48. J. R. Peterson, *J. Alloys & Compounds*, 1995, *223*, 180–184.
49. J. R. Peterson, J. P. Young, R. G. Haire and U. Benedict, *Physica B+C*, 1986, *144*, 85–90.
50. J. R. Peterson and R. D. Baybarz, *Inorg. Nucl. Chem. Lett.*, 1972, *8*, 423–431.
51. J. F. Wild, E. K. Hulet, R. W. Lougheed, W. N. Hayes *et al.*, *J. Inorg. Nucl. Chem.*, 1978, *40*, 811–817.
52. (a) G. R. Willey, T. J. Woodman and M. G. B. Drew, *Polyhedron*, 1997, *16*, 3385–3393; (b) S. Wu, Z. Ding and X. Li, *Polyhedron*, 1994, *13*, 2679–2681; (c) C. Wenqi, J. Zhongsheng, X. Yuguo and Y. Guangdi, *Inorg. Chim. Acta*, 1987, *130*, 125–129.
53. T. Aitasalo, J. Hölsä, M. Lastusaari, J. Legendziewicz *et al.*, *J. Alloys & Compounds*, 2004, *380*, 296–302.
54. S. Hosokawa, S. Iwamoto and M. Inoue, *J. Alloys & Compounds*, 2006, *408–412*, 529–532.
55. T. Schleid and G. Meyer, *J. Less-Common Met.*, 1987, *127*, 161–166.
56. R. Wilson, *Inorg. Chem.*, 2015, *54*, 10208–10213.
57. G. Ionova, C. Madic and R. Guillaumont, *Radiochimica Acta*, 1997, *78*, 83–90.
58. D. Brown, *Inorganic Chemistry, Series Two*, 1975, *7*, 111–150.
59. J. H. Forsberg and T. Moeller, *Inorg. Chem.*, 1969, *8*, 883.

60. K. E. Mironov, A. P. Popov, V. Y. Vorob'eva and Z. A. Grankina, *Russ. J. Inorg. Chem.*, 1971, *16*, 1476.
61. E. E. Kriss and Z. A. Sheka, *Radiochimica*, 1962, *4*, 312–322.
62. E. S. Pilkington and A. W. Wylie, *J. Applied Chem.*, 1954, *4*, 568–580.
63. Z. Yu, X. Li, K. Wang, B. Yan *et al.*, *Faming Zhuanli Shenqing*. CN 106480329 A, March 8, 2017.
64. S. Topin and J. Aupiais, *J. Environ. Rad.*, 2016, *153*, 237–244.
65. B. N. Ivanov-Emin, Z. K. Odinets, B. E. Zaitsev, V. M. Akimov *et al.*, *Zhurnal Neorganicheskoi Khimii*, 1974, *19*, 2319–2322.
66. P. Charpin, G. Chevrier, M. Lance, M. Nierlich *et al.*, *Acta Cryst.*, 1987, *C43*, 1239–1241.
67. J. Habash and A. J. Smith, *Acta Cryst.*, 1983, *C39*, 413–415.
68. P. Kierkegaard, *Acta Chem. Scand.*, 1956, *10*, 599–616.
69. B. M. Casari and V. Langer, *J. Solid State Chem.*, 2007, *180*, 1616–1622.
70. J. Zink, *Inorg. Chem.*, 1975, 14, 555–558.
71. K. Hughes and P. C. Burns, *Acta Cryst.*, 2003, *C59*, i7–i8.
72. M. S. Wickleder, *Chem. Rev.*, 2002, *102*, 2011–2088.
73. A. D. Burns, B. O. Patrick, A. E. Lam and D. Dreisinger, *Acta Cryst.*, 2014, *C70*, 726–731.
74. J. I. Bullock, M. F. C. Ladd, D. C. Povey and A. E. Storey, *Inorg. Chim. Acta*, 1980, *43*, 101–108.
75. C. Hennig, A. Ikeda-Ohno, S. Tsushima and A. C. Scheinost, *Inorg. Chem.*, 2009, *48*, 5350–5360.
76. (a) J. Habash and A. J. Smith, *J. Cryst. & Spectr. Res.*, 1992, *22*, 21–24; (b) D. D. Schnaars and R. E. Wilson, *Inorg. Chem.*, 2012, *51*, 9481–9490; (c) R. Wilson, *Inorg. Chem.*, 2011, *50*, 5663–5370.
77. K. E. Knope, S. Skanthakumar and L. Soderholm, *Inorg. Chem.*, 2015, *54*, 10192–10196.
78. R. A. Lybrand, J. G. Bockheim, W. Ge, R. C. Graham *et al.*, *Chem. Geol.*, 2016, 442, 174–186.
79. R. Wolfram and G. Irmer, *Dalton Trans.*, 2017, *46*, 4235–4244.
80. F. Favier and J. Pascal, *J. Chem. Soc., Dalton Trans.*, 1992, *13*, 1997–2002.
81. F. Favier, J. L. Pascal, A. Fitch and G. Vaughan, *Applied Cryst.*, 1998, *17*, 93–96.
82. M. S. Grigor'ev, E. V. Gubanov and A. M. Fedoseev, *Radiochemistry*, 2013, *55*, 472–475.
83. V. Philippini, T. Vercouter, A. Chausśe and P. Vitorge, *J. Solid State Chem.*, 2008, *181*, 2143–2154.
84. G. Brauer, K. Wetzel, *Manual of Preparative Inorganic Chemistry*, 3rd edn., *2*, Stuttgart, pp. 1066–1116, 1978.

85. T. L. Webb and J. E. Krűger, *Diff. Therm. Anal.*, 1970, *1*, 303–341.
86. H. Ehrhardt, H. Schweer and H. Z. Seidel, *J. Inorg. Gen. Chem. Lett.*, 1967, *3*, 403.
87. M. Altmaier, V. Neck, R. Mueller and T. Fanghaenel, *Radiochim. Acta.*, 2005, *93*, 83–92.
88. R. Marquart, *Uber Carbonate and Oxide (V)-und (VI)-wertiger Actiniden*, 1981, Ph.D. Thesis, Ludwig-Maximilians-Uniersitat, Műchen, Germany.
89. R. J. Finch, M. A. Cooper and F. C. Hawthorne, *Can. Min.*, 1999, *37*, 929–938.
90. C. E. Bamberger, *Topics in f-Element Chemistry*, 1991, *2*, 279–320.
91. T. Thevenin, J. Jove and C. Madic, *J. Less-Common Met.*, 1986, *121*, 477.
92. L. E. Drabkina, *Russ J. Inorg. Chem.*, 1958, *3*, 63.
93. S. D. Reilly, W. Runde and M. P. Neu, *Geochim. Cosmochim. Acta.*, 2007, *71*, 2672–2679.
94. R. Janicki, P. Starynowicz and A. Mondry, *Eur. J. Inorg. Chem.*, 2011, 3601–3616.
95. D. M. Ananias, M. Kostova, F. A. Paz, A. Ferreira *et al.*, *J. Am. Chem. Soc.*, 2004, *126*, 10410–10417.
96. J. Felsche, The Crystal Chemistry of the Rare-Earth Silicates, in *Rare-Earths: Structure and Bonding*, 1973, Berlin, Springer.
97. G. Morrison, M. Smith and H. C. zur Loye, *J. Am. Chem. Soc.*, 2016, *138*, 7121–7129.
98. M. F. Tang, P. Y. Chiang, Y. H. Su, Y. C. Jung *et al.*, *Inorg. Chem.*, 2008, 8985–8989.
99. G. Morrison and H. C. zur Loye, *Cryst. Growth Des.*, 2016, *16*, 1294–1299.
100. K. Popa, P. E. Raison, L. Martel, P. M. Martin *et al.*, *J. Solid State Chem.*, 2015, *230*, 169–174.
101. K. Linthout, *Can. Min.*, 2007, *45*, 503–508.
102. (a) M. J. Polinski, J. N. Cross, E. M. Villa, J. Lin *et al.*, *Inorg. Chem.*, 2013, *52*, 8099–8105; (b) M. A. Silver and T. E. Albrecht-Schmitt, *Coord. Chem. Rev.*, 2016, *323*, 36–51.
103. J. Peterson and W. Xu, *J. Rad. Nucl. Chem.*, 1996, *203*, 301–307.
104. J. Huang and D. Chen, *Applied Phys. A.*, 2011, *102*, 239–243.
105. A. A. Kaminskii, K. Ueda, H. J. Eichler, Y. Kuwano *et al.*, *Laser Phys.*, 2001, *11*, 1124–1133.
106. A. A. Kaminskii, K. Ueda, H. Eichler, Y. Kuwano *et al.*, *J. Opt. Comm.*, 2001, *194*, 201–206.
107. Y. Oka, T. Yao and N. Yamamoto, *J. Solid State Chem.*, 2000, *152*, 486–491.

108. Y. Wang, Y. Xuemiao, Y. Zhao, Y. Gao *et al.*, *Inorg. Chem.*, 2015, *54*, 8449–8455.
109. (a) F. Shi, J. Meng and Y. Ren, *J. Solid State Chem.*, 1996, *121*, 236–239; (b) N. Taira and Y. Hinatsu, *J. Mat. Chem.*, 2002, *12*, 148–152.
110. F. Weigel and S. Volker, *Radiochim. Acta.*, 1970, *13*, 6–10.
111. H. Naruke and T. Yamase, *Inorg. Chem.*, 2002, *41*, 6514–6520.
112. E. Banks and M. Nemiroff, *Inorg. Chem.*, 1974, *13*, 2715–2718.
113. (a) S. R. Bharadwaj, M. S. Chandrasekharaiah and S. R. Dharwadkar, *J. Mat. Science Lett.*, 1984, *3*, 840–842; (b) T. Nagai, N. Sata, S. Kitawaki and A. Uehara *et al.*, *J. Nucl. Mat.*, 2013, *433*, 397–403.
114. T. L. Cremers, P. G. Eller and R. A. Penneman, *Acta Cryst.*, 1983, *39*, 1165–1167.
115. Z. Assefa, J. Ling, R. G. Haire, T. E. Albrecht-Schmitt and R. E. Sykora, *J. Solid State Chem.*, 2006, *179*, 3653–3663.
116. (a) R. E. Sykora, Z. Assefa, R. G. Haire and T. E. Albrecht-Schmitt, *Inorg. Chem.*, 2006, *45*, 475–477; (b) R. E. Sykora, Z. Assefa, R. G. Haire and T. E. Albrecht-Schmitt, *Inorg. Chem.*, 2005, *44*, 5667–5676.
117. A. Bean, B. Scott, T. E. Albrecht-Schmitt and W. Runde, *Inorg. Chem.*, 2003, *42*, 5632–5636.

https://doi.org/10.1142/9781800610163_0004

Chapter 4

Coordination Chemistry of Lanthanides

Yaofeng Chen
State Key Laboratory of Organometallic Chemistry,
Shanghai Institute of Organic Chemistry,
Chinese Academy of Sciences, 345 Lingling Road,
Shanghai 200032, P. R. China
yaofchen@mail.sioc.ac.cn

4.1 Introduction

The coordination chemistry of lanthanides is more complicated than that of d-block metals. The lanthanides have varying coordination numbers, and prefer to form complexes with high coordination numbers. The coordination geometry of lanthanides is less regular and predictable than that of d-block metals. The bonding between lanthanide ion and ligand is predominantly ionic and 4f-orbitals of lanthanide ions are efficiently shielded by electrons in 5s and 5p orbitals. The lanthanide ions are hard Lewis acids, a large number of lanthanide complexes with ligands containing O or N donor atoms have been reported while those with ligands containing S or P donor atoms are much less so. While this chapter mainly focuses on recent developments in coordination chemistry of lanthanide amides, imides, phosphides, phosphindenes and boryl complexes, some basic contents, such as coordination numbers, aqua ions, hydrated salts, β-diketonates, nitrates and halides, are also covered.

4.2 Coordination Numbers[1,2]

The lanthanide complexes do not obey the "18 electron rule" in classic transition-metal chemistry, and have varying coordination numbers from 2 to 12. Ionic radii and oxidation state of the lanthanide ion and the steric properties of the ligands mainly determine the coordination numbers of lanthanide complexes. Larger lanthanide ions and less bulky ligands usually result in higher coordination numbers. In the absence of adverse steric factors the lanthanide ions tend to have coordination numbers of eight or higher. The high and variable coordination numbers of lanthanide complexes are due to the following facts: the ionic radii of lanthanide ions are larger than d-block metal ions at a comparable oxidation state, and bonding interactions between lanthanide ions and ligands are predominantly ionic and not directional.

The lanthanide complexes of coordination number 2 or 3 are very limited, and these complexes contain very bulky ligands. Representative examples are $[Yb(II)\{C(SiMe_3)_3\}_2]$ (CN = 2),[3] $[Ln\{N(SiMe_3)_2\}_3]$ (Ln = all lanthanides except Pm(III)) (CN = 3),[4] $[Ln\{CH(SiMe_3)_2\}_3]$ (Ln = La, Ce, Pr, Nd, Sm, Er, Yb, Lu, Y) (CN = 3)[5] and $[(Me_3Si)_2NLn\{\mu_2\text{-}N(SiMe_3)_2\}_2Na]$ (Ln = Yb(II), Eu(II)) (CN = 3).[6] In these low coordinate lanthanide complexes, $Ln\cdots H\text{-}C$ agostic type interactions commonly exist due to the unsaturation of coordination sphere around the lanthanide ions. The neutral Lewis base containing lanthanide amides, such as $[\{(Me_3Si)_2N\}_2Ln(THF)_2]$ (Ln = Sm(II), Eu(II), Yb(II))[7] and $[\{(Me_3Si)_2N\}_3Ln(OPPh_3)]$ (Ln = La, Sm, Eu, Er, Lu),[8,9] have coordination numbers of 4. Some "ate" complexes, such as $[\{(Me_3Si)_2N\}_3Ln(\mu_2\text{-}Cl)Li(THF)_3]$ (Ln = Pr, Nd, Sm, Eu, Yb),[10] and ion-pair complexes, such as $[Li(THF)_4][Ln(NPh_2)_4]$ (Ln = Er, Yb),[11] also have coordination numbers of 4. Addition of two neutral Lewis base molecules to the three-coordinate $Ln[N(SiMe_3)_2]_3$ complexes provides five-coordinate lanthanide complexes, such as $[\{(Me_3Si)_2N\}_3Ln(CNC_6H_{11})_2]$ (Ln = Nd, Sm, Er).[9] The trialkyl complexes $[Ln\{CH_2(SiMe_3)\}_3(THF)_2]$ (Ln = Tb, Er, Yb, Lu, Y, Sc) represent another important type of five-coordinate lanthanide complexes,[12] which are very useful precursors for synthesis of other lanthanide complexes via alkane ($SiMe_4$) elimination. To pursue low coordinate lanthanide complexes as catalysts or catalyst precursors for organic reactions and/or polymerizations, an increasing number of five-coordinate lanthanide complexes have emerged in the past two decades. These complexes are usually supported by a bulky bidentate or tridentate monoanionic ligands. β-diketiminate and amidinate are two of the mostly-used bidentate monoanionic ligands, representative complexes include [{ArNC(Me)CHC(Me)NAr}Ln(NPh_2)$_2$(THF)] (Ln = Nd, Yb; Ar = Xyl),[13] [{ArNC(Me)CHC(Me)NAr)}Lu(CH_2SiMe_3)$_2$(THF)] (Ar = Ph, Xyl, Mes)[14] and [{(Dipp)NC(Ph)N(Dipp)}Ln(CH_2SiMe_3)$_2$ (THF)] (Ln = Lu, Y, Sc).[15] Lanthanide complexes having coordination numbers between 6 and 10 are very common, and the most popular coordination number for lanthanide complexes is 8. The 11 and 12-coordinate lanthanide complexes are much less common than the 10-coordinate ones, due to increasing steric congestion around the metal center. The complexes of coordinate number >10 usually contain small size nitrate anion, and the nitrate anion acts as a bidentate ligand with a small "bite angle" in the complexes. Such examples can be founded in $[Ln(NO_3)_3(H_2O)_5]\cdot H_2O$ (Ln = La, Ce) (CN = 11), [Ln(15-crown-5)(NO_3)$_3$] (Ln = Nd-Lu) (CN = 11) and [Ln(18-crown-6)(NO_3)$_3$] (Ln = La, Nd) (CN = 12).[1]

4.3 Complexes

4.3.1 *Aqua ions*

Lanthanide(III) ions are able to form aqua ions of coordination number eight and nine in aqueous solution. The formation of eight and nine water-coordinated species can be presented as the following equilibrium:

$$[Ln(H_2O)_8]^{3+} + H_2O \leftrightharpoons [Ln(H_2O)_9]^{3+}$$

For early lanthanides, the nona-aqua ion is energetically more favourable (from an alternative perspective, the large ionic radii allow them to adopt more water molecules around them), while for late lanthanides, the octa-aqua ion is the preferable one. Upon moving across the lanthanide series from La^{3+}, which is of the largest ionic radius, to Lu^{3+}, which is of the smallest ionic radius, the equilibrium shifts more and more to the left-hand side, i.e. the octa-aqua ion is more preferable against the nona-aqua ion upon decreasing of ionic radius of Ln^{3+}. For the mid-series lanthanides, nona- or octa-aqua ions are in equilibrium, and the water exchange rate of these lanthanides is faster than that of those Ln^{3+} ions where the equilibrium is totally shifted to either the nona- or octa-aqua ion. An octa-aqua ion adopts a square antiprismatic geometry, whereas the water molecules of a nona-aqua ion have a strong tendency to form a tricapped trigonal prismatic geometry.[16,17]

4.3.2 *Hydrated salts*[1,2]

Lanthanide hydrated salts $LnX_3 \cdot nH_2O$ (X = Cl, Br, I, ClO_4, NO_3, SO_4, CF_3SO_3, CH_3COO) are usually prepared by reactions of lanthanide oxides, hydroxides or carbonates with certain acids. The number of water molecules in lanthanide hydrated salts ranges from 1 to 11, and the small size of X^- ligands and strong oxophilicity of lanthanide ions allow water molecules to be accommodated in the coordination sphere. In $[La(NO_3)_3(H_2O)_5] \cdot H_2O$, three nitrate ions and five water molecules coordinate to the La^{3+} ion and the complex has coordination number 11.[18] In $[Pr_2(SO_4)_3(H_2O)_8]$, each Pr^{3+} ion is coordinated by one oxygen from each of the four sulfate ions (one of the sulfate ions belongs to an adjoining unit cell) and four water molecules; two sulfate ions are bonded to three Pr^{3+} ions and the third sulfate ions is bonded to two Pr^{3+} ions.[19] If the coordination between the X^- ligands and the Ln^{3+} ions is relatively weak, the X^- ligands can be partially or completely displaced by water molecules from the complexes' coordination sphere forming a separated ion-pair structure. For example, $LnX_3{\cdot}7H_2O$ (X = Cl, Ln = La-Pr; X = Br, Ln = La, Ce) exist as the dimeric species $[(H_2O)_7Ln(\mu_2\text{-}X)_2Ln(H_2O)_7]X_4$, where each Ln^{3+} ion is coordinated by two μ_2-Cl^-/Br^- ions and seven water molecules, four remaining chlorides (or bromides) exist in the lattice and are not coordinated to the Ln^{3+} ion;[20,21] $LnX_3{\cdot}6H_2O$ (X = Cl, Ln = Nd-Lu, Y; X = Br, Ln = Pr-Dy,) exist as monomers $[LnX_2(H_2O)_6]X$, the Ln^{3+} ion is coordinated by two Cl^- (or Br^-) ions and six water molecules, the remaining halide discretely exists in the lattice.[21,22] The cations in the compounds $[Ln(H_2O)_8]Br_3$ (Ln = Ho-Lu, Y), $[Ln(H_2O)_n]I_3$ (n = 8 or 9), $[Ln(H_2O)_6](ClO_4)_3$ and $[Ln(H_2O)_9](CF_3SO_3)_3$ are the pure aqua ions $[Ln(H_2O)_n]^{3+}$ (n = 6–9), and the X^- ligands (X = Br, I, ClO_4, CF_3SO_3) are no longer coordinated to the metal ions.

4.3.3 *Phosphines*

Since lanthanide ions are among the hardest Lewis acids, and phosphines are soft Lewis bases, lanthanide metal–phosphine coordination is thus mismatched based on the Pearson's HSAB principle. Although the first example of the lanthanide phosphine, $[(C_5H_5)_3Yb(PPh_3)]$, was reported in 1966,[23] examples of lanthanide phosphines are still very sparse. In one case, weakly coordinated diethyl ether in $[\{(Me_3Si)_2N\}_2Yb(OEt_2)_2]$ can be displaced by a bidentate phosphine $Me_2PCH_2CH_2PMe_2$ to afford the lanthanide phosphine $[\{(Me_3Si)_2N\}_2Yb(Me_2PCH_2CH_2PMe_2)]$, in which the phosphine coordinates to the Yb^{2+} ion through both of the two phosphorus atoms.[24] $[\{(Me_3Si)_2N\}_2Yb(OEt_2)_2]$ also reacts with a monodentate phosphine P^nBu_3 to give $[\{(Me_3Si)_2N\}_2Yb(P^nBu_3)_2]$.[24] Reactions of $[(C_5Me_5)_2Ln(OEt_2)_2]$ (Ln = Yb(II), Eu(II)) with $Me_2PCH_2CH_2PMe_2$ give the polymeric compound $[(C_5Me_5)_2Ln(Me_2PCH_2CH_2PMe_2)]_n$ with the bidentate phosphine coordinating to two Ln^{2+} ions through each phosphorus atom respectively.[25] $[(C_5Me_5)_2Ln(Me_2PCH_2CH_2PMe_2)]_n$ is insoluble in non-coordinating solvents. When one carbon tethered $Me_2PCH_2PMe_2$ phosphine is used, the reactions give the monomeric complex $[(C_5Me_5)_2Ln(Me_2PCH_2PMe_2)]$, which are hydrocarbon-soluble. The phosphine in $[(C_5Me_5)_2Ln(Me_2PCH_2PMe_2)]$ can be reversibly displaced by diethyl ether to give $[(C_5Me_5)_2Ln(OEt_2)_2]$; while for the polymeric $[(C_5Me_5)_2Ln(Me_2PCH_2CH_2PMe_2)]_n$, the phosphine ligand is immune from exchange with Et_2O, although tetrahydrofuran can displace the phosphine from the complexes. The divalent ytterbium complex $[(C_5Me_5)_2Yb(\kappa^2\text{-}P,P'\text{-}Me_2PCH_2PMe_2)]$ reacts with $YbCl_3$ to give a trivalent ytterbium complex $[(C_5Me_5)_2YbCl(\eta^1\text{-}P\text{-}Me_2PCH_2PMe_2)]$, in which the phosphine acts a monodentate ligand. Lanthanide ion-pair complexes $[\{MeC(NDipp)CHC(Me)(NCH_2CH_2PPh_2)LnOC_6H_3\text{-}2,6\text{-}{}^tBu_2]$ $[B(C_6F_5)_4]$ (Ln = Sc, Y, Lu) which contain a phosphine pendant arm were recently synthesized, and the internal $Ln\cdots P$ interactions in these complexes were revealed by solution ^{31}P

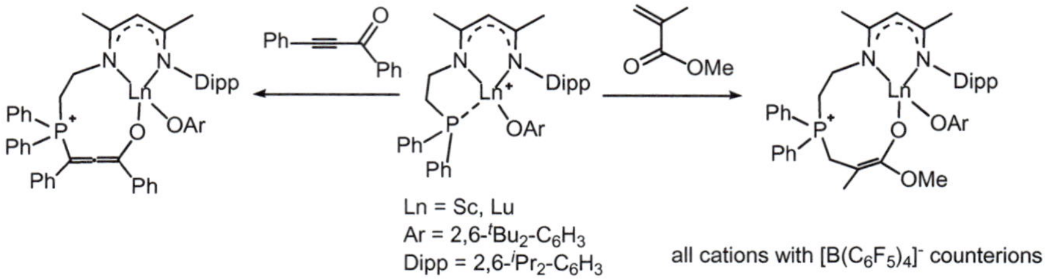

Scheme 4.1. Lewis pair reactivity of lanthanide phosphines.

NMR spectra.[26] These interesting complexes display Ln/P Lewis pair reactivity, undergoing 1,4-addition reactions toward conjugated carbonyl substrates (Scheme 4.1). Furthermore, these complexes catalyse polymerizations of conjugated polar alkene monomers (methyl methacrylate, α-methylene-γ-butyrolactone and γ-methyl-α-methylene-γ-butyrolactone).

4.3.4 *β-diketonates*[27,28]

β-diketones ($RCOCH_2COCR'$) are readily deprotonated to form the monoanionic ligands $[RCOCHCOCR']^-$, which coordinate to Ln^{3+} ions as bidentate ligands to give the lanthanide β-diketonates. Due to the strong Ln–O bonds and the chelating structure, the lanthanide β-diketonates are usually highly stable. Lanthanide β-diketonates represent one of the most important classes of lanthanide complexes and have wide applications, such as light conversion molecular devices, nuclear resonance shift reagents, organic electroluminescent devices, stains and labels for immunoassays and the imaging of biological cells, diode lasers and optical fibers.

The R and R′ substituents of $[RCOCHCOCR']^-$ can be alky, aryl and fluoro-substituted alkyl, as shown in Figure 4.1. The introduction of fluoro-substituted alkyls enhances thermal stability and volatility of the lanthanide β-diketonates, and they also have better photoluminescence and electroluminescence properties in comparison with their carbon-hydrogen congeners. Lanthanide β-diketonates may contain one, two, three or four β-diketonato ligands (Figure 4.2), and those with three β-diketonato ligands are the most common. $Ln(acac)_3(H_2O)_n$ (acac = $[CH_3COCHCOCCH_3]^-$, n = 1 or 2) are the simplest lanthanide β-diketonates. These

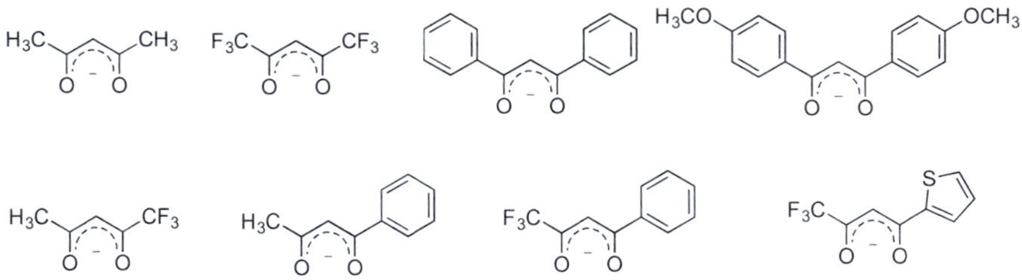

Figure 4.1. Some representative β-diketonato ligands commonly used.

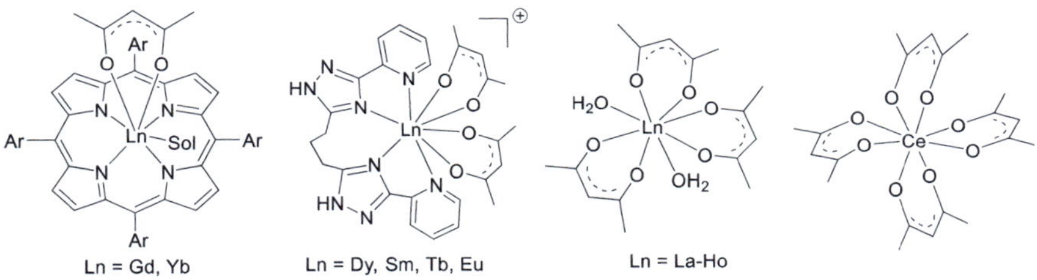

Figure 4.2. Lanthanide β-diketonates containing one, two, three or four β-diketonato ligands.

complexes are difficult to dehydrate, even *in vacuo*; they decompose upon being heated, and on dehydration at room temperature tend to oligomerize to involatile materials. Anhydrous lanthanide β-diketonates, such as $Ln(dmp)_3$ (dmp = $[Me_3CCOCHCOCMe_3]^-$), can be prepared by using the bulkier β-diketonato ligand under anhydrous conditions; they are coordinatively unsaturated and tend to hydrate to give $Ln(dmp)_3(H_2O)$. Ce(IV)$(acac)_4$, $[C_5H_{10}NH_2][Eu(PhCOCHCOCCH_3)_4]$ and $NaLn(hfac)_4$ (Ln = Y, Eu, Er; hfac = $[F_3CCOCHCOCF_3]^-$) represent examples of lanthanide complexes with four β-diketonato ligands. Ce(IV)$(acac)_4$ is synthesized by reaction of $[NH_4]_4[Ce(IV)(SO_4)_4]$ with acetylacetone or oxidation of Ce(III)$(acac)_3$ in the presence of acetylacetone. Ce(IV)$(acac)_4$ has two forms: α- and β-forms, which are dependent on the solvents used for recrystallization.[29] $[C_5H_{10}NH_2][Eu(PhCOCHCOCCH_3)_4]$ can be prepared from reaction of europium chloride with benzoylacetone and piperdine in ethanol.[30] $NaLn(hfac)_4$ are readily synthesized on a large scale from reactions of Na(hfac) and anhydrous $LnCl_3$, and these complexes can be used as volatile single-source

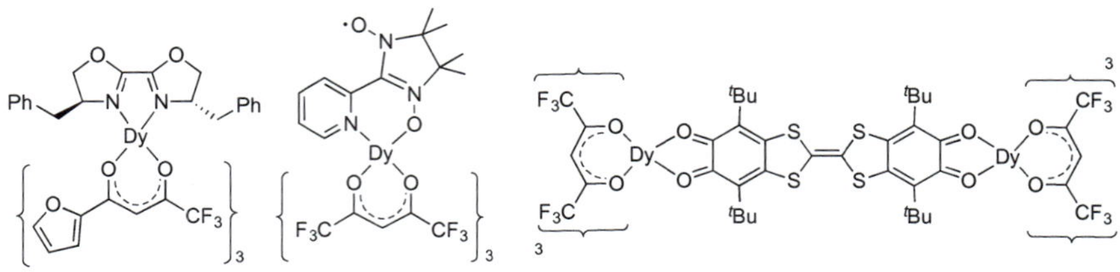

Figure 4.3. SMMs of dysprosium β-diketonates.

precursors for low-temperature preparation of sodium-rare earth metal fluorides.[31] Introducing other multidentate ligands may result in mono- and bis(β-diketonato) lanthanide complexes, such as Ln(acac)(L)(sol.) (Ln = Gd, Yb; L = porphyrins or porpholactones)[32] and [Ln(acac)$_2$(L′)]Cl (Ln = Dy, Sm, Tb, Eu; L′ = bis(5-(pyridine-2-yl)-1,2,4-triatriazol-3-yl)propane).[33] Water molecules in hydrated lanthanide β-diketonates Ln(RCOCCOCR′)$_3$(H$_2$O)$_n$ can be replaced by strong Lewis bases (L), such as the oxygen or nitrogen containing organic molecules, to form a variety of new lanthanide β-diketonates Ln(RCOCCOCR′)$_3$(L)$_x$(H$_2$O)$_y$ with different properties. Actually, the ability to coordinate organic ligands is one of the facts that make lanthanide β-diketonates useful as nuclear resonance shift reagents. Reactions of Dy(RCOCCOCR′)$_3$ (H$_2$O)$_n$ [RCOCCOCR′ = 2-furyl-trifluoro-acetonate or hfac] with (*S*,*S*)-2,2′-bis(4-benzyl-2-oxazoline),[34] 2-(2′-pyridyl)-4,4,5,5-tetramethylimidazoline-1-oxyl-3-oxide[35] or 4,4′,7,7′-tetra-*tert*-butyl-2,2′-bi-1,3-benzo-dithiole-5,5′,6,6′-tetrone[36] provide the new Single Molecule Magnets (SMMs) of dysprosium complexes (Figure 4.3).

4.3.5 *Nitrates*[37]

Lanthanide nitrates can be readily prepared from reactions of nitric acid with lanthanide oxides, hydroxides, carbonates or lanthanide metals. Lanthanide nitrates are commonly the hydrated ones with the formula of $Ln(NO_3)_3 \cdot nH_2O$ where n is usually 5 or 6, depending on the lanthanide ion and preparative conditions. Lanthanide ions La-Nd tend to form the hexahydrates while for Eu–Lu the pentahydrates are favourable. The nitrate anion is a bidentate ligand, and its small size allows four or five water molecules to stay in the coordination sphere of the Ln^{3+} ion. Therefore, lanthanide nitrates $Ln(NO_3)_3 \cdot nH_2O$ have a high coordination number (10–11). Due to their easy accessibility and high solubility in both aqueous and polar organic phases, lanthanide nitrates are widely used as starting materials for the synthesis of other lanthanide complexes. Anhydrous lanthanide nitrates $Ln(NO_3)_3$ have been observed as intermediate phases during thermal decomposition of the hydrated nitrates for all lanthanides. By contrast, the hydrated scandium nitrate tends to decompose directly to the oxynitrate phase. Tetravalent cerium nitrate contains four nitrate ions with a formula of $Ce(NO_3)_4 \cdot 5H_2O$. Lanthanide pentanitrates and hexanitrates, such as $M_2Ln(III)(NO_3)_5 \cdot nH_2O$ (M = ammonium ion or alkali Li-Cs ion), $[NH_4]_2[Ce(IV)(NO_3)_6] \cdot 5H_2O$ (cerium ammonium nitrate) and $M'Ce(NO_3)_6 \cdot 8H_2O$ (M' = Mg(II), Zn(II)), are also known.

4.3.6 *Halides*[1,38–40]

Lanthanide halides are the most widely used as starting materials for the synthesis of other lanthanide complexes. Hydrated lanthanide halides $LnX_3 \cdot nH_2O$ are generally prepared by reactions of lanthanide oxides or carbonates with corresponding hydrogen halides in water. The solubility of hydrated lanthanide fluorides in water is very low, while other hydrated lanthanide halides are highly soluble in water. Anhydrous lanthanide halides LnX_3 cannot be directly obtained by dehydrating $LnX_3 \cdot nH_2O$; the dehydration of $LnX_3 \cdot nH_2O$ under elevated temperature leads to the formation of LnOX. Alternatively, anhydrous LnX_3 can be prepared by reactions of lanthanide metals with X_2 or HgX_2, but the ammonium chloride route is probably the most popular one because it is inexpensive and straightforward even for large-scale preparations.[41,42] $[NH_4]_2[MCl_5]$(M = La–Gd) or $[NH_4]_3[MCl_6]$(M = Tb–Lu) is readily prepared by reactions of lanthanide oxides Ln_2O_3 or hydrated lanthanide chlorides with NH_4Cl, and then the thermal decomposition of $[NH_4]_2[MCl_5]$ or $[NH_4]_3[MCl_6]$ by slowly raising the temperature to above 350–400°C under dynamic vacuum produces the anhydrous lanthanide chlorides $LnCl_3$ in high yields. The anhydrous lanthanide bromides and iodides ($LnBr_3$ and LnI_3) can be prepared by the same method by using $LnBr_3 \cdot nH_2O$ and NH_4Br or $LnI_3 \cdot nH_2O$ and NH_4I as the starting materials. Anhydrous lanthanide halides LnX_3 are very hygroscopic and readily form the hydrates $LnX_3 \cdot nH_2O$ or coordinate with Lewis bases (L: THF, ether, alcohol, amine, phosphine oxides etc.) to give the lanthanide halide-Lewis base adducts $LnX_3 \cdot n(L)$. Divalent lanthanide iodides LnI_2 (Ln = La, Ce, Pr, Nd, Sm, Gd, Dy, Tm, Yb) can be prepared by a high-temperature (600°C) comproportionation reaction of a stoichiometric mixture of trivalent lanthanide iodides LnI_3 and Ln powder.[43] LnI_2 (Ln = Eu, Yb, Sm) are soluble and stable in THF. LnI_2 (Ln = Nd, Dy, Tm) are also soluble in THF, but the solutions of DyI_2 and NdI_2 quickly decompose at room temperature. $LnI_2(sol.)_n$ (Ln = Eu(II), Yb(II), Sm(II); sol. = THF, DME) can be readily prepared by oxidation of lanthanide metals with 1,2-diiodoethane or molecular iodine in THF or DME. $SmI_2(sol.)_n$ is now a family of the most useful reducing reagents in organic synthesis.[44,45] $TmI_2(DME)_3$ can be prepared by reduction of TmI_3 with thulium metal in refluxing DME.[46] In the solid state and in solution, $TmI_2(DME)_3$ slowly decomposes at room temperature, and the decomposition is noticeably accelerated by sunlight. Unlike the preparation of $TmI_2(DME)_3$, metallic neodymium or dysprosium does not reduce NdI_3 or DyI_3 at 80°C and ultrasonic conditions. Alternatively, $LnI_2(THF)_5$ and $LnI_2(DME)_3$ (Ln = Dy, Nd) are synthesized by reaction of neodymium or dysprosium metal with iodine at high temperature (>1500°C), and crystallization of the reaction products in THF or DME at low temperature.[47] The divalent lanthanide bromides, chlorides and fluorides also exist. Staying with the same lanthanide metal, the stability of the divalent lanthanide halides decreases from the iodides to the fluorides, therefore the fluorides are much less common than the iodides. All the divalent lanthanide halides are sensitive to air, and need to be kept under a dinitrogen or argon atmosphere. Tetravalent lanthanide halides are very rare. CeF_4 and TbF_4 are prepared by reactions of fluorine, xenon difluoride or chlorine trifluoride with CeO_2 or CeF_3, and Tb_4O_7 or TbF_3, respectively. PrF_4 is prepared by extracting NaF from a complex alkali metal fluoride of tetravalent praseodymium (Na_2PrF_6 or $Na_7Pr_6F_{30}$) with anhydrous hydrofluoric acid or by the solid-state reaction of krypton difluoride with Pr_6O_{11} or PrO_2. Upon heating, the tetrafluorides decompose slowly to form the corresponding trifluorides. The thermostability of LnF_4 decreases in the order $CeF_4 > TbF_4 > PrF_4$. At room temperature, all the tetrafluorides are more or less sensitive to moist air. Structurally, the most popular geometry of anhydrous LnX_3 is a nine-coordinate cation geometry. The structures of some trivalent hydrated lanthanide halides have been determined by single crystal X-ray diffraction. $GdCl_3 \cdot 6H_2O$ exits as $[GdCl_2 \cdot 6H_2O] \cdot Cl$, the Gd^{3+} ion is eight-coordinated by two chlorides and six water molecules, one-third of the chlorides are not coordinated to the gadolinium ion.[48] On the contrary, the structure of $LaCl_3 \cdot 7H_2O$ can be described as a dimeric complex $[(H_2O)_7La(\mu\text{-}Cl)_2La(H_2O)_7]Cl_4$ where each La^{3+} ion is nine-coordinated by two μ-chlorides and seven water molecules.[21] The solvated divalent complexes $SmI_2(THF)_5$, $SmI_2(THF)_3(DME)$ and $TmI_2(DME)_3$ are all neutral species, the metal ion is seven-coordinated by five oxygen atoms and two iodides, displaying a pentagonal-bipyramidal coordination geometry.[46]

4.3.7 *Amides*[49,50]

A large number of lanthanide amides have been synthesized, most of them are thermally stable, and some can even be sublimed. However, it also should be noted that the Ln–N(amide) bond is much weaker than the Ln–O(alkoxide) bond, and comparable to the Ln–C(alkyl) bond. For example, the absolute bond disruption enthalpies of $Cp^*_2Sm\text{-}O^tBu$, $Cp^*_2Sm\text{-}NMe_2$ and $Cp^*_2Sm\text{-}CH(SiMe_3)_2$ are 82.4(3.5), 48.2(1.8) and 47.0(1.5) kcal/mol, respectively.[51] There are several synthetic routes to lanthanide amides, and the salt metathesis of rare-earth metal halides with alkali metal (and alkali-earth metal) amides is the most common method for synthesizing a variety of homoleptic and heteroleptic lanthanide amides. Reactions of liquid ammonia with europium or ytterbium metal readily produce divalent lanthanide primary amides $Ln(NH_2)_2$ (Ln = Eu, Yb). On the other hand, trivalent lanthanide primary amides $Ln(NH_2)_3$ are synthesized under much harsher conditions, e.g. in supercritical ammonia up to 5,000 atm, 550°C and 5–50 days.[52] $Ln(NH_2)_x$ (X = 2, 3) are inorganic compounds and related to the solid-state chemistry. In the crystal structure of $La(NH_2)_3$, there are two kinds of crystallographically independent lanthanum ions. Each La^{3+} cation is surrounded by eight $[NH_2]^-$ anions, one La^{3+} ion has a distorted square antiprismatic geometry and the other has a distorted trigonal prismatic geometry.[53] Introducing alkyl, aryl or silyl substituents to the nitrogen of amide ligands significantly increase the steric congestion around metal centers leading to the low coordination lanthanide amides of good solubility in organic solvents. $[(^iPr_2N)_3Ln(THF)_n]$ (n = 0 or 1),[54–56] $[Ln\{N(SiMe_3)_2\}_3]$,[4b] $[\{(Me_3Si)_2N\}_2Ln(THF)(\mu\text{-}Cl)]_2$[57] and "ate" complexes $[(^iPr_2N)_2Ln\ (\mu\text{-}N^iPr_2)_2Li(THF)]$,[55,56] $[\{(Me_3Si)_2N\}_3Ln(\mu\text{-}Cl)Li(THF)_3]$,[10] $[\{(Me_3Si)_2N\}_2Ln(\mu\text{-}Cl)Li(THF)_2(\mu^3\text{-}Cl)]_2$ (Ln = Eu, Ho) and $[\{(Me_3Si)_2N\}Ln(\mu\text{-}Cl)_2Li(THF)_2(\mu\text{-}Cl)]_2$ (Ln = Nd, Sm, Eu, Ho, Yb)[58] can be prepared from reactions of $LnCl_3$ with LiN^iPr_2, $KN(SiMe_3)_2$ or $LiN(SiMe_3)_2$ in THF. In the synthesis of $[Ln\{N(SiMe_3)_2\}_3]$, a workup of extraction with hexane and sublimation under vacuum is required to remove LiCl and THF from the crude products. $[Ln\{N(SiMe_3)_2\}_3]$ and $[\{(Me_3Si)_2N\}_3Ln(\mu\text{-}Cl)Li(THF)_3]$ now have been used as the versatile precursors for synthesis of other lanthanide complexes via amine elimination with protic compounds such as alcohols, phenols, amines, phosphines, acetylenes, cyclopentadienes and pyrroles. Salt metathesis of divalent lanthanide iodides with $MN(SiMe_3)_2$ (M = Na, K) provides divalent lanthanide bis(trimethylsilyl)amides.[7,59] These divalent lanthanide bis(trimethylsilyl)amides are generally obtained as solvent adducts, $[\{(Me_3Si)_2N\}_2Ln(L)_x]$ (Ln = Sm, Eu, Yb; L = THF, Et_2O, DME). The solvent-free complex $[Yb\{N(SiMe_3)_2\}_2]_2$ is synthesized by removal of Et_2O from $[Yb\{N(SiMe_3)_2\}_2(Et_2O)_2]$ in toluene. This complex exists as a dimer, and each Yb^{2+} ion is coordinated by one terminal and two bridging $[N(SiMe_3)_2]^-$ ligands.[60] A mononuclear solvent-free divalent samarium complex $[Sm\{N(Si^iPr_3)_2\}_2]$ was recently prepared by salt metathesis of $[SmI_2(THF)_2]$ with $KN(Si^iPr_3)_2$ in toluene.[61] The bulky Si^iPr_3 substituents are vital for the formation of this mononuclear and Lewis base-free complex. The complex has a near-linear N-Sm-N angle (175.52(18)°), and displays four agostic type Sm $\cdots$ C(methine) interactions. Tetravalent cerium amides are also known. The heteroleptic complexes $[\{N(CH_2CH_2NSiMe^t_2Bu)_3\}Ce(IV)I]$[62] and $[\{(Me_3Si)_2N\}_3Ce(IV)Cl]$[63–65] are obtained from oxidation of $[Ce(III)\{N(CH_2CH_2NSiMe^t_2Bu)_3\}]$ with I_2 in pentane and $[Ce(III)\{N(SiMe_3)_2\}_3]$ with $TeCl_4$, Ph_3CCl or $PhICl_2$ in toluene, respectively, and the homoleptic $[Ce(IV)\{N(SiHMe_2)_2\}_4]$ is synthesized by treatment of $[\{(Me_2HSi)_2N\}_3Ce(III)(THF)_2]$ with $PhICl_2$,

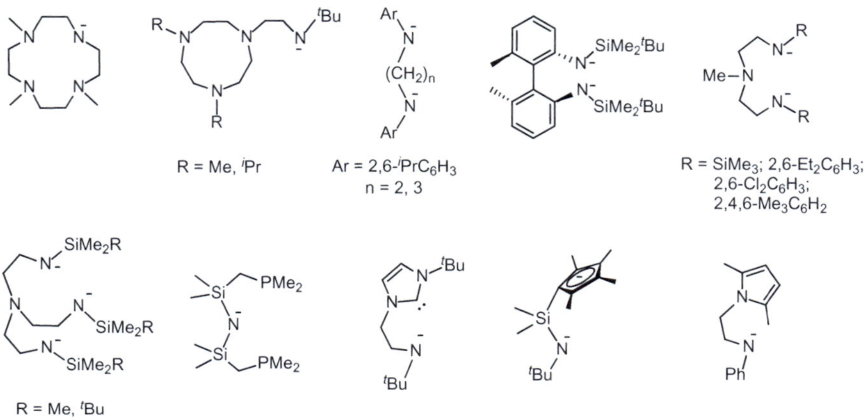

Figure 4.4. Donor-functionalized amide ligands for lanthanide complexes.

Ph_3CCl or C_2Cl_6 in THF.[66] A variety of lanthanide complexes containing donor-functionalized amido ligands (Figure 4.4), such as 1,4,7-trimethyl-1,4,7,10-tetraazacyclododecane,[67] 1,4,7-triazacyclononane-amide,[68] chelating aryl diamide,[69] chiral bis (silylamido)biphenyl,[70] diamidoamine,[71] triamidoamine[62,72] diphosphine amides,[73] linked *N*-heterocyclic carbene amide,[74] linked cyclopentadienyl amides[75] and linked pyrrole amides,[76] also have been reported.

4.3.8 *Imides*[77–79]

Despite burgeoning development in the last decade, lanthanide imides are still rare, and most of them contain bridging imido ligands. The representative synthesis and coordination modes are given in Scheme 4.2 and Figure 4.5, respectively. A redox reaction of divalent ytterbium complex $[Yb(C_{10}H_8)(THF)_3]$ with azobenzene in THF provides a trivalent ytterbium imide $[Yb_4(\mu\text{-}\eta^2{:}\eta^2\text{-}N_2Ph_2)_4(\mu_3\text{-}NPh)_2(THF)_4]$.[80] In the complex, four Yb^{3+} ions are connected by four $\eta^2{:}\eta^2$-bridging azobenzene ligands and two μ_3-bridging imido ligands. This redox method is also applicable for the synthesis of samarium imides. Treatment of divalent samarium in $[Sm(C_{10}H_8)(THF)_3]$ or $[Me_2Si\{NC(Ph)N(Dipp)\}_2SmI_2Li_2(THF)(Et_2O)_2]$ with azobenzene affords the trivalent samarium imide $[Sm_4(\mu\text{-}\eta^2{:}\eta^2\text{-}N_2Ph_2)_4(\mu_3\text{-}NPh)_2(THF)_6]$ or $[(Me_2Si\{NC(Ph)N(Dipp)\}_2)_2Sm_4(\mu_3\text{-}NPh)_4]$,

4 [Ln(C$_{10}$H$_8$)(THF)$_3$] + 5 PhN=NPh ⟶ [Ln$_4$(μ-η^2:η^2-N$_2$Ph$_2$)$_4$(μ_3-NPh)$_2$(THF)$_n$] + 4 C$_{10}$H$_8$
(Ln = Yb(II), Sm(II)) (Ln = Yb(III), Sm(III))

2[{(Dipp)NH}$_2$Yb{μ_2-NH(Dipp)}$_2$Na(THF)] + 4 *n*-BuLi ⟶ [(THF)Li{(Dipp)N}{(Dipp)NH}Yb{μ_2-N(Dipp)}Na(THF)]$_2$ + 4 *n*-BuH + 2 LiNH(Dipp)

[Sm{NH(Dipp)}$_3$]$_2$ + 4 AlMe$_3$ ⟶ [{μ_2-(Dipp)N}Sm{μ_2-NH(Dipp)}(μ_2-Me)AlMe$_2$]$_2$ + 2 CH$_4$ + [Me$_2$AlNH(Dipp)]$_2$

2 Ln(AlMe$_4$)$_3$ + 2 KNH(Mes*) ⟶ [(AlMe$_4$)Ln{μ_2-N(Mes*)}]$_2$ + 2 [K(AlMe$_4$)] + 2 CH$_4$ + 2 AlMe$_3$
(Ln = Y, Nd, Lu)

4 NdI$_3$ + 4 [PhNMg(THF)] ⟶ [Nd$_4$(NPh)$_4$I$_4$(THF)$_n$] + 4 MgI$_2$

[{(C$_5$Me$_4$SiMe$_3$)Ln(μ-H)$_2$}$_4$(THF)] + 4 PhCN ⟶ [(C$_5$Me$_4$SiMe$_3$)Ln(μ_3-NCH$_2$Ph)$_4$] + THF
(Ln = Y, Lu)

[(NNN)Sc(Me)NH(Dipp)] + DMAP ⟶ [(NNN)(DMAP)Sc=N(Dipp)] + CH$_4$

Scheme 4.2. Representative synthesis of lanthanide imides.

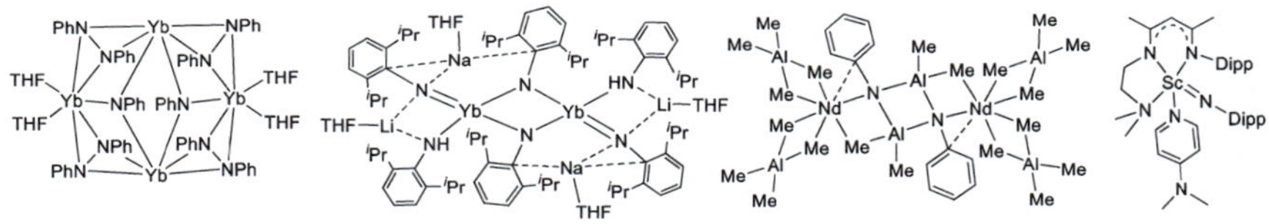

Figure 4.5. Representative coordination modes of the imides with lanthanide ions.

respectively.[81,82] These two samarium complexes also contain the μ_3-imido ligands. Deprotonation of primary or secondary amides with a strong base provides another efficient method to the imides, and several different types of lanthanide imides are prepared via this method. Reaction of $Me_2Si(C_9H_7)(C_2B_{10}H_{11})$ with 4 equivalents of $NaNH_2$ in THF, followed by treatment with 1 equivalent of $LnCl_3$, gives $[\{(\eta^5\text{-}\mu_2\text{-}C_9H_6SiMe_2NH)Ln\}_2(\mu_3\text{-Cl})(THF)]_2(\mu_4\text{-NH})(THF)_n$ ($n = 1$, Ln = Gd, Er; $n = 0$, Ln = Dy).[83] In these complexes, the imido ligands adopt a μ_4-ligation mode. Treatment of ytterbium amide [{(Dipp)NH}$_2$Yb{μ_2-NH(Dipp) }$_2$Na(THF)] with 2 or 4 equivalents of ^{n}BuLi give a ytterbium complex containing both amido and imido ligands [(THF)Li{(Dipp)N}{(Dipp)NH}Yb{μ_2-N(Dipp)}Na(THF)]$_2$ or a ytterbium imido complex [(THF)Li{(Dipp)N}$_2$Yb{μ_2-N(Dipp)}Li{Na(THF)}]$_2$.[84] These ytterbium complexes bear bridging imido ligands and feature very short Yb-N(imido) distances and large Yb-N(imido)-C angles. Deprotonation of the amides with

$AlMe_3$ provides a variety of Ln-Al bimetallic imido complexes. For example, reactions of $[Nd(NHPh)_3(KCl)_3]$ or $[Sm\{NH(Dipp)\}_3]_2$ with excess $AlMe_3$ yield $[\{Me_2Al(\mu_2\text{-}Me_2)\}_2Nd(\mu_3\text{-}NPh)(\mu_2\text{-}Me)AlMe]_2$ or $[\{\mu_2\text{-}(Dipp)N\}Sm\{\mu_2\text{-}NH(Dipp)\}(\mu_2\text{-}Me)AlMe_2]_2$, respectively;[85,86] Reactions of homoleptic tetramethylaluminates $Ln(AlMe_4)_3$ with KNH(Mes*) or LiNH(Dipp) afford $[(AlMe_4)Ln\{\mu_2\text{-}N(Mes^*)\}]_n$ (n = 2, Ln = Y, Nd, Lu; n = 4, Ln = La) or $[\{(AlMe_4)Ln\}_2\{\mu_2\text{-}N(Dipp)\}\{\mu_3\text{-}N(Dipp)\}\{(\mu_2\text{-}Me)_2AlMe\}]$ (Ln = Y, La, Ce, Nd), respectively.[87,88] Lanthanide imides can also be prepared by protolysis of lanthanide dialkyl complexes with amines. Reactions of $[\{(NCN)Ln(CH_2C_6H_4NMe_2\text{-}o)_2]$ (NCN = $[PhC\{N(Dipp)_2\}_2]^-$) or $[\{Bo^MCp^{tet}\}Lu(CH_2Ph)_2]$ (Bo^MCp^{tet} = $MeC(Ox^{Me2})_2C_5Me_4$; Ox^{Me2} = 4,4-dimethyl-2-oxazoline) with $(Dipp)NH_2$ or $(1\text{-}C_{10}H_7)CH_2NH_2$ give $[(NCN))Ln\{\mu_2\text{-}N(Dipp)\}]_2$ (Ln = Sc, Lu) or $[\{Bo^MCp^{tet}\}Lu\{\mu_2\text{-}NCH_2(1\text{-}C_{10}H_7)\}]_2$.[89,90] Two polyimido clusters of neodymium, $[Nd_4(NPh)_4I_4(THF)_n]$ (n = 3, 4) and $[Nd_6(NPh)_8I_2(py)_6]$, are obtained straightforwardly by salt metathesis of NdI_3 with 1 or 4/3 equivalents of [PhNMg(THF)] in refluxing THF or pyridine, respectively.[91] Slow diffusion of Et_2O into a solution of a 1:1 mixture of NdI_3 and [PhNMg(THF)] in pyridine/THF or into a solution of $[Nd_6(NPh)_8I_2(py)_6]$ in pyridine afforded single crystals of $[Nd_4(\mu_3\text{-}NPh)_4I_4(py)_5(THF)_2]\cdot py$ or $[Nd_6(\mu)3\text{-}NPh)_8I_2(py)_8]\cdot 5py$. The former displays a cubane core structure, while the later one has a rhombododecahedron core structure. Yttrium and lutetium polyhydrides $[\{(C_5Me_4SiMe_3)Ln(\mu\text{-}H)_2\}_4(THF)]$ (Ln = Y, Lu) reacts with benzonitrile to give tetranuclear imides $[(C_5Me_4SiMe_3)Ln(\mu_3\text{-}NCH_2Ph)_4]$ by double addition of the Ln–H bond across the C≡N bond.[92] Interestingly, it was reported that methylidene complexes $[\{(NCN)_3Ln_3(\mu_2\text{-}Me)_3(\mu_3\text{-}Me)(\mu_3\text{-}CH_2)]$ (NCN = $[PhC\{N(Dipp)\}_2]^-$, Ln = Y, Lu) are able to react with azobenzene, imine or aniline to generate the imides in high yields via N=N, C=N or N–H bond transformations.[93] Due to HOMO/LUMO orbital energy mismatch between the lanthanide ions and the imido groups, the formation of lanthanide terminal imido complexes is unfavourable. The lanthanide terminal imido species once formed can easily assemble into more stable bridged species,[94] or undergo reaction with solvent via C–H bond activation.[95] Therefore, lanthanide complexes with terminal bonded imido ligands are very rare. Only six scandium,[96] one yttrium, one lutetium[97] and one cerium (IV)[98] complexes with terminal bonded imido ligands have been reported so far. For stabilisation of the lanthanide terminal imido complexes, bulky and/or multidentate supporting ligands, such as the *tert*-butyl substituted β-diketiminato ligand, β-diketiminato based tri- or tetradentate ligands and the tris(pyrazolyl)borato ligand, are used. All the terminal complexes except the cerium(IV) one are synthesized from the alkyl-amido precursors via alkane elimination. The cerium(IV) terminal imido complex [Cs(2.2.2-cryptand)][(TriNO$_x$)Ce=NC$_6$H$_3$(3,5-(CF$_3$)$_2$)] (TriNO$_x$ = [{2-^{t}BuN(O)C$_6$H$_4$CH$_2$}$_3$N]$^{3-}$) is prepared from a one-pot reaction of the cerium(IV) amido complex, CsN(SiMe$_3$)$_2$, and 2.2.2-cryptand; the Cs$^+$ ion is encapsulated by 2.2.2-cryptand to provide the terminal imido structure. The Ln–N (imido) bond lengths of the terminal imido complexes are significantly shorter than the Ln–N(amido) bond lengths of the corresponding amido complexes. For example, the Sc–N(imido) bond length of [(NNN)(DMAP)Sc=N(Dipp)] (1.881(5) Å) is 0.16 Å shorter than the Sc–N(amido) bond length of [(NNN)(Me)Sc-NH(Dipp)] (NNN = [MeC(N(Dipp))CHC(Me)(NCH$_2$CH$_2$NMe$_2$)]$^-$) (2.047(3) Å).[96a] Besides the short Ln–N(imido) bond lengths, the imido ligands are nearly linear with large Ln-N-C angles (167.90(17)° ~ 175.8(5)°; the Ln-N-C angle in the cerium (IV) complex [Cs(2.2.2-cryptand)][(TriNO$_x$)Ce=NC$_6$H$_3$(3,5-(CF$_3$)$_2$)] is somewhat smaller 157.3(3)°). In agreement with the structural data, DFT studies on the imido complex [(NNN)(DMAP)Sc=N(Dipp)] and the amido complex [(NNN)(Me)Sc-NH(Dipp)] show two *p* orbitals of the N(imido) atom form two bonds with two *d* orbitals of the scandium ion while one *p* orbital of the N(amido) atom overlaps with a *d* orbital of the scandium ion.[96a] The scandium terminal imido complexes are capable of activating elemental selenium,[96b] undergoing cycloaddition with CO$_2$ and a series of unsaturated organic small molecules,[99] activating the Si–H bond of phenylsilane and the C–H bond of terminal alkenes in 1,2-addition,[96f,100] and reacting with metal halides,[101] demonstrating the nucleophilicity of the nitrogen atom and the Lewis acidity of the scandium in the Sc=N double bond.

4.3.9 *Phosphides*[102,103]

The number of reported lanthanide phosphides is much less than that of lanthanide amides. This can be partially ascribed to the weaker Ln–P bond compared to the Ln–N bond. The compounds $[Cp_2Ln(P^tBu_2)]$ (Ln = Tb, Ho, Er, Tm, Yb, Lu)[104] and $[Cp_2Yb(P(C_6H_{11})_2)]$[105] are the first lanthanide phosphides, however these complexes are only characterized by elemental analysis and spectroscopic methods. The first structurally well-characterized lanthanide phosphide, $[Cp_2Lu(\mu_2\text{-}PPh_2)_2Li(TMEDA)]$ (TMEDA = N,N,N′,N′-tetramethylethylenediamine), has not been reported until 1986.[106] The lanthanide phosphides are generally synthesized by salt metathesis of lanthanide halides or triflates with alkali metal phosphides or alkane (or amine) elimination of lanthanide alkyls (or amides) with phosphines. Reactions of $LnI_3(THF)_x$ (Ln = Tm(III), Nd(III); X = 3.5) with 3 equivalents of $KP(SiMe_3)_2$ in THF provide the homoleptic phosphides $[Ln\{P(SiMe_3)_2\}_3(THF)_2]$.[107] The products of reactions between lanthanide halides/triflates and LiP^tBu_2 depend on the ionic radii of the Ln^{3+} ion, giving the "ate" complex $[(^tBu_2P)_2La(\mu_2\text{-}P^tBu_2)_2Li(THF)]$[108] with the largest La^{3+} ion and the neutral complexes $[Ln(P^tBu_2)_3(THF)_x]$ (x = 0 or 2) with the smaller Ho^{3+}, Er^{3+}, Tm^{3+}, Lu^{3+} or Y^{3+} ions.[109] Interestingly, $Ln(OSO_2CF_3)_3$ (Ln = Sm, Yb, Eu) reacts with 5 equivalent of LiP^tBu_2 to produce divalent $[Ln\{(\mu_2\text{-}P^tBu_2)_2Li(THF)\}_2]$ and 0.5 equivalent of $^tBu_2P\text{-}P^tBu_2$, where reduction of the metal ion occurs during the reactions.[108,110] A THF-free homoleptic yttrium phosphide $[Y\{P(SiMe_3)_2\}_3]_2$ is prepared by alkane elimination of $[Y\{CH(SiMe_3)_2\}_3]$ with $HP(SiMe_3)_2$ in toluene.[111] Reaction of YCl_3 with 1 equivalent of K[PH(Mes)] in THF gives a dinuclear yttrium phosphido dichloride $[YCl_2\{\mu_2\text{-}PH(Mes)\}(THF)_3]_2$,[112] while that of $YI_3(THF)_{3.5}$ with $K[P(SiMe_3)(Dipp)]$ in toluene provides a mononuclear complex $[Y\{P(SiMe_3)Dipp\}I_2(THF)_3]$.[113] In the latter case, a very bulky phosphide ligand is employed. Divalent $[Ln(PPh_2)_2(THF)_4]$ (Ln = Sm, Yb) can be prepared by three different synthetic routes: (1) Salt metathesis of $[LnI_2(THF)_2]$ with $KPPh_2$; (2) Salt metathesis of $[LnI_3(THF)_3]$ with $KPPh_2$ accompanied by reduction of the metal ion with $Ph_2P\text{-}PPh_2$ elimination; (3) Amine elimination of $[Ln\{N(SiMe_3)_2\}_2(THF)_2]$ with Ph_2PH.[114]

Salt metathesis of [$SmI_2(THF)_2$] with 2 equivalents of $KP(SiMe_3)_2$ produces a dinuclear divalent complex, [{$(Me_3Si)_2P$}Sm{μ_2-$P(SiMe_3)_2$}$_3$Sm$(THF)_3$], in which two Sm^{2+} ions are in different coordination environments.[115] An *N*-heterocyclic carbene (NHC) ytterbium phosphide complex [$(IMe_4)_3Yb(PPh_2)_2$] (IMe_4 = 1,3, 4,5-tetramethylimidazol-2-ylidene) is synthesized by reaction of [$(IMe_4)_2Yb\{N(SiMe_3)_2\}_2$] with Ph_2PH or reaction of [$(THF)_4Yb(PPh_2)_2$] with IMe_4.[116] Some lanthanide phosphides containing other anionic organic ancillary ligands, such as the trivalent complexes [$Ln\{N(SiMe_3)_2\}_2(PPh_2)$] (Ln = La, Eu) and [$Ln\{N(SiMe_3)_2\}_2(PPh_2)(Ph_3PO)_2$] (Ln = La, Eu and Y),[117] [$Cp_2^{Me}Sm(PPh_2)$],[$Cp^*{}_2Sm(PPh_2)$] and [$Cp^*{}_2Sm(PPh_2)(THF)$],[118] [{1,3-$(Me_3Si)_2C_5H_3$}$_2$Y(μ_2-P(H)Si^{i}Pr$_3$)$_2$M(sol.)] (M = Li, sol. = THF; M = K, sol. = benzene) and [{1,3-$(Me_3Si)_2C_5H_3$}$_2$Y(μ_2-PH$_2$)$_2$Li$_2$(μ_2-Cl)(TMEDA)$_2$],[119] [$(Tp^{Me2})(Cp)Y(PPh_2)(THF)$] ($Tp^{Me2}$ = tris(3,5-dimethylpyrazolyl)borate),[120] and divalent complexes [Cp*Sm(THF){μ_2-P(H)Ar}KCp*(THF)]$_\infty$,[121] [{^{t}BuC(NC$_6$H$_3$-2,6-^{i}Pr$_2$)$_2$}Yb(μ_2-H)(μ_2-PPh$_2$)Yb{^{t}BuC(NC$_6$H$_3$-2,6-^{i}Pr$_2$)$_2$}][122] have also been prepared. A mixed-valent Sm(II/III) complex [Cp*$_2$Sm(μ_2-PPh$_2$)SmCp*$_2$] is obtained from a redox reaction of [Cp*$_2$Sm] with Ph_2P-PPh_2 or reaction of [Cp*$_2$Sm] with [Cp*$_2$Sm(PPh$_2$)].[118b]

Diffusion of P_4 vapor into a toluene solution of the divalent samarium complex [Cp*$_2$Sm] gives the first molecular rare-earth metal polyphosphide [{Cp*$_2$Sm}$_4$P$_8$] (Figure 4.6(a)).[123] The structure of this complex can be regarded as a realgar-type $P_8{}^{4-}$ ligand trapped in a cage of four samarocenes. Redox reactions of rare-earth metal naphthalene complexes [{(1,1′-fc(NSi^{t}BuMe$_2$)$_2$)Ln(THF)$_n$}$_2$(μ_2-C$_{10}$H$_8$)] (fc = ferrocenylene; Ln = Sc, $n = 0$; Ln = Y, $n = 1$) with P_4 produce the $P_8{}^{4-}$ complex [{(1,1′-fc(NSi^{t}BuMe$_2$)$_2$)Sc}$_4$P$_8$] and the P_7^{3-} complexes [{(1,1′-fc(NSi^{t}BuMe$_2$)$_2$)Ln(THF)$_n$}$_3$P$_7$] (Ln = Sc, $n = 0$; Ln = Y, $n = 1$) (Figure 4.6(b)).[124] Reaction of [(Dipp$_2$pyr)SmI(THF)$_3$](Dipp$_2$pyr = 2,5-bis{*N*-(2,6-diisopropylphenyl)iminomethyl}pyrrolyl) with [Cp*Fe(η^5-P$_5$)] in THF in the presence of potassium-naphthalene yields the cyclo-P_5 bridging Sm/Fe complexes.[125] Recrystallization from THF/toluene or toluene/pentane gives the monomeric or dimeric complex, [Cp*FeP$_5$Sm(Dipp$_2$pyr)(THF)$_2$] (Figure 4.6(c)) or [Cp*FeP$_5$Sm(Dipp$_2$pyr)]$_2$. Reaction of [Y{P(SiMe$_3$)Dipp}I$_2$(THF)$_3$] with K[PH(Dipp)] in toluene results in an unusual polymetallic yttrium phosphinidene phosphide

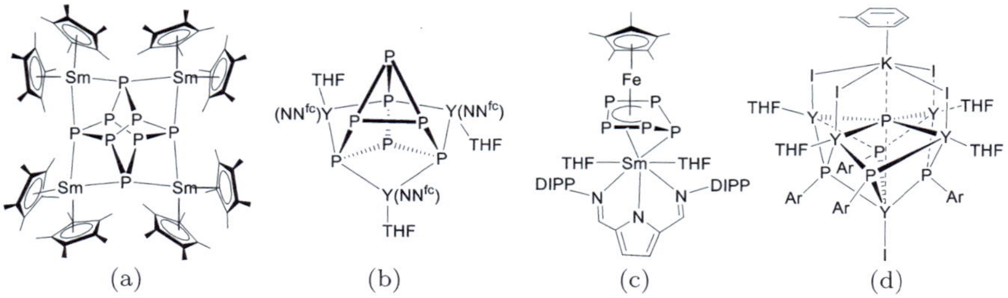

Figure 4.6. The P_8^{4-}, P_7^{3-}, cyclo-P_5^{3-} and P^{3-} complexes.

(Figure 4.6(d)) via P–Si (or H) and P–C bond cleavage, which is the first well-defined soluble P^{3-} rare-earth metal complex.[113] This complex can be transferred into other P^{3-} yttrium complexes by salt metathesis.

4.3.10 *Phosphinidenes*

Lanthanide phosphinidene complexes had not been synthesized until recently. Reaction of $[\{2\text{-}(^iPr_2P)4\text{-}(Me)C_6H_3\}_2NLu(CH_2SiMe_3)_2]$ with $MesPH_2$ at 80°C leads to $SiMe_4$ elimination providing a lutetium phosphinidene complex $[\{2\text{-}(^iPr_2P)4\text{-}(Me)C_6H_3\}_2NLu\{\mu_2\text{-}P(Mes)\}]_2$ (Figure 4.7).[126] Reaction of $NdI_3(THF)_{3.5}$ with $K[P(SiMe_3)(Dipp)]$ at room temperature produces a neodymium phosphinidene halide $[(THF)_3(I)Nd\{\mu_2\text{-}P(Dipp)\}]_2$ via salt metathesis and silyl redistribution.[127] $[(THF)_3(I)Nd\{\mu_2\text{-}P(Dipp)\}]_2$ undergoes salt metathesis with KCp* to give a neodymium pentamethylcyclopentadienyl phosphinidene complex $[(THF)(Cp^*)Nd\{\mu_2\text{-}P(Dipp)\}]_2$ or with potassium hydrotris(pyrazolyl)borate to afford a neodymium hydrotris (pyrazolyl)borate phosphinidene complex $[(THF)(Tp^{Ph*})Nd\{\mu_2\text{-}P(Dipp)\}]_2$.[128] The geometries of the phosphinidene phosphorus atoms in $[\{2\text{-}(^iPr_2P)4\text{-}(Me)C_6H_3\}_2NLu\{\mu_2\text{-}P(Mes)\}]_2$, $[(THF)_3(I)Nd\{\mu_2\text{-}P(Dipp)\}]_2$ and $[(THF)(Tp^{Ph*})Nd\{\mu_2\text{-}P(Dipp)\}]_2$ are trigonal planar; while that in $[(THF)(Cp^*)Nd\{\mu_2\text{-}P(Dipp)\}]_2$ is somewhat tetrahedral. $[\{2\text{-}(^iPr_2P)4\text{-}(Me)C_6H_3\}_2NLu\{\mu_2\text{-}P(Mes)\}]_2$ and $[(THF)_3(I)Nd\{\mu_2\text{-}P(Dipp)\}]_2$ show phospha-Wittig chemistry toward ketones to yield the corresponding phosphaalkenes. $[\{2\text{-}(^iPr_2P)4\text{-}(Me)C_6H_3\}_2NSc(Me)Br]$ reacts with Li[PH(Trip)] to give a bis(μ_2-phosphinidene) discandium complex $[\{2\text{-}(^iPr_2P)4\text{-}(Me)C_6H_3\}_2NSc\{\mu_2\text{-}P(Trip)\}]_2$ or with Li[P(H)DMP]

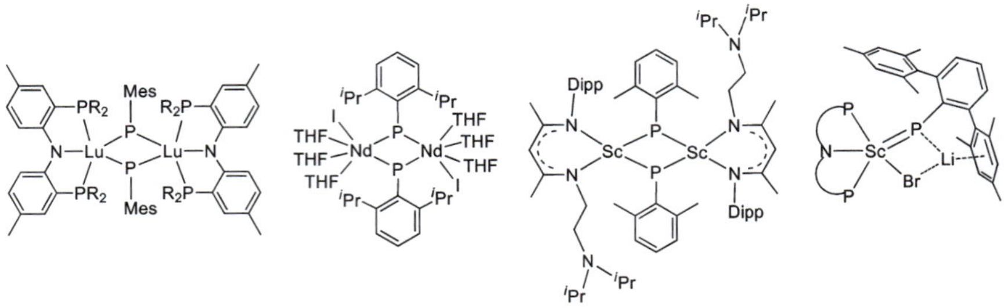

Figure 4.7. Representative lanthanide phosphinidene complexes.

to provide an alkali metal-capped phosphinidene complex [{2-(^{i}Pr$_2$P)4-(Me)C$_6$H$_3$}$_2$NSc{μ_2-P(DMP)}(μ_2-Br)Li]).[129] The reactivity of the bis(μ_2-phosphinidene)discandium complex [{2-(^{i}Pr$_2$P) 4-(Me)C$_6$H$_3$}$_2$NSc{μ_2-P(Trip)}]$_2$ is very low, but the alkali metal-capped phosphinidene complex [{2-(^{i}Pr$_2$P)4-(Me)C$_6$H$_3$}$_2$ NSc{μ_2-P(DMP)}(μ_2-Br)Li] reacts with ketones and phosphorous dichlorides to yield the corresponding phosphaalkenes and diphosphenes. [{2-(^{i}Pr$_2$P)4-(Me)C$_6$H$_3$}$_2$NSc{μ_2-P(DMP)}(μ_2-Br)Li]) can also deliver the [P(DMP)]$^{2-}$ unit to metal-halide precursors to provide other transition metal phosphinidene complexes, such as [Cp$_2$Zr=P(DMP)(P(CH$_3$)$_3$)]. A four-coordinate scandium phosphinidene complex [{MeC(NDipp)CHC(Me)(NCH$_2$ CH$_2$N(^{i}Pr)$_2$}Sc{μ_2-P(Xyl)}]$_2$ is synthesized by reaction of [{MeC(NDipp)CHC(Me)(NCH$_2$CH$_2$N(^{i}Pr)$_2$}Sc(Me)Cl] with K[PH (Xyl)].[130] Although the complex has a bis(μ-phosphinidene) discandium structural unit, this coordinatively unsaturated complex shows high and versatile reactivity toward a variety of substrates. It undergoes two electron reduction with 2,2′-bipyridine, elemental selenium, elemental tellurium, Me$_3$P=S, or Ph$_3$P=E (E = S, Se), resulting in oxidative coupling of two phosphinidene ligands 2[PAr]$^{2-}$ into a diphosphene ligand [ArP-PAr]$^{2-}$. It also undergoes nucleophilic addition reactions with unsaturated substrates (benzylallene, benzonitrile, *tert*-butyl isocyanide and CS$_2$) and initiates homologation of CO and coordinated CO in Mo(CO)$_6$. Methane elimination reactions of methylidene complexes [{(NCN)$_3$Ln$_3$(μ_2-Me)$_3$(μ_3-Me)(μ_3-CH$_2$)] with PhPH$_2$ provide trinuclear phosphinidene complexes [{(NCN)$_3$Ln$_3$(μ_2-Me)$_3$(μ_3-Me)(μ_3-PPh)] (NCN = [PhC{N(Dipp)}$_2$]$^-$, Ln = Y, Lu).[131] These trinuclear phosphinidene complexes react with CS$_2$ to give unusual thione-dianion-ligated rare-earth sulfide complexes $[\{(NCN)_3Ln_3(\mu_2\text{-}Me)_2(\mu_3\text{-}S)(P,C,S\text{-}Ph(Me)PC(Me)S)]$. Deprotonation of the P–H bonds in a dysprosium phosphide $[Cp_2^{Me}Dy\{\mu_2\text{-}P(H)Mes\}]_3$ by ^{n}BuLi in THF produces heterometallic phosphinidene complexes $[Li(THF)_4]_2[(Cp_2^{Me}Dy)_3(\mu_3\text{-}PMes)_3Li]$.[132] $[Cp_2^{Me}Dy\{\mu_2\text{-}P(H)Mes\}]_3$ exhibits promising single-molecule magnet (SMM) properties, while $[Li(THF)_4]_2[(Cp_2^{Me}Dy)_3(\mu_3\text{-}PMes)_3Li]$ does not.

4.3.11 *Boryls*

The first lanthanide boryl complexes are reported in 2011. To date, there are only three scandium, one yttrium, one gadolinium, and one lutetium boryl complexes reported. Attempts to synthesize boryl complexes through salt metathesis of boryl lithium $[Li\{B(N(Dipp)CH)_2\}(THF)_2]$ with metal halides ($ScI_3(THF)_3$, $YCl_3(THF)_{3.5}$, $CpYCl_2$) or deprotonation of a hydroborane with $[Ln(CH_2SiMe_3)_3(THF)_2]$ (Ln = Sc, Gd) were not successful. The boryl complexes $[Ln\{B(N(Dipp)CH)_2\}(CH_2SiMe_3)_2(THF)_n]$ (Ln = Sc, n = 1; Ln = Y, Gd, Lu, n = 2) were obtained by salt metathesis of $[Li\{B(N(Dipp)CH)_2\}(THF)_2]$ with $[Ln(CH_2SiMe_3)_2(THF)_x][BPh_4]$.[133,134] Density functional theory analysis reveals that the Ln–B bonds in these complexes are predominantly ionic, with covalent character in the σ-bonding Ln–B HOMO. The scandium complex contains the least ionic Ln–B bond and the yttrium complex the most ionic one. Salt metathesis of $[Li\{B(N(Dipp)CH)_2\}(THF)_2]$ with $[\{Me_2Si(C_5Me_4)N^tBu\}ScCl(THF)]$ does not provide isolable products, but that with $[\{Me_2Si(C_5Me_4)NPh\}ScCl(THF)]_2$ gives an "ate"-type boryl complex, $[\{Me_2Si(C_5Me_4)NPh\}Sc\{B(N(Dipp)CH)_2\}(\mu_2\text{-}Cl)Li(THF)_3]$, in a high yield. Treatment of this complex with 1,2-dimethoxyethane (DME) yields an ion-pair complex, $[\{Me_2Si(C_5Me_4)NPh\}Sc\{B(N(Dipp)CH)_2\}Cl][Li(DME)_3]$.[135] Reactivity of these boryl complexes has been briefly studied. Reaction of $[Sc\{B(N(Dipp)CH)_2\}(CH_2SiMe_3)_2(THF)]$ with N,N'-diisopropylcarbodiimide gives a bis(amidinato) scandium boryl complex, which further reacts with N,N'-diisopropylcarbodiimide to produce a tris(amidinato) scandium complex (Scheme 4.3). Therefore, the Sc-C(alkyl) bond in the complex is more reactive than the Sc–B(boryl) bond, this is

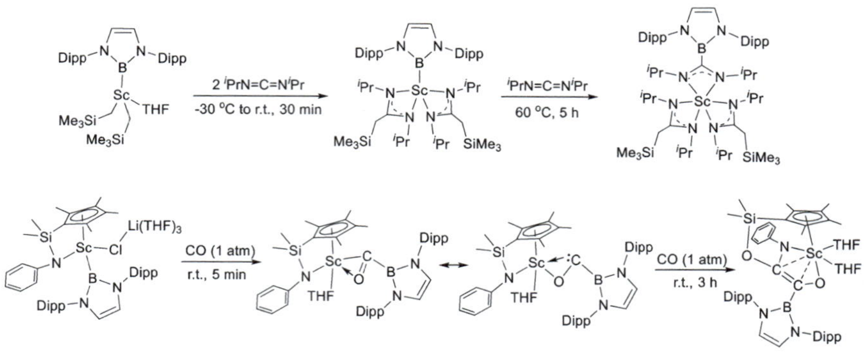

Scheme 4.3. Reactivity of scandium boryl complexes.

possibly due to the bulky substituents on the boron atom. $[\{Me_2Si(C_5Me_4)NPh\}Sc\{B(N(Dipp)CH)_2\}(\mu_2\text{-}Cl)Li(THF)_3]$ reacts with CO to give a novel scandium boryl oxycarbene with CO insertion into Sc–B(boryl) bond, and the generated scandium boryl oxycarbene further reacts with another CO molecule to produce a phenylamido- and boryl-substituted enediolate complex through C–C coupling of CO and the oxycarbene ligand and cleavage and rearrangement of the Si–N bond.[136]

References

1. S. Cotton, *Lanthanide and Actinide Chemistry*, John Wiley & Sons Ltd., New York, 2006.
2. C. H. Huang (ed.), *Rare Earth Coordination Chemistry: Fundamentals and Applications*, John Wiley & Sons (Asia) Pte Ltd., New York, 2010.
3. C. Eaborn, P. B. Hitchcock, K. Izod and J. D. Smith, *J. Am. Chem. Soc.*, 1994, *116*, 12071–12072.
4. (a) J. S. Ghotra, M. B. Hursthouse and A. J. Welch, *J. Chem. Soc. Chem. Commun.*, 1973, 669–670; (b) D. C. Bradley, J. S. Ghotra and F. A. Hart, *J. Chem. Soc., Dalton Trans.*, 1973, 1021–1023.
5. (a) P. B. Hitchcock, M. F. Lappert, R. G. Smith, R. A. Bartiett and P. P. Power, *J. Chem. Soc. Chem. Commun.*, 1988, 1007–1009; (b) C. J. Schaverien and A. G. Orpen, *Inorg. Chem.*, 1991, *30*, 4968–4978; (c) C. Guttenberger, H.-D. Amberger, *J. Organomet. Chem.*, 1997, *545–546*, 601–606; (d) S. Tian, V. M. Arredondo, C. L. Stern and T. J. Marks, *Organometallics*, 1999, *18*, 2568–2570; (e) H. Reddmann,

C. Guttenberger and H.-D. Amberger, *J. Organomet. Chem.*, 2000, *602*, 65–71; (f) A. G. Avent, C. F. Caro, P. B. Hitchcock, M. F. Lappert, Z. N. Li and X.-H. Wei, *Dalton Trans.*, 2004, 1567–1577.
6. T. D. Tilley, R. A. Andersen and A. Zalkin, *Inorg. Chem.*, 1984, *23*, 2271–2276.
7. (a) T. D. Tilley, A. Zalkin, R. A. Andersen and D. H. Templeton, *Inorg. Chem.*, 1981, *20*, 551–554; (b) Y. F. Rad'kov, E. A. Fedorova, S. Y. Khorshev, G. S. Kalinina, M. N. Bochkarev and G. A. Razuvaev, *Russ. J. Gen. Chem. (Engl. Transl.)*, 1985, *55*, 1911; (c) W. J. Evans, D. K. Drummond, H. M. Zhang and J. L. Atwood, *Inorg. Chem.*, 1988, *27*, 575–579.
8. D. C. Bradley, J. S. Ghotra, F. A. Hart, M. B. Hursthouse and P. R. Raithby, *J. Chem. Soc., Dalton Trans.*, 1977, 1166–1172.
9. (a) S. Jank, C. Guttenberger, H. Reddmann, J. Hanss and H.-D. Amberger, *Z. Anorg. Allg. Chem.*, 2006, *632*, 2429–2438; (b) S. Jank, H. Reddmann, L. Zhang and H.-D. Amberger, *Z. Anorg. Allg. Chem.*, 2012, *638*, 1159–1166.
10. (a) F. T. Edelmann, A. Steiner, D. Stalke, J. W. Gilje, S. Jagner and M. Håkansson, *Polyhedron*, 1994, *13*, 539–546; (b) S. L. Zhou, S. W. Wang, G. S. Yang, X. Y. Liu, E. H. Sheng, K. H. Zhang, L. Cheng and Z. X. Huang, *Polyhedron*, 2003, *22*, 1019–1024; (c) E. H. Sheng, S. W. Wang, G. S. Yang, S. L. Zhou, L. Cheng, K. H. Zhang and Z. X. Huang, *Organometallics*, 2003, *22*, 684–692; (d) M. H. Xie, X. Y. Liu, S. W. Wang, L. Liu, Y. Y. Wu, G. S. Yang, S. L. Zhou, E. H. Sheng and Z. X. Huang, *Chin. J. Chem.*, 2004, *22*, 678–682.
11. W.-K. Wong, L. L. Zhang, F. Xue and T. C. W. Mak, *Polyhedron*, 1997, *16*, 2013–2020.
12. (a) M. F. Lappert and R. Pearce, *J. Chem. Soc., Chem. Commun.*, 1973, 126; (b) J. L. Atwood, W. E. Hunter, R. D. Rogers, J. Holton, J. McMeeking, R. Pearce and M. F. Lappert, *J. Chem. Soc., Chem. Commun.*, 1978, 140–142; (c) S. Arndt, P. Voth, T. P. Spaniol, and J. Okuda, *Organometallics*, 2000, *19*, 4690–4700.
13. M. Q. Xue, Y. M. Yao, Q. Shen and Y. Zhang, *J. Organomet. Chem.*, 2005, *690*, 4685–4691.
14. (a) Z. C. Zhang, D. M. Cui and X. L. Liu, *J. Polym. Sci., Part A: Polym. Chem.*, 2008, *46*, 6810–6818; (b) D. T. Liu, C. G. Yao, R. Wang, M. Y. Wang, Z. C. Wang, C. J. Wu, F. Lin, S. H. Li, X. H. Wan and D. M. Cui, *Angew. Chem. Int. Ed.*, 2015, *54*, 5205–5209.
15. S. Bambirra, M. W. Bouwkamp, A. Meetsma and B. Hessen, *J. Am. Chem. Soc.*, 2004, *126*, 9182–9183.
16. Th. Kowall, F. Foglia, L. Helm and A. E. Merbach, *J. Phys. Chem.*, 1995, *99*, 13078–13087.

17. Th. Kowall, F. Foglia, L. Helm and A. E. Merbach, *Chem. Eur. J.*, 1996, *2*, 285–294.
18. (a) B. Eriksson, L. O. Larsson and L. Niinistö, *J. Chem. Soc., Chem. Commun.*, 1978, 616–617; (b) B. Eriksson, L. O. Larsson, L. Niinistö and J. Valkonen, *Inorg. Chem.*, 1980, *19*, 1207–1210.
19. E. G. Sherry, *J. Solid State Chem.*, 1976, *19*, 271–279.
20. A. Habenschuss and F. H. Spedding, *Cry. Struct. Commun.*, 1979, *8*, 511–516.
21. V. V. Bakakin, R. F. Klevstova and L. P. Solov'eva, *J. Struct. Chem.*, 1974, *15*, 723–732.
22. C. J. Kepert, B. W. Skelton and A. H. White, *Aust. J. Chem.*, 1994, *47*, 385–390.
23. E. O. Fischer and H. Fischer, *J. Organomet. Chem.*, 1966, *6*, 141–148.
24. T. D. Tilley, R. A. Andersen and A. Zalkin, *J. Am. Chem. Soc.*, 1982, *104*, 3725–3727.
25. T. D. Tilley, R. A. Andersen and A. Zalkin, *Inorg. Chem.*, 1983, *22*, 856–859.
26. P. F. Xu, Y. M. Yao and X. Xu, *Chem. Eur. J.*, 2017, *23*, 1263–1267.
27. K. Z. Wang, β-Diketonate Lanthanide Complexes, in *Rare Earth Coordination Chemistry: Fundamentals and Applications* (Ed. C. H. Huang), John Wiley & Sons (Asia) Pte Ltd., New York, 2010.
28. K. Binnemans, Rare-Earth β-Diketonates, in *Handbook on the Physics and Chemistry of Rare Earths* (Eds., K. A. Gschneidner, Jr., J.-C. G. Bünzli, V. K. Pecharsky), Vol. 35, Elsevier, Amsterdam, 2005.
29. (a) B. Matković and D. Grdenić, *Acta Cryst.*, 1963, *16*, 456–461; (b) H. Titze, *Acta Chem. Scand. A*, 1974, *28*, 1079–1088; (c) H. Titze, *Acta Chem. Acand.*, 1969, *23*, 399–408; (d) T. Behrsing, A. M. Bond, G. B. Deacon, C. M. Forsyth, M. Forsyth, K. J. Kamble, B. W. Skelton and A. H. White, *Inorg. Chim. Acta*, 2003, *352*, 229–237.
30. L. R. Melby, N. J. Rose, E. Abramson and J. C. Caris, *J. Am. Chem. Soc.*, 1964, *86*, 5117–5125.
31. M. C. Barry, Z. Wei, T. Y. He, A. S. Filatov and E. V. Dikarev, *J. Am. Chem. Soc.*, 2016, *138*, 8883–8887.
32. X.-S. Ke, B.-Y. Yang, X. Cheng, S. L.-F. Chan and J.-L. Zhang, *Chem. Eur. J.*, 2014, *20*, 4324–4333.
33. A. N. Gusev, M. Hasegawa, T. Shimizu, T. Fukawa, S. Sakurai, G. A. Nishchymenko, V. F. Shul'gin, S. B. Meshkova and W. Linert, *Inorg. Chim. Acta*, 2013, *406*, 279–284.
34. D.-P. Li, T.-W. Wang, C.-H. Li, D.-S. Liu, Y.-Z. Li and X.-Z. You, *Chem. Commun.*, 2010, *46*, 2929–2931.

35. X.-L. Wang, L.-C. Li and D.-Z. Liao, *Inorg. Chem.*, 2010, *49*, 4735–4737.
36. F. Pointillart, S. Klementieva, V. Kuropatov, Y. L. Gal, S. Golhen, O. Cador, V. Cherkasov and L. Ouahab, *Chem. Commun.*, 2012, *48*, 714–716.
37. M. Leskelä and L. Niinistö, Inorganic Complex Compounds I, in *Handbook on the Physics and Chemistry of Rare Earths* (Eds. K. A. Gschneidner, Jr. and L. Eyring), Vol. 8, Elsevier, Amsterdam, 1986.
38. J. M. Haschke, Halides, in *Handbook on the Physics and Chemistry of Rare Earths* (Ed. K. A. Gschneidner, Jr. and L. Eyring), Vol. 4, Elsevier, 1979.
39. O. Greis and J. M. Haschke, Rare Earth Fluorides, in *Handbook on the Physics and Chemistry of Rare Earths* (Eds. K. A. Gschneidner, Jr., L. Eyring), Vol. 5, Elsevier, 1982.
40. G. Meyer and M. S. Wickleder, Simple and Complex Halides, in *Handbook on the Physics and Chemistry of Rare Earths* (Ed. K. A. Gschneidner, Jr., L. Eyring), Vol. 28, Elsevier, 2000.
41. M. D. Taylor, *Chem. Rev.*, 1962, *62*, 503–511.
42. G. Meyer, *Inorg. Synth.*, 1989, *25*, 146–150.
43. F. Nief, *Dalton Trans.*, 2010, *39*, 6589–6598.
44. H. B. Kagan, *Tetrahedron*, 2003, *59*, 10351–10372.
45. R. A. Flowers II and E. Prasad, Samarium(II) Based Reductants, in *Handbook on the Physics and Chemistry of Rare Earths* (Eds. K. A. Gschneidner, Jr., J.-C. G. Bünzli, V. K. Pecharsky), Vol. 36, Elsevier, 2006.
46. M. N. Bochkarev, I. L. Fedushkin, A. A. Fagin, T. V. Petrovskaya, J.W. Ziller, R. N. R. Broomhall-Dillard and W. J. Evans, *Angew. Chem., Int. Ed. Engl.*, 1997, *36*, 133–135.
47. M. N. Bochkarev and A. A. Fagin, *Chem. Eur. J.*, 1999, *5*, 2990–2992.
48. M. Marezio, H. A. Plettinger and W. H. Zachariasen, *Acta Crystallogr.*, 1961, *14*, 234–236.
49. R. Anwander, *Lanthanide Amides*, in *Topics Current Chem* (Ed. W. A. Herrmann), Vol. 179, Springer, 1996.
50. M. F. Lappert, P. P. Power, A. Protchenko and A. Seeber, *Metal Amide Chemistry*, John Wiley & Sons Ltd., 2009.
51. S. P. Nolan, D. Stern and T. J. Marks, *J. Am. Chem. Soc.*, 1989, *111*, 7844–7853.
52. R. Juza, H. Jacobs and H. Gerke, *Ber. Bunsenges. Phys. Chem.*, 1966, *70*, 1103–1105.
53. V. C. Hadenfeldt, B. Gieger and H. Jacobs, *Z. Anorg. Allg. Chem.*, 1974, *408*, 27–36.

54. D. C. Bradley, J. S. Ghotra and F. A. Hart, *Inorg. Nucl. Chem. Lett.*, 1976, *12*, 735–737.
55. H. C. Aspinall and M. R. Tillotson, *Polyhedron*, 1994, *13*, 3229–3234.
56. W. J. Evans, R. Anwander, J. W. Ziller and S. I. Khan, *Inorg. Chem.*, 1995, *34*, 5927–5930.
57. H. C. Aspinall, D. C. Bradley, M. B. Hursthouse, K. D. Sales, N. P. C. Walker and B. Hussain, *J. Chem. Soc. Dalton Trans.*, 1989, 623–626.
58. H.-X. Li, Q.-F. Xu, J.-X. Chen, M.-L. Cheng, Y. Zhang, W.-H. Zhang, J.-P. Lang and Q. Shen, *J. Organomet. Chem.*, 2004, *689*, 3438–3448.
59. T. D. Tilley, J. M. Boncella, D. J. Berg, C. J. Burns and R. A. Andersen, *Inorg. Synth.*, 1990, *27*, 146–150.
60. T. D. Tilley, PhD Thesis, University of California, Berkeley, 1982.
61. N. F. Chilton, C. A. P. Goodwin, D. P. Mills and R. E. P. Winpenny, *Chem. Commun.*, 2015, *51*, 101–103.
62. C. Morton, N. W. Alcock, M. R. Lees, I. J. Munslow, C. J. Sanders and P. Scott, *J. Am. Chem. Soc.*, 1999, *121*, 11255–11256.
63. O. Eisenstein, P. B. Hitchcock, A. G. Hulkes, M. F. Lappert and L. Maron, *Chem. Commun.*, 2001, 1560–1561.
64. P. L. Arnold, Z. R. Turner, N. Kaltsoyannis, P. Pelekanaki, R. M. Bellabarba and R. P. Tooze, *Chem. Eur. J.*, 2010, *16*, 9623–9629.
65. P. Dröse, A. R. Crozier, S. Lashkari, J. Gottfriedsen, S. Blaurock, C. G. Hrib, C. Maichle-Mössmer, C. Schädle, R. Anwander and F. T. Edelmann, *J. Am. Chem. Soc.*, 2010, *132*, 14046–14047.
66. A. R. Crozier, A. M. Bienfait, C. Maichle-Mössmer, K. W. Törnroos and R. Anwander, *Chem. Commun.*, 2013, *49*, 87–89.
67. (a) M. Ohashi, M. Konkol, I. D. Rosal, R. Poteau, L. Maron and J. Okuda, *J. Am. Chem. Soc.*, 2008, *130*, 6920–6921; (b) J.-C. Buffet and J. Okuda, *Dalton Trans.*, 2011, *40*, 7748–7754.
68. (a) S. Bambirra, D. van Leusen, A. Meetsma, B. Hessen and J. H. Teuben, *Chem. Commun.*, 2001, 637–638; (b) S. Bambirra, A. Meetsma, B. Hessen and A. P. Bruins, *Organometallics*, 2006, *25*, 3486–3495.
69. (a) F. G. N. Cloke, B. R. Elvidge, P. B. Hitchcock and V. M. E. Lamarche, *J. Chem. Soc., Dalton Trans.*, 2002, 2413–2414; (b) P. W. Roesky, *Organometallics*, 2002, *21*, 4756–4761; (c) A. G. Avent, F. G. N. Cloke, B. R. Elvidge and P. B. Hitchcock, *Dalton Trans.*, 2004, 1083–1096.
70. (a) T. I. Gountchev and T. D. Tilley, *Organometallics*, 1999, *18*, 2896–2905; (b) T. I. Gountchev and T. D. Tilley, *Organometallics*, 1999, *18*, 5661–5667.

71. (a) A. Dumitrescu, B. Martin-Vaca, H. Gornitzka, J.-B. Cazaux, D. Bourissou and G. Bertrand, *Eur. J. Inorg. Chem.*, 2002, 1948–1951; (b) K. C. Hultzsch, F. Hampel and T. Wagner, *Organometallics* 2004, *23*, 2601–2612.
72. (a) H. C. Aspinall and M. R. Tillotson, *Inorg. Chem.*, 1996, *35*, 2163–2164; (b) T. G. Wetzel and P. W. Roesky, *Z. Anorg. Allg. Chem.*, 1999, *625*, 1953–1954.
73. (a) M. D. Fryzuk and T. S. Haddad, *J. Am. Chem. Soc.*, 1988, *110*, 8263–8265; (b) M. D. Fryzuk, T. S. Haddad and S. J. Rettig, *Organometallics*, 1991, *10*, 2026–2036.
74. (a) P. L. Arnold, S. A. Mungur, A. J. Blake and C. Wilson, *Angew. Chem. Int. Ed.*, 2003, *42*, 5981–5984; (b) S. T. Liddle and P. L. Arnold, *Organometallics*, 2005, *24*, 2597–2605; (c) P. L. Arnold and I. J. Casely, *Chem. Rev.*, 2009, *109*, 3599–3611.
75. (a) P. J. Shapiro, E. E. Bunel, W. P. Schaefer and J. E. Bercaw, *Organometallics*, 1990, *9*, 867–869; (b) J. Okuda, *Dalton Trans.*, 2003, 2367–2378.
76. F. H. Wang, S. W. Wang, X. C. Zhu, S. L. Zhou, H. Miao, X. X. Gu, Y. Wei and Q. B. Yuan, *Organometallics*, 2013, *32*, 3920–3931.
77. G. R. Giesbrecht and J. C. Gordon, *Dalton Trans.*, 2004, 2387–2393.
78. Z. X. Liu and Y. F. Chen, *Sci. Sin. Chim.*, 2011, *41*, 304–313.
79. O. T. Summerscales and J. C. Gordon, *RSC Adv.*, 2013, *3*, 6682–6692.
80. A. A. Trifonov, M. N. Bochkarev, H. Schumann and J. Loebel, *Angew. Chem. Int. Ed. Engl.*, 1991, *30*, 1149–1151.
81. N. S. Emelyanova, M. N. Bochkarev, H. Schumann, J. Loebel and L. Esser, *Koord. Khim.*, 1994, *20*, 789–793.
82. C. L. Pan, W. Chen, S.Y. Song, H. J. Zhang and X. W. Li, *Inorg. Chem.*, 2009, *48*, 6344–6346.
83. (a) Z. W. Xie, S. W. Wang, Q. C. Yang and T. C. W. Mak, *Organometallics*, 1999, *18*, 1578–1579; (b) S. W. Wang, Q. C. Yang, T. C. W. Mak and Z. W. Xie, *Organometallics*, 1999, *18*, 5511–5517.
84. H.-S. Chan, H.-W. Li and Z. W. Xie, *Chem. Commun.*, 2002, 652–653.
85. W. J. Evans, M. A. Ansari, J. W. Ziller and S. I. Khan, *Inorg. Chem.*, 1996, *35*, 5435–5444.
86. J. C. Gordon, G. R. Giesbrecht, D. L. Clark, P. J. Hay, D. W. Keogh, R. Poli, B. L. Scott and J. G. Watkin, *Organometallics*, 2002, *21*, 4726–4734.
87. D. Schädle, C. Schädle, K. W. Törnroos and R. Anwander, *Organometallics*, 2012, *31*, 5101–5107.
88. D. Schädle, C. Schädle, D. Schneider, C. Maichle-Mössmer and R. Anwander, *Organometallics*, 2015, *34*, 4994–5008.

89. J. Q. Hong, L. X. Zhang, K. Wang, Z. X. Chen, L. M. Wu and X. G. Zhou, *Organometallics*, 2013, *32*, 7312–7322.
90. N. L. Lampland, J. Zhu, M. Hovey, B. Jana, A. Ellern and A. D. Sadow, *Inorg. Chem.*, 2015, *54*, 6938–6946.
91. J.-C. Berthet, P. Thuéry and M. Ephritikhine, *Eur. J. Inorg. Chem.*, 2008, 5455–5459.
92. D. M. Cui, O. Tardif and Z. M. Hou, *J. Am. Chem. Soc.*, 2004, *126*, 1312–1313.
93. J. Q. Hong, L. X. Zhang, K. Wang, Y. Zhang, L. H. Weng and X. G. Zhou, *Chem. Eur. J.*, 2013, *19*, 7865–7873.
94. D. J. Beetstra, A. Meetsma, B. Hessen and J. H. Teuben, *Organometallics*, 2003, *22*, 4372–4374.
95. J. Scott, F. Basuli, A. R. Fout, J. C. Huffman and D. J. Mindiola, *Angew. Chem. Int. Ed.*, 2008, *47*, 8502–8505.
96. (a) E. L. Lu, Y. X. Li and Y. F. Chen, *Chem. Commun.*, 2010, *46*, 4469–4471; (b) E. L. Lu, J. X. Chu, Y. F. Chen, M. V. Borzov and G. Y. Li, *Chem. Commun.*, 2011, *47*, 743–745; (c) Z. Jian, W. Rong, Z. Mou, Y. Pan, H. Xie and D. Cui, *Chem. Commun.*, 2012, *48*, 7516–7518; (d) T. Chu, W. E. Piers, J. L. Dutton and M. Parvez, *Organometallics*, 2013, *32*, 1159–1165; (e) W. F. Rong, J. H. Cheng, Z. H. Mou, H. Y. Xie and D. M. Cui, *Organometallics*, 2013, *32*, 5523–5529; (f) J. X. Chu, X. H. Han, C. E. Kefalidis, J. L. Zhou, L. Maron, X. B. Leng and Y. F. Chen, *J. Am. Chem. Soc.*, 2014, *136*, 10894–10897.
97. D. Schädle, M. Meermann-Zimmermann, C. Schädle, C. Maichle-Mössmer and R. Anwander, *Eur. J. Inorg. Chem.*, 2015, 1334–1339.
98. L. A. Solola, A. V. Zabula, W. L. Dorfner, B. C. Manor, P. J. Carroll and E. J. Schelter, *J. Am. Chem. Soc.*, 2017, *139*, 2435–2442.
99. J. X. Chu, E. L. Lu, Z. X. Liu, Y. F. Chen, X. B. Leng and H. B. Song, *Angew. Chem. Int. Ed.*, 2011, *50*, 7677–7680.
100. J. X. Chu, E. L. Lu, Y. F. Chen and X. B. Leng, *Organometallics*, 2013, *32*, 1137–1140.
101. E. L. Lu, Q. H. Zhou, Y. X. Li, J. X. Chu, Y. F. Chen, X. B. Leng and J. Sun, *Chem. Commun.*, 2012, *48*, 3403–3405.
102. F. Nief, *Coord. Chem. Rev.*, 1998, *178–180*, 13–81.
103. T. S. Li, S. Kaercher and P. W. Roesky *Chem. Soc. Rev.*, 2014, *43*, 42–57.
104. (a) H. Schumann and H. Jarosch, *Z. Anorg. Allgem. Chem.*, 1976, *426*, 127–130; (b) H. Schumann and G. M. Frisch, *Z. Naturforsch.*, 1981, *36b*, 1244–1246.
105. G. Bielang and R. D. Fischer, *J. Organomet. Chem.*, 1978, *161*, 335–346.

106. H. Schumann, E. Palamidis, G. Schmid and R. Boese, *Angew. Chem. Int. Ed. Engl.*, 1986, *25*, 718–719.
107. (a) G. W. Rabe, J. Riede and A. Schier, *J. Chem. Soc., Chem. Commun.*, 1995, 577–578; (b) G. W. Rabe and J. W. Ziller, *Inorg. Chem.*, 1995, *34*, 5378–5379.
108. G. W. Rabe, J. Riede and A. Schier, *Inorg. Chem.*, 1996, *35*, 40–45.
109. H. Schumann and G. M. Frisch, *Z. Naturforsch.*, 1979, *34b*, 748–749.
110. (a) G. W. Rabe, J. Riede and A. Schier, *Inorg. Chem.*, 1996, *35*, 2680–2681; (b) G. W. Rabe, G. P. A. Yap and A. L. Rheingold, *Inorg. Chem.*, 1997, *36*, 3212–3215.
111. M. Westerhausen, S. Schneiderbauer, M. Hartmann, M. Warchhold and H. Nöth, *Z. Anorg. Allg. Chem.*, 2002, *628*, 330–332.
112. S. Krieck, H. Görls and M. Westerhausen, *Inorg. Chem. Commun.*, 2009, *12*, 409–411.
113. Y. D. Lv, X. Xu, Y. F. Chen, X. B. Leng and M. V. Borzov, *Angew. Chem. Int. Ed.*, 2011, *50*, 11227–11229.
114. G. W. Rabe, G. P. A. Yap and A. L. Rheingold, *Inorg. Chem.*, 1995, *34*, 4521–4522.
115. G. W. Rabe, J. Riede and A. Schier, *Organometallics*, 1996, *15*, 439–441.
116. J. Yuan, H. F. Hu and C. M. Cui, *Chem. Eur. J.*, 2016, *22*, 5778–5785.
117. (a) H. C. Aspinall, D. C. Bradley and K. D. Sales, *J. Chem. Soc. Dalton Trans.*, 1988, 2211–2213; (b) H. C. Aspinall, S. R. Moore and A. K. Smith, *J. Chem. Soc. Dalton Trans.*, 1992, 153–156.
118. (a) W. J. Evans, I. Bloom, W. E. Hunter and J. L. Atwood, *Organometallics*, 1983, *2*, 709–714; (b) W. J. Evans, J. T. Leman, J. W. Ziller and S. I. Khan, *Inorg. Chem.*, 1996, *35*, 4283–4291.
119. M. Westerhausen, S. Schneiderbauer, N. Makropoulos, M. Warchhold, H. Nöth, H. Piotrowski and K. Karaghiosoff, *Organometallics*, 2002, *21*, 4335–4341.
120. W. Y. Yi, J. Zhang, L. C. Hong, Z. X. Chen and X. G. Zhou, *Organometallics*, 2011, *30*, 5809–5814.
121. Z. M. Hou, Y. G. Zhang, H. Tezuka, P. Xie, O. Tardif, T.-A. Koizumi, H. Yamazaki and Y. Wakatsuki, *J. Am. Chem. Soc.*, 2000, *122*, 10533–10543.
122. I. V. Basalov, D. M. Lyubov, G. K. Fukin, A. V. Cherkasov and A. A. Trifonov, *Organometallics*, 2013, *32*, 1507–1516.
123. S. N. Konchenko, N. A. Pushkarevsky, M. T. Gamer, R. Köppe, H. Schnöckel and P. W. Roesky, *J. Am. Chem. Soc.*, 2009, *131*, 5740–5741.

124. W. L. Huang and P. L. Diaconescu, *Chem. Commun.*, 2012, *48*, 2216–2218.
125. T. S. Li, J. Wiecko, N. A. Pushkarevsky, M. T. Gamer, R. Köppe, S. N. Konchenko, M. Scheer and P. W. Roesky, *Angew. Chem. Int. Ed.*, 2011, *50*, 9491–9495.
126. J. D. Masuda, K. C. Jantunen, O. V. Ozerov, K. J. T. Noonan, D. P. Gates, B. L. Scott and J. L. Kiplinger, *J. Am. Chem. Soc.*, 2008, *130*, 2408–2409.
127. P. Cui, Y. F. Chen, X. Xu and J. Sun, *Chem. Commun.*, 2008, 5547–5549.
128. P. Cui, Y. F. Chen and M. V. Borzov, *Dalton Trans.*, 2010, *39*, 6886–6890.
129. B. F. Wicker, J. Scott, J. G. Andino, X. F. Gao, H. Park, M. Pink and D. J. Mindiola, *J. Am. Chem. Soc.*, 2010, *132*, 3691–3693.
130. Y. D. Lv, C. E. Kefalidis, J. L. Zhou, L. Maron, X. B. Leng and Y. F. Chen, *J. Am. Chem. Soc.*, 2013, *135*, 14784–14796.
131. K. Wang, G. Luo, J. Q. Hong, X. G. Zhou, L. H. Weng, Y. Luo and L. X. Zhang, *Angew. Chem. Int. Ed.*, 2014, *53*, 1053–1056.
132. T. Pugh, F. Tuna, L. Ungur, D. Collison, E. J. L. McInnes, L. F. Chibotaru and R. A. Layfield, *Nature Commun.*, 2015, *6*, 7492–7497.
133. L. M. A. Saleh, K. H. Birjkumar, A. V. Protchenko, A. D. Schwarz, S. Aldridge, C. Jones, N. Kaltsoyannis and P. Mountford, *J. Am. Chem. Soc.*, 2011, *133*, 3836–3839.
134. S. H. Li, J. H. Cheng, Y. H. Chen, M. Nishiura and Z. M. Hou, *Angew. Chem., Int. Ed.*, 2011, *50*, 6360–6363.
135. B. L. Wang, M. Nishiura, J. H. Cheng and Z. M. Hou, *Dalton Trans.*, 2014, *43*, 14215–14218.
136. B. L. Wang, X. H. Kang, M. Nishiura, Y. Luo and Z. M. Hou, *Chem. Sci.*, 2016, *7*, 803–809.

https://doi.org/10.1142/9781800610163_0005

Chapter 5

Coordination Chemistry of Actinides

Grégory Nocton[*,‡] *and Marinella Mazzanti*[†,§]

[*]*LCM, CNRS, Ecole Polytechnique,*
Institut Polytechnique de Paris, Route de Saclay,
91120 Palaiseau, France
[†]*Institut des Sciences et Ingénierie Chimiques,*
Ecole Polytechnique Fédérale de Lausanne (EPFL),
CH-1015 Lausanne, Switzerland
[‡]*gregory.nocton@polytechnique.edu*
[§]*marinella.mazzanti@epfl.ch*

5.1 Introduction

The coordination chemistry of the actinides has seen a vigorous development in the last decades especially since good, reliable starting materials and synthetic procedures are more accessible and since the traditional oxidation states have been challenged when the solution chemistry has been extended to non-aqueous environments. Moreover, the technology for handling very sensitive and radioactive material has opened new opportunities for the growth of this discipline and the large scope of their potential interest gathered in several other chapters of this book has encouraged coordination chemists to exploit all kinds of new strategies for the synthesis of coordination compounds containing actinides. Nonetheless, the radioactive character and the inherent toxicity of most actinides do not allow large-scale synthesis and radiation damages often prevent the use of typical spectroscopic tools available in ordinary laboratories. Because of this,

Table 5.1. Common (**bold**) and other accessible oxidation states of actinides (S for solid-state only).

Oxidation state/element	Th	Pa	U	Np	Pu	Am	Cm
7				X	X		
6			**X**	X	X	X	
5		**X**	X	**X**	X		
4	**X**	X	**X**	X	**X**	X	X
3	X	S	X	X	X	**X**	**X**
2	X		X			S	

Oxidation state/element	Bk	Cf	Es	Fm	Md	No	Lr
7							
6							
5							
4	X	S					
3	**X**	**X**	**X**	**X**	**X**	X	**X**
2		S	S	X	X	**X**	

aside uranium and thorium chemistry that represent the vast majority of the molecules made, the coordination chemistry of actinides is still very challenging and is limited to few research groups in the world. This chapter aims at gathering a selection of the important modern aspects of this area with a strong focus on the coordination chemistry in non-aqueous environments.

Due to the larger spatial extension of 5f orbitals compared to 4f orbitals and due to the low energy gap between 5f and 6d orbitals, the removal of several electrons from the outer shell is possible and a variable range of oxidation states can be observed for the earlier actinides (An = Th–Pu). Table 5.1 gathers the known oxidation states (the most common appear in bold) for most actinides. A quick look at this table shows that U, Np and Pu are the actinide ions that present the highest number of accessible oxidation states. The +IV oxidation state is by far the most stable oxidation state for Th, with only few examples of complexes containing Th in the +3 oxidation state isolated in anaerobic and non-aqueous conditions and one unique example of a molecular compound containing Th^{2+} recently reported.[1] For high-valent U^{6+}, and Np^{5+}, the actinyl form $AnO_2{}^{+/2+}$ is the stable form in aerated water. Coordination complexes of non-uranyl U^{VI} and U^{V} have been reported, but are less common. Complexes of U^{IV} and to a lesser extent U^{III} can be prepared in anhydrous anaerobic conditions. Two molecular complexes of formally U^{II} are also known to date. Plutonium as Pu^{4+} is the most stable oxidation state in aqueous solution, but all the first four oxidation states (+III, +IV, +V, +VI) can coexist in aqueous solution and their concentration depends on the pH. Complexes of Pu^{III} are more easily formed in anhydrous media compared to Pu^{IV} complexes. Most reported coordination complexes of neptunium contain Np in the oxidation states +IV to +VI. A few compounds of Np^{III} have been prepared in non-aqueous anaerobic conditions. The +III oxidation state is the common and most stable oxidation state for minor actinides (Am–Lr).

5.2 Aqua Ions

The most well-characterised aqua ion is the uranyl(VI) complex $[UO_2(OH_2)_5]^{2+}$, **1** (Figure 5.1). Uranyl(V) can be generated in water in the absence of oxygen but, unlike neptunyl(V) and plutonyl(V), uranium is only briefly stable and readily disproportionates to form tetravalent uranium and uranyl(VI) species. The hydrated tetravalent U^{IV} and Th^{IV} cations are nine-coordinated in aqueous solution. Thorium(IV) extensively hydrolyses from pH 4 affording polynuclear hydroxide species. In strongly acidic solutions, Pu^{III} and Pu^{IV} exist as the nona-aqua and octa-aqua ions, respectively. Plutonium in the

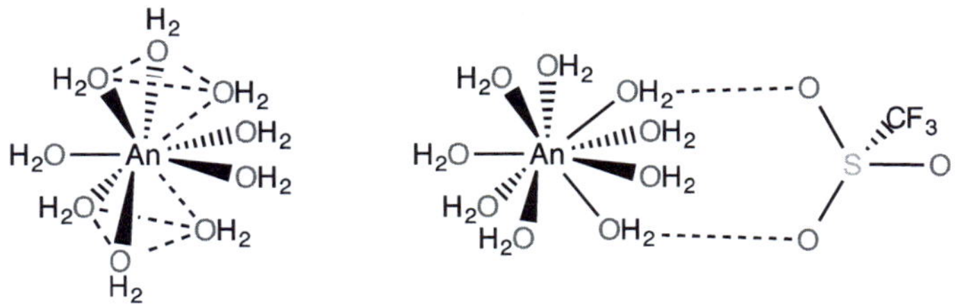

An = U, 3; Np, 4; Pu, 5; Am, 6; Cm, 7: Cf, 8

Figure 5.1. Representations of the solid state structures of actinide(III) aqua ions.

oxidation states +V and +VI are also present in water as penta-aquo trans dioxo cations. In contrast Pu^{VII}, prepared by the ozonation of Pu^{VI} in alkaline solutions, is present in water as the tetra oxo bis-hydroxo species $PuO_4(OH)_2$, **2** (Figure 5.1). The cation, $NpO_2{}^+$ is the most stable species in aqueous solution over a wide range of pH values. High-energy X-ray scattering data of $NpO_2{}^+$ and $NpO_2{}^{2+}$ show that the neptunyl cations coordinate five water molecules in the equatorial plane in both cases. In the case of neptunyl(V), T-shaped cation–cation interactions (the term often used to define the oxo ligand of one neptunyl(V) coordinated in the equatorial plane of a neighbouring neptunyl group) are also observed.[2]

The aqueous chemistry of later actinides in the oxidation state +III is very similar to lanthanide(III) chemistry. The decrease of the ionic radii along the An^{3+} series leads to a change in the hydration number from nine to eight (between Cm^{3+} and Es^{3+}) as observed for Ln^{3+} ions. The molecular structure of the triflate salts of nona-aqua complexes of An^{III}, $[An(H_2O)_9](CF_3SO_3)_3$, **3–8**, have been crystallographically characterised for An = U, Np, Pu, Am, Cm, Cf.[3] The U^{3+} and Np^{3+} cation are extremely sensitive to oxidation. They show a tricapped trigonal prismatic coordination geometry analogous to the corresponding lanthanide (III) aqua ions. In the solid-state structure, each of the nine water molecules forms hydrogen bonds with the neighbouring triflate anion leading to structure with C_{3h} symmetry (Figure 5.1). Particularly remarkable is the isolation of the $[U(H_2O)_9](CF_3SO_3)_3$ complex when considering the ability of the U^{III} cation and of the $[U(CF_3SO_3)_3]$ complex in particular to reduce water.[4]

5.3 Common Precursors

Generally the most common precursors for the synthesis of actinide coordination compounds in various oxidation states are halide, nitrate or acetate salts. In anhydrous conditions, amide compounds are also convenient precursors that allow the controlled synthesis of low coordinate actinide complexes via the protonolysis route. The bulky bis(trimethylsilyl)amido anion, $[N(SiMe_3)_2]^-$ is largely used because it acts as a strong base and a good leaving group in the reaction with protonated ligands.

The most common precursors used for the synthesis of uranyl(VI) coordination complexes both in water and non-anhydrous organic solvents are the commercially available (at least in the USA) hexahydrate nitrate salt $UO_2(NO_3)_2 \cdot 6H_2O$, **9**, and the analogous acetate $UO_2(OAc)_2 \cdot 2H_2O$, **10**. $Th(NO_3)_4(H_2O)_5$, **11**, is the most commonly used commercial starting material for the development of thorium chemistry. No commercial sources exist for the heavier actinide ions. This, together with their higher radioactivity compared to the naturally available isotopes of uranium and thorium, confines the development of the chemistry of these ions to specialised laboratories and nuclear facilities.

The chemistry of actinide ions has been for a long time dominated by studies in aqueous solution or in wet organic solvents. However, the development of new anhydrous precursors led to a greater expansion of actinide chemistry rendering the coordination of a much wider variety of ligands possible. Notably, the coordination chemistry of actinide cations in aqueous solutions is dominated by electrostatic metal–ligand interactions and limited to hard donor ligands (according to the Pearson classification). In anhydrous solvents, softer ligands can be used to preferentially bind actinides leading to more covalent M–L bonds and unexpected structure and reactivity.

The non-aqueous chemistry of uranyl(VI) has been developed using the dimeric compound $[UO_2Cl_2(thf)_2]$, **12** and monomeric compound $[UO_2Cl_2(thf)_3]$, **13**[5] as precursors. The dehydration of $[UO_2Cl_2(H_2O)_n]$, **14**, with $SiMe_3Cl$ provides a convenient route to the $[UO_2Cl_2(thf)_3]$ (**13**) complex (Scheme 5.1).

$$UO_3 + HCl_{(aq)} \xrightarrow{\Delta} \underset{\mathbf{14}}{[UO_2(Cl)_2(H_2O)_n]} \xrightarrow{\text{thf, SiMe}_3\text{Cl}} \underset{\mathbf{13}}{[UO_2Cl_2(thf)_3]}$$

Scheme 5.1. Synthetic route to the anhydrous uranyl(VI) starting material $[UO_2Cl_2(thf)_3]$.

The triflate analogue $[UO_2(OTf)_2(thf)_3]$, **15**, is easily obtained from the reaction of the chloride complex with AgOTf. The iodide complexes $[UO_2I_2(py)_3]$, **16** and $[UO_2I_2(thf)_3]$, **17**, are easily obtained from the metathesis reaction of the triflate salt with Me_3SiI or from the oxidation of $[UI_3(thf)_4]$, **18**, with two equivalents of pyridine *N*-oxide (PyNO).

In all these complexes, the uranium adopts a pentagonal bipyramidal geometry which is the most common geometry for uranyl(VI). The reaction of $[UO_2Cl_2(thf)_2]$, **13**, with $K[N(SiMe_3)_2]$ allows the synthesis of the bis-amido precursor $[UO_2(N(SiMe_3)_2)_2(thf)_2]$, **19**. The presence of the bulky amide results in an unusual six coordinate uranyl(VI) complex.

The $[NpO_2Cl_2(thf)]$, **20** and $[PuO_2Cl_2(thf)_3]$, **21**, analogues have also been prepared by adding HCl to a THF solution of $NpO_2(OH)_2 \cdot xH_2O$, **22** and $[PuO_2(CO_3)_2]$, **23**, respectively (Eq. (5.1)). The $[PuO_2Cl_2(thf)_3]$ (**21**) complex crystallises as a dimer with a pentagonal bipyramidal geometry isostructural to the uranyl analogue while $[NpO_2Cl_2(thf)]$ (**20**) crystallises in a polymeric form with the neptunyl cations linked by two bridging chloride anions.[6] The redox chemistry of neptunyl is difficult to control in organic solution. Notably, a mixed-valent Np(VI)/Np(V) trimer $([\{Np^{VI}O_2Cl_2\}\{Np^{V}O_2Cl(thf)_3\}_2])$, **24** (Figure 5.2), was obtained after leaving the $[NpO_2Cl_2(thf)]$ complex in THF for several days.

$$\underset{\mathbf{23}}{[PuO_2(CO_3)]} \xrightarrow[\text{thf}]{\text{HCl/Et}_2\text{O}} \underset{\mathbf{21}}{[PuO_2Cl_2(thf)_3]_2} + 2CO_2(g) + 2H_2O \quad (5.1)$$

A stable uranyl(V) iodide complex $[UO_2(py)_5(KI_2Py_2)]_\infty$, **25**, was first prepared from the reaction of $[UI_3(thf)_4]$ (**18**) with one equivalent of PyNO and 1 equivalent of H_2O followed by addition of KI (Eq. (5.2)).[7] The same complex can also be prepared by reduction of $[UO_2I_2(THF)_3]$ (**17**) with KC_5R_5 (R = H, Me).[8] An isostructural neptunyl(V) complex $[(NpO_2Py_5)(KI_2Py_2)]_n$, **26**, has been prepared

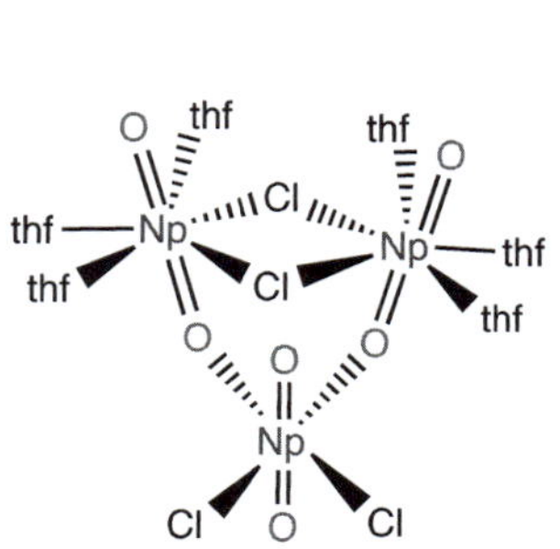

24

Figure 5.2. Mixed valent $NpO_2{}^+/NpO_2{}^{2+}$ trimeric chloride complex **24**, which exhibits cation-cation interactions in the solid state.

from dry "NpO_2Cl" by anion exchange with potassium iodide in pyridine.[9]

$$\underset{\mathbf{18}}{[UI_3(thf)_4]} + PyNO + H_2O \xrightarrow{\text{pyridine}} \underset{\mathbf{25}}{[UO_2(py)_5(KI_2py_5)]_n} \qquad (5.2)$$

High oxidation states are rare beyond plutonium, $AmO_2{}^+$ and $AmO_2{}^+$ species are known but are difficult to stabilise and there are no reported precursors for their anhydrous chemistry.[10]

While for a long time UCl_4[11] (Scheme 5.2) was the only entry into U(IV) chemistry, a wide variety of other convenient precursors have now been reported such as $[UX_4(MeCN)_4]$ (Scheme 5.2), (X = Cl, **26**; I, **27**; Br, **28**), $[UI_4(PhCN)_4]$, **29** (Scheme 5.2), $[UI_4(OEt_2)_2]$, **30**, $[UI_4(1,4\text{-dioxane})_2]$, **31**.[12]

$$UO_3 + CCl_3C(Cl){=}CCl_2 \xrightarrow{\Delta} UCl_4$$

$$U + 2\,I_2 \xrightarrow{PhCN} \underset{\mathbf{29}}{[UI_4(PhCN)_4]}$$

Scheme 5.2. Synthetic routes to synthetically useful anhydrous U(IV) precursors.

$ThCl_4(dme)_2$ (**32**), $ThCl_4(1,4\text{-dioxane})_2$ (**33**) and $ThCl_4(thf)_{3.5}$ (**34**)[13] are very convenient anhydrous starting materials for the development of thorium chemistry. These precursors have been recently prepared from commercial thorium sources, $Th(NO_3)_4(H_2O)_5$, **11**, using HCl and the dehydrating agent $SiMe_3Cl$. These precursors render the anhydrous chemistry of this element more accessible to non-specialised laboratories (Scheme 5.3, Figure 5.3).

$$\underset{\mathbf{11}}{[Th(NO_3)_4(H_2O)_4]} + HCl_{(aq)} \xrightarrow{\Delta} [Th(Cl)_4(H_2O)n] \xrightarrow{\text{dme, }SiMe_3Cl} \underset{\mathbf{32}}{[ThCl_4(dme)_2]}$$

Scheme 5.3. Synthesis of the organic soluble starting material $[ThCl_4(dme)_2]$, **32**.

The precursors $[UI_3(S)_4]$, $[NpI_3(S)_4]$ and $[PuI_3(S)_4]$ (S = thf; U, **35**; Np, **36**; Pu, **37**, or pyridine; U, **38**; Np, **39**; Pu, **40**)) and $[UI_3(S)_{1.5}]$ (S = 1,4-dioxane), **41**, are conveniently prepared from the reaction of uranium or plutonium metal with iodine in coordinating solvents (S = thf or pyridine).[14] The synthesis of $[UCl_3(py)_4]$,

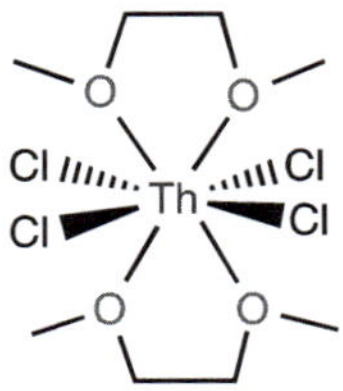

32

Figure 5.3. Representation of the molecular structure of $[ThCl_4(dme)_2]$, **32**.

42, by reduction of UCl_4 has also been recently reported (Eq. 5.3). This synthesis renders low valent chemistry of uranium more accessible since UCl_4 is prepared from UO_3, a precursor easier to purchase than uranium metal.[15]

$$UCl_4 + 3\,NaBH_4 \xrightarrow[\text{2. hexane/pyridine}]{\text{1. Mg, dioxane, 100°C, 3d}} \underset{\mathbf{42}}{[UCl_3(py)_4]_2} \qquad (5.3)$$

The oxidation of plutonium metal with bromine leads to $[PuBr_3(thf)_4]$, **43**. The $[UI_3(thf)_4]$ (**18**) and $[PuI_3(thf)_4]$ complexes are isostructural with seven coordinate Pu^{3+} and U^{3+} cations displaying a distorted pentagonal bipyramidal geometry. Both $[UI_3(thf)_4]$ (**18**) and $[PuI_3(thf)_4]$ (**43**) can be reacted with $Na[N(SiMe_3)_2]$ to yield the tris-amide complexes $[An\{N(SiMe_3)_2\}_3]$ (An = U, **44**; Np, **45**; Pu, **46**).[14a] The single crystal X-ray structures of the isostructural complexes $[U\{N(SiMe_3)_2\}_3]$ (**44**)[16] and $[Pu(N(SiMe_3)_2)_3]$ (**46**)[17] show that the metals adopt a pyramidal geometry. The use of a bulkier amide leads to a trigonal planar geometry for the complex $[U[N(SiMe_2tBu)_2]_3$, **47**.[18] The tris-amide complex $[U\{N(SiMe_3)_2\}_3]$ (**44**) has become a great precursor for the synthesis of monodentate and polydentate alkoxide and amide complexes. By contrast, the use of the Pu and Np precursors is less developed due to the lower availability of these actinide ions and the difficulty in handling them.

Finally the reproducible synthesis of the $[U(BH_4)_3(thf)_2]$, **48**, from UI_3 and $NaBH_4$ has recently provided another entry into the chemistry of U(III) (Eq. 5.4).[19]

$$\underset{\mathbf{18}}{[UI_3(thf)_4]} + 3\,NaBH_4 \xrightarrow[-3\,NaI]{\text{thf}} \underset{\mathbf{48}}{[U(BH_4)_3(thf)_2]} \qquad (5.4)$$

5.4 Uranyl(VI) Complexes

The UO_2^{2+} cation, the *trans*-dioxo di-cation, is by far the oxidation state of uranium with the most developed coordination chemistry as a result of the high stability of uranyl(VI) complexes in ambient conditions. Several important exceptions are to be noted, in particular the halides UF_6, **49**, (an important compound in the nuclear industry) and UCl_6 and several alkoxy derivatives whose first syntheses were reported in the 1950s by Gilman.[20] Since then, several others have been described and structurally characterised.[21] Another important class of U(VI) complexes do not possess the UO_2 motif and are the *trans*-(bis)imido analogues UNR_2^{2+}, ($UN^tBu_2I_2(thf)_2$, **50**), developed in 2005 by Boncella and his group (Figure 5.4).[22] The uranyl di-cation is easily obtained from any low-valent uranium complex in the presence of air.

Unlike in transition metal di-oxo chemistry, in the UO_2^{2+} moiety, the two oxo groups are found *trans* to each other with short U–O distances (range from 1.7 to 1.9 Å) that give rise to typical bands in vibrational spectroscopy; a strong one around 920–980 cm^{-1} in IR (ν_3 asymmetric O–U–O stretch) and one around 860 cm^{-1} in Raman spectroscopy (ν_3 total symmetric O–U–O stretch). The linear geometry of the uranyl moiety ($D_{\infty h}$) is possible because of the involvement in bonding of both the 4f and the 5d valence shells that can form σ and π bonds with oxygen (σ_g, σ_u, π_g and π_u) resulting in a bond order of three for both U–O bonds.[23] Because of the strong axial oxo bonding, the remaining ligands bind in the equatorial plane. The coordination number in the equatorial positions ranges from 3 to 6, i.e. a total coordination number of 5–8 for the uranium ion.

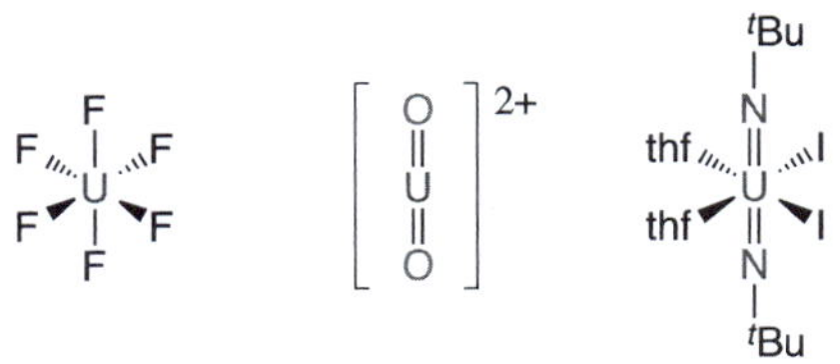

49 Uranyl dication UO_2^{2+} **50**

Figure 5.4. Molecular structures of the U(VI)F_6, U(VI)-bisimido complexes and of the uranyl(VI) dication.

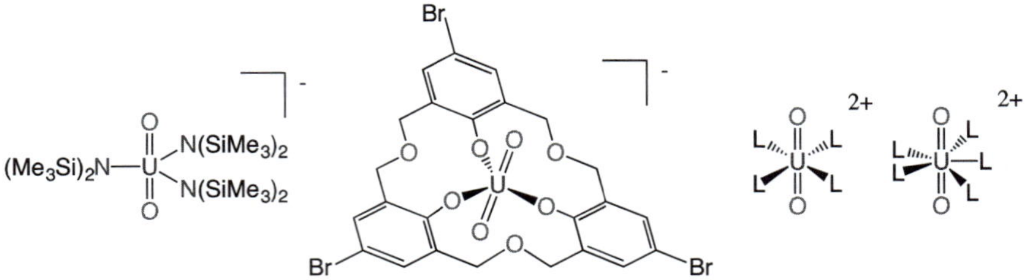

51 **52** $[UO_2L_4]^{2+}$ $[UO_2L_5]^{2+}$

Figure 5.5. Examples where the uranyl(VI) ion coordinates 3–5 donor atoms in the equatorial plane (here, L = neutral monodentate ligand).

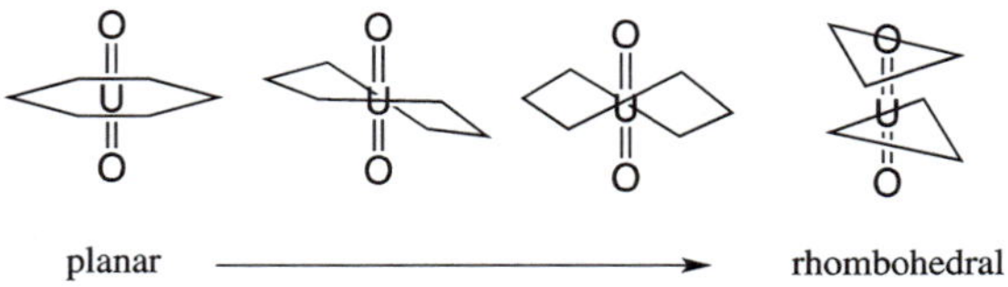

Figure 5.6. Coordination geometries of complexes of uranyl.

With sterically bulky ligands such as the anionic $N(SiMe_3)_2^-$ ligand ($[UO_2(N(SiMe_3)_2)_3]^-$, **51**) or the calixarene ligand (($[UO_2(calix^{2-}]^-$, **52**, Figure 5.5), only three donors wrap up around the $UO_2{}^{2+}$ moiety while monodentate donors favour the coordination of four to five ligands and multidentate ligands may allow the coordination of up to six donors.

The coordination geometry varies depending upon the nature of the equatorial ligands. Depending on how the ligand lays in the equatorial plane it will give rise to perfect bipyramidal geometries (the coordination number in the equatorial plane defines the nature of the bipyramid, i.e. trigonal, square planar, tetragonal, pentagonal and hexagonal) or to distorted bipyramidal geometries (above and below the equatorial plane) up to a rhombohedral coordination geometry (see Figure 5.6).[24] The aqua ion of the uranyl dication $[UO_2(OH_2)_5]^{2+}$, **1**, possesses a pentagonal bipyramidal geometry and this constitutes the most common geometry for the uranyl dication.

Because of the nuclear industry, the coordination chemistry of the uranyl ion has attracted considerable interest for both nuclear processing (extraction and fuel) and decontamination. In this context, the carbonate[25] and the nitrate complexes are particularly relevant.

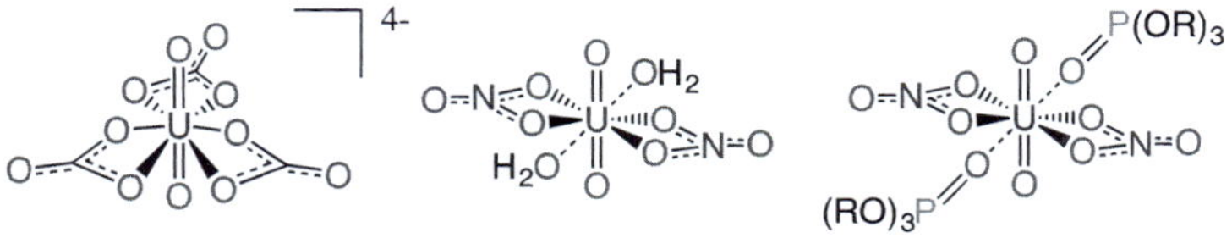

53 **54** **55**

Figure 5.7. Solution structures of uranyl(VI) carbonate (**53**), nitrate (**54**) and nitrate bis-trialkylphosphate complexes (**55**, $R = {}^{n}$Bu, Et).

The eight coordinate complex $[UO_2(CO_3)_3]^{4-}$ **53** (Figure 5.7), forms after carbonate leaching of uranium from contaminated solids, while the reaction of uranium oxides with nitric acid (used in the dissolution of spent nuclear fuel rods) leads to hydrated nitrate complexes $[UO_2(NO_3)_2(H_2O)_2].xH_2O$, (x = 2; **54**) where the hydration number depends upon the acid concentration. The substitution of the water molecules in the nitrate complex by a variety of neutral ligands based on phosphines, phosphates (**55**), oxides, amides and sulfoxides has been largely studied. These studies were in part motivated by the search of optimal systems for the separation of uranium from other elements in nuclear reprocessing (see Figure 5.7).[26]

Linear poly(amino)carboxylate ligands, that are commercially available, provide good chelates for the uranyl complexation but have limited selectivity with respect to other cations present in large amounts in the human body or in the environment. Therefore, the design of ligands with increased selectivity towards uranyl has motivated coordination chemistry studies due to their relevance in uranium sequestration and decorporation. Bulky anionic ligands were used to produce the anionic $[UO_2(carbox^1)_3][Et_3NH]$, **56**, complex with $carbox^1$ as 2,6-terphenyl carboxylate. This ligand allows edge to face π-stacking contributing to the selective complexation of the uranyl complex allowing its extraction from aqueous solutions even in the presence of considerable quantities of NaCl.[27]

The rational design of ligands using secondary interactions (stereognostic coordination chemistry) of the ligand with the uranyl oxo group was first introduced by Raymond's report of a podand amine ligand bearing carboxylate groups ($carbox^3$). The protonation of the apical amine enhances the stability of the complex with the uranyl cation due to hydrogen bonding to the oxo group in $[UO_2(carbox^3)]$, **58**.[28]

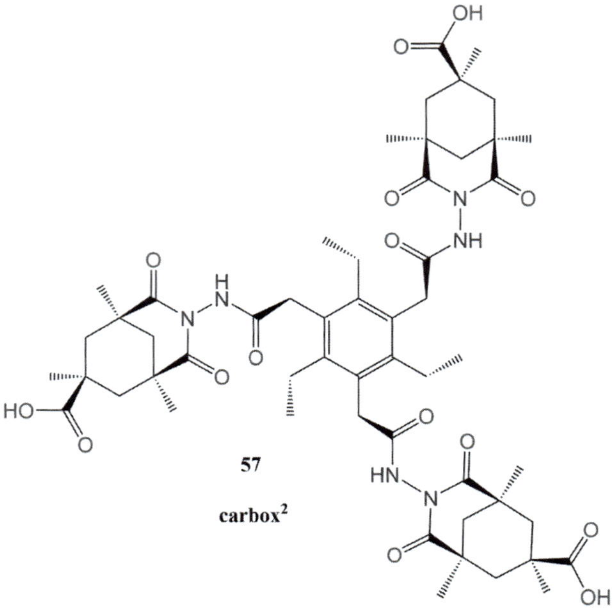

57

carbox²

Figure 5.8. Structure of tripodal tris-carboxylate ligand carbox^2.

More recently, the first crystallographic structure of a uranyl(VI) complexes with a tripodal tris-carboxylate ligand has been reported. In the polydentate carbox^2 ligand (carbox^2, Figure 5.8) designed by Rebek *et al.* the triethylbenzene platform is used to connect, through amide moieties, three bulky carboxylate groups that bind the uranyl group to afford the complex $[UO_2(\text{carbox}^2)]^-$, **57** (Figure 5.8). The coordination of the carboxylate groups, as well as the secondary interactions of the amide group in the chelating ligand provides a very good selectivity of carbox^2 for the uranyl dication in presence of seawater concentration of alkali and alkali-earth cations and of numerous anionic competitor ligands (Cl^-, Na^+, Mg^{2+}, Ca^{2+}, K^+ and SO_4^{2-}) (Figure 5.9).[29]

Many other ligand design strategies have been employed to selectively coordinate the uranyl dication. Among them, the coordination to phenolate groups has been largely studied with multidentate phenolate ligands and with calixarenes bearing phenoxides and ether oxygen atoms. The use of these macrocycles leads to high stability of their uranyl complexes that is reinforced by potential secondary

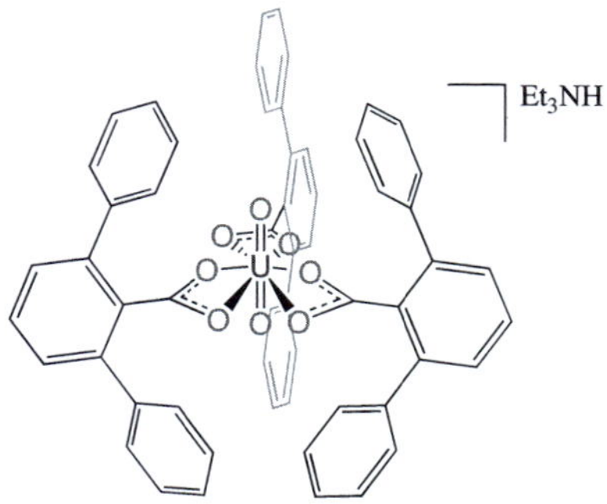

56

Figure 5.9. Molecular structure of $[UO_2(carbox^1)_3][Et_3NH]$, **56**, ($carbox^1$ = 2, 6-terphenyl carboxylate).

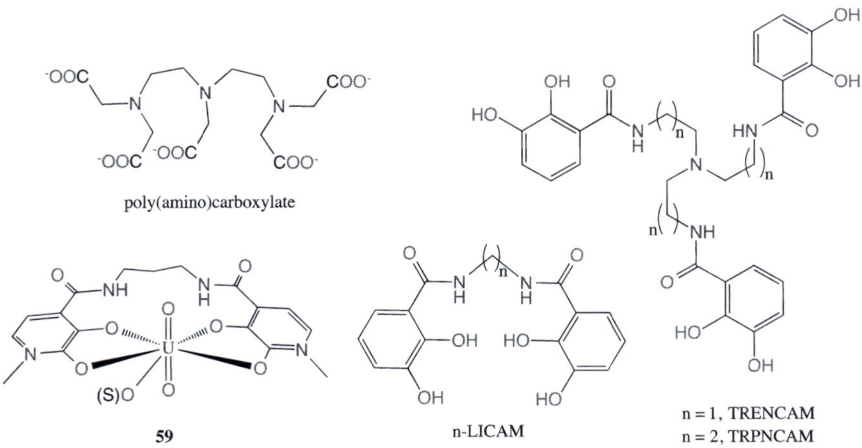

Figure 5.10. Polyaminocarboxylate and siderophore inspired multidentate ligands for the selective coordination of uranyl.

interactions with oxygen donor atoms (**52**, Figure 5.5). Parallel to this, the series of hydroxypyridonate ligands (HOPO; $[UO_2$(3LI-Me-3,2-HOPO)(dmso)], **59**), the catecholamide ligands (CAM) and the terephthalamide ligands (TAM), inspired by biologically relevant siderophores, have been successfully applied to the selective coordination of the uranyl dication and of other actinides by the Raymond group and many ligands with different architectures (multidentate, podand, macrocyclic and mixed ligands) have been proposed (Figure 5.10).[30]

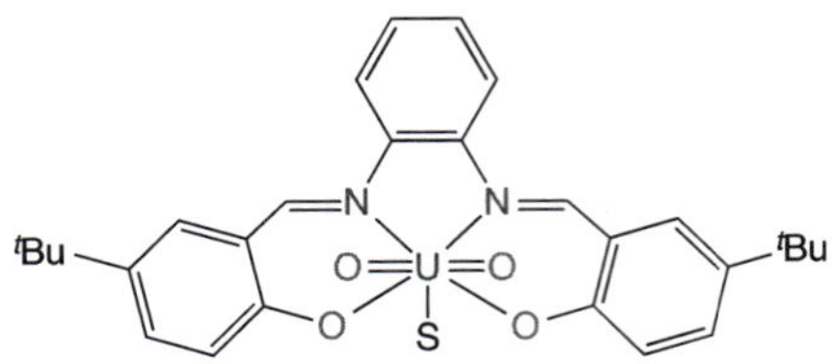

60

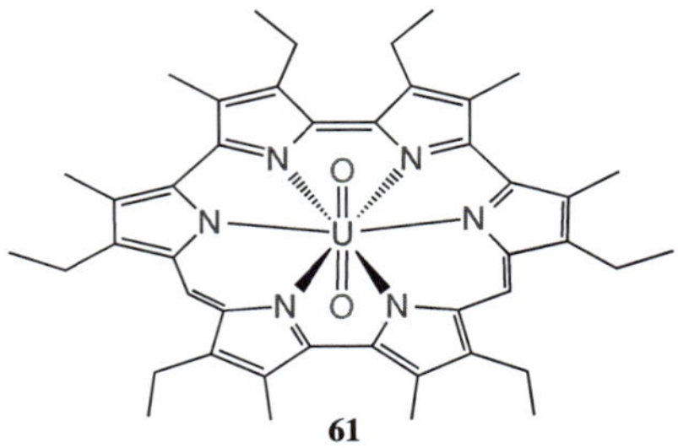

61

Figure 5.11. Examples of uranyl(VI) Schiff base (**60**) and hexaphyrin (**61**) complexes (S = coordinated solvent molecule).

Schiff bases constitute another class of ligands that have been extensively used to coordinate the linear uranyl cation. The presence of strong σ-donor oxygen atoms allows strong coordination. In general the tetradentate Schiff bases give heptacoordinated complexes of pentagonal bipyramid coordination geometry with the two oxo groups in axial positions; the tetradendate ligand being in the equatorial plane and a solvent molecule (e.g. in [UO_2(^{t}Bu_2salophen)(S)], **60**).[31] Some of these complexes have been used for molecular recognition of several ions such as fluoride or lithium.[32]

The coordination of uranyl with extended porphyrins, such as hexaphyrins (for example isoamethyrin, [UO_2(isoamethyrin)], **61**, Figure 5.11),[31] also leads to the formation of a planar hexagonal bipyramidal complex. These ligands are able to bind the uranyl cation selectively over many transition metal ions.

The coordination of uranyl with the larger, expanded macrocycle (H_4L^1) leads to compartmental coordination ([$UO_2(H_2L^1)$], **62**, Figure 5.12), which provides a route for the synthesis of heterometallic complexes. This strategy has been developed by Arnold and Love and has allowed the functionalisation[33] and reductive silylation[34] of the highly unreactive uranyl(VI) oxygen.

5.5 Actinyl(V) and Cation–Cation Interactions

Neptunyl(V) is the most stable AnO_2^+ species and is the form under which neptunium would be present in most environmental conditions. NpO_2^+ does not readily undergo disproportionation (unlike

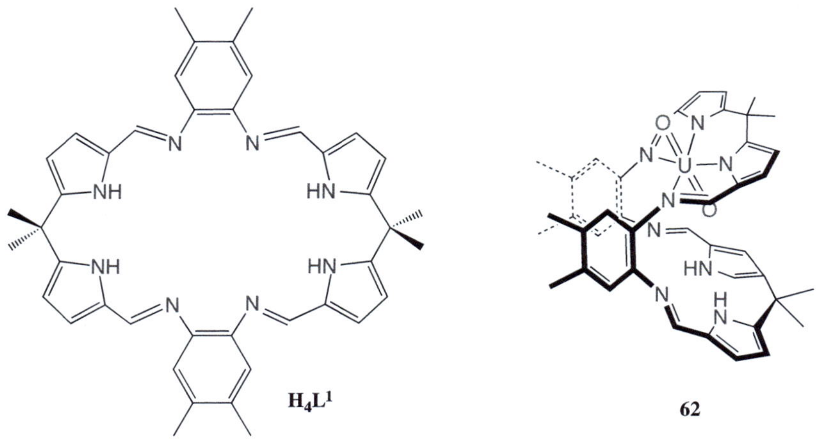

Figure 5.12. Compartmental coordination of uranyl(VI) in one pocket of the expanded tetrapyrrole macrocyclic ligand (H_4L^1).

UO_2^+ and PuO_2^+) but is easily reduced by Fe^{2+} to Np^{4+}. The coordination chemistry of NpO_2^+ neptunyl(V) is quite developed and many compounds have been isolated and characterised in the solid state.

Carbonate is an important ligand in neptunyl(V) chemistry because it forms stable complexes and as such, plays a key role in the transport and fate of Np in the environment, a relevant problem in nuclear fuel storage underground.[35]

Mono-carbonate and bis-carbonate solids of formula $MAnO_2(CO_3)$ (An = Np, Pu, Am and M = Li, Na, K, Rb, Cs) and M_3AnO_2 $(CO_3)_2$ (An = Np, Pu, Am and M = Na, K, Rb) are known in the solid state for actinyl(V). EXAFS studies of neptunyl(V) carbonate in solution, indicate the presence of seven-coordinate mono-carbonato $[NpO_2(CO_3)(H_2O)_3]^-$, **63**, and eight-coordinate bis- and tris-carbonato $[NpO_2(CO_3)_2(H_2O)_2]^{2-}$, **64** and $[NpO_2(CO_3)_3(H_2O)_2]^{3-}$, **65**, species.[36]

The stability of NpO_2^+ complexes in solution with respect to ligand dissociation is quite low due to the weak charge of the cation. The use of the O-donor [18]crown-6 ligand and of the N-donor macrocyclic ligand hexaphyrin has allowed the isolation of stable mononuclear complexes of neptunyl(V) ($NpO_2([18]\text{-}C\text{-}6)^+$, **66,** Figure 5.13).[37] In these complexes, the Np cation is eight coordinate, with a hexagonal bipyramidal coordination environment, with Np–O ligand bond lengths of 1.762(1)–1.826(1) Å which are shorter than those found in simple metal salts (1.85 Å). The coordination of the large neptunyl ion leads to significant ligand distortion.

The structural chemistry of neptunyl(V) in particular is dominated by the so-called "cation–cation" interaction (CCI). CCI describes the interaction of the Lewis basic $O{=}An{=}O^+$ unit with a metal cation and the mutual interaction of two actinyl groups.

In neptunyl(V) chemistry, T-shaped and diamond-shaped cation–cation interactions lead to polymeric chains and tri-dimensional arrays (Figure 5.14).[38] Examples of discrete polynuclear assemblies containing CCIs are rarer and are limited to the diamond-shaped mellitate dimer $Na_4[NpO_2(C_6COO_6)_2]_2$, **67**, the mixed-valent trimeric complex $([\{Np^{VI}O_2Cl_2\}\{Np^{V}O_2Cl(THF)_3\}_2]$, **68**,[39] and the T-shaped tetramer $[\{NpO_2(salen)\}_4(\mu_8\text{-}K)_2][K([18]C\text{-}6)Py]_2$, **69**.[40]

The $UO_2{}^+$ species undergoes disproportionation reaction in protic media leading to uranyl(VI) and U(IV) species. Experimental and computational studies suggest that the disproportionation mechanism involves a dimeric cation–cation intermediate. The use of bulky polydentate ligands, pentadentate ligands or macrocyclic ligands that prevent the cation–cation interaction allow the synthesis and the isolation of stable complexes of uranyl(V).[12,41] Stable complexes of uranyl(V) can be prepared either by reduction of uranyl(VI)

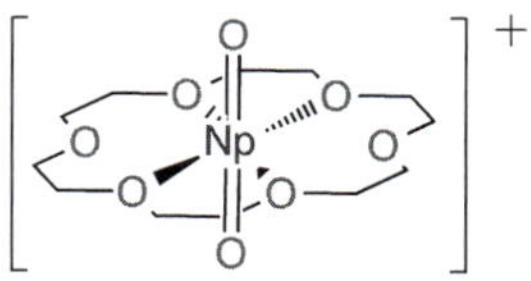

66

Figure 5.13. Molecular structure of NpO_2^+ encapsulated in [18] crown-6.

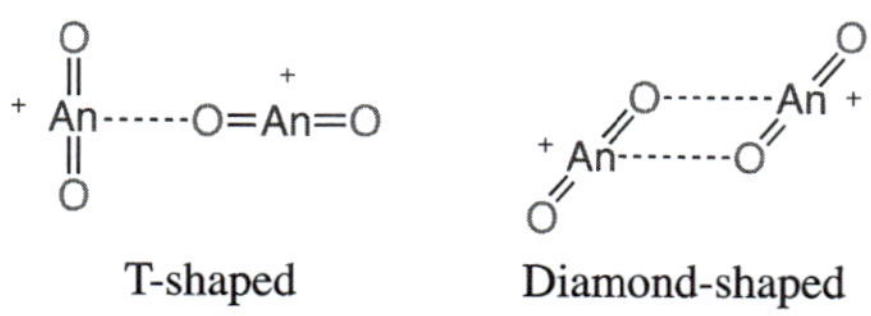

Figure 5.14. T-shaped and diamond shaped CCI interactions observed in neptunyl(V) chemistry.

analogous complexes using suitable reducing agents (usually cobaltocene or pentamethylcobaltocene is used) or by salt metathesis reactions using the iodide precursor $[(UO_2Py)_5(KI_2Py_2)]_n$ (**25**). Bulky bidentate dianionic diketonate or diketiminate ligands such as dibenzoylmethanate (Kdbm) and KAr_2nacnac allow the preparation and characterisation of complexes of pentavalent uranyl but their stability in solution is limited.[42] By contrast, bulky diamine bis-phenolate ligands such as salan^{t}Bu$_2$H$_2$(N,N'-bis(2-hydroxybenzyl-3,5-ditertbutyl)-1,2-dimethylaminomethane) bulky Schiff base ligands such as H_2salophen-^{t}Bu$_2$ (H_2salophen-^{t}Bu$_2$ = N,N'-phenylene-bis-(3,5-di-*tert*-butylsalicylideneimine), pentadentate Schiff bases such as H_2Mesaldien (H_2Mesaldien = N,N'-(2-aminomethyl)diethylene bis-(salicylideneimine) or macrocyclic dinucleating pyrrole-imine ligands (pacman, $\mathbf{H_4L^1}$) form complexes that are stable in organic solvents and in some cases even in the presence of up to more than 100 equivs. of water.[43] Solid and solution state structural studies show that in all these complexes the uranium atom is seven coordinated with pentagonal bipyramidal coordination geometries (Figure 5.15).

When the ligands do not provide five coordination sites, the presence of co-ligands or coordinating solvents such as pyridine lead to higher stability. The increased Lewis basicity of the UO_2^+ oxo groups compared to the UO_2^{2+} ones renders it susceptible to form cation–cation interactions similarly to those found in Np(V) compounds. In anionic complexes of bulky tetradentate, pentadentate or macrocyclic Schiff bases where there are no accessible coordination sites at the uranium centres, $UO_2^+ \cdots UO_2^+$ CCI interactions are

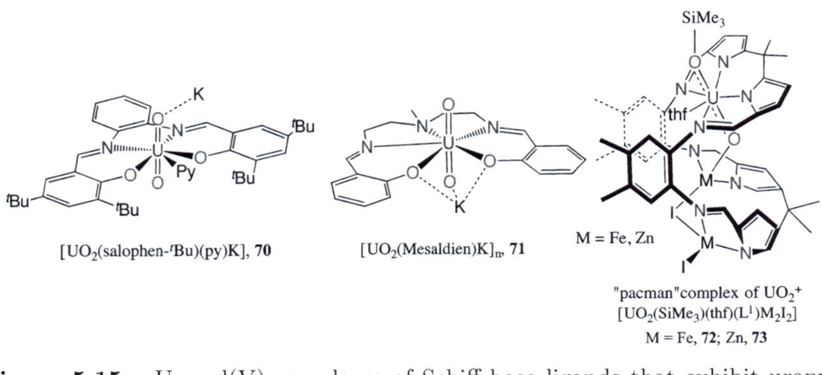

Figure 5.15. Uranyl(V) complexes of Schiff base ligands that exhibit uranyl-oxo-metal CCIs.

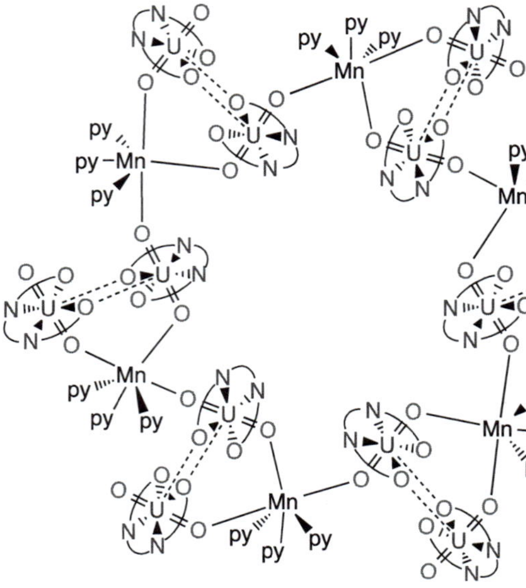

74

Figure 5.16. A uranyl(V)-Mn(II) CCI assembly that exhibits single molecule magnetic behaviour.

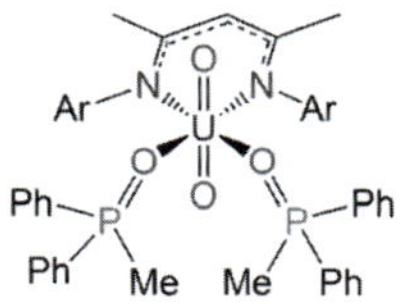

75

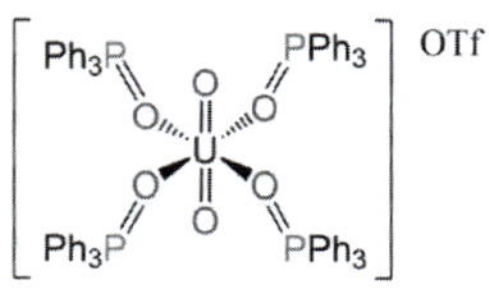

76

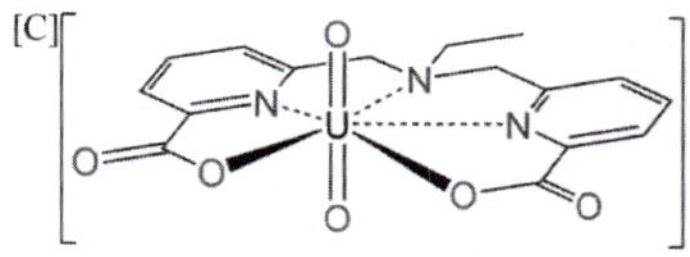

A; C = $Co(Cp^*)_2$, K(2.2.2-cryptand)

Figure 5.17. Mononuclear uranyl(V) phosphine oxide and carboxylate complexes.

not observed. However, the uranyl oxygen easily binds other cations such as alkali, alkaline earth, d-block metals such as Zn^{2+}, Ni^{2+}, Fe^{2+}, Mn^{2+} or other actinides such as U(IV),[44] leading to polymeric chains, or polynuclear discrete assemblies. Several of these complexes show magnetic coupling between the uranium and the d-block metal leading to, in some case to single molecule magnet/single chain magnet behaviour (Figure 5.16).[45]

A relatively stable neutral mononuclear uranyl(V) complex with an unusual six-coordinate uranium centre was obtained combining a bulky bidentate monoanionic ligand and neutral very bulky O-donor ligands in the [UO_2(Ar_2nacnac)(Ph_2MePO)$_2$] complex, **75** (Figure 5.17).[42b] Six-coordination of the uranium centre is also found in the cationic complex [UO_2(Ph_3PO)$_4$]OTf, **76**, serendipitously isolated from a solution of the hexavalent analogue complex [UO_2(Ph_3PO)$_4$]OTf_2. A lengthening (0.06–0.08 Å) of the An^{V} =O bonds compared to the An^{VI} =O bond is observed for all mononuclear complexes in agreement with theoretical calculations.[46] In 2018, a polydentate aminocarboxylate ligand was shown to stabilise uranyl(V) in aqueous solution around neutral pH; [C][UO_2(dpaea)] (**A**; dpaeaH_2 = bis(pyridyl-6-methyl-2-carboxylate)-ethylamine; C = [$CoCp_2$]$^+$ or [K(2.2.2.cryptand)]$^+$) were both characterised in this work (Figure 5.17).[47]

When one coordination site becomes available at the uranium centre, for example, when using non-bulky tetradentate ligands to bind UO_2^+, polynuclear complexes exhibiting $UO_2^+ \cdots UO_2^+$ interactions are obtained. The $UO_2^+ \cdots UO_2^+$ interaction can adopt different geometries depending on the ligand denticity and charge. Complexes with nuclearities of 2–4 have been isolated that present diamond-shaped, **77**, T-shaped, **78**, **79**, or triangular CCI, **80** (Figure 5.18).[48] In these complexes, magnetic coupling between the

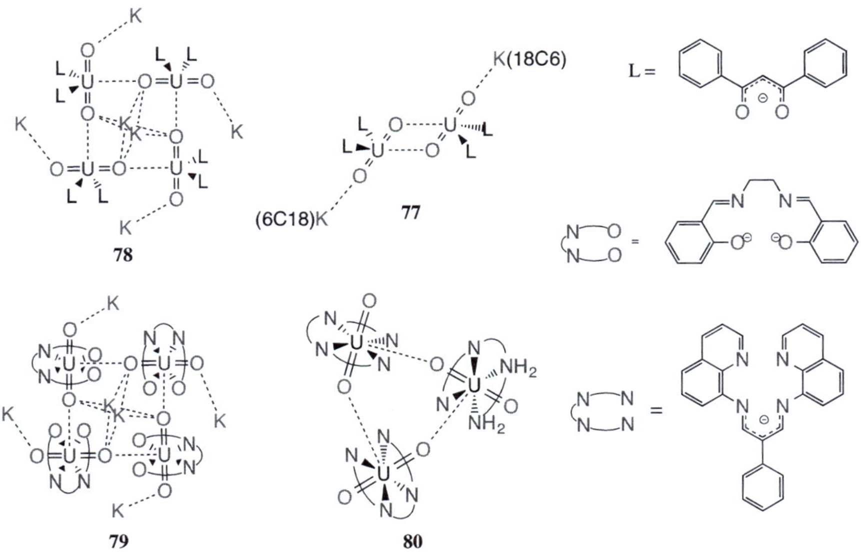

Figure 5.18. Complexes **77–80**.

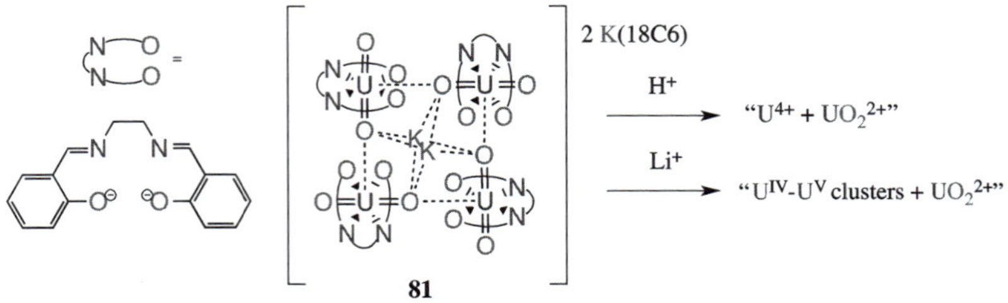

Scheme 5.4. Reactivity of **81**.

$5f^1$ centres is observed. The strength of coupling (as determined by maxima in plots of magnetic susceptibility against temperature between 5 and 12 K) varies both with the geometry of the CCI and with the nature of the supporting ligand.

Isolated CCI complexes can be fully stable towards metal disproportionation in aprotic organic solutions. NMR spectroscopy shows that the CCI is retained in solution. Electronic and structural parameters control the solution stability of these cation–cation complexes. Notably, the tetranuclear complex $[UO_2(\text{salen})]_4[\mu_8\text{-}K]_2\}(K(18C6)Py)_2$, **81**, (Scheme 5.4) prepared from the reaction of $[(UO_2Py)_5(KI_2Py_2)]_n$ (**25**) with K_2 salen is fully stable in pyridine solution, but disproportionates immediately when protons are added, affording U(VI) and U(IV) species. In the salen tetramer, the uranyl(V) groups are also involved in CCIs with the potassium cations, and the cation clearly plays an important role in the stability of this complex. Indeed, if potassium is replaced by Li^+ then disproportionation occurs, leading to U(IV)/U(V) oxo clusters.[49] The structural parameters of cation–cation complexes show that the $UO_2{}^+ \cdots UO_2{}^+$ interaction results in a significant lengthening of the U=O bonds involved in the CCI with respect to the unbound oxygen, with a mean difference between the two U=O bonds of 0.1 Å.

The isostructural $[\{NpO_2(\text{salen})\}_4(\mu_8\text{-}K)_2][K(18C6)Py]_2$, **82**, has also been prepared by reacting $[(NpO_2Py_5)(KI_2Py_2)]_n$ with K_2 salen.[40] NMR studies again confirmed the presence of $NpO_2{}^+ \cdots NpO_2{}^+$ CCI in solution. A significant lengthening of the Np=O bond involved in the CCI is observed with respect to the unbound oxygen, with a mean difference of about 0.05 Å. However, the mean Np–(O=Np) distance in **82** is 0.1 Å longer than the analogous U–(O=U) distance in **81**, suggesting the presence of a weaker cation–cation interaction.

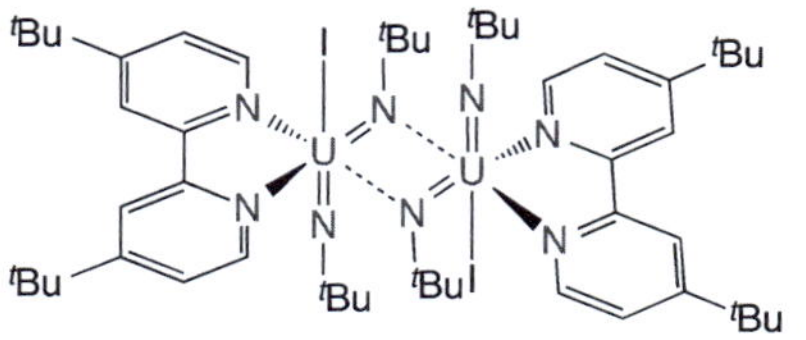

83

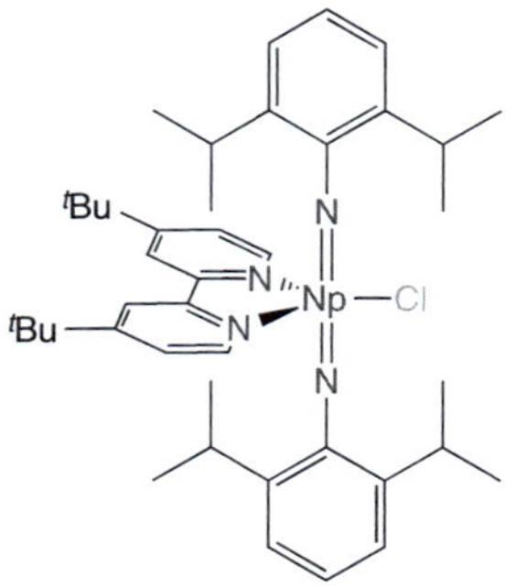

84

Figure 5.19. Complexes **83** and **84**.

Imido analogues of the UO_2^+ and UO_2^{2+} species are also known. The complex $[U^{VI}(N^tBu)_2I_2(thf)_2]$, **50**, is easily prepared by the oxidation of $UI_3(thf)_4$ with iodine in the presence of *tert*-butylamine.[22] DFT studies suggest the presence of two triple bonds in the linear fragment $R_2N{\equiv}U{\equiv}NR_2$. A pentavalent uranium bis(imido) analogue of UO_2^+ was prepared by reduction of the U^{VI} complex $[U^{VI}(N^tBu)_2(I)_2(^tBu_2bpy)]$, **82**. In the resulting dimer $[U^V(N^tBu)_2(I)(^tBu_2bpy)]_2$, **83** (Figure 5.19), two imido groups are involved in a diamond-shaped CCI that results in a strong antiferromagnetic coupling between the two $5f^1$ uranium centres, with a Néel temperature of 13 K.[50] The closely related mononuclear Np(V) *trans*-bis(imido) complex, $[U^V(NDipp)_2(Cl)(^tBu_2bpy)]$, **84** (Figure 5.19), was isolated in low yield by reacting $NpCl_4(dme)_2$ with 2 equiv. of $^tBu_2bipy$, 4 equiv. of LiNHDipp, and deliberate addition of excess CH_2Cl_2. Theoretical calculations show the participation of both 5f and 6d orbitals of Np(V) to the multiple bonding in the $[N{=}Np^V{=}N]^+$ unit, although to a smaller extent than that found for the U(V) analogue.[51]

5.6 Non-Actinyl Complexes of Actinides in High Oxidation State (VI and V)

Non-actinyl complexes of actinides in high oxidations state are quite rare. However, an increasing number of imido, nitride and oxo complexes of high valent uranium are being prepared from redox

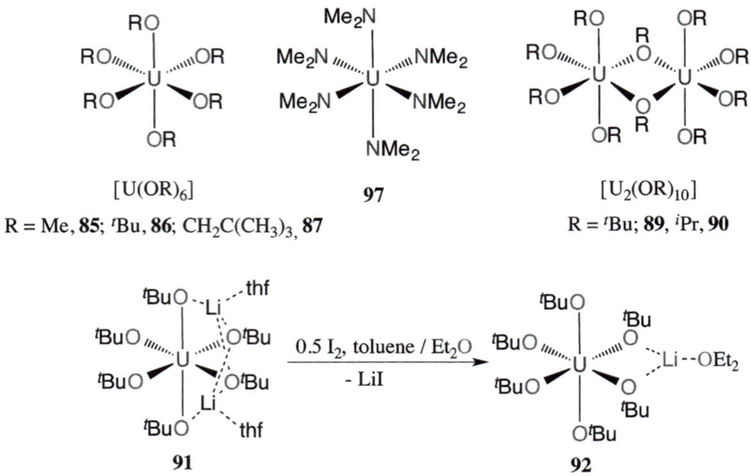

Figure 5.20. Complexes **85–87**, **89–92** and **97**.

reactivity studies of low-valent actinides. These complexes will be presented in dedicated sections.

Homoleptic octahedral complexes have been isolated with alkoxide, amide and ketimide ligands, which are all strong σ- and π-donor ligands. These complexes are of interest for studying the nature of metal–ligand bonding and the contribution of 5f and 6d orbitals to bonding. The largest number of non-uranyl uranium(VI) compounds are the alkoxides, $[U(OR)_6]$, which have been known for more than 60 years. Initial interest in these volatile complexes stemmed from their potential application in uranium isotope separation.[52]

Several U(VI) alkoxide complexes have been structurally characterised; $[U(OMe)_6]$, **85**, $[U(OCH_2C(CH_3)_3)]_6$, **86**, $[U(OtBu)_6]$, **87** (U–O 2.001–2.002 Å) (Figure 5.20). The large U-O–C bond angles in these complexes have been interpreted in terms of strong π-donation of electron density from the alkoxide oxygens to the metal centre. U^{VI} alkoxides are usually monomeric, in contrast to U^{V} alkoxides, which are usually polynuclear. U^{V} alkoxides were among the first examples of stable complexes isolated for the pentavalent oxidation state. $[U(OEt)_5]$, **88**, (a liquid) was the first prepared from UCl_5 and was used to prepare other alkoxides by protonolysis.

$[U_2(O^iPr)_{10}]$, **90**, was the first crystallographically characterised U^{V} alkoxide. In this dimeric complex, the uranium atoms are arranged as two edge-sharing octahedra with a U$\cdots$U distance of

3.789(1) Å.[53] These complexes were considered as potential candidates for U–U bond formation but the strong π-donation from the oxygen to the metal dominates instead. Alkoxides of high valent uranium can also be prepared by oxidation of U^{IV} compounds. Notably, the homoleptic mononuclear U^{V} alkoxide $[Li(Et_2O)][U(O^tBu)_6]$, **91**, has been prepared by oxidation of $[Li(THF)]_2[U(O^tBu)_6]$, **92**.[21] Heteroleptic $[U(O^tBu)_5(OC_6F_5)]$, **93** and $[U(O^tBu)_4(OC_6F_5)_2]$, **94**, have also been prepared by protonolysis reactions from $[U(OtBu)_6]$. Homoleptic U^{V} $[Li(dme)_3][U(OC_6F_5)_6]$, **95**, was also prepared, but attempts to prepare homoleptic aryloxide complexes of U(VI) have not succeeded, probably due to the lower π-donating ability of aryloxides, which fail to stabilise U(VI).

High valent uranium amides have been known since the Manhattan project.[20] but are still quite rare. $[U(NEt_2)_5]$, **96** and $[U(NMe)_6]$, **97** are prepared by oxidation of the respective U^{IV} complexes $[Li(THF)]_2[U(NEt_2)_5]$, **98** and $[Li(THF)][U(NMe)_6]$, **99**, with $TlBPh_4$ or AgI; the precursors were synthesised by salt metathesis reactions from UCl_4 and the parent lithium amide salt.[54] More recently homoleptic $[Li(DME)_3][U^{V}(NC_5H_{10})_6]$, **100**, was prepared from the U^{IV} amide, $[Li(DME)][U(NC_5H_{10})_5]$, **101**, by oxidation with iodine followed by addition of $Li(NC_5H_{10})$. Further oxidation with iodine leads to the U^{VI} amide $[U^{VI}(NC_5H_{10})_6]$, **102** (Figure 5.21).[55] These complexes were all crystallographically characterised.

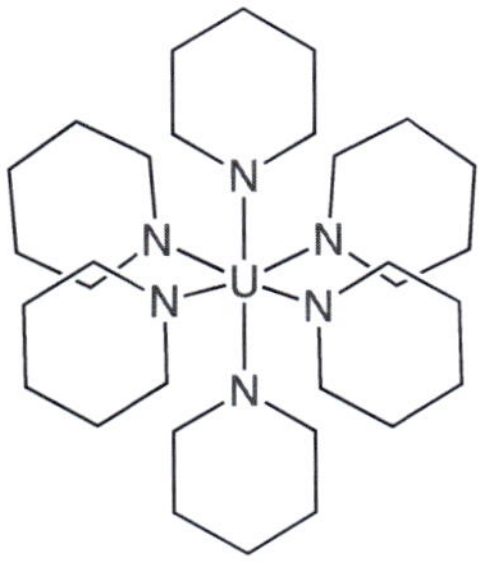

102

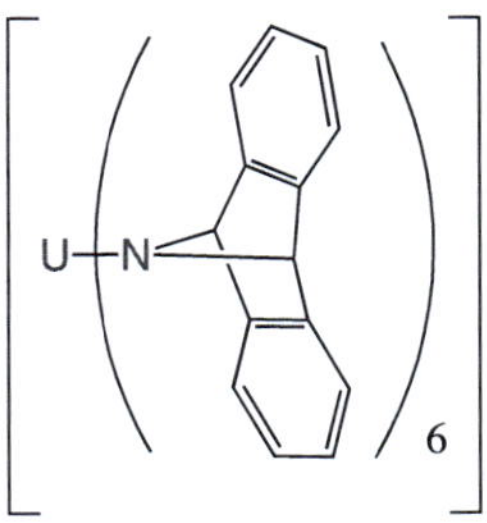

104

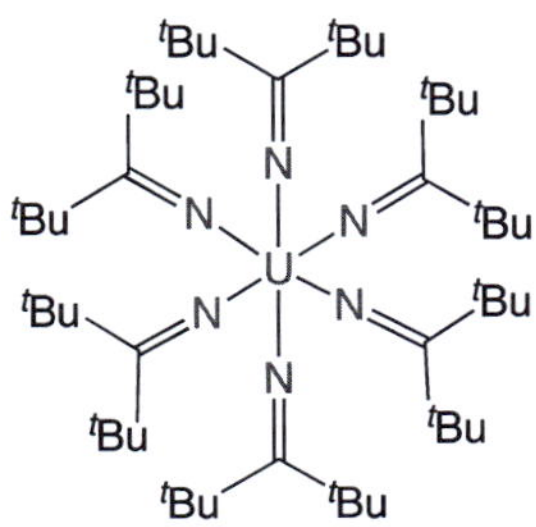

106

Figure 5.21. Complexes **102**, **104** and **106**.

An alternative preparation of high valent uranium amides used the redox active ligand dbabh (Hdbabh = 2,3:5,6-dibenzo-7-azabicyclo[2.2.1]hepta-2,5-diene), which effects two-electron oxidation of U^{III} to U^{V} by elimination of anthracene. The U^{V} hexakis(amido) complex $[Li(THF)_x][U^{V}(dbabh)_6]$, **103**, was isolated from the reaction of $UI_3(THF)_4$ with Li(dbabh), and it can be further oxidised to $[U^{VI}(dbabh)_6]$, **104**.[56]

The ketimide complexes $[Li][U^{V}(N{=}C^{t}BuPh)_6]$, **105** and $[U^{VI}(N{=}C^{t}BuPh)_6]$, **106**, were prepared by oxidation of the respective U^{IV} ketimide complexes $[Li(THF)_2]_2[U^{IV}(N{=}C^{t}BuPh)_5]$, **107** and $[Li(THF)][U(N{=}C^{t}Bu_2)_5]$, **108**.[57] In these complexes the strong uranium–ketimide interaction was postulated to be due to the participation of the uranium 6d orbitals.

5.7 Complexes of An^{II}, An^{III} and An^{IV}

Polydentate ligands containing negative oxygen donor groups such as carboxylates, phenolate or 3-hydroxy-2(1H)-pyridinone have been extensively used in the chemistry of low valent actinides.[30a] Molecular complexes of An^{IV} (An = U, Th, Pu, Np) can be isolated with a large variety of ligands including soft N- or S-donor ligands if water and oxygen are avoided. Trivalent uranium is highly reducing and therefore only redox innocent ligands can be used to prepare uranium(III) complexes. Moreover, uranium(III) tends to disproportionate to uranium(0) and uranium(IV), and therefore very bulky monodentate monoanionic ligands or bulky tripodal neutral, monoanionic or trianionic ligands appear to be the best choice to stabilise uranium in the +3 oxidation state. Several of these ligands are also effective supporting ligands for the synthesis of Pu^{III} and Np^{III} complexes.

Heavier actinides are stable in air and water in the +3 oxidation state. Stable complexes of these cations have been prepared up to berkelium using polydentate anionic oxygen donor ligands.[58] The +2 oxidation state is known for all actinides in the solid state. Molecular compounds of Th^{II}, U^{II}, Np^{II} and Pu^{II} have only recently been characterised and are mostly cyclopentadienyl-based so are beyond the scope of this chapter, though an example of a U^{II} amide is included below.

5.7.1 Neutral N-donor ligands

Polydentate ligands containing heterocyclic N-donors (with denticity 2–6) have been extensively used in the coordination chemistry of actinides and more particularly in comparative studies with lanthanides.[31] Interest in these ligands arises from the ability of neutral N-donor ligands to selectively extract actinides with respect to lanthanides from acidic mixtures of these elements obtained during spent nuclear fuel recycling processes.[59] Notably, one possible solution for the long-term management of used nuclear fuel involves the separation of plutonium and other minor actinides from these mixtures by solvent extraction as a key step. The selective complexation of actinides compared to lanthanides is thought to arise from the different nature of M–L bonding in 5f elements compared to the 4f elements. The more radially diffuse 5f orbitals can participate in covalent interactions, but the bonding in 4f elements complexes is essentially electrostatic.

Homoleptic and heteroleptic An^{III} and An^{IV} complexes of chelating ligands containing neutral N-donors (see Scheme 5.1) have been characterised both in solution and the solid state. Structural comparisons between Ln^{III} and An^{III} complexes indicate the presence of a more covalent An–N interaction. Heteroleptic An^{III} complexes of the tripodal tpza ligand, $[An(tpza)I_3(MeCN)]$, (An = U, **109**; Pu, **110**), have been obtained. Homoleptic tris-cationic complexes $[ML_2]^{3+}$ (M= U^{III}, Ln^{III}) and $[ML_3]^{3+}$ (M= U^{III}, Ln^{III}) have been prepared using weakly coordinating anions (iodide or triflate salts) and have been crystallograpically characterised for the tetradentate tripodal ligands tpa (see example in Scheme 5.5; $[U(tpa)_2]I_3$ **111**)

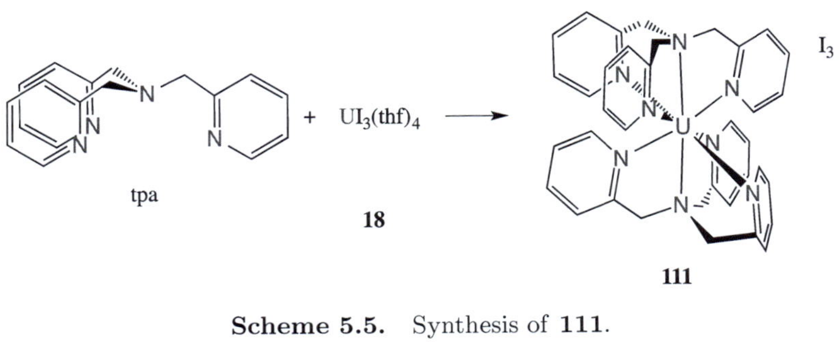

Scheme 5.5. Synthesis of **111**.

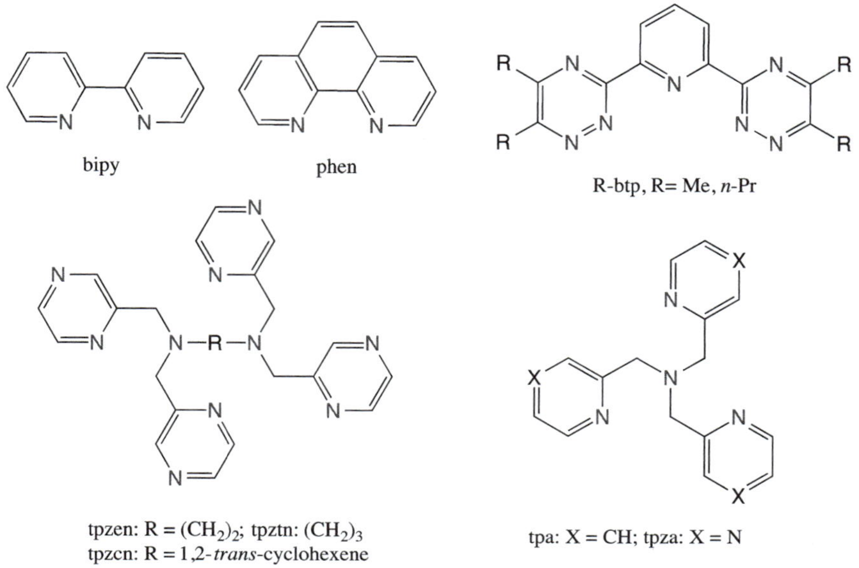

Figure 5.22. Selection of N-donor ligands.

and ntb and tridentate btp, respectively (Figure 5.22). The complexation of Am^{III} and Cm^{III} by heterocyclic N-donor ligands has been extensively studied in solution by TRFLS, EXAFS and UV-Visible spectroscopic studies.[59] The high selectivity of BTP ligands in the extraction of An^{III} from An^{III}/Ln^{III} mixtures can be correlated to the high stability of the resultant $[AnL_3]^{3+}$ complexes in organic solution.

5.7.2 *Anionic N-donor ligands*

Amides are the most common anionic N-donor ligands in non-aqueous actinide chemistry. The synthesis and structures of $[An\{N(SiMe_3)_2\}_3]$ (**44**–**46**) have already been discussed. In 2019, Evans and co-workers reported the KC_8 reduction of **44** in the presence of 2.2.2-cryptand to give the anionic U^{II} amide $[K(2.2.2\text{-cryptand})][U\{N(SiMe_3)_2\}_3]$ (**44-A**).[60] U^{IV} amides are also easily synthesised by salt metathesis reactions. $[U(NEt_2)_4]$, **112**, is a useful and versatile starting material for the synthesis of amide, alkoxide and thiolate complexes as $NEt_2{}^-$ is a good proton acceptor. Complex **112** has a dimeric structure with two amides bridging the two pentacoordinate uranium centres. $[U(NPh_2)_4]$, **113**, and $[U\{N(SiMe_3)_2\}_4]$, **114**, are monomers with a *pseudo*-tetrahedral coordination environment. The U^{IV} tetra-amido complex **114**[61] can be reduced with potassium without or with crown ether to afford the U^{III} separated ion pair complexes $[K(THF)_6][U\{N(SiMe_3)_2\}_4]$, **115**, (Scheme 5.6) and $[K(18C6)]$ $[U\{N(SiMe_3)_2\}_4]$, **116** that can be oxidized back to **114** using CuI. The latter complex reacts with CO_2 to afford the U^{IV} bis-cyanate complex $[K(18C6)]$ $[U\{N(SiMe_3)_2\}_3(NCO)_2]_n$, **117**, and unidentified products of CO_2 reduction.[62] While Pu^{IV} is the most stable oxidation state of Pu in aqueous conditions, the oxidation of Pu^{III} to Pu^{IV} is difficult in non-aqueous conditions.[63] The reaction of $[Pu\{N(SiMe_3)_2\}_3]$ (**46**) with 0.25 equivalents of $TeCl_4$ in toluene yields Pu^{IV} $[Pu\{N(SiMe_3)_2\}_3Cl]$, **118**. In this complex, the metal centre is four coordinate with a distorted tetrahedral geometry.

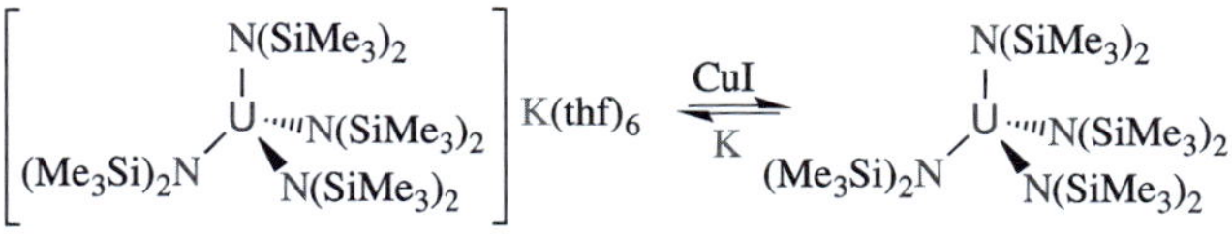

115 **114**

Scheme 5.6. Interconversion of **114** and **115**.

Poly(pyrazolyl)borate ligands have been largely used as supporting ligands in actinide coordination chemistry because their steric and electronic properties can easily be modified by changing the number of pyrazolyl rings coordinated to the boron atom and by introducing different substituents on the pyrazolyl ring. Monoanionic tridentate polypyrazolyl borate or "scorpionate" ligands were among the first polydentate ligands used in uranium coordination chemistry to stabilise low-valent uranium. The tripodal nature of the ligand and the low charge both contribute to the stabilisation properties. The complex $[(Tp^*)_2UI]$, **119**, (Tp^* = hydrotris(3,5-dimethylpyrazolyl)borate) was first prepared by Takats (Figure 5.23) in 1994 by the salt metathesis reaction of $UI_3 \cdot thf$ with 2 $KTpMe_2$.[64] The Tp^* ligand is able to support the reactivity at the metal centre, leading for example to the characterisation of rare U^{III} alkyl derivatives that could easily undergo CO_2 insertion into the U–C bond.[65]

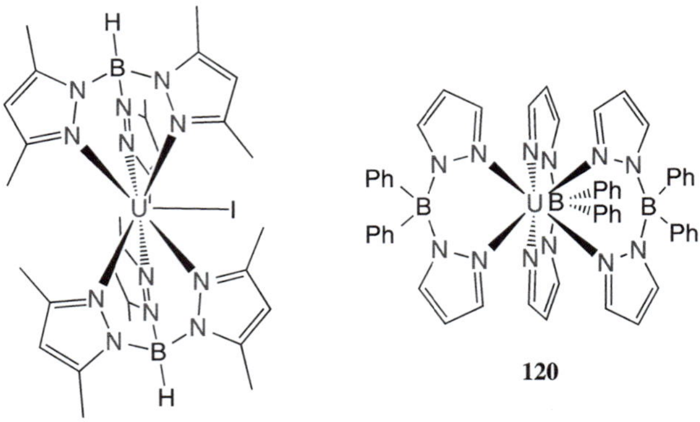

Figure 5.23. Complexes **119** and **120**.

A neutral homoleptic hexacoordinated U^{III} complex with a D_{3h} symmetric trigonal prismatic co-ordination geometry was obtained using the monoanionic bidentate bis-pyrazolate ligand Ph_2BPz_2. $[U(Ph_2BPz_2)_3]$, **120**, was the first reported example of a uranium complex showing single molecule magnet behaviour.[66]

Actinide complexes of porphyrins have also been reported. Due to their relatively small core, the coordinated Th^{IV} or U^{IV} cations are situated outside of the ring plane. Alternatively, larger ligands have also been identified as suitable candidates in actinide coordination chemistry: calix[4]pyrrole, expanded porphyrins, (macrocycles containing four or more pyrroles and 17 or more atoms in their inner core), pyrrole-containing Schiff-base macrocycles containing both iminic and pyrrolic nitrogens, and corroles.[31] The coordination geometries of these macrocycles vary since they can coordinate η^1- by the nitrogen atom or η^5- *via* the pyrrole ring. U^{III} complexes with calix[4]pyrrole are stable in a dimeric form as lithium salts ({[(calix[4]pyrrole)U][Li(thf)]}$_2$, **121**), but are very unstable as the potassium salt ({[(calix[4]pyrrole)U(thf)][K(thf)$_3$]}, **122**, Figure 5.24).

The trivalent, mononuclear complex [(Et_8–calix[4]pyrrole) U(dme)][K(dme)], **123**, was obtained by reacting $UI_3(dme)_2$ with the calixpyrrole salt [(Et_8–calix[4]pyrrole)K_4(dme)$_2$]. Reduction of this complex gave a bis-nitride U^V/U^{IV} complex that provided the

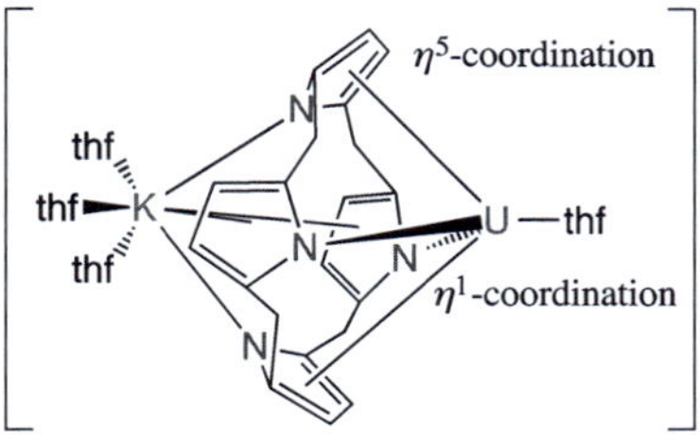

122

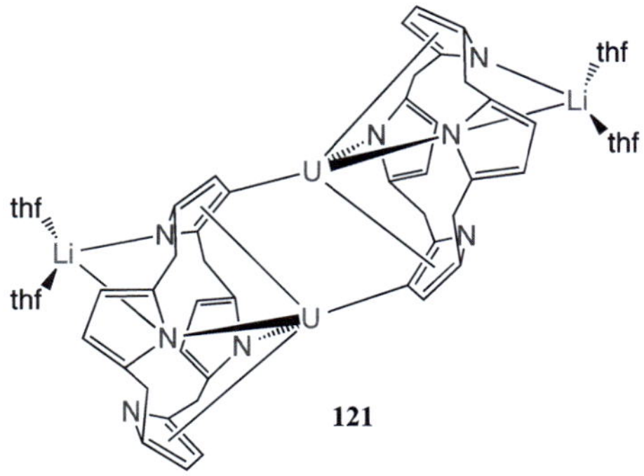

121

Figure 5.24. Complexes **121** and **122**.

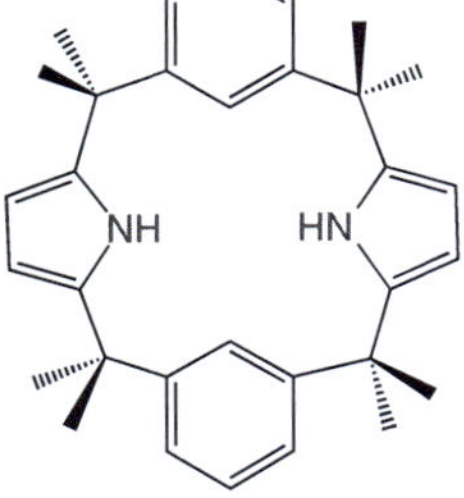

H_2L^{Ar}

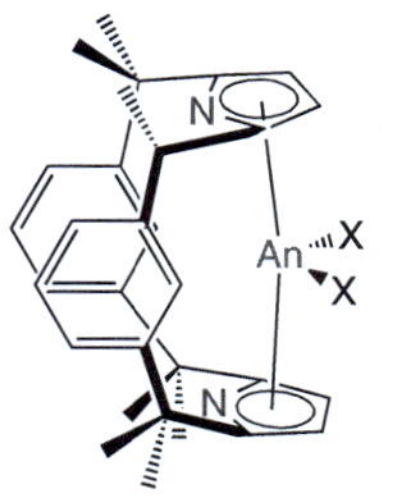

An = U, X = I, **124**; Th, X = Cl, **125**

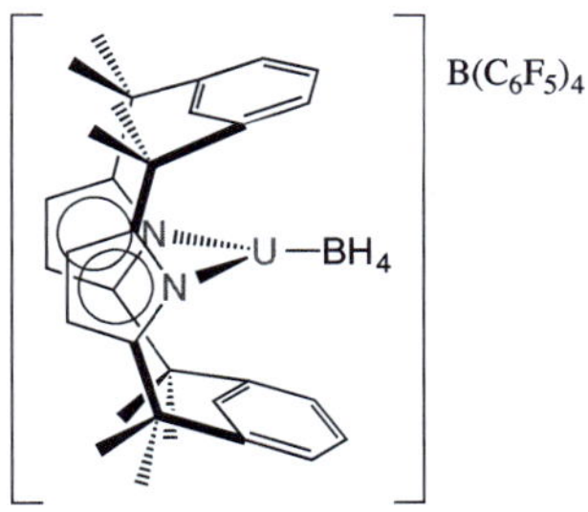

126

Figure 5.25. Complexes **124–126**.

first example of dinitrogen activation by a molecular uranium compound.[67] More recently Arnold and co-workers used the flexible macrocycle trans-calix[2]benzene[2]pyrrole (H_2L^{Ar}) to prepare the An^{IV} complexes $[AnX_2(L^{Ar})]$ (An = U, **124** and Th, **125**) where the pyrrolides adopt an unusual κ^5:κ^5 binding mode.[19] The U^{IV} complex is easily reduced by KC_8 to yield the stable U^{III} analogue where the ligand switches to a different binding mode, a κ^1-pyrrolide η^6-arene sandwich; $[UBH_4L^{Ar})]$, **126**, is represented in Figure 5.25. An analogous Np^{III} complex showing the same η^6:κ^1:η^6:κ^1 binding mode was also prepared both from $Np^{III}Cl_3.THF_x$ and $Np^{IV}Cl_4$ precursors.[68]

Dinuclear complexes of U^{IV} and U^{III} have also been prepared by the same group using large pacman-type pyrrole-containing Schiff-base macrocycles.[19] Thorium(IV) and uranium(IV) macrocyclic complexes of Mes_2(p-OMePh)corrole were prepared in high yields via salt metathesis with the corresponding lithium corrole yielding

[{(Mes$_2$(p-OMePh)corrole)An(dme)}$_2$(Cl$_2$)] (An = U, **127**; Th, **128**, Scheme 5.7).[69]

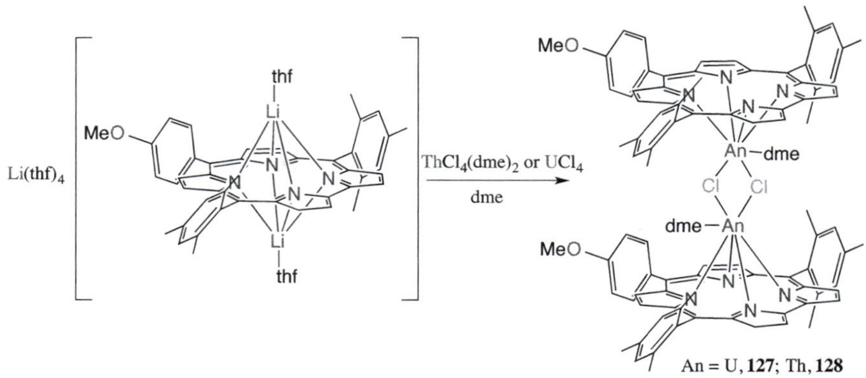

Scheme 5.7. Synthesis of **127** and **128**.

Tripodal trianionic N-donor ligands such as N-anchored tetradentate ligands with different substituents ([tren^{DMBS} = {N(CH$_2$CH$_2$NSiMe$_2$tBu)$_3$}$^{3-}$, tren^{TMS} = {N(CH$_2$CH$_2$NSiMe$_3$}$^{3-}$, tren^{TIPS} = {N(CH$_2$CH$_2$NSi^{i}Pr$_3$)$_3$}]) or tacn anchored hexadentate ligands ((SiMe$_2$NPh)$_3$tacn) have proven very effective in stabilising uranium in the +3 oxidation state (U(tren^{DMBS}), **129**, U(tren^{TMS}), **130**, U(tren^{TIPS}), **131** and U((SiMe$_2$NPh)$_3$tacn, **132**) (Figure 5.26) and in controlling the redox reactivity at the metal centres. The tripodal nature of the ligand prevents undesirable disproportionation reactions leading to U^{IV} and U^0, while the strong donor character of the anionic amido groups give highly reducing metal centres. A variety of U^{IV} complexes supported by tripodal tris-amido ligands have also been obtained from redox reactivity studies of U^{III} complexes or from salt metathesis reactions of the U^{IV} chloride complexes.

The U^{IV} complexes [U(tren^R)(Cl)], (R = DMBS, **133**; TMS, **134**; TIPS, **135**) of the tren^{R3-} ligands are easily prepared from UCl$_4$ by salt metathesis reactions. The Pu^{IV} and Np^{IV} complexes [An(tren^{TIPS})(Cl)] (An = U, **136**; Np, **137**; Pu, **138**) were also prepared by reacting [AnCl$_4$(dme)$_2$] (An = U, Np, Pu) with one equivalent of [Li$_3$(tren^{TIPS})] in thf (Scheme 5.8).[70]

The reduction of [U(tren^{DMBS})(Cl)] (**133**) to the U^{III} product was first reported by Scott, who also reported N$_2$ binding to afford

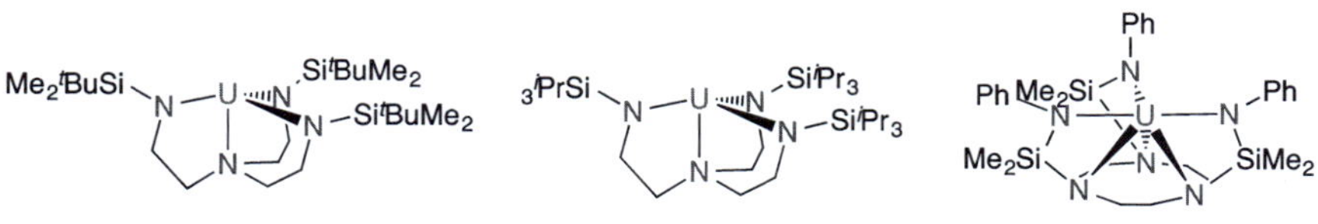

129 **131** **132**

Figure 5.26. Complexes **129**, **131** and **132**.

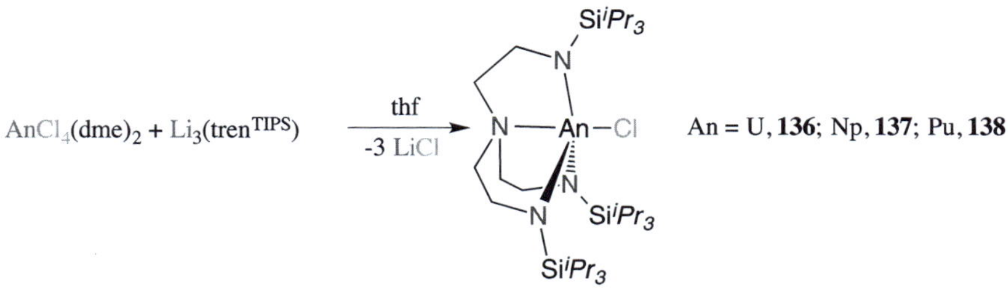

Scheme 5.8. Synthesis of **136–138**.

the side-on bridged dinitrogen complex $[\{U(tren^{DMBS})\}_2(\mu\text{-}\eta_2{:}\eta_2\text{-}N_2)]$, **139**. Reduction of $[U(tren^{TIPS})(Cl)]$, **135**, with potassium also affords the U^{III} product cleanly. The steric bulk of $[U(tren^{TIPS})]$, **131**, provides control of the reactivity at the metal centre and allowed the isolation of terminal U^V nitrido, phosphidene, oxido and imido complexes. $[U(tren^{DMBS})]$ is able to promote the homologation of CO.

5.7.3 *Anionic O donor ligands*

An^{IV} (An = Th, U, Pu, Np) ions all form stable complexes with anionic oxygen donor ligands; in 2017, Abergel and co-workers reported the only example to date of a stable Bk^{IV} complex through the use of a siderophore ligand.[71] Water-stable tetracatecholate complexes of formula $Na_4[An(C_6H_4O_2)_4]$ form for Th, **140**, U, **141** and Pu, **142**, and have been crystallographically characterised to reveal an octacoordinated metal centre for An = Th and U.[30a] Polyaminocarboxylate ligands such as EDTA and DTPA strongly bind An^{IV} ions and are currently used in chelating therapy for the removal of these ions from contaminated patients. However, the low selectivity of these ligands compared to alkali ions has inspired the design of more specific ligands based on multidentate catecholate

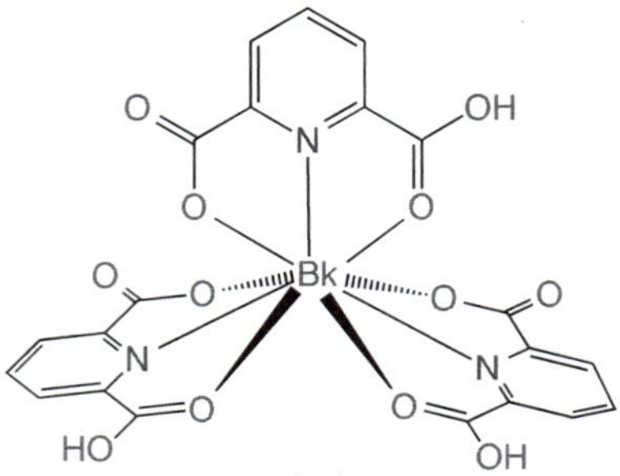

147

Figure 5.27. Complex **147**.

and hydroxypyridinone ligands among others (see examples in Figure 5.10).[30a] The coordination chemistry of Pu^{IV}, the most stable oxidation state of Pu in aerobic aqueous media, is dominated by hard ligands and in particular by negative O-donor ligands. $Pu(CO_3)_5^{6-}$, **143** and $Pu(NO_3)_6^{2-}$, **144**, have been structurally characterised and they show 10-coordinate and 12-coordinate Pu^{IV}, respectively. The coordination chemistry of the later actinides (An = Am–Lr) is less developed due to their low availability and high radioactivity. These ions are mostly found in the +3 oxidation state and as such their chemistry is very similar to that of Ln^{III} cations. These ions form stable complexes in aqueous solution with anionic multidentate O-donor ligands. Crystallographically characterised complexes of these ions are rare. However, isostructural tris-dipicolinate complexes have been synthesised for An = Am, **145**, Cm, **146** and Bk, **147** (Figure 5.27) and recently their structures have been determined up to the Cf^{3+} ion (**148**).[58b,71b] The structures of these complexes are similar to analogous Ln^{III} complexes, with a nine-coordinate tricapped trigonal prismatic environment about the An^{III} cations. Magnetic studies carried out on the Cf complex suggest the presence of more important ligand-field effects compared to Cm^{3+} and Am^{3+}, and this helps explain the higher stability of the +2 oxidation state for Cf and later actinides.

Homoleptic An^{IV} alkoxide and aryloxide complexes $[An(OR)_4]$ (An =U, **149**; Th, **150**) are prepared in non-aqueous conditions by protonolysis or salt metathesis protocols.[72] The synthesis of homoleptic U^{III} aryloxides is more difficult due to the relatively low stability of U^{III}. Stable U^{III} aryloxide complexes were first obtained by Sattelberger in 1989 from the reaction of $[U\{N(SiMe_3)_2\}_3]$ (**45**) with

$HO\text{-}2,6\text{-}R_2C_6H_3$ ($R = {}^tBu$ or ${}^iPr$) in hexane solutions to give $[U(O\text{-}2,6\text{-}R_2C_6H_3)_3]$, $R = {}^tBu$, **151** and ${}^iPr$, **152** (Scheme 5.9). The crystal structures reveal dimeric complexes $[U(OAr)_3]_2$ where the coordinatively unsaturated uranium ion is bound by the ligand arene ring.[73] More recently the activation of N_2 and CO by these complexes has been reported.[74] An analogous $[Np(OAr)_3]$ complex was also prepared from $[Np\{N(SiMe_3)_2\}_3]$ (**46**). The Thorium(IV) complex $[Th(O\text{-}2,6\text{-}{}^tBu2C_6H_3)_4]$, **153**, was prepared by salt metathesis protocols from $ThI_4(thf)_4$, **154**. The protonolysis route does not lead to the desired tetraphenoxide, even at high temperature (Scheme 5.9).[75]

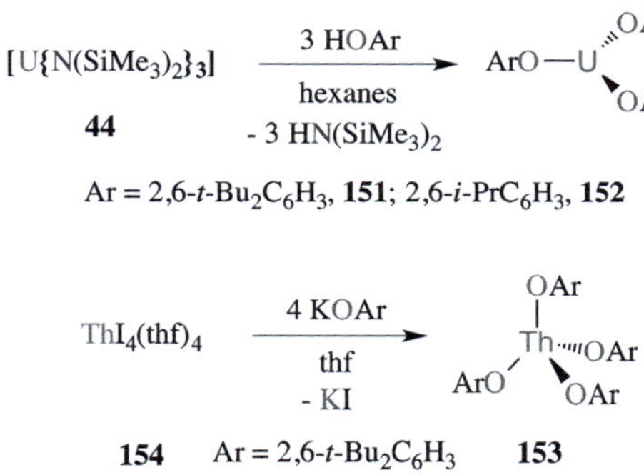

Scheme 5.9. Synthesis of **152** and **153**.

Siloxide ligands have been less studied then aryloxides. However, the reaction of $[U\{N(SiMe_3)_2\}_3]$ (**46**)[76] with three equivalents of tris-tertbutoxysilanol $HOSi(O^tBu)_3$ in hexane at −40°C afforded the dinuclear uranium(III) complex $[U(OSi(O^tBu)_3)_2(\mu\text{-}OSi(O^tBu)_3)]_2$, **155**, which is stable at −40°C (Scheme 5.10).[77]

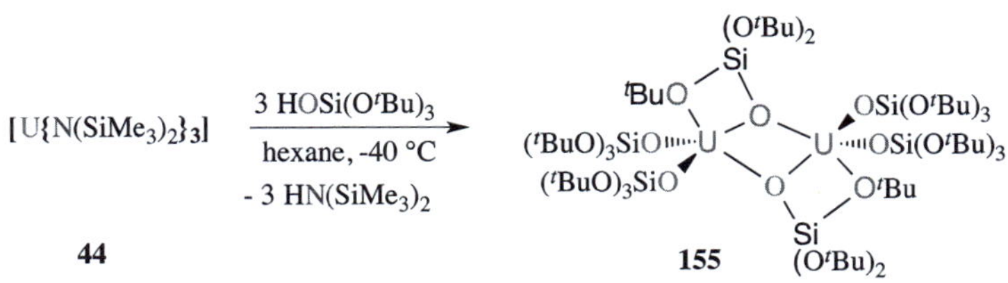

Scheme 5.10. Synthesis of **155**.

Complex **155** was found to reduce a wide range of substrates such as CO_2, N_3^-, toluene and CS_2.[78] Moreover, thermal decomposition of the tris-siloxide complex provides a route to the heteroleptic U^{IV} siloxy/silandiolate complex $[U(OSi(O^tBu)_3)_2(\mu\text{-}O_2Si(O^tBu)_2)]_2$, **156**, via C–O bond cleavage and isobutene elimination.[79] Further reduction of the decomposition product afforded the U^{III}/U^{III} complex $[K(THF)U(OSi(O^tBu)_3)_2(\mu\text{-}O_2Si(O^tBu)_2)]_2$, **157**. The U^{IV} complex $[U(OSi(O^tBu)_3)_4]$, **158**, was also prepared from UI_4 and $KOSi(O^tBu)_3$ by a salt metathesis reaction and can be easily reduced to afford the thermally stable (up to 100°C) but highly reactive heterobimetallic U^{III} complex $[K\{U(OSi(O^tBu)_3)_4\}]$, **159** (Scheme 5.11). The multidentate binding mode of the tris-tertiobutoxysiloxide ligand is particularly effective in promoting the formation of multimetallic species.

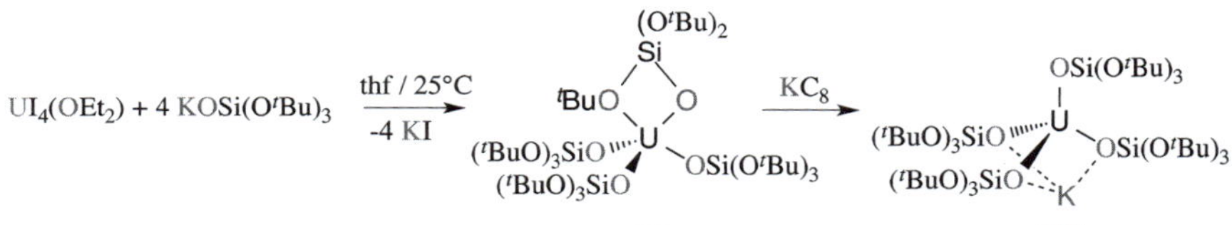

Scheme 5.11. Synthesis of **158** and **159**.

A molecular coordination complex of U^{II} has been isolated using a tripodal chelating tris(aryloxide) arene ligand (Scheme 5.12).[80] The U^{II} complex ($[\{(^{Ad,Me}ArO)_3mes\}U]^-$, **160**) was prepared by reduction of the U^{III} precursor $[\{(^{Ad,Me}ArO)_3mes\}U]$, **161**, with potassium in the presence of cryptand at low temperature (−40°C). Complex **160** decomposes rapidly at room temperature. The δ-backbonding between uranium and the arene of the chelating ligand is key in the stabilisation of U^{II} in this molecular compound.

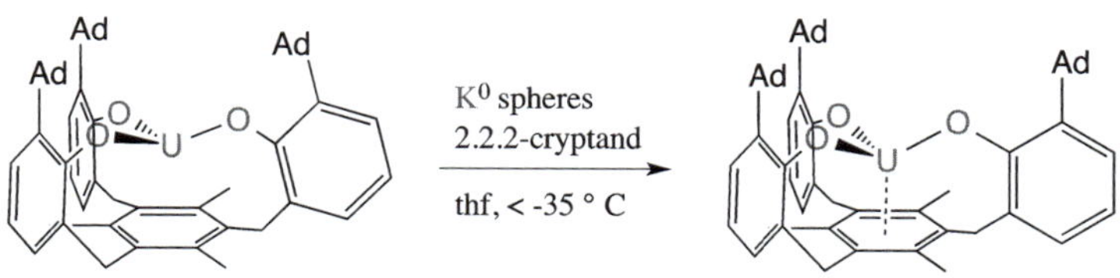

Scheme 5.12. Synthesis of **161** by reduction of **160**.

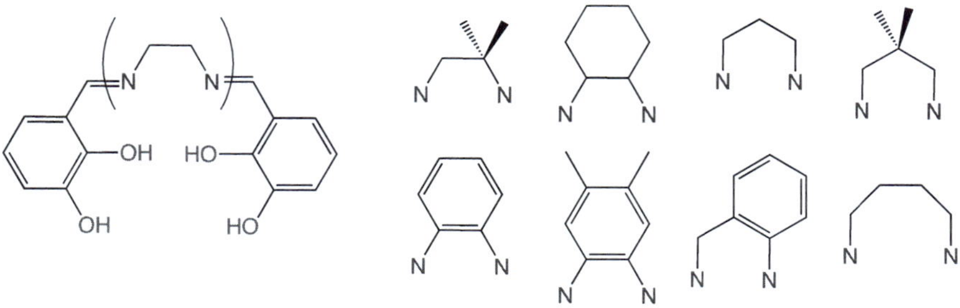

Figure 5.28. Selection of Schiff base ligands.

U^{IV} calixarene complexes are also readily accessed from U^{IV} chloride, acac and triflate precursors. Notably, the reaction of UCl_4 with calix[n]arene (n = 4, 6, 8) afforded mononuclear, bis-binuclear and trinuclear calixarene complexes, respectively.

5.7.4 *O,N donor ligands*

Tetradentate Schiff bases such as salophen^{2-} and salen^{2-} (Figure 5.28) form heteroleptic [An(L)X_2] and homoleptic [An$(L)_2$] complexes (An = U^{IV} and Th^{IV}).

Hexadentate compartmental Schiff bases presenting two different binding sites were used to promote the formation of trinuclear heterobimetallic complexes where only oxygen atoms of the Schiff base ligands bind the actinide ion, while two oxygen and two nitrogen atoms are coordinated to a 3d transition metal ion. Comparative magnetic studies of Cu_2U, Zn_2U (**163** and **166**) and Cu_2Zr trimers suggest that the U^{IV} ion mediates ferromagnetic interactions between the three metal centres in the Cu_2U trimer (**162**) (Figure 5.29).[81]

The tripodal trianionic ligands [U{(^{R}ArO)$_3$tacn}][{(^{R}ArO)$_3$ tacn} = {$C_6H_{12}N_3(CH_2C_6H_2$-3-^{t}Bu-5-R-6-O$)_3\}^{3-}$; R =^{t}Bu, **167**, Ad, **168**, diamantyl (Dia), **169**] and [U{(^{R}ArO)$_3$N}], **170**, [{(^{R}ArO)$_3$N} = {N($CH_2C_6H_2$-3-Me-5-Ad-6-O$)_3\}^{3-}$] have been used by the Meyer group to prepare highly reactive U^{III} complexes.[122] The bulk of the ligand can be tuned by varying the R groups, leading to different reactivity with small molecules such as CO_2, CS_2 and S_8 (Figure 5.30).

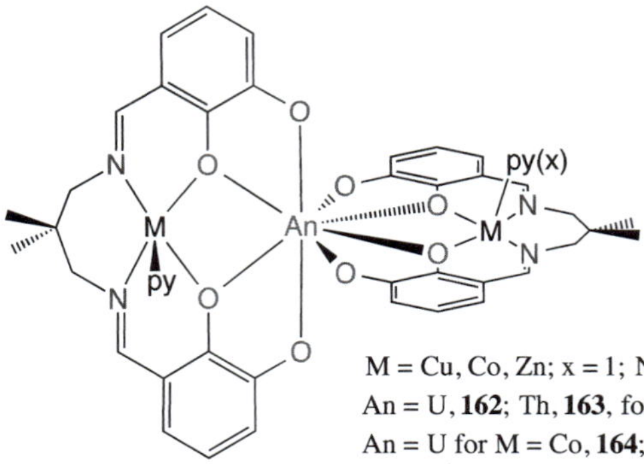

M = Cu, Co, Zn; x = 1; Ni, x = 0
An = U, **162**; Th, **163**, for M = Cu
An = U for M = Co, **164**; Ni, **165**; Zn, **166**

Figure 5.29. Complexes **162–166**.

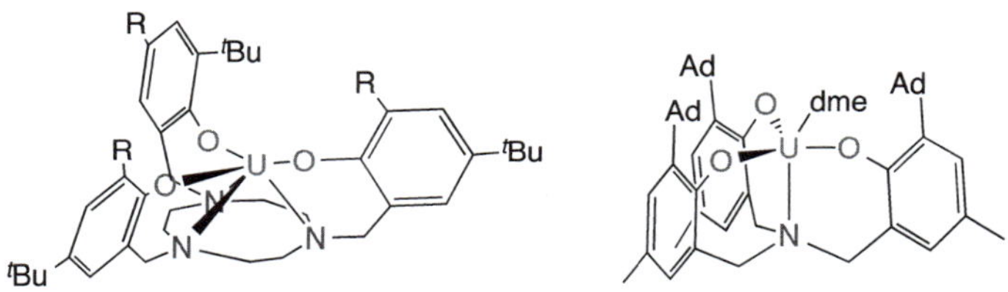

R = ᵗBu, **167**; Ad, **168**; Dia, **169**

170

Figure 5.30. Complexes **167–170**.

Several dinuclear U^{IV} oxo-, sulfido-, persulfido-, carbonato-, thiocarbonato-bridged complexes have been obtained from these reactions and characterised. Examples of terminal U^{IV} sulfido and hydrosulfido complexes have also been prepared.

N,O-donor β-ketoiminates have proved to be suitable ligand systems for the structural comparison of U^{IV} and Pu^{IV} cations in a similar environment. Acnac complexes of U^{IV} $[UCl_2(^{Ar}acnac)_2]$, **171** and $[UI_2(^{Ar}acnac)_2]$, **172**, ($Ar = 3,5\text{-}^tBu_2C_6H_3$) were obtained from the reactions of UCl_4 or $[UI_4(OEt_2)]$ with $Na(^{Ar}acnac)$, respectively. The Pu^{IV} complex $[PuI_2(^{Ar}acnac)_2]$, **173**, was prepared by oxidation of Pu^0 metal with iodine in the presence of $Na(^{Ar}acnac)$. Comparison of U–N vs. Pu–N bonding in these complexes show the expected shortening from the actinide contraction across the 5f series.[82]

5.7.5 *S-donor ligands*

Thermally stable $[U(SR)_4]$ compounds were first prepared by protonolysis of $[U(NEt_2)_4]$ by RSH (R = Et, **174**; ^{i}Pr, **175**; ^{n}Bu, **176**; ^{t}Bu, **177**) by Gilman in 1954. These complexes are highly sensitive to moisture and air. Anionic homoleptic complexes of dithiolate ligands such as ddt^{2-} (dddt = 5,6-dihydro-1,4-dithiine-2,3-dithiolate, **178**) or edt^{2-}, **179**, are prepared by salt metathesis protocols from UCl_4. The oxidative reaction of uranium metal with dichalcogenides is an alternative method for isolating molecular uranium(IV) thiolates and selenolates such as $[U(EPh)_4(py)_3]$ (E = S, **180**, Se, **181**, see Scheme 5.13).[83]

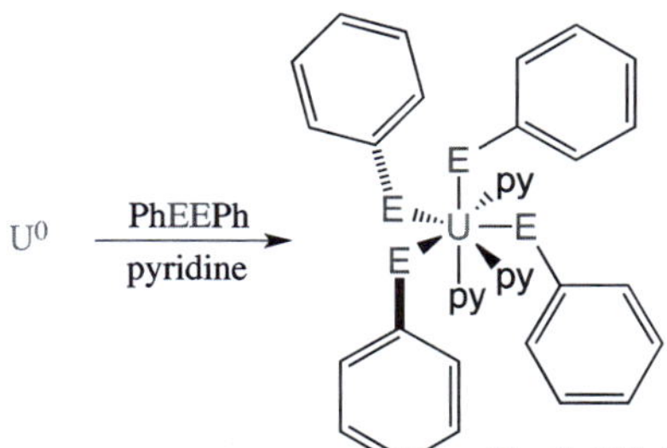

Scheme 5.13. Synthesis of **180** and **181**.

The first neutral homoleptic uranium(IV) thiolate characterised by X-ray diffraction, $[U(SMes^*)_4]$, **182**, was prepared from the reaction of $[U(BH_4)_4]$ with KSMes* (HSMes* = HS-2,4,6-$^t Bu_3C_6H_2$) (Scheme 5.14), and the first homoleptic thiolate complex of uranium(III), $[U(SMes^*)_3]$, **183**, was prepared by the protonolysis of $[U\{N(SiMe_3)_2\}_3]$ with HSMes* in cyclohexane.[83]

Anionic imidodiphosphinochalcogenide ligands of the type $[N(EPPh_2)_2]^-$, which present a delocalised negative charge, were used to prepare trivalent complexes $[An\{N(EPR_2)_2\}_3]$ (R = *i*Pr, Ph; An = U, Pu; E = S, **184–187**; Te **188–191**; or Se, **192–195**) (Figure 5.31), by salt metathesis (E = Te) or by protonolysis from $[An\{N(SiMe_3)_2\}_3]$ (E = S, Se). The An–E bond lengths are shorter than the Ln–E bond lengths for lanthanide ions of similar ionic radii, suggesting that the An–E bonds are stronger. DFT calculations indicate that the observed increase in covalency for the An–E bonding is

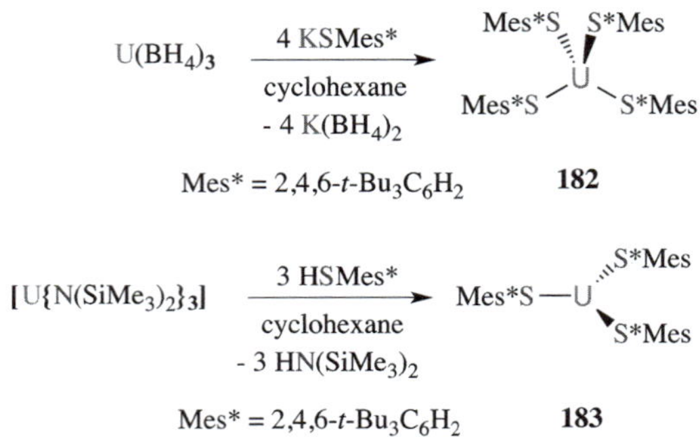

Scheme 5.14. Synthesis of **182** and **183**.

R = Ph, ^{i}Pr; An = U, Pu; E = S, Te, Se

184-195

Figure 5.31. Complexes **184–195**.

due to 5f-orbital participation in bonding, whilst the increased covalency of An–Te compared to An–S arises from increased participation of the An 6d-orbitals.[84]

Neutral crown-thioethers have also been used to incorporate actinides: [AnI_3(9-thiacrown-3)$(MeCN)_2$] (An = U, **196**; Pu, **197**) were obtained by reacting [$AnI_3(thf)_4$] and 1,4,7-trithiacyclononane in acetonitrile and [$U(BH_4)_2$(18-thiacrown-6)][BPh_4], **198**, which was obtained from [$U(BH_4)_2(thf)_5$][BPh_4], **199** and 18-thiacrown-6. Structural comparisons of the M–S bonding in 9-thiacrown complexes of U^{III} and La^{III} also point to the presence of covalent U–S bonding.[85]

5.7.6 *Redox active ligands*

Actinide complexes of the redox-active ligand $(dpp\text{-}BIAN)^{2-}$ (dpp-BIAN=1,2-bis(2,6-diisopropylphenylimino)acena-phthylene), U(dpp-BIAN)$_2$, **200**, U(dpp-BIAN)$_2$(thf), **201** and Th(dpp-BIAN)$_2$(thf),

202, have been prepared. Reversible intramolecular electron transfer was demonstrated for these systems (Scheme 5.15).[86]

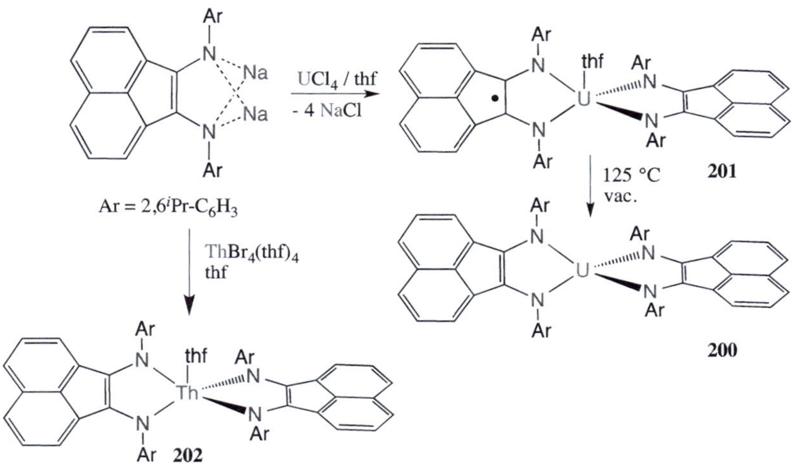

Scheme 5.15. Synthesis of **200–202**.

Non-innocent ligands have been more recently used to effect ligand and metal centred multi-electron reactions at the uranium centre. Pyridine(diimine) (PDI) ligands have been used in uranium chemistry because of their ability to store up to three electrons in their conjugated π^* orbitals. A U^{IV} complex of the mono-reduced PDI ligand $[U(^{Mes}PDI^{Me})I_3(thf)]$, **203**, ($^{Mes}PDI^{Me}$ = 2,6-((Mes)N=CMe)$_2C_5H_3N$) was prepared by reaction of $^{Mes}PDI^{Me}$ with $UI_3(thf)_4$. Subsequent reduction of this complex leads to U^{IV} complexes (**204–206**) containing up to four electrons stored on the ligand (Scheme 5.16). The U^{IV} complex $[U(^{Mes}PDI^{Me})^{4-}]_2$, **206**, can transfer six electrons to an organic azide using the electrons stored in the ligand framework, leading to a tris-imido complex of U^{VI}.[87]

Schiff base ligands provide another class of non-innocent ligands. Reduction of U^{IV} complexes with tetradentate of tridentate Schiff bases (salophen^{2-} for example) does not lead to reduction of the metal centre but to the reduction of the imine function and formation of intramolecular or intermolecular C–C bonds (Scheme 5.17). The electrons stored in these C–C bonds can be used to oxidise substrates.

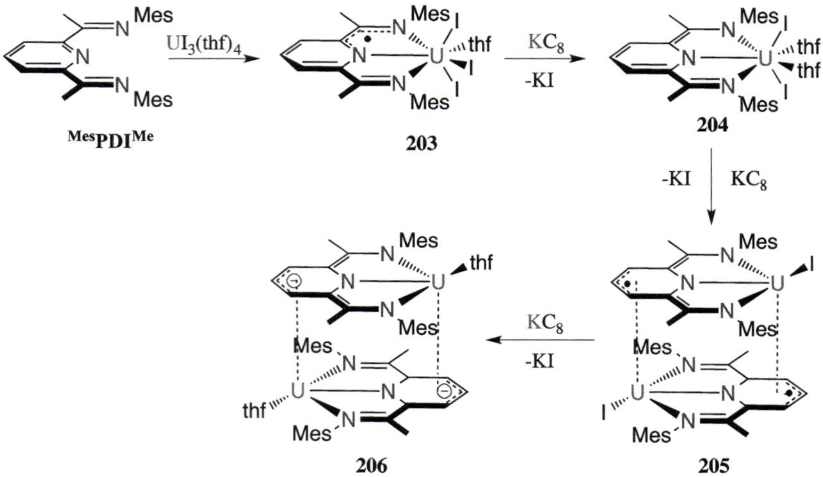

Scheme 5.16. Synthesis of **203**–**206**.

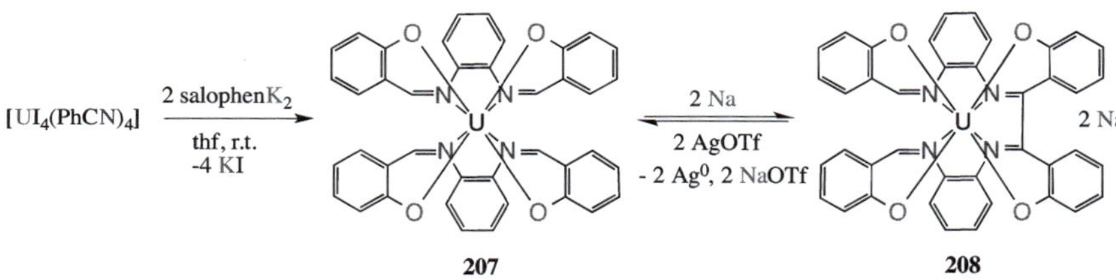

Scheme 5.17. Synthesis of **207** and **208**.

5.8 Actinide Complexes Containing Multiply Bonded Atoms

Due to the electrostatic nature of bonding in actinides, the formation of An-heteroatom bonds are much less common than in d-block transition metals. Usually, reactions with oxo, nitride or sulphur transfer reagents result in the formation of sulfido, oxo or nitrido bridged complexes.

In recent years, an increasing number of An complexes containing terminally bonded atoms have been prepared through the development of alternative synthetic routes or through the design of bulky ligands capable of preventing the formation of dinuclear complexes. The study of the bonding in these complexes has clearly shown the presence of multiple bonds and in some cases the participation of

5f orbitals and 6d orbitals to the bond has been identified by DFT studies.

5.8.1 *Imides*

The first imido complexes of uranium were reported in 1984 and concerned organometallic species of uranium. A typical route involves reacting organoazides with low valent uranium complexes to provide imido moieties as the result of the two electron reduction of azides. The U^{III} complex $[U\{N(SiMe_3)_2\}_3]$ (**44**) was reacted with several organic azides to yield the U^{V} complexes $[U\{N(SiMe_3)_2\}_3(NR)]$ (R = Ph, **209** and $SiMe_3$, **210**). These complexes exhibit short U=N distances (1.910(16) Å for $[\{N(SiMe_3)_2\}_3U{=}NSiMe_3]$) (Scheme 5.18).[88] Further oxidation with $AgPF_6$ yields the U^{VI} complexes $[U\{N(SiMe_3)_2\}_3(NR)F]$ (R = Ph, **211** and $SiMe_3$, **212**).[89]

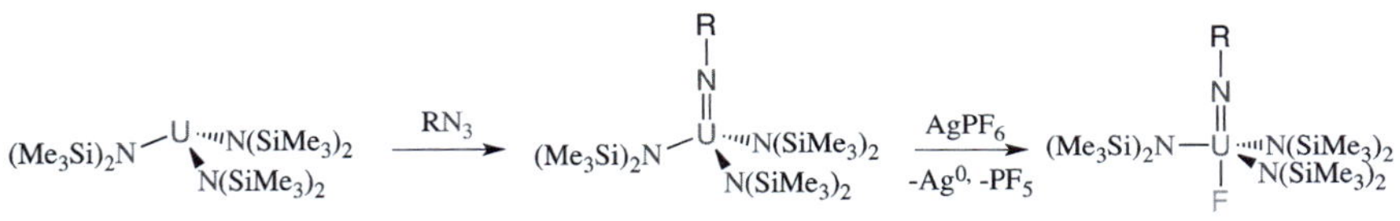

Scheme 5.18. Synthesis of **209**–**212**.

This synthetic route was successfully extended to U^{III} complexes containing different supporting ligands presented in previous chapters, such as $[U\{(^{R}ArO)_3tacn\}]$ and $[U\{(^{R}ArO)_3N\}]$ (see Figure 5.30) to yield the U^{V} imido complexes $[U\{(^{R}ArO)_3tacn\}NSiMe_3]$ (R = Ad, **213**) and $[U\{(^{R}ArO)_3N\}(OEt_2)NMes]$ (R = Ad, **214**) (Scheme 5.19). Complex **213** can undergo σ-bond metathesis in the presence of acetonitrile and CO.[90] Complex **214** does not have the expected C_3 symmetry but rather exhibits C_s symmetry, suggesting the presence of the inverse-*trans* influence, an effect mainly observed in actinide compounds.[89]

As described earlier, oxidative strategies were also used with uranium complexes containing redox non-innocent ligands and notably a tris-imido complex was isolated from the reaction of $[U^{IV}(^{Mes}PDI^{Me})^{4-}]_2$ (**206**) with organic azides, resulting in a formal

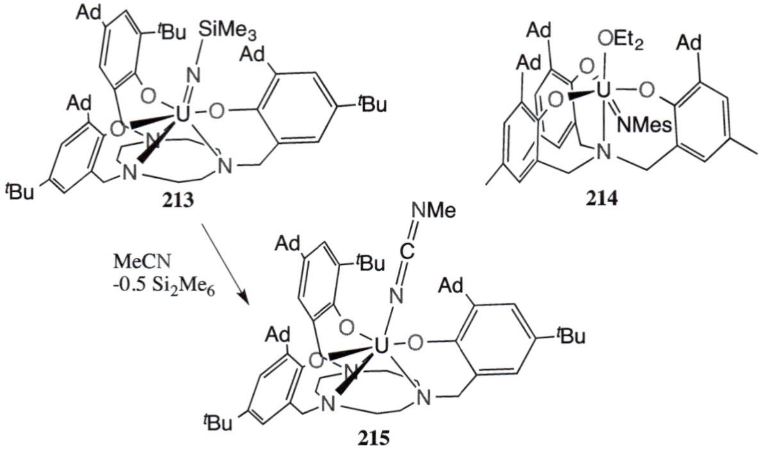

Scheme 5.19. Synthesis of **214** and **215** from **213**.

6-electron reduction per uranium centre.[87] In the previous examples, bulky ancillary ligands are present to stabilise imido complexes of U^{V}. In the absence of ancillary ligands a bulky imido ligand such as NDipp (Dipp= 2,6-di-isopropylphenyl) also allowed the formation of a mono-imido complex, $[UI_2(thf)_3NDipp]$, **216**. The formation of such complexes is very sensitive to the reaction conditions and the use of LiN(H)Dipp instead of KN(H)Dipp led only to the dimeric form $[UCl_2(thf)_2NDipp]_2$, **217** (Figure 5.32). A uranium(IV) imido complex, $[UI_2(thf)_2(^{t}Bu_2bipy)N\text{-}^{t}Bu]$, **218**, was prepared from the comproportionation reaction of $[UI_3(thf)_4]$ with the U^{V} dimer $[U(N^{t}Bu)_2(I)(^{t}Bu_2bpy)]_2$.[91] The imido group on actinides is only rarely stabilised by small R substituents such as H. One example of a complex containing a NH moiety, $[U(tren^{TIPS})NH][K(15C6)]$, **219**, was prepared by the deprotonation of $[U(tren^{TIPS})NH_2]$, **220**, with K and subsequent abstraction of K^+ by [15]C-5.[92]

Trans-bis(imido) uranium complexes have been synthesised by the oxidation of uranium turnings in presence of *tert*-butylamine, and were described earlier in this chapter (**50**, Figure 5.4). Np analogues were also reported in 2015 (**84**, Figure 5.19). Treatment of the *trans*-bis(imido) complex **50** by $B(C_6F_5)_3{\cdot}H_2O$ led to the exchange of one imido ligand by an oxo group, leading to a rare mixed terminal imido/oxo complex.[93]

When the mononuclear anionic uranium tetrasiloxide complex **159** reacts with bulky organic azides (RN_3, R = Ad, $SiMe_3$), mononuclear imido complexes are obtained, $[K(18C6)][(OSi(O^tBu)_3)_4(SiMe_3)_2)_3U{=}NR]$ (R = Ad, **221**; $SiMe_3$, **222**; Scheme 5.20).

5.8.2 *Nitrides*

Similarly to imido complexes, the few reported uranium nitride complexes have been mostly prepared from the reaction of low-valent uranium complexes with azides.[94] This approach has resulted in the formation of dinuclear and multinuclear nitrido-bridged complexes,[95] borane-capped nitrides, stable[96] or highly reactive terminal nitrides that insert into the supporting ligand C–H bonds.[97] However, the first example of molecular uranium nitride was reported in 2002 from the reaction of $[(Et_8\text{–calix}[4]\text{pyrrole})U(dme)][K(dme)]$ (**123**) with molecular dinitrogen to yield a bis-nitride U(V)/U(IV) complex.[67] In the absence of bulky supporting ligands the reactions of ionic or metallic azides with low-valent uranium leads to large assemblies of uranium in which the nitride does not necessarily engage in multiple bonding. For example, a transient uranium(IV) heptaazide reacts with $[UI_3(thf)_4]$ to yield a tetrameric uranium(IV) assembly where four uranium are bridged by azide ligands and a nitride ligand located in the centre of the cluster (**223**) (Figure 5.33).[95b]

An azide salt complexed by a simple Lewis acid has been used by Cummins to deliver a protected nitride ligand to the U^{III} complex $[U\{N(^tBu)Ar\}_3(thf)]$ (Ar = 3,5-$Me_2C_6H_3$), **224**, in order to prevent the formation of bimetallic complexes.[98] Structural data (U–N

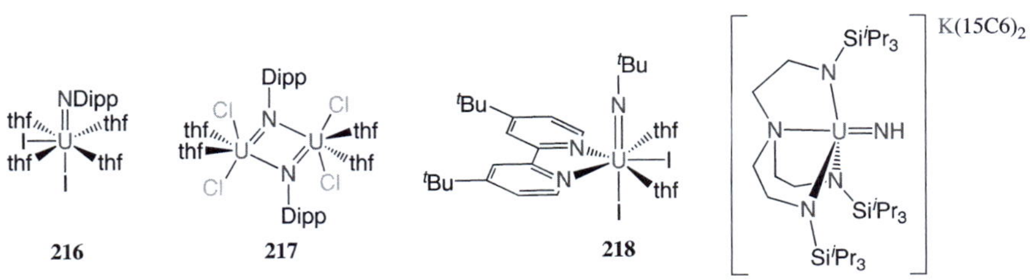

216 217 218 219

Figure 5.32. Complexes **216–219**.

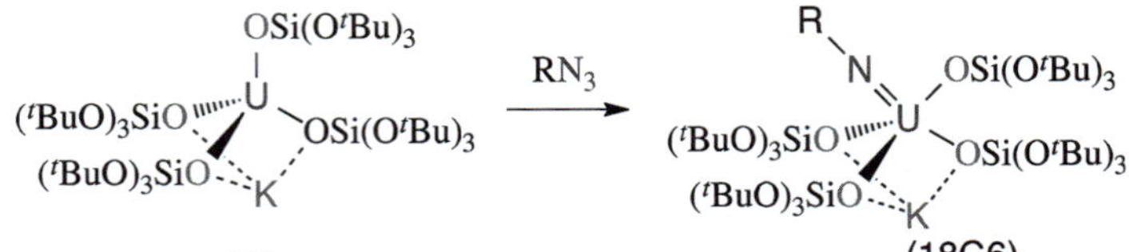

159

R = Ad, **221**; $SiMe_3$, **222**

Scheme 5.20. Synthesis of **221** and **222**.

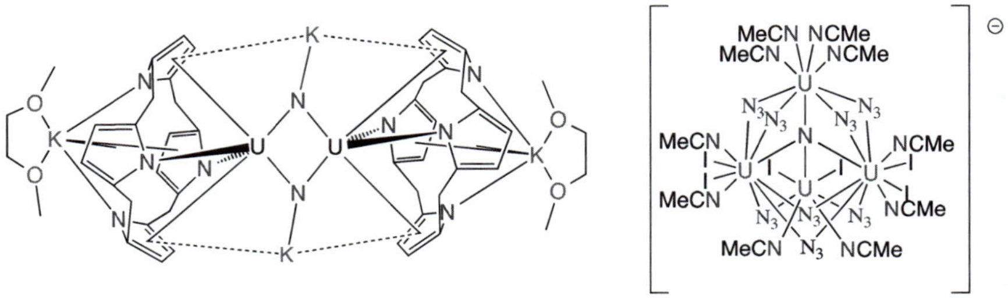

Figure 5.33. Bridged nitride and azide uranium complexes.

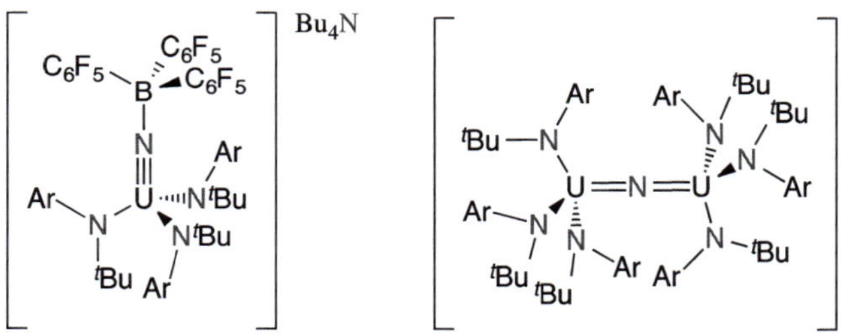

225 **227**

Figure 5.34. Complexes **225** and **227**.

distance of 1.880(4) Å) and DFT calculations of the resulting U^V nitridoborate complex, $[N(^nBu)_4][\{N(^tBu)Ar\}_3U{\equiv}N\text{-}B(C_6F_5)_3]$, **225** (Figure 5.34), indicate that this complex can be described as a borane adduct of a multiply bonded terminal nitride with a U–N bond similar to other formally triply bonded uranium(VI) organoimido complexes. Deprotection of the nitride group in **225** to afford a non-capped terminal nitride has not been accomplished to date. One electron oxidation of this U^{V} complex gave the neutral uranium(VI) nitridoborate derivative $[\{N(^{t}Bu)Ar\}_3U{\equiv}N\text{-}B(C_6F_5)_3]$, **226**.

The reaction of the same U^{III} amide precursor **224** with NaN_3 or $[N(^{n}Bu)_4][N_3]$ afforded the nitrido-bridged anionic diuranium(IV) complex $Na[(\mu\text{-}N)(U(N[^{t}Bu]Ar)_3)_2]$, **227** (Figure 5.34), which could be further oxidised step-wise to afford the analogous neutral $U^{IV}\text{-}U^{V}$, **228** and cationic $U^{V}\text{–}U^{V}$ complexes, **229**. All of these complexes were crystallographically characterised, showing a linear U=N=U core with equivalent U=N bond distances.[99] In this series, the successive one-electron oxidation of the linear U–N–U core results in a decrease of the U–N nitride distance of ~0.03 Å that was explained in terms of a mainly electrostatic interaction in the multiply bonded U=N=U core. The $U^{V}\text{–}U^{V}$ complex was found to react as a metallonitrene with NaCN, suggesting that the nitride group might be available for N-transfer reactions, even in these dinuclear compounds.

Similar results were obtained from the reactions of the neutral tris-siloxide dimer $[U(OSi(O^{t}Bu)_3)_3]_2$ (**155**) with CsN_3, affording the diuranium(IV) nitrido bridged complex $[Cs\{[(\mu\text{-}N)[U(OSi(O^{t}Bu)_3)_3]_2\}]$, **230**, which also contains a linear U=N=U core (Figure 5.35).[77] However, this complex is not an ion pair, but the Cs cation is bound in the complex to both the nitride group and the siloxide oxygens. The presence of Cs in **230** is crucial to effect the reduction by Cs metal of the $U^{IV}{=}N{=}U^{IV}$ core to the analogous $U^{IV}{=}N{=}U^{III}$ and $U^{III}{=}N{=}U^{III}$ compounds $[Cs_2\{[(\mu\text{-}N)[U(OSi(O^{t}Bu)_3)_3]_2\}]$, **231** and $[Cs_3\{[(\mu\text{-}N)[U(OSi(O^{t}Bu)_3)_3]_2\}]$, **232** (Figure 5.35). In the structure of the latter complex three Cs cations are firmly bound to the nitride group and to the siloxide oxygens.[100] The multimetallic nature of these complexes and the high nucleophilicity of the bridging nitride results in unusual N transfer reactivity with CO_2, CO and CS_2, leading to N–C bond formation and CO cleavage.[101] Reduction of **230** with excess KC_8 gave $[K_3\{[(\mu\text{-}N)[U(OSi(O^{t}Bu)_3)_3]_2\}]$ (**232-K**); this complex was shown to activate N_2 to and effect the four electron reduction of N_2 to yield the diuranium(V) complex $[K_3\{[(\mu\text{-}N)(\mu\text{-}\eta^2{:}\eta^2\text{-}N_2)\}[U(OSi(O^{t}Bu)_3)_3]_2\}]$ (**232-K-N$_2$**).[100b]

The reaction of $[U\{N(SiMe_3)_2\}_3]$ (**44**) with NaN_3 also leads to a nitrido bridged diuranium(IV) complex, **233**. However, in this case

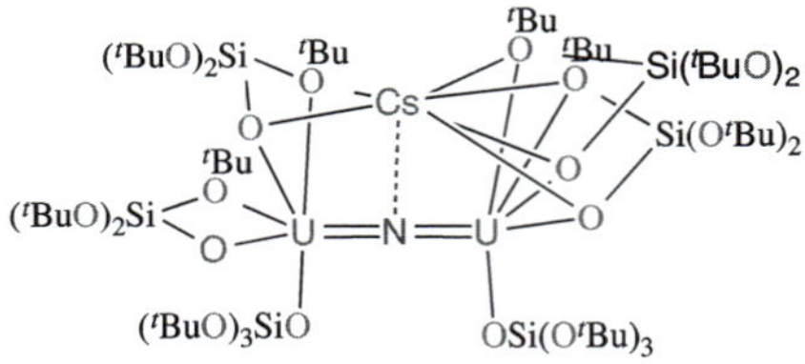

230

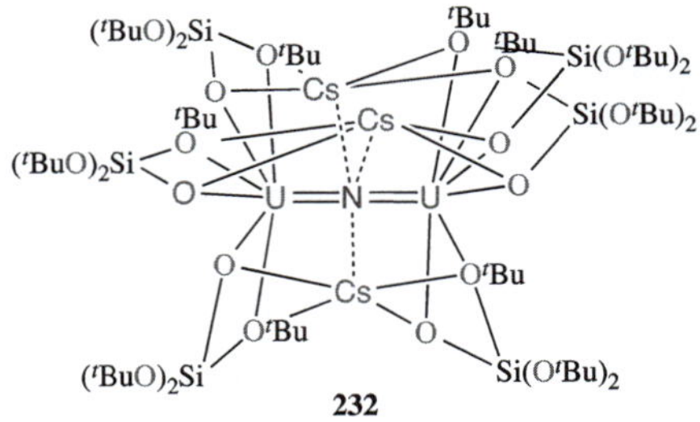

232

Figure 5.35. Complexes **230** and **232**.

deprotonation of a methyl group of $\{N(SiMe_3)_2\}$-occurs, yielding a μ-CH_2 moiety and $H\{N(SiMe_3)_2\}$, and resulting in a bent U–N–U moiety and delocalised U–N=U bonding.[102]

Finally, in 2012, the trivalent complex [U(tren^{TIPS})], **131**, which contains a very bulky polydentate amide, was used in a series of reactions with sodium azide to yield dimeric [{U(tren^{TIPS})(NNa)$_2$}$_2$], **234**, which exhibits a short U–N bond length of 1.883(4) Å. Abstraction of the sodium atom by [12]C-4 yielded the first terminal pentavalent uranium nitride complex, [U(tren^{TIPS})N][Na([12]C-4)], **235**, in which the U–N bond length is 1.825(15) Å, in agreement with a triple U–N bond (Figure 5.36).[96] This triple bonding interaction was confirmed by theoretical calculations. Oxidation of **235** afforded an analogous hexavalent terminal nitride.[103]

A similar approach based on the use of the sterically demanding Tren^{TIPS} tripodal ligand to stabilise weak U-As and U–P bonds has led to the synthesis of a rare terminal uranium phosphide U-PH_2, **236**, phosphinidene U=PH,[104] **237**, arsenide U-AsH_2, **238**, arsinidene [U=AsH], **239** and arsenido [U≡AsK_2], **240**, which all exhibit U–X multiple bonds (Figure 5.37).[105] The parent phosphide and arsenide complexes were prepared by reacting [U(tren^{TIPS})] (**131**) with $NaPH_2$ and [$KAsH_2$], respectively, and subsequent deprotonation yielded the multiply bonded U=PH, [U=AsH] and [U≡AsK_2] species.

5.8.3 *Chalcogenides*

Mono-oxo complexes have been investigated since the 1970s in the presence of simple chloride and fluoride ligands ($[OUCl_5]^{2-}$, **241**,

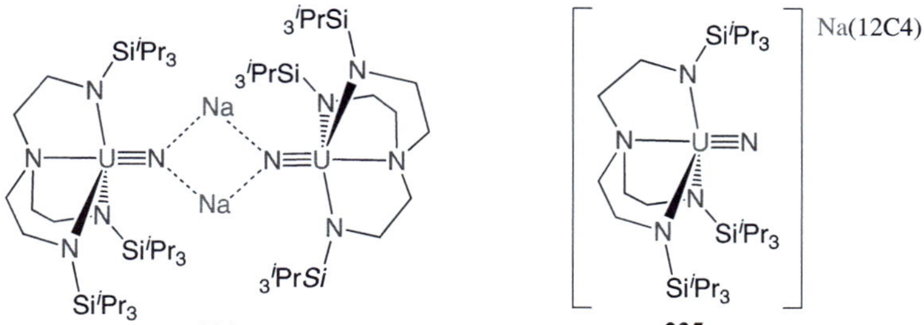

234 **235**

Figure 5.36. Complexes **234** and **235**.

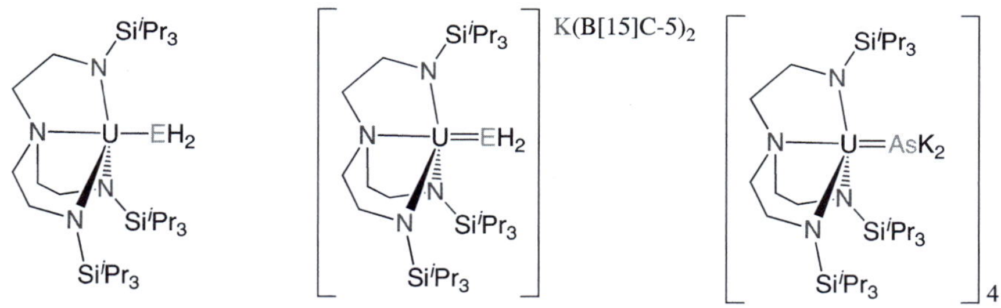

E = P, 236; As, 238 E = P, 237; As, 239 **240**

Figure 5.37. Complexes **236**–**240**.

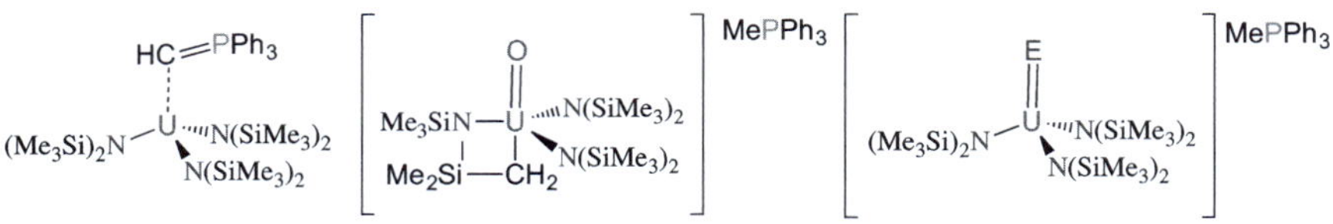

244 **245** E = S, **246**; Se, **247**; Te, **248**

Figure 5.38. Complexes **244**–**248**.

$[OUCl_5]^-$, **242** and $[OUF_4]$, **243**). Later, in 2011, it was shown that $[U\{N(SiMe_3)_2\}_3(H_2C{=}PPh_3)]$, **244**, can react with TEMPO to yield $[U\{N(SiMe_3)_2\}_2(CH_2SiMe_2NSiMe_3)(O)][MePPh_3]$, **245**, and can be oxidised to the equivalent neutral complex (Figure 5.38).[106] Addition of other chalcogenides in elemental forms (S_8, Se, Te) to **244** afforded a series of terminal chalcogenide complexes, $[U\{N(SiMe_3)_2\}_3E]$ (E = S, **246**; Se, **247**; Te, **248**) (Figure 5.38).[107]

The treatment of $[U\{N(SiMe_3)_2\}_3R]$ (R = Me, **249**, CCPh, **250**) with 4-morpholine N-oxide yielded $[U(N(SiMe_3)_2)_3R(O)]$ ((R =

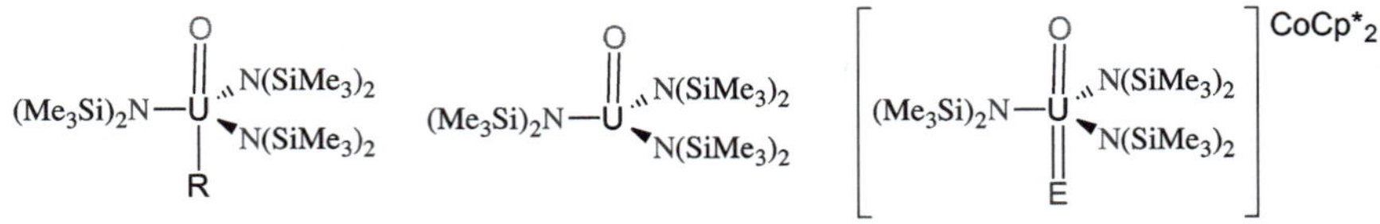

R = Me, **251**; CCPh, **252** **253** E = S, **254**; Se, **255**

Figure 5.39. Complexes **251–255**.

Me, **251**, CCPh, **252**, Figure 5.39)[108] and the direct oxidation of $[U\{N(SiMe_3)_2\}_3]$ (**44**) with TEMPO afforded $[U\{N(SiMe_3)_2\}_3(O)]$ (**253**, Figure 5.39). An important feature in **253** is its trigonal pyramidal geometry, which is much less common than the expected tetrahedral geometry.[109] Reduction of $[U(N(SiMe_3)_2)_3(O)]$ with decamethylcobaltocene, followed by oxidation with chalcogenides such as S_8 and Se gave the mixed U=O/E complexes $[U(N(SiMe_3)_2)_3(O)(E)][CoCp^*{}_2]$ (E = S, **254**; Se, **255**).[110]

Terminal oxo An complexes have been prepared from low-valent uranium complexes using bulky monodentate or polydentate ligands that prevent the more favourable formation of bridging oxo dimers. Most reported terminal oxos contain high valent uranium (+V or

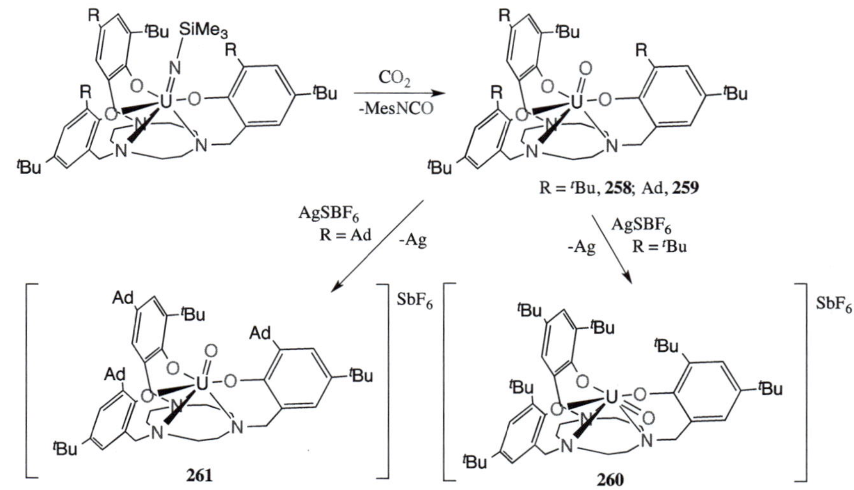

Scheme 5.21. Synthesis of **258–261**.

+VI). A rare U^{IV} terminal oxo complex [$(Tp^*)_2U(O)$], **256**, was prepared from the reaction of [$(Tp^*)_2U(bipy)$], **257**, (a complex that contains a U^{III} centre bound by a $bipy^{\bullet-}$ radical anion) with pyridine N-oxide.[111] The bulky tris-aryloxide ligand $(^{R}ArO)_3$tacn (R = Ad, ^{t}Bu) also supported the synthesis of U^{V} terminal mono-oxo complexes, [(O)U{$(^{R}ArO)_3$tacn}] (R =^{t}Bu, **258**; Ad, **259**), but through a different pathway involving a metathesis reaction of the U^{V} imido complex [$(NSiMe_3)$U{$(^{R}ArO)_3$tacn}] with CO_2 (Scheme 5.21).[112] The one-electron oxidation of **258** affords a U^{VI} terminal oxo complex, **260**, in which the oxo group is found in the equatorial plane *trans*- to one of the aryloxide groups (C_s symmetry), as the result of inverse *trans*-influence (Scheme 5.21). In contrast, when one of the *tert*-butyl aryloxide substituents is replaced by more bulky adamantyl groups, the oxidation of the U^{III} tris-aryloxide complex **259** leads to a C_3 symmetric complex, **261**.[113]

The sterically hindered and highly reducing U^{III} tetrasiloxide complex [$KU((OSi(O^tBu)_3)_4)$] (**159**) reacted with CO_2 to yield a terminal U^{V} oxide by reductive deoxygenation of CO_2 (Scheme 5.22), where cooperative binding of CO_2 by both the U^{III} and the K^+ centres leads to a rare complete cleavage of the C=O bond in CO_2 to form a U^{V} mono-oxo complex, **262**.[114]

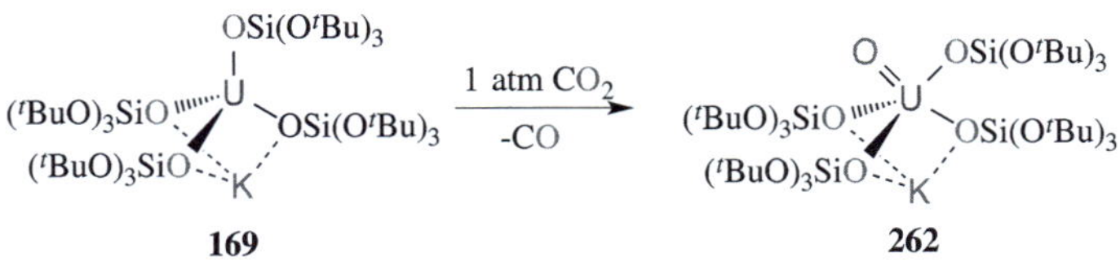

Scheme 5.22. Synthesis of **262**.

The reduction of S or Se with low-valent uranium complexes has led to many different complexes in which the chalcogen element is bridged between two uranium centres or its reduction is incomplete and features the $(E\text{-}E)^{2-}$ fragment (E=S, Se and Te).[115] However, the terminal sulphide complex, **263**, could be isolated as it has sufficient steric hindrance in the metal coordination sphere and an alternative ligand transfer agent $KSCPh_3$ was used to oxidise [$U\{N(SiMe_3)_2\}_3$] (**44**) (Scheme 5.23).[107b]

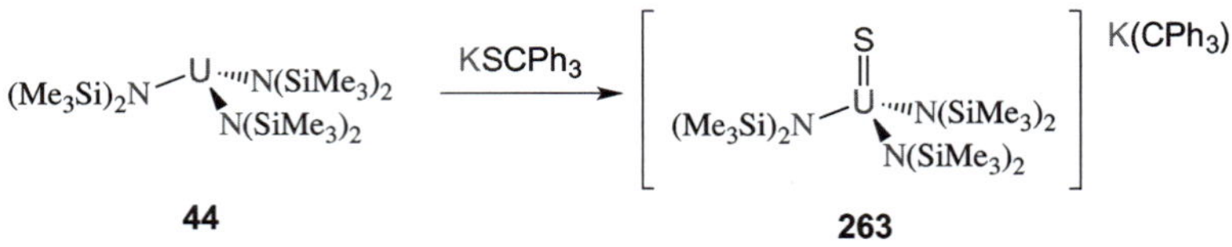

Scheme 5.23. Synthesis of **263**.

The siloxide complex [KU((OSi(O^{t}Bu)$_3$)$_4$)] (**169**) reacts with Ph_3PS to yield a dimeric complex in which potassium bridges between U≡S bonds, **264**. The use of a cryptand allowed potassium abstraction to yield monomolecular [KU((OSi(O^{t}Bu)$_3$)$_4$)(S)] [Kcryptand], **265**, in which a short U≡S bond of 2.5220(14) Å is reported (Scheme 5.24).[116] Complex **265** reacts with proton sources to yield the U^{IV} hydrosulfido complex [KU((OSi(O^{t}Bu)$_3$)$_4$)(SH)] [K18C6], **266**. On the other hand, [U($^{R,R'}$ArO)$_3${tacn}] (**267**, R= Me, R′=Ad) reacts with H_2S to yield the hydrosulfido U^{IV} complex [U{($^{R,R'}$ArO)$_3$tacn}(SH)], **268**. This complex can be deprotonated with amide salts in the presence of 18C6 or cryptand to afford the terminal sulphide U^V complex [U{($^{R,R'}$ArO)$_3$tacn}(S)], **269**, in which the U≡S bond distance is 2.536(2) Å (Scheme 5.25).[117]

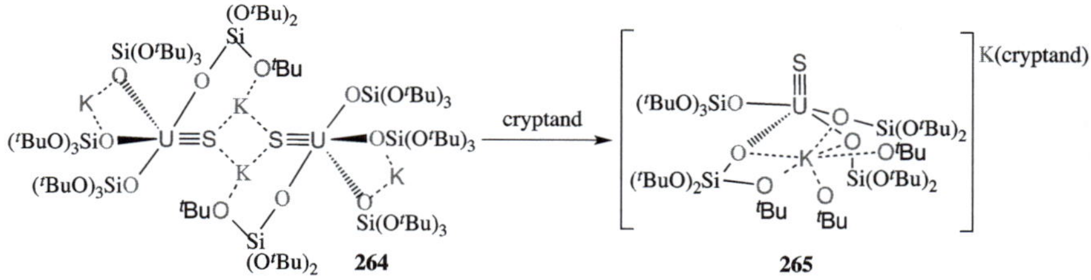

Scheme 5.24. Synthesis of **265** from **264**.

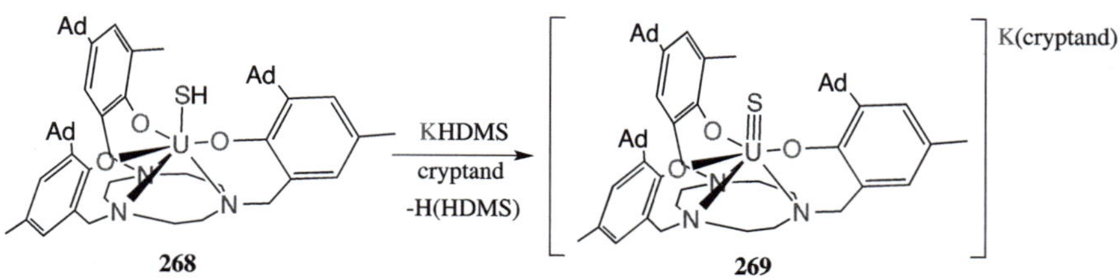

Scheme 5.25. Synthesis of **269** from **268**.

In 2015, $[Th\{N(SiMe_3)_2\}_3(S)][K18C6]$, **270**, was also synthesised using the reductive trityl deprotection strategy, while the reactions of $[Th\{N(SiMe_3)_2\}_3(E–E)][K18C6]$ (E=Se, **271**; Te, **272**) with Hg or Et_3P yielded the terminal chalcogenide complexes $[Th(N(SiMe_3)_2)_3(E)][K18C6]$ (E=Se, **273**; Te, **274**) by the trapping and removal of one of the E atoms (Scheme 5.26).[118]

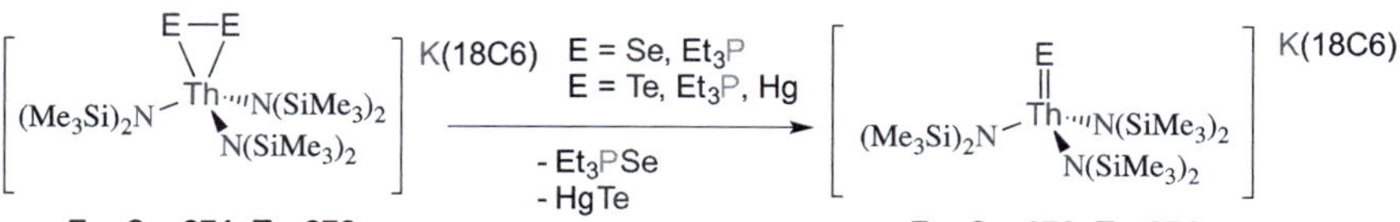

E = Se, **271**; Te, **272** E = Se, **273**; Te, **274**

Scheme 5.26. Synthesis of **273** and **274** from **271** and **272**.

5.9 Conclusions

The previous chapters clearly show that the last 20 years have witnessed a remarkable development in the chemistry of actinides. Compounds that were generally thought impossible to isolate have been prepared and new original reaction pathways as well as physical properties that are significantly different from those of d-block and 4f-elements have been identified that might lead to the development of new technological applications. A large number of coordination chemistry studies have focused on the design of polydentate O,N ligands for the selective complexation of actinides in aqueous solution. New ligands have been identified for the selective complexation of uranyl(VI) that could lead to applications in environmental remediation, metallurgical extraction, and water purification and could open the possibility of extracting uranyl(VI) from water and therefore provide a larger availability of this feedstock.

In another direction, the use of anaerobic and anhydrous conditions has expanded the number of accessible oxidation states and has widened the range of reactivity accessible to actinide ions. In this context, two key steps were ligand design and the development of the synthesis of suitable precursors, such as $[UI_3(S)_4]$, $[NpI_3(S)_4]$ and $[PuI_3(S)_4]$ or $[U^VO_2(py)_5(KI_2Py_2)]_n$ (with S being typical solvents, such as thf, pyridine or dme) for the study of coordination chemistry in anhydrous conditions. Ligand design has proved crucial for the synthesis of stable complexes of uranyl(V), uranium(III) and more recently uranium(II). For example, complexes of uranyl(V) that are stable in organic media can be obtaining using polydentate ligands that prevent the formation of dimeric species through cation–cation interactions, that eventually could result in disproportionation. Future development of this chemistry should continue to develop ligands capable of stabilising uranyl(V) in water, thus providing important information on the role of this species in the environmental conditions.

Careful tuning of the electronics of the ligands can result in the synthesis of stable homo- and hetero-polymetallic complexes, and cation–cation interactions in these complexes provides a pathway for magnetic communication between actinide ions or actinide ions and d-block transition metals. It has been shown that such interaction can lead to single-molecule magnet behaviour. Thus, the use of actinides could provide a route to molecular magnets with optimised properties for future applications. To this aim more information needs to be gathered on the parameters leading to single molecule magnet behaviour in these systems.

Ligand design has also been key for the development of An^{III} and An^{II} chemistry. In particular, tripodal trianionic ligands have allowed the isolation of stable U^{III} complexes and the second example of a molecular complex of U^{II}. These ligands have also proved to be convenient for the expansion of the coordination chemistry of low-valent Pu and Np. Tripodal ligands are able to control reactivity at the actinide centre, leading to the activation of unreactive molecules of great concern in the current energetic panorama such as CO_2 and CO. Tripodal ligands have also supported the characterisation of multiply bonded N, O, S, As, P atoms, providing insights on the ability of 5f orbitals to participate in bonding. In parallel to the development of carefully tuned sophisticated ligands, studies involving simpler or commercially available bulky monodentate ligands such as phenoxides, amides or siloxides have also resulted in stable complexes that can support reactivity at actinide centres in high and low oxidation states, showing that very new chemistry can be obtained with "old" ligands provided that the correct conditions are identified and that we are ready to question scientific assumptions.[119]

Among the many challenges that remain to be tackled in the chemistry of coordination complexes of low-valent actinides some obvious targets come to mind: C–H and N_2 activation remain still highly challenging targets, together with the identification of possible catalytic cycles for small molecule transformations.[120] The association of actinide to redox-active ligands is a recent and very promising area of research for the development of multi-electron redox reactions at actinide centres that should see a successful expansion in the future years. A still unsolved fundamental challenge is the design of ligands capable of promoting, if at all possible, An–An bonds.[121]

It can be anticipated that the chemistry of low-valent Th, Pu and Np will grow significantly in future years, leading to an increased overall knowledge of bonding in these elements with important technological and environmental repercussions.

References

1. R. R. Langeslay, M. E. Fieser, J. W. Ziller, F. Furche and W. J. Evans, *Chem. Sci.*, 2015, *6*, 517–521.
2. S. Skanthakumar, M. R. Antonio and L. Soderholm, *Inorg. Chem.*, 2008, *47*, 4591–4595.
3. (a) C. Apostolidis, B. Schimmelpfennig, N. Magnani, P. Lindqvist-Reis, O. Walter, R. Sykora, A. Morgenstern, E. Colineau, R. Caciuffo, R. Klenze, R. G. Haire, J. Rebizant, F. Bruchertseifer and T. Fanghaenel, *Angew. Chem. Int. Ed. Engl.*, 2010, *49*, 6343–6347; (b) P. Lindqvist-Reis, C. Apostolidis, J. Rebizant, A. Morgenstern, R. Klenze, O. Walter, T. Fanghaenel and R. G. Haire, *Angew. Chem. Int. Ed. Engl.*, 2007, *46*, 919–922; (c) J. H. Matonic, B. L. Scott and M. P. Neu, *Inorg. Chem.*, 2001, *40*, 2638–2639.
4. L. Natrajan, M. Mazzanti, J.-P. Bezombes and J. Pécaut, *Inorg. Chem.*, 2005, *44*, 6115–6121.
5. M. P. Wilkerson, C. J. Burns, R. T. Paine and B. L. Scott, *Inorg. Chem.*, 1999, *38*, 4156–4158.
6. S. M. Cornet, L. J. L. Haller, M. J. Sarsfield, D. Collison, M. Helliwell, I. May and N. Kaltsoyannis, *Chem. Commun.*, 2009, 917–919.
7. L. Natrajan, F. Burdet, J. Pecaut and M. Mazzanti, *J. Am. Chem. Soc.*, 2006, *128*, 7152–7153.
8. J. C. Berthet, G. Siffredi, P. Thuery and M. Ephritikhine, *Chem. Commun.*, 2006, 3184–3186.

9. R. Copping, V. Mougel, S. Petit, C. Den Auwer, P. Moisy and M. Mazzanti, *Chem. Commun.*, 2011, *47*, 5497–5499.
10. W. H. Runde and B. J. Mincher, *Chem. Rev.*, 2011, *111*, 5723–5741.
11. (a) I. A. Khan and H. S. Ahuja, *Inorg. Synth.*, 1982, *21*, 187; (b) J. L. Kiplinger, D. E. Morris, B. L. Scott and C. J. Burns, *Organometallics*, 2002, *21*, 5978–5982.
12. A. J. Gaunt and M. B. Jones, *Chem. Rev.*, 2013, *113*, 1137–1198.
13. T. Cantat, B. L. Scott and J. L. Kiplinger, *Chem. Commun.*, 2010, *46*, 919–921.
14. (a) L. R. Avens, S. G. Bott, D. L. Clark, A. P. Sattelberger, J. G. Watkin and B. D. Zwick, *Inorg. Chem.*, 1994, *33*, 2248–2256; (b) M. J. Monreal, R. K. Thomson, T. Cantat, N. E. Travia, B. L. Scott and J. L. Kiplinger, *Organometallics*, 2011, *30*, 2031–2038.
15. H. S. La Pierre, F. W. Heinemann and K. Meyer, *Chem. Commun.*, 2014, *50*, 3962–3964.
16. R. A. Andersen, *Inorg. Chem.*, 1979, *18*, 1507–1509.
17. A. J. Gaunt, A. E. Enriquez, S. D. Reilly, B. L. Scott and M. P. Neu, *Inorg. Chem.*, 2008, *47*, 26–28.
18. C. A. P. Goodwin, F. Tuna, E. J. L. McInnes, S. T. Liddle, J. McMaster, I. J. Vitorica-Yrezabal and D. P. Mills, *Chem. Eur. J.*, 2014, *20*, 14579–14583.
19. P. L. Arnold, C. J. Stevens, J. H. Farnaby, M. G. Gardiner, G. S. Nichol and J. B. Love, *J. Am. Chem. Soc.*, 2014, *136*, 10218–10221.
20. R. G. Jones, G. Karmas, G. A. Martin and H. Gilman, *J. Am. Chem. Soc.*, 1956, *78*, 4285–4286.
21. S. Fortier, G. Wu and T. W. Hayton, *Inorg. Chem.*, 2008, *47*, 4752–4761.
22. T. W. Hayton, J. M. Boncella, B. L. Scott, P. D. Palmer, E. R. Batista and P. J. Hay, *Science*, 2005, *310*, 1941–1943.
23. R. G. Denning, *J. Phys. Chem. A*, 2007, *111*, 4125–4143.
24. J. C. Berthet, M. Nierlich and M. Ephritikhine, *Chem. Commun.*, 2003, 1660–1661.
25. D. L. Clark, D. E. Hobart and M. P. Neu, *Chem. Rev.*, 1995, *95*, 25–48.
26. M. B. Jones and A. J. Gaunt, *Chem. Rev.*, 2013, *113*, 1137–1198.
27. S. Beer, O. B. Berryman, D. Ajami and J. Rebek, *Chem. Sci.*, 2010, *1*, 43–47.
28. T. S. Franczyk, K. R. Czerwinski and K. N. Raymond, *J. Am. Chem. Soc.*, 1992, *114*, 8138–8146.
29. A. C. Sather, O. B. Berryman and J. Rebek, *J. Am. Chem. Soc.*, 2010, *132*, 13572–13574.

30. (a) A. E. V. Gorden, J. D. Xu, K. N. Raymond and P. Durbin, *Chem. Rev.*, 2003, *103*, 4207–4282; (b) G. Szigethy and K. N. Raymond, *J. Am. Chem. Soc.*, 2011, *133*, 7942–7956.
31. J. L. Sessler, P. J. Melfi and G. D. Pantos, *Coord. Chem. Rev.*, 2006, *250*, 816–843.
32. M. Cametti, M. Nissinen, A. D. Cort, L. Mandolini and K. Rissanen, *J. Am. Chem. Soc.*, 2007, *129*, 3641–3648.
33. P. L. Arnold, D. Patel, A. J. Blake, C. Wilson and J. B. Love, *J. Am. Chem. Soc.*, 2006, *128*, 9610–9611.
34. P. L. Arnold, D. Patel, C. Wilson and J. B. Love, *Nature*, 2008, *451*, 315–318.
35. D. L. Clark, D. E. Hobart and M. P. Neu, *Chem. Rev.*, 1995, *95*, 25–48.
36. D. L. Clark, S. D. Conradson, S. A. Ekberg, N. J. Hess, M. P. Neu, P. D. Palmer, W. Runde and C. D. Tait, *J. Am. Chem. Soc.*, 1996, *118*, 2089–2090.
37. (a) D. L. Clark, D. W. Keogh, P. D. Palmer, B. L. Scott and C. D. Tait, *Angew. Chem. Int. Ed. Engl.*, 1998, *37*, 164–166; (b) J. L. Sessler, D. Seidel, A. E. Vivian, V. Lynch, B. L. Scott and D. W. Keogh, *Angew. Chem. Int. Ed. Engl.*, 2001, *40*, 591–594.
38. N. N. Krot and M. S. Grigoriev, *Russ. Chem. Rev.*, 2004, *73*, 89–100.
39. N. Magnani, E. Colineau, R. Eloirdi, J. C. Griveau, R. Caciuffo, S. M. Cornet, I. May, C. A. Sharrad, D. Collison and R. E. P. Winpenny, *Phys. Rev. Lett.*, 2010, *104*, 197202(197204).
40. R. Copping, V. Mougel, C. Den Auwer, C. Berthon, P. Moisy and M. Mazzanti, *J. Chem. Soc. Dalton Trans.*, 2012, *41*, 10900–10902.
41. (a) P. L. Arnold, J. B. Love and D. Patel, *Coord. Chem. Rev.*, 2009, *253*, 1973–1978; (b) S. Y. Kim, T. Asakura, Y. Morita, G. Uchiyama and Y. Ikeda, *Radiochimica Acta*, 2005, *93*, 75–81.
42. (a) F. Burdet, J. Pecaut and M. Mazzanti, *J. Am. Chem. Soc.*, 2006, *128*, 16512–16513; (b) T. W. Hayton and G. Wu, *J. Am. Chem. Soc.*, 2008, *130*, 2005–2014.
43. (a) K. Mizuoka, S. Tsushima, M. Hasegawa, T. Hoshi and Y. Ikeda, *Inorg. Chem.*, 2005, *44*, 6211–6218; (b) K. Takao, S. Tsushima, S. Takao, A. C. Scheinost, G. Bernhard, Y. Ikeda and C. Hennig, *Inorg. Chem.*, 2009, *48*, 9602–9604; (c) G. Nocton, P. Horeglad, V. Vetere, J. Pecaut, L. Dubois, P. Maldivi, N. M. Edelstein and M. Mazzanti, *J. Am. Chem. Soc.*, 2010, *132*, 495–508.
44. V. Mougel, J. Pecaut and M. Mazzanti, *Chem. Commun.*, 2012, *48*, 868–870.
45. (a) V. Mougel, L. Chatelain, J. Pecaut, R. Caciuffo, E. Colineau, J. C. Griveau and M. Mazzanti, *Nature Chem.*, 2012, *4*, 1011–1017; (b) L. Chatelain, J. P. S. Walsh, J. Pecaut, F. Tuna and M. Mazzanti, *Angew.*

Chem. Int. Ed., 2014, *53*, 13434–13438; (c) L. Chatelain, J. Pecaut, F. Tuna and M. Mazzanti, *Chem. Eur. J.*, 2015, *21*, 18038–18042; (d) V. Mougel, L. Chatelain, J. Hermle, R. Caciuffo, E. Colineau, F. Tuna, N. Magnani, A. de Geyer, J. Pecaut and M. Mazzanti, *Angew. Chem. Int. Ed.*, 2014, *53*, 819–823.

46. J. C. Berthet, M. Nierlich and M. Ephritikhine, *Angew. Chem. Int. Ed. Engl.*, 2003, *42*, 1952–1954.
47. R. Faizova, R. Scopelliti, A.-S. Chauvin and M. Mazzanti, *J. Am. Chem. Soc.*, 2018, *140*, 13554–13557.
48. (a) G. Nocton, P. Horeglad, J. Pécaut and M. Mazzanti, *J. Am. Chem. Soc.*, 2008, *130*, 16633–16645; (b) L. Chatelain, V. Mougel, J. Pecaut and M. Mazzanti, *Chem. Sci.*, 2012, *3*, 1075–1079.
49. V. Mougel, P. Horeglad, G. Nocton, J. Pecaut and M. Mazzanti, *Chem. Eur. J.*, 2010, *16*, 14365–14377.
50. L. P. Spencer, E. J. Schelter, P. Yang, R. L. Gdula, B. L. Scott, J. D. Thompson, J. L. Kiplinger, E. R. Batista and J. M. Boncella, *Angew. Chem. Int. Ed.*, 2009, *48*, 3795–3798.
51. J. L. Brown, E. R. Batista, J. M. Boncella, A. J. Gaunt, S. D. Reilly, B. L. Scott and N. C. Tomson, *J. Am. Chem. Soc.*, 2015, *137*, 9583–9586.
52. E. A. Cuellar, S. S. Miller, T. J. Marks and E. Weitz, *J. Am. Chem. Soc.*, 1983, *105*, 4580–4589.
53. A. F. Cotton, O. D. Marler and W. Schwotzer, *Inorg. Chem.*, 1984, *23*, 4211–4215.
54. J. C. Berthet and M. Ephritikhine, *Coord. Chem. Rev.*, 1998, *178*, 83–116.
55. L. A. Seaman, S. Fortier, G. A. Wu and T. W. Hayton, *Inorg. Chem.*, 2011, *50*, 636–646.
56. K. Meyer, D. J. Mindiola, T. A. Baker, W. M. Davis and C. C. Cummins, *Angew. Chem. Int. Ed. Engl.*, 2000, *39*, 3063–3066.
57. L. A. Seaman, G. Wu, N. Edelstein, W. W. Lukens, N. Magnani and T. W. Hayton, *J. Am. Chem. Soc.*, 2012, *134*, 4931–4940.
58. (a) M. A. Silver *et al.*, *Science* 2016, *353*, 888; (b) T. Albrecht-Schmitt, *Nature Chem.*, 2014, *6*, 840–840.
59. (a) P. J. Panak and A. Geist, *Chem. Rev.*, 2013, *113*, 1199–1236; (b) M. A. Denecke, P. J. Panak, F. Burdet, M. Weigl, A. Geist, R. Klenze, M. Mazzanti and K. Gompper, *Comptes Rendus Chimie.*, 2007, *10*, 872–882.
60. A. J. Ryan, M. A. Angadol, J. W. Ziller and W. J. Evans, *Chem. Commun.*, 2019, *55*, 2325–2327.
61. A. J. Lewis, U. J. Williams, P. J. Carroll and E. J. Schelter, *Inorg. Chem.*, 2013, *52*, 7326–7328.

62. C. Camp, L. Chatelain, C. E. Kefalidis, J. Pecaut, L. Maron and M. Mazzanti, *Chem. Commun.*, 2015, *51*, 15454–15457.
63. A. J. Gaunt and M. P. Neu, *Comptes Rendus Chimie.*, 2010, *13*, 821–831.
64. Y. Sun, R. McDonald, J. Takats, V. Day and W. T. A. Eperspacher, *Inorg. Chem.*, 1994, *33*, 4433–4434.
65. E. M. Matson, W. P. Forrest, P. E. Fanwick and S. C. Bart, *J. Am. Chem. Soc.*, 2011, *133*, 4948–4954.
66. J. D. Rinehart and J. R. Long, *J. Am. Chem. Soc.*, 2009, *131*, 12558–12559.
67. I. Korobkov, S. Gambarotta and G. P. A. Yap, *Angew. Chem. Int. Ed.*, 2002, *41*, 3433–3436.
68. M. S. Dutkiewicz, J. H. Farnaby, C. Apostolidis, E. Colineau, O. Walter, N. Magnani, M. G. Gardiner, J. B. Love, N. Kaltsoyannis, R. Caciuffo and P. L. Arnold, *Nature Chem.* 2016, *8*, 797–802.
69. A. L. Ward, H. L. Buckley, W. W. Lukens and J. Arnold, *J. Am. Chem. Soc.*, 2013, *135*, 13965–13971.
70. J. L. Brown, A. J. Gaunt, D. M. King, S. T. Liddle, S. D. Reilly, B. L. Scott and A. J. Wooles, *Chem. Commun.*, 2016, *52*, 5428–5431.
71. (a) G. J.-P. Deblonde, M. Sturzbecher-Hoehne, P. B. Rupert, D. D. An, M.-C. Illy, C. Y. Ralston, J. Brabec, W. A. de Jong, R. K. Strong and R. J. Abergel, *Nature Chemistry*, 2017, *9*, 843–849; (b) S. K. Cary, M. A. Silver, G. K. Liu, J. C. Wang, J. A. Bogart, J. T. Stritzinger, A. A. Arico, K. Hanson, E. J. Schelter and T. E. Albrecht-Schmitt, *Inorg. Chem.*, 2015, *54*, 11399–11404.
72. W. G. Vandersluys and A. P. Sattelberger, *Chem. Rev.*, 1990, *90*, 1027–1040.
73. W. G. Vandersluys, C. J. Burns, J. C. Huffman and A. P. Sattelberger, *J. Am. Chem. Soc.*, 1988, *110*, 5924–5925.
74. S. M. Mansell, N. Kaltsoyannis and P. L. Arnold, *J. Am. Chem. Soc.*, 2011, *133*, 9036–9051.
75. J. M. Berg, D. L. Clark, J. C. Huffman, D. E. Morris, A. P. Sattelberger, W. E. Streib, W. G. Vandersluys and J. G. Watkin, *J. Am. Chem. Soc.*, 1992, *114*, 10811–10821.
76. L. R. Avens, S. G. Bott, D. L. Clark, A. P. Sattelberger, J. G. Watkin and B. D. Zwick, *Inorg. Chem.*, 1994, *33*, 2248–2256.
77. V. Mougel, C. Camp, J. Pecaut, C. Coperet, L. Maron, C. E. Kefalidis and M. Mazzanti, *Angew. Chem. Int. Ed.*, 2012, *51*, 12280–12284.
78. C. Camp, J. Pecaut and M. Mazzanti, *J. Am. Chem. Soc.*, 2013, *135*, 12101–12111.
79. C. Camp, C. E. Kefalidis, J. Pecaut, L. Maron and M. Mazzanti, *Angew. Chem. Int. Ed.*, 2013, *52*, 12646–12650.

80. H. S. La Pierre, A. Scheurer, F. W. Heinemann, W. Hieringer and K. Meyer, *Angew. Chem. Int. Ed.*, 2014, *53*, 7158–7162.
81. T. Le Borgne, E. Riviere, J. Marrot, J. J. Girerd and M. Ephritikhine, *Angew. Chem. Int. Ed.*, 2000, *39*, 1647–1649.
82. D. D. Schnaars, E. R. Batista, A. J. Gaunt, T. W. Hayton, I. May, S. D. Reilly, B. L. Scott and G. Wu, *Chem. Commun.*, 2011, *47*, 7647–7649.
83. M. Ephritikhine, *Coord. Chem. Rev.*, 2016, *319*, 35–62.
84. A. J. Gaunt, B. L. Scott and M. P. Neu, *Angew. Chem. Int. Ed.*, 2006, *45*, 1638–1641.
85. L. Karmazin, M. Mazzanti and J. Pécaut, *Chemical Commun.*, 2002, 654–655.
86. E. J. Schelter, R. L. Wu, B. L. Scott, J. D. Thompson, T. Cantat, K. D. John, E. R. Batista, D. E. Morris and J. L. Kiplinger, *Inorg. Chem.*, 2010, *49*, 924–933.
87. N. H. Anderson, S. O. Odoh, Y. Yao, U. J. Williams, B. A. Schaefer, J. J. Kiernicki, A. J. Lewis, M. D. Goshert, P. E. Fanwick, E. J. Schelter, J. R. Walensky, L. Gagliardi and S. C. Bart, *Nature Chem.*, 2014, *6*, 919–926.
88. A. Zalkin, J. G. Brennan and R. A. Andersen, *Acta Cryst. Sect. C*, 1988, *44*, 1553–1554.
89. C. J. Burns, W. H. Smith, J. C. Huffman and A. P. Sattelberger, *J. Am. Chem. Soc.*, 1990, *112*, 3237–3239.
90. I. Castro-Rodriguez, H. Nakai and K. Meyer, *Angew. Chem. Int. Ed.*, 2006, *45*, 2389–2392.
91. R. E. Jilek, L. P. Spencer, D. L. Kuiper, B. L. Scott, U. J. Williams, J. M. Kikkawa, E. J. Schelter and J. M. Boncella, *Inorg. Chem.*, 2011, *50*, 4235–4237.
92. D. M. King, J. McMaster, F. Tuna, E. J. L. McInnes, W. Lewis, A. J. Blake and S. T. Liddle, *J. Am. Chem. Soc.*, 2014, *136*, 5619–5622.
93. T. W. Hayton, J. M. Boncella, B. L. Scott and E. R. Batista, *J. Am. Chem. Soc.*, 2006, *128*, 12622–12623.
94. D. M. King and S. T. Liddle, *Coord. Chem. Rev.*, 2014, *266*, 2–15.
95. (a) W. J. Evans, S. A. Kozimor and J. W. Ziller, *Science*, 2005, *309*, 1835–1838; (b) G. Nocton, J. Pecaut and M. Mazzanti, *Angew. Chem. Int. Ed.*, 2008, *47*, 3040–3042.
96. D. M. King, F. Tuna, E. J. L. McInnes, J. McMaster, W. Lewis, A. J. Blake and S. T. Liddle, *Science*, 2012, *337*, 717–720.
97. R. K. Thomson, T. Cantat, B. L. Scott, D. E. Morris, E. R. Batista and J. L. Kiplinger, *Nature Chem.*, 2010, *2*, 723–729.
98. A. R. Fox and C. C. Cummins, *J. Am. Chem. Soc.*, 2009, *131*, 5716–5717.
99. A. R. Fox, P. L. Arnold and C. C. Cummins, *J. Am. Chem. Soc.*, 2010, *132*, 3250–3251.

100. (a) L. Chatelain, R. Scopelliti and M. Mazzanti, *J. Am. Chem. Soc.*, 2016, *138*, 1784–1787; (b) M. Falcone, L. Chatelain, R. Scopelliti, I. Živković and M. Mazzanti, *Nature*, 2017, *547*, 332–335.
101. (a) M. Falcone, L. Chatelain and M. Mazzanti, *Angew. Chem. Int. Ed.*, 2016, *55*, 4074–4078; (b) M. Falcone, C. E. Kefalidis, R. Scopelliti, L. Maron and M. Mazzanti, *Angew. Chem. Int. Ed.*, 2016, *55*, 12290–12294.
102. S. Fortier, G. Wu and T. W. Hayton, *J. Am. Chem. Soc.*, 2010, *132*, 6888–6889.
103. D. M. King, F. Tuna, E. J. L. McInnes, J. McMaster, W. Lewis, A. J. Blake and S. T. Liddle, *Nature Chem.*, 2013, *15*, 482–488.
104. B. M. Gardner, G. Balazs, M. Scheer, F. Tuna, E. J. L. McInnes, J. McMaster, W. Lewis, A. J. Blake and S. T. Liddle, *Angew. Chem. Int. Ed.*, 2014, *53*, 4484–4488.
105. B. M. Gardner, G. Balazs, M. Scheer, F. Tuna, E. J. L. McInnes, J. McMaster, W. Lewis, A. J. Blake and S. T. Liddle, *Nature Chem.*, 2015, *7*, 582–590.
106. S. Fortier, N. Kaltsoyannis, G. Wu and T. W. Hayton, *J. Am. Chem. Soc.*, 2011, *133*, 14224–14227.
107. (a) J. L. Brown, S. Fortier, R. A. Lewis, G. Wu and T. W. Hayton, *J. Am. Chem. Soc.*, 2012, *134*, 15468–15475; (b) D. E. Smiles, G. Wu and T. W. Hayton, *J. Am. Chem. Soc.*, 2014, *136*, 96–99.
108. A. J. Lewis, P. J. Carroll and E. J. Schelter, *J. Am. Chem. Soc.*, 2013, *135*, 13185–13192.
109. S. Fortier, J. L. Brown, N. Kaltsoyannis, G. Wu and T. W. Hayton, *Inorg. Chem.*, 2012, *51*, 1625–1633.
110. J. L. Brown, S. Fortier, G. Wu, N. Kaltsoyannis and T. W. Hayton, *J. Am. Chem. Soc.*, 2013, *135*, 5352–5355.
111. S. J. Kraft, J. Walensky, P. E. Fanwick, M. B. Hall and S. C. Bart, *Inorg. Chem.*, 2010, *49*, 7620–7622.
112. S. C. Bart, C. Anthon, F. W. Heinemann, E. Bill, N. M. Edelstein and K. Meyer, *J. Am. Chem. Soc.*, 2008, *130*, 12536–12546.
113. B. Kosog, H. S. La Pierre, F. W. Heinemann, S. T. Liddle and K. Meyer, *J. Am. Chem. Soc.*, 2012, *134*, 5284–5289.
114. O. Cooper, C. Camp, J. Pecaut, C. E. Kefalidis, L. Maron, S. Gambarelli and M. Mazzanti, *J. Am. Chem. Soc.*, 2014, *136*, 6716–6723.
115. S. T. Liddle, *Angew. Chem. Int. Ed.*, 2015, *54*, 8604–8641.
116. J. Andrez, J. Pecaut, R. Scopelliti, C. E. Kefalidis, L. Maron, M. W. Rosenzweig, K. Meyere and M. Mazzanti, *Chem. Sci.*, 2016, *7*, 5846–5856.

117. M. W. Rosenzweig, A. Scheurer, C. A. Lamsfus, F. W. Heinemann, L. Maron, J. Andrez, M. Mazzanti and K. Meyer, *Chem. Sci.*, 2016, *7*, 5857–5866.
118. (a) D. E. Smiles, G. Wu, N. Kaltsoyannis and T. W. Hayton, *Chem. Sci.*, 2015, *6*, 3891–3899; (b) D. E. Smiles, G. Wu, P. Hrobarik and T. W. Hayton, *J. Am. Chem. Soc.*, 2016, *138*, 814–825.
119. W. J. Evans, *Inorg. Chem.* 2007, *46*, 3435–3449.
120. D. P. Halter, F. W. Heinemann, J. Bachmann and K. Meyer, *Nature*, 2016, *530*, 317–321.
121. L. Gagliardi and B. O. Roos, *Nature*, 2005, *433*, 848–851.
122. S. C. Bart and K. Meyer, Highlights in uranium coordination chemistry. *In Struct. Bond.*, 2008, *127*, 119–176.

https://doi.org/10.1142/9781800610163_0006

Chapter 6

Organometallic Chemistry of Lanthanides

Wenliang Huang[*,†,‡] *and Paula L. Diaconescu*[*,§]

[*]*Beijing National Laboratory for Molecular Sciences,*
College of Chemistry and Molecular Engineering, Peking University,
Beijing 100871, P. R. China
[†]*Department of Chemistry & Biochemistry, University of California,*
Los Angeles, Los Angeles, CA 90095-1569, USA
[‡]*wlhuang@pku.edu.cn*
[§]*pld@chem.ucla.edu*

6.1 Introduction

Lanthanides are a group of elements that mainly differ by their size (Table 6.1) and share similar chemical properties. The group 3 metals (scandium and yttrium) also have similar chemical properties to the lanthanides and are often used as diamagnetic analogues for the ease of study via nuclear magnetic resonance spectroscopy and other characterization methods. They are usually grouped together with a collective name, rare-earth metals, a name given because of the difficulty of their mining from the earth crust, despite their relative abundance.[1] The organometallic chemistry of lanthanides has witnessed a renaissance in recent years together with that of their closely related cousins, actinides. This wave of discovery is driven by pure scientific curiosity as well as enthusiasm for applications in the fields of magnetism and optics. For instance, it has been long thought that lanthanides, except europium, ytterbium and samarium, do not exist in the divalent state in solution because the redox

Table 6.1. Effective ionic radii of Ln^{3+}[30] and electronic configurations of Ln^{3+} (in the order from smallest to largest ions, radioactive promethium excluded).

Ln	**Radius (Å)**	**Ln^{3+} electron configuration**	**Ln**	**Radius (Å)**	**Ln^{3+} electron configuration**
Sc	0.74	[Ar]	Tb	0.92	[Xe]$4f^8$
Lu	0.86	[Xe]$4f^{14}$	Gd	0.94	[Xe]$4f^7$
Yb	0.87	[Xe]$4f^{13}$	Eu	0.95	[Xe]$4f^6$
Tm	0.88	[Xe]$4f^{12}$	Sm	0.96	[Xe]$4f^5$
Er	0.89	[Xe]$4f^{11}$	Nd	0.98	[Xe]$4f^3$
Y	0.90	[Kr]	Pr	0.99	[Xe]$4f^2$
Ho	0.90	[Xe]$4f^{10}$	Ce	1.02	[Xe]$4f^1$
Dy	0.91	[Xe]$4f^9$	La	1.03	[Xe]

potential for Ln^{3+}/Ln^{2+} is so negative that Ln(II) would react with solvents, and thus resulting in decomposition.[2–5] However, a decade ago, molecular complexes containing lanthanide ions (lanthanum and cerium) in the +2 oxidation state were synthesized and characterized by M. F. Lappert and co-workers at the University of Sussex (Figure 6.1(a)).[6,7] Later on, W. J. Evans and co-workers at the University of California, Irvine, employed a similar strategy with optimized reaction conditions to expand this chemistry to most rare-earth metals (except for the smallest one, scandium)[8–12] and even actinides, uranium(II)[13–15] and thorium(II)[16] (Figure 6.1(b)). On the other hand, the pursuit of magnetic[17–20] and optical[21] materials with superior performance also calls for a burst in lanthanide chemistry. The intrinsic nature of lanthanide ions, i.e. their high single-ion spins and f-f transitions,[5] shows an unparalleled potential in areas ranging from quantum computing[17,19] to magnetic resonance imaging (MRI).[21] The two driving forces occasionally meet. An inspiring case is the isolation of lanthanide stabilized $(N_2)^{3-}$ radical trianionic complexes. Evans and co-workers have been studying dinitrogen activation by lanthanide complexes under strong reducing conditions for over two decades, with the first example being dinitrogen activation by $Cp^*{}_2Sm$ ($Cp^* = \eta^5$-C_5Me_5) to afford a bimetallic samarium dinitrogen complex, $(Cp^*{}_2Sm)_2(\mu$-$\eta^2{:}\eta^2$-$N_2)$ (Figure 6.1(c)).[22] Over the years, this group accomplished the synthesis of a series of $(Z_2Ln)_2(\mu$-$\eta^2{:}\eta^2$-$N_2)$ complexes using the combination of Z_3Ln/K (Z is a

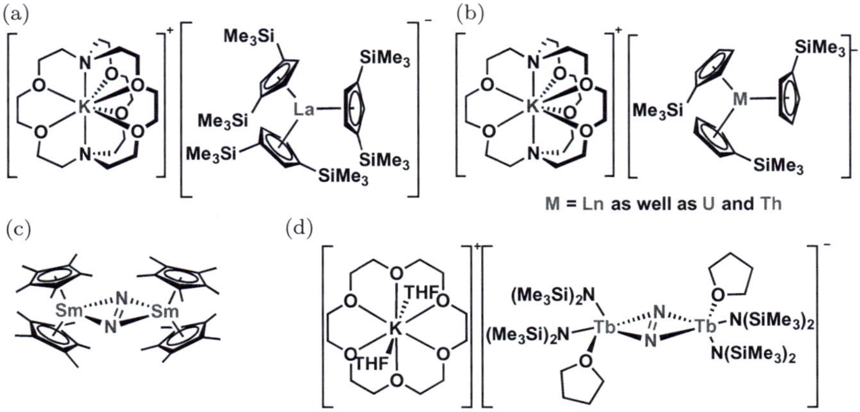

Figure 6.1. Recent milestones of organometallic lanthanide chemistry: (a) First crystallographically characterized molecular lanthanum(II) complex [K(crypt-222)][La(C_5H_3-1,3-($SiMe_3$)$_2$)$_3$]; (b) Molecular lanthanide(II) and uranium(II) and thorium(II) complexes [K(crypt-222)][M($C_5H_4SiMe_3$)$_3$]; (c) First f-element dinitrogen complex ($Cp^*{}_2Sm$)$_2$(μ-η^2:η^2-N_2); (d) Dinitrogen radical trianion terbium complex [K(18-crown-6)$_2$(THF)$_2$][(([(Me_3Si)$_2$N]$_2$Tb(THF))$_2$(μ-η^2:η^2-N_2)] as a single molecule magnet with a blocking temperature of 14 K.

monoanionic ligand such as N($SiMe_3$)$_2$ or cyclopentadienyl and its derivatives; Ln = most rare-earth metals except Eu and Yb).[4,23,24] In 2009, they achieved a new stage of dinitrogen activation, the radical trianion, $(N_2)^{3-}$, in [K(THF)$_6$][(([(Me_3Si)$_2$N]$_2$Y)$_2$(μ-η^2:η^2-N_2)] (Figure 6.1(d)).[25–27] Interestingly, through collaboration with the Long group at the University of California, Berkeley, the dysprosium and terbium analogues were found to be single-molecule magnets with the highest blocking temperature of 14 K at the time.[28,29] This synergistic effort shows the power of combining two fundamental scientific approaches: one driven by the desire of synthetic chemists to push the boundary of knowledge, and the other motivated by the pursuit of potential applications.

This chapter will highlight the recent accomplishments achieved by synthetic chemists on advancing the knowledge of organometallic lanthanide chemistry. The content will be categorized in four sections. In Section 6.2, we will show recent breakthroughs in stabilizing molecular complexes of lanthanide ions in previously considered unlikely oxidation states, such as the crystallographically characterized Ln(II) complexes. In Section 6.3, organometallic lanthanide complexes containing Ln—C σ-bonds will be summarised with the emphasis on their use as synthetic precursors. In Section 6.4, lanthanide complexes supported by π-ligands will be discussed, including carbon-based cyclic (including cyclopentadienyl, cyclooctatetraene dianion and reduced arenes) and acyclic (including alkene, diene and alkyne) π-ligands. In Section 6.5, we will present another emerging field, that of lanthanide carbene complexes, including both Fischer and Schrock-type carbenes. Overall, we intend to present a comprehensive discussion of the rapidly evolving organometallic lanthanide chemistry with a focus on fundamental aspects, such as the nature of bonding interactions and periodic trends, i.e. generality and individuality across the lanthanide series.

6.2 Organometallic Chemistry of Lanthanide Ions in Unconventional Oxidation States

6.2.1 *Conventional oxidation states for organometallic lanthanide chemistry*

The rare-earth metals, scandium, yttrium and lanthanides, have long been considered redox inert. Due to the lanthanide contraction, 4f orbitals are much lower in energy than 5d and 6s orbitals, except for La and Ce, for which 5d orbitals are lower in energy than 4f orbitals.[2,3] This leads to a general electronic configuration for gas phase lanthanide atoms of [Xe]$4f^n6s^2$, except for La, [Xe]$5d^16s^2$, Ce, [Xe]$4f^15d^16s^2$ and Gd, [Xe]$4f^75d^16s^2$ (the former two exceptions are the result of the relative energy levels of the 4f and 5d orbitals, while the latter is due to Hund's rule of half-shell). When forming molecular complexes, lanthanides readily lose three electrons to form a +3 cation, while the fourth electron in the 4f orbital resists oxidation. Therefore, only cerium has a relatively accessible +4 oxidation state with the redox potential of the Ce^{4+}/Ce^{3+} pair at +1.61 V (vs. NHE);[31] cerium ammonia nitrate (CAN) is a widely used oxidant in organic synthesis.[32] Praseodymium and terbium do exist in higher oxidation states in their oxides, such as the commercially available oxides, Pr_4O_9 and Tb_4O_7. However, the redox potentials of Pr^{4+}/Pr^{3+} and Tb^{4+}/Tb^{3+} are over +3.0 V[3] and, as such, no example of organometallic praseodymium(IV) and terbium(IV) complexes are known. Even for cerium, few examples of organometallic cerium(IV) complexes have been reported; they are usually coordination complexes with electronegative nitrogen, oxygen or halogen ligands. The difficulty in developing the organometallic chemistry of lanthanides in a higher oxidation state, i.e. tetravalent, also lies in the fact that carbon-based ligands in organometallic chemistry are usually so electron rich that they will be oxidized by Ln(IV) ions spontaneously with the concomitant formation of Ln(III) species.[33] Nonetheless, some recent advances in the stabilization of cerium(IV) ions show the potential for developing high valent (+4) organometallic lanthanide chemistry.[34–37]

On the low valent side, europium and ytterbium have relatively accessible +2 oxidation states because Eu(II) and Yb(II) ions meet Hund's rule. As a result, Eu(II) and Yb(II) are common in the organometallic chemistry of those elements; in fact, attempts to synthesize organometallic Eu(III) complexes containing Ln–C σ-bonds proved to be challenging.[38] The redox potential for Sm^{3+}/Sm^{2+} at −1.55 V (vs. NHE) is also reasonably accessible.[3] In fact, an organometallic Sm(II) complex was first synthesized in 1969,[39] merely 15 years after the first organometallic lanthanide complexes were reported.[40] However, the next in line, Tm(II), is further away from Sm(II) according to the estimated redox potential for the Ln^{3+}/Ln^{2+} pairs (Table 6.2). Despite a report of a proposed cerium(II) complex, $[K(DME)_n]_2[Ce(COT)_2]$ (DME = dimethoxyethane, COT = cyclooctatetraenyl dianion), whose structure has not been determined,[41] it was considered impossible to isolate any stable molecular Ln(II) complexes beyond Sm(II) even as late as the 1990s.[4]

6.2.2 *Synthesis of molecular divalent thulium, dysprosium and neodymium complexes*

Nothing can stop fearless synthetic chemists (a nice story behind this discovery can be found in Ref. [3]): the first molecular thulium(II) complex was the product of a collaboration between the group of Mikhail N. Bochkarev in Nizhny Novgorod, Russia, and the Evans group in California, USA.[42] The molecule, with a formula as simple as $TmI_2(DME)_3$, was characterized by X-ray crystallography (Scheme 6.1(a)). The isolation of this Tm(II) complex was truly

Table 6.2. Calculated redox potentials of the Ln^{3+}/Ln^{2+} pairs vs. NHE[3] and electronic configurations of Ln^{2+} (in the order from the least negative to the most negative, NA = not available).

Ln	Ln^{3+}/Ln^{2+} vs. NHE (V)	Ln^{2+} electron configuration	Ln	Ln^{3+}/Ln^{2+} vs. NHE (V)	Ln^{2+} electron configuration
Eu	−0.35	$[Xe]4f^7$	Er	−3.1	$[Xe]4f^{12}$
Yb	−1.15	$[Xe]4f^{14}$	La	−3.1	$[Xe]4f^1$
Sm	−1.55	$[Xe]4f^6$	Ce	−3.2	$[Xe]4f^2$
Tm	−2.3	$[Xe]4f^{13}$	Tb	−3.7	$[Xe]4f^9$
Dy	−2.5	$[Xe]4f^{10}$	Gd	−3.9	$[Xe]4f^8$
Nd	−2.6	$[Xe]4f^4$	Lu	NA	$[Xe]4f^{14}5d^1$
Pr	−2.7	$[Xe]4f^3$	Y	NA	$[Kr]4d^1$
Ho	−2.9	$[Xe]4f^{11}$	Sc	NA	$[Ar]3d^1$

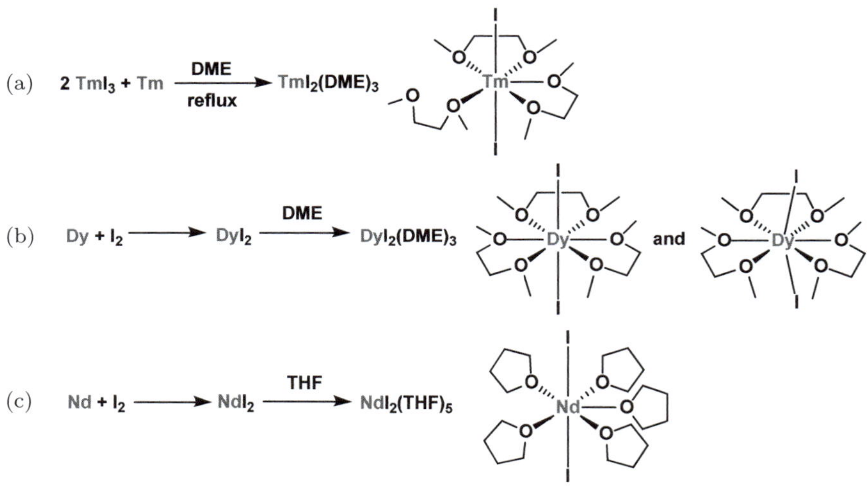

Scheme 6.1. Synthesis of non-classical divalent molecular lanthanide complexes of (a) thulium, (b) dysprosium and (c) neodymium.

encouraging: since the huge gap between Sm and Tm had been overcome, the relatively small difference between Tm, Dy and Nd was likely not to be much of an obstacle, as turned out to be the case: in 2000 and 2001, Evans and Bochkarev's groups published the synthesis and characterization of $DyI_2(DME)_3$[43] and $NdI_2(THF)_5$[44] (Scheme 6.1(b),(c)), respectively. More examples emerged since then,[45–49] summarized in a review by F. Nief in 2010, in which divalent thulium, dysprosium and neodymium complexes were called non-classical divalent lanthanide complexes in order to differentiate them from those considered classical, i.e. europium, ytterbium and samarium.[50] It is interesting that the success of synthesizing divalent thulium, dysprosium, and neodymium complexes can be traced back to solid-state chemistry. Gerd Meyer at Köln, Institut für Anorganische Chemie der Universität, had studied subvalent lanthanide halides for decades and discovered that albeit LnI_2 can be made for all rare-earth metals, they exist in two different forms: one behaves like a normal salt, for Eu, Yb, Sm, Tm, Dy and Nd; the other behaves like a metal, for the rest of lanthanides, scandium and yttrium. The different physical and chemical properties of the two types of LnI_2 rise from their difference in the electronic configuration of the corresponding lanthanide ions: while the former represents a genuine Ln^{2+} with a $[Xe]4f^{n+1}$ electron configuration, the latter has a $[Xe]4f^{n}5d^{1}$ electron configuration and is better described as $(Ln^{3+})(e^{-})(I^{-})_2$. The electron in the 5d orbitals forms a conduction band and thus results in metal-like properties.[51–54] The solid-state chemistry of LnI_2 echoes the developments in molecular divalent lanthanide chemistry.[4,11,46,50,55,56] However, since the solid-state chemistry suggests that the line separating genuine $Ln(II)I_2$ salts and metallic $Ln(III)(e)I_2$ be drawn at neodymium, does that mean that other lanthanides are excluded from forming stable divalent molecular complex? Before answering this question, an important method, reductive pathway, should be discussed, because this method is the key to opening new territories in synthetic low-valent lanthanide chemistry.[4,50,55,56]

6.2.3 *The reductive pathway and completion of molecular divalent complexes for the whole lanthanide series*

There are two ways for entry to molecular divalent lanthanide chemistry. One approach is by salt metathesis of inorganic Ln(II) starting materials, such as YbI_2, SmI_2 and LnI_2 as shown in Scheme 6.2, with anionic organic ligands, e.g. cyclopentadienyl. Although this method is straightforward, it has several limitations. First, only genuine Ln(II) halides can be employed in the salt metathesis reaction

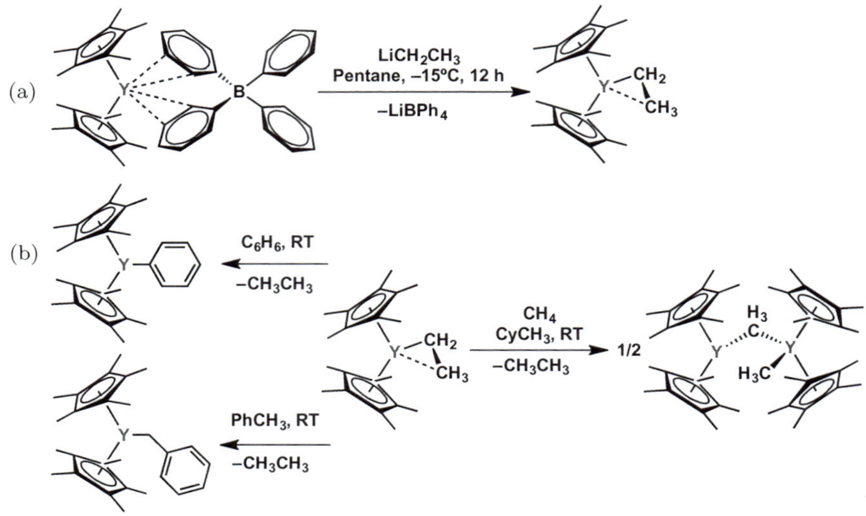

Scheme 6.2. (a) Synthesis of $Cp^*{}_2Y(CH_2CH_3)$; (b) Reactivity of $Cp^*{}_2Y(CH_2CH_3)$ toward hydrocarbons.

because the metallic $Ln(III)(e)I_2$ are not soluble in organic solvents.[50] Second, upon dissolving them in solution, non-classical Ln(II) ions are so reactive that they may react with solvents or dinitrogen (!) to form Ln(III) complexes instead of undergoing salt metathesis to form stable organometallic divalent lanthanide complexes.[43,45] Therefore, another approach, called reductive pathway, became practical in such cases. In the reductive pathway, an organometallic trivalent lanthanide precursor is first synthesized and then treated with strong reductants, such as alkali metals or their alternatives, e.g. potassium graphite (KC_8) or lithium naphthalenide ($LiC_{10}H_8$). The logic behind the reductive pathway is that although a Ln(II) complex is thermally unstable, its kinetic stability can be realized by a pre-established coordination environment. One way to achieve this goal is to encapsulate the lanthanide ion with sterically demanding ligands. Since the most commonly used ancillary ligand for large lanthanide ions is cyclopentadienyl (Cp), it is natural that the first choice of ligands is substituted cyclopentadienyls, among which methyl, *tert*-butyl, and trimethylsilyl are the most common substituents. Some widely used ligands are: Cp* (C_5Me_5), C_5Me_4H, Cp^{tt} ($C_5H_3(t\text{-butyl})_2$-1,3),

Cp′ ($C_5H_4SiMe_3$), and Cp″ ($C_5H_3(SiMe_3)_2$-1,3). Larger ligands, like Cp^{ttt} (C_5H_2(*t*-butyl)$_3$-1,2,4) and Cp‴ ($C_5H_2(SiMe_3)_3$-1,2,4), have also been used in reductive chemistry but rarely.[48] $(Cp^X)_3Ln$ (Cp^X represents any of these Cp derivatives) can be synthesized by salt metathesis of LnX_3 (X = halide) and MCp^X (M = alkali metal). In these complexes, the lanthanide ion is completely shielded by three bulky Cp^X ligands. Lappert was a pioneer in exploiting the reductive pathway. In 1998, his group reported the isolation of a bimetallic lanthanum complex, $[(Cp^{tt})_2La(\mu\text{-}C_6H_6)La(Cp^{tt})_2][K(18\text{-crown-6})(C_6H_6)_2]$, by reacting Cp^{tt}_3La with potassium mirror in benzene in the presence of a cationic trapping agent, 18-crown-6.[57] In the report, the authors suggested that the compound be formulated as two La(II) ions bridged by a benzene radical monoanion since the alternative formulation of two La(III) ions bridged by a benzene radical trianion was considered less plausible. Unfortunately, due to the non-innocence of the benzene ligand (which will be discussed in Section 6.4), the oxidation state of lanthanum could not be assigned rigorously. A similar case can be found in inverse sandwich actinide benzene complexes.[58–60] Ten years later, after numerous attempts, Lappert *et al.* eventually synthesized the first unambiguous molecular complex of divalent lanthanum.[6] By using a different ligand, Cp″, and a different solvent, tetrahydrofuran (THF), in the presence of crypt-222 (4,7,13,16,21,24-hexaoxa-1,10-diazabicyclo[8.8.8]hexacosane), they were able to isolate [K(crypt-222)][$Cp_3''La$]. Similarly, [K(18-crown-6)(Et_2O)][$Cp_3''La$], as well as co-crystals of [K(18-crown-6)(Et_2O)][$Cp_3''Ce$] · [$Cp_3''Ce$], could be obtained by this method. The X-ray crystallography study of [K(crypt-222)][$Cp_3''La$] clearly supported the formulation of a divalent lanthanum compound (Figure 6.2(a)). Surprisingly, the average distance of La-$C_{centroid}$ is 2.62 Å, barely longer than that of the trivalent precursor, $Cp_3''La$. DFT calculations unfolded that La(II) had a [Xe]$5d^1$ electron configuration rather than [Xe]$4f^1$. This discovery is important because all previous molecular divalent lanthanide species have a [Xe]$4f^n$ electron configuration. Thus, those findings opened up a whole new world for divalent lanthanide chemistry: now that La(II) was achieved, in principle, a lot of other lanthanides should exist in the +2 oxidation state as well, since according to Table 6.2, lanthanum is not right after dysprosium but rather in the middle of the series.

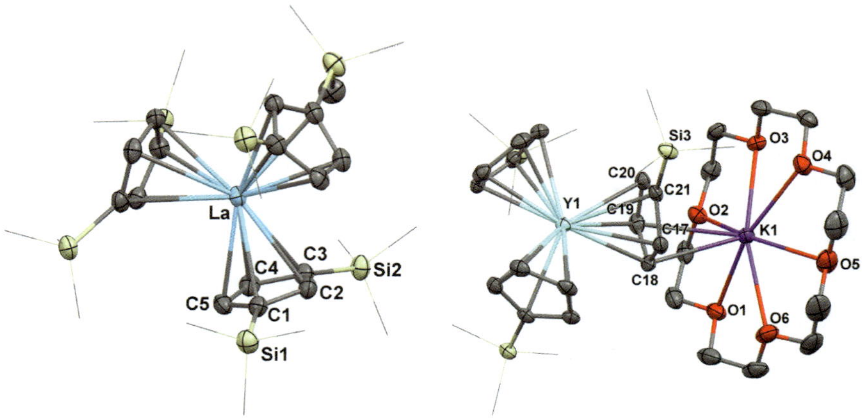

(a) (b)

Figure 6.2. Molecular structures of $[K(crypt\text{-}222)][Cp''_3La]$ (a) and $[K(18\text{-}crown\text{-}6)][Cp'_3Y]$ (b). Adapted with permission from Ref. [6], Copyright (2008) WILEY-VCH Verlag GmbH & Co. KGaA, Weinheim. And Ref. [8], Copyright (2011) American Chemical Society.

That turned out to be the case. Evans's group has since carried out this chemistry and expanded the scope of divalent lanthanides to almost all rare-earth metals except scandium and radioactive promethium.[8–10,12] In addition, this method could be applied to actinides to afford the first genuine uranium(II) and thorium(II) complexes, $[K(crypt\text{-}222)][Cp'_3U]$[13,14] and $[K(crypt\text{-}222)][Cp''_3Th]$.[16] It is important to note that uranium(II) and thorium(II) ions in those complexes have unusual electronic configurations, $[Rn]5f^36d^1$ and $[Rn]6d^2$, but these configurations are similar to those of the lanthanide series. Recently, this chemistry was further extended to plutonium(II) through the synthesis of $[K(crypt\text{-}222)][Cp''_3Pu]$.[61] Details of the divalent actinide chemistry will be disclosed in the actinide sections. Instead of using the Cp″ ligand, Evans *et al.* chose a less sterically demanding ligand, Cp′, to fit lanthanides of all sizes, from La to Lu. They also found that the choice of cyclopentadienyl ligand is critical to the result of the reaction. In the case of lanthanum, $[Cp''_3La]^-$ could be synthesized even under a dinitrogen atmosphere, meaning that the divalent lanthanum in $[Cp''_3La]^-$ is stable under N_2. This was attributed to a kinetic barrier instead of thermodynamic stability since (1) $[Cp''_2Tm]_2(\mu\text{-}\eta^2{:}\eta^2\text{-}N_2)$ and $[Cp''_2Dy]_2(\mu\text{-}\eta^2{:}\eta^2\text{-}N_2)$ were known;[45] (2) the lanthanum dinitrogen complex $[Cp^*{}_2La(THF)]_2$

(μ-η^2:η^2-N_2) could be obtained by KC_8 reduction of the trivalent precursor [$Cp^*{}_2La$][(μ-Ph)$_2BPh_2$] under a dinitrogen atmosphere;[24] (3) La^{2+} is more reducing than Dy^{2+} and Tm^{2+}, as shown by the redox potential values in Table 6.2. This proposal was tested experimentally and, indeed, when adding 2 equiv. of KCp^* to a $[Cp''_3La]^-$ solution under a dinitrogen atmosphere, the known $[Cp^*{}_2La(THF)]_2(\mu$-η^2:η^2-$N_2)$ was obtained.[8] Therefore, the ligand choice among cyclopentadienyl derivatives does make a big difference in reductive chemistry.[62] Cp′ seemed to be a good fit for stabilizing divalent lanthanide ions because (1) the trimethylsilyl substituent can provide not only steric hindrance but also stabilize the cyclopentadienyl anion through a p-σ^* interaction;[63] (2) unlike the more sterically demanding Cp*[64] and Cp″,[8] Cp'_3Ln can be made for most lanthanides from the largest lanthanum to the smallest lutetium, allowing a systematic study for the entire lanthanide series.[11,65] Evans *et al.* first synthesized a divalent yttrium complex, [K(18-crown-6)][Cp'_3Y] (Figure 6.2(b)),[8] and later expanded the chemistry to other lanthanides, while finding that crypt-222 is a better ligand to encapsulate the potassium cation.[10] In the process of completing the whole series of divalent lanthanides, it was found that they all have a $4f^n5d^1$ electronic configuration, consistent with the electronic configuration of the lanthanide ions in the solid-state LnI_2. This triggered the authors to explore the barrier between conventional ($4f^{n+1}$) and unconventional ($4f^n5d^1$) divalent lanthanide ions in molecular lanthanide complexes by a systematic study of Cp'_3Ln, including those already known to form stable molecular divalent complexes, i.e. Eu, Yb, Sm, Tm, Dy and Nd. Besides DFT calculations, another criterion to determine the ground state of the lanthanide ion is the elongation of the Ln–$Cp_{centroid}$ distance in anionic $[Cp'_3La]^-$ compared to neutral Cp'_3La (Figure 6.3). For conventional divalent lanthanides, Eu, Yb and Sm, this elongation is significant (>0.10 Å) and comparable to the difference of the ionic radii between Ln^{2+} and Ln^{3+};[30] for lanthanides having the Ln^{3+}/Ln^{2+} redox potential more negative than that of neodymium, the elongation is minimal (<0.05 Å). The same small difference is found for metal-ligand distances of d-block transition metal complexes. For instance, Cp_2TiCl_2[66] and $[Cp_2Ti(\mu - Cl)]_2$[67] have an average Ti-$Cp_{centroid}$ distance of 2.370 and 2.350 Å, respectively. Similarly, Cp_4Zr[68] and Cp_3Zr[69] both have the same average Zr-$Cp_{centroid}$

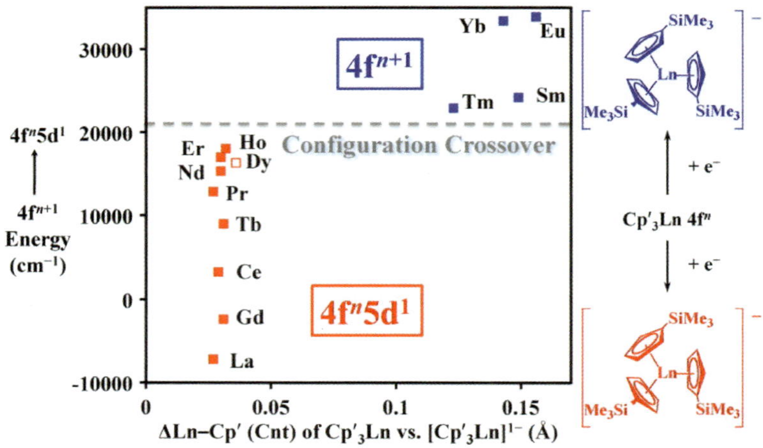

Figure 6.3. Plot of the $4f^n$ to $4f^n5d^1$ promotion energies (only an estimated energy is available for Dy) vs. the differences in the Ln–($Cp'_{centroid}$) distances of $[Cp'_3Ln]^-$ and $[Cp'_3Ln]$. The gray dashed line indicates the barrier in promotion energies to reduce the $4f^n$ $[Cp'_3Ln]$ to the $4f^{n+1}$ (blue squares on right) or $4f^n5d^1$ (red squares on left) configuration of $[Cp'_3Ln]^-$. Reprinted with permission from Ref. [11]. Copyright (2015) American Chemical Society.

distance of 2.58 Å. While 4f orbitals are contracted toward the nucleus and not impacted by the ligand field, 5d orbitals are diffused and subjected to ligand field as well as bonding interactions. Therefore, the small difference of the Ln–$Cp'_{centroid}$ distance upon reduction suggests that the extra electron goes into 5d instead of 4f orbitals. DFT calculations also support this view: the energy required for the $4f^{n+1}$ to $4f^n5d^1$ transition is small for lanthanides with a highly negative Ln^{3+}/Ln^{2+} redox potential but large for conventional divalent lanthanides (Figure 6.3). An interesting thing to note is the presence of a subgroup of lanthanides having a medium negative value of the Ln^{3+}/Ln^{2+} redox potential, namely thulium, dysprosium, and neodymium. In the solid state, they all behave similarly and as genuine $Ln(II)I_2$ salts. However, in the molecular divalent complexes $[Cp'_3Ln]^-$, dysprosium and neodymium showed a small difference in distances upon reduction and a $4f^n5d^1$ electron configuration in the ground state that is distinct from that of the solid-state $Ln(II)I_2$ salts. In addition, even for Tm^{2+}, the difference in distances is smaller when compared to conventional Ln^{2+}, indicating some mixing of $4f^{13}$ and $4f^{12}5d^1$ configurations in the ground state.[11]

Evans *et al.* carried out an investigation of the Cp^x ligand effects,[62] the magnetic properties,[70] as well as the chemical reactivity of $[Cp'_3Ln]^-$ toward arenes, including naphthalene, biphenyl and benzene.[71,72] Those preliminary studies showed a great potential of divalent lanthanide chemistry and physics. The arene complexes obtained by reduction with $[Cp'_3Ln]^-$ will be discussed in Section 6.4. Another class of molecular subvalent lanthanide complexes, zero-valent arene sandwich complexes,[73–77] credited to F. G. Cloke and co-workers at the University of Sussex, United Kingdom, will also be included in Section 6.4 since they can be classified as lanthanide arene complexes.

In addition to the isolation of stable divalent lanthanide complexes, the reductive pathway can be used to achieve masked Ln(II) chemistry by combining a trivalent lanthanide precursor and strong reducing agents such as alkali metals. This strategy was most widely used in the lanthanide mediated dinitrogen reduction developed by Evans's group.[4,23,24] Since this content is best described as small molecule activation by lanthanides, it will not be elaborated here.

6.2.4 *Outlook for organometallic lanthanide chemistry in unconventional oxidation states*

The isolation of stable molecular divalent lanthanide complexes for all lanthanides greatly expands the breadth and depth of organometallic lanthanide chemistry. As G. Meyer stated in the highlight titled "Superbulky Ligands and Trapped Electrons: New Perspectives in Divalent Lanthanide Chemistry" in 2008: "Furthermore, as lanthanum(I) is known in the solid state as LaI,[78,79] why should it not be possible to realize molecular lanthanum(I) compounds in solution? And, with the aforementioned cerium(II) compound as well as cerium's well-established oxidation state +4 in mind, two electron reduction processes, which are so important in transition-metal chemistry, are perhaps in reach."[7] In addition to low-valent lanthanides, organometallic cerium(IV) chemistry may soar, while terbium(IV) and praseodymium(IV) chemistries are potentially accessible with a sophisticated ligand design. All of these may open a brand new territory for synthetic lanthanide chemistry, as well as lanthanide mediated catalysis, and lanthanide-based magnetic and luminescent materials.[70,80]

6.3 Lanthanide Complexes Containing Ln-C σ-Bonds

6.3.1 *General considerations on Ln-C σ-bonds*

By the classical definition, organometallic lanthanide compounds should contain at least one Ln–C bond. The Ln–C interaction is mainly electrostatic in nature and usually delocalised over a multi-atom π system. For example, unlike transition metals, which form a covalent interaction with the cyclopentadienyl ligand, e.g. ferrocene, the Ln–Cp interaction is mainly ionic and there is little if any orbital interaction between the lanthanide d or f orbitals and the π orbitals of the Cp ligand. Furthermore, the Ln–C σ-bond is considered to be a highly polarized covalent bond, similar to organolithium (RLi) or Grignard reagents (RMgBr),[81] since the electronegativity of lanthanides (from 1.36 of scandium to 1.1 of lanthanum) is in between that of lithium (0.98) and magnesium (1.31, all by the Pauling scale).[82,83] Therefore, the Ln–C σ-bond is truly a mismatch between the electropositive lanthanide ion and carbanions and subject to decomposition processes such as protonation or oxidation. Organometallic lanthanide complexes containing Ln–C σ-bonds were once considered beyond reach,[84] until chemists developed multiple approaches to stabilize them.[85–87] These approaches include: (1) Forming Lewis base adducts that can lead to mononuclear species and improve solubility in organic solvents (important for crystallization); the addition of a Lewis base also saturates the coordination sphere of lanthanide ions. (2) Introducing large silyl substituents at the α-position, such as CH_2SiMe_3, $CH(SiMe_3)_2$ and $C(SiMe_3)_3$, which can provide steric protection around the lanthanide ion; at the same time, silyl substituents can further stabilize the carbanion through p-σ^* interactions and prevent potential decomposition pathways such as β-H or β-Me elimination. (3) Adding an adjacent π system such as benzene not only stabilizes the carbanion by delocalising the negative charge through conjugation but also forms η^2 or η^3-interactions resulting in bending of the Ln–C–C_{ipso} angle (in order to allow a better overlap with the π system). (4) Exploiting chelating alkyl ligands with a tethered neutral donor, e.g. NMe_2, OMe or $SiMe_3$ (agostic interaction), can endow thermal as well as kinetic stability; (5) Forming adducts with Lewis acids like MR_3 (M = group 13 elements, R = alkyl); the resulting lanthanide aluminate complexes behave similarly as lanthanide alkyl complexes and the latter can be generated *in situ* from the aluminate by the addition of a stoichiometric amount of a Lewis base to remove MR_3; (6) Forming mono or dialkyl lanthanide complexes supported by π ligands like Cp* leads to far more stable complexes than homoleptic trialkyls. The organization of this section will be based on the nature of the alkyl groups: (1) simple alkyl, such as methyl, *tert*-butyl and *neo*-pentyl; (2) alkyl with silyl substituents; (3) benzyl and its derivatives; (4) tetraalkylaluminates; (5) aryl; (6) alkynide. We also want to direct readers to some excellent comprehensive reviews on Ln–C σ bond complexes for further details and references.[88,89]

6.3.2 *Simple alkyl complexes*

6.3.2.1 *Methyl*

Lanthanide methyl complexes are well known for their reactivity in the σ-bond metathesis reaction[90,91] and considered to be the most reactive among lanthanide alkyl complexes;[89] however, homoleptic lanthanide trimethyl complexes proved difficult to elucidate. No X-ray structure of $LnMe_3$ or its polymeric form $[LnMe_3]_n$ has been reported.[89] Schumann and co-workers were pioneers in the search of permethylated lanthanide complexes and succeeded in synthesizing lanthanide methyl ate complexes, $[Li(donor)_x]_3[LnMe_6]$.[87] The stability of ate methyl complexes over the corresponding neutral counterpart is due to both electronic and steric reasons: electronically, the high Lewis acidity of the lanthanide ion is quenched by the additional donation from the methyl anion; sterically, the fulfilment of the octahedral coordination sphere around a large lanthanide ion prevents an attack from external molecules. About two decades later, Anwander and co-workers succeeded in preparing quantitatively pure $[LnMe_3]_n$ (Ln = Y and Lu) by treating homoleptic $[Ln(AlMe_4)_3]$ with an equimolar amount of tetrahydrofuran (per aluminum) and characterized them by elemental analysis, IR (Infrared), and ^1H and ^{13}C MAS (magic-angle-spinning) NMR spectroscopy, despite their polymeric nature in the solid state and insolubility in non-coordinating organic solvents.[92] Although neutral $LnMe_3$ and $LnMe_3(donor)_x$ have not been characterized crystallographically, the corresponding cationic lanthanide dimethyl and dicationic lanthanide methyl species could

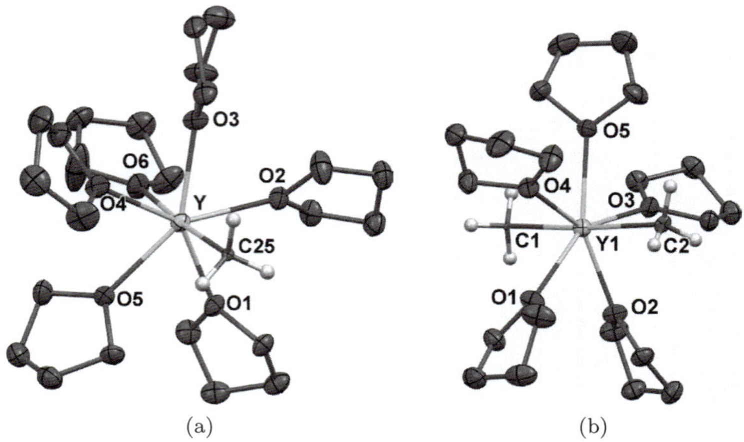

Figure 6.4. Molecular solid-state structure of cation $[YMe_2(THF)_5]^+$ in $[YMe_2(THF)_5][BPh_4]$ (a) and dication $[YMe(THF)_6]^{2+}$ in $[YMe(THF)_6][BPh_4]_2$ (b). Adapted with permission from Ref. [93]. Copyright (2005) Wiley-VCH Verlag GmbH & Co. KGaA, Weinheim and with permission from Ref. [95]. Copyright (2003) Wiley-VCH Verlag GmbH & Co. KGaA, Weinheim.

be obtained and structurally characterized.[93,94] $[YMe_2(THF)_5]^+$ in $[YMe_2(THF)_5][BPh_4]$ (Figure 6.4(a))[93] and $[YMe(THF)_6]^{2+}$ in $[YMe(THF)_6][BPh_4]_2$ (Figure 6.4(b))[95] are depicted as examples. Due to its instability, $LnMe_3$ has not been exploited as a synthetic precursor for mono- or dimethyl complexes as much as other homoleptic lanthanide alkyl complexes. Lanthanide tetramethyl aluminate complexes, $Ln(AlMe_4)_3$, are used as an alternative; lanthanide methyl complexes are obtained by treating the tetramethylaluminate complexes with a Lewis base to release $AlMe_3$(donor). The merit of lanthanide tetraalkylaluminate complexes will be discussed in Section 6.3.7.

6.3.2.2 *Other simple alkyls*

Lanthanide complexes of linear alkyls are even more rare than methyl complexes because they usually undergo fast β-H elimination or other decomposition reactions and are only stable in solutions of coordinating solvents.[96] For example, an early attempt by Watson *et al.* to isolate a stable metallocene lutetium ethyl complex resulted in obtaining a hydride species with the concomitant loss of ethylene.[90] Some crystallographically characterized lanthanide ethyl

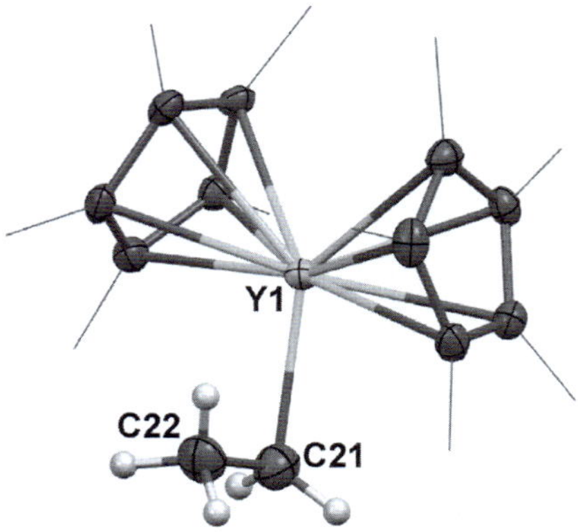

Figure 6.5. Molecular structure of $Cp^*{}_2Y(CH_2CH_3)$. Hydrogen atoms on Cp* rings are omitted for clarity. Adapted with permission from Ref. [96]. Copyright (2015) American Chemical Society.

or diethyl complexes are known when supported by bulky chelating non-metallocene ligands.[97–99] Recently, Evans *et al.* reported the synthesis, structure and reactivity of the donor-free ethyl yttrium metallocene complex, $Cp^*{}_2Y(CH_2CH_3)$.[96] The synthesis required a suitable starting material, $Cp^*{}_2Y(\mu\text{-}Ph)_2BPh_2$, and the right reaction conditions, such as low temperature, $LiCH_2CH_3$ and *n*-pentane as a solvent (Scheme 6.2(a)). The molecular structure of $Cp^*{}_2Y(CH_2CH_3)$ showed a β-agostic interaction between the coordinatively unsaturated yttrium ion and the terminal methyl group that can be considered an intermediate for β-H elimination (Figure 6.5). The yttrium ethyl complex is extremely reactive toward hydrocarbons, such as methane, benzene and toluene (Scheme 6.2(b)). The success in the synthesis of the ethyl yttrium metallocene complex demonstrated the potential of synthetic lanthanide chemistry to afford highly reactive and once considered elusive species.

One strategy to stabilize the Ln–C σ-bond is to use steric demanding alkyl groups. Therefore, in principle, though containing β-hydrogens, lanthanide *tert*-butyl complexes are relatively stable.[100] Despite the fact that the neutral Ln^tBu_3 remains elusive, the lithium ate complexes $[Li(donor)_x][Ln^tBu]_4$ could be obtained by salt metathesis between $LnCl_3$ and Li^tBu.[100,101] With a sterically demanding ^tBu ligand, the coordination number of the lanthanide ion is down to 4 from 6 in $[Li(donor)_x]_3[LnMe_6]$.

Another obvious choice is *neo*-pentyl, a sterically bulky alkyl without any β hydrogen. Lappert and Pearce first reported the synthesis of $Ln(CH_2^tBu)_3(THF)_2$ (Ln = Sc, Y) together with the *neo*-silyl analogue $Ln(CH_2SiMe_3)_3(THF)_2$, which later became popular as a synthetic precursor for lanthanide alkyl complexes (see below).[102] $Ln(CH_2^tBu)_3(THF)_2$ was prepared by reacting $LnCl_3$ with $LiCH_2^tBu$ and the structure was anticipated as trigonal bipyramidal according to the ^1H NMR spectrum. However, a solid-state structure determination only took place 30 years later for $Yb(CH_2^tBu)_3(THF)_2$, which was prepared directly from ytterbium metal and ICH_2^tBu.[103] It turned out that lanthanide *neo*-pentyl complexes suffer from low thermal stability, especially when compared to their *neo*-silyl counterpart, $Ln(CH_2SiMe_3)_3(THF)_x$. β-Methyl elimination in $Ln(CH_2^tBu)_3(THF)_2$ was found to be a facile decomposition pathway, which limited their application.

6.3.3 *Lanthanide complexes of silyl-substituted alkyls*

6.3.3.1 CH_2SiMe_3

One of the most popular synthetic precursors for lanthanide alkyl chemistry is $Ln(CH_2SiMe_3)_3(THF)_x$. It combines three advantages: steric bulk, the silyl stabilizing effect of the carbanion through a p-σ^* interaction, and lack of a β-hydrogen. However, due to a limited sterically shielding effect, homoleptic $Ln(CH_2SiMe_3)_3(THF)_x$ complexes are restricted to small-to-medium size lanthanide ions, with the largest reported to be samarium.[104] Lappert and Schumann are the major contributors to the development of this chemistry.[102,104,105] Donors like THF (tetrahydrofuran) are usually required to stabilize a monomeric compound and the number of donor molecules depends on the size of lanthanide ions, e.g. two in $Lu(CH_2SiMe_3)_3(THF)_2$ and three in $Sm(CH_2SiMe_3)_3(THF)_3$. The change in number of donor molecules will also affect the geometry of the solid-state structure (Figure 6.6(a)/(b)).

$Ln(CH_2SiMe_3)_3(THF)_x$ are widely used as synthetic precursors for lanthanide mono- or dialkyl complexes. In most cases, the isolation of $Ln(CH_2SiMe_3)_3(THF)_x$ was not necessary and it could be generated *in situ* (Eq. (6.1)). This provides a convenient and high yielding entry for organometallic lanthanide alkyl chemistry[89] and

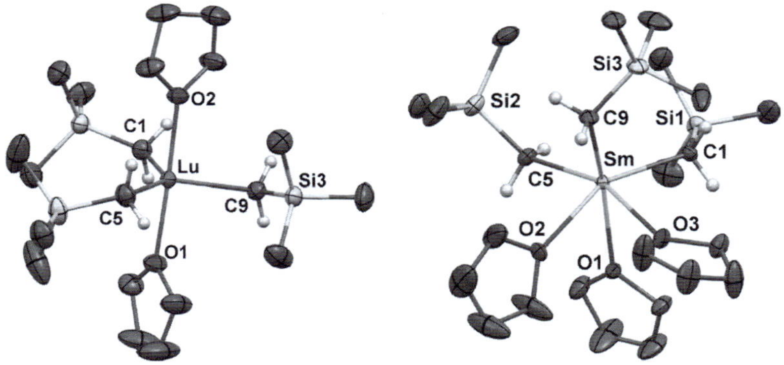

(a) (b)

Figure 6.6. Molecular structure of $Lu(CH_2SiMe_3)_3(THF)_2$ (a) and $Sm(CH_2SiMe_3)_3(THF)_3$ (b). Adapted with permission from Ref. [104]. Copyright 2002 WILEY-VCH Verlag GmbH & Co. KGaA, Weinheim.

avoids the isolation of a thermally unstable $Ln(CH_2SiMe_3)_3(THF)_x$ complex, which can readily degrade through γ-H elimination and other decomposition pathways.[106] Unfortunately, the lack of sufficient steric protection for large-size lanthanide ions limits its scope in synthesis.[107]

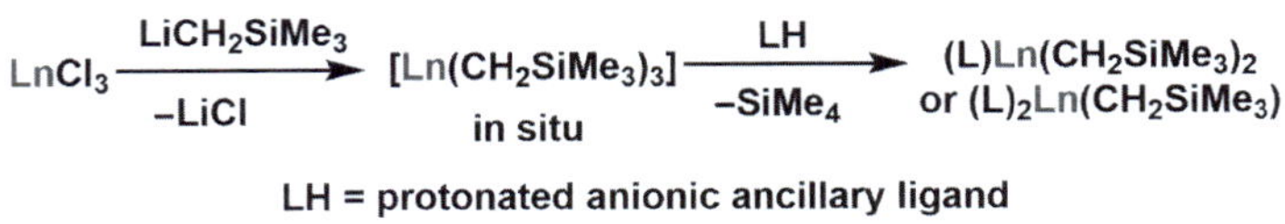

(6.1)

6.3.3.2 CH_2SiMe_2Ph

In the pursuit to increase the thermal stability of the lanthanide alkyl species, a more sterically demanding CH_2SiMe_2Ph group was introduced by replacing a methyl by a phenyl substituent. Piers *et al.* reported the synthesis of $Ln(CH_2SiMe_2Ph)_3(THF)_2$ through the salt-metathesis reaction of $LnCl_3(THF)_x$ and $LiCH_2SiMe_2Ph$.[108] Albeit CH_2SiMe_2Ph provided a slightly higher thermal stability when compared to CH_2SiMe_3, the fact that $LiCH_2SiMe_2Ph$ is not commercially available and the low volatility of the by-product of the alkane-elimination reaction, $SiMe_3Ph$, limited its application in lanthanide alkyl chemistry.

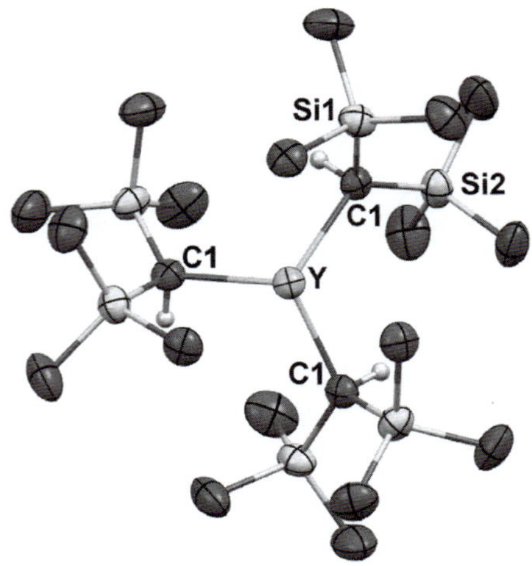

Figure 6.7. Molecular structures of $Y[CH(SiMe_3)_2]_3$ as representative of $Ln[CH(SiMe_3)_2]_3$. Reproduced from Ref. [109] with permission from the Royal Society of Chemistry.

6.3.3.3 *CH(SiMe*$_3$*)*$_2$

It was mentioned previously that the formation of a Lewis base adduct is crucial to the stability of lanthanide alkyl complexes; while that is generally the case, when sufficient steric protection is provided, donor-free homoleptic lanthanide trialkyl complexes can be obtained and are stable enough for structure determination and reactivity studies. That was the case for the $CH(SiMe_3)_2$ group. In 1974, Barker and Lappert reported the synthesis of $Ln[CH(SiMe_3)_2]_3$ and found no size limitation for this sterically hindered alkyl group, i.e. $Ln[CH(SiMe_3)_2]_3$ could be obtained from the smallest scandium to the largest lanthanum.[109] The molecular structure of $Ln[CH(SiMe_3)_2]_3$ clearly showed multiple agostic interactions between lanthanide ions and silicon methyl groups (Figure 6.7).[110,111] Divalent lanthanide $Yb[CH(SiMe_3)_2]_2$ could also be obtained by salt metathesis between YbI_2 and an alkali metal salt of $CH(SiMe_3)_2$[112,113]; when excess $KCH(SiMe_3)_2$ was used, ate complexes $[Yb[CH(SiMe_3)_2]_3K]$ were formed instead.[114] However, for the larger divalent lanthanides such as samarium and europium, no such complexes could be obtained, probably due to insufficient steric protection of the lanthanide(II) ions.

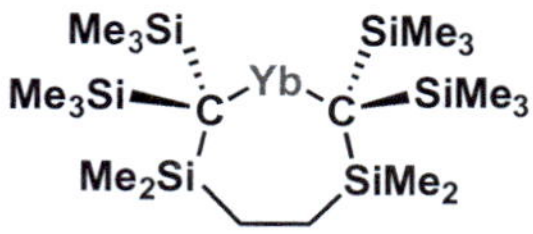

Figure 6.8. Structure of $Yb[(Me_3Si)_2CSiMe_2CH_2CH_2SiMe_2C(SiMe_3)_2]$. Due to the low quality of the data, detailed structural information was not available.

6.3.3.4 $C(SiMe_3)_3$

The steric bulk can be pushed to an extreme by introducing a third trimethylsilyl substituent. These extreme bulky alkyl ligands are found to be too large for forming trialkyl complexes with trivalent lanthanide but proved to be excellent for stabilizing divalent lanthanide dialkyl complexes.[115] With this largest ligand available, $Eu[C(SiMe_3)_3]_2$, $Sm[C(SiMe_3)_3]_2$, as well as $Yb[C(SiMe_3)_3]_2$[116] could be obtained. A structurally related bidentate ligand, $[(Me_3Si)_2CSiMe_2CH_2CH_2SiMe_2C(SiMe_3)_2]$ (Figure 6.8), is also worth mentioning here since it furnished a rare case of a 1:1 lanthanide:alkyl complex $Yb[(Me_3Si)_2CSiMe_2CH_2CH_2SiMe_2C(SiMe_3)_2]$.[117]

6.3.4 *Lanthanide benzyl complexes*

6.3.4.1 *Non-chelating benzyls*

Due to the limitation of obtaining $Ln(CH_2SiMe_3)_3(THF)_3$ complexes for all lanthanides, another class of alkyls, benzyls, was investigated. It turned out that even a simple benzyl group could tolerate all metal sizes ranging from the smallest, scandium, to the largest, lanthanum. Salt metathesis between LnX_3 (X = halide) and potassium benzyl (KBn) is effective for obtaining salt-free neutral $Ln(CH_2Ph)_3(THF)_3$ complexes (Ln = La, Nd, Y and Sc).[110,118,120] Derivatives of benzyl, such as $CH_2C_6H_4Me$-4, $CH_2C_6H_4{}^tBu$-4 and $CH_2C_6H_3Me_2$-3,5, were applied in some cases to improve the solubility of the resulting lanthanide tribenzyl complexes in non-polar hydrocarbon solvents.[121,122] The use of potassium alkyl reagents seemed to be advantageous over the corresponding lithium reagents because it prevented the contamination of the product with an alkali metal salt, since potassium halides are much less soluble

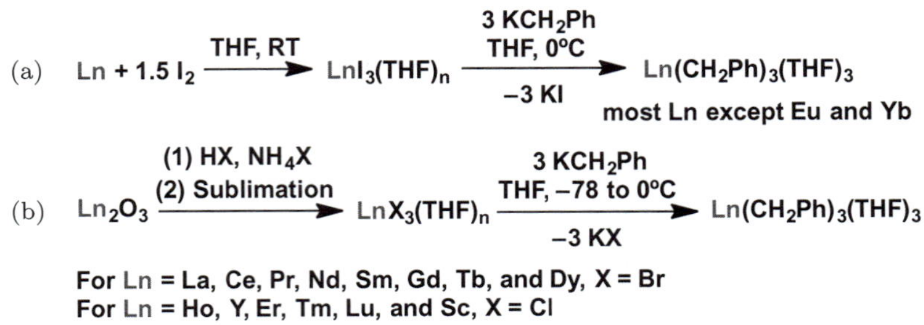

Scheme 6.3. Synthesis of $Ln(CH_2Ph)_3(THF)_3$ for the whole lanthanide series; (a) from Ln metal and (b) from the oxide.

than lithium halides in polar organic solvents (such as tetrahydrofuran and diethyl ether) and thus drive the salt-metathesis reaction to completion. Recently, the groups of Liddle and Diaconescu independently synthesized and structurally characterized $Ln(CH_2Ph)_3(THF)_3$ for the whole lanthanide series through different synthetic routes.[38,123] Liddle *et al.* conducted salt metathesis reactions of LnI_3 (obtained from Ln metal and iodine)[124] and KCH_2Ph to afford $Ln(CH_2Ph)_3(THF)_3$ (Scheme 6.3(a)).[123] Diaconescu *et al.* reported the use of $LnBr_3(THF)_x$ or $LnCl_3(THF)_x$ (obtained from aqueous HBr or HCl treatment of lanthanide oxides in the presence of ammonium bromide or chloride followed by step-wise vacuum sublimation)[125] as starting materials. It was found that the choice of lanthanide halide was critical to the outcome of the salt metathesis reaction: for larger lanthanide ions, bromide gave a moderate to high yield for the lanthanide tribenzyl product, while chloride gave intractable oily products; for the smaller lanthanide ions, on the contrary, the metal chloride afforded higher yields than the metal bromide (Scheme 6.3(b)).[38] Although this phenomenon has not been fully understood yet, it was suggested that it should be related to the relative Lewis acidity of the lanthanide ions as well as the form of the lanthanide halide salt (monomeric or polymeric). For larger lanthanides, the THF adduct of their tribromide is monomeric but the corresponding trichloride is polymeric. In addition to the choice of lanthanide halides, the stoichiometry was also crucial to the results: it was found that dropwise addition of exactly 3 equiv. of KBn in a cold THF solution to a cold slurry of $LnX_3(THF)_x$ ($X = Cl$ or Br) with stirring gave the best results. The salt metathesis between lanthanide

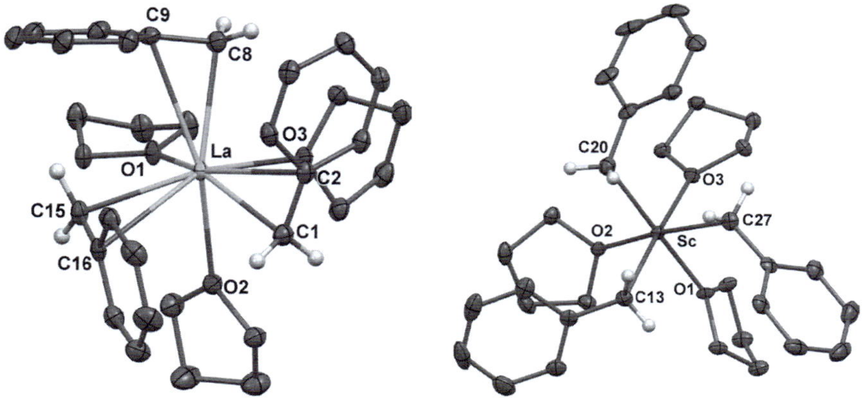

(a) (b)

Figure 6.9. Molecular structures of $La(CH_2Ph)_3(THF)_3$ (a) and $Sc(CH_2Ph)_3(THF)_3$ (b). Hydrogen atoms are omitted for clarity. Adapted with permission from Ref. [118], Copyright (2006) American Chemical Society. And Ref. [119], Copyright (2008) American Chemical Society.

halides and KBn is likely to be kinetically controlled. In order to prevent formation of polymeric species or ate complexes,[119,126] it was important to use a monomeric form of the THF adduct of lanthanide trihalides and to prevent excess of KBn (even locally). The recent advancements in the synthesis of homoleptic lanthanide trisbenzyl complexes for the entire lanthanide series make these compounds a top choice as synthetic precursors for lanthanide alkyl complexes.

The solid-state molecular structures of $La(CH_2Ph)_3(THF)_3$ and $Sc(CH_2Ph)_3(THF)_3$ are shown in Figure 6.9 as representatives of the series. Apparently, the hapticity of the benzyl ligand is variable: for the large La(III) ion, all three benzyls adopt an η^2-coordination mode through both the benzylic and *ipso* carbons; for the small Sc(III) ion, all three benzyls bind only through the benzylic carbon. This flexibility of its coordination mode explains why benzyl can serve as a universal alkyl ligand for the whole size range of lanthanides and group 3 metals. Similarly to $Ln(CH_2SiMe_3)_3(THF)_x$, $Ln(CH_2Ph)_3(THF)_3$ can be generated *in situ* and used directly to obtain ancillary ligand supported lanthanide benzyl complexes with an improved yield and ease of handling compared to the step-wise synthesis.[38,123]

A non-chelating benzyl ligand is, however, insufficient to stabilize large divalent lanthanide dibenzyl complexes. A notable exception is the mixed-valent dinuclear ytterbium complex, $[Yb(II)(CH_2Ph)(THF)_5]^+[Yb(III)(CH_2Ph)_4(THF)_2]^-$ (Eq. (6.2)), synthesized in low yield by the salt-metathesis reaction of $YbI_3(THF)_{3.5}$ and KCH_2Ph in THF at either 0°C or −78°C.[123]

$$YbI_3(THF)_{3.5} + 3\ KCH_2Ph \xrightarrow[-3\ KI]{THF} [Yb(II)(CH_2Ph)(THF)_5]^+[Yb(III)(CH_2Ph)_4(THF)_2]^- + \text{other products} \tag{6.2}$$

6.3.4.2 *Chelating benzyls*

In the early age of organometallic lanthanide chemistry, when homoleptic lanthanide trialkyl complexes were still considered beyond reach due to their poor stability and high reactivity, Manzer introduced the *ortho*-dimethylamino benzyl ligand for scandium after having success with other early transition metals like titanium and chromium. $Sc(CH_2C_6H_4\text{-}o\text{-}NMe_2)_3$ was obtained by the salt metathesis reaction between anhydrous $ScCl_3$ and $Li(CH_2C_6H_4\text{-}o\text{-}NMe_2)$ as an extremely air-sensitive crystalline solid but further purification was too difficult at the time.[127,128] Three decades later, Harder revisited this chemistry and was able to synthesize and structurally characterize solvent-free $Y(CH_2C_6H_4\text{-}o\text{-}NMe_2)_3$ and $La(CH_2C_6H_4\text{-}o\text{-}NMe_2)_3$ with a modified procedure and using $K(CH_2C_6H_4\text{-}o\text{-}NMe_2)$ instead of its lithium analogue (Scheme 6.4(a)).[129] Soon after, this strategy was found to be universal for the whole lanthanide series.[121] Independently, Hou *et al.* published a different synthetic route to access $Ln(CH_2C_6H_4\text{-}o\text{-}NMe_2)_3$ through the reaction of a lanthanide trihalide and $Li(CH_2C_6H_4\text{-}o\text{-}NMe_2)$ (Scheme 6.4(b)).[130] It is interesting to note that Hou *et al.* also found that the use of a pre-formed THF adduct of lanthanide tribromide and $K(CH_2C_6H_4\text{-}o\text{-}NMe_2)$ instead of lanthanide trichloride and $Li(CH_2C_6H_4\text{-}o\text{-}NMe_2)$ is advantageous for the larger lanthanide ions (Scheme 6.4(c)). This echoes the findings from the preparation of $Ln(CH_2Ph)_3(THF)_3$.[38] The chelating nature of $CH_2C_6H_4\text{-}o\text{-}NMe_2$ provides extra stability to homoleptic $Ln(CH_2C_6H_4\text{-}o\text{-}NMe_2)_3$ complexes: their solutions can be stored at room temperature under an inert atmosphere for months without significant degradation.[121] They can be used as synthetic precursors for half-sandwich lanthanide dialkyl complexes that can be highly reactive catalysts for olefin polymerization

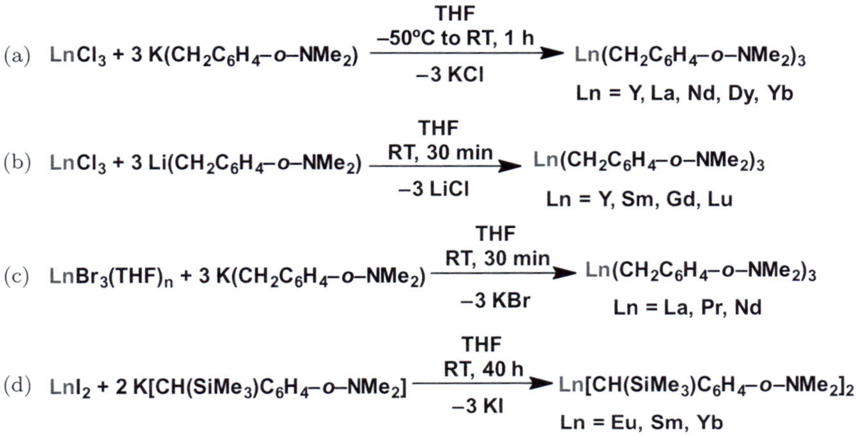

Scheme 6.4. Synthesis of $Ln(CH_2C_6H_4\text{-}o\text{-}NMe_2)_3$ (a to c) and $Ln[CH(SiMe_3)C_6H_4\text{-}o\text{-}NMe_2]_2$ (d).

and copolymerization upon converting to a cationic species in the presence of a strong Lewis acid.[131]

Parallel to trivalent lanthanide chemistry, Harder *et al.* also designed a more sterically demanding chelating benzyl ligand, $CH(SiMe_3)C_6H_4\text{-}o\text{-}NMe_2$, for divalent lanthanide ions. The corresponding complexes were obtained by a salt-metathesis reaction between LnI_2 and $K[CH(SiMe_3)C_6H_4\text{-}o\text{-}NMe_2]$ in THF at room temperature (Scheme 6.4(d)). Interestingly, the solid-state molecular structure of $Ln[CH(SiMe_3)C_6H_4\text{-}o\text{-}NMe_2]_2$ (Ln = Eu, Sm and Yb) revealed diastereomers with an R and S configuration at the two benzylic carbons.

6.3.5 *Lanthanide aryl complexes*

6.3.5.1 *Phenyl*

Lanthanide phenyls were among the first organometallic complexes containing a Ln–C σ-bond. Hart and Saran reported the first synthesis of homoleptic lanthanide phenyl complexes by reacting $LnCl_3$ and LiPh.[85] For small rare-earth metals, i.e. scandium and yttrium, neutral $LnPh_3$ could be obtained (Scheme 6.5(a)), while for large lanthanides, i.e. lanthanum and praseodymium, regardless of the stoichiometry, the ate complex, $Li[LnPh_4]$, was formed

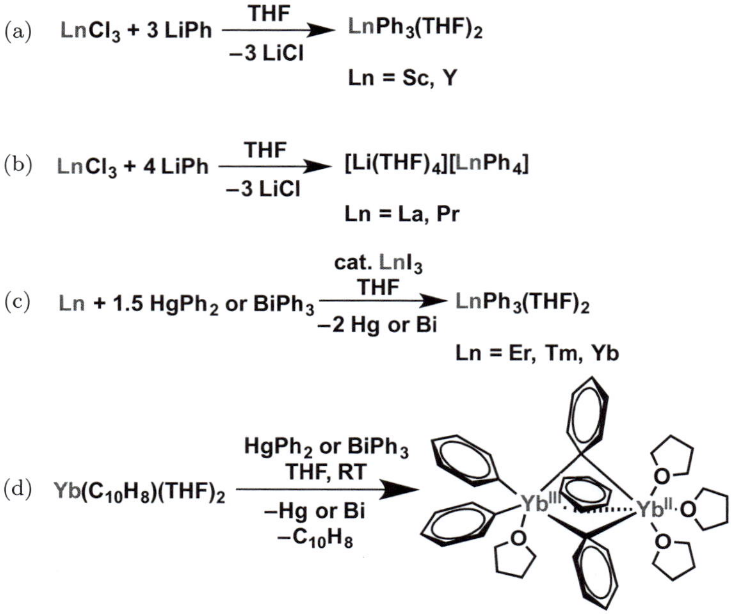

Scheme 6.5. Synthesis of $LnPh_3(THF)_3$ by salt-metathesis ((a) and (b) and redox transmetallation (c) and (d)).

exclusively (Scheme 6.5(b)). An analogue of the ate complexes, $[Li(THF)_4][Lu(C_6H_3Me_2\text{-}2,6)_4]$, was structurally characterized two years later;[86] however, it took 30 years before the structural determination of neutral $LnPh_3$ was finally provided for monomeric $ScPh_3(THF)_2$ (Figure 6.10).[132,133] Bochkarev developed redox transmetallation reactions between a lanthanide metal and $HgPh_2$ or $BiPh_3$ in the presence of a catalytic amount of lanthanide iodide to afford $LnPh_3(THF)_3$ (Ln = Er, Tm, Yb, Scheme 6.5(c)).[134,135] Due to the insufficient steric shielding provided by the phenyl ligand, neutral $LnPh_3(THF)_3$ is still unknown for large-size lanthanides ions. Nonetheless, for small to middle size lanthanides, their $LnPh_3(THF)_3$ complexes were rather thermally stable and some could even tolerate moderate heating without decomposition.[132,134] For the redox-active ytterbium, when using $Yb(C_{10}H_8)(THF)_2$ as the metal source, the oxidation reaction with $HgPh_2$ or $BiPh_3$ yielded a mixed-valent dinuclear ytterbium complex as the main product, $Ph_2Yb(III)(THF)(\mu\text{-}Ph)_3Yb(II)(THF)_3$ (Scheme 6.5(d)).[136]

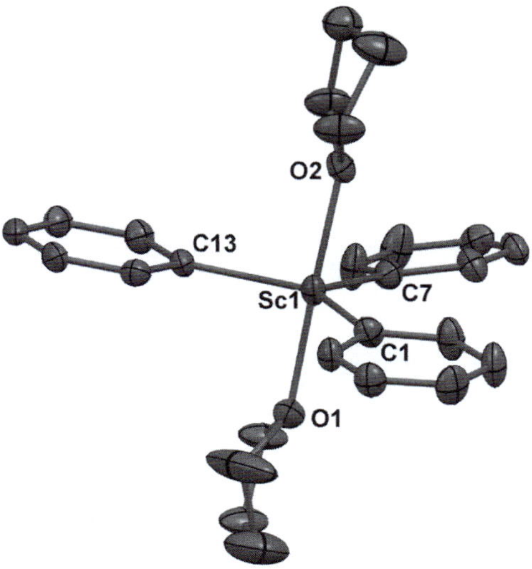

Figure 6.10. Molecular structure of $ScPh_3(THF)_2$. Hydrogen atoms are omitted for clarity. Adapted with permission from Ref. [132]. Copyright (1999) Wiley-VCH Verlag GmbH & Co. KGaA, Weinheim.

Similarly to chelating benzyl ligands, a chelating aryl ligand, [*ortho*-(dimethylamino)methyl]phenyl (o-$C_6H_4CH_2NMe_2$), was introduced to stabilize lanthanide ions at the same time as their benzyl counterparts by Manzer.[128] Wayda *et al.* extended this synthetic protocol to other lanthanides and were able to obtain good yields and high purity for $Ln(o\text{-}C_6H_4CH_2NMe_2)_3$ for small size lanthanides (Ln = Lu, Y, Er).[137] However, attempts to isolate their counterparts for large to middle-size lanthanides failed,[138] in contrast to the success in the preparation of $Ln(CH_2C_6H_4\text{-}o\text{-}NMe_2)_3$ for the whole lanthanide series.[121,129,130] The ability of benzyl to coordinate to the metal with a higher hapticity through the *ipso*-carbon may explain this difference.

6.3.5.2 *Other aryls*

Since simple phenyl falls short in terms of steric protection for large divalent lanthanide ions, substituted aryl ligands (Ar) were developed in order to obtain $Ln(II)Ar_2$ species. The perfluorophenyl ligand (C_6F_5) was found suitable to stabilise the corresponding europium(II) and ytterbium(II) complexes.

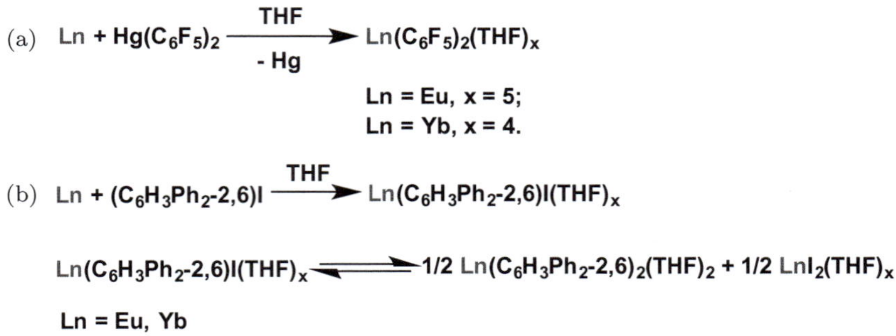

Scheme 6.6. Synthesis of $Ln(C_6F_5)_2(THF)_x$ (a) and $Ln(C_6H_3Ph_2\text{-}2,6)_2(THF)_2$ (b).

Deacon and co-workers used redox transmetallation to synthesize $Ln(C_6F_5)_2(THF)_x$ (Scheme 6.6(a)). Both $Yb(C_6F_5)_2(THF)_4$ and $Eu(C_6F_5)_2(THF)_5$ were structurally characterized and no agostic interaction between the *o*-fluorine atoms and the metal was observed.[139,140] Adding large substituents in the *o*-position of the phenyl ligand is another approach to stabilize divalent lanthanide aryl complexes. Heckmann and Niemeyer introduced *m*-terphenyl ligands ($C_6H_3Ph_2$-2,6) to divalent lanthanides through the oxidation of the metal by *m*-terphenyl iodide (Scheme 6.6(b)). $Ln(C_6H_3Ph_2\text{-}2,6)_2(THF)_2$ was obtained for europium and ytterbium.[141]

6.3.6 *Lanthanide alkynide complexes*

The strength of the metal–carbon bond follows the trend M–*sp*C > M–*sp*^{2}C > M–*sp*^{3}C;[142] as a result, lanthanide alkynide complexes are anticipated to have the highest thermal stability among all lanthanide complexes containing Ln–C σ-bonds. However, the lanthanide alkynide chemistry is far less developed compared to the corresponding alkyl chemistry because the high acidity of the *sp*C–H bond inhibits the common acid-base-type ligand exchange reaction that allows lanthanide alkyl complexes to serve as synthetic precursors through alkane elimination. The acidity of the *sp*C–H bond of acetylene or substituted alkynes is much higher than that of the corresponding bond in alkanes, and even higher than in amines.[143] Nonetheless, the synthesis and characterization of lanthanide alkynide complexes have been studied to some

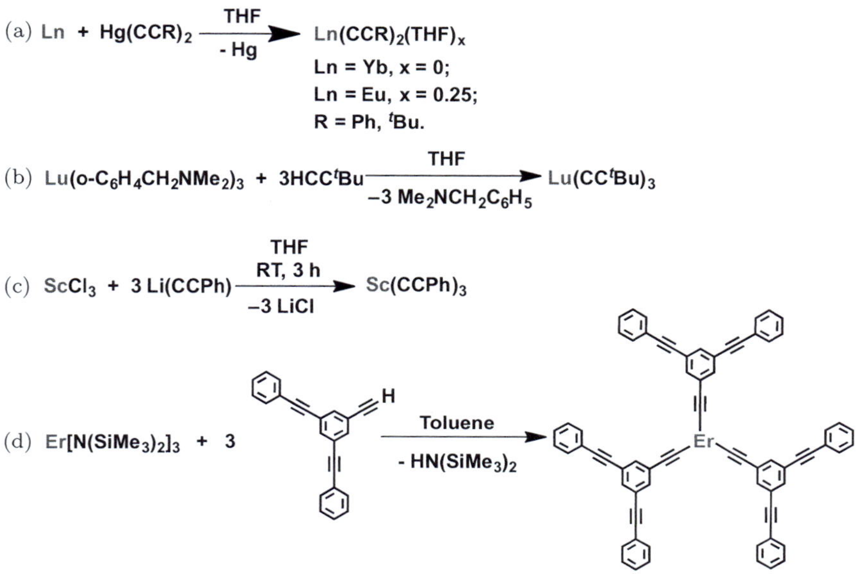

Scheme 6.7. Synthesis of lanthanide alkynide complexes; (a) from Ln metal, (b) from a tris-aryl precursor, (c) from the anhydrous chloride precursor, (d) via amine elimination.

extent. Divalent compounds are readily accessible through redox transmetallation between lanthanide metal and mercury alkynide (Scheme 6.7(a)).[144] However, this method failed to give trivalent lanthanide alkynide complexes. Since the alkyne can behave as a relatively strong acid in organometallic chemistry, amine elimination, alkane elimination and salt metathesis are valid approaches to obtain lanthanide alkynide complexes.[89] Examples include the alkane elimination from triphenyl Lu(o-$C_6H_4CH_2NMe_2$)$_3$ with HCC^tBu, to form solvent-free homoleptic $Lu(CC^tBu)_3$ (Scheme 6.7(b)),[138] and the salt metathesis between $ScCl_3$ and LiCCPh, to give $Sc(CCPh)_3$ (Scheme 6.7(c)).[85] Bochkarev *et al.* published the synthesis and characterization of a structurally interesting organoerbium dendrimer. By reacting $Er[N(SiMe_3)_3]_3$ with 3 equiv. of the dendritic ligand $H[CCC_6H_3(CCPh)_2\text{-}3,5]$, they were able to obtain $Er[CCC_6H_3(CCPh)_2\text{-}3,5]_3$ in good yield (Scheme 6.7(d)). A second generation dendrimer $Er(CCC_6H_3[CC(C_6H_3(CCPh_2)\text{-}3,5]_2\text{-}3,5)_3$ could be made in a similar fashion.[145] As such, that report opened up the study of supramolecular organometallic lanthanide chemistry.

6.3.7 *Lanthanide tetraalkylaluminate and tetraalkylgallate complexes*

A rapidly advancing field in lanthanide alkyl chemistry is their complexation with Lewis acids, namely group 13 metal trialkyls to form multinuclear metal compounds with much improved stability.[89,146,147] This approach may seem counterintuitive at first since lanthanide ions are highly electrophilic and so are the group 13 metals. However, the stability of the Lewis acid–Lewis acid adduct over that of their free forms can be explained as such: first, the highly polarized Ln–C σ-bond and orbital mismatch between lanthanide and carbon results in the negative charge being localized on the carbon atom that can be significantly stabilized by coordination to a highly Lewis acidic group 13 metal center; second, the formation of an adduct usually adds two bridging alkyl ligands (for large lanthanide ions, up to three bridging alkyl ligands),[148] which not only partially quench the Lewis acidity of both the lanthanide and the group 13 metal centers but also provide additional steric shielding around the large lanthanide ion (for instance, the coordination numbers of Ln in $Lu(AlMe_4)_3$ and $La(AlMe_4)_3$ are six and seven, respectively).[148] Among group 13 metals, lanthanide tetraalkylaluminate complexes are the most studied; lanthanide tetraalkylgallate complexes have also been studied and found to behave similarly to the analogous tetraalkylaluminate complexes except for minor structural differences.[146,149] Anderson *et al.* and Evans *et al.* are among the early investigators in this field. Anwander's group has been the major contributor in the 21st century.

Lanthanide tetraalkylaluminate complexes can be synthesized in multiple ways. For divalent lanthanides, treating lanthanide diamide or diaryloxide precursors with excess trialkylaluminum reagents can afford the corresponding $[Ln(AlR_4)_2]_n$ species (Scheme 6.8 (a),(b)).[150–152] Unlike the methyl version, $[Ln(AlMe_4)_2]_n$, which could not be structurally characterized due to its low solubility in non-coordinating solvents, the ethyl version, $[Ln(AlEt_4)_2]_n$, was structurally characterized and showed a polymeric structure with two distinct units: the anionic $[Ln(AlEt_4)_3]^-$ and the cationic $[Ln(AlEt_4)]^+$. The two are interconnected through ethyl bridges (Figure 6.11).[153] The synthesis of trivalent lanthanide tetraalkylaluminate complexes follows similar routes to

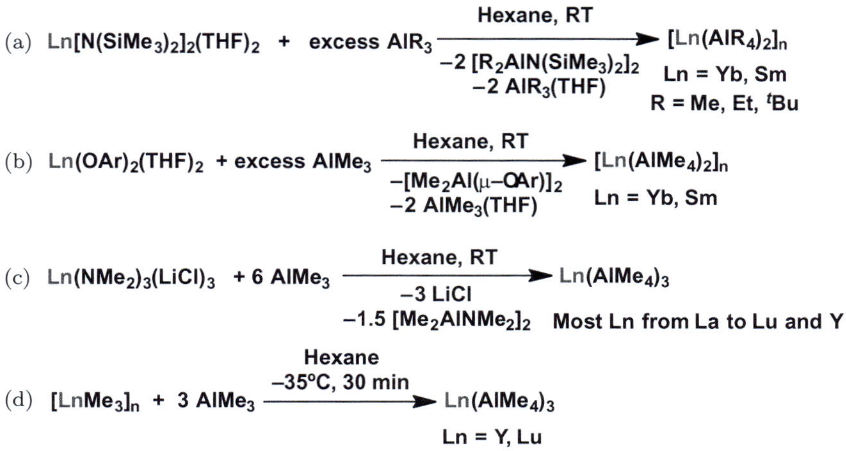

Scheme 6.8. Synthesis of $[Ln(AlR_4)_2]_n$ ((a) and (b)) and $Ln(AlR_4)_3$ ((c) and (d)).

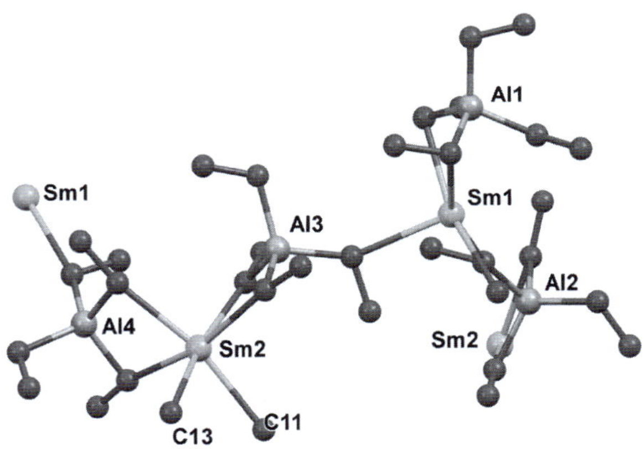

Figure 6.11. Molecular structures of $[Sm(AlEt_4)]^+$ and $[Sm(AlEt_4)_3]^-$. Hydrogen atoms are omitted for clarity. Reproduced from Ref. [153] with permission from the Royal Society of Chemistry.

that of divalent lanthanides. Evans *et al.* reported the first synthesis of $Ln(AlMe_4)_3$ from amide-aluminate exchange between $Ln(NMe_2)_3(LiCl)_3$ and excess $AlMe_3$.[146] That synthetic protocol was later adapted and optimized by the group of Anwander and extended to the whole lanthanide series (Scheme 6.8(c)).[148] Although reacting a lanthanide amide or alkoxide with trialkylaluminum reagents can afford the corresponding lanthanide tetraalkylaluminate products, their separation from by-products proved to be difficult because of the similar solubility of the desired product and that of the by-products (usually $R_2AlNR'_2$ or R_2AlOR').[89] Anwander *et al.* succeeded in preparing $Ln(AlMe_4)_3$ in high yield and purity by reacting $AlMe_3$ with polymeric $[LnMe_3]_n$ species (Scheme 6.8(d)).[92] Recently, the same group systematically synthesized and studied the stability of homoleptic lanthanide(III) tetramethylaluminate complexes for the entire lanthanide series,[147] as well as expanded this chemistry to the first synthesis of an perethylated lanthanide aluminum multi-nuclear complex, $LaAl_3Et_{12}$.[154] It is worth noting that lanthanide tetraalkylaluminate complexes can undergo multiple C–H bond activations at the methyl site to afford methylidene, R_2C^{2-}, methylidyne, RC^{3-} and even carbide, C^{4-}, which are, in turn, stabilized by multi-metal centers in the formed clusters.[155] This part of chemistry will be discussed in Section 5.4, together with lanthanide carbene complexes.

The molecular structures of $Ln(AlMe_4)_3$ are shown in Figure 6.12, using $Lu[(\mu\text{-}Me)_2AlMe_2]_3$ and $La[(\mu\text{-}Me)_2AlMe_2]_2[(\mu\text{-}Me)_3AlMe]$ as representatives. It is apparent that the coordination mode of tetraalkylaluminate is flexible and the hapticity can change from two, for small lanthanides, to three, for large lanthanides.[92] This structural flexibility seems to be crucial for a single ligand to fit the whole lanthanide series. Another case is the synthesis of lanthanide tribenzyl complexes.[38,123] On the contrary, "coordination-rigid" ligands, such as (trimethylsilyl)methyl and chelating phenyls, despite their considerable steric and electronic stabilization effects, still fail to accomplish the completion of the whole lanthanide series. The similar chemical properties but rather different sizes of lanthanides (from the smallest Sc(III) of 0.74 Å to the largest La(III) of 1.03 Å, there is a 39% increase; even for the narrower definition of lanthanides, i.e. excluding scandium and yttrium, the difference between Lu(III) and La(III) is 0.17 Å, a 20% difference, see Table 6.1 for details) require the ligand to adjust itself when binding to large or small lanthanide ions. This observation is not only true for lanthanide alkyl complexes but also holds for other ligands, including cyclopentadienyl derivatives[64] and amide-based ligands.[156]

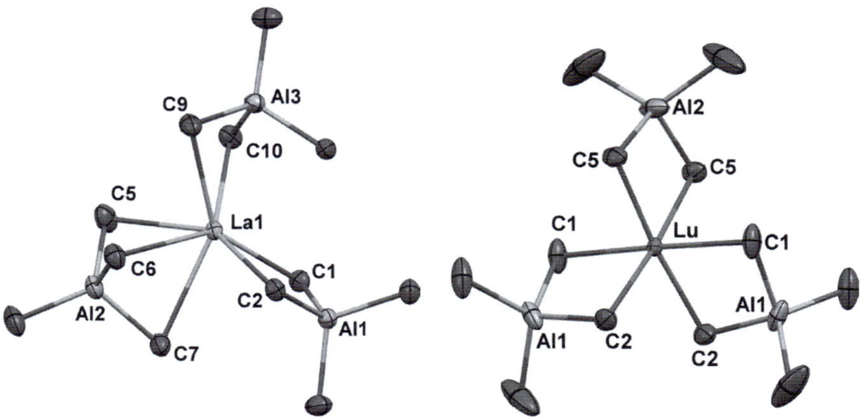

(a) (b)

Figure 6.12. Molecular structures of Lu[(μ-Me)$_2$AlMe$_2$]$_3$ (a) and La[μ-Me)$_2$AlMe$_2$]$_2$[(μ-Me)$_3$AlMe] (b). Adapted with permission from Ref. [184]. Copyright (2007) Wiley-VCH Verlag GmbH & Co. KGaA, Weinheim.

6.4 Lanthanide π-Complexes Containing Carbocyclic and Acyclic π-Ligands

Part of this section is adapted from Ref. [171].

6.4.1 *Overview of organometallic lanthanide π-complexes*

Organometallic chemistry can be traced back to the 19th century with Zeise's epoch-making synthesis of the eponymous salt, whose structure was correctly suggested by its original discoverer[157] and eventually determined by X-ray crystallography in the 20th century.[158] The next major accomplishment in organometallic chemistry was the discovery of ferrocene, its synthesis[159,160] and, more importantly, its structural interpretation and determination.[161,162] Interestingly, Zeise's salt and ferrocene both contain organic π-ligands, ethylene and cyclopentadienyl, respectively. This reflects the importance of π-ligands in organometallic chemistry, also true for organometallic lanthanide chemistry. Indeed, the first organometallic lanthanide complexes, tricyclopentadienyl complexes Cp_3Ln, were reported in 1954, only three years after

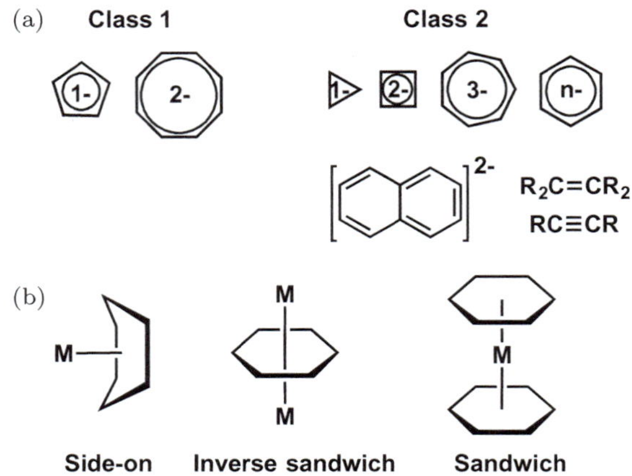

Figure 6.13. (a) Class 1 and 2 π-ligands; (b) Typical coordination modes for lanthanides and π-ligands: side-on, inverse sandwich, and sandwich.

the synthesis of ferrocene.[40] Shortly after, the di- and monocyclopentadienyl lanthanide complexes were isolated.[163,164] While the chemistry with cyclopentadienyl and its derivatives as supporting ligands has continued to soar and dominates organometallic lanthanide chemistry,[64,165,166] other π-ligands, e.g. the cyclooctatetraenyl dianion[167,168] and arenes,[169] also emerged and enjoyed a rapid growth in recent years.[170,171] In this section, we will review organometallic lanthanide complexes containing carbocyclic and acyclic π-ligands. There are mainly two classes of π-ligands: the first class includes cyclopentadienyl and its derivatives (C_5), as well as the cyclooctatetraenyl dianion (C_8), which are usually used as supporting ligands; the second class includes carbocylic rings other than C_5 and C_8 and carbo-acyclic ligands such as alkenes and dienes (Figure 6.13(a)). Typical coordination modes include side-on, inverse sandwich, and sandwich (Figure 6.13(b)). For the first class, we will emphasize their applications in catalysis and materials; while for the second class, we will discuss the bonding interactions and electronic structures of those lanthanide π-complexes.

6.4.2 *Lanthanide cyclopentadienyl complexes*

The number of lanthanide cyclopentadienyl complexes is enormous and it would be impossible to include even one-tenth of organometallic lanthanide complexes containing cyclopentadienyl or its derivatives as ancillary ligands by listing them in this chapter. In 1995, Shumann *et al.* wrote a comprehensive review on organometallic π-complexes of rare-earth metals, in which 96 pages out of 111 pages were about lanthanide complexes supported by cyclopentadienyl and its derivatives.[165] Since then, the non-cyclopentadienyl chemistry of lanthanides has grown rapidly;[172–174] however, that did not slow down the pace of lanthanide cyclopentadienyl chemistry. In the 21st century, because the volume of literature existent on lanthanide cyclopentadienyl complexes has exceeded the normal length of a journal review, reviews are usually dedicated to one of the following sub-fields: tricyclopentadienyl Cp^X_3Ln, dicyclopentadienyl Cp^X_2LnR (or metallocene), and monocyclopentadienyl Cp^XLnR_2 (or half-sandwich).[64,166,175] Herein, we intend to provide a general overview of each sub-field and direct the interested readers to journal review articles for specific topics or details.

6.4.2.1 Cp^X_3Ln

Cp_3Ln are the first organometallic lanthanide complexes ever been characterized as the result of the race to expand ferrocene chemistry to the whole periodic table.[40] They are stable complexes under regular conditions. In the solid state, the solvent free Cp_3Ln form chain-like structures with a change of the hapticity of the bridging Cp rings[165] from one to three or five. Due to the lack of steric bulkiness of the parent Cp ring, Cp_3Ln can form adducts with Lewis bases such as pyridine, phosphine, tetrahydrofuran and organic nitriles. Cp_3Ln can even form water adducts by absorbing traces of water from the solvent or atmosphere;[176] this is in sharp contrast to Ln–C σ-bonds, which will be protonated immediately by the water molecule. As organic chemists changed the substituents on phosphorus to tune the cone angles of the corresponding phosphine ligands, organometallic chemists also explored whether it is possible to achieve similar effects by adding substituents to cyclopentadienyl rings. As such, a good number of cyclopentadienyl derivatives has been developed and exploited in organometallic lanthanide chemistry. Some widely used ligands are listed below in the order of steric bulkiness from low to high: C_5H_4Me, C_5HMe_4, $C_5H_4(SiMe)_3$ (Cp′), C_5H_4(*tert*-butyl) (Cp^t), C_5Me_5 (Cp*), C_5H_3-1,3-$(SiMe_3)_2$ (Cp″), and

C_5H_3-1,3-(*tert*-butyl)$_2$ (Cp^{tt}). Extremely bulky examples such as Cp^{ttt} (C_5H_2-1,2,4-(*tert*-butyl)$_3$) are also available but reports of their complexes are rare.[48] Among them, the chemistry of $Cp^*{}_3Ln$ and Cp'_3Ln is the most studied. For $Cp^*{}_3Ln$, due to the steric hindrance caused by the Cp* ligand, the average $Ln{-}Cp^*{}_{centroid}$ distance increases ca. 0.10 Å compared to Cp_3Ln. The elongation of the Ln−Cp* centroid distance is interpreted as a release of steric strain.[64] It also results in the weakening of the bond between the lanthanide ion and the Cp* rings. A change of hapticity can occur and, for the smaller rare-earth metals, a $(\eta^5\text{-}Cp^*)_3Ln$ could not be prepared; when using the smaller ligand C_5HMe_4, a lower hapticity was found for one of the three C_5HMe_4 rings.[177] This unique coordination mode led to the discovery of new reactivity, which was named sterically induced reduction (Eq. (6.3)).[64,178,179] Later, it was found that those sterically crowded complexes could be photochemically activated to reduce small molecules, e.g. N_2,[180] or initiate polymerization reactions.[181] Cp'_3Ln was used as a trivalent lanthanide precursor for the isolation of divalent lanthanide complexes [K(crypt-222)][Cp'_3Ln] as discussed in Section 6.2.

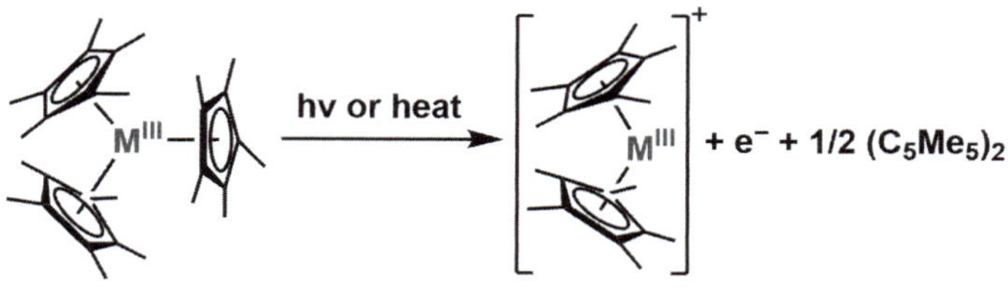

(6.3)

6.4.2.2 Cp^X_2LnR

Metallocene complexes with a formula of Cp^X_2LnR are widely used in small molecule activation as well as catalysis. A general starting point for lanthanide metallocene chemistry is dicyclopentadienyl lanthanide halides.[163] The first example of an early transition metal mediated C−H bond activation is the methane exchange reaction catalysed by $Cp^*{}_2LuMe$ (Eq. (6.4)).[90] Later, this transformation was found to be general for hydrocarbons containing sp^3C−H, sp^2C−H, or *sp*C−H bonds with $Cp^*{}_2LnR$ (R = hydride or alkyl) and was termed σ-bond metathesis by J. E. Bercaw of California Institute of Technology.[91] Similarly to the analogous group 4 metal complexes, lanthanide metallocenes have found use in catalysis, especially olefin polymerization.[182] Because they cannot form a more reactive cationic species and maintain the crucial Ln–C σ-bond, which is necessary for catalysis,[183] the half-sandwich complexes Cp^XLnR_2 were targeted instead of $Cp^*{}_2LnR$.

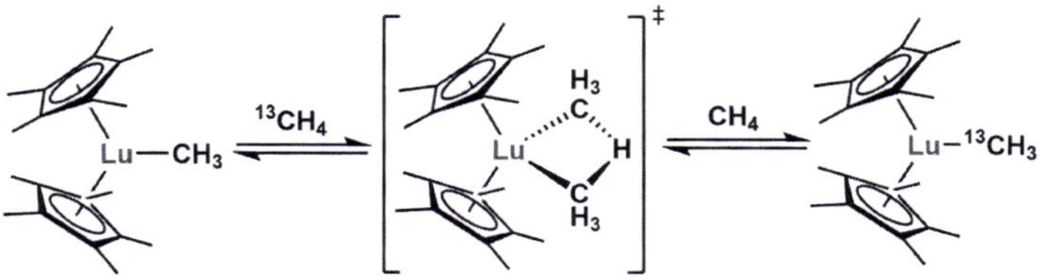

(6.4)

6.4.2.3 $Cp^X LnR_2$

Because the lanthanide ions are only complexed by only one Cp ring (usually necessary to be Cp* because of its strong electron donating capacity and large steric hindrance), Cp^XLnR_2 are usually called half-sandwich complexes. Half-sandwich complexes can be made from common precursors, monocyclopentadienyl lanthanide halides.[164] They have been extensively exploited as catalysts for olefin polymerization, carbometalation and hydroarylation. Hou's group at Rikagaku Kenkyūsho in Japan is leading the research in these areas.[175] In addition to R = alkyl species, when R = H, these complexes can form lanthanide polyhydride species that can serve as active catalysts for regio-selective olefin polyermization.[184,185] A structurally related trinuclear titanium polyhydride complex was found to be able to perform dinitrogen cleavage to form ammonia using H_2 as the hydrogen source.[186] In recent years, non-cyclopentadienyl mono-anionic ligands have been developed in order to achieve variable steric and electronic properties.[187–190]

Overall, despite the emergence of non-cyclopentadienyl supporting ligands, cyclopentadienyl and its derivatives still dominate the synthetic lanthanide chemistry, as well as lanthanide based catalysts for organic transformations or polymerizations. The privilege of cyclopentadienyl ligands originates from their good match with lanthanide ions in both electronic and steric properties. Chemists will likely continue to rely on this class of ligands in exploring new reactivity and properties of organometallic lanthanide complexes.

6.4.3 *Lanthanide cyclooctatetraenyl dianion complexes*

The cyclooctatetraenyl dianion (COT) is an intriguing ligand in organometallic chemistry. Neutral 1,3,5,7-cyclooctatetraene is a non-aromatic, non-planar molecule with alternating C–C distances typical of C–C double and single bonds; upon a two-electron reduction, it becomes planar with uniform C–C distances. Derivatives of COT ligands, e.g. COT″ (1,4-bis(trimethylsilyl)cyclooctatetraenyl) and COT‴ (1,3,6-tris(trimethylsilyl)cyclooctatetraenyl), were also developed in order to provide steric bulkiness. Similarly to the cyclopentadienyl anion, COT forms a strong interaction with lanthanides and actinides that is mainly ionic. However, it carries a higher negative charge density and has a larger size than cyclopentadienyl. Therefore, there is more orbital overlap between the metal and the cyclooctatetraenyl ligand, i.e. a stronger covalent interaction is found between COT and the metal ion. The most famous COT complex is probably uranocene, $U(COT)_2$, which is a sandwich complex.[191,192] Lanthanides also form sandwich complexes with COT, but because of their predominant trivalent oxidation state, anionic $[Ln(COT)_2]^-$ species are usually obtained,[172] except for Eu and Yb, which prefer to form the neutral complexes Eu(COT) and Yb(COT),[193,194] and Ce, which can form the neutral $Ce(COT)_2$ complex.[168,195] The bonding interaction between U(IV) and COT in $U(COT)_2$ is a textbook example of a δ covalent interaction between uranium 5f and 6d orbitals and the π^*-orbitals of the COT ligand.[196] Lanthanide COT sandwich complexes were found to exhibit a similar covalent interaction but to a lesser extent.[197] It is interesting to note that some lanthanide sandwich COT complexes or mixed Cp* and COT complexes, such as [K(18-crown-6)][$Er(COT)_2$][198,199] and (Cp*)Er(COT) (Figure 6.14(a)),[200] are among the best performing single-molecule magnets reported. Owing to its flat geometry and high hapticity, the COT ligand can be used as a building block for multi-decker complexes, such as the tetradecker sandwich complex [(Cp*)Yb(μ-η^8:η^8-COT‴)Yb(μ-η^8,

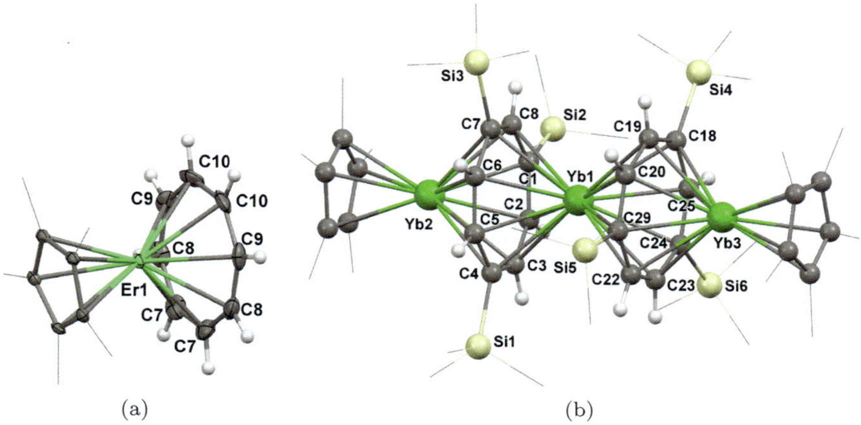

Figure 6.14. Molecular structures of (a) (Cp*)Er(COT) and (b) [(Cp*)Yb(μ-η^8:η^8-COT‴)Yb(μ-η^8,η^8-COT‴)Yb(Cp*)]. Adapted with permission from Ref. [200], Copyright (2011) American Chemical Society. And Ref. [201], Copyright (2008) Wiley-VCH Verlag GmbH & Co. KGaA, Weinheim.

η^8-COT‴)Yb(Cp*)] (Figure 6.14(b)),[201] which can eventually lead to one-dimensional organometallic polymers that may find use in materials applications.[202]

6.4.4 *Lanthanide complexes with C_3, C_4 and C_7 rings*

6.4.4.1 *C_3 rings*

The simplest example of a reduced three-membered ring is the cyclopropenyl anion, $C_3H_3^-$, which is considered to be either antiaromatic[203] or nonaromatic.[204] A substituted cyclopropenyl, such as triphenylcyclopropenyl and tributylcyclopropenyl, forms complexes with late transition metals by reacting $C_3R_3^+X^-$ (X = halogen) and metal carbonyls. However, in those complexes, the cyclopropenyl is best described as an aromatic cation but not as an anion. No cyclopropenyl lanthanide complexes have been reported, likely due to two reasons: first, Ln(III) ions are highly Lewis acidic and lack the necessary π-basicity to stabilize an interaction with the cyclopropenyl cation; second, although Ln(III) ions could stabilize the negative charge of the cyclopropenyl anion by an electrostatic interaction, its antiaromaticity[203] or non-aromaticity[204] makes its isolation elusive.

To date, there is no example of a cyclopropenyl anion metal complex reported.

6.4.4.2 *C_4 rings*

Cyclobutadiene has been long sought after by synthetic chemists. Despite the fact that free cyclobutadiene C_4H_4 has not been isolated, its dianion, $(C_4H_4)^{2-}$, can be stabilized by coordination to a metal and fulfills the Hückel rule for aromaticity. Although lanthanide dianionic cyclobutadienyl complexes are still unknown, a uranium analogue was recently synthesized by reacting the inverse sandwich uranium toluene complex $[U(Ts^{Xy})]_2(\mu-\eta^6{:}\eta^6\text{-}C_7H_8)$ (Ts^{Xy} = $HC(SiMe_2NAr)_3$; Ar = 3,5-$Me_2C_6H_3$) with excess diphenylacetylene for a prolonged period of time to form an inverse sandwich uranium dianionic cyclobutadienyl complex, $[U(Ts^{Xy})]_2(\mu-\eta^5{:}\eta^5\text{-}C_4Ph_4)$.[205] Considering the similar chemical properties of uranium and lanthanides and the presence of inverse sandwich lanthanide arene complexes,[170] a lanthanide analogue will likely be made in the near future.

6.4.4.3 *C_7 rings*

Despite the absence of compounds with reduced C_3 and C_4 rings, lanthanides do form stable complexes with cycloheptatrienyl. There are two ways for cycloheptatrienyl (C_7H_7) to achieve aromaticity: by losing an electron to form the 6π-electron tropylium cation, $C_7H_7^+$, or by accepting three electrons to form the 10π-electron trianion, $C_7H_7^{3-}$. While the tropylium cation is more common in organic chemistry, metal complexes show ambiguous assignments of the charge of the cycloheptatrienyl ligand. In the case of late transition metal complexes, it is usually considered to be +1 and the metal-ring bonding interaction to be covalent.[206] For early transition metals, the situation becomes more complicated since the metal is highly electropositive and the line between complete electron transfer from metal to ligand and partial electron transfer, i.e. back-donation, becomes blurry. For example, the oxidation state of titanium in $CpTi(\eta^7\text{-}C_7H_7)$ is either 0 if $C_7H_7^+$ or +4 if $C_7H_7^{3-}$. In both cases, the seven-membered ring satisfies Hückel's rule. Initially, the first assignment was suggested but it was later found that the latter formulation, Ti(IV) and $C_7H_7^{3-}$, was more appropriate.[206] It is important to note that there

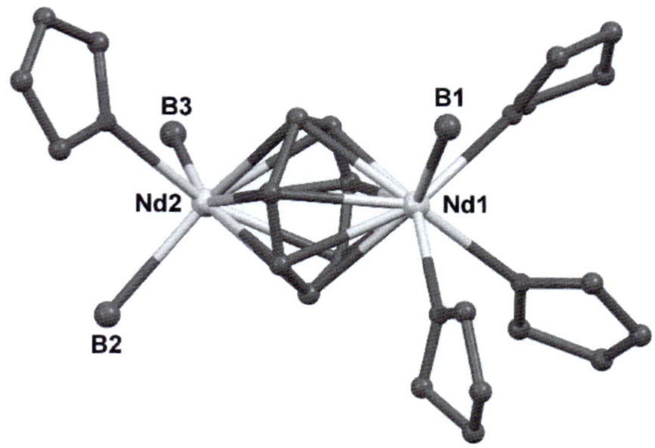

Figure 6.15. Solid-state molecular structure of $[(THF)(BH_4)_2Nd(\mu-\eta^7{:}\eta^7\text{-}C_7H_7)Nd(BH_4)(THF)_3]$ with hydrogen atoms omitted for clarity. Reproduced from Ref. [211] with permission from the Royal Society of Chemistry.

is a significant orbital overlap between the 3d orbitals of titanium and the π-orbitals of the cycloheptatrienyl ring indicating covalent bonding.[206]

Despite a good number of trianionic cycloheptatrienyl complexes of early transition metals and actinides,[207–211] only one example of a trianionic cycloheptatrienyl complex of a lanthanide is known,[211] an inverse sandwich complex of neodymium with the formula $[(THF)(BH_4)_2Nd(\mu-\eta^7{:}\eta^7\text{-}C_7H_7)Nd(BH_4)(THF)_3]$ (THF = tetrahydrofuran, Figure 6.15). Assigning a +3 oxidation state for neodymium, the cycloheptatrienyl ligand carries a 3- charge and fulfils Hückel's rule. The cycloheptatrienyl ring was found to be planar, however, due to high disorder in the crystal structure, the C–C distance within the ring could not be determined accurately. Nonetheless, the interactions between the two Nd(III) ions and the cycloheptatrienyl ligand are almost identical despite the fact that the two Nd(III) ions differ in their coordination sphere by the number of borate and tetrahydrofuran ligands. This neodymium complex was synthesized by reacting 2 equiv. of $Nd(BH_4)_3(THF)_2$ and 3 equiv. of potassium cycloheptadienyl, $K(C_7H_9)$. Disproportionation of $K(C_7H_9)$ led to *in situ* formation of $K_3(C_7H_7)$, which then underwent a salt metathesis reaction to form the inverse sandwich complex with the concomitant formation of KBH_4 as a by-product. Despite this facile synthesis, the neodymium complex remained the sole example of a structurally characterized cycloheptatrienyl lanthanide complex until recently, when a series of inverse sandwich lanthanide trianionic cycloheptatrienyl complexes was reported for Ln = Dy, Gd, Er. Those complexes were found to exhibit intriguing magnetic properties featuring ferromagnetic exchange interactions between the lanthanide ions bridged by the $(C_7H_7)^{3-}$.[212]

6.4.5 *Lanthanide arene complexes*

Six-membered ring arenes, classic aromatic molecules, with one or more phenyl rings either linked or sharing one side (fused), are common ligands in organometallic chemistry. Arenes can serve as both neutral and anionic ligands in complexes with highly electropositive metals. The first lanthanide arene complex, $Sm(\eta^6\text{-}C_6Me_6)(\eta^2\text{-}AlCl_4)_3$, was reported in 1986 by Cotton and Schwotzer.[213] Between 1987 and 1993, Cloke *et al.* reported a series of sandwich rare-earth metal arene complexes with a formal oxidation state of zero for the metal.[73,74,76] They resembled benzene sandwich complexes of transition metals, such as $Cr(C_6H_6)_2$; surprisingly, the metal-arene bond was found to be even stronger than that of $Cr(C_6H_6)_2$.[75] Lappert and co-workers isolated and structurally characterized several lanthanide benzene complexes with various degrees of benzene reduction.[57,214,215] Fryzuk *et al.* reported the synthesis of inverse sandwich dianionic biphenyl complexes of lanthanides.[216] Recently, Diaconescu *et al.* synthesized the first rare-earth metal arene complexes in which the phenyl ring sandwiched by two metal ions is quadruply reduced.[217,218] Evans *et al.* also reported lanthanide reduced benzene, biphenyl or naphthalene complexes through reduction of arenes by divalent lanthanide precursors.[71,72] For fused arenes, numerous examples of rare-earth metal fused-arene complexes are known and have been summarized in a comprehensive review.[169] In most cases, the fused arenes are doubly reduced and the negative charge is delocalized over the whole π-electron system.

6.4.5.1 *Anionic benzene or alkyl substituted benzene complexes*

6.4.5.1.1 C_6 radical anion

The benzene radical anion was proposed as an intermediate in Birch reductions.[219] However, its isolation and characterization is difficult

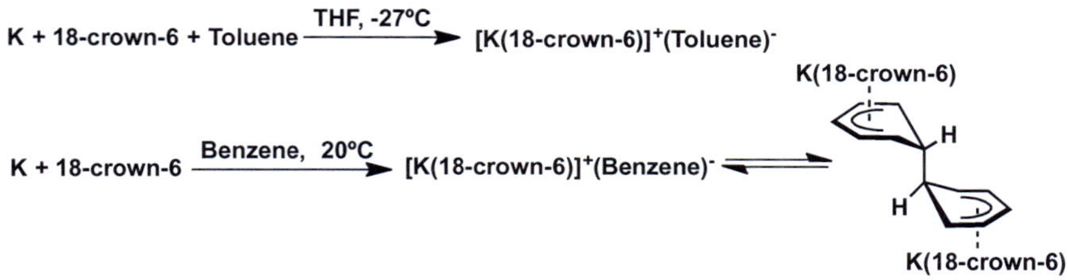

Scheme 6.9. Reduction of toluene and benzene by potassium metal in the presence of 18-crown-6.

due to its radical nature as well as the extremely negative reduction potential of benzene.[220] Although no rare-earth metal complex of the benzene radical anion is known, Lappert and co-workers succeeded in isolating and structurally characterizing the potassium crown ether salt of the toluene radical anion and benzene radical anion dimer by treating potassium metal with the corresponding arenes in the presence of crown ethers, which encapsulate the potassium ion (Scheme 6.9).[221] The crystal structure of the toluene radical anion showed that the C−C distances within the ring vary and that the coordination of potassium to individual ring carbon atoms also differs significantly (K−C distances range from 3.04 to 3.31 Å). The unsymmetrical nature of the toluene radical anion could rise from the methyl substitution or its high polarizability.

6.4.5.1.2 C_6 dianion

The benzene dianion forms stable complexes with various rare-earth metals as the 1,4-cyclohexa-2,5-dienyl ligand (benzene 1,4-dianion, Eq. (6.5)).[214,215] In all cases, the benzene dianion is sandwiched by a rare-earth metal ion and a potassium ion; the latter is coordinated by a crown ether ligand. The benzene 1,4-dianion in [K(18-crown-6)][$Cp''_2La(\eta^2-C_6H_6)$] ($Cp'' = \eta^5-C_5H_3(SiMe_3)_2$-1,3, Figure 6.16) adopts a boat-type configuration with the two out of plane carbon atoms showing the closest contact to the lanthanide ion. The negative charges are localized on these carbon atoms, and the alternating C−C distances also reflect this localization (C2−C3 is 1.35 Å, close to a C−C double bond; C1−C2 is 1.47 Å, close to a C−C single bond). Therefore, the benzene 1,4-dianion is best described as an interrupted π-system, with a strong σ-bond character between C1 or

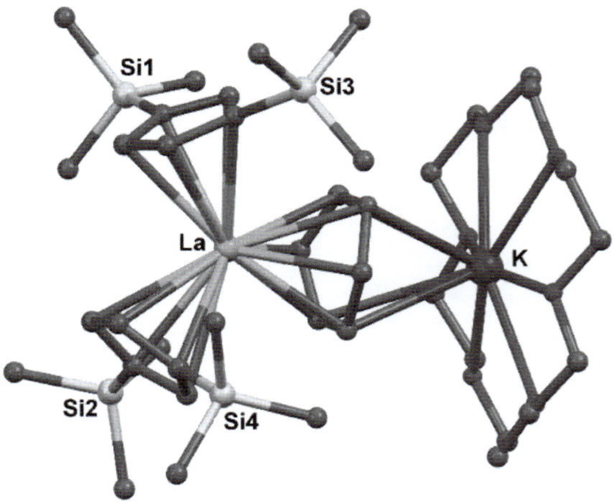

Figure 6.16. Molecular structure of $[K(18\text{-crown-6})][Cp''_2La(\eta^2\text{-}C_6H_6)]$ with hydrogen atoms omitted for clarity. Reproduced from Ref. [215] with permission from the Royal Society of Chemistry.

C4 and La(III).

$$Cp''_3Ln \xrightarrow[C_6H_6]{2\ K,\ 2\ 18\text{-crown-6}\ \ -\ K(18\text{-crown-6})Cp''} [K(18\text{-crown-6})][Cp''_2Ln(\eta^2\text{-}C_6H_6)]$$

$$Ln = La, Ce, Pr, Nd;\ Cp'' = \eta^5\text{-}C_5H_3(SiMe_3)_2\text{-}1,3 \qquad (6.5)$$

6.4.5.1.3 C_6 radical trianion

The benzene trianion has not been identified in any transition metal complexes. Nonetheless, a structurally characterized compound reported in the literature likely contains a trianionic benzene ligand. Lappert *et al.* reported the synthesis and characterization of $[K(18\text{-crown-6})(\eta^2\text{-}C_6H_6)_2][(Cp^{tt}_2La)_2(\mu\text{-}\eta^6\text{:}\eta^6\text{-}C_6H_6)]$ and proposed two possible formulations: two La(II) ions with a benzene monoanion or two La(III) ions with a trianionic benzene ligand (Figure 6.17).[57]

The authors preferred the former because, at the time, a benzene trianion seemed implausible. However, an examination of its geometrical parameters and other anionic radical or dianionic benzene compounds reveals that the formulation of two La(III) ions coordinated to a trianionic benzene ligand is more appropriate. For example, the average C–C distance in reduced benzene complexes is typically used to determine the extent of reduction of the benzene ligand since

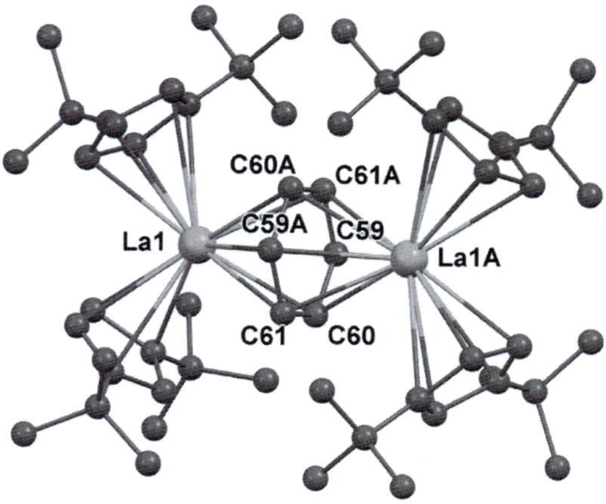

Figure 6.17. Molecular structure of $[K(18\text{-crown-}6)(\eta^2\text{-}C_6H_6)_2][(Cp_2^{tt}La)_2(\mu\text{-}\eta^6{:}\eta^6\text{-}C_6H_6)]$ with hydrogen atoms omitted for clarity. Adapted with permission from Ref. [57]. Copyright (1998) American Chemical Society.

the extra electrons are added to the anti-bonding π-orbitals of benzene and, therefore, the weakened π-bond results in elongated C–C distances. In the case of $[K(18\text{-crown-}6)(\eta^2\text{-}C_6H_6)_2][(Cp_2^{tt}La)_2(\mu\text{-}\eta^6{:}\eta^6\text{-}C_6H_6)]$, the average C–C distance is longer than the average C–C distance of the benzene dianion.[215] The almost planar benzene ring and rather short La–C distances suggest a strong interaction between lanthanum and the reduced benzene ligand. In view of the recent advances in the organometallic chemistry of f-elements, especially the isolation and characterization of tetraanionic biphenyl complexes (see below),[217,218,222] the formulation of a benzene trianion in $[K(18\text{-crown-}6)(\eta^2\text{-}C_6H_6)_2][(Cp_2^{tt}La)_2(\mu\text{-}\eta^6{:}\eta^6\text{-}C_6H_6)]$ seems appropriate.

6.4.5.1.4 C_6 tetraanion

Recently, Evans *et al.* reported synthesis of a structurally intriguing inverse sandwich lanthanide complex with the formula $[K(\text{crypt-}222)]_2[(Cp'_2Ln)_2(\mu\text{-}\eta^6{:}\eta^6\text{-}C_6H_6)]$ (Ln = La or Ce, Eq. (6.6), Figure 6.18).[72] The ^1H NMR spectrum of the lanthanum compound is diamagnetic. Instead of interpreting it as $La(III)_2(C_6H_6)^{4-}$, the authors preferred an alternative formulation as $La(II)_2(C_6H_6)^{2-}$ based on structural parameters and DFT calculations. The structural

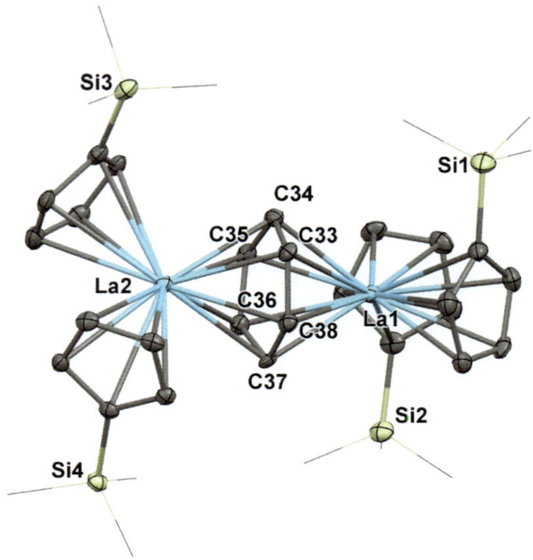

Figure 6.18. Molecular structure of $[K(crypt\text{-}222)]_2[(Cp'_2La)_2(\mu\text{-}\eta^2{:}\eta^2\text{-}C_6H_6)]$ with hydrogen atoms omitted for clarity. Reproduced from Ref. [72] with permission from the Royal Society of Chemistry.

parameters showed a significant elongation of the $La{-}Cp'_{centroid}$ distance and DFT calculations suggested that the HOMO and HOMO-1 have both lanthanide 5d and benzene π^*-orbital character. The elongation of the $La{-}Cp'_{centroid}$ distance was explained to be the consequence of the larger ionic radius of La(II) compared to La(III). However, in our opinion, a $La(III)_2(C_6H_6)^{4-}$ formulation is more appropriate for the following reasons. In $[K(crypt\text{-}222)]_2[(Cp'_2La)_2(\mu\text{-}\eta^6{:}\eta^6\text{-}C_6H_6)]$, unlike $[Cp'_3La]^-$, the HOMO and HOMO-1 are not localized on the metal but rather delocalized between the two metal ions and the bridging benzene ligand. Indeed, according to the DFT calculation results,[72] over 60% of HOMO and HOMO-1 consist of benzene π^*-orbitals with less than 40% from the two metal centers. Therefore, lanthanum is indeed trivalent, i.e. it uses all three of its valence electrons in forming bonds, in which one could be viewed as ionic and the other two as covalent. For comparison, the Mulliken population analysis of the water molecule showed ca. 60% orbital contribution from the 2p orbitals of oxygen and ca. 40% orbital contribution from 1s orbitals of hydrogen in the bonding orbital $1b_1$.[223] The oxidation states of H and O in H_2O are no doubt H(I) and O(-II). In addition, when comparing [K(18-crown-6)(η^2-C_6H_6)$_2$][($Cp_2^{tt}La$)$_2$(μ-η^6:η^6-C_6H_6)] and [K(crypt-222)]$_2$[(Cp'_2La)$_2$(μ-η^6:η^6-C_6H_6)] to the biphenyl tetraanion examples stabilized by lanthanides as well as the inverse sandwich uranium benzene complexes, a similarity in the bonding interaction can be found among them, featuring a three-center (two metal ions and one C_6 ring) and four-electron bond with δ symmetry (for a detailed discussion see below).[59,60,217,224–226] Normally, the bonding interactions in organometallic lanthanide π-complexes are mainly ionic; however, under extreme circumstances like the presence of a highly reducing arene anion, covalent interactions start to play an important role.[171] Furthermore, [K(crypt-222)]$_2$[(Cp'_2La)$_2$(μ-η^6:η^6-C_6H_6)] could act as a four-electron reducing agent to reduce two naphthalene molecules to the known compound of [K(crypt-222)][(Cp'_2La)$_2$(η^4-$C_{10}H_8$)].[71]

$$Cp'_3Ln \xrightarrow[C_6H_6]{2\ K,\ 2\ [2.2.2]crpt} [K(crypt\text{-}222)]_2[(Cp'_2Ln)_2(\mu\text{-}\eta^6\text{:}\eta^6\text{-}C_6H_6)]$$

$$Ln = La, Ce;\ Cp' = \eta^5\text{-}C_5H_3(SiMe_3) \qquad (6.6)$$

6.4.5.1.5 Miscellaneous

Between the late 1980s and early 1990s, Cloke and co-workers published a series of papers on sandwich benzene complexes of lanthanide and group 3 metals.[73–77] Typically, sterically demanding benzene derivatives, such as 1,3,5-tri-*tert*-butylbenzene, were required to obtain stable products characterized as Ln(1,3,5-C_6H_3-$^tBu_3$)$_2$. The sandwich complexes were synthesized by co-condensation of arene and atomized lanthanide vapors (Eq. (6.7)). It was found that the bonding interaction between some lanthanide ions and arene is exceptionally strong, even stronger than the interaction between chromium and benzene in $Cr(C_6H_6)_2$.[75] The authors formulated the sandwich complexes as two neutral arenes sandwiching a zero valent metal ion. However, when taking into account the strong metal-arene bonding interaction and the electronic structure of the inverse sandwich lanthanide arene complexes reported afterward, it seems appropriate to formulate this class of complexes as containing an M(III) ion sandwiched by two reduced arenes. The observed magnetic moment and EPR signal could be explained by the presence of unpaired electrons in partially filled anti-bonding π-orbitals with a strong interaction with metal-based d orbitals. In some respect, the sandwich lanthanide benzene complexes could be viewed as molecular analogues of the graphite intercalation compounds such as C_6Yb, which was found to show superconductivity.[227]

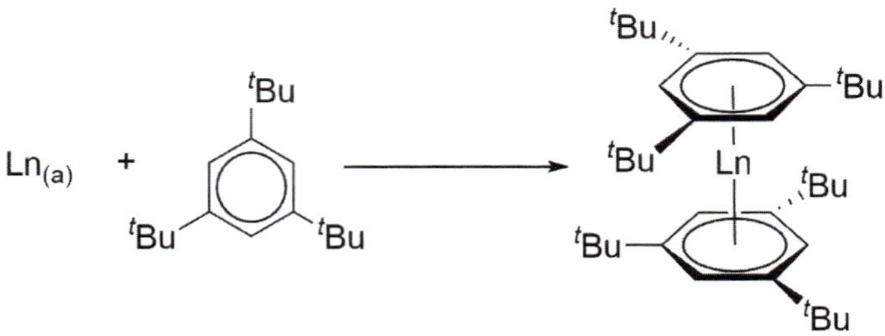

Stable for Ln = Sc, Y, Pr, Nd, Gd, Tb, Dy, Ho, Er, Lu
dec > 40°C Pr, dec > 0°C La, dec > -30°C Sm
Unstable for Ln = Ce, Eu, Tm, Yb (6.7)

Besides isolated compounds, a lanthanide anionic benzene complex was also proposed to be the key intermediate in a new type of C–H bond activation. Recently, Diaconescu *et al.* reported a bimetallic cleavage of aromatic C–H bonds by rare-earth metal complexes.[228] When reacting $(NN^{TBS})ScI(THF)_2$ (NN^{TBS} = $fc(NSi^tBuMe_2)_2$, fc = 1,1′-ferrocenediyl) and KC_8 in benzene under a dinitrogen atmosphere, instead of forming reduced dinitrogen complexes as Evans's Z_3Ln/K system, equal molar (per scandium) amounts of $(NN^{TBS})ScPh(THF)$ and $[(NN^{TBS})Sc(THF)]_2(\mu\text{-}H)_2$ were formed and characterized by X-ray crystallography and multinuclear NMR spectroscopy (Scheme 6.10). The mechanism of this unprecedented C–H bond activation was studied both experimentally and computationally. A benzene dianion bridging two $(NN^{TBS})Sc$ moieties was found to be the most likely intermediate according to intramolecular and intermolecular kinetic isotope effects, regio-selectivity of toluene C–H bond activation, and DFT calculations of reaction coordinates. This result is encouraging since it shows that lanthanide arene complexes not only provide fascinating molecular structures and intriguing bonding interactions but also provoke new reactivity that may have fundamental importance and practical uses.

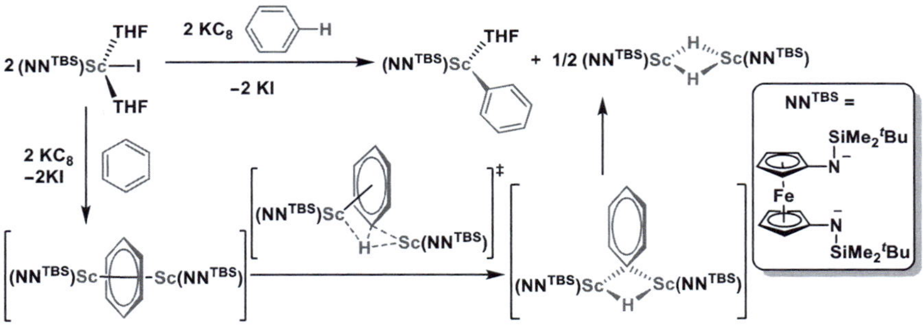

Scheme 6.10. Bimetallic cleavage of aromatic C–H bond by scandium complexes and proposed mechanism.

6.4.5.2 *Anionic aryl substituted benzene complexes*

Biphenyl, *p*-terphenyl and 1,3,5-triphenylbenzene are typical aryl substituted benzenes. Unlike fused arenes, the phenyl rings linked by the $C_{ipso}-C_{ipso}$ bond are only weakly conjugated and considered separate π-systems. Therefore, their anions can behave similarly to those of benzene. However, in some cases, the shortening of the $C_{ipso}-C_{ipso}$ bond could result in strong conjugation of the neighbouring phenyl rings, delocalization of the π-system, and sharing of the negative charge. These lead to two different coordination modes of biphenyl: coordination of rare-earth metal ions to the same phenyl ring, i.e. localization of negative charges, and coordination of rare-earth metal ions to two different phenyl rings, i.e. delocalization of negative charges.

6.4.5.2.1 Dianionic ligands

Fryzuk *et al.* reported the synthesis and characterization of dianionic 4,4'-dimethylbiphenyl complexes of yttrium and lutetium.[216] Albeit an unusual synthetic route from a metal phenyl intermediate was proposed, the biphenyl dianion was a typical π-ligand (Scheme 6.11(a)). Two coordination modes were observed in the solid-state molecular structures: two yttrium ions coordinated to different phenyl rings in $[(P_2N_2)Y]_2[\mu-\eta^6{:}\eta^{6\prime}\text{-}(C_6H_5)_2]$ (Figure 6.19) and two yttrium ions coordinated to the same phenyl ring in $[(P_2N_2)Y]_2[\mu-\eta^6{:}\eta^6\text{-}C_6H_4Me\text{-}4(C_6H_4Me\text{-}4')_2]$ $((P_2N_2) = [PhP(CH_2SiMe_2NSiMe_2CH_2)_2PPh])$.

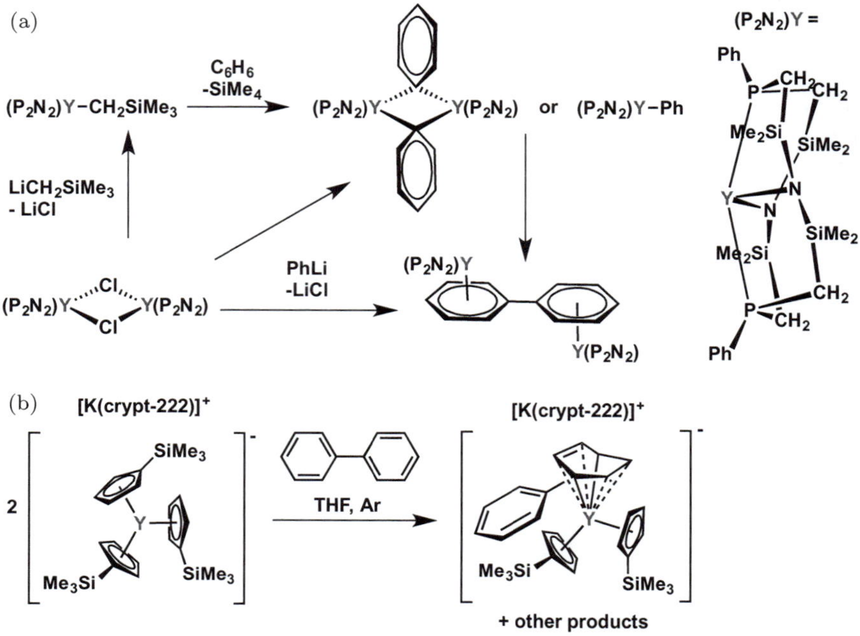

Scheme 6.11. (a) Synthesis of $[(P_2N_2)Y]_2[\eta^6,\eta^{6\prime}\text{-}(C_6H_5)_2]$ by two different methods; (b) Reaction of [K(crypt-222)][Cp'_3Y] and biphenyl to form [K(crypt-222)][$Cp'_2Y(\eta^6\text{-}C_6H_5\text{-}Ph)$] as one of the products.

Despite the difference in their solid-state structures, both compounds showed fluxional behaviour in solution and, in both cases, the phenyl rings were indistinguishable on the ^1H NMR time scale in solution at ambient temperature. This suggests that the negative charges are equally distributed between the phenyl rings and the interaction between yttrium and the dianionic biphenyl ligand is mainly electrostatic in nature. Recently, Evans *et al.* obtained a monometallic lanthanide complex of the biphenyl dianion, [K(crypt-222)][$Cp'_2Y(\eta^6\text{-}C_6H_5\text{-}Ph)$], utilizing a divalent lanthanide starting material, [K(crypt-222][(Cp'_3Y] (Scheme 6.11(b)).[71] This reaction demonstrated the reducing power of a molecular yttrium(II) species. The crystal structure of the anion $[Cp'_2Y(\eta^6\text{-}C_6H_5\text{-}Ph)]^-$ showed coordination of the Cp (Cp'_2Y) moiety to one of the phenyl rings. The structural data argued for the localization of the two negative charges on the coordinating phenyl ring, which is structurally similar

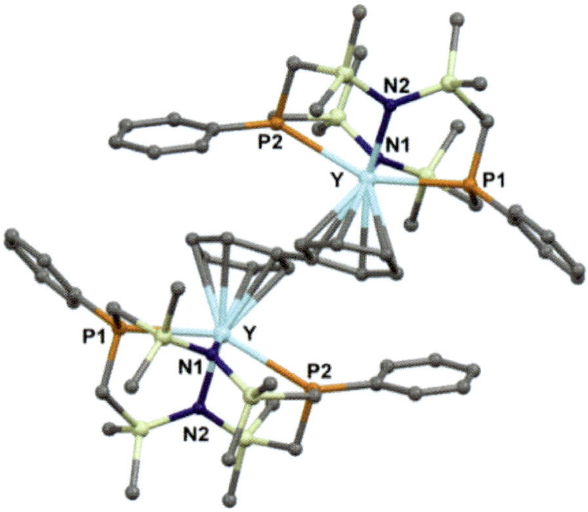

Figure 6.19. Molecular structure of $[(P_2N_2)Y]_2[\eta^6{:}\eta^{6\prime}\text{-}(C_6H_5)_2]$ with hydrogen atoms omitted for clarity. Adapted with permission from Ref. [216]. Copyright (1997) American Chemical Society.

to the benzene 1,4-dianion described by Lappert and co-workers in $[K(18\text{-crown-6})][Cp''_2La(\eta^2\text{-}C_6H_6)]$.[215]

6.4.5.2.2 Tetraanionic ligands

In 2013, Diaconescu *et al.* reported the synthesis and characterization of inverse sandwich rare-earth metal complexes of biphenyl, *p*-terphenyl, and 1,3,5-triphenylbenzene with a general formula of $[(NN^{TBS})Ln]_2(\mu-\eta^6{:}\eta^6\text{-arene})[K(solvent)]_2$ (Scheme 6.12).[217] Despite the difference in arenes, all bimetallic complexes share the same coordination mode as evidenced by their respective solid-state structures: two $(NN^{TBS})Ln$ (NN^{TBS} = fc$(NSi^tBuMe_2)_2$, fc = 1,1′-ferrocenediyl, Ln = Sc, Y, La and Lu) moieties bind to opposite sides of one of the phenyl rings (the end ring for *p*-terphenyl and 1,3,5-triphenylbenzene). A variable temperature ^{1}H NMR spectroscopic study confirmed that this asymmetric structure was maintained in solution and no fluxional behaviour was observed, in contrast to previously reported rare-earth metal complexes of biphenyl or naphthalene dianions.[216,229–231] This unusual structural rigidity pointed to the possibility of an unprecedented benzene tetraanion stabilized by

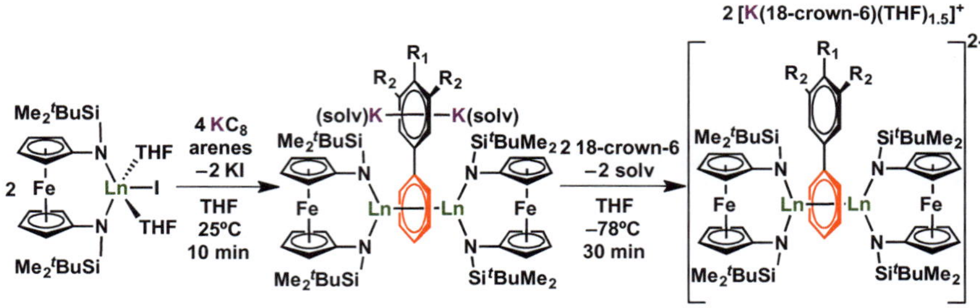

Ln: Sc, Y, La, Lu, Gd, Dy, Er
Arenes: $R_1 = R_2 = H$, biphenyl; $R_1 = Ph$, $R_2 = H$, *p*-terphenyl; $R_1 = H$, $R_2 = Ph$, 1,3,5-triphenylbenzne
solv: Et_2O or toluene

Scheme 6.12. Synthesis of $[(NN^{TBS})Ln]_2(\mu\text{-}\eta^6\text{:}\eta^6\text{-}C_6H_5Ar)[K(solv)]_2$ and $[(NN^{TBS})Ln]_2(\mu\text{-}\eta^6\text{:}\eta^6\text{-}C_6H_5Ar)[K(18\text{-crown-}6)(THF)_{1.5}]_2$ ($Ar = C_6H_5$ for biphenyl, 4-PhC_6H_4 for *p*-terphenyl, and 3,5-$Ph_2C_6H_3$ for 1,3,5-triphenylbenzene).

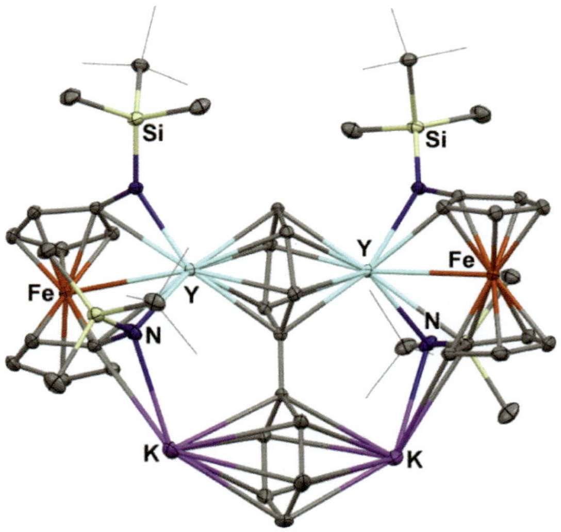

Figure 6.20. Molecular structure of $[(NN^{TBS})Y]_2[K(toluene)]_2(\mu\text{-}\eta^6\text{:}\eta^6\text{-}C_6H_5C_6H_5)$ with hydrogen atoms and toluene molecules omitted for clarity.

coordination to rare-earth metal ions along with a weak conjugation to the other phenyl ring(s).

Compounds $\mathbf{Y_2K_2}$**-biph**, $[(NN^{TBS})Y]_2[K(toluene)]_2(\mu\text{-}\eta^6\text{:}\eta^6\text{-}C_6H_5C_6H_5)$ (Figure 6.20), and $\mathbf{Y_2K_2}$**-biph-crown**$_2$, $[(NN^{TBS})Y]_2(\mu\text{-}\eta^6\text{:}\eta^6\text{-}C_6H_5Ar)[K(18\text{-crown-}6)(THF)_{1.5}]_2$, were singled out for detailed studies. X-ray crystallographic data of $\mathbf{Y_2K_2}$**-biph** showed that the two phenyl rings of the biphenyl ligand are distinct from each other: the phenyl ring sandwiched by two $(NN^{TBS})Y$ moieties has a longer average C–C distance, of 1.46 Å, than the average C–C distance of 1.41 Å of the adjacent phenyl ring and also longer than that found for the aforementioned benzene dianions.[57,215] This phenomenon is more prominent upon removal of the potassium ions by crown ether in **Y_2K_2-biph-crown$_2$**. The longer the C–C distance, the more reduced the phenyl ring is. Other evidence supporting the formulation of a benzene tetraanion included ^{89}Y NMR chemical shifts, reflecting shielding effects from the aromatic ring current of the 6C, 10π benzene tetraanion, and DFT calculations, favouring the localization of the four negative charges on one phenyl ring. The successful isolation of a series of rare-earth metal complexes containing the aryl-substituted benzene tetraanion demonstrates the predicting power of Hückel's rule in addition to the capacity of lanthanides to stabilize such highly charged species.

Another important information obtained from DFT calculations is that the covalent interaction between the arene ligand and the rare-earth metal ions also plays an important role in stabilizing the highly negatively charged benzene ring. The HOMO and HOMO-1 of **Y_2K_2-biph** showed significant δ overlap of the benzene π^* orbitals and metal based orbitals. Rare-earth metal complexes are usually considered to engage in ionic interactions due to the high electropositive character. However, in **Y_2K_2-biph**, the energy level of the benzene π^* orbitals is close enough to the energy level of the empty 4d orbitals of the yttrium(III) ion to induce a covalent bonding interaction and stabilize the high electron density of the benzene tetraanion. Later on, the dysprosium and erbium analogues of these inverse sandwich biphenyl complexes were found to show single-molecule magnet properties.[218]

Inverse sandwich arene complexes of uranium are structurally similar to the rare-earth metal stabilized benzene tetraanion. The strong interaction between uranium ions and benzene usually involves four electrons and δ overlap between uranium 5f or 6d orbitals and the π^* orbitals of benzene.[58,224,232,233] It is interesting to note that, for uranium, this binding mode (two uranium moieties coordinated to the same phenyl ring) is preferred over other possibilities: a series of μ-η^6:η^6-arene-bridged diuranium hexakisketimide complexes of a variety of arenes, including naphthalene, biphenyl, (*E*)-stilbene, and *p*-terphenyl, all showed the same coordination motif through the entire series.[224] Besides f-elements, Westerhausen *et al.*[222] reported an enthralling inverse sandwich calcium complex of 1,3,5-triphenylbenzene, $[(THF)_3Ca]_2(\mu\text{-}\eta^6{:}\eta^6\text{-}C_6H_3\text{-}1,3,5\text{-}Ph_3)$ (Eq. (6.8)). The complex was formulated as two calcium(I) ions and a 1,3,5-triphenylbenzene dianion. However, the structural data support the formulation of two calcium(II) ions and a 1,3,5-triphenylbenzene tetraanion: the average C–C distance in the central benzene ring was 1.46 Å, same as for $\mathbf{Y_2K_2}$**-biph**, although DFT calculations reported in the original paper provided different results. Nonetheless, it is clear that all four electrons from two calcium atoms are involved in bonding interactions in the complex.

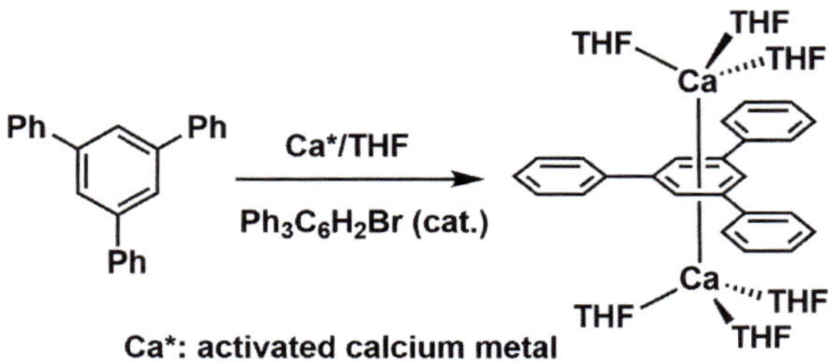

(6.8)

6.4.6 *Lanthanide fused arene complexes*

Fused arenes, such as naphthalene and anthracene, are more prone to reduction than benzene because the larger the π-electron system is, the better the negative charge is stabilized. As a result, more rare-earth metal complexes of reduced fused arene are known than of reduced benzene. Bochkarev wrote a comprehensive review on rare-earth metal arene complexes and a large portion of the review was about fused arene complexes.[169] Lanthanide fused arene complexes outnumber lanthanide arene complexes and due to the length limit of the present chapter, we are not able to discuss individual examples but rather summarize their general features. Readers with a particular interest should refer to Bochkarev's review for detailed information. First, most fused arene complexes of rare-earth metals contain a doubly reduced arene ligand. Second, in the case of bimetallic compounds, the two metal ions (two rare-earth metal ions or a rare-earth metal ion and an alkali metal ion) coordinate to opposite faces of different phenyl rings. Third, alternating C–C distances were observed, indicating a localization of π-bonds and

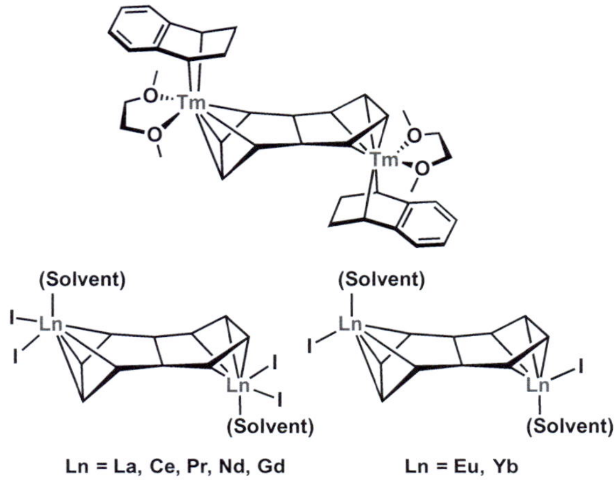

Figure 6.21. Examples of lanthanide fused arene complexes.

negative charges, i.e. interruption of the π-electron system. Indeed, the geometric parameters are almost identical for dianionic naphthalene complexes of lithium[234] and dianionic naphthalene complexes of rare-earth metals.[230,231] Figure 6.21 presents some representative lanthanide fused arene complexes. Therefore, all three features point to an ionic interaction in these complexes.

6.4.7 *Lanthanide alkene, diene and alkyne complexes*

Lanthanide alkene complexes are much less abundant compared to their corresponding transition metal counterparts. The lack of ability for back-donation of Ln(III) ions prevents the formation of stable neutral alkene complexes. However, fully reduced alkenes or partially reduced dienes can form stable complexes with lanthanides. Evans *et al.* reported several examples of samarium complexes of doubly reduced alkenes, including (*E*)-stilbene and styrene, by reacting the solvent free $Cp^*{}_2Sm$ and alkenes (Scheme 6.13(a)),[235] a bimetallic yttrium/alkali metal complex of the tetraphenylethylene dianion, $[M(THF)_x][Y(Ph_2C = CPh_2)]$, by reacting $YCl_3(THF)_3$ and $M_2(Ph_2C{=}CPh_2)$ (M = Na, K; x = 4-6, Scheme 6.13(b)),[236] and samarium complexes of doubly reduced dienes, such as isoprene and myrcene, by reacting $Cp^*{}_2Sm$ and the corresponding substituted butadienes (Scheme 6.13(c)).[237] Unlike yttrium tetraphenylethylene complexes, the formation of

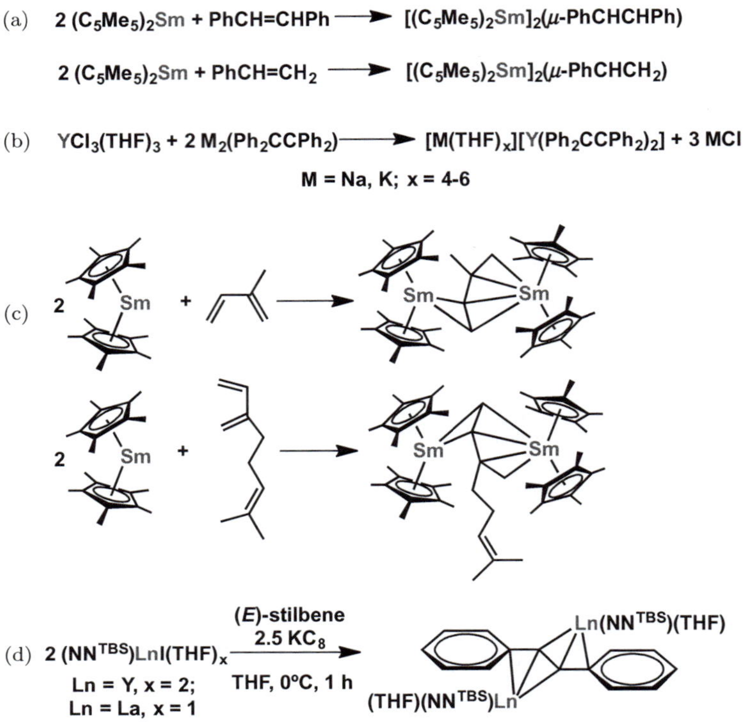

Scheme 6.13. (a) Synthesis of samarium alkene complexes; (b) Synthesis of sandwich yttrium tetraphenylethylene anion; (c) Synthesis of samarium diene complexes; (d) Synthesis of yttrium and lanthanum complexes of (E)-stilbene dianion supported by a ferrocene diamide ligand.

samarium alkenes or dienes complexes was reversible: the addition of THF generated $Cp^{*}{}_2Sm(THF)_2$ and free alkenes or dienes.[237] Mashima and co-workers reported the crystal structure of a diene-bridged dilanthanum complex, $[LaI_2(THF)_3(\mu-\eta^4{:}\eta^4$-PhCH=CHCH=CHPh)$LaI_2(THF)_3]$.[238] Recently, Diaconescu *et al.* reported the synthesis of yttrium and lanthanide complexes of the (E)-stilbene dianion. Treating $(NN^{TBS})LnI(THF)_2$ (NN^{TBS} = $fc(NSi^tBuMe_2)_2$, fc = 1,1′-ferrocenediyl, Ln = Y and La) with KC_8 in the presence of (E)-stilbene in THF resulted in the formation of $[(NN^{TBS})Ln(THF)]_2[\mu-\eta^2{:}\eta^2$-($E$)-stilbene] (Scheme 6.13(d)). The use of additional KC_8 did not lead to further reduction but instead to the scrambling of one $(NN^{TBS})Ln(THF)$ moiety by a potassium ion, i.e. formation of $[(NN^{TBS})Ln(THF)][\mu-\eta^2{:}\eta^{6'}$-($E$)-stilbene][K(THF)].[239] All aforementioned lanthanide complexes of alkenes and dienes adopted an inverse sandwich structure with the metal ion coordinated to the π electrons of the alkene or diene ligand. Albeit examples of lanthanide reduced alkene complexes exist, rare-earth metal complexes of reduced alkynes remain elusive to synthetic chemists. The stronger π bond in alkynes is probably responsible for this absence. Indeed, alkyne complexes of f-elements are essentially unknown, despite examples of dianionic species derived from the dimerization of terminal alkynes.[240–244] In these cases, C–C homo-coupling reactions were achieved using lanthanocene(III) complexes with terminal alkynes as substrates. For example, phenylacetylene reacted with different mono- (in the +2 or +3 oxidation state) or bimetallic (in the +3 oxidation state) samarocene complexes to afford a new class of trienediyl complexes of samarium(III), $(Cp^*{}_2Sm)_2(\mu\text{-}\eta^2\text{:}\eta^2\text{-Ph–C=C=C=C–Ph})$.[241,242]

6.4.8 *Summary and outlook of organometallic lanthanide π-complexes*

Historically, rare-earth metals are assumed to form ionic complexes because they are highly electropositive metals (electronegativity ranging from 1.36 for scandium to 1.1 for lanthanum, Pauling scale).[82,83] However, in organometallic chemistry, this assumption may not always hold. The 5d orbitals of lanthanides (3d for scandium or 4d for yttrium) can participate in bonding interactions if matched energetically and symmetrically. The π^* orbitals of some carbocyclic or acyclic ligands have the right symmetry, π or δ, and are high enough in energy. For example, the reduction potential of benzene, biphenyl and naphthalene was measured to be -3.42,[220] -2.69 and -2.50 eV (vs. NHE),[31] respectively. These values are comparable to the reduction potential of most Ln^{3+}/Ln^{2+} pairs; therefore, the frontier orbitals of these π-ligands can match the d-orbitals of Ln(III). From Figure 6.22, we can conclude that there are two types of bonding interactions, π or δ, that can occur between π-ligands and lanthanide ions based on the symmetry of the frontier orbitals of the π-ligands. It was found that δ-interactions are usually stronger than π-interactions and have a significant covalent character.[217,239]

Compared to actinides, lanthanides are less involved in covalent bonding. This is partially because they are more electropositive, however, probably more important, because of orbital mismatch. For actinides, 6d and 5f orbitals can both participate in bonding

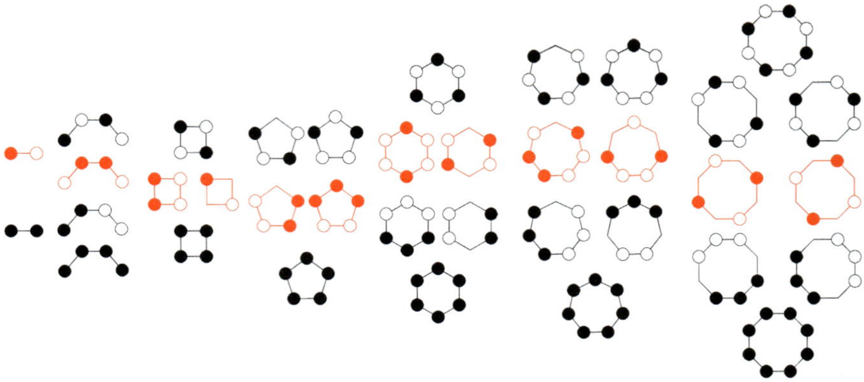

Figure 6.22. Frontier molecular orbitals for C_2, diene and C_3–C_8 rings. The orbitals highlighted in red are the HOMOs of the corresponding π-ligands that can participate in π or δ interactions with metal-based d-orbitals.

interactions, and, benefiting from a more acute angle, 5f orbitals are dominant in covalent interactions. In the case of lanthanides, 4f orbitals are core-like and do not participate in bonding, which leaves only the 5d orbitals for bonding. Besides their high energy level, d orbitals are less efficient in covalent bonding interactions than f orbitals. However, despite the lower tendency to form covalent interactions, lanthanides can still form δ-interactions with highly reduced π-ligands such as the quadruply reduced benzene.[72,217] Taking into account the recent advances in low valent lanthanide chemistry,[6,8,9,11] as well as lanthanide arene complexes[72,225] and their reactivity,[228] it is likely that the organometallic chemistry of rare-earth metals will witness another era of blossoming with an emphasis on ligand field effects[11] and covalent bonding interactions.[245]

6.5 Lanthanide Carbene Complexes

6.5.1 *Common aspects for lanthanide carbene complexes*

Carbene, a neutral carbon atom with a valence of two and two unshared valence electrons, is probably one of the most intriguing molecules or ligands in organic and organometallic chemistry. Carbenes can be either triplet or singlet: in the case of the parent

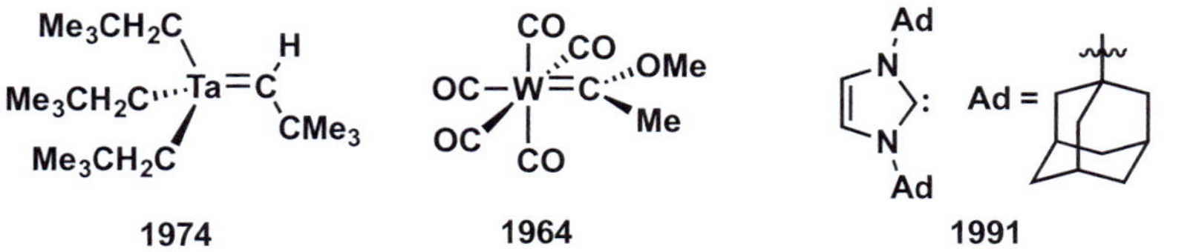

(a) **Schrock carbene** (b) **Fischer carbene** (c) **N-heterocyclic carbene**

Figure 6.23. First examples of (a) Schrock and (b) Fischer carbene metal complexes as well as of a stable (c) *N*-heterocyclic carbene.

carbene, CH_2, the triplet state is about 8 kcal/mol lower in energy than the singlet state. However, substituents on the carbon atom can significantly shift the relative energy level of singlet and triplet carbenes: for instance, π-donating ligands, such as alkoxy groups and fluorine, can greatly stabilize the singlet state (for CF_2, the ground state is singlet, which is over 56 kcal/mol more stable in energy than the first excited triplet state).[246] Recent studies also showed that certain groups are able to stabilize a triplet carbene, resulting in a wider energy gap between singlet and triplet carbene when compared to the parent CH_2 carbene.[247] Both triplet and singlet carbenes are short-lived but can serve as ligands to form metal complexes to retain stability. In organometallic chemistry, the triplet and singlet carbenes correspond to Schrock alkylidenes and Fischer carbenes. Schrock alkylidenes are considered to be in the triplet state and form a covalent double bond with a high oxidation state metal center.[248] The first metal alkylidene complex, $(Me_3CCH_2)_3Ta{=}C0HCMe_3$ (Figure 6.23(a)), was discovered in an attempt to synthesize a penta-alkyl tantalum(V) compound.[249] On the contrary, Fischer carbenes are a class of carbenes having a singlet ground state: when forming metal complexes, they act as a σ-donor, by donating a lone pair of electrons, and a π-acceptor, since it has a vacant p orbital to accept back-bonding from Lewis basic metal centers. In addition to the metal acting as a π-base, the carbon usually has one or more π-donating substituents that also donate to the empty π-orbital of the carbon center.[250] The first metal carbene complex, $(OC)_5W{=}C(OMe)(Me)$ (Figure 6.23(b)), was characterized as a Fischer-type carbene.[251] Despite the non-saturated nature of carbenes, when certain electronic and steric conditions are satisfied,

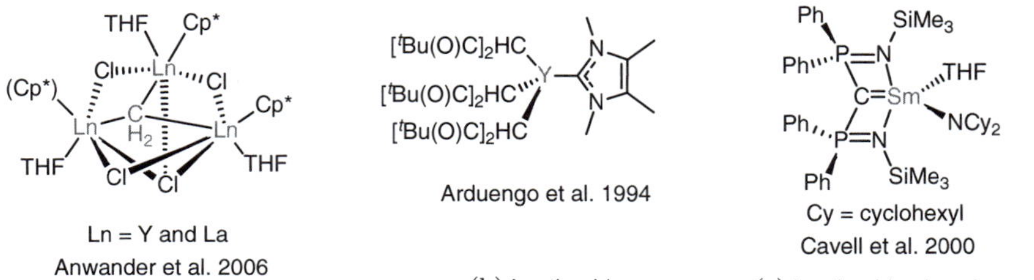

(a) Lanthanide methylidene (b) Lanthanide N-heterocyclic carbene (c) Lanthanide phosphorus-stabilized carbene

Figure 6.24. The first examples of lanthanide carbene complexes of a (a) methylidene, (b) *N*-heterocyclic carbene and (c) phosphorus-stabilized carbene.

they can become stable enough to be isolated and stored as intact species. *N*-heterocyclic carbenes (NHCs) are the best known and most developed subgroup of persistent carbenes (Figure 6.23(c)).[252] They are highly σ-donating ligands with a weak π-acidity due to two strong π-donating amino substituents on the carbon center. Since their discovery, NHCs have been widely studied and found great use as ligands for transition metals as well as p-block elements and as organocatalysts.[253] All types of carbenes, Schrock and Fischer, with the subgroup of *N*-heterocyclic carbenes, can form stable transition metal complexes. It is worth noting that the two most widely used olefin metathesis (awarded Nobel Prize in Chemistry in 2005) catalysts, group six (molybdenum and tungsten) and group eight ruthenium catalysts are transition metal carbenes containing ligands of both types.[254,255] However, for lanthanides, only Schrock carbenes and NHCs have been reported.

In this section, we will categorize lanthanide carbene complexes into three classes: first, lanthanide methylidene (CH_2^{2-}) complexes and related methylidyne (CH^{3-}) and carbide (C^{4-}) complexes; second, lanthanide *N*-heterocyclic carbene complexes; and third, a special subgroup of alkylidenes, lanthanide phosphorus-stabilized carbene complexes. All three classes, methylidene,[155] N-heterocyclic carbene[256,257] and phosphorus-stabilized carbene,[258] have been reviewed recently. Readers with specific interests may turn to those comprehensive reviews for details. The first examples of structurally characterized lanthanide carbene complexes in those three classes are depicted in Figure 6.24.

6.5.2 *Lanthanide methylidene complexes*

Metal alkylidene complexes are common for early transition metals from group 4 to group 6 metals. Usually, the alkylidenes serve as a terminal ligand and form a multiple bond with the metal center.[248] The d-orbitals of transition metals are capable of forming stable σ- and π-bonds with the 2p orbitals of carbon. However, this is not the case for lanthanides and group 3 metals. At the left end of d-block metals, they are highly electropositive metals and tend to form ionic rather than covalent bonds. Therefore, rare-earth metal alkylidene complexes are much more rare than their transition metal counterparts, and even in the known literature examples, they are multi-nuclear species with the alkylidene (actually in most cases, methylidene, CH_2^{2-}) bridging two or more metal centers.[155] Lanthanide methylidene complexes are usually formed by C–H activation of a methyl ligand in a multi-nuclear metal species; thus, the formed methylidene complexes are always multi-nuclear complexes, and, in a lot of cases, even large clusters. Due to the non-controlled nature of their formation, multiple C–H activation can occur either at one or different methyl ligands, which may result in the formation of methylidyne CH^{3-} or carbide C^{4-} complexes/clusters.

6.5.2.1 *CH_2^{2-}*

In 1979, Schumann *et al.* described the first likely lanthanide alkylidene complexes, an anionic lutetium complex, $[Li(donor)]^+[(Me_3SiCH_2)_2Lu{=}CHSiMe_3]^-$, and a neutral erbium complex, $[(Me_3SiCH_2)Er{=}CHSiMe_3]_n$, through decomposition of the homoleptic alkyl species $[Li(donor)]^+[Lu(CH_2SiMe_3)_4]$ and $[Er(CH_2SiMe_3)_3]$.[259] However, these complexes have never been structurally characterized. With the exception of this early report, all other lanthanide alkylidene (no hetero-atom supported) complexes are essentially methylidene complexes. In 2006, Anwander *et al.* published the first structurally characterized rare-earth metal methylidene complexes, $([Cp^*Ln(THF)]_3(\mu_2\text{-}Cl)_3(\mu_3\text{-}Cl)(\mu_3\text{-}CH_2))$ (Ln = Y, La) from the slow decomposition of mixed chloride tetramethylaluminate lanthanide complexes, $[Cp^*Ln(AlMe_3)_xCl_y]_z$ (Ln = Y: z = 2, y = x = 1; Ln = La: z = 6, y = 4/3, x = 2/3, Scheme 6.14(a)). The lanthanide aluminate complexes are common precursors for lanthanide methylidene complexes. The strong Lewis acidity of aluminum to

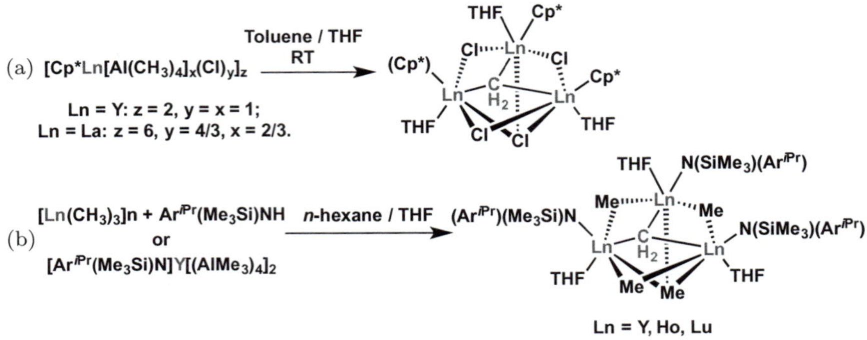

Scheme 6.14. Formation of ([Cp*Ln(THF)]$_3$(μ_2-Cl)$_3$(μ_3-Cl)(μ_3-CH$_2$)) (a) and [(([Ar^{iPr}(SiMe$_3$)N]Ln(THF)]$_3$(μ_2-CH$_3$)$_3$(μ_3-CH$_3$)(μ_3-CH$_2$)] (Ar^{iPr} = C$_6$H$_3$ (^{i}Pr)$_2$-2,6) (b).

facilitate C–H activation and its ability to stabilize the product are crucial here. A structurally similar yttrium complex was obtained by Lewis base induced cleavage of a homoleptic yttrium tetramethylaluminate or polymeric [LnMe$_3$]$_n$ species (Scheme 6.14(b)).[260]

The structures of both complexes feature a trinuclear core: the lanthanide ions are each capped by an anionic ligand (Cp* or amide), a neutral THF donor and three μ_2-Cl or μ_2-CH$_3$ ligands to form a pseudo plane, which is capped by a μ_3-CH$_2$ and a μ_3-Cl or μ_3-CH$_3$ ligand on opposite sides (Figure 6.25, [(([Ar^{iPr}(Me$_3$Si)N]Y(THF)]$_3$(μ_2-CH$_3$)$_3$(μ_3-CH$_3$)(μ_3-CH$_2$)] used as a representative of the series).[260] Regarding the Ln–C bonds, there are three distinct Ln–C distances in [(([Ar^{iPr}(Me$_3$Si)N]Y(THF)]$_3$(μ_2-CH$_3$)$_3$(μ_3-CH$_3$)(μ_3-CH$_2$)], Y-CH$_2$, Y-(μ_2-CH$_3$), and Y-(μ_3-CH$_3$), averaging 2.39 Å, 2.55 Å and 2.79 Å, respectively. The shortest Y–C distance with the methylidene unit clearly demonstrates a stronger bonding interaction compared to the methyl ligand. However, the Y–CH$_2$ distance was not much different from the terminal Y–CH$_3$ distance of 2.42 Å in [Y(CH$_3$)(THF)$_6$]$^{2+}$[BPh$_4$]$^{-}$[95], indicating little if any double bond character between Y and the methylidene ligand. This trinuclear yttrium methylidene complex was found to show nucleophilic reactivity similarly to other Schrock carbenes.[260]

It is not always necessary to pre-form lanthanide tetraalkylaluminate complexes in order to obtain lanthanide methylidene complexes. By treating a lanthanide dibenzyl precursor with trimethylaluminum in toluene, Zhou *et al.* was able to obtain a trinuclear lanthanide

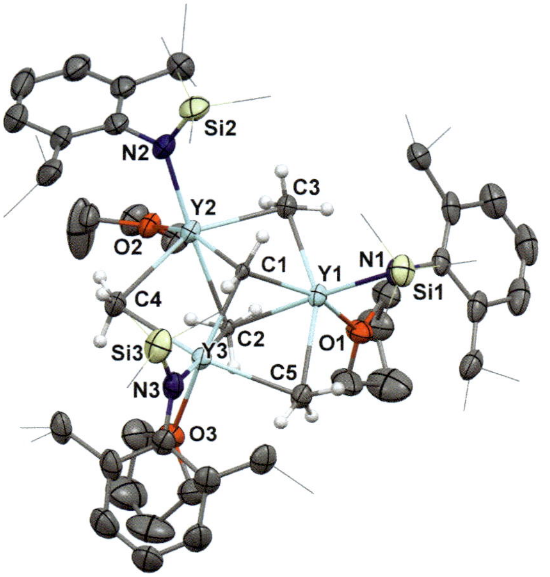

Figure 6.25. Molecular structure of $[([Ar^{iPr}(Me_3Si)N]Y(THF)]_3(\mu_2\text{-}CH_3)_3(\mu_3\text{-}CH_3)(\mu_3\text{-}CH_2)]$. Hydrogen atoms are omitted for clarity. Reproduced from Ref. [260] with permission from the Royal Society of Chemistry.

methylidene complex, which has the same core, $Ln_3(\mu_2\text{-}CH_3)_3(\mu_3\text{-}CH_3)(\mu_3\text{-}CH_2)$, as the examples previously reported by Anwander *et al.* (Scheme 6.15(a)).[261] Hou *et al.* reported a similar thulium complex, $[(Cp'Tm)_3(\mu_2\text{-}CH_3)(\mu_3\text{-}CH_3)(\mu_3\text{-}CH_2)]$, by the reaction of $[Cp'Tm(CH_3)_2]_3$ with 1 equiv. of $AlMe_3$ (Scheme 6.15(b)). The repeated appearance of the trinuclear core with various supporting ligands, different lanthanide ions, as well as distinct reaction conditions, suggests that this may be a thermodynamic well in the decomposition pathway of lanthanide alkyl species. Moreover, Hou *et al.* was able to identify a tetranuclear complex $[Cp'Tm(\mu_3\text{-}CH_2)]_4$ from heating the trinuclear complex in benzene. This is the only known example of a lanthanide methylidene complex composed of merely "$LLnCH_2$" unit and can be considered a tetramer. The structure of $[Cp'Ln(\mu_3\text{-}CH_2)]_4$ is symmetric and resembles cubane (Figure 6.26). The average Tm−C distance is 2.38 Å, which is comparable to the Y−CH_2 distance in the trinuclear core of $Y_3(\mu_2\text{-}CH_3)_3(\mu_3\text{-}CH_3)(\mu_3\text{-}CH_2)$, if accounting for the ionic size difference between Tm(III) and Y(III).

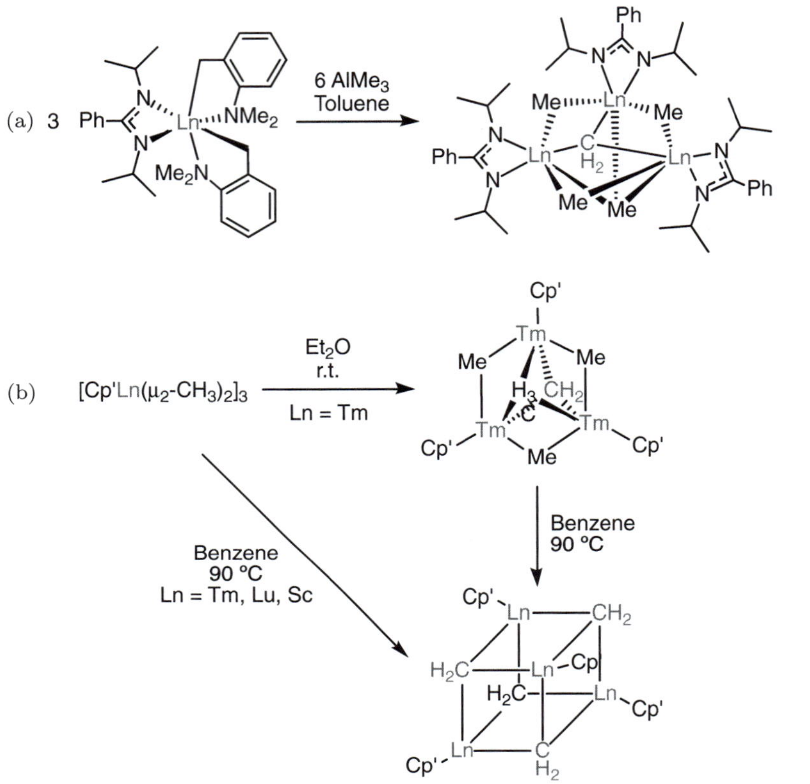

Scheme 6.15. Synthesis of $[(\mathrm{LLn})_3(\mu_2\text{-}\mathrm{CH_3})_3(\mu_3\text{-}\mathrm{CH_3})(\mu_3\text{-}\mathrm{CH_2})]$ (L = *N*,*N'*-diisopropylbenzamidinate, Ln = Sc, Lu) (a) and $[(\mathrm{Cp'Tm})_3(\mu_2\text{-}\mathrm{CH_3})(\mu_3\text{-}\mathrm{CH_3})(\mu_3\text{-}\mathrm{CH_2})]$ as well as $[\mathrm{Cp'Tm}(\mu_3\text{-}\mathrm{CH_2})]_4$ (b).

Since the formation of lanthanide methylidene complexes occurs usually in the presence of $\mathrm{AlMe_3}$, in many cases, heterometallic lanthanide aluminum methylidene complexes could be isolated and characterized. Numbers of examples are known with various ancillary ligands (Figure 6.27), including a bulky trispyrazolylborate ($\mathrm{Tp}^{t\mathrm{Bu,Me}}$ = tris(3-*tert*-butyl-5-methylpyrazolyl)borate) by Anwander *et al.*,[262] a pincer-type PNP ligand (PNP = N[2-P-(^{i}Pr)$_2$-4-methylphenyl]$_2$) by Mindiola *et al.*,[263] a flexible ferrocene diamide ligand ($\mathrm{NN^{TBS}}$ = 1,1′-fc(NSi^{t}BuMe$_2$)$_2$) by Diaconescu *et al.*,[264] and a neutral tri-amino ligand (TMTAC = 1,3,5-trimethyl-1,3,5-triazacyclohexane or T*i*PTAC = 1,3,5-tri-isopropyl-1,3,5-triazacyclohexane) by Mitzel *et al.*[265–268] In these cases, the methylidene is bridging one lanthanide and two

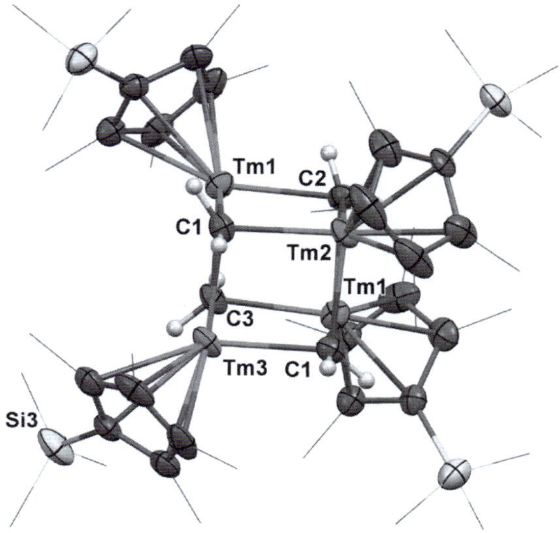

Figure 6.26. Molecular structure of tetranuclear $[Cp'Tm(\mu_3\text{-}CH_2)]_4$. Hydrogen atoms are omitted for clarity. Adapted with permission from Ref. [131]. Copyright (2011) American Chemical Society.

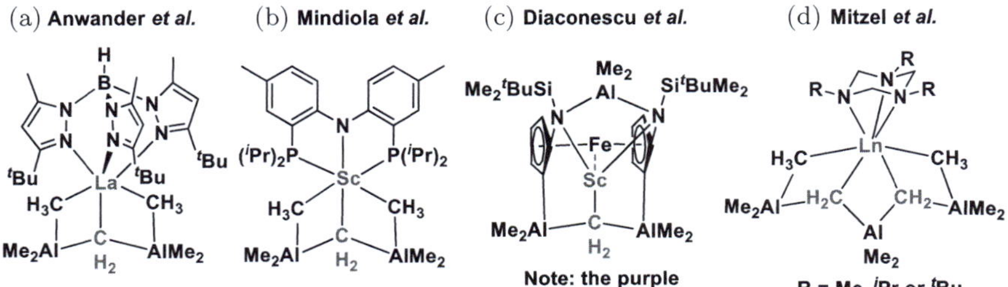

Figure 6.27. Heterometallic lanthanide aluminum complexes with various supporting ligands as reported by the groups of (a) Anwander, (b) Mindiola, (c) Diaconescu and (d) Mitzel.

aluminum centers in a T-shape with CH_2 placed in the center as a common Tebbe like unit, $((\mu_3\text{-}CH_2)[(\mu_2\text{-}CH_3)Al(CH_3)_2]_2)^{2-}$. Not surprisingly, they all show similar reactivity to the Tebbe reagent.[269] For example, $[(PNP)Sc(\mu_3\text{-}CH_2)[(\mu_2\text{-}CH_3)Al(CH_3)_2]_2)]$ can undergo methylidene-oxo exchange with Ph_2CO to form $[(PNP)Sc(\mu_3\text{-}O)[(\mu_2\text{-}CH_3)Al(CH_3)_2]_2)]$ and $Ph_2C{=}CH_2$.[263]

Besides lanthanide aluminum methylidene complexes/clusters, Hou *et al.* reported a rare case of a lanthanide transition metal heterometallic methylidene complex in the attempt to reduce transition metal carbonyl complexes by yttrium polyhydride species (Eq. (6.9)).[270] A Y_4Ir cluster was obtained from the reaction of $Cp^*Ir(CO)_2$ and $[(Cp'Y)_4(\mu_2\text{-}H)_6(\mu_3\text{-}H)_2(THF)]$ as $[(Cp'Y)_4(Cp^*Ir)(\mu_2\text{-}H)_2(\mu_3\text{-}H)(\mu_3\text{-}CH_3)(\mu_3\text{-}CH_2)(\mu_3\text{-}O)_2]$. The presence of both oxo and methylidene ligands as well as hydride and methyl ligands is remarkable.

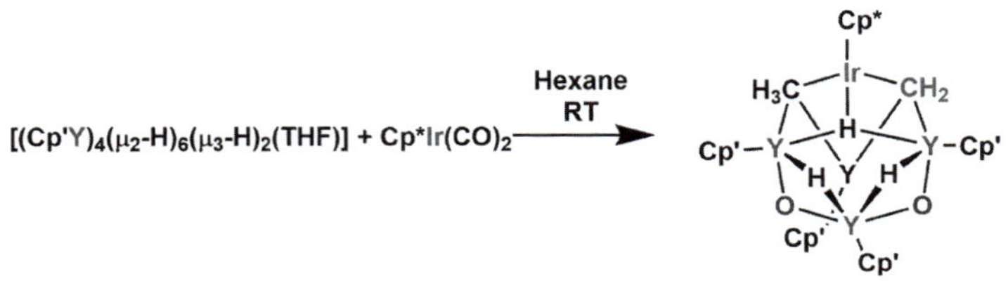

(6.9)

Recently, Chen *et al.* reported a rare case of a bimetallic scandium complex containing both bridging methylidene and phosphinidene ligands from the intermolecular C–H and P–H bond activation of a scandium methyl/phosphide precursor (Eq. (6.10)).[271] The reactivity study showed that the methylidene unit is prone to migratory insertion reactions with unsaturated molecules, such as CS_2, CO_2, ^tBuNC and PhCN.

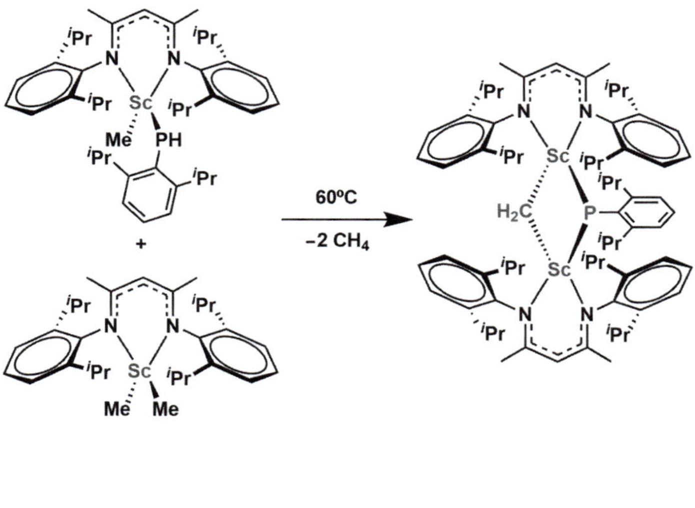

(6.10)

6.5.2.2 CH^{3-} and C^{4-}

Methylidyne CH^{3-} (or methine) and carbide C^{4-} are rare ligands in organometallic chemistry; however, they may play an important role in industrial[272] as well as biologic processes, as their existence in important systems such as nitrogenase has been revealed recently.[273] Actually, despite the underdevelopment of lanthanide alkylidene chemistry in general, lanthanide methylidyne and carbide complexes are known. They are usually the ultimate thermal decomposition product of lanthanide methylidene complexes after multiple C–H bond activation steps. Right before the publication of the first structurally characterized lanthanide methylidene complexes, Anwander's group reported the synthesis of an Y_4Al_4 cluster with two μ_4-bridging methylidyne ligands.[274] Instead of treating a half-sandwich yttrium bis(tetramethylaluminate) with 2 equiv. of the Lewis base diethyl ether to obtain the trinuclear $[Cp^*Y(\mu\text{-}CH_3)_2]_3$ complex, using only 1 equiv. diethyl ether, a Y_4Al_4 cluster was precipitated out as a crystalline solid, reproducibly but in low yield, with the chemical formula of $([(Cp^*Y)_2[(CH_3)Al(\mu_2\text{-}CH_3)_2]_2(\mu_4\text{-}CH)]_2(\mu_2\text{-}CH_3)_2)$ (Scheme 6.16). The molecular structure of the Y_4Al_4 cluster is depicted in Figure 6.28. The structure can be viewed as a $[(Me_3Al)_2(CH)]$ unit bridging two Cp^*Y fragments

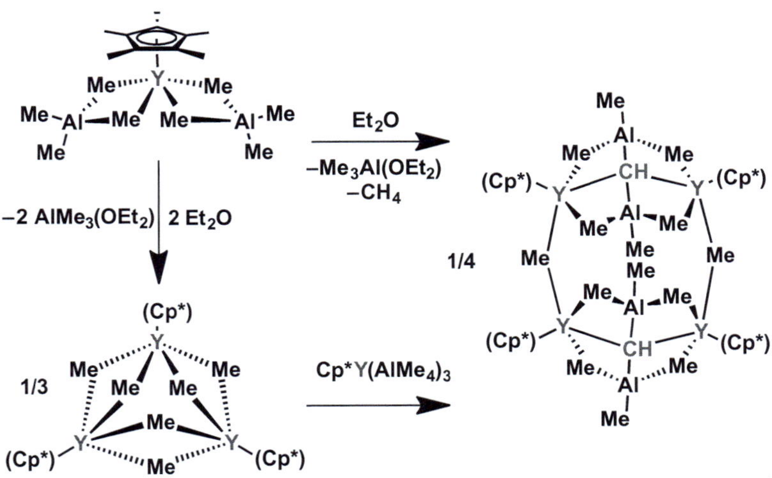

Scheme 6.16. Formation of an Y_4Al_4 cluster from Lewis base induced C–H bond activation of an yttrium tetramethylaluminate complex.

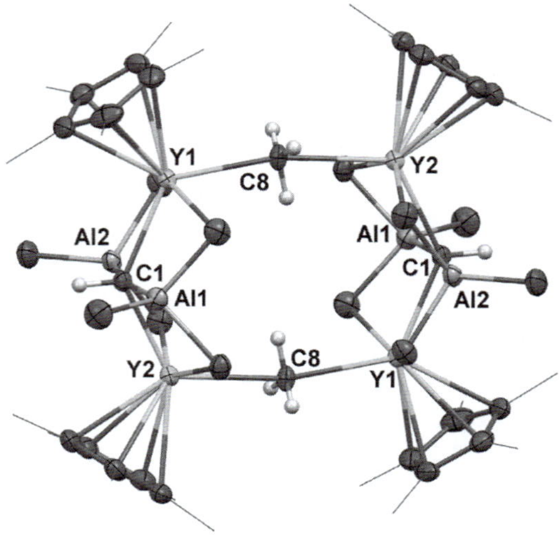

Figure 6.28. Molecular structure of ([$(Cp^*Y)_2[(CH_3)Al(\mu_2\text{-}CH_3)_2]_2(\mu_4\text{-}CH)]_2(\mu_2\text{-}CH_3)_2$). Hydrogen atoms are omitted for clarity. Adapted with permission from Ref. [275]. Copyright (2006) American Chemical Society.

through four μ_2-CH_3 and one μ_4-CH ligands; two of this moieties are linked by two μ_2-CH_3 bridging ligands. The average Y−CH distance is 2.45 Å, close to the average Y−CH_2 distance of 2.39 Å found in $[([Ar^{iPr}(SiMe_3)N]Y(THF)]_3(\mu_2\text{-}CH_3)_3(\mu_3\text{-}CH_3)(\mu_3\text{-}CH_2)]$ and the Y−CH_3 distance of 2.42 Å in $[Y(CH_3)(THF)_6]^{2+}[BPh_4]^{2-}$, but much shorter than the Y−CH_3 distance in Y_4Al_4 of 2.58–2.70 Å. The comparison of Y−C distance indicates that the bond strength of one Y−CH bond is close to that of a Y−C bond of a terminal methyl ligand. The Y_4Al_4 cluster is proposed to form through intermolecular C−H bond activation of the bridging methyl group in $Cp^*Y(AlMe_4)_2$ by an *in situ* generated $[Cp^*Y(\mu\text{-}CH_3)_2]_3$ species, since reacting equimolar amounts of $Cp^*Y(AlMe_4)_2$ and $[Cp^*Y(\mu\text{-}CH_3)_2]$ yielded the Y_4Al_4 cluster in a moderate yield (47%).

Later, Anwander's group went even deeper into the world of cluster chemistry driven by the pursuit to characterize homoleptic lanthanide trimethyl species of large lanthanide ions. When treating $La(AlMe_4)_3$ with 1 equiv. of PMe_3, two La_4Al_8 clusters and one La_5Al_9 cluster, with multiple methylidene, methylidyne and carbide ligands, were obtained. Those large heterometallic clusters were regarded as intermediates in the degradation of $La(AlMe_4)_3$ to

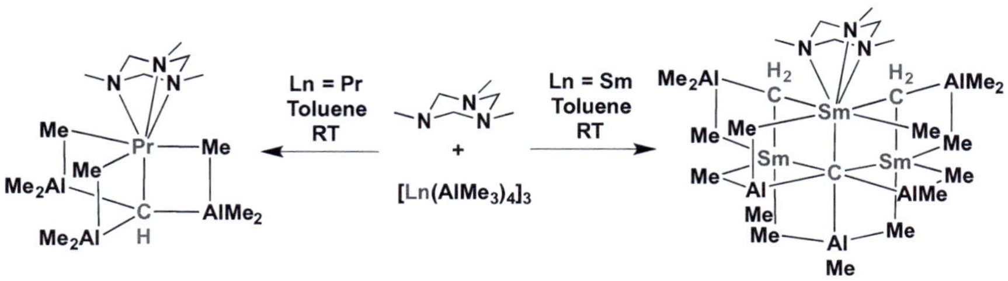

Scheme 6.17. Formation of praseodymium methylidyne complexes and samarium methylidene-carbide cluster by treating lanthanide tetramethylaluminate precursors with the Lewis base TMTAC (1,3,5-trimethyl-1,3,5-triazacyclohexane).

polymeric $[LaMe_3]_n$ species.[275] The tendency of lanthanum to form multinuclear clusters, in contrast to small lanthanides to form discrete small molecules, may be attributed to its high coordination number and more electropositive nature. The formation of methylidyne or carbide species could also be induced by a Lewis base with a 1,3,5-triazacyclohexane core. Mitzel *et al.* reported isolation of a praseodymium methylidyne complex supported by three aluminum centers and a samarium methylidene-carbide cluster, in which the carbide acts as a μ_6-ligand to coordinate three samarium and three aluminum ions (Scheme 6.17).[265,266] In these reactions, multiple products with a different degree of clustering were obtained. It is interesting to note that the presence of aluminum seems to be essential for multiple C–H bond activations since no methylidyne or carbide have been made without aluminum, and aluminum always binds to the methylidyne or carbide center in the product. However, little is known about the mechanism of CH^{3-} and C^{4-} formation so far. It will be of great importance to uncover the mechanism in order to synthesize those rare species in a controlled manner; in addition, this may shine light on developing a synthetic model for biological systems since synthetic molecular metal carbide complexes are rare.

6.5.3 *Lanthanide N-heterocyclic carbene complexes*

Since the first synthesis of persistent *N*-heterocyclic carbene, these compounds evolved into a large class composed of derivatives differing by the substituents on nitrogen as well as on the backbone of the cyclic ring. *N*-heterocyclic carbenes can act as a catalyst by itself or form metal complexes that can catalyse important transformations with various advantages over other common ligand classes such as organic phosphines.[253] Their strong σ-donating ability allows the binding to electropositive metals despite the mismatch between the soft carbene ligand and hard electropositive metal centers. Therefore, a good number of lanthanide *N*-heterocyclic carbene complexes are known in literature and this number is growing rapidly. Arnold *et al.* has written two reviews recently to summarise f-block *N*-heterocyclic carbene complexes.[256,257] Readers with specific interests are recommended to refer to these reviews for details and comprehensive references in literature. Herein, we are going to present some representative lanthanide *N*-heterocyclic carbene (NHC) complexes divided into two groups: the first group is that of simple adducts of a lanthanide and a neutral monodentate NHC, while the second group is composed of a lanthanide and a chelating NHC with a tethered monoanionic anchor to strengthen the binding.

6.5.3.1 *Simple lanthanide-NHC adducts*

As mentioned earlier, despite being a soft base, NHC is a strong σ-donor capable to bind electropositive lanthanide ions. Almost all rare-earth metals can form stable NHC adducts. The first lanthanide-NHC complex was reported by Arduengo *et al.* in 1994, three years after the publication of the first persistent NHC by the same group, as the monoadduct $Y(THD)_3[C(NMeCMe)_2]$ (THD = tetramethylheptanedioate, Figure 6.29(a)).[276] Anwander's group later published the synthesis and characterization of a series of mono- and bis-NHC yttrium adducts (Figure 6.29(b),(c)) and used the ^{89}Y-^{13}C coupling constant $^1J_{YC}$ as a probe to interpret the strength of NHC binding.[277] Analogous lanthanide(III)-NHC adducts are known for other lanthanides, prepared by a similar synthetic protocol, including lanthanum,[277] cerium,[278] europium,[276] erbium[277,279] and lutetium.[279] Selected examples are depicted in Figure 6.29(d),(e),(f). Divalent samarium(II)-NHC complexes were among the first examples of lanthanide-NHC complexes and were published at the same time as the yttrium example.[276,280,281] They could be made by simply replacing a solvent molecule in $Cp^*{}_2Sm$(solvent) with free NHCs. Both mono- and bis-NHC adducts are known (Figure 6.29(g),(h)).

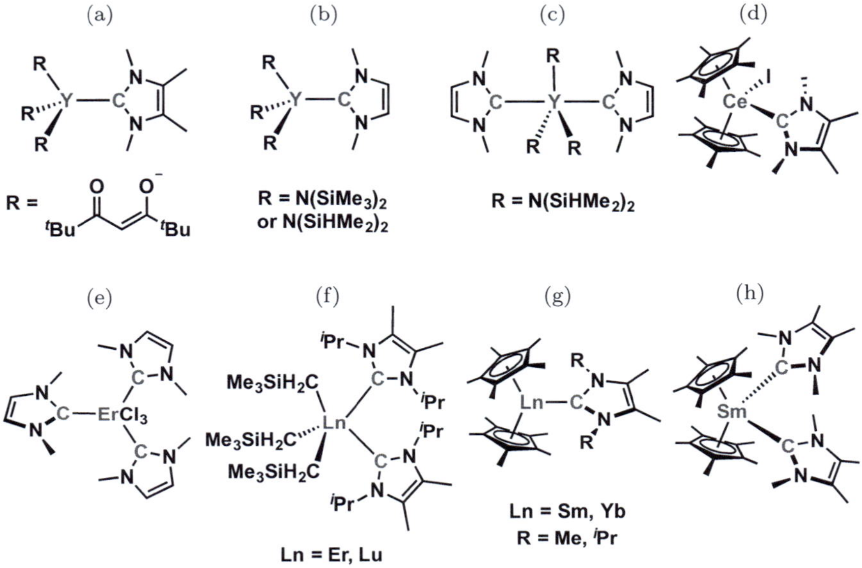

Figure 6.29. (a)–(h) Selected examples of simple lanthanide-NHC adduct complexes.

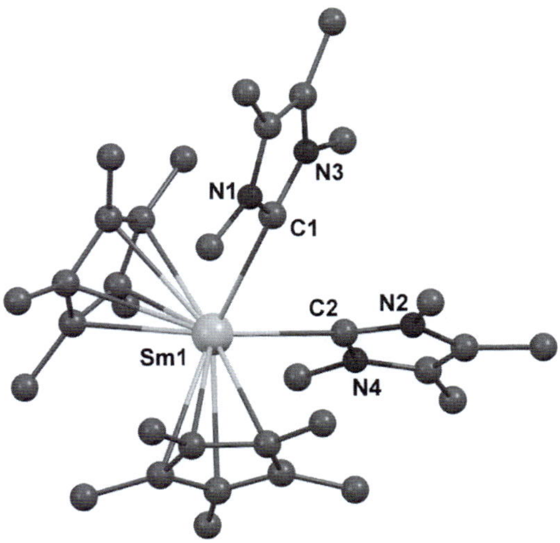

Figure 6.30. Molecular structure of a samarocene bis-NHC complex. Hydrogen atoms are omitted for clarity. Adapted with permission from Ref. [272]. Copyright (1994) American Chemical Society.

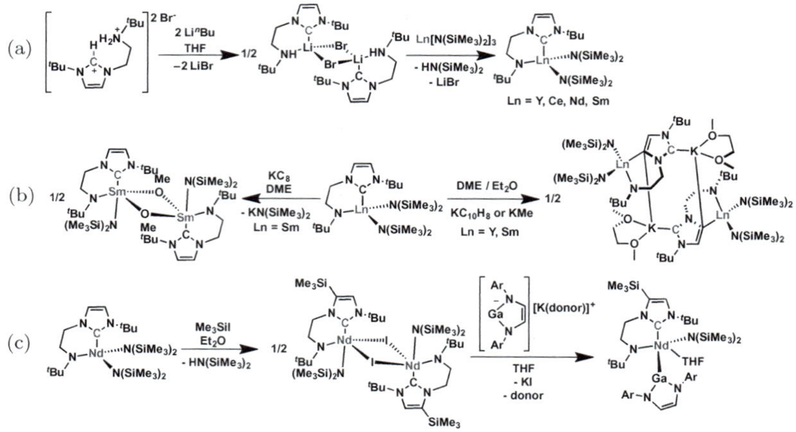

Scheme 6.18. Synthesis (a) and reactivity (b) (c) of lanthanide complexes of amide-NHC ligands.

The solid-state structure of bis-NHC adducts is shown in Figure 6.30. Similarly, ytterbium(II)-NHC complexes could be made in a similar fashion and reported by Schumann *et al.* The ytterbium(II)-NHC complexes with different cyclopentadienyl derivatives displayed a similar Yb–$C_{carbene}$ distance and the ^{13}C chemical shifts for the carbene center were within a narrow range of 198–205 ppm, typical of diamagnetic metal-NHC complexes.[280,282]

6.5.3.2 *Chelating-NHC ligands with an anionic anchor*

Besides the simple adducts of lanthanide-NHC complexes, another class of NHC ligands contains a tethered anionic anchor that can stabilize the lanthanide-NHC complexes and also introduce unusual reactivity. Because the tethered arm is usually stored by *N*-functionalisation, these are called *N*-functionalized NHCs. Based on the nature of the anionic tethered arm, we classify them into following four categories: (1) amide-NHC; (2) alkoxide/aryloxide-NHC; (3) indenyl/fluorenyl-NHC; (4) miscellaneous NHCs.

6.5.3.3 *Amide-NHCs*

Amide-functionalized NHCs are the most commonly used chelating NHC ligands because of the ease of synthesis and the strong σ-donating ability of the amide. Polly L. Arnold *et al.* reported

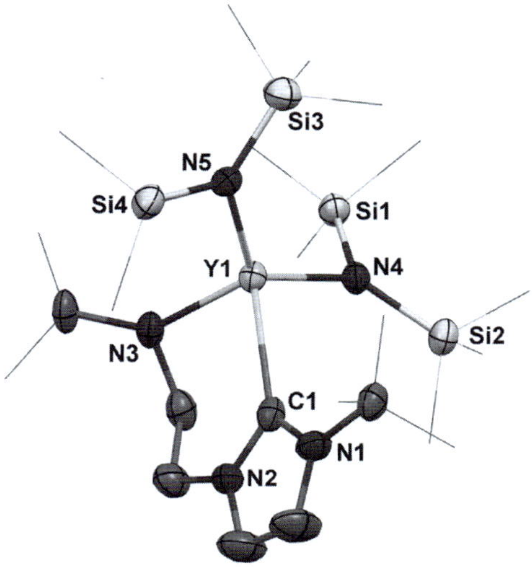

Figure 6.31. Molecular structure of the first *N*-functionalized NHC yttrium complex. Adapted with permission from Ref. [283]. Copyright (2003) Wiley-VCH Verlag GmbH & Co. KGaA, Weinheim.

the first *N*-functionalized NHC yttrium complex, which was synthesized by reacting a lithium bromide adduct of an amine-NHC with $Y[N(SiMe_3)_2]_3$ (Scheme 6.18(a)).[282] The solid-state structure of the amide-NHC yttrium complex is shown in Figure 6.31. The $Y{-}C_{carbene}$ distance was found to be 2.50 Å, shorter than the $Y{-}C_{carbene}$ distance of 2.55 Å in the simple yttrium-NHC adduct.[277] In addition, the $^1J_{YC}$ was found to be 54.7 Hz, the largest in the literature for any yttrium-NHC complex. These two features indicated a strong interaction between yttrium and the carbene, which was supported by a reactivity study showing that common Lewis bases such as THF or PPh_3 could not perturb the Y–NHC bond, while stronger donors like $OPPh_3$ could. The yttrium complex was found to be a bifuncational catalyst for the ring-opening polymerization of racemic lactide to afford poly(racemic-lactide) with low polydispersity and high heterotacticity. It was proposed that the bifuncational nature of the yttrium catalyst should be attributed to the labile NHC ligand: while the Lewis acidic yttrium center binds the lactide for activation, the NHC ligand dissociated after lactide coordination can act as a Lewis base to ring open the lactide substrate through a nucleophilic mechanism.[283] Using the same amide-NHC, Arnold's group expanded the chelating-NHC chemistry to other lanthanides, including cerium(III),[284] neodymium(III)[285] and samarium(III).[282] The reactivity of those lanthanide-NHC complexes was explored. The yttrium and samarium complexes were treated with strong reducing agents such as potassium naphthalenide ($KC_{10}H_8$) or potassium graphite (KC_8). For the redox-inactive yttrium, deprotonation at the NHC backbone was the only outcome when using $KC_{10}H_8$ or the more straightforward base KMe, while no reaction took place with KC_8; for the redox-active samarium, deprotonation took place with $KC_{10}H_8$ but reduction to a Sm(II) species happened when KC_8 was used. However, no Sm(II) product could be characterized and, instead, heating the reaction mixture resulted in the cleavage of the DME solvent molecule (Scheme 6.18(b)).[286] The neodymium-NHC complex was used to prepare the first example of an f-element complex containing a metal-gallium bond (Scheme 6.18(c)).

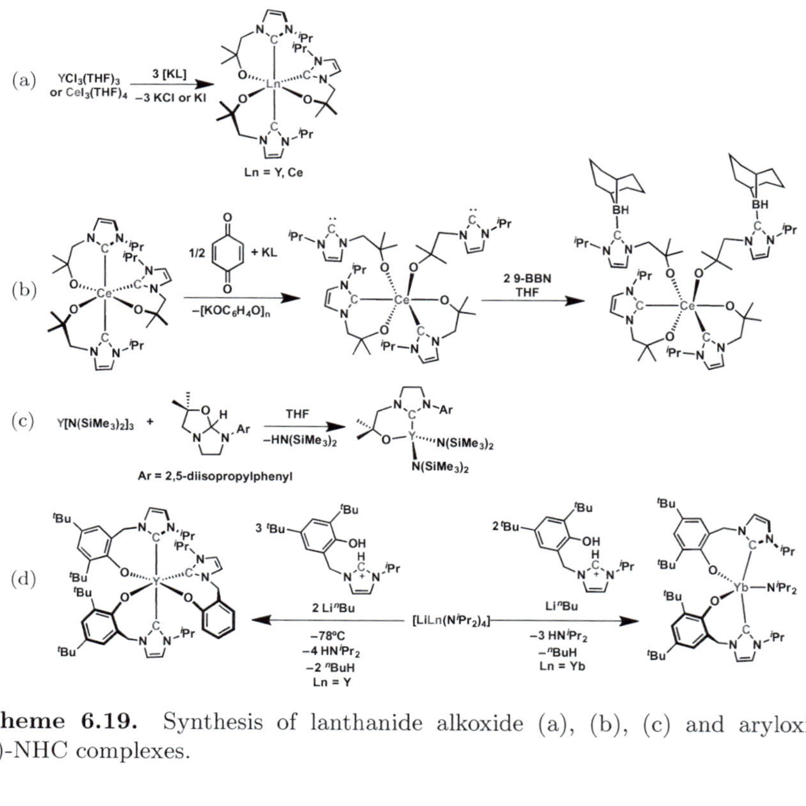

Scheme 6.19. Synthesis of lanthanide alkoxide (a), (b), (c) and aryloxide (d)-NHC complexes.

6.5.3.4 *Alkoxide/aryloxide-NHCs*

Lanthanides are strongly oxophilic so it is not surprising that an alkoxide/aryloxide-functionalized NHC will be a good ancillary ligand for lanthanides. Arnold *et al.* introduced an alkoxide-NHC to yttrium by the salt-metathesis reaction between $YCl_3(THF)_3$ and the potassium salt of the alkoxide-NHC.[287] The resulting tris-alkoxide-NHC yttrium complex has a pseudo-octahedral geometry. The same group also synthesized the cerium(III) analogue and was able to oxidize the cerium(III) tris-alkoxide-NHC complex to cerium(IV) using benzoquinone as an oxidant (Scheme 6.19(a)).[288] In the cerium(IV) complex, four alkoxide-NHC ligands bind to cerium(IV), but only two carbene donors actually coordinate to the metal center, while the other two are pendant and could be quenched by the addition of the Lewis acid 9-BBN (9-borabicylco[3.3.1]nonane) to afford a borane adduct (Scheme 6.19(b)). The flexibility of the coordination mode of the NHC ligand, either bidentate or monodentate, offers potential in bifuncational reactivity or even catalysis. Arnold's group also reported an yttrium-NHC complex with a saturated backbone and a tethered alkoxide arm (Scheme 6.19(c)). The $Y{-}C_{carbene}$ distance was found to be longer than that of other unsaturated Y-NHC complexes and so was the smaller $^1J_{YC}$ value. Shen *et al.* introduced aryloxide-NHC ligand to lanthanide chemistry. By treating $[LiY(N^iPr_2)_4]$ with the corresponding NHC chloride salt in the presence of Li^nBu with the stoichiometry Y:L:Li = 1:3:2, an yttrium tris-aryloxide-NHC complex was obtained (Scheme 6.19(d)).[289] The same group also reported the synthesis of an ytterbium(III) bis-aryloxide-NHC amide complex around the same time, using a similar strategy but a different stoichiometry (Yb:L:Li = 1:2:1, Scheme 6.19(d)).[290]

6.5.3.5 *Indenyl/fluorenyl-NHCs*

Indenyl and fluorenyl can be considered as fused cyclopentadienyl derivatives and they have been tethered on NHCs to create a multi-dentate ligand, which can coordinate to a metal at both cyclopentadienyl and carbene sides. Cui *et al.* reported the synthesis of indenyl/fluorenyl complexes of several rare-earth metals, including scandium, yttrium, holmium and lutetium.[291,292] The synthesis is straightforward using alkane elimination between a protonated indenyl/fluorenyl-NHC and $Ln(CH_2SiMe_3)_3(THF)_2$ (Eq. (6.11)). The lanthanide fluorenyl-NHC complexes were found to catalyse a highly 3,4-selective living polymerization of isoprene.[292] Independently, Danopoulos reported a structurally similar yttrium indenyl-NHC complex bearing one bis(trimethylsilyl)methyl and one bridging bromide ligand.[293]

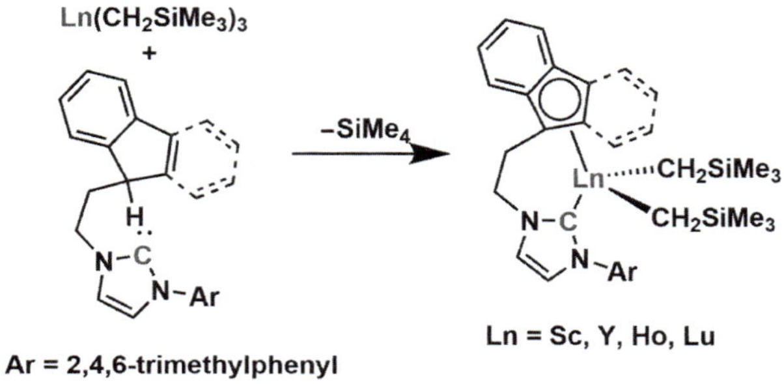

(6.11)

6.5.3.6 *Miscellaneously functionalized NHCs*

The N-functionalisation of an NHC can be versatile and some functionalized NHC ligands cannot be simply included in any of the previous three classes. Cui *et al.* managed to coordinate a tridentate CCC-pincer type bis-carbene ligand to scandium to afford an NHC-metal dibromide complex by *in situ* deprotonation of the ligand precursor with Li^nBu in the presence of $ScCl_3$ (Scheme 6.20(a)).[294] The solid-state structure of the pincer-type bis-NHC scandium complex is shown in Figure 6.32. Arnold *et al.* published the synthesis of a structurally related yttrium complex of a CNC-pincer type bis-carbene ligand from $Y[N(SiMe_3)_2]_3$ and the lithium chloride adduct of the amino-bis-carbene (Scheme 6.20(b)). Shen *et al.* reported the use of a salicylaldimine-functionalized NHC to synthesize a neodymium bis-NHC monobromide complex by using $[LiNd(^iPr_2)_4]$ as the metal precursor and the bromide salt of the functionalized NHC (Scheme 6.20(c)),[295] the same method exploited as in the previous preparation of the lanthanide aryloxide-NHC complex.[289] All these multi-dentate *N*-functionalized carbene or bis-carbene complexes share similar structural and NMR spectroscopic data with their bidentate analogous.[257]

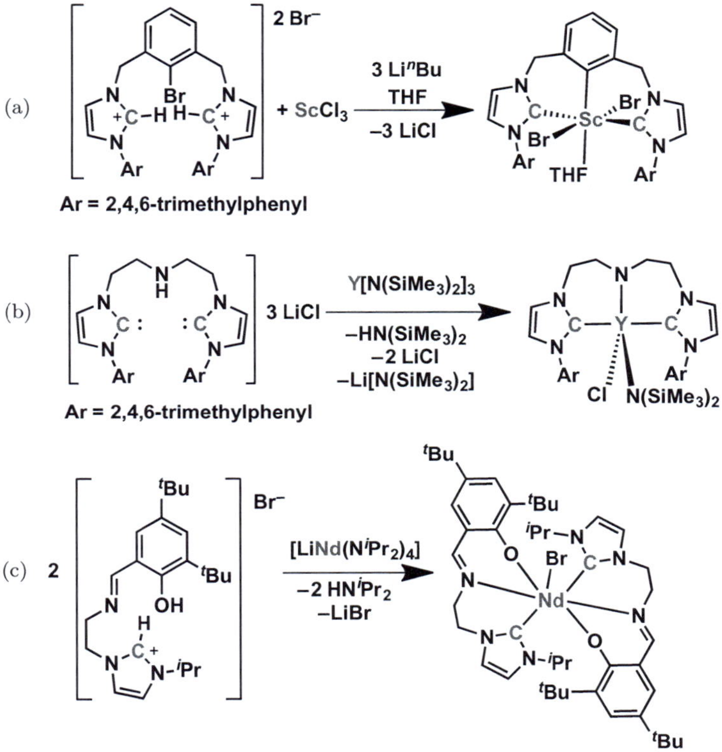

Scheme 6.20. Synthesis of miscellaneously functionalized NHC complexes; (a) aryl bridged NHC, (b) N-bridged NHC, (c) Schiff base NHC.

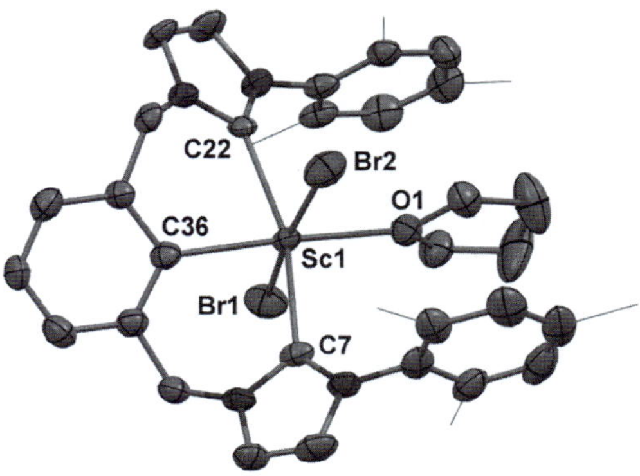

Figure 6.32. Molecular structure of CCC-pincer type NHC scandium complex. Adapted with permission from Ref. [295]. Copyright (2008) American Chemical Society.

(a)

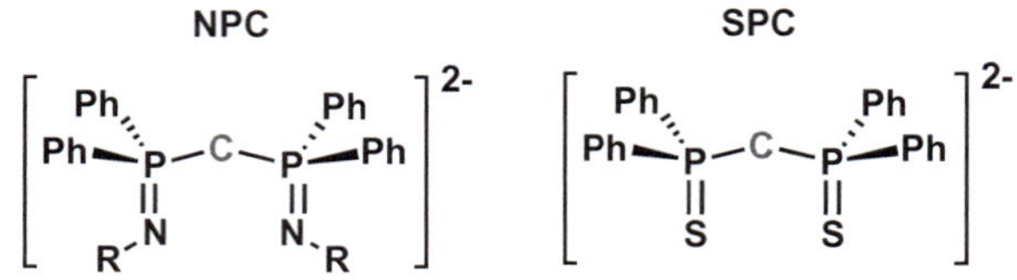

(b)

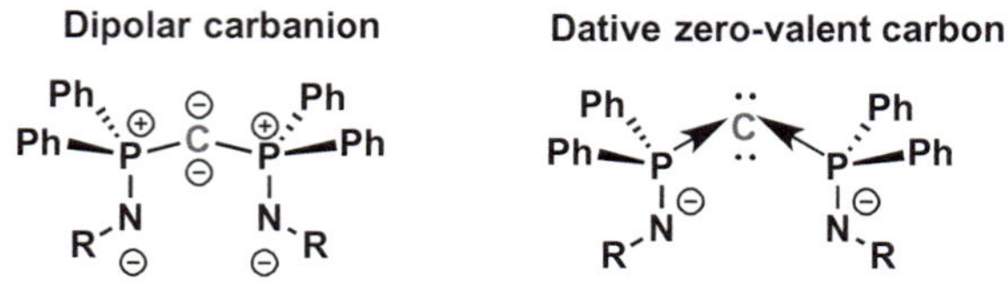

Figure 6.33. (a) Bis(diphenyliminophosphorano)carbenes (NPCs) and bis (diphenylthiophosphinoyl)carbenes (SPCs); (b) Dominant resonance structures for an NPC.

6.5.4 *Lanthanide phosphorus-stabilized carbene complexes*

Aside from NHCs, a new class of carbene-like ligands, bis(phosphorus-stabilized)carbene, has emerged in recent years. Instead of stabilizing the carbene carbon by a lone pair from the neighbouring nitrogen atoms in NHC, in phosphorus-stabilized carbenes, the carbene carbon is best described as an eight electron carbanion stabilized through back-donation to high-valent phosphorus(V). Two most commonly used classes of bis(phosphorus-stabilized)carbenes are depicted in Figure 6.33(a), bis(diphenyliminophosphorano) carbenes (NPCs) and bis(diphenylthiophosphinoyl) carbenes (SPCs). Because of the presence of the high-valent phosphorus, several resonance structures could be drawn for NPCs and SPCs. DFT calculations on the NPC moiety suggested that the dominant resonance form is the dipolar N^{-}-P^{+}-C^{2-}-P^{+}-N^{-} structure, while the possibility of a zero-valent carbon coordinated by two amido-functionalized P(III) ligands cannot be ruled out (Figure 6.33(b)).[296] Liddle *et al.* have recently reviewed early metal bis(phosphorus-stabilized)carbene chemistry for group 1–4 metals as well as the f-elements.[258] In this section, we will summarize the major advances in this field and highlight important aspects of this class of complexes, including their synthesis and electronic structures. Also recently, a scandium mono(phosphorous-stabilized)carbene (non-pincer type)

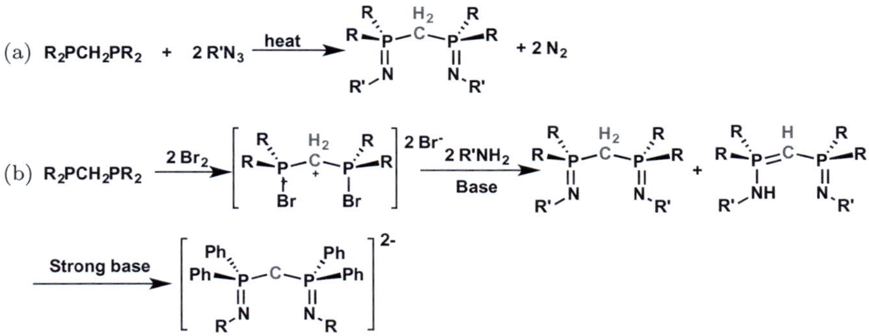

Scheme 6.21. Synthetic protocols for the preparation of an NPC: (a) Phospha-Staudinger method; (b) Kirsanov method.

complex was isolated and studied in detail with regard to its bonding and reactivity.[297] This is a big step forward to realize the synthesis of a non-stabilized, non-bridging alkylidene complex of rare-earth metals.

6.5.4.1 *NPC*

The Phospha-Staudinger method was used to prepare bis(iminophosphorano)methane, the neutral precursor of NPC. Reacting a bis(phosphino)methane and an organic azide will afford the desired bis(iminophosphorano)methane product with concomitant elimination of dinitrogen (Scheme 6.21(a)).[298] The reaction has a wide substrate scope for both bis(phosphino)methanes and organic azides, where R could be Me, Ph, or cyclohexyl, and R′ could be silyl, alkyl, or aryl. In the case when the alkyl azide is unavailable or extremely unstable, and thus dangerous to handle, the Kirsanov method serves as an attractive alternative: oxidation of bis(phosphino) methane by bromine followed by treatment with a primary amine (Scheme 6.21(b)).[299] Despite the formation of a mixture of bis-imino and imino-amine tautomers that could be generated from the reaction, subsequent deprotonation of the mixture will generate a single product of a dianionic NPC.

Cavell *et al.* reported the first lanthanide NPC complex, $Sm[C(PPh_2NSiMe_3)_2](NCy_2)(THF)$ (Cy = cyclohexyl) in 2000. The complex was synthesized from the reaction of bis(diphenyliminophosphorano)methane and the homoleptic samarium tris

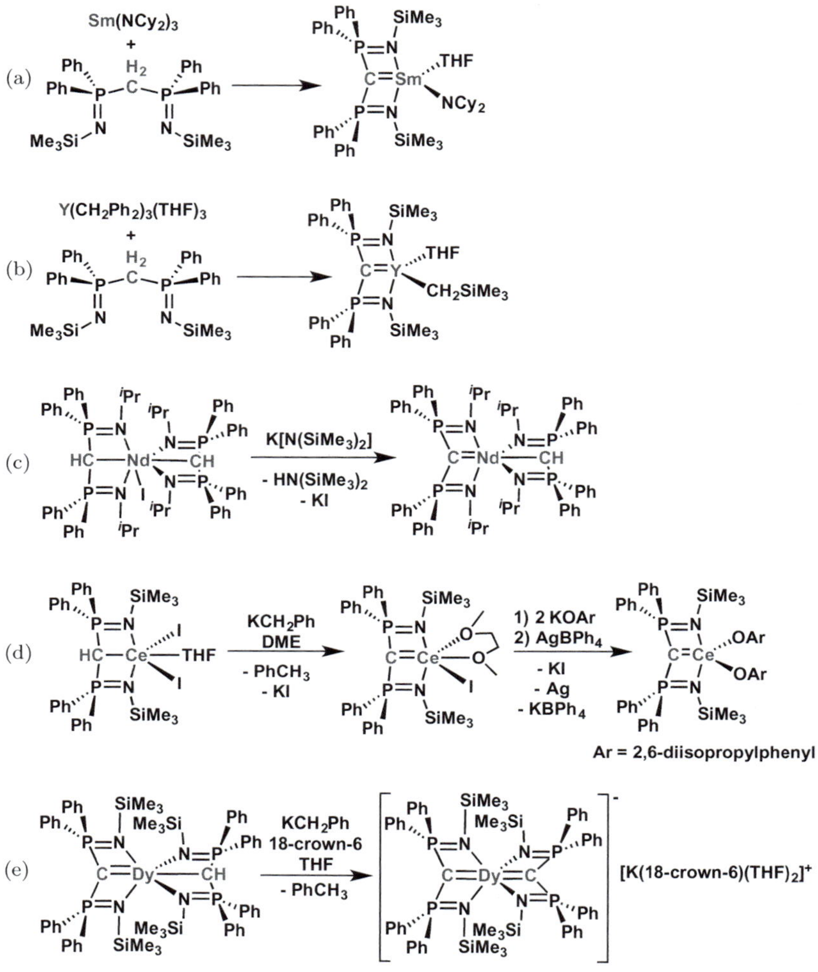

Scheme 6.22. Synthesis of lanthanide-NPC complexes: (a) $Sm[C(PPh_2NSiMe_3)_2](NCy_2)(THF)$; (b) $Y[C(PPh_2NSiMe_3)_2]$ $(CH_2Ph)(THF)$; (c) $Nd[C(PPh_2N^iPr)_2][CH(PPh_2N^iPr)_2]$; (d) $Ce[C(PPh_2NSiMe_3)_2](OAr)_2$; (e) $[K(18\text{-crown-6})(THF)_2](Dy[C(PPh_2NSiMe_3)_2]_2)$.

(dicyclohexylamide) complex (Scheme 6.22(a)).[300] The solid-state structure of $Sm[C(PPh_2NSiMe_3)_2](NCy_2)(THF)$ is shown in Figure 6.34(a). In 2008, Liddle *et al.* reported a related yttrium NPC complex, $Y[C(PPh_2NSiMe_3)_2]$ $(CH_2SiMe_3)(THF)$, obtained by the alkane elimination reaction between yttrium

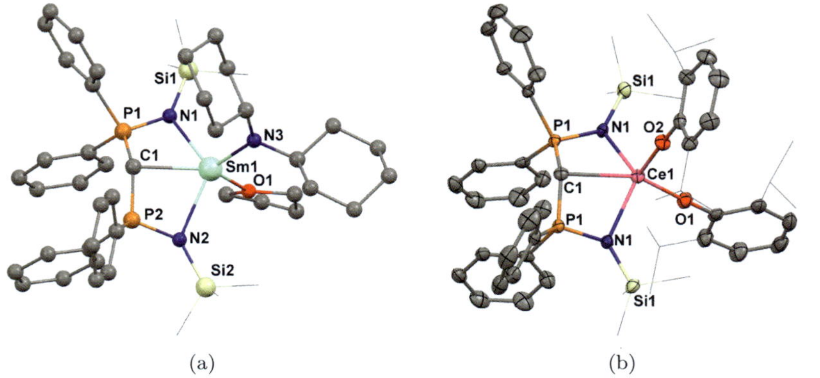

Figure 6.34. Molecular structures of (a) the first lanthanide-NPC complex $Sm[C(PPh_2NSiMe_3)_2](NCy_2)(THF)$ and (b) cerium(IV)-NPC complex. Adapted with permission from Ref. [301], copyright (2000) American Chemical Society. And Ref. [307]. Copyright (2013) Wiley-VCH Verlag GmbH & Co. KGaA, Weinheim.

tris(trimethylsilyl)methyl and the parent NPC.[301] Albeit both complexes seemed to contain a metal-carbon double bond at a first glance, a detailed spectroscopic and computational study suggested that the bonding between lanthanide and carbon is best described as a highly polarized interaction with the electron density predominantly localized on carbon and the nitrogens of the NPC ligand, with minimal metal contributions. Liddle's group subsequently reported the benzyl analogue $Y[C(PPh_2NSiMe_3)_2](CH_2Ph)(THF)$ from $Y(CH_2Ph)_3$ (Scheme 6.22(b))[302] and later published the *in situ* method using $LnCl_3$, KCH_2Ph, and the parent NPC for the synthesis of the corresponding NPC complexes for the whole lanthanide series: while for the small size lanthanide ions, this method was sufficient to afford mono-NPC lanthanide alkyl products, for the middle to large size lanthanides, the alkane elimination reaction was less selective and resulted in the formation of a mixed carbene-methanide complex. The carbene and methanide ligands could be differentiated by their geometry: the former has an almost planar CP_2N_2M core, while the latter has a boat conformation of the core. The same group also used $Ln(CH_2Ph)_2I$ as starting materials to afford the synthetically useful $Ln[C(PPh_2NSiMe_3)_2(I)(THF)_2$ (Ln = Y, Er) complexes.[303] When using LnI_3 and the cesium salt of a monoanionic NPC, lanthanide methanide diiodide complexes were obtained, but subsequent deprotonation still resulted in ligand scrambling to end up in a mixed carbene-methanide complex.[304] Independently, Le Floch *et al.* published the synthesis of an analogous neodymium-NPC complex $Nd[C(PPh_2N^iPr)_2][CH(PPh_2N^iPr)_2]$ around the same time (Scheme 6.22(c)).[305] Recently, Liddle *et al.* reported the synthesis of the first cerium(IV)-NPC complex $Ce[C(PPh_2NSiMe_3)_2](OAr)_2$ (Ar = 2,6-diisopropylphenyl) by the oxidation of the corresponding cerium(III) potassium ate complex (Scheme 6.22(d)).[306] The solid-state structure of $Ce[C(PPh_2NSiMe_3)_2](OAr)_2$ is shown in Figure 6.34(b). Experimental and computational studies of the Ce(IV)-NPC complex by comparison with the structurally similar Th(IV) and U(IV) analogues revealed an appreciable covalent character in the Ce(IV)$-C_{carbene}$ bond, in line with that for U(IV) and more prominent than that for Th(IV).[307] When treating the lanthanide mixed carbene-methanide complex with the strong base KCH_2Ph in the presence of 18-crown-6, the potassium salt of the bis-NPC lanthanide anion could be obtained (Scheme 6.22(e)). The dysprosium bis-NPC complex $[K(18\text{-crown-}6)(THF)_2](Dy[C(PPh_2NSiMe_3)_2]_2)$ was found to behave as a single molecule magnet.[308] The reactivity of the yttrium-NPC complex was explored and showed surprisingly rich chemistry including a cascade C–H activation/insertion reaction with benzophenone,[309] C–F and C–O bond activation[310] and other non-typical transformations.[311] The reactivity of yttrium–NPC complexes varied based on the other ligands coordinated to yttrium, such as alkoxide, iodide and alkyl. This suggested that NPC could serve as an excellent non-innocent supporting ligand.

6.5.4.2 *SPC*

The bis(diphenylthiophosphinoyl)carbene (SPC) is stabilized through a similar dipolar/dative interaction as that found in bis(diphenyliminophosphorano)carbene (NPC). Its synthesis is very straightforward and it is obtained by the direct oxidation of bis(dipenylphosphino)methane by excess elemental sulfur (Scheme 6.23(a)). Le Floch, Nief and Mézailles introduced the SPC ligand $[C(PPh_2S)_2]^{2-}$ in rare-earth metal chemistry. They reported the synthesis of $[Ln[C(PPh_2S)_2](THF)_2(\mu\text{-}I)]_2$ and $[Li(THF)_4](Ln[C(PPh_2S)_2])_2$ by reacting LnI_3 with the lithium salt

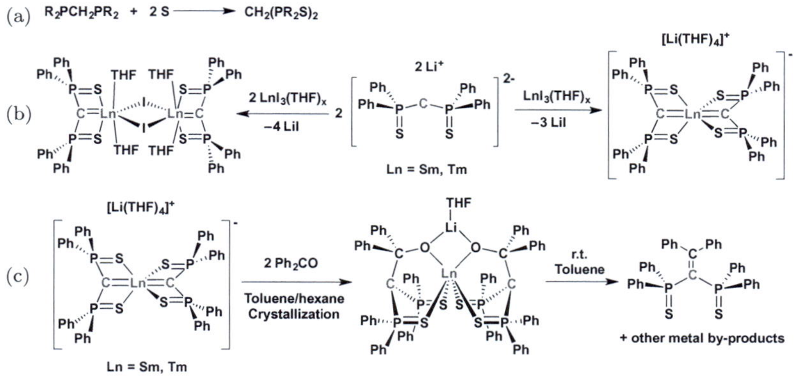

Scheme 6.23. (a) Synthesis of bis(diphenylthiophosphinoyl)methane; (b) Synthesis of $[Ln[C(PPh_2S)_2](THF)_2(\mu\text{-}I)]_2$ and $[Li(THF)_4](Ln[C(PPh_2S)_2])_2$; (c) Intermediate metallo-oxetane complex isolated in the reaction between $[Li(THF)_4](Ln[C(PPh_2S)_2])_2$ and benzophenone (Ln = Sm, Tm).

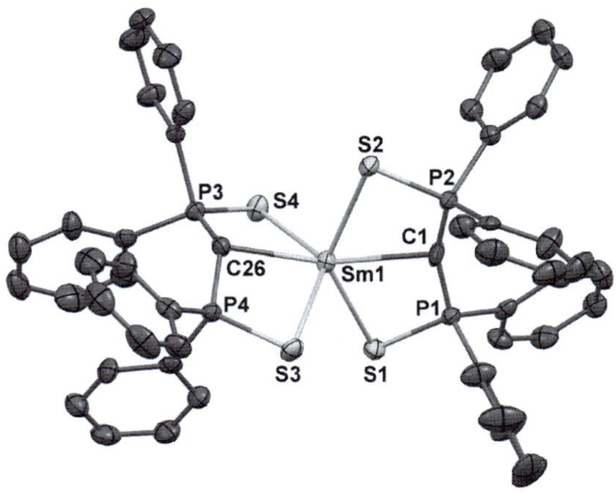

Figure 6.35. Molecular structure of $[Li(THF)_4](Sm[C(PPh_2S)_2])_2$. Hydrogen atoms omitted for clarity. Reproduced from Ref. [313] with permission from the Royal Society of Chemistry.

of $[C(PPh_2S)_2]^{2-}$ (Ln = Sm, Tm) with the correct stoichiometry 2:1 and 1:1, respectively (Scheme 6.23(b)).[312,313] In the solid-state structures (Figure 6.35), all lanthanide-SPC complexes have a planar CP_2S_2M core. The lanthanide-SPC complexes exhibited Tebbe-type reactivity with benzophenone to generate metal oxide products and the organic by-product $Ph_2C{=}C(PPh_2S)_2$. However, the so-called "open" metallo-oxetane species could be isolated when reacting

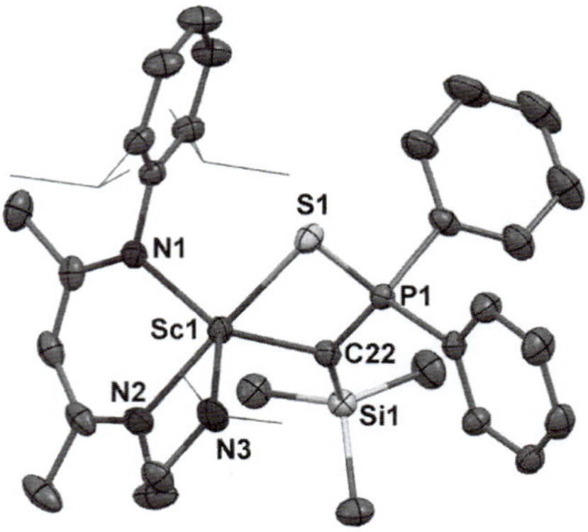

Figure 6.36. Molecular structure of a mononuclear non-pincer type scandium carbene complex. Hydrogen atoms omitted for clarity. Adapted with permission from Ref. [298]. Copyright (2013) Wiley-VCH Verlag GmbH & Co. KGaA, Weinheim.

$[Li(THF)_4](Ln[C(PPh_2S)_2])_2$ with benzophenone (Scheme 6.23(c)). This species was proposed to be an intermediate in the reactions of early metal carbene complexes with molecules containing a carbonyl group but were rarely isolated due to their instability and transient nature. Recently, Mézailles employed a similar strategy to afford a series of scandium-SPC complexes with different monoanionic ligands to study the bonding interaction between scandium and the SCP ligand systematically.[314,315]

6.5.4.3 *Non-pincer type carbenes*

Recently, Chen *et al.* reported a non-pincer type mononuclear scandium alkylidene complex by using a tridentate monoanionic supporting ligand and a (silyl/diphenylthiophosphinoyl)carbene (Eq. (6.12)).[317] It is important to note that the silyl group also has a significant stabilization effect on the neighbouring carbanion,[318] similar to the examples discussed earlier.

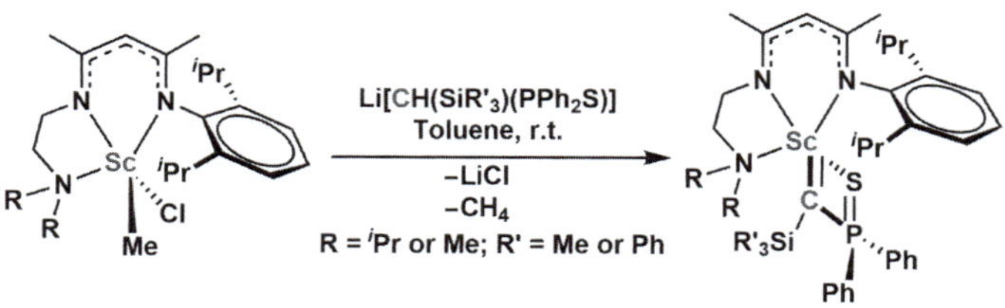

(6.12)

The molecular structures of the scandium carbene complexes featured short Sc-$C_{carbene}$ distances (Figure 6.36). The shortest Sc–$C_{carbene}$ distance in the series was found to be 2.11 Å, much shorter than those in the pincer-type SPC scandium complexes.[312,313] Moreover, this is also the shortest Sc–C distance reported to date. The authors carried DFT calculations on these scandium carbene complexes and found a strong Schrock-type carbene character. Besides the polarized Sc–C σ-bond, the lone pair is mainly localized on the carbon atom but interacts strongly with an empty 3d orbital of scandium. This donor-acceptor interaction has a π-symmetry and was calculated to be very strong with a bonding energy of 82 kcal/mol, much stronger than previously reported for the pincer-type scandium-SPC complex.[314] The preliminary reactivity study showed that the Sc=C double bond is subject to formal [2 + 2] cycloadditions with unsaturated molecules to form pincer-type products. The basicity of the carbene carbon led to C–H activation of the moderately acidic phenylacetylene. Albeit this scandium carbene complex still requires one thiophosphinoyl group for stabilization, it is a step forward to synthesizing genuine rare-earth metal alkylidene complexes. It is worth noting that a phosphorano-stabilized U(IV)-carbene was reported by Hayton *et al.*, in which the carbene is a monodentate ligand.[319]

6.6 Summary and Outlook for the Organometallic Chemistry of Lanthanides

In the last two decades, the organometallic chemistry of lanthanides witnessed a renaissance together with that of the actinides. The trend seems to move in two directions. One is the individuality of each lanthanide ions despite the traditional sense that lanthanides only differ by their sizes but not much in reactivity or properties. This movement is driven by the desire to separate individual lanthanide elements efficiently for production and their unique magnetic and optic properties. The other is the generality of the whole lanthanide series. That is to develop a universal platform for all lanthanide ions. At first glance, the two directions seem to be contrary to each other; however, they are actually organically connected and rely on each other for future advancement. Only systematic studies can reveal the individuality of each lanthanide and a good knowledge of each individual lanthanide is essential for designing a universal platform intended for systematic studies. The key here is the supporting ligand, or to be more specific, the ligand field. For transition metals, crystal field theory and later ligand field theory define their coordination chemistry and lead to rational design for new catalysts or materials. In f-element chemistry, it has been long thought that the ligand field has little if any impact on the electronic configuration of the metal center; thus, no general theory can be sought for design and prediction. However, with efforts led by synthetic chemists around the world, the organometallic chemistry of f-elements now looks more like that of the d-block transition metals. Due to the participation of 5f orbitals in bonding interactions, actinides are found to be more like transition metals than lanthanides. The recent discovery of the non-conventional lanthanide(II) chemistry suggests that 5d orbitals can be occupied and therefore participate in bonding interactions. This opens up a whole new world for lanthanide chemistry since, with electrons in 5d orbitals, those complexes might behave similarly to the nearby d-block elements such as group 4 metals. Rational ligand design to enhance reactivity or physical properties is already on the way, in fields like lanthanide single-molecule magnets.

A very successful approach in organometallic lanthanide chemistry is to exploit ligands with flexible coordination capacity. Notable examples are: cyclopentadienyl ligands capable to adopt η^1, η^3, and η^5-coordination modes, tetraalkylaluminate ligands capable of delivering two or three bridging methyl donors, benzyl ligands that can bend the Ln–C_{benzylic}–C_{ipso} angle to achieve a Ln–C_{ipso} interaction, NN^{TBS} ligand that bears a Lewis basic iron(II) center that can form a weak donor-acceptor type interaction with Lewis acidic lanthanide ions, *N*-functionalized NHC ligands that can have "on" and "off" mode for an NHC carbene donor, and other ancillary ligands that have labile tethered donors such as tertiary amine. The flexibility of the ligands can not only stabilize the lanthanide ions of different sizes but also allow high reactivity by providing open coordination sites for small molecule activation as well as catalysis. Ligand design will continue to be at the center stage of organometallic lanthanide chemistry and is probably the only way to meet the need of both individuality and generality.

References

1. K. Hans Wedepohl, *Geochim. Cosmochim. Acta*, 1995, *59*, 1217–1232.
2. L. J. Nugent, R. D. Baybarz, J. L. Burnett and J. L. Ryan, *J. Phys. Chem.*, 1973, *77*, 1528–1539.
3. L. R. Morss, *Chem. Rev.*, 1976, *76*, 827–841.
4. W. J. Evans, *Inorg. Chem.*, 2007, *46*, 3435–3449.
5. D. A. Atwood Ed., *The Rare Earth Elements: Fundamentals and Applications*, Wiley, Chichester, 2012.
6. P. B. Hitchcock, M. F. Lappert, L. Maron and A. V. Protchenko, *Angew. Chem. Int. Ed.*, 2008, *47*, 1488–1491.
7. G. Meyer, *Angew. Chem. Int. Ed.*, 2008, *47*, 4962–4964.
8. M. R. MacDonald, J. W. Ziller and W. J. Evans, *J. Am. Chem. Soc.*, 2011, *133*, 15914–15917.
9. M. R. MacDonald, J. E. Bates, M. E. Fieser, J. W. Ziller, F. Furche and W. J. Evans, *J. Am. Chem. Soc.*, 2012, *134*, 8420–8423.
10. M. R. MacDonald, J. E. Bates, J. W. Ziller, F. Furche and W. J. Evans, *J. Am. Chem. Soc.*, 2013, *135*, 9857–9868.
11. M. E. Fieser, M. R. MacDonald, B. T. Krull, J. E. Bates, J. W. Ziller, F. Furche and W. J. Evans, *J. Am. Chem. Soc.*, 2015, *137*, 369–382.
12. W. J. Evans, *Organometallics*, 2016, *35*, 3088–3100.
13. M. R. MacDonald, M. E. Fieser, J. E. Bates, J. W. Ziller, F. Furche and W. J. Evans, *J. Am. Chem. Soc.*, 2013, *135*, 13310–13313.
14. G. Meyer, *Angew. Chem. Int. Ed.*, 2014, *53*, 3550–3551.
15. C. J. Windorff, M. R. MacDonald, K. R. Meihaus, J. W. Ziller, J. R. Long and W. J. Evans, *Chem. Eur. J.*, 2016, *22*, 772–782.
16. R. R. Langeslay, M. E. Fieser, J. W. Ziller, F. Furche and W. J. Evans, *Chem. Sci.*, 2015, *6*, 517–521.
17. D. N. Woodruff, R. E. P. Winpenny and R. A. Layfield, *Chem. Rev.*, 2013, *113*, 5110–5148.
18. R. A. Layfield, *Organometallics*, 2014, *33*, 1084–1099.
19. R. A. Layfield and M. Murugesu (Eds.), *Lanthanides and Actinides in Molecular Magnetism*, John Wiley & Sons, Hoboken, 2015.
20. P. Zhang, L. Zhang and J. Tang, *Dalton Trans.*, 2015, *44*, 3923–3929.
21. J.-C. G. Bünzli, *Chem. Rev.*, 2010, *110*, 2729–2755.
22. W. J. Evans, T. A. Ulibarri and J. W. Ziller, *J. Am. Chem. Soc.*, 1988, *110*, 6877–6879.
23. W. J. Evans, D. S. Lee and J. W. Ziller, *J. Am. Chem. Soc.*, 2004, *126*, 454–455.
24. W. J. Evans, D. S. Lee, C. Lie and J. W. Ziller, *Angew. Chem. Int. Ed.*, 2004, *43*, 5517–5519.

25. W. J. Evans, M. Fang, G. Zucchi, F. Furche, J. W. Ziller, R. M. Hoekstra and J. I. Zink, *J. Am. Chem. Soc.*, 2009, *131*, 11195–11202.
26. M. Fang, J. E. Bates, S. E. Lorenz, D. S. Lee, D. B. Rego, J. W. Ziller, F. Furche and W. J. Evans, *Inorg. Chem.*, 2011, *50*, 1459–1469.
27. M. Fang, D. S. Lee, J. W. Ziller, R. J. Doedens, J. E. Bates, F. Furche and W. J. Evans, *J. Am. Chem. Soc.*, 2011, *133*, 3784–3787.
28. J. D. Rinehart, M. Fang, W. J. Evans and J. R. Long, *Nature Chem.*, 2011, *3*, 538–542.
29. J. D. Rinehart, M. Fang, W. J. Evans and J. R. Long, *J. Am. Chem. Soc.*, 2011, *133*, 14236–14239.
30. R. Shannon, *Acta Cryst.*, 1976, *A32*, 751–767.
31. N. G. Connelly and W. E. Geiger, *Chem. Rev.*, 1996, *96*, 877–910.
32. V. Nair, L. Balagopal, R. Rajan and J. Mathew, *Acc. Chem. Res.*, 2004, *37*, 21–30.
33. K. Jacob and K. H. Thiele, *Z. Anorg. Allg. Chem.*, 1986, *543*, 192–198.
34. J. A. Bogart, A. J. Lewis, S. A. Medling, N. A. Piro, P. J. Carroll, C. H. Booth and E. J. Schelter, *Inorg. Chem.*, 2013, *52*, 11600–11607.
35. J. E. Kim, P. J. Carroll and E. J. Schelter, *New J. Chem.*, 2015, *39*, 6076–6084.
36. J. R. Levin, W. L. Dorfner, P. J. Carroll and E. J. Schelter, *Chem. Sci.*, 2015, *6*, 6925–6934.
37. J. E. Kim, P. J. Carroll and E. J. Schelter, *Chem. Commun.*, 2015, *51*, 15047–15050.
38. W. Huang, B. M. Upton, S. I. Khan and P. L. Diaconescu, *Organometallics*, 2013, *32*, 1379–1386.
39. G. W. Watt and E. W. Gillow, *J. Am. Chem. Soc.*, 1969, *91*, 775–776.
40. G. Wilkinson and J. M. Birmingham, *J. Am. Chem. Soc.*, 1954, *76*, 6210.
41. A. Greco, S. Cesca and W. Bertolini, *J. Organomet. Chem.*, 1976, *113*, 321–330.
42. M. N. Bochkarev, I. L. Fedushkin, A. A. Fagin, T. V. Petrovskaya, J. W. Ziller, R. N. R. Broomhall-Dillard and W. J. Evans, *Angew. Chem. Int. Ed.*, 1997, *36*, 133–135.
43. W. J. Evans, N. T. Allen and J. W. Ziller, *J. Am. Chem. Soc.*, 2000, *122*, 11749–11750.
44. M. N. Bochkarev, I. L. Fedushkin, S. Dechert, A. A. Fagin and H. Schumann, *Angew. Chem. Int. Ed.*, 2001, *40*, 3176–3178.
45. W. J. Evans, N. T. Allen and J. W. Ziller, *Angew. Chem. Int. Ed.*, 2002, *41*, 359–361.
46. M. N. Bochkarev, *Coord. Chem. Rev.*, 2004, *248*, 835–851.
47. A. A. Fagin, M. N. Bochkarev, S. A. Kozimor, J. W. Ziller and W. J. Evans, *Z. Anorg. Allg. Chem.*, 2005, *631*, 2848–2853.

48. F. Jaroschik, F. Nief, X.-F. Le Goff and L. Ricard, *Organometallics*, 2007, *26*, 1123–1125.
49. F. Jaroschik, F. Nief, X.-F. Le Goff and L. Ricard, *Organometallics*, 2007, *26*, 3552–3558.
50. F. Nief, *Dalton Trans.*, 2010, *39*, 6589–6598.
51. G. Meyer, *Chem. Rev.*, 1988, *88*, 93–107.
52. G. Meyer and H. J. Meyer, *Chem. Mater.*, 1992, *4*, 1157–1168.
53. G. Meyer and N. Gerlitzki, *J. Alloys Compd.*, 2004, *380*, 71–78.
54. G. Meyer, *Z. Anorg. Allg. Chem.*, 2007, *633*, 2537–2552.
55. W. J. Evans, *Coord. Chem. Rev.*, 2000, *206–207*, 263–283.
56. F. Nief, Molecular Chemistry of the Rare-Earth Elements in Uncommon Low-Valent States, in *Handbook on the Physics and Chemistry of Rare Earths* (Eds. K. A. Gschneidner Jr. and J.-C. G. Bünzli, K. P. Vitalij), Vol. 40, Ch. 246, Elsevier, Amsterdam, 2010, pp. 241–300.
57. M. C. Cassani, D. J. Duncalf and M. F. Lappert, *J. Am. Chem. Soc.*, 1998, *120*, 12958–12959.
58. P. L. Diaconescu, P. L. Arnold, T. A. Baker, D. J. Mindiola and C. C. Cummins, *J. Am. Chem. Soc.*, 2000, *122*, 6108–6109.
59. W. J. Evans, S. A. Kozimor, J. W. Ziller and N. Kaltsoyannis, *J. Am. Chem. Soc.*, 2004, *126*, 14533–14547.
60. B. Vlaisavljevich, P. L. Diaconescu, W. L. Lukens Jr., L. Gagliardi and C. C. Cummins, *Organometallics*, **2013**, *32*, 1341–1352.
61. C. J. Windorff, G. P. Chen, J. N. Cross, W. J. Evans, F. Furche, A. J. Gaunt, M. T. Janicke, S. A. Kozimor and B. L. Scott, *J. Am. Chem. Soc.*, 2017, *139*, 3970–3973.
62. J. F. Corbey, D. H. Woen, C. T. Palumbo, M. E. Fieser, J. W. Ziller, F. Furche and W. J. Evans, *Organometallics*, 2015, *34*, 3909–3921.
63. A. Sekiguchi, K. Ebata, C. Kabuto and H. Sakurai, *J. Am. Chem. Soc.*, 1991, *113*, 1464–1465.
64. W. J. Evans and B. L. Davis, *Chem. Rev.*, 2002, *102*, 2119–2136.
65. J. K. Peterson, M. R. MacDonald, J. W. Ziller and W. J. Evans, *Organometallics*, 2013, *32*, 2625–2631.
66. A. Clearfield, D. K. Warner, C. H. Saldarriaga-Molina, R. Ropal and I. Bernal, *Can. J. Chem.*, 1975, *53*, 1622–1629.
67. R. Jungst, D. Sekutowski, J. Davis, M. Luly and G. Stucky, *Inorg. Chem.*, 1977, *16*, 1645–1655.
68. R. D. Rogers, R. V. Bynum and J. L. Atwood, *J. Am. Chem. Soc.*, 1978, *100*, 5238–5239.
69. W. L. Lukens Jr. and R. A. Andersen, *Organometallics*, 1995, *14*, 3435–3439.

70. K. R. Meihaus, M. E. Fieser, J. F. Corbey, W. J. Evans and J. R. Long, *J. Am. Chem. Soc.*, 2015, *137*, 9855–9860.
71. C. M. Kotyk, M. R. MacDonald, J. W. Ziller and W. J. Evans, *Organometallics*, 2015, *34*, 2287–2295.
72. C. M. Kotyk, M. E. Fieser, C. T. Palumbo, J. W. Ziller, L. E. Darago, J. R. Long, F. Furche and W. J. Evans, *Chem. Sci.*, 2015, *6*, 7267–7273.
73. J. G. Brennan, F. G. N. Cloke, A. A. Sameh and A. Zalkin, *J. Chem. Soc. Chem. Commun.*, 1987, 1668–1669.
74. D. M. Anderson, F. G. N. Cloke, P. A. Cox, N. Edelstein, J. C. Green, T. Pang, A. A. Sameh and G. Shalimoff, *J. Chem. Soc., Chem. Commun.*, 1989, 53–55.
75. W. A. King, T. J. Marks, D. M. Anderson, D. J. Duncalf and F. G. N. Cloke, *J. Am. Chem. Soc.*, 1992, *114*, 9221–9223.
76. F. G. N. Cloke, *Chem. Soc. Rev.*, 1993, *22*, 17–24.
77. P. L. Arnold, M. A. Petrukhina, V. E. Bochenkov, T. I. Shabatina, V. V. Zagorskii, G. B. Sergeev and F. G. N. Cloke, *J. Organomet. Chem.*, 2003, *688*, 49–55.
78. J. D. Martin and J. D. Corbett, *Angew. Chem. Int. Ed.*, 1995, *34*, 233–235.
79. M. Ryazanov, L. Kienle, A. Simon and H. Mattausch, *Inorg. Chem.*, 2006, *45*, 2068–2074.
80. M. Xémard, A. Jaoul, M. Cordier, F. Molton, O. Cador, B. Le Guennic, C. Duboc, O. Maury, C. Clavaguéra and G. Nocton, *Angew. Chem. Int. Ed.*, 2017, *56*, 4266–4271.
81. F. M. Bickelhaupt, M. Solà and C. Fonseca Guerra, *J. Chem. Theory Comput.*, 2006, *2*, 965–980.
82. L. Pauling, *J. Am. Chem. Soc.*, 1932, *54*, 3570–3582.
83. A. L. Allred, *J. Inorg. Nucl. Chem.*, 1961, *17*, 215–221.
84. F. A. Cotton, *Chem. Rev.*, 1955, *55*, 551–594.
85. F. A. Hart, A. G. Massey and M. S. Saran, *J. Organomet. Chem.*, 1970, *21*, 147–154.
86. S. A. Cotton, F. A. Hart, M. B. Hursthouse and A. J. Welch, *J. Chem. Soc. Chem. Commun.*, 1972, 1225–1226.
87. H. Schumann and J. Müller, *Angew. Chem. Int. Ed.*, 1978, *17*, 276–276.
88. S. A. Cotton, *Coord. Chem. Rev.*, 1997, *160*, 93–127.
89. M. Zimmermann and R. Anwander, *Chem. Rev.*, 2010, *110*, 6194–6259.
90. P. L. Watson, *J. Am. Chem. Soc.*, 1983, *105*, 6491–6493.
91. M. E. Thompson, S. M. Baxter, A. R. Bulls, B. J. Burger, M. C. Nolan, B. D. Santarsiero, W. P. Schaefer and J. E. Bercaw, *J. Am. Chem. Soc.*, 1987, *109*, 203–219.

92. H. M. Dietrich, G. Raudaschl-Sieber and R. Anwander, *Angew. Chem. Int. Ed.*, 2005, *44*, 5303–5306.
93. S. Arndt, K. Beckerle, P. M. Zeimentz, T. P. Spaniol and J. Okuda, *Angew. Chem. Int. Ed.*, 2005, *44*, 7473–7477.
94. M. U. Kramer, D. Robert, S. Arndt, P. M. Zeimentz, T. P. Spaniol, A. Yahia, L. Maron, O. Eisenstein and J. Okuda, *Inorg. Chem.*, 2008, *47*, 9265–9278.
95. S. Arndt, T. P. Spaniol and J. Okuda, *Angew. Chem. Int. Ed.*, 2003, *42*, 5075–5079.
96. M. R. MacDonald, R. R. Langeslay, J. W. Ziller and W. J. Evans, *J. Am. Chem. Soc.*, 2015, *137*, 14716–14725.
97. M. D. Fryzuk, G. Giesbrecht and S. J. Rettig, *Organometallics*, 1996, *15*, 3329–3336.
98. T. I. Gountchev and T. D. Tilley, *Organometallics*, 1999, *18*, 2896–2905.
99. P. G. Hayes, W. E. Piers, L. W. M. Lee, L. K. Knight, M. Parvez, M. R. J. Elsegood and W. Clegg, *Organometallics*, 2001, *20*, 2533–2544.
100. A. L. Wayda and W. J. Evans, *J. Am. Chem. Soc.*, 1978, *100*, 7119–7121.
101. H. Schumann, J. Mueller, N. Bruncks, H. Lauke, J. Pickardt, H. Schwarz and K. Eckart, *Organometallics*, 1984, *3*, 69–74.
102. M. F. Lappert and R. Pearce, *J. Chem. Soc. Chem. Commun.*, 1973, 126.
103. M. Niemeyer, *Z. Anorg. Allg. Chem.*, 2000, *626*, 1027–1029.
104. H. Schumann, D. M. M. Freckmann and S. Dechert, *Z. Anorg. Allg. Chem.*, 2002, *628*, 2422–2426.
105. H. Schumann and J. Müller, *J. Organomet. Chem.*, 1978, *146*, C5–C7.
106. K. A. Rufanov, D. M. M. Freckmann, H.-J. Kroth, S. Shutte and H. Z. Shumann, *Naturforsch. B: Chem. Sci.*, 2005, *60b*, 533.
107. S. Bambirra, F. Perazzolo, S. J. Boot, T. J. J. Sciarone, A. Meetsma and B. Hessen, *Organometallics*, 2008, *27*, 704–712.
108. D. J. H. Emslie, W. E. Piers, M. Parvez and R. McDonald, *Organometallics*, 2002, *21*, 4226–4240.
109. G. K. Barker and M. F. Lappert, *J. Organomet. Chem.*, 1974, *76*, C45–C46.
110. P. B. Hitchcock, M. F. Lappert, R. G. Smith, R. A. Bartlett and P. P. Power, *J. Chem. Soc., Chem. Commun.*, 1988, 1007–1009.
111. A. G. Avent, C. F. Caro, P. B. Hitchcock, M. F. Lappert, Z. Li and X.-H. Wei, *Dalton Trans.*, 2004, 1567–1577.
112. P. B. Hitchcock, S. A. Holmes, M. F. Lappert and S. Tian, *J. Chem. Soc., Chem. Commun.*, 1994, 2691–2692.

113. J. R. van den Hende, P. B. Hitchcock, S. A. Holmes, M. F. Lappert and S. Tian, *J. Chem. Soc., Dalton Trans.*, 1995, 3933–3939.
114. P. B. Hitchcock, A. V. Khvostov and M. F. Lappert, *J. Organomet. Chem.*, 2002, *663*, 263–268.
115. C. Eaborn, P. B. Hitchcock, K. Izod and J. D. Smith, *J. Am. Chem. Soc.*, 1994, *116*, 12071–12072.
116. C. Eaborn, P. B. Hitchcock, K. Izod, Z.-R. Lu and J. D. Smith, *Organometallics*, 1996, *15*, 4783–4790.
117. C. Eaborn, M. S. Hill, P. B. Hitchcock, J. D. Smith, S. Zhang and T. Ganicz, *Organometallics*, 1999, *18*, 2342–2348.
118. S. Bambirra, A. Meetsma and B. Hessen, *Organometallics*, 2006, *25*, 3454–3462.
119. N. Meyer, P. W. Roesky, S. Bambirra, A. Meetsma, B. Hessen, K. Saliu and J. Takats, *Organometallics*, 2008, *27*, 1501–1505.
120. S. Ge, A. Meetsma and B. Hessen, *Organometallics*, 2009, *28*, 719–726.
121. S. Harder, C. Ruspic, N. N. Bhriain, F. Berkermann and M. Z. Schürmann, *Naturforsch. B: Chem. Sci.*, 2008, *63b*, 267.
122. C. T. Carver, M. J. Monreal and P. L. Diaconescu, *Organometallics*, 2008, *27*, 363–370.
123. A. J. Wooles, D. P. Mills, W. Lewis, A. J. Blake and S. T. Liddle, *Dalton Trans.*, 2010, *39*, 500–510.
124. K. Izod, S. T. Liddle and W. Clegg, *Inorg. Chem.*, 2004, *43*, 214–218.
125. G. G. Meyer and J. D. Corbett, *Inorg. Synth.*, John Wiley & Sons, Inc., New York, 2007, p. 146.
126. N. Meyer, P. W. Roesky, S. Bambirra, A. Meetsma, B. Hessen, K. Saliu and J. Takats, *Organometallics*, 2013, *32*, 3427–3427.
127. L. E. Manzer, *J. Organomet. Chem.*, 1977, *135*, C6–C9.
128. L. E. Manzer, *J. Am. Chem. Soc.*, 1978, *100*, 8068–8073.
129. S. Harder, *Organometallics*, 2005, *24*, 373–379.
130. W.-X. Zhang, M. Nishiura, T. Mashiko and Z. Hou, *Chem. Eur. J.*, 2008, *14*, 2167–2179.
131. F. Jaroschik, T. Shima, X. Li, K. Mori, L. Ricard, X.-F. Le Goff, F. Nief and Z. Hou, *Organometallics*, 2007, *26*, 5654–5660.
132. M. A. Putzer and G. P. Bartholomew, *Z. Anorg. Allg. Chem.*, 1999, *625*, 1777–1778.
133. M. A. Putzer, J. S. Rogers and G. C. Bazan, *J. Am. Chem. Soc.*, 1999, *121*, 8112–8113.
134. L. N. Bochkarev, T. A. Stepantseva, L. N. Zakharov, G. K. Fukin, A. I. Yanovsky and Y. T. Struchkov, *Organometallics*, 1995, *14*, 2127–2129.

135. L. N. Bochkarev, T. A. Zheleznova, A. V. Safronova, M. S. Drozdov, S. F. Zhil'tsov, L. N. Zakharov, G. K. Fukin and S. Y. Khorshev, *Russ. Chem. Bull.*, 1998, *47*, 165–168.
136. M. N. Bochkarev, V. V. Khramenkov, Y. F. Rad'kov, L. N. Zakharov and Y. T. Struchkov, *J. Organomet. Chem.*, 1992, *429*, 27–39.
137. A. L. Wayda, J. L. Atwood and W. E. Hunter, *Organometallics*, 1984, *3*, 939–941.
138. A. L. Wayda and R. D. Rogers, *Organometallics*, 1985, *4*, 1440–1444.
139. C. M. Forsyth and G. B. Deacon, *Organometallics*, 2000, *19*, 1205–1207.
140. G. B. Deacon and C. M. Forsyth, *Organometallics*, 2003, *22*, 1349–1352.
141. G. Heckmann and M. Niemeyer, *J. Am. Chem. Soc.*, 2000, *122*, 4227–4228.
142. P. E. M. Siegbahn, *J. Phys. Chem.*, 1995, *99*, 12723–12729.
143. F. G. Bordwell, G. E. Drucker, N. H. Andersen and A. D. Denniston, *J. Am. Chem. Soc.*, 1986, *108*, 7310–7313.
144. G. Deacon, A. Koplick and T. Tuong, *Aust. J. Chem.*, 1982, *35*, 941–949.
145. M. N. Bochkarev, M. A. Katkova, S. Y. Khorshev and N. P. Makarenko, *Russ. Chem. Bull.*, 1998, *47*, 349–351.
146. W. J. Evans, R. Anwander, R. J. Doedens and J. W. Ziller, *Angew. Chem. Int. Ed. Engl.*, 1994, *33*, 1641–1644.
147. G. Occhipinti, C. Meermann, H. M. Dietrich, R. Litlabø, F. Auras, K. W. Törnroos, C. Maichle-Mössmer, V. R. Jensen and R. Anwander, *J. Am. Chem. Soc.*, 2011, *133*, 6323–6337.
148. M. Zimmermann, N. Å. Frøystein, A. Fischbach, P. Sirsch, H. M. Dietrich, K. W. Törnroos, E. Herdtweck and R. Anwander, *Chem. Eur. J.*, 2007, *13*, 8784–8800.
149. H. M. Dietrich, C. Meermann, K. W. Törnroos and R. Anwander, *Organometallics*, 2006, *25*, 4316–4321.
150. J. M. Boncella and R. A. Andersen, *Organometallics*, 1985, *4*, 205–206.
151. M. G. Klimpel, R. Anwander, M. Tafipolsky and W. Scherer, *Organometallics*, 2001, *20*, 3983–3992.
152. M. G. Schrems, H. M. Dietrich, K. W. Tornroos and R. Anwander, *Chem. Commun.*, 2005, 5922–5924.
153. H.-M. Sommerfeldt, C. Meermann, M. G. Schrems, K. W. Tornroos, N. A. Froystein, R. J. Miller, E.-W. Scheidt, W. Scherer and R. Anwander, *Dalton Trans.*, 2008, 1899–1907.
154. H. M. Dietrich, K. W. Törnroos and R. Anwander, *Angew. Chem. Int. Ed.*, 2011, *50*, 12089–12093.

155. J. Kratsch and P. W. Roesky, *Angew. Chem. Int. Ed.*, 2014, *53*, 376–383.
156. P. L. Diaconescu, *Comments Inorg. Chem.*, 2010, *31*, 196–241.
157. W. C. Zeise, *Annalen der Physik*, 1831, *97*, 497–541.
158. M. Black, R. H. B. Mais and P. G. Owston, *Acta Cryst. Sect. B*, 1969, *25*, 1753–1759.
159. T. J. Kealy and P. L. Pauson, *Nature*, 1951, *168*, 1039–1040.
160. S. A. Miller, J. A. Tebboth and J. F. Tremaine, *J. Chem. Soc.*, 1952, 632–635.
161. G. Wilkinson, M. Rosenblum, M. C. Whiting and R. B. Woodward, *J. Am. Chem. Soc.*, 1952, *74*, 2125–2126.
162. J. D. Dunitz, L. E. Orgel and A. Rich, *Acta Cryst.*, 1956, *9*, 373–375.
163. R. E. Maginn, S. Manastyrskyj and M. Dubeck, *J. Am. Chem. Soc.*, 1963, *85*, 672–676.
164. S. Manastyrskyj, R. E. Maginn and M. Dubeck, *Inorg. Chem.*, 1963, *2*, 904–905.
165. H. Schumann, J. A. Meese-Marktscheffel and L. Esser, *Chem. Rev.*, 1995, *95*, 865–986.
166. S. Arndt and J. Okuda, *Chem. Rev.*, 2002, *102*, 1953–1976.
167. R. G. Hayes and J. L. Thomas, *J. Am. Chem. Soc.*, 1969, *91*, 6876–6876.
168. M. D. Walter, C. H. Booth, W. W. Lukens and R. A. Andersen, *Organometallics*, 2009, *28*, 698–707.
169. M. N. Bochkarev, *Chem. Rev.*, 2002, *102*, 2089–2118.
170. W. Huang and P. L. Diaconescu, Rare Earth Arene-bridged Complexes Obtained by Reduction of Organometallic Precursors, in *Handbook on the Physics and Chemistry of Rare Earths* (Eds. J.-C. Bünzli and V. Pecharsky), Vol. 45, Elsevier, Amsterdam, 2014, pp. 261–329.
171. W. Huang and P. L. Diaconescu, *Dalton Trans.*, 2015, *44*, 15360–15371.
172. F. T. Edelmann, D. M. M. Freckmann and H. Schumann, *Chem. Rev.*, 2002, *102*, 1851–1896.
173. N. Marques and A. Sella, J. Takats, *Chem. Rev.*, 2002, *102*, 2137–2160.
174. P. L. Diaconescu, *Acc. Chem. Res.*, 2010, *43*, 1352–1363.
175. M. Nishiura, F. Guo and Z. Hou, *Acc. Chem. Res.*, 2015, *48*, 2209–2220.
176. H. Schumann, F. H. Görlitz, F. E. Hahn, J. Pickardt, C. Qian and Z. Xie, *Z. Anorg. Allg. Chem.*, 1992, *609*, 131–138.
177. W. J. Evans, D. S. Lee, M. A. Johnston and J. W. Ziller, *Organometallics*, 2005, *24*, 6393–6397.
178. W. J. Evans, J. M. Perotti, S. A. Kozimor, T. M. Champagne, B. L. Davis, G. W. Nyce, C. H. Fujimoto, R. D. Clark, M. A. Johnston and J. W. Ziller, *Organometallics*, 2005, *24*, 3916–3931.

179. W. J. Evans, B. L. Davis, T. M. Champagne and J. W. Ziller, *Proc. Natl. Acad. Sci. U.S.A.* 2006, *103*, 12678–12683.
180. M. E. Fieser, J. E. Bates, J. W. Ziller, F. Furche and W. J. Evans, *J. Am. Chem. Soc.*, 2013, *135*, 3804–3807.
181. M. E. Fieser, C. W. Johnson, J. E. Bates, J. W. Ziller, F. Furche and W. J. Evans, *Organometallics*, 2015, *34*, 4387–4393.
182. W. J. Evans, T. M. Champagne, D. G. Giarikos and J. W. Ziller, *Organometallics*, 2005, *24*, 570–579.
183. P. M. Zeimentz, S. Arndt, B. R. Elvidge and J. Okuda, *Chem. Rev.*, 2006, *106*, 2404–2433.
184. X. Li, J. Baldamus, M. Nishiura, O. Tardif and Z. Hou, *Angew. Chem. Int. Ed.*, 2006, *45*, 8184–8188.
185. Z. Hou, M. Nishiura and T. Shima, *Eur. J. Inorg. Chem.*, 2007, 2535–2545.
186. T. Shima, S. Hu, G. Luo, X. Kang, Y. Luo and Z. Hou, *Science*, 2013, *340*, 1549–1552.
187. P. G. Hayes, W. E. Piers and R. McDonald, *J. Am. Chem. Soc.*, 2002, *124*, 2132–2133.
188. S. Bambirra, M. W. Bouwkamp, A. Meetsma and B. Hessen, *J. Am. Chem. Soc.*, 2004, *126*, 9182–9183.
189. S. Bambirra, A. Meetsma, B. Hessen and A. P. Bruins, *Organometallics*, 2006, *25*, 3486–3495.
190. X. Li, M. Nishiura, K. Mori, T. Mashiko and Z. Hou, *Chem. Commun.*, 2007, 4137–4139.
191. A. Streitwieser and U. Mueller-Westerhoff, *J. Am. Chem. Soc.*, 1968, *90*, 7364–7364.
192. A. Zalkin and K. N. Raymond, *J. Am. Chem. Soc.*, 1969, *91*, 5667–5668.
193. C. J. Schaverien, Organometallic Chemistry of the Lanthanides, in *Advances in Organometallic Chemistry* (Eds. F. G. A. Stone and W. Robert), Vol. 36, Academic Press, Cambridge, 1994, pp. 283–362.
194. F. T. Edelmann, *Angew. Chem. Int. Ed. Engl.*, 1995, *34*, 2466–2488.
195. N. M. Edelstein, P. G. Allen, J. J. Bucher, D. K. Shuh, C. D. Sofield, N. Kaltsoyannis, G. H. Maunder, M. R. Russo and A. Sella, *J. Am. Chem. Soc.*, 1996, *118*, 13115–13116.
196. A. H. H. Chang and R. M. Pitzer, *J. Am. Chem. Soc.*, 1989, *111*, 2500–2507.
197. N. Hosoya, K. Yada, T. Masuda, E. Nakajo, S. Yabushita and A. Nakajima, *J. Phys. Chem. A*, 2014, *118*, 3051–3060.
198. K. R. Meihaus and J. R. Long, *J. Am. Chem. Soc.*, 2013, *135*, 17952–17957.
199. L. Ungur, J. J. Le Roy, I. Korobkov, M. Murugesu and L. F. Chibotaru, *Angew. Chem. Int. Ed.*, 2014, *53*, 4413–4417.

200. S.-D. Jiang, B.-W. Wang, H.-L. Sun, Z.-M. Wang and S. Gao, *J. Am. Chem. Soc.*, 2011, *133*, 4730–4733.
201. A. Edelmann, S. Blaurock, V. Lorenz, L. Hilfert and F. T. Edelmann, *Angew. Chem. Int. Ed.*, 2007, *46*, 6732–6734.
202. K. Miyajima, M. B. Knickelbein and A. Nakajima, *J. Phys. Chem. A*, 2008, *112*, 366–375.
203. R. Breslow, J. Brown and J. J. Gajewski, *J. Am. Chem. Soc.*, 1967, *89*, 4383–4390.
204. S. R. Kass, *J. Org. Chem.*, 2013, *78*, 7370–7372.
205. D. Patel, J. McMaster, W. Lewis, A. J. Blake and S. T. Liddle, *Nature Commun.*, 2013, *4*, 2323.
206. M. L. H. Green and K. D. P. Ng, *Chem. Rev.*, 1995, *95*, 439–473.
207. J. T. Miller and C. W. Dekock, *J. Organomet. Chem.*, 1981, *216*, 39–48.
208. T. Arliguie, M. Lance, M. Nierlich, J. Vigner and M. Ephritikhine, *J. Chem. Soc. Chem. Commun.*, 1994, 847–848.
209. T. Arliguie, M. Lance, M. Nierlich, J. Vigner and M. Ephritikhine, *J. Chem. Soc. Chem. Commun.*, 1995, 183–184.
210. J. Li and B. E. Bursten, *J. Am. Chem. Soc.*, 1997, *119*, 9021–9032.
211. T. Arliguie, M. Lance, M. Nierlich and M. Ephritikhine, *J. Chem. Soc., Dalton Trans.*, 1997, *14*, 2501–2504.
212. K. L. M. Harriman, J. J. Le Roy, L. Ungur, R. J. Holmberg, I. Korobkov and M. Murugesu, *Chem. Sci.*, 2017, *8*, 231–240.
213. F. A. Cotton and W. Schwotzer, *J. Am. Chem. Soc.*, 1986, *108*, 4657–4658.
214. M. C. Cassani, Y. K. Gun'ko, P. B. Hitchcock, M. F. Lappert and F. Laschi, *Organometallics*, 1999, *18*, 5539–5547.
215. M. C. Cassani, Y. K. Gun'ko, P. B. Hitchcock and M. F. Lappert, *Chem. Commun.*, 1996, 1987–1988.
216. M. D. Fryzuk, J. B. Love and S. J. Rettig, *J. Am. Chem. Soc.*, 1997, *119*, 9071–9072.
217. W. Huang, F. Dulong, T. Wu, S. I. Khan, J. T. Miller, T. Cantat and P. L. Diaconescu, *Nature Commun.*, 2013, *4*, 1448.
218. W. Huang, J. J. Le Roy, S. I. Khan, L. Ungur, M. Murugesu and P. L. Diaconescu, *Inorg. Chem.*, 2015, *54*, 2374–2382.
219. H. E. Zimmerman, *Acc. Chem. Res.*, 2012, *45*, 164–170.
220. J. Mortensen and J. Heinze, *Angew. Chem. Int. Ed. Engl.*, 1984, *23*, 84–85.
221. P. B. Hitchcock, M. F. Lappert and A. V. Protchenko, *J. Am. Chem. Soc.*, 2001, *123*, 189–190.
222. S. Krieck, H. Görls, L. Yu, M. Reiher and M. Westerhausen, *J. Am. Chem. Soc.*, 2009, *131*, 2977–2985.

223. P. Hunt, 2008, "Molecular Orbitals and Population Analysis." http://www.huntresearchgroup.org.uk/teaching/teaching_comp_chem_year4/L7_bonding.pdf.
224. P. L. Diaconescu and C. C. Cummins, *Inorg. Chem.*, 2012, *51*, 2902–2916.
225. D. Patel, F. Tuna, E. J. L. McInnes, J. McMaster, W. Lewis, A. J. Blake and S. T. Liddle, *Dalton Trans.*, 2013, *42*, 5224–5227.
226. S. T. Liddle, *Coord. Chem. Rev.*, 2015, *293–294*, 211–227.
227. T. E. Weller, M. Ellerby, S. S. Saxena, R. P. Smith and N. T. Skipper, *Nature Phys.* 2005, *1*, 39–41.
228. W. Huang, F. Dulong, S. I. Khan, T. Cantat and P. L. Diaconescu, *J. Am. Chem. Soc.*, 2014, *136*, 17410–17413.
229. W. Huang and P. L. Diaconescu, *Eur. J. Inorg. Chem.*, 2013, 4090–4096.
230. W. Huang, S. I. Khan and P. L. Diaconescu, *J. Am. Chem. Soc.*, 2011, *133*, 10410–10413.
231. M. D. Fryzuk, L. Jafarpour, F. M. Kerton, J. B. Love and S. J. Rettig, *Angew. Chem. Int. Ed.*, 2000, *39*, 767–770.
232. B. Vlaisavljevich, P. L. Diaconescu, W. W. Lukens, L. Gagliardi and C. C. Cummins, *Organometallics*, 2013, *32*, 1341–1352.
233. P. L. Diaconescu and C. C. Cummins, *J. Am. Chem. Soc.*, 2002, *124*, 7660–7661.
234. C. Melero, A. Guijarro and M. Yus, *Dalton Trans.*, 2009, 1286–1289.
235. W. J. Evans, T. A. Ulibarri and J. W. Ziller, *J. Am. Chem. Soc.*, 1990, *112*, 219–223.
236. D. M. Roitershtein, J. W. Ziller and W. J. Evans, *J. Am. Chem. Soc.*, 1998, *120*, 11342–11346.
237. W. J. Evans, D. G. Giarikos, C. B. Robledo, V. S. Leong and J. W. Ziller, *Organometallics*, 2001, *20*, 5648–5652.
238. K. Mashima, H. Sugiyama and A. Nakamura, *J. Chem. Soc., Chem. Commun.*, 1994, 1581–1582.
239. W. Huang, P. M. Abukhalil, S. I. Khan and P. L. Diaconescu, *Chem. Commun.*, 2014, *50*, 5221–5223.
240. C. E. Kefalidis, L. Perrin and L. Maron, *Dalton Trans.*, 2014, *43*, 4520–4529.
241. W. J. Evans, R. A. Keyer and J. W. Ziller, *Organometallics*, 1990, *9*, 2628–2631.
242. W. J. Evans, R. A. Keyer and J. W. Ziller, *Organometallics*, 1993, *12*, 2618–2633.
243. H. J. Heeres, J. Nijhoff, J. H. Teuben and R. D. Rogers, *Organometallics*, 1993, *12*, 2609–2617.

244. C. M. Forsyth, S. P. Nolan, C. L. Stern, T. J. Marks and A. L. Rheingold, *Organometallics*, 1993, *12*, 3618–3623.
245. M. W. Löble, J. M. Keith, A. B. Altman, S. C. E. Stieber, E. R. Batista, K. S. Boland, S. D. Conradson, D. L. Clark, J. Lezama Pacheco, S. A. Kozimor, R. L. Martin, S. G. Minasian, A. C. Olson, B. L. Scott, D. K. Shuh, T. Tyliszczak, M. P. Wilkerson and R. A. Zehnder, *J. Am. Chem. Soc.*, 2015, *137*, 2506–2523.
246. D. L. S. Brahms and W. P. Dailey, *Chem. Rev.*, 1996, *96*, 1585–1632.
247. A. Nemirowski and P. R. Schreiner, *J. Org. Chem.*, 2007, *72*, 9533–9540.
248. R. R. Schrock, *Chem. Rev.*, 2002, *102*, 145–180.
249. R. R. Schrock, *J. Am. Chem. Soc.*, 1974, *96*, 6796–6797.
250. M. A. Sierra, *Chem. Rev.*, 2000, *100*, 3591–3638.
251. E. O. Fischer and A. Maasböl, *Angew. Chem. Int. Ed. Engl.*, 1964, *3*, 580–581.
252. A. J. Arduengo, R. L. Harlow and M. Kline, *J. Am. Chem. Soc.*, 1991, *113*, 361–363.
253. M. N. Hopkinson, C. Richter, M. Schedler and F. Glorius, *Nature*, 2014, *510*, 485–496.
254. R. R. Schrock, *Acc. Chem. Res.*, 1986, *19*, 342–348.
255. T. M. Trnka and R. H. Grubbs, *Acc. Chem. Res.*, 2001, *34*, 18–29.
256. P. L. Arnold and S. T. Liddle, *Chem. Commun.*, 2006, 3959–3971.
257. P. L. Arnold and I. J. Casely, *Chem. Rev.*, 2009, *109*, 3599–3611.
258. S. T. Liddle, D. P. Mills and A. J. Wooles, *Chem. Soc. Rev.*, 2011, *40*, 2164–2176.
259. H. Schumann and J. Müller, *J. Organomet. Chem.*, 1979, *169*, C1–C4.
260. M. Zimmermann, D. Rauschmaier, K. Eichele, K. W. Tornroos and R. Anwander, *Chem. Commun.*, 2010, 5346–5348.
261. J. Hong, L. Zhang, X. Yu, M. Li, Z. Zhang, P. Zheng, M. Nishiura, Z. Hou and X. Zhou, *Chem. Eur. J.*, 2011, *17*, 2130–2137.
262. R. Litlabø, M. Zimmermann, K. Saliu, J. Takats, K. W. Törnroos and R. Anwander, *Angew. Chem. Int. Ed.*, 2008, *47*, 9560–9564.
263. J. Scott, H. Fan, B. F. Wicker, A. R. Fout, M.-H. Baik and D. J. Mindiola, *J. Am. Chem. Soc.*, 2008, *130*, 14438–14439.
264. W. Huang and C. T. Carver, P. L. Diaconescu, *Inorg. Chem.*, 2011, *50*, 978–984.
265. A. Venugopal, I. Kamps, D. Bojer, R. J. F. Berger, A. Mix, A. Willner, B. Neumann, H.-G. Stammler and N. W. Mitzel, *Dalton Trans.*, 2009, 5755–5765.
266. D. Bojer, B. Neumann, H.-G. Stammler and N. W. Mitzel, *Eur. J. Inorg. Chem.*, 2011, 3791–3796.

267. D. Bojer, B. Neumann, H.-G. Stammler and N. W. Mitzel, *Chem. Eur. J.*, 2011, *17*, 6239–6247.
268. D. Bojer, A. Venugopal, A. Mix, B. Neumann, H.-G. Stammler and N. W. Mitzel, *Chem. Eur. J.*, 2011, *17*, 6248–6255.
269. F. N. Tebbe, G. W. Parshall and G. S. Reddy, *J. Am. Chem. Soc.*, 1978, *100*, 3611–3613.
270. Y. Takenaka, T. Shima, J. Baldamus and Z. Hou, *Angew. Chem. Int. Ed.*, 2009, *48*, 7888–7891.
271. J. Zhou, T. Li, L. Maron, X. Leng and Y. Chen, *Organometallics*, 2015, *34*, 470–476.
272. N. N. Greenwood and A. Earnshaw, *Chemistry of the Elements*, Oxford, Pergamon Press, 1984.
273. K. M. Lancaster, M. Roemelt, P. Ettenhuber, Y. Hu, M. W. Ribbe, F. Neese, U. Bergmann and S. DeBeer, *Science*, 2011, *334*, 974–977.
274. H. M. Dietrich, H. Grove, K. W. Törnroos and R. Anwander, *J. Am. Chem. Soc.*, 2006, *128*, 1458–1459.
275. L. C. H. Gerber, E. Le Roux, K. W. Törnroos and R. Anwander, *Chem. Eur. J.*, 2008, *14*, 9555–9564.
276. A. J. Arduengo, M. Tamm, S. J. McLain, J. C. Calabrese, F. Davidson and W. J. Marshall, *J. Am. Chem. Soc.*, 1994, *116*, 7927–7928.
277. W. A. Herrmann, F. C. Munck, G. R. J. Artus, O. Runte and R. Anwander, *Organometallics*, 1997, *16*, 682–688.
278. T. Mehdoui, J.-C. Berthet, P. Thuery and M. Ephritikhine, *Chem. Commun.*, 2005, 2860–2862.
279. H. Schumann, D. M. M. Freckmann, S. Schutte, S. Dechert and M. Hummert, *Z. Anorg. Allg. Chem.*, 2007, *633*, 888–892.
280. H. Schumann, M. Glanz, J. Winterfeld, H. Hemling, N. Kuhn and T. Kratz, *Chem. Ber.*, 1994, *127*, 2369–2372.
281. M. Glanz, S. Dechert, H. Schumann, D. Wolff and J. Springer, *Z. Anorg. Allg. Chem.*, 2000, *626*, 2467–2477.
282. H. Schumann, M. Glanz, J. Winterfeld, H. Hemling, N. Kuhn and T. Kratz, *Angew. Chem. Int. Ed. Engl.*, 1994, *33*, 1733–1734.
283. P. L. Arnold, S. A. Mungur, A. J. Blake and C. Wilson, *Angew. Chem. Int. Ed.*, 2003, *42*, 5981–5984.
284. D. Patel, S. T. Liddle, S. A. Mungur, M. Rodden, A. J. Blake and P. L. Arnold, *Chem. Commun.*, 2006, 1124–1126.
285. S. T. Liddle and P. L. Arnold, *Organometallics*, 2005, *24*, 2597–2605.
286. P. L. Arnold and S. T. Liddle, *Chem. Commun.*, 2005, 5638–5640.
287. P. L. Arnold and S. T. Liddle, *Organometallics*, 2006, *25*, 1485–1491.

288. P. L. Arnold, S. Zlatogorsky, N. A. Jones, C. D. Carmichael, S. T. Liddle, A. J. Blake and C. Wilson, *Inorg. Chem.*, 2008, *47*, 9042–9049.
289. I. J. Casely, S. T. Liddle, A. J. Blake, C. Wilson and P. L. Arnold, *Chem. Commun.*, 2007, 5037–5039.
290. Z.-G. Wang, H.-M. Sun, H.-S. Yao, Q. Shen and Y. Zhang, *Organometallics*, 2006, *25*, 4436–4438.
291. Z.-G. Wang, H.-M. Sun, H.-S. Yao, Y.-M. Yao, Q. Shen and Y. Zhang, *J. Organomet. Chem.*, 2006, *691*, 3383–3390.
292. B. Wang, D. Wang, D. Cui, W. Gao, T. Tang, X. Chen and X. Jing, *Organometallics*, 2007, *26*, 3167–3172.
293. B. Wang, D. Cui and K. Lv, *Macromolecules*, 2008, *41*, 1983–1988.
294. S. P. Downing, S. C. Guadaño, D. Pugh, A. A. Danopoulos, R. M. Bellabarba, M. Hanton, D. Smith and R. P. Tooze, *Organometallics*, 2007, *26*, 3762–3770.
295. K. Lv and D. Cui, *Organometallics*, 2008, *27*, 5438–5440.
296. J. Zhang, H. Yao, Y. Zhang, H. Sun and Q. Shen, *Organometallics*, 2008, *27*, 2672–2675.
297. M. Alcarazo, C. W. Lehmann, A. Anoop, W. Thiel and A. Fürstner, *Nature Chem.* 2009, *1*, 295–301.
298. C. Wang, J. Zhou, X. Zhao, L. Maron, X. Leng and Y. Chen, *Chem. Eur. J.*, 2016, *22*, 1258–1261.
299. R. Appel and I. Ruppert, *Z. Anorg. Allg. Chem.*, 1974, *406*, 131–144.
300. A. V. Kirsanov, *Izv. Akad. Nauk SSSR*, 1950, 426.
301. K. Aparna, M. Ferguson and R. G. Cavell, *J. Am. Chem. Soc.*, 2000, *122*, 726–727.
302. S. T. Liddle, J. McMaster, J. C. Green and P. L. Arnold, *Chem. Commun.*, 2008, 1747–1749.
303. D. P. Mills, O. J. Cooper, J. McMaster, W. Lewis and S. T. Liddle, *Dalton Trans.*, 2009, 4547–4555.
304. D. P. Mills, A. J. Wooles, J. McMaster, W. Lewis, A. J. Blake and S. T. Liddle, *Organometallics*, 2009, *28*, 6771–6776.
305. G. W. Marshall, A. J. Wooles, D. P. Mills, W. Lewis, A. J. Blake and S. T. Liddle, *Inorganics*, 2013, *1*, 46–49.
306. A. Buchard, A. Auffrant, L. Ricard, X.-F. Le Goff, R. H. Platel, C. K. Williams and P. Le Floch, *Dalton Trans.*, 2009, 10219–10222.
307. M. Gregson, E. Lu, J. McMaster, W. Lewis, A. J. Blake and S. T. Liddle, *Angew. Chem. Int. Ed.*, 2013, *52*, 13016–13019.
308. M. Gregson, E. Lu, F. Tuna, E. J. L. McInnes, C. Hennig, A. C. Scheinost, J. McMaster, W. Lewis, A. J. Blake, A. Kerridge and S. T. Liddle, *Chem. Sci.*, 2016, *7*, 3286–3297.

309. M. Gregson, N. F. Chilton, A.-M. Ariciu, F. Tuna, I. F. Crowe, W. Lewis, A. J. Blake, D. Collison, E. J. L. McInnes, R. E. P. Winpenny and S. T. Liddle, *Chem. Sci.*, 2016, *7*, 155–165.
310. D. P. Mills, L. Soutar, W. Lewis, A. J. Blake and S. T. Liddle, *J. Am. Chem. Soc.*, 2010, *132*, 14379–14381.
311. D. P. Mills, W. Lewis, A. J. Blake and S. T. Liddle, *Organometallics*, 2013, *32*, 1239–1250.
312. D. P. Mills, L. Soutar, O. J. Cooper, W. Lewis, A. J. Blake and S. T. Liddle, *Organometallics*, 2013, *32*, 1251–1264.
313. T. Cantat, F. Jaroschik, F. Nief, L. Ricard, N. Mézailles and P. Le Floch, *Chem. Commun.*, 2005, 5178–5180.
314. T. Cantat, F. Jaroschik, L. Ricard, P. Le Floch, F. Nief and N. Mézailles, *Organometallics*, 2006, *25*, 1329–1332.
315. M. Fustier, X.-F. Le Goff, P. Le Floch and N. Mézailles, *J. Am. Chem. Soc.*, 2010, *132*, 13108–13110.
316. M. Fustier, X.-F. Le Goff, M. Lutz, J. C. Slootweg and N. Mézailles, *Organometallics*, 2015, *34*, 63–72.
317. C. Wang, J. Zhou, X. Zhao, L. Maron, X. Leng and Y. Chen, *Chem. Eur. J.*, 2016, *22*, 1258–1261.
318. B. Römer, G. G. Gatev, M. Zhong and J. I. Brauman, *J. Am. Chem. Soc.*, 1998, *120*, 2919–2924.
319. S. Fortier, J. R. Walensky, G. Wu and T. W. Hayton, *J. Am. Chem. Soc.*, 2011, *133*, 6894–6897.

https://doi.org/10.1142/9781800610163_0007

Chapter 7

Organoactinide Chemistry

David J. H. Emslie, Nicholas R. Andreychuk, and Carlos A. Cruz*
Department of Chemistry and Chemical Biology, McMaster University, 1280 Main Street West, Hamilton, Ontario, L8S 4M1, Canada
**emslied@mcmaster.ca*

7.1 Introduction

Actinides are the group of elements from actinium (element 89) to nobelium (element 102), with lawrencium (element 103) typically considered a group 3 transition metal.[1] Of these elements, only thorium and uranium have substantial natural abundances, similar to those of tantalum, tin, boron and lead in the earth's crust (2–14 ppm). Thorium consists almost exclusively of ^{232}Th with a half-life of 14.1 billion years. By contrast, natural-abundance uranium consists of a mixture of ^{238}U ($t_{1/2}$ 4.47 billion years), ^{235}U (704 million years), and ^{234}U (246 thousand years), with the latter formed on the decay series from ^{238}U. Protactinium and actinium also have non-zero natural abundances, since ^{234}Pa ($t_{1/2}$ 6.7 hours) is formed in the "uranium series" decay chain originating at ^{238}U, ^{231}Pa ($t_{1/2}$ 33 thousand years) and ^{227}Ac ($t_{1/2}$ 21.8 years) are formed in the "actinium series" decay chain that starts at ^{235}U, and ^{228}Ac ($t_{1/2}$ 6.1 hours) is formed in the "thorium series" originating from ^{232}Th. These decay series are also responsible for the trace natural abundance of ^{234}Th, ^{231}Th, ^{230}Th, ^{228}Th and ^{227}Th, with half lives of 24 days, 26 hours, 75 thousand years, 1.9 years and 19 days, respectively.[2]

Anthropogenic neptunium and plutonium also have several fairly long-lived isotopes, including ^{237}Np ($t_{1/2}$ 2.14 million years), ^{239}Pu ($t_{1/2}$ 24.1 thousand years), ^{242}Pu ($t_{1/2}$ 373 thousand years) and ^{244}Pu ($t_{1/2}$ 80.8 million years).[3] Chemical studies are most often conducted with ^{237}Np and ^{239}Pu, although research with these highly toxic elements is only possible in highly regulated facilities; typically government facilities, utilizing specialized equipment (e.g. negative atmosphere gloveboxes) with a variety of measures to guard against, and monitor for, any accidental release. Consequently, only a limited number of organometallic Np and Pu complexes have been reported. Additionally, a very small number of organometallic Ac, Pa, Am, Cm, Cf and Bk compounds have also been prepared, including $Pa(COT)_2$,[4] $PaCp_4$,[5] $AnCp_3$ (An = Am,[6] Cm,[7] Cf, Bk).[8] However, the organometallic chemistry of these man-made elements has not been more extensively investigated due to a combination of very low availability and high or very high radioactivity (i.e. short or very short half-lives) of all isotopes of these elements.

The vast majority of organoactinide chemistry involves thorium and uranium, but the field is not as well developed as that of organolanthanide chemistry. In addition, while the organometallic chemistry of lanthanide elements has focused more on diamagnetic compounds of trivalent Sc, Y, Lu and La, paramagnetic non-uranium(VI) organometallic chemistry is better developed than diamagnetic thorium(IV) organometallic chemistry, as evidenced by over 300 compounds with U–C bonds in Cambridge Structural Database at the time of writing (few of these are uranium(VI) complexes), vs. less than 120 with Th–C bonds. Greater interest in uranium likely stems from the increased covalency of uranium compounds relative to thorium compounds, including greater participation of the 5f-orbitals in bonding, combined with a rich redox chemistry; uranium provides routine access to organometallic compounds in oxidation states III–VI, whereas almost all organothorium chemistry involves thorium(IV). The appreciable covalency of uranium compounds is apparent from the volatility of UF_6, $[U(NMe_2)_4]$ and $[U(BH_4)_4]$, the accessibility of higher oxidation states, and may also be responsible for the increased solubility of most uranium organometallic compounds vs. thorium analogues in non-polar solvents such as hexane.

The covalency of most uranium–ligand bonds is believed to be significantly lower than that in related transition metal complexes (groups 4–11), but is generally far greater than that in trivalent rare earth complexes, and so uranium is uniquely positioned as a high natural-abundance f-element with certain properties in common with lanthanides (large size and electropositivity) and certain properties more in common with mid-transition metals (appreciable covalency and a rich redox chemistry), combined with unique availability of the f-orbitals for participation in bonding (due to greater radial extension of early actinide 5f orbitals vs. lanthanide 4f orbitals). Np and Pu, and to a lesser extent Pa and Am, share many of these properties, whereas the late actinide elements (Cm–No) are more lanthanide-like, generally forming highly ionic compounds, with one primary oxidation state and a second less-common oxidation state; as with the lanthanide elements, the last member of the actinide series, nobelium, has the most readily accessible divalent oxidation state, with an f^{14} configuration.

The ionic radii for Th(IV) and U(IV) are 0.94 and 0.89 Å, respectively (for a coordination number of 6), which is smaller than that of early trivalent lanthanide ions such as La(III) (1.03 Å), but is comparable with later members of the lanthanide series and yttrium (e.g. the ionic radii for Sm(III), Y(III) and Yb(III) are 0.96, 0.90 and 0.87 Å, respectively), and is significantly larger than that of the group 4 transition metals Ti, Zr and Hf (0.61–0.72 Å). By contrast, the ionic radius of U(III) is 1.03 Å, which is nearly identical to that of lanthanum(III). The ionic radii of U(V) and U(VI) are 0.76 and 0.73 Å, respectively, which are significantly larger than those of Ta(V) (0.64 Å) and W(VI) (0.60 Å). The Pauling electronegativities of Th and U are 1.3 and 1.4, respectively, which are on par with those of Sc, Y and Lu (1.4, 1.2 and 1.3, respectively).

7.2 Anhydrous Halide, Triflate, Amido and Aryloxide Starting Materials

The availability of suitable anhydrous starting materials has played a critical role in the development of actinide organometallic chemistry. However, none of these compounds are commercially available, so in this section, synthetic routes to anhydrous halide, triflate, amido and aryloxide compounds are outlined, with a focus on compounds with demonstrated or potential utility as starting materials for the preparation of organometallic derivatives. Base-free and diethylether-, dme-, THF- or 1,4-dioxane-coordinated compounds are of the most general utility, since stronger nitrogen donor ligands are not easily displaced, and nitriles and pyridines are incompatible with many strong nucleophiles.

The most common halide starting materials in organothorium chemistry are $ThCl_4$ and square antiprismatic [$ThCl_4$(dme)$_2$]. $ThCl_4$ has not been commercially available for many years, but can be prepared by passing N_2 containing CCl_4 vapors over ThO_2 at 750°C.[9] A more common starting point is [$ThCl_4$(dme)$_2$], since it can be accessed using standard wet-chemistry techniques; [$Th(NO_3)_4(H_2O)_n$] ($n = 4$–6) is boiled in concentrated HCl until no more NO_2 is evolved, and the solvent is then removed under reduced pressure to afford [$ThCl_4(H_2O)_x$]; reduced pressure is required because hydrated thorium(IV) chloride decomposes to a mixed hydroxide-chloride between 100°C and 160°C.[10] The resulting [$ThCl_4(H_2O)_4$] is converted to square antiprismatic [$ThCl_4$(dme)$_2$] either by: (a) stirring in $SOCl_2$ to remove H_2O, yielding $ThCl_4(OSCl_2)$,[11] followed by Soxhlett extraction in 1,2-dimethoxyethane (dme),[12] or (b) reaction with excess Me_3SiCl in dme.[13] [$ThCl_4(H_2O)_4$] can also be refluxed in 1,4-dioxane with excess Me_3SiCl and anhydrous HCl/OEt_2 to form [$ThCl_4$(1,4-dioxane)$_2$], and reaction of this product with THF yielded [$ThCl_4$(THF)$_{3.5}$].[13] Furthermore, [$ThCl_4$(dme)$_2$] can be converted to [ThX_4(dme)$_2$] (X = Br or I) by treatment with Me_3SiX.[13,14]

In organouranium(IV) and (III) chemistry, the most common halide starting materials are UCl_4 and [$UI_3(OR_2)_x$]. By contrast, simple uranium(V) and (VI) halide complexes such as UCl_5 and UF_6 are rarely used as entry points to high valent uranium chemistry, since they are highly oxidizing, and UCl_5 is also prone to disproportionation. Instead, higher oxidation state complexes are often accessed by initial ligand attachment to uranium(III) or (IV) and subsequent chemical oxidation, or alternatively, uranyl precursors such as [{UO_2Cl_2(THF)$_n$}$_x$] (n = 2, x = 2; n = 3, x = 1) are employed.[15]

Uranium tetrachloride may be accessed by passing CCl_4 vapors over UO_2 in a tube furnace at 400 °C,[16] or by cautious[17] slow addition of solid UO_3 to hexachloropropene at 190 °C.[a] Analogous syntheses of UCl_4 starting from U_3O_8, $UO_2Cl_2 \cdot xH_2O$ or $UO_2(NO_3)_2 \cdot 6H_2O$ were also recently reported.[18] Additionally, reaction of UCl_4 with Me_3SiI in diethylether or acetonitrile afforded $[UI_4(OEt_2)_2]$[19] and $[UI_4(NCMe)_4]$,[20] respectively; these uranium(IV) tetraiodo complexes are stable at room temperature, in contrast to base-free UI_4 which eliminates I_2 to form UI_3.[21] $[UI_4(OEt_2)_2]$ has also been prepared by reaction of UH_3[20] or uranium turnings[22] with 2 equiv. of I_2 in OEt_2, although the reaction with UH_3 has been reported to proceed more cleanly than that with uranium metal.[19] Reaction of UH_3 with 4 equiv. of AgBr, AgCl, $CuCl_2$ or AgOTf in dme also yields $[UX_4(dme)_2]$ (X = Br, Cl or OTf),[19] and base-free $U(OTf)_4$ is readily accessible via the reaction of UCl_4 with HOTf.[23]

Base-free UI_3 can be prepared via solvent-free reactions between uranium turnings and HgI_2 (1.5 equiv.)[24] or I_2 (1.5 equiv.) at high temperature,[25] or more conveniently via the reaction of uranium turnings with 1.5 equiv. of I_2 in diethylether.[22] Alternatively, $[UX_3(THF)_4]$ (X = I or Br), $[UI_3(dme)_2]$ or $[UI_3(pyridine)_4]$ can be prepared via the reactions of amalgamated uranium turnings with 1.5 equiv. of I_2 or Br_2 in the appropriate donor solvent, although it has been noted that the THF-coordinated compounds are prone to decomposition involving THF ring-opening.[26,27] Along similar lines, $[UBr_3(dme)_2]$ was prepared from UH_3 with 3 equiv. of AgBr,[19] $U(OTf)_3$ can be prepared from uranium turnings and HOTf at 120°C,[23] and $[U(OTf)_3(NCMe)_3]_n$ can be prepared by reaction with UI_3 with KOTf in acetonitrile.[28] However, uranium turnings, and by extension UH_3, are not readily accessible to many research groups, so the recent synthesis of $[UCl_3(pyridine)_4]$ from UCl_4, by reduction with Mg turnings in 1,4-dioxane (100°C) followed by reaction with pyridine, provides an alternative pathway into low-valent uranium chemistry.[29] This compound is a well-defined uranium(III) chloro

---

[a]For the synthesis of UCl_4 from UO_3 with hexachloropropene, it is recommended to add UO_3 via a solid addition funnel placed at the top of a reflux condenser, and the use of silicone grease rather than hydrocarbon-based H-grease is required in order to obtain a forest green product.

compound, in contrast to $UCl_3(THF)_x$ ($x = 1$–2), which is prepared from UCl_4 and excess NaH in THF.[30]

Actinide halide precursors such as $[ThCl_4(dme)_2]$, UCl_4 and UI_3 can also be converted to useful amido starting materials such as $[An(NEt_2)_4]$[31] and $[U\{N(SiMe_3)_2\}_3]$[27] by reaction with $LiNEt_2$, or $KN(SiMe_3)_2$, and $[U(OAr)_3]$ (Ar = $C_6H_3{}^iPr_2$-2,6) can be synthesized by reaction of $[U\{N(SiMe_3)_2\}_3]$ with 2,6-diisopropylphenol.[32]

Anhydrous UO_2Cl_2 can be prepared via several high temperature routes including passing O_2 over UCl_4 at 300°–350°C.[33] However, $[\{UO_2Cl_2(THF)_2\}_2]$ is a more straightforwardly accessible reagent, prepared by initial reaction of $UO_2(NO_3)_2 \cdot 6H_2O$ with concentrated HCl to form $UO_2Cl_2 \cdot xH_2O$,[34] followed by dehydration using Me_3SiCl in THF.[15] Related $[UO_2(OTf)_2]$ can also be prepared from UO_3 in neat triflic anhydride (TfOTf), or via the reaction of UO_3 with HOTf in water to form $UO_2(OTf)_2 \cdot xH_2O$, followed by dehydration under vacuum at 200°C.[35]

7.3 Homoleptic Hydride, Borohydride, Aluminohydride, and Aminodiboronate Complexes and Their Lewis Base Adducts

Binary actinide hydrides include ThH_2, Th_4H_{15},[36] UH_3,[19] PaH_3 (two polymorphs)[37] and AnH_x (An = Np,[38] Pu,[39] or Am;[40] $x \approx 2$–3 as a solid solution), all of which are prepared by reaction of elemental metal with H_2. The thorium and uranium hydride products have been used as precursors to finely divided actinide metal powders by heating to around 900°C and 400°C in vacuo, respectively.[19,36]

Colorless $[Th(BH_4)_4]$,[41] red-orange $[Pa(BH_4)_4]$,[42] dark green $[U(BH_4)_4]$,[43] blue-green $[Np(BH_4)_4]$ and purple-black $[Pu(BH_4)_4]$,[42] published between 1949 and 1978, were synthesized via the solvent-free reactions between AnF_4 and explosively pyrophoric $Al(BH_4)_3$.[44] Alternatively, $[Th(BH_4)_4]$ is reported to be accessible from $ThCl_4$ and 4 equiv. of $LiBH_4$ in diethylether, initially forming $[Th(BH_4)_4(OEt_2)_2]$[45] from which the base-free compound can be obtained by sublimation at 140°–150°C in high vacuum.[46] The Th and U complexes are polymeric in the solid state, and only slowly hydrolyze in air, which contrasts the much greater air-sensitivity of molecular $[Zr(BH_4)_4]$ and $[Hf(BH_4)_4]$.[47] The Pa complex is also polymeric, and of the Th, Pa and U complexes, the latter is the

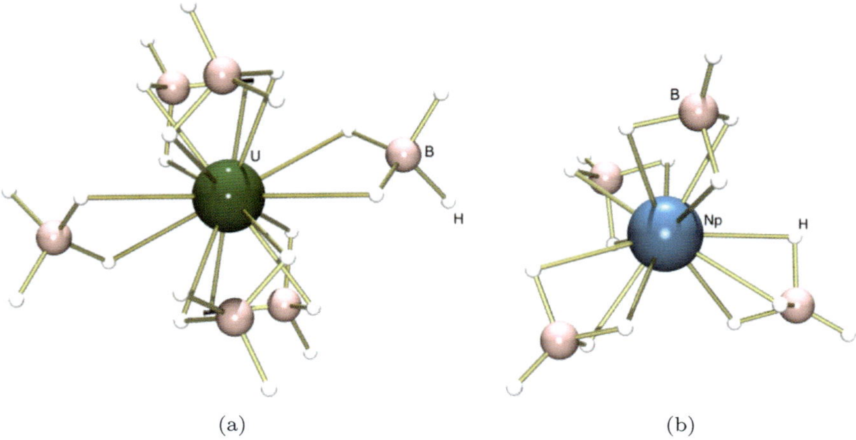

Figure 7.1. (a) Neutron structure of polymeric $[U(BH_4)_4]$ (the three κ^2-coordinated BH_4 anions bridge (μ-κ^2:κ^2) between uranium atoms)[48] and (b) X-ray structure of monomeric $[Np(BH_4)_4]$.[50]

most volatile, subliming even at room temperature at 0.1 mTorr (vapor pressure ~0.3 mm Hg at 34°C).[42] The uranium compound was studied by X-ray and neutron diffraction ((a) in Figure 7.1), revealing that each uranium center is coordinated to two terminal and four bridging borohydride ligands; the terminal BH_4 anions are κ^3-coordinated, whereas the bridging BH_4 anions are μ-κ^2:κ^2-coordinated.[48,49] By contrast, the Np ((b) in Figure 7.1) and Pu complexes are monomeric with four κ^3-coordinated BH_4 anions. These transuranium complexes are thermally unstable at room temperature, and have low melting points (14°C) and remarkably high vapor pressures (~10 mm Hg at 25°C).[42,50]

$[U(BH_4)_4]$ reacted with 3 equiv. of BMe_3 to form monomeric $[U(H_3BMe)_4]$, and with less than half an equivalent of BMe_3 to form highly volatile $[U(BH_4)_3(H_3BMe)]$.[51] $[U(H_3BMe)_4]$ was also prepared more straightforwardly from UCl_4 and $LiBH_3Me$ in chlorobenzene, and monomeric $[An(H_3BMe)_4]$ (An = Th, Pa and Np) were prepared analogously from $AnCl_4$ (Th, Pa, Np) or $PaCl_5$.[52] The thorium H_3BMe compound is significantly more volatile than the BH_4 analogue, subliming in vacuo at 40°–50°C vs. 120°–150°C. Reaction of $[U(BH_4)_4]$ with $Me_2C{=}CMe_2$ in THF provided a route to $[U(H_3BCMe_2CHMe_2)_4(THF)_2]$, and treatment of $[U(BH_4)_4]$ with various Lewis bases afforded $[U(BH_4)_4(OR_2)]$

(R = Me, Et or ^{n}Pr),[53] [$U(BH_4)_4$(dme)] and [$U(BH_4)_4(L)_2$] (L = THF, HMPA or Ph_3PO).[b,54,55] Several neutral Lewis base adducts of [$U(H_3BMe)_4$] have also been reported, including THT (tetrahydrothiophene), dme, btme ($MeSCH_2CH_2SMe$), tmeda ($Me_2NCH_2CH_2NMe_2$) and dmpe ($Me_2PCH_2CH_2PMe_2$) adducts.[56,57] Similarly, reactions of [$Th(BH_4)_4(OEt_2)_2$] with PMe_3, PEt_3 and dmpe afford [$Th(BH_4)_4(PR_3)_2$] and [$Th(BH_4)_4$(dmpe)$_2$];[45] these reactions contrast the reactions of [$Zr(BH_4)_4$] and [$Hf(BH_4)_4$] with phosphines, which yield mixed hydrido/borohydride products with elimination of $R_3P(BH_3)$.[58]

Solid [$U(BH_4)_4$] is unstable at elevated temperatures, eliminating B_2H_6 and H_2 at 100°C to form an explosively pyrophoric red-brown compound thought to be [$U(BH_4)_3$]. Furthermore, heating [$U(BH_4)_4$] above 150°C affords a silver mirror of UB_4 (or U with 4B) and H_2 contaminated with only traces of volatile boron compounds.[43] [$U(BH_4)_3$] can also be prepared by heating a toluene solution of [$U(BH_4)_4$] at 125°C in a sealed vessel, with periodic removal of H_2 and B_2H_6 in vacuo.[59] By contrast, heating [$U(BH_4)_4$] at 150°C in mesitylene afforded [$U(BH_4)_3(\eta^6$-mesitylene)] which reacts with hexamethylbenzene in toluene to form [$U(BH_4)_3(\eta^6\text{-}C_6Me_6)$] ((a) in Figure 7.2).[60] Along similar lines, the brown-black uranium(III) dmpe complex, [$U(H_3BMe)_3$(dmpe)$_2$], was prepared in low yield by heating green [$U(H_3BMe)_4$(dmpe)] in toluene at 80°C in the presence of excess dmpe.[57]

Synthetically useful [$U(BH_4)_3(THF)_x$] ($x = 2$ or 3 depending on crystallization conditions) is accessible by reaction of UI_3[63] or less ideally $UCl_3(THF)_x$ ($x = 1$–2)[30] with 3 equiv. of $NaBH_4$ in THF. By contrast, the related reaction between $UCl_3(THF)_x$ (generated *in situ* by reaction of UCl_4 with 1 equiv. of $LiAlH_4$ in THF) and 3 equiv. of $LiAlH_4$ resulted in precipitation of a pyrophoric grey solid presumed to be [$U(AlH_4)_3(THF)_x$], which was reported to decompose to aluminium metal and an unidentified uranium-containing solid above −20°C.[63] Reactions of [$U(BH_4)_3(THF)_x$] with dmpe and $PyPPh_2$ ($NC_5H_4(PPh_2)$-2) afforded structurally characterized [$U(BH_4)_3$(dmpe)$_2$] and [$U(BH_4)_3(PyPPh_2)_2$], respectively.[64]

---

[b] [$U(BH_4)_4$(dme)] was reported to eliminate BH_3 to form [{(dme)$U(BH_4)_3(\mu$-H)}$_2$], but re-examination of the X-ray structure of the product shows it to be [{(dme)$U(BH_4)_3$}$_2(\mu$-O)].

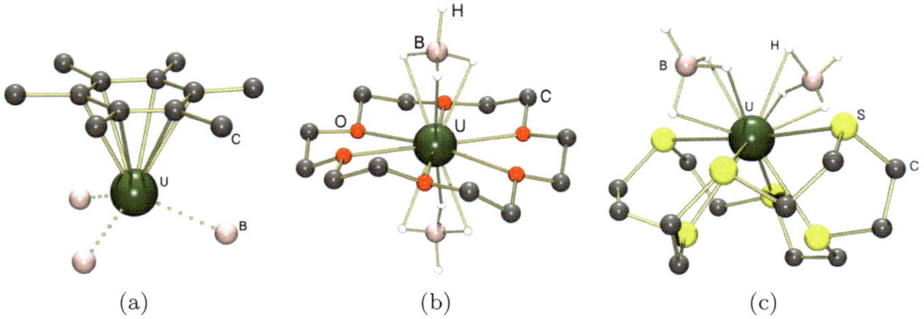

Figure 7.2. X-Ray crystal structures of uranium(III) borohydride complexes (a) $[U(BH_4)_3(\eta^6\text{-}C_6Me_6)]$,[60] (b) the cationic portion of $[U(BH_4)_2$(18-crown-6)][BPh_4], and (c) the cationic portion of $[U(BH_4)_2$(18-thiacrown-6)][BPh_4].[61]

Cationic $[U(BH_4)_2(THF)_5][BPh_4]$ was also prepared by reaction of $[U(BH_4)_3(THF)_3]$ with $[HNEt_3][BPh_4]$, releasing H_2 and $Et_3N(BH_3)$, and reaction of the THF-coordinated cation with 18-crown-6 in THF or 18-thiacrown-6 in THT provided $[U(BH_4)_2$ (18-crown-6)][BPh_4] and $[U(BH_4)_2$(18-thiacrown-6)][BPh_4] in which the crown ether is meridionally coordinated, while the larger thiacrown is facially-coordinated ((b) and (c) in Figure 7.2).[61]

Related to borohydride compounds are the aminodiboronate ($H_3BNMe_2BH_3^-$) complexes developed by Girolami *et al.* Reaction of $ThCl_4$ with 4 equiv. of $Na(H_3BNMe_2BH_3)$ afforded the first 15-coordinate complex, $[Th(H_3BNMe_2BH_3)_4]$ in which seven of the eight boron atoms bridge to thorium via two hydrogen atoms, and heating $[Th(H_3BNMe_2BH_3)_4]$ to 100°C afforded 14-coordinate $[Th(H_3BNMe_2BH_3)_2(BH_4)_2]$ in quantitative yield.[65] By contrast, reaction of UCl_4 with 4 equiv. of $Na(H_3BNMe_2BH_3)$ in diethylether afforded the polymeric uranium(III) complex, $[U(H_3BNMe_2BH_3)_3]$, which crystallized as different polymorphs from pentane and toluene with coordination numbers of 13 and 14 ((a) in Figure 7.3) respectively.[66] Conducting the reaction between UCl_4 with $Na(H_3BNMe_2BH_3)$ in THF or dme afforded $[U(H_3BNMe_2BH_3)_3(THF)]$ ((b) in Figure 7.3) and $[U(H_3BNMe_2BH_3)_3(dme)]$, and treatment of the THF complex with PMe_3 provided $[U(H_3BNMe_2BH_3)_3(PMe_3)_2]$ which slowly decomposes with release of $H_3BNMe_2BH_2(PMe_3)$.[67] Unlike the lanthanide $[Ln(H_3BNMe_2BH_3)_3(THF)]$ analogues which lose THF at elevated temperatures to form readily-sublimed $[Ln(H_3BNMe_2BH_3)_3]$ complexes, attempts to sublime the uranium

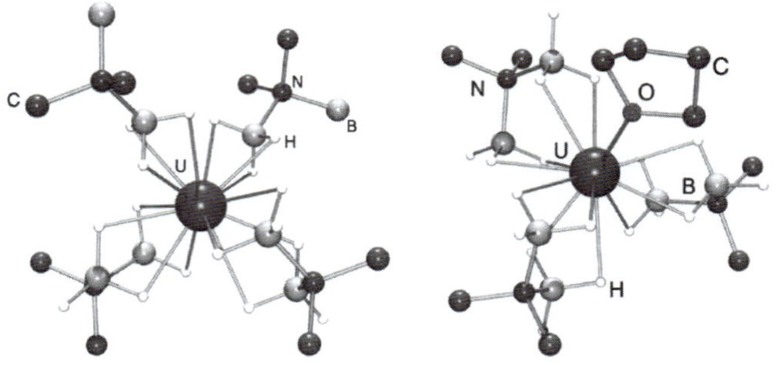

(a) (b)

Figure 7.3. X-Ray crystal structures of (a) the polymorph of $[U(H_3BNMe_2BH_3)_3]$ with a coordination number of 14 (polymeric; the two dangling BH_3 groups are κ^3-coordinated to a neighbouring uranium center)[66] and (b) monomeric $[U(H_3BNMe_2BH_3)_3(THF)]$.[67]

analogue, $[U(H_3BNMe_2BH_3)_3(THF)]$, at 10 mTorr resulted only in decomposition.[68]

7.4 Homoleptic Acyclic Hydrocarbyl Compounds and Their Lewis Base Adducts

7.4.1 *Homoleptic alkyl complexes*

Interest in the development of homoleptic actinide alkyl complexes dates back to the Manhattan project in the 1940s,[69] where the search for stable and readily volatilized compounds for isotope separation resulted in numerous attempted syntheses of σ-bonded uranium alkyl compounds. However, these efforts proved unsuccessful due to thermal instability which prevented isolation of these neutral peralkyl species. This thermal instability was attributed by some authors to coordinative unsaturation of the metal,[70] and while it precluded the use of actinide peralkyl compounds for isotope separation, they remain of fundamental interest and importance as potential precursors to new organoactinide complexes via alkane elimination.

The Marks group, having recognized the problems associated with actinide peralkyl syntheses, proceeded to employ excess amounts of alkyl lithium reagents to produce isolable actinide(IV) "ate" compounds of the type $[Li(OR_2)_4]_2[UR_6]$ (OR_2 = THF, Et_2O; R = CH_3, C_6H_5, CH_2SiMe_3)[72] and $[Li(tmeda)]_3[Th(CH_3)_7]$,[72] and since

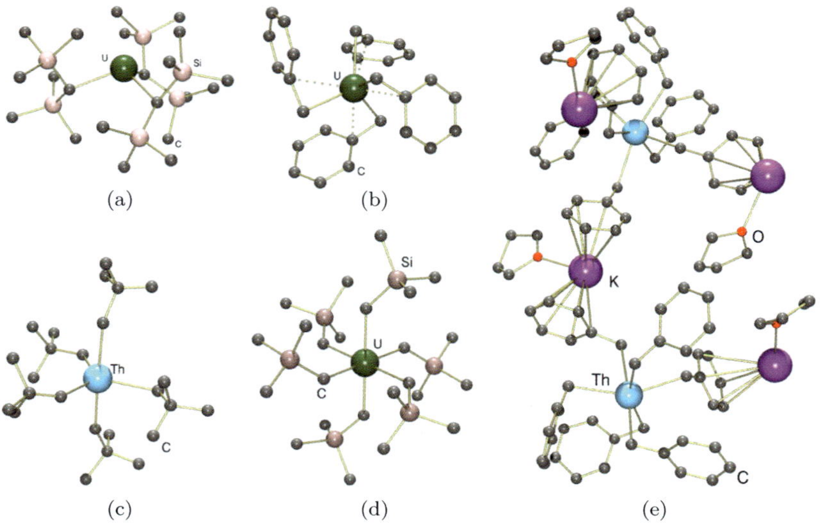

Figure 7.4. X-Ray crystal structures of (a) $[U\{CH(SiMe_3)_2\}_3]$ bearing 3 alkyl groups,[77] (b) $[U(CH_2Ph)_4]$ bearing 4 benzyl groups,[78] (c) the anionic portion of $[Li(THF)_4][Th(CH_2{}^tBu)_5]$ featuring 5-coordinate thorium,[75] (d) the anionic portion of $[Li(THF)_4][U(CH_2SiMe_3)_6]$ featuring 6-coordinate uranium,[79] and (e) part of the extended structure of $[K(THF)]_2[Th(CH_2Ph)_6]$.[75]

2000, this series has been extended to include $[Li_2(py)_3][U(Fc)_3]$ (Fc $= 1,1'$-ferrocenediyl),[73] $[Li(dme)_3][U(CH_2SiMe_3)_5]$, $[Li(THF)_4][U(CH_2{}^tBu)_5]$, $[Li(tmeda)]_2[UMe_6]$, $\{[K(THF)]_3[K(THF)_2][U(CH_2Ph)_6]_2\}_x$,[74] $[Li(THF)_4][Th(CH_2{}^tBu)_5]$ ((c) in Figure 7.4), $[Li(dme)_2][Th(CH_2SiMe_3)_5]$, $[K(THF)]_2[Th(CH_2Ph)_6]$ ((e) in Figure 7.4),[75] $[Li(dme)_3]_2[ThPh_6]$, and $[Li(THF)(12crown4)]_2[ThPh_6]$.[76]

Neutral actinide alkyl compounds are relatively rare. The thorium(IV) tetraalkyl complex $[Th(CH_2SiMe_3)_4(dme)_x]$,[80] formed from $[ThCl_4(dme)_2]$ with 4 equiv. of $LiCH_2SiMe_3$, has been proposed based on its alkane elimination reactions with protonated ligand precursors (*vide infra*), but was not isolated. Additionally, the reaction of benzyllithium with thorium tetrachloride has been reported to yield $Th(CH_2Ph)_4$, but characterization is limited to IR spectroscopy.[81] However, the more sterically protected tetrabenzyl derivative $[Th(CH_2C_6H_3Me_2\text{-}3,5)_4]$, synthesized by the reaction of thorium tetrachloride with $LiCH_2C_6H_3Me_2$-3,5 in THF, has been thoroughly characterized via NMR spectroscopy and elemental analysis.[82] Related $[U(CH_2Ph)_4(MgCl_2)]$ was also reported to form as a finely-crystalline red-brown solid from reaction of $[UCl_4(THF)_3]$ with $Mg(CH_2Ph)_2$, but was only characterized by elemental analysis.[83] More recently, Bart *et al.* reported the synthesis of a family of tetrabenzyluranium(IV) compounds, $[U(CH_2Ar)_4]$ {Ar = Ph ((b) in Figure 7.4), C_6H_4Me-*p*, $C_6H_3Me_2$-*m*, $C_6H_4{}^iPr$-*p*, $C_6H_4{}^tBu$-*p*, $C_6H_4(NMe_2)$-*p*, $C_6H_4(SMe)$-*p*, $C_6H_4(OMe)$-*p*, $C_6H_4(OMe)$-*o*, 2-pyridinyl}, via reactions of UCl_4 with 4 equiv. of KCH_2Ar in THF, and all but the *p*-NMe_2 and *p*-SMe derivatives are stable in the solid state at room temperature. The benzyl groups in these complexes are polyhapto coordinated with short U–C_{ipso} distances in the solid state, except in the latter two compounds where uranium–heteroatom coordination is observed.[78] Along similar lines, reaction of $[ThCl_4(dme)_2]$ with Li{$C_6H_4(CH_2NMe_2)$-*o*} in cold THF afforded the homoleptic aryl complex, [Th{$C_6H_4(CH_2NMe_2)$-*o*}$_4$] stabilized by thorium-amine interactions,[84] and reaction of $[UI_3(THF)_4]$ or UCl_4 with $KCH(NMe_2)Ph$ (3 or 4 equiv., respectively) provided [U{$CH(NMe_2)Ph$}$_3$] in which each amine-substituted benzyl ligand is $\eta^4 NC_3$-coordinated.[85]

Trialkylphosphine-stabilized tetralkyl compounds have also been accessed via the reaction of $[(dmpe)_2AnCl_4]$ (An = Th, U) with 4 equiv. of methyllithium[86] or benzyllithium[87] to produce $[(dmpe)_xAnR_4]$ {R = CH_3, x = 2; R = CH_2Ph, x = 1}. These compounds were characterized by elemental analysis, and X-ray diffraction in the case of the methyl compound, and reactions with phenol provided the corresponding $[(dmpe)An(OPh)_4]$ compounds. The related mixed methyl/benzyl derivative, $[(dmpe)An(CH_2Ph)_3Me]$, was obtained by reaction of $[(dmpe)_2AnCl_4]$ with 3 equiv. of $PhCH_2Li$ and 1 equiv. of MeLi.[87]

Based on their alkane elimination reactions with protonated ligand precursors (*vide infra*), the *in situ*-generated uranium(III) trialkyl complexes, $[U(CH_2R)_3(THF)_x]$ (R = Ph, $SiMe_3$ or CMe_3),[88] have been proposed. However, the only isolated homoleptic trialkyluranium(III) complex is royal blue [U{$CH(SiMe_3)_2$}$_3$] ((a) in Figure 7.4) prepared by Sattelberger *et al.* via the reaction of $[U(OAr)_3]$ (Ar = $C_6H_3{}^tBu_2$-2,6) with 3 equiv. of $LiCH(SiMe_3)_2$ in hexanes. By contrast, the reaction of $[UCl_3(THF)_x]$ with 3 equiv. of $LiCH(SiMe_3)_2$ in THF afforded green $[Li(THF)_3]$[UCl{$CH(SiMe_3)_2$}$_3$]; an "ate" complex resulting from LiCl occlusion. In the solid state, room temperature-stable [U{$CH(SiMe_3)_2$}$_3$] is trigonal pyramidal with C–U–C angles of

108°;[77] this was initially attributed to γ-agostic U–H–C interactions on the more open face of the molecule, but based on computational studies on [Ln{CH(SiMe$_3$)$_2$}$_3$] (Ln = La and Sm), pyramidalization may well be a consequence of U–(β-C–Si) interactions.[89] Yellow-brown [Pu{CH(SiMe$_3$)$_2$}$_3$] was also prepared via the reaction of [Pu(OAr)$_3$] (Ar = 2,6-^{t}Bu$_2$C$_6$H$_3$) with 3 equiv. of LiCH(SiMe$_3$)$_2$, and [Np{CH(SiMe$_3$)$_2$}$_3$] was accessible from [NpI$_3$(THF)$_4$], although characterization was limited to IR spectroscopy.[90]

High-valent homoleptic alkyl compounds are particularly rare. Addition of excess LiR to [U$_2$(OEt)$_{10}$] in 1,4-dioxane was reported by Wilkinson *et al.* to yield 8-coordinate uranium(V) complexes, [Li(dioxane)]$_3$[UR$_8$] (R = Me, CH$_2$SiMe$_3$; CH$_2$^{t}Bu), but these compounds have not been structurally characterized.[70] In 2011, Hayton reported the first well-characterized U(V) alkyl complex, octahedral [Li(THF)$_4$][U(CH$_2$SiMe$_3$)$_6$] ((d) in Figure 7.4), via the reaction of [Li(dme)$_3$][U(CH$_2$SiMe$_3$)$_5$] with half an equivalent of I$_2$, followed by rapid addition of LiCH$_2$SiMe$_3$. Cyclic voltammetry of [Li(THF)$_4$][U(CH$_2$SiMe$_3$)$_6$] revealed a reversible $U^{V/VI}$ wave at -1.22 V vs. [FeCp$_2$]$^{0/+}$ in THF, and reaction with [U(O^{t}Bu)$_6$] ($U^{V/VI}E_{1/2} = -1.12$ V) afforded [U(CH$_2$SiMe$_3$)$_6$] and [Li(THF)$_4$][U(O^{t}Bu)$_6$]. However, isolation of [U(CH$_2$SiMe$_3$)$_6$] was prevented by high solubility combined with rapid decomposition above -25°C.[79,91] Hayton *et al.* also recently isolated and structurally characterized the uranium(VI) alkyl complex, [Li(dme)$_{1.5}$]$_2$[UO$_2$(CH$_2$SiMe$_3$)$_4$];[91] a dianionic relative of the thermally unstable neutral uranyl [UO$_2$(R)$_2$(THF)$_x$] (R = Me, Et, CH=CH$_2$, ^{i}Pr, ^{n}Bu, ^{t}Bu, Ph) complexes generated *in situ* in the early 1980s by Seyam.[92]

7.4.2 *Homoleptic allyl complexes*

Allyl anions may be η^1-coordinated like alkyl ligands, or they may be η^3-coordinated via a π-system with 2 filled MOs (with 0 and 1 node) and 1 empty MO (with 2 nodes), depending on the requirements of the metal center. This flexible bonding situation bears some resemblance to the variable hapticity of benzyl ligands, although the extent of delocalization is greater in η^3-allyl complexes than η^3-benzyl complexes. The first example of a homoleptic actinide allyl complex, [Th(C$_3$H$_5$)$_4$], was first mentioned by Wilke in 1966,[93] and a detailed description was published by Marks in 1992.[82] This complex was prepared by reaction of $[ThCl_4(THF)_3]$ with $(C_3H_5)MgBr$, and since it decomposes readily at 0°C, it is best handled at temperatures of –20°C or lower. Homoleptic uranium(IV) allyl analogues, $[U(C_3H_5)_4]$ and $[U(C_3H_4Me\text{-}2)_4]$ were prepared similarly via reactions of UCl_4 with $(C_3H_4R)MgBr$ (R = H or Me) at –30°C,[94] and as in the case of thorium, both complexes are thermally unstable, decomposing above −20°C.[95]

Almost 40 years later, sterically stabilized analogues of $[Th(C_3H_5)_4]$ were reported by Hanusa and Brennessel,[96] who increased the steric bulk of the allyl ligands through single or double $SiMe_3$ substitution. $[\{1,3\text{-}(SiMe_3)_2C_3H_3\}_4Th]$ and $[\{1\text{-}(SiMe_3)C_3H_4\}_4Th]$ were prepared by reaction of $K[1,3\text{-}(SiMe_3)_2C_3H_3]$ or $K[1\text{-}(SiMe_3)C_3H_4]$ with $[ThBr_4(THF)_4]$, and are remarkably thermally robust, melting with decomposition at temperatures of 124°C and 90°C, respectively.

7.5 Ligand Attachment Protocols for the Synthesis of Heteroleptic Compounds

7.5.1 *Salt metathesis*

Ancillary ligand attachment in actinide chemistry is frequently achieved by salt metathesis (transmetallation), typically utilizing an appropriate alkali-metal or thallium(I) reagent in combination with an actinide halide or triflate. In a minority of cases, especially in donor solvents, this results in halide products containing occluded alkali metal halide salts. However, these can still serve as precursors for further elaboration, and often yield salt-free products upon substitution of remaining halide anions with bulkier and more electron donating organometallic ligands (Scheme 7.1).[97,98] In fact, "ate" complexes may in some cases offer synthetic benefits. For example, Evans *et al.* have reported substantial differences in reactivity between anionic $[NBu_4][Cp^*{}_2UCl_3]$ and neutral $[Cp^*{}_2UCl_2]$; the former reacted in minutes, rather than hours or days, with 1 equiv. of KL {L = hpp (1,3,4,6,7,8-hexahydro-2H-pyrimido[1,2-*a*]-pyrimidinate) or NC_4Me_4} to afford $[Cp^*{}_2UCl(L)]$, and reaction of $[NBu_4][Cp^*{}_2UCl_3]$ with 3 equiv. of K(hpp) afforded $[Cp^*U(hpp)_3]$ (via KCl, $[NBu_4]Cl$ and KCp^* elimination), which was not observed as a product in the reaction of neutral $[Cp^*{}_2UCl_2]$ with 3 equiv. of K(hpp).

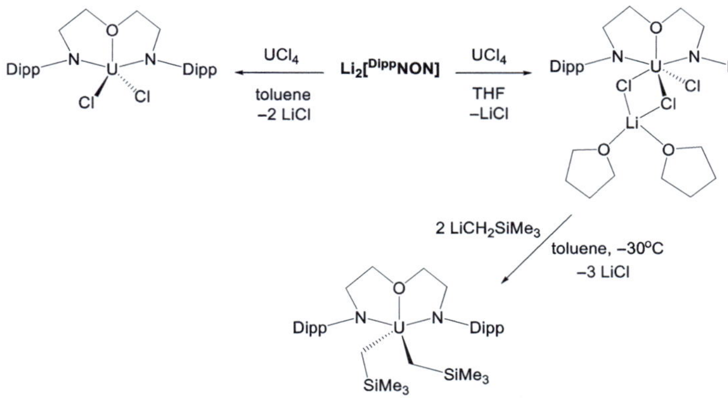

Scheme 7.1. Ancillary ligand attachment by salt metathesis, illustrating solvent-dependent "ate" complex formation, and subsequent derivatization to yield a salt-free dialkyl complex (Dipp = 2,6-diisopropylphenyl).[97]

Problems have in some cases been encountered using lithium alkyl reagents in combination with actinide iodide precursors; for example, reaction of [Tp*UI_2(THF)$_2$] {Tp* = tris(3,5-dimethylpyrazolyl)borate} with 2 equiv. of $LiCH_2SiMe_3$ in THF yielded [Li(THF)$_4$][Tp*UI_3] in over 60% yield, and the same triiodide "ate" complex was formed in reactions of [Tp*$_2$UI] with $LiCH_2SiMe_3$ or MeLi. However, sodium alkyl reagents (NaR; R = CH_2SiMe_3, Me or ^{n}Bu) proved to be much more effective in the latter reaction, cleanly yielding the desired [Tp*$_2$UR] compounds and poorly soluble NaI as a non-interfering by-product.[99]

Magnesium reagents (e.g. Mg($C_5H_2{}^tBu_3$-1,2,4)$_2$, Mg(CH_2CR=$CRCH_2$)(THF)$_2$, $MgMe_2$, Cp*MgCl or MeMgBr) have also been utilised to install organometallic ligands, although in rare cases this has resulted in competing ancillary ligand transfer to magnesium,[100,101] or halide exchange reactivity,[102,103] rather than the expected salt metathesis (Scheme 7.2); halide exchange presumably occurs via Grignard adducts similar to that in Figure 7.5.[100]

Actinide borohydride and tetraarylborate compounds can also be utilized as salt metathesis precursors, eliminating $LiBH_3R$ or $LiBAr_4$ salts rather than a lithium halide. For example, reaction of [U(BH_4)$_4$] with KCp* (2 equiv.) in toluene afforded [Cp*$_2$U(BH_4)$_2$], and the reactions of [(TMP)U(BH_4)$_3$] (TMP = tetramethylphospholyl) with KCH_2Ph (3 equiv.), or [(TMP)$_2$U(BH_4)$_2$] with LiR (2 equiv.), afforded [(TMP)U(CH_2Ph)$_3$] (Scheme 7.3) and [(TMP)$_2$UR_2] (R = Me or CH_2SiMe_3) in substantially higher yields than the analogous

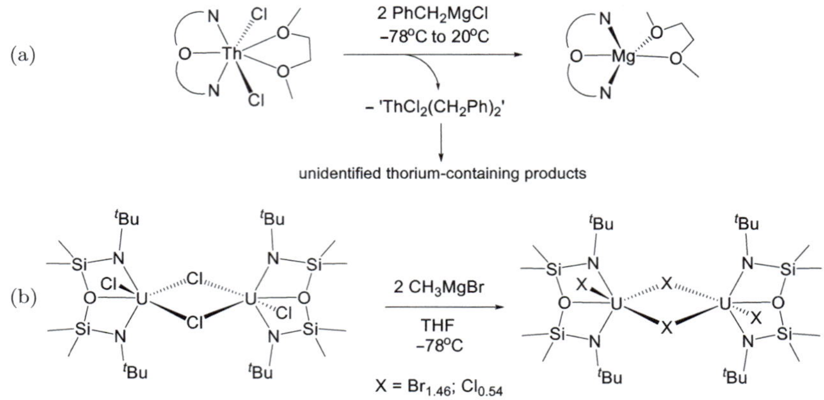

Scheme 7.2. Reactions between actinide halide precursors and RMgBr Grignard reagents that do not yield the expected alkylated products: (a) transfer of a dianionic NON-donor ligand {4,5-bis(2,6-diisopropylanilido)-2,7-di-*tert*-butyl-9,9-dimethylxanthene}from thorium to magnesium[100] and (b) halide exchange converting $[\{(^{tBu}NON)UCl(\mu\text{-}Cl)\}_2]$ $\{^{tBu}NON = O(SiMe_2N^tBu)_2\}$ to a mixed chloride/bromide analogue.[103]

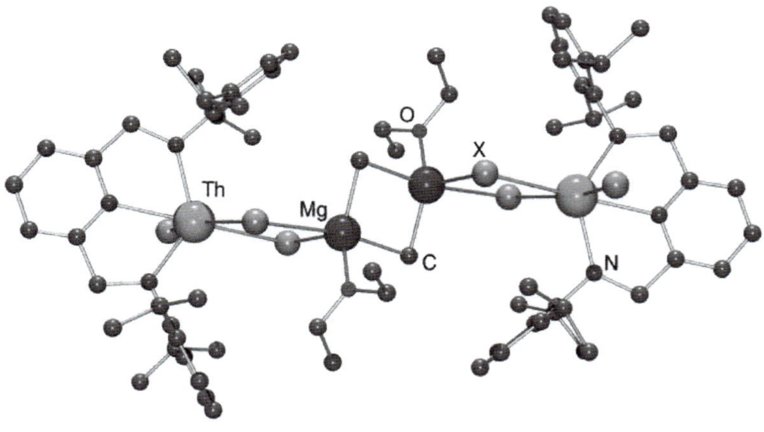

Figure 7.5. X-ray crystal structure of $[\{(BDPP)ThX(\mu\text{-}X)_2Mg(OEt_2)(\mu\text{-}Me)\}_2]$ (X = $Br_{0.73-0.87}/Cl_{0.13-0.27}$; BDPP = 2,6-bis(2,6-diisopropylanilidomethyl)pyridine).[100]

reactions with $[(TMP)UCl_3(OR_2)_2]$ and $[(TMP)_2UCl_2]$.[104] Additionally, the reaction of $[Cp^*{}_2U\{(\mu\text{-}Ph)_2BPh_2\}]$ with KX (X = Cp* or NC_4Me_4) in non-coordinating solvents is synthetically valuable as a means to access base-free $[Cp^*{}_2UX]$ (Scheme 7.3),[105,106] and

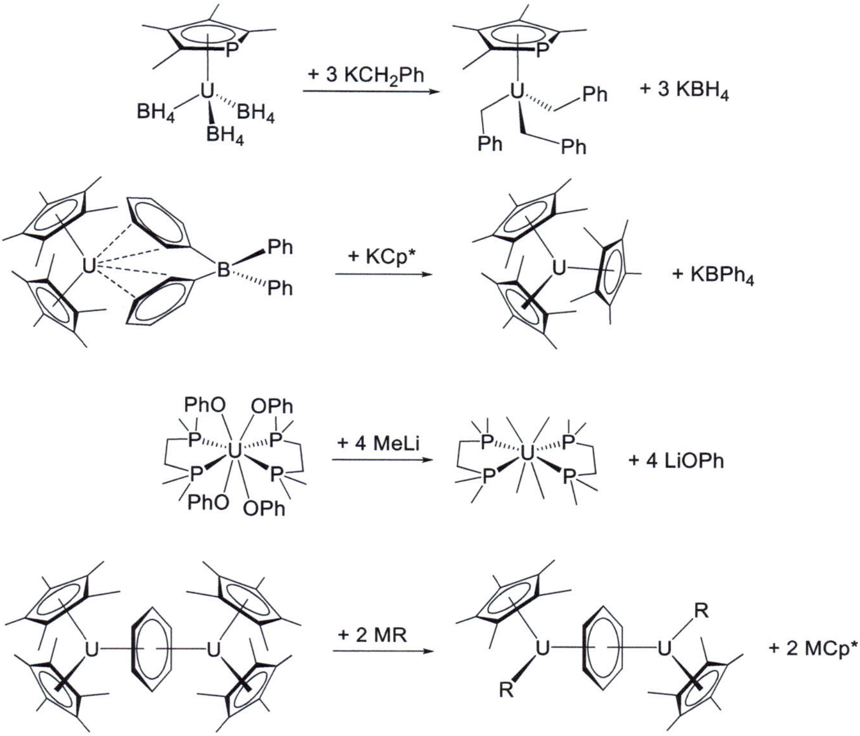

(M = Li or K; R = CH(SiMe$_3$)$_2$, N(SiMe$_3$)$_2$, OAr {Ar = C$_6$H$_2$(^{t}Bu-*o*)$_2$(Me-*p*)}, ^{i}PrNCMeN^{i}Pr)

Scheme 7.3. Selected salt metathesis-like reactions starting with borohydride,[104] tetraphenylborate,[106] aryloxide,[86] and sterically very hindered pentamethylcyclopentadienyl precursors.[109,110]

reactions of [Cp*$_2$UMeX′] (X′ = Me or Cl) with BPh$_3$, followed by addition of K(C$_5$HMe$_4$) provided [Cp*$_2$(C$_5$HMe$_4$)UX′], presumably via [Cp*$_2$UX′][MeBPh$_3$] (isolated for X′ = Me).[107]

Along similar lines, actinide alkoxide or aryloxide compounds have be utilized as alternative salt metathesis precursors, eliminating LiOR salts rather than a lithium halide. For example, [U(CH(SiMe$_3$)$_2$)$_3$] was prepared by reaction of [U(OC$_6$H$_3$^{t}Bu$_2$-2,6)$_3$] with 3 equiv. of LiCH(SiMe$_3$)$_2$,[108] [Li(dioxane)]$_3$[UR$_8$] formation is reported in the reactions of [U$_2$(OEt)$_{10}$] with excess LiR (R = Me, CH$_2$SiMe$_3$; CH$_2$^{t}Bu),[70] [UMe$_4$(dmpe)$_2$] was prepared from [U(OPh)$_4$(dmpe)$_2$] with MeLi (Scheme 7.3),[86] and [Li$_{14}$(O^{t}Bu)$_{12}$Cl][U(CH$_2$SiMe$_3$)$_5$] was obtained in the reaction of

$[Li(THF)]_2[U(O^tBu)_6]$ with $LiCH_2SiMe_3$.[74] Furthermore, in very sterically hindered complexes such as $[UCp^*_3]$ and $[Cp^*_2U(\mu\text{-}\eta^6{:}\eta^6\text{-}C_6H_6)UCp^*_2]$, the Cp* ligands become unusually vulnerable to replacement by less sterically hindered κ^1- or κ^2-coordinating anions such as $[N(SiMe_3)_2]^-$, $[CH(SiMe_3)_2]^-$, $[OAr]^-$ {Ar = $C_6H_2(^tBu\text{-}o)_2(Me\text{-}p)$} and $[MeC(N^iPr)_2]^-$ (Scheme 7.3).[109,110]

7.5.2 *Alkane elimination*

Alkane elimination is a frequently employed ligand attachment protocol in the chemistry of the lanthanides, facilitated by the ready accessibility of trialkyl $[Ln(CH_2R)_3(THF)_n]$ (R = $SiMe_3$ or Ph) and related starting materials.[111] However, this approach has rarely been employed to install multidentate ligands on actinide metals, partly due to the low thermal stability of homoleptic trimethylsilylmethyl thorium and uranium compounds and only recent availability of well-defined homoleptic benzyl uranium complexes (Section 7.4.1). Notable examples of alkane elimination from a homoleptic alkyl actinide precursor include the reactions of (a) $H_2[XA_2]$, $H_2[BDPP]$ and $H_2[NN^{Fc}]$ {XA_2 = 4,5-bis(2,6-diisopropylanilido)-2,7-di-tert-butyl-9,9-dimethylxanthene; BDPP = 2,6-bis(2,6-diisopropylanilidomethyl)pyridine; NN^{Fc} = $Fe(\eta^5\text{-}C_5H_4NSiR_3)_2$} with 1 equiv. of *in situ* generated "$Th(CH_2SiMe_3)_4(dme)_x$"[80] or "$U(CH_2R)_3(THF)_x$" (R = Ph, $SiMe_3$ and/or CMe_3);[89] Section 7.4.1, (b) the reaction of thermally unstable $[U(C_3H_5)_4]$ with 2 equiv. of ^iPrOH at –20°C to afford $[\{U(allyl)_2(O^iPr)_2\}_2]$,[95] and reaction of $[U\{CH(NMe_2)Ph\}_3]$ with 3 equiv. of $HS_2C(C_6H_3Mes_2\text{-}2,6)$ in THF to produce $[U\{S_2C(C_6H_3Mes_2\text{-}2,6)\}_4(THF)]$.[85]

By comparison, alkane elimination from non-homoleptic precursors such as $[Cp^*AnMe_2]$, in combination with protic reagents such as terminal alkynes, primary or secondary amines and phosphines, alcohols, and thiols is common and also constitutes a key step in hydroelementation catalysis proceeding via the amido mechanism. Alkane or alkylsilane elimination is also a common strategy for the synthesis of heteroleptic actinide hydride complexes (Section 7.6.3), via σ-bond metathesis between a heteroleptic alkyl complex and H_2 or a hydrosilane (most commonly $PhSiH_3$).[112–114] These reactions have typically been carried out in solution, but in the case of H_2, Evans *et al.* recently reported a range of reactions which proceed more cleanly in the absence of solvent.[115]

7.5.3 H_2 elimination

Ligand attachment by H_2 elimination from a hydride, borohydride or aluminohydride precursor is relatively uncommon. Starting from a hydride precursor, examples include (a) the reactions of [$Cp^*{}_2UH$(hpp)] (hpp = 1,3,4,6,7,8-hexahydro-2*H*-pyrimido[1,2-*a*]pyrimidinato) with HC_2Ph, HSPh or H_2NPh to afford [$Cp^*{}_2U$(X)(hpp)] (X = C_2Ph, SPh or HNPh),[116] (b) the reaction of the tuck-in tuck-over metallocene [$Cp^*U\{\mu\text{-}\eta^5{:}\eta^1{:}\eta^1\text{-}C_5Me_3(CH_2)_2\}(\mu\text{-}H)_2UCp^*{}_2$] with 4 equiv. of phenol to generate [$Cp^*{}_2U(OPh)_2$],[117] and (c) the reaction of [$\{(1,3\text{-}C_5H_3{}^tBu_2)_2U\}_2(\mu\text{-}H)_2$] with 2 equiv. of water to form [$\{(1,3\text{-}C_5H_3{}^tBu_2)_2U\}_2(\mu\text{-}OH)_2$].[118]

Examples of ligand attachment by H_2 (with EH_3; E = B or Al) elimination from borohydride or aluminohydride precursors include the reaction of [$Cp_3U(AlH_4)$] with 1 equiv. of ROH (R = Et or ^{n}Bu) to afford [$Cp_3U(OR)$],[119] reaction of [(COT)$U(BH_4)_2$] with excess HSR (R = ^{n}Bu and ^{i}Pr) to form [(COT)$U(SR)_2$],[120] and the reaction of [$U(BH_4)_4$] with HCp to afford [$CpU(BH_4)_3$].[121] The latter product is also formed via reaction of [$U(BH_4)_4$] with TlCp, and this type of salt metathesis-like reactivity is employed much more commonly for ligand attachment using borohydride precursors, compared with H_2/EH_3 elimination. In fact, in some cases only the salt metathesis route has proven effective. For example, [$LU(BH_4)$] (L = *trans*-calix[2]benzene[2]pyrrolide) reacted with KOAr (Ar = $C_6H_3{}^tBu_2$-2,6) at room temperature to eliminate THF-insoluble KBH_4, whereas the reaction with HOAr did not proceed, even at elevated temperatures.[122]

7.5.4 Amine elimination

Ligand attachment can be achieved by amine elimination from an amido precursor, although this protocol has not been used widely in actinide chemistry. Examples include the reaction of [$An(NEt_2)_4$] (An = Th or U) with 2 or 3 equiv. of HCp to afford [$Cp_2An(NEt_2)_2$] or [$Cp_3An(NEt_2)$],[123,124] and reaction of [$An(NEt_2)_3$][BPh_4] with 3 equiv. of HCp in THF to provide [Cp_3An(THF)][BPh_4]. Similarly, reaction of [$An(NMe_2)_4$] (An = Th or U) with H_2[CGC] (CGC = ^{t}BuN-$SiMe_2$-C_5Me_4) afforded [(CGC)$An(NMe_2)_2$] in high yield, and subsequent treatment with Me_3SiX (X = Cl or I) provided readily derivatized [(CGC)AnX_2].[125]

7.5.5 *Less-common ligand attachment protocols*

7.5.5.1 *Trimethylsilyl- or trialkyltin-halide elimination*

Multidentate ancillary ligand attachment to actinide elements via trimethylsilyl halide or trialkyltin halide elimination has rarely been reported, although it is a common ligand attachment protocol in the chemistry of the lanthanides; one example in actinide chemistry is the synthesis of $[(C_5Me_4Et)_2UCl_2]$ from UCl_4 and 2 equiv. of $C_5Me_4Et(SnBu_3)$.[126] Elimination of Me_3SiF has also been employed for the synthesis of cyanide and azide compounds using Me_3SiCN or Me_3SiN_3 in combination with $[Cp^*{}_2UF_2(Py)]$.[127] However, in the chemistry of uranium, organic azides, including Me_3SiN_3,[108,109] are more commonly employed for the synthesis of imido compounds via N_2 extrusion,[128] or triazenido (RNNNR) compounds by insertion.[129]

7.5.5.2 *Insertion reactions*

Anionic ancillary ligands have in some instances been installed by insertion, for example, $[\{Cp^*{}_2UH\}_2(\mu\text{-}H)_2]$ reacted with neutral $Me_4C_5{=}CH_2$ to afford $[UCp^*{}_3]$, presumably via 1,2-insertion followed by H_2 reductive elimination.[106] Other insertion reactions (e.g. those with CO, RNC, RCN, CO_2, CS_2, $C(NR)_2$, RN_3, RNCO, C_2R_4, C_2R_2) are illustrated elsewhere in the chapter.

7.5.5.3 *Addition of (a) X^- with concurrent oxidation, or (b) an $X\cdot$ radical*

Neutral $[(COT)CpU(NEt_2)_2]$ was isolated from the reaction of $[(COT)U(NEt_2)_2]$ with TlCp,[130] and $[Cp^*{}_2U(am)]$ (am = amidinate) reacted with 2 equiv. of TlCp to form $[Cp^*Cp_2U(am)]$, with elimination of Tl metal and TlCp*.[131] This reactivity is closely related to a range of reactions of uranium(III), (IV) and (V) complexes with CuX, AgX, PbX_2 or HgX_2 salts, or $[(Ph_3P)AuX]$, to form U–X complexes in a higher oxidation state. For example, $[U(C_5R_5)_3]$ reacted with $PbCl_2$,[132] HgX_2 (X = Cl or I),[25] HgF_2,[132] or AgF[133] to afford uranium(IV) $[(C_5R_5)_3UX]$ (X = halide) complexes, and various bis-Cp* uranium(III) and (IV) complexes reacted with $[(Ph_3P)AuX]$ (X = Me, CF_3, C≡CR or N_3)[134] to form uranium(IV) and (V) methyl, trifluoromethyl, acetylide and azide compounds, respectively.

7.5.5.4 *Reductive elimination, alkyl radical extrusion and sterically induced reduction*

Reaction of $[U(CH_2Ph)_4]$ with the neutral α-diimine, MesN=CMe–CMe=NMes ($^{Mes}DAB^{Me}$), resulted in 1,2-diphenylethane reductive elimination and formation of [(MesN–CMe=CMe–NMes) $U(CH_2Ph)_2$] containing a dianionic enediamide ligand.[78] Similarly, $[U(CH_2Ph)_4]$ reacted with an *ortho*-iminoquinone ligand (^{dipp}ap) to eliminate 1,2-diphenylethane and generate $[(^{dipp}ap)U(CH_2Ph)_2]$, which contains a dianionic *ortho*-amidophenolate ligand.[135] However, the former reaction proceeded via concerted reductive elimination while the latter proceed via benzyl radical extrusion (Section 7.7).

Thorium and uranium hydride complexes can also eliminate H_2, allowing the hydride ligands to serve as reducing equivalents. For example, $[\{Cp^*{}_2ThH\}_2(\mu\text{-}H)_2]$, $[\{Cp^*{}_2UH\}_2]$ and $[Cp^*{}_2UH(dmpe)]$ (2 equiv.) reacted with COT (3 equiv. per An) to form $[\{Cp^*(COT)An\}_2(\mu\text{-}COT)]$, eliminating both 1.0 H_2 and 0.5 $(C_5Me_5)_2$ per thorium center.[106,136] Additionally, reaction of $[Cp^*{}_2UH(dmpe)]$ with 0.5 equiv. of $PbCp^*{}_2$ provided $[UCp^*{}_3]$;[106] this reaction bears resemblance to the reactions of $[(C_5R_5)_2UMe_2]$ and $[Cp^*{}_2UMe(am)]$ (am = amidinate) with AgX (X = OTf)[107,137] or CuX (X = Br or I),[137,138] to yield $[(C_5R_5)_2UX_2]$ and $[Cp^*{}_2UX(am)]$ complexes.

Elimination of $(C_5Me_5)_2$ from extremely sterically hindered pentamethylcyclopentadienyl complexes has also been observed. For example, $[UCp^*{}_3]$ reacted with COT (1.5 equiv.), N_2Ph_2 or C_2Ph_2 (2 equiv.) to form $[\{Cp^*(COT)U\}_2(\mu\text{-}\eta^3{:}\eta^3\text{-}COT)]$, $[Cp^*{}_2U(=NPh)_2]$ and $[Cp^*{}_2U(\kappa^2\text{-}C_4Ph_4)]$, respectively. This unusual sterically induced reduction (SIR) reactivity has been developed primarily by the Evans group, and is only observed in the most hindered complexes, indicated by U–Cp* centroid distances that are approximately 0.1 Å longer than normal, and Cp* methyl substituents that are displaced 0.48–0.54 Å out of the plane of the cyclopentadienyl ring.[139] Other examples of SIR include the reactions of $[\{Cp^*{}_2U\}_2(\mu\text{-}\eta^6{:}\eta^6\text{-}C_6H_6)]$ with COT to form $[\{Cp^*(COT)U\}_2(\mu\text{-}C_8H_8)]$, benzene and $(C_5Me_5)_2$,[109] and the reaction of $[\{Cp^*(C_5Me_4H)U\}_2(COT)]$ with phenazine to afford $[\{Cp^*(COT)U\}_2(\mu\text{-}N_2C_{12}H_8)]$ and $(C_5Me_4H)_2$.[140]

7.5.5.5 *Reactions with low-valent precursors and synthons*

A range of ligand attachment reactions make use of low-valent actinide precursors or low-valent synthetic equivalents (synthons). These reactions are discussed at multiple points in the chapter, and include reactions with COT, alkynes, P_4, As_4, S_8, azobenzenes, azides and pyridine-*N*-oxides. Additionally, a small number of ligand attachment reactions starting with elemental uranium metal have been described. For example, uranium powder (in the presence of a small amount of iodine to activate the surface of the metal) reacted with E_2Ph_2 in pyridine, to form $[An(EPh)_4(Py)_3]$ (E = S or Se).[141] Furthermore, $[U(COT)_2]$ and $[Cp_3UX]$ (X = Cl, Br or I) were prepared by reaction of uranium powder (prepared electrolytically as an amalgam followed by removal of Hg by high-temperature distillation) with COT, or HCp with CCl_4, EtBr or EtI.[142]

7.6 Cyclopentadienyl Actinide Complexes

Among carbocyclic ligands, a particularly robust and versatile system is the cyclopentadienyl anion (C_5R_5 anion; Cp^x), and as in the chemistry of the transition metals, the cyclopentadienyl ligand has played a dominant role in the field of organoactinide chemistry since its inception. The most commonly employed cyclopentadienyl anions in organoactinide chemistry are C_5H_5 (Cp), C_5H_4Me (Cp′), $C_5H_4(SiMe_3)$ (Cp^{Si}), 1,3-$C_5H_3(SiMe_3)_2$ (Cp^{Si2}), 1,3-$C_5H_3(^tBu)_2$ (Cp^{t2}), 1,2,4-$C_5H_2(SiMe_3)_3$ (Cp^{Si3}), 1,2,4-$C_5H_2(^tBu)_3$ (Cp^{t3}), C_5HMe_4 (Cp^{Me4}) and C_5Me_5 (Cp*), and these ligands may bind to metal centers in an η^1-, η^3- or η^5-coordination mode, with η^5-coordination being observed almost exclusively in actinide chemistry, although lower hapticities are more favourable for related indenyl and fluorenyl anions. An interesting feature of the cyclopentadienyl ligand system, that highlights its flexibility as a supporting ancillary, is the possibility of forming mono-, bis-, tris- and tetrakis-η^5-cyclopentadienyl actinide complexes with actinide elements in a range of oxidation states. Representative tetravalent examples of each of these types of complexes will be described in the following sections, although the number of cyclopentadienyl complexes, especially bis- and tris-cyclopentadienyl complexes, makes it impossible to include more than a fraction of the known complexes.

7.6.1 [*AnCp*$_4^x$] *complexes*

Some of the earliest successes in the field of organoactinide chemistry involved the synthesis of homoleptic tetrakis (cyclopentadienyl)actinide complexes, [$AnCp_4$], which have been prepared for thorium,[143] uranium,[144] protactinium[5] and neptunium[145] through the reactions of MCl_4 (M = Th, U or Np) with 4 KCp, or $PaCl_4$ with 2 $BeCp_2$. From a combination of infrared, powder and single crystal X-ray data, these four complexes were confirmed to be pseudo-tetrahedral with η^5-coordination of all four cyclopentadienyl rings ((a) in Figure 7.6). This contrasts the bonding situation in group 4 transition metal analogues, which adopt [$(\eta^5\text{-Cp})_2M(\eta^1\text{-Cp})_2$] (M = Ti or Hf)[146] and [$(\eta^5\text{-Cp})_3Zr(\eta^1\text{-Cp})$] structures ((b) and

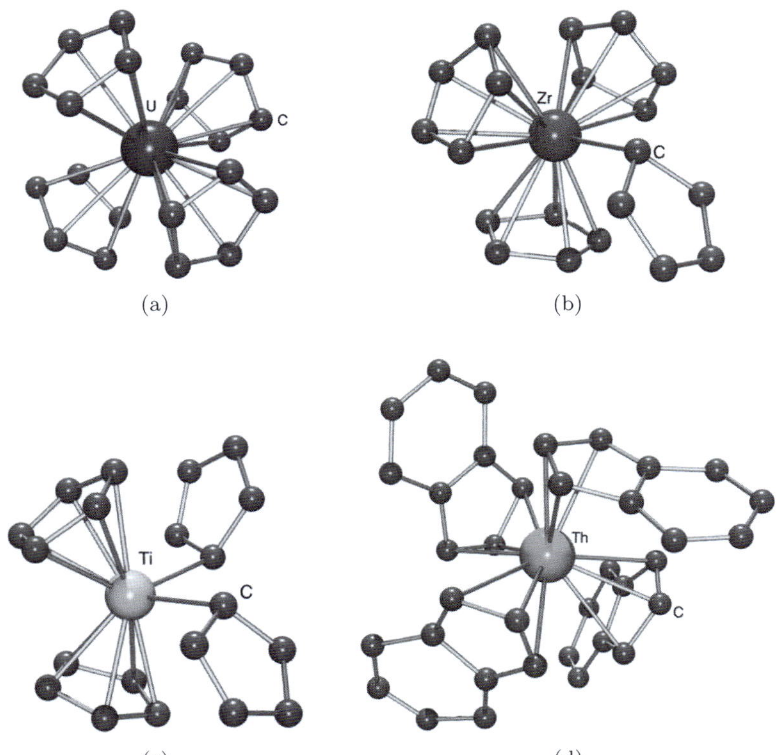

(a) (b) (c) (d)

Figure 7.6. X-ray crystal structures of (a) [UCp_4], (b) [$ZrCp_4$], (c) [$TiCp_4$], and (d) [$Th(ind)_4$] illustrating the effects of steric and electronic influences on π-ligand hapticity.

(c) in Figure 7.6)[147] in the solid state and in solution. A related thorium(IV) complex bearing four indenyl ($C_9H_7^-$) ancillaries has also been prepared by salt metathesis (*vide infra*) using $K(C_9H_7)$ and $ThCl_4$ in THF ((d) in Figure 7.6).[148] However, while compositionally analogous to the thorium(IV) Cp complex, each indenyl ring adopts an η^3-coordination mode, with longer M–C bond lengths as a result of increased steric crowding.

^{237}Np Mössbauer spectroscopy on [$NpCp_4$] yielded a substantially higher isomer shift than in other neptunium(IV) compounds, including $NpCl_4$, indicative of increased covalency in the organometallic compound.[149] The [$AnCp_4$] (An = Th–Cm) series was also investigated computationally (DFT and ADF) by Kaltsoyanis *et al.* who noted that while MOs with larger Cp π-MO and 5f-contributions are observed as the series is traversed from Th to Am, this orbital mixing is due to an increasingly close match between the Cp π-orbital and 5f-orbital energies (Figure 7.7), and does not in fact represent an increase in extent of overlap between the metal and ligand orbitals; i.e. it does not indicate that [$AmCp_4$] is more covalent, in the sense of charge buildup in the bonding region, than early members of the series such as [UCp_4].[150,151] This conclusion is in keeping with the substantial decrease in 5f-orbital radial extension as the actinide series is traversed. Similar arguments also apply to [$CmCp_4$], but in this case the 5f-based orbitals are lower in energy than the Cp π-MO-based orbitals, resulting in a $5f^7$ configuration that is better classified as Cm(III) than Cm(IV) (Figure 7.7).[150] While structurally remarkable and valuable for investigation of actinide(IV)–organometallic ligand bonding, the lack of readily accessible reactive valences in [$AnCp_4$] places limitations on the scope of subsequent reactivity.

7.6.2 [Cp^x_3AnR] complexes

In the mid-1950s, Birmingham and Wilkinson were the first to employ the cyclopentadienyl ligand to prepare [MCp_3] compounds of lanthanide elements.[152] Their results strongly suggested that cyclopentadienyl ligands would be suitable for the synthesis of various organoactinide complexes, due to similarities in the size of lanthanide and actinide elements. Isolation of [$(C_5H_5)_3UCl$], the first reported organoactinide complex, as a red-black crystalline

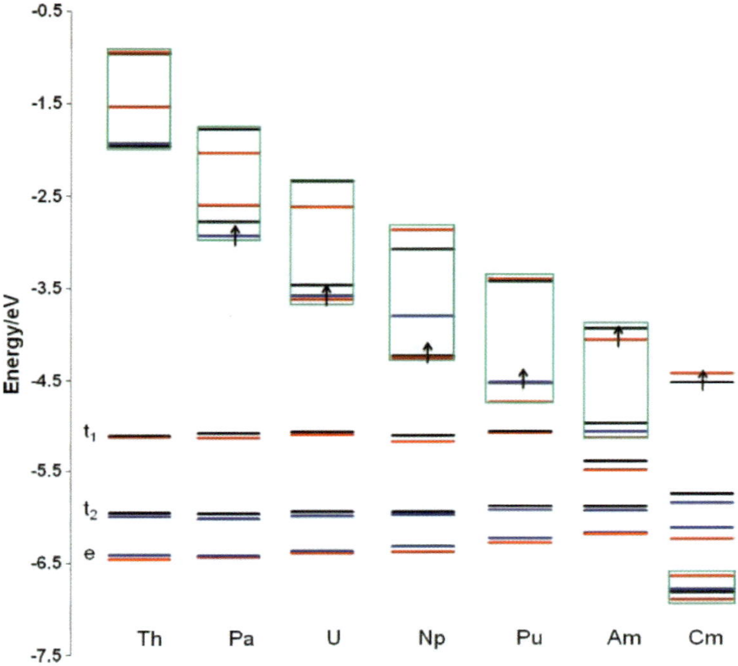

Figure 7.7. Molecular orbital energy level diagram for $[AnCp_4]$ showing the e, t_2 and t_1 Cp π-orbital based levels, and the metal 5f-based orbitals. MOs spanning the a irreducible representation (in the S_4 point group) are given in red, b in blue and e in black. Green boxes surround the 5f-based orbitals. The highest occupied orbital is indicated by an arrow, except for $[ThCp_4]$, where the t_1 Cp π-orbital based level is the HOMO). Reproduced from reference 150 with permission from the Royal Society of Chemistry.

solid followed soon thereafter.[153] The synthesis of $[(C_5H_5)_3UCl]$ was achieved by the reaction of UCl_4 with 3 equiv. of NaCp.

Tetravalent actinide compounds bearing three cyclopentadienyl or indenyl ligands make up a fairly extensive class of organoactinide complexes, and further derivatization of $[(C_5H_5)_3AnX]$ halide and hydrocarbyl complexes may be readily accomplished through salt metathesis and protonation or σ-bond metathesis routes to yield compounds containing a variety of donors, including alkyl, allyl, aryl, vinyl and alkynyl,[154–157] hydrido,[158] borohydride,[159,160] aluminohydride,[161] silyl, germyl, stannyl,[162,163] amido and alkoxide,[164,165] phosphido,[165] thiolate[159] ligands (Scheme 7.4), or a different cyclopentadienyl ligand.[123,154] $[Cp_3U(PPh_2)]$ is one example of a complex

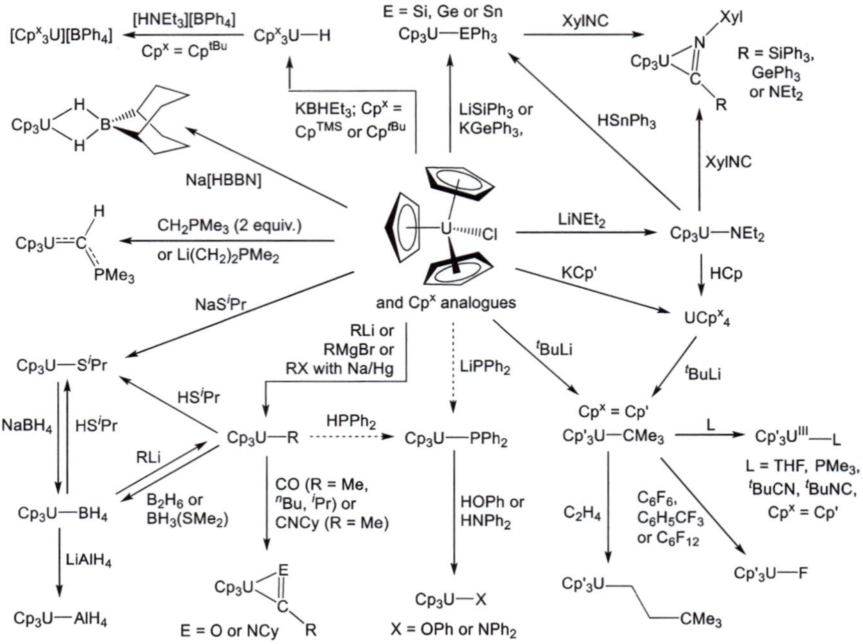

Scheme 7.4. Representative reactivity stemming from $[Cp^x_3UCl]$ (Cp^x = Cp unless noted below). In reactions converting Cp_3UCl to Cp_3UR, RLi or RMgBr reagents include those with R = Me, ^{n}Bu, ^{i}Pr, Ph, C_6F_5, ferrocenyl, vinyl, allyl, C_2Ph,[155,156] and for reactions with RX and Na/Hg, RX = MeI, ^{n}BuBr or ^{i}PrCl.[157] BBN = 9-borabicyclo[3.3.1]-nonane.

that has been prepared either by σ-bond metathesis or salt metathesis, although neither of these reactions proceeds rapidly to completion, and $[Cp_3U(PPh_2)]$ is more straightforwardly prepared via the unusual amine elimination/ligand redistribution reaction between $Cp_2U(NEt_2)_2$ and 2 equiv. of $HPPh_2$.[165] 1,1-Insertion reactions between actinide alkyl, silyl, germyl or amido complexes and isonitriles have also yielded η^2-coordinated iminoacyl and iminocarbamoyl complexes.[162,166] Additionally, reactions between $[Cp_3AnR]$ (An = Th or U; R = Me, ^{n}Bu, ^{i}Pr, ^{t}Bu or CH_2SiMe_3) and CO provided η^2-acyl complexes, which in some cases rearranged via H or SiR_3 migration due to significant contributions from both acyl An–C(=O)R and carbene An–O–C–R resonance structures.[167] The unusual tertiary alkyl complex, $[Cp'_3U^tBu]$, also underwent monoinsertion with ethylene to form $[Cp'_3U(CH_2CH_2{}^tBu)]$.[154] However, with isonitriles and several other neutral Lewis bases, reduction of $[Cp'_3U^tBu]$ to uranium(III) was observed yielding $[Cp'_3U(L)]$ complexes.[154] Furthermore, $[Cp'_3U^tBu]$ reacted with C_6F_6 in toluene, neat $C_6H_5CF_3$, or C_6F_{12} in *o*-xylene to afford $[Cp'_3UF]$ in quantitative yield *via* reactions considered to involve ^tBu radicals (Scheme 7.4).[133,168]

The reactions of $[Cp_3UCl]$ with $Li(CH_2)_2PMe_2$ (1 equiv.) or $H_2C{=}PMe_3$ (2 equiv., eliminating $[PMe_4]Cl$) are also included in Scheme 7.4, affording $[Cp_3U\{CH(PMe_3)\}]$ which features a short U–C bond (2.27(1) Å) and an elongated C–PMe_3 bond (1.70(1) Å), indicative of significant U=C multiple bond character.[169] $[Cp_3U\{CH(PMeRPh)\}]$ (R = Me and Ph) analogues were also accessible using $Li(CH_2)_2PRPh$. However, with 2 or 3 equiv. of $Li(CH_2)_2PPh_2$, $[\{Cp_2U(\mu_3\text{-}CH)(\mu\text{-}CH_2)PPh_2\}_2]$ and $[CpU\{(\mu\text{-}CH_2)_2PPh_2\}_3]$ were formed, and in the case of $[Cp_3ThCl]$, reactions with $Li(CH_2)_2PPh_2$ only yielded $[Cp_2Th\{(\mu\text{-}CH_2)_2PPh_2\}_2]$.[170]

Uranium $[Cp_3U\{CH(PMeRPh)\}]$ complexes have been shown to undergo a range of insertion reactions with organic substrates such as CO,[171] CyNC,[172] MeCN[173] and PhNCO,[174] generating reaction products analogous to those of alkyl complexes. Furthermore, insertion reactivity was observed with the CO ligands in organometallic compounds such as $[W(CO)_6]$,[175] $[CpMn(CO)_3]$[176] and $[CpCo(CO)_2]$[177] to yield $[Cp_3U\text{–}O\text{–}C({=}CH\text{–}PMeRPh)\text{–}ML_x]$ $\{ML_x$ = $W(CO)_5$, $CpMn(CO)_2$ or $CpCo(CO)\}$ products which can be described by several alternative resonance structures including $[Cp_3U\text{–}O{=}C(\text{–}CH{=}PMeRPh)\text{–}ML_x]$. At elevated temperatures, these complexes undergo unique decomposition reactions to afford $[Cp_3U\text{–}O\text{–}CH{=}CH\text{–}PRPh\text{–}CH_2\text{–}W(CO)_5]$, $[Cp(CO)_2Mn\text{–}C{\equiv}C\text{–}PMeRPh]$ and $[CpCo(CO)(PMeRPh)]$, respectively (Scheme 7.5).

X-ray diffraction studies of several unsubstituted triscyclopentadienyl actinide halide complexes show that they share a similar pseudo-tetrahedral structure (cent–U–Cl ≈ 100°; cent = the centroid of each Cp ring), with the halide residing on a threefold axis of symmetry, and very similar An–C bond lengths regardless of the nature of halide. By contrast, longer actinide–halide distances are observed in sterically hindered Cp^*_3UX (X = Cl or F) complexes, and these complexes, as well as Cp^*_3UMe and Cp^*_3ThH, are pseudo-trigonal pyramidal with the centroids of the three cyclopentadienyl rings and uranium located in a plane.[132,178,179] The extent to which even three unsubstituted cyclopentadienyl anions saturate the

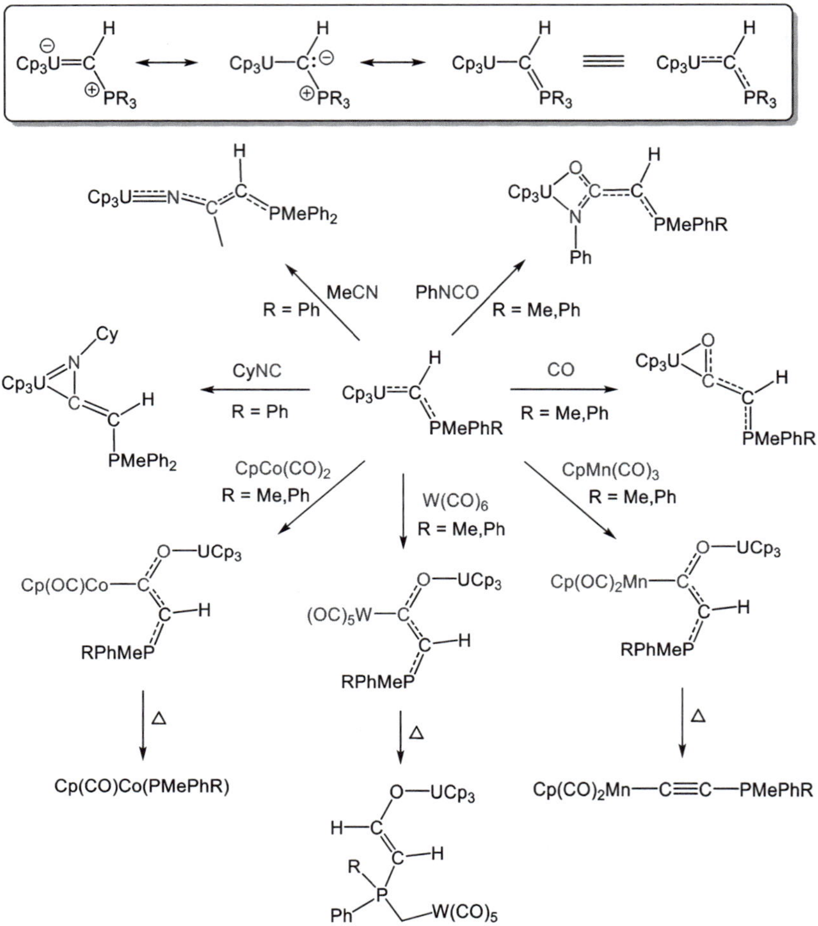

Scheme 7.5. Resonance structures- and representitive reactivity of $[Cp_3U\{CH(PMeRPh)\}]$ (R = Me, Ph).

coordination environment around the metal center is illustrated in the following examples: (a) addition of large excesses of alkyl lithium (LiR) reagents to $[Cp_3AnR]$ complexes does not result in formation of $[Cp_3AnR_2]^-$ derivatives,[155] and (b) in $[(C_5H_5)_3U(C_3H_4R\text{-}2)]$ the allyl ligand is η^1-coordinated in the solid state (R = Me),[180] and η^1-coordination is maintained in solution (although for R = H at room temperature, exchange of the α and γ allyl carbon atoms is observed, presumably occurring via an unobserved η^3-coordinated intermediate).[155]

Another effect of substantial steric protection is a high degree of thermal stability. For example, the β-hydrogen-containing primary and tertiary alkyl complexes [Cp_3U^nBu] and [Cp_3U^tBu] are only 50% decomposed after heating at 97 °C for 47 days and 11 days, respectively. Slow thermolysis of [Cp_3Th^nBu] at 170 °C also afforded [{$Cp_2Th(\mu$-η^5:η^1-$C_5H_4)$}$_2$] and n-butane in near quantitative yield, indicative of decomposition via σ-bond metathesis rather than β-hydride elimination.[181,182] By contrast, room temperature UV photolysis of [Cp_3Th^iPr] or [Cp_3Th^nBu] yielded a dark-green compound formulated as [Cp_3Th] and an approximate 1:1 mixture of alkane (propane or n-butane) and alkene (propene or 1-butene), presumably by β-hydride elimination to form "Cp_3ThH" and alkene, followed by bimolecular alkane reductive elimination via rapid reaction of the hydride complex with remaining [Cp_3ThR] (R =^{i}Pr or ^{n}Bu). Photolysis of the uranium [Cp_3UR] analogues also afforded low to moderate yields of [UCp_3] accompanied by alkane and alkene by-products. However, these reactions yielded significantly more alkane than alkene, with the additional hydrogen atoms derived from the Cp ligands and solvent, indicating that for [Cp_3UR] photolysis, β-H elimination is a minor pathway, and is accompanied by a major σ-bond metathesis or homolysis pathway.[183]

7.6.3 [$Cp^x{}_2AnR_2$] *complexes*

Tetravalent organoactinide complexes of the general formula [Cp_2AnX_2] initially proved difficult to synthesize due to instability of the metallocene complex with respect to ligand redistribution to yield mono- and tris-ligand species. For example, the first reported synthesis of [Cp_2UCl_2] by reaction of 2 equiv. of TlCp with UCl_4 in 1971 was later shown to actually yield a mixture of [Cp_3UCl] and [$CpUCl_3$(dme)].[184] However, complexes supported by the sterically undemanding cyclopentadienyl ancillaries can be stabilized by the coordination of strong Lewis bases such as dmpe, as is the case for [Cp_2ThR_2(dmpe)] (R = Me, CH_2Ph).[185]

An alternative strategy for the synthesis of stable [$Cp^X{}_2AnX_2$] complexes is the use of a sterically bulkyl cyclopentadienyl ligand, such as Cp* which has become one of the most widely employed ancillaries in organoactinide chemistry, typically yielding complexes that exhibit high thermal stability, solubility in a variety of solvents, and

Figure 7.8. X-ray crystal structures illustrating the differences in Cent–An–Cent (Cent = cyclopentadienyl ring centroid) angles in (a) $[\{Me_2Si(C_5Me_4)_2\}Th(CH_2SiMe_3)_2]$,[192] (b) $[Cp^*{}_2U(Me)_2]$,[197] and (c) the dicationic portion of $[Cp^*{}_2U(NCMe)_5][BPh_4]_2$.

appreciable crystallinity.[228,229] Complexes of the form $[Cp^*{}_2AnCl_2]$ (An = Th, U) are pseudo-tetrahedral with bent-metallocene geometries, and are readily amenable to dialkylation, typically by treatment with an alkyl lithium reagent.[71,112,113,186,187] Other substituted cyclopentadienyl ligand sets have also had various degrees of success for the stabilization of tetravalent actinide metallocenes; among the most successful ligands are the Cp″, Cp^{t2} and Cp^{t3} anions, which have allowed the synthesis of a range of monomeric $[Cp^x{}_2UR_2]$ bis-hydrocarbyl complexes.[188–190] Furthermore, $[Me_2Si(C_5Me_4)_2]^{2-}$ dianions have proven suitable for the synthesis of *ansa* actinide metallocenes (e.g. $[\{Me_2Si(C_5Me_4)_2\}An\{(\mu\text{-}Cl)_2\{Li(dme)\}_2]$) with particularly acute centroid–metal–centroid angles (~115° vs. ~140° in unlinked metallocenes; Figure 7.8). This results in a more open metal coordination environment yielding complexes capable of accommodating more than two equatorial ligands, and increasing the accessibility of substrates to the metal center for potential reactivity.[191]

Ansa-metallocenes have also been employed successfully for the formation and isolation of organoactinide dialkyl complexes bearing CH_2SiMe_3, $CH_2{}^tBu$, C_6H_5, *n*–butyl and benzyl ligands,[192] and related $[(^tBuN)SiMe_2(C_5Me_4)]^{2-}$ ligands have been utilized to generate sterically open "constrained-geometry" catalysts (CGCs) such as $[\{(^tBuN)SiMe_2(C_5Me_4)\}An(NMe_2)_2]$ for intramolecular alkene hydroamination[193] and alkyne hydroalkoxylation.[194] At the other end of the spectrum of Cp–An–Cp angles, linear actinide metallocenes were accessed by coordination of a

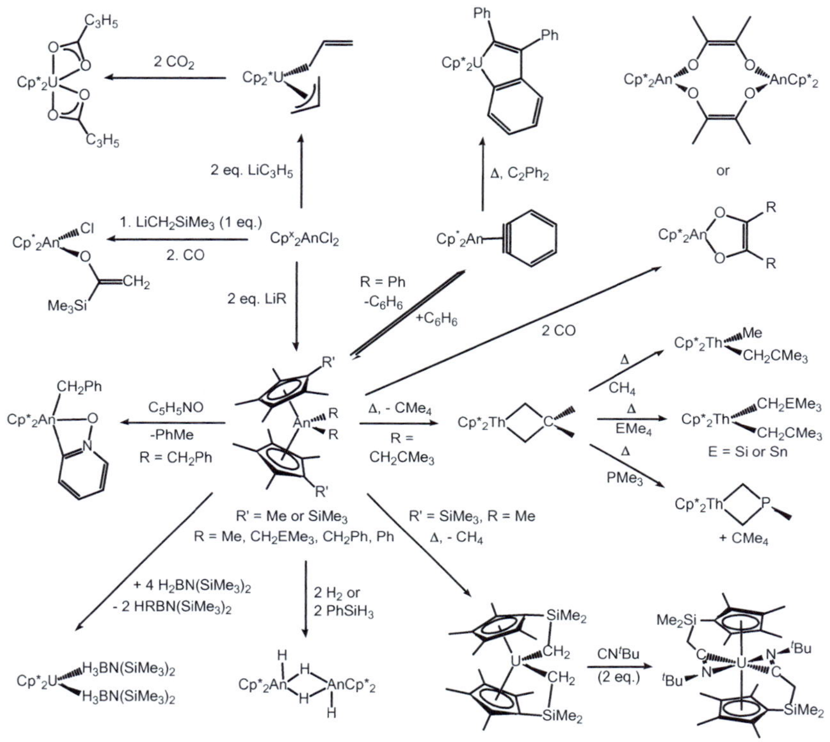

Scheme 7.6. Synthesis and selected reactions of alkyl, allyl and aryl actinide metallocene complexes bearing Cp* and $C_5Me_4(SiMe_3)$ ancillary ligands.

dicationic $[Cp^*_2U]^{2+}$ core to five neutral or anionic donor atoms; example complexes include dicationic $[Cp^*_2U(NCMe)_5][BPh_4]_2$ (Figure 7.8) and $[Cp^*_2U(phen)(NCMe)_3][BPh_4]_2$ (phen = 1,10-phenanthroline),[195] and trianionic $[NEt_4]_3[Cp^*_2U(CN)_5]$.[196]

Selected reactivity of alkyl[112] and allyl[187] actinide metallocene complexes is highlighted in Scheme 7.6, including insertion reactions with CO_2 and CN^tBu,[187,198] insertion of CO followed by rearrangement (due to significant contributions from both acyl An–C(=O)R and carbene An–O–C–R resonance structures),[199] reversible benzene elimination from the diphenyl complex to generate a benzyne complex which can be trapped with diphenylacetylene,[113] unusual cyclometallation rather than oxygen-atom transfer reactivity with pyridine-*N*-oxide,[200] cyclometallation reactions leading to metallacyclobutane products which are particularly capable of σ-bond metathesis with the C–H bonds in substrates including methane, $SiMe_4$, $SnMe_4$ and PMe_3,[201] double cyclometallation of $[\{C_5Me_4(SiMe_3)\}_2UMe_2]$ to form a double tuck-in complex,[198] reaction of dialkyl complexes with H_2 or $PhSiH_3$ to form a dimetallic tetrahydride species (in equilibrium with a uranium(III) hydride species for An = U),[112–115] and reaction of $[Cp^*{}_2UMe_2]$ with the dihydroborane $H_2BN(SiMe_3)_2$ (2 or 4 equiv.) to form $[Cp^*{}_2UMe\{H_3BN(SiMe_3)_2\}]$ and $[Cp^*{}_2U\{H_3BN(SiMe_3)_2\}_2]$.[202]

The reactivity of actinide metallocenes bearing bulky Cp^{t3} ancillary ligands has also been explored in some detail, especially due to the accessibility of monomeric $[(Cp^{t3})_2An{=}NR]$ complexes through various routes,[189] including reaction of dialkyl $[(Cp^{t3})_2AnR_2]$ complexes[188] with 1 equiv. of RNH_2 (for An = Th, R = *p*-tolyl; for An = U, R = Me or *p*-tolyl; Scheme 7.7).[190] Both thorium and uranium $[(Cp^{t3})_2An{=}NR]$ (for An = U, R = Me) complexes were shown to react with $Ph_2C{=}O$ to afford $[(Cp^{t3})_2An{=}O]$ and $Ph_2C{=}NR$.[189] Furthermore, $[(Cp^{t3})_2U{=}NR]$ reacted with internal alkynes ($C_2R'_2$, R′ = Me or Ph) to form the 2 + 2 cycloaddition product,[190]

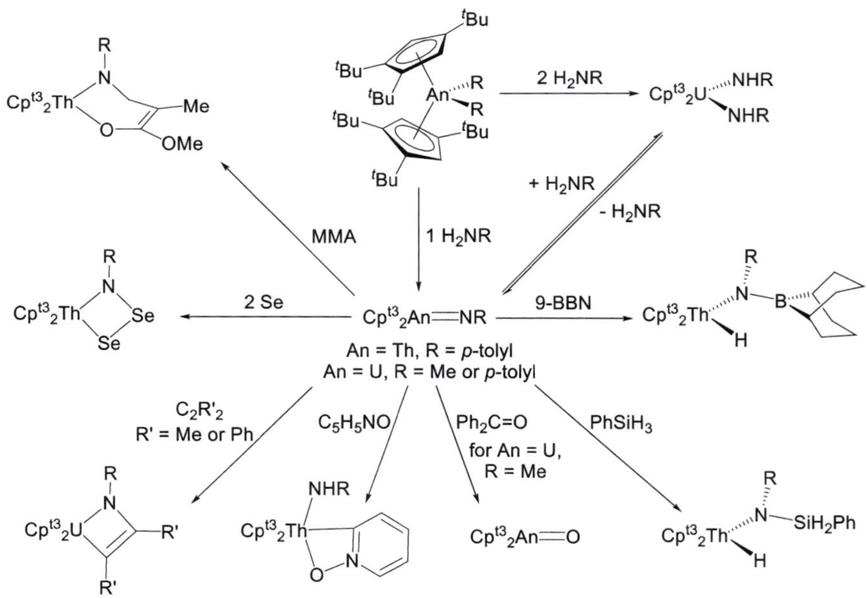

Scheme 7.7. Synthesis and selected reactions of $[(Cp^{t3})_2AnR_2]$ and $[(Cp^{t3})_2An{=}NR]$ complexes (MMA = methyl methacrylate).

and $[(Cp^{t3})_2Th{=}N(p\text{-tolyl})]$ reacted with Se (2 equiv.) and methyl methacrylate (MMA) to form 2 + 2 and 2 + 4 cycloaddition products, respectively. The thorium imido complex also reacted with with pyridine-*N*-oxide, $PhSiH_3$ and BBN (9-borabicyclo[3.3.1]nonane) to afford organometallic $[(Cp^{t3})_2U(\kappa^2 OC\text{-}C_5H_4NO)\{NH(p\text{-tolyl})\}]$ and $[(Cp^{t3})_2UH\{N(X)(p\text{-tolyl})\}]$ (X = SiH_2Ph or BC_8H_{14}) products in which a C–H, Si–H or B–H bond has added across the U=N moiety.[203,204]

7.6.4 $[Cp^x AnR_3]$ complexes

Complexes of the general formula $[Cp^xAnX_3L_x]$ (L = neutral donor ligand) employing a single cyclopentadienyl ligand are quite rare. The synthesis of $[(C_5H_5)UCl_3(dme)]$ was first described in 1972 via the reaction of UCl_4 with the thallium cyclopentadienyl salt in 1,2-dimethoxyethane (dme).[205] In this complex and in other mono-cyclopentadienyl or mono-indenyl halide derivatives, the low level of steric protection at the metal center requires coordination of external neutral bases (e.g. THF, dme) to yield isolable and stable metal complexes. Thorium and neptunium $[CpAnX_3L_x]$ complexes have also been prepared, but as the ionic radius decreases from Th to Np, the number of coordinated neutral donors switches from 3 to 2, and in THF solution the neptunium complexes show a strong tendency to undergo ligand redistribution to form $[Cp_3NpCl]$ and $[NpCl_4L_2]$ (L = THF, Me_2N-COMe, or R_3PO).[206] More sterically hindered and electron-rich $[CpU(BH_4)_3]$ can be prepared in base-free form.[121] However, in the presence of Lewis bases (L = THF, dme or HMPA), this complex also undergoes ligand redistribution to form $[Cp_3U(BH_4)]$ and $[U(BH_4)_4L_x]$.[207]

Various substituted cyclopentadienyl ligands have been investigated for the synthesis of mono-cyclopentadienyl actinide complexes in order to impart greater stability to the resulting complexes, and have met with some success. For example, a single bulky tris-trimethylsilyl substituted cyclopentadienyl ligand $\{Cp^{Si3}$ = 1,2,4-$(Me_3Si)_3C_5H_2\}$ may be coordinated to uranium through the reaction of UCl_4 with $LiCp^{Si3}$ in the presence of THF, leading to the formation of a mononuclear uranate complex $[(Cp^{Si3})UCl_2(THF)(\mu\text{-}Cl)_2Li(THF)_2]$.[208] However, formation of a closely analogous thorium complex was not observed; reaction of $ThCl_4$ with $NaCp^{Si3}$ in

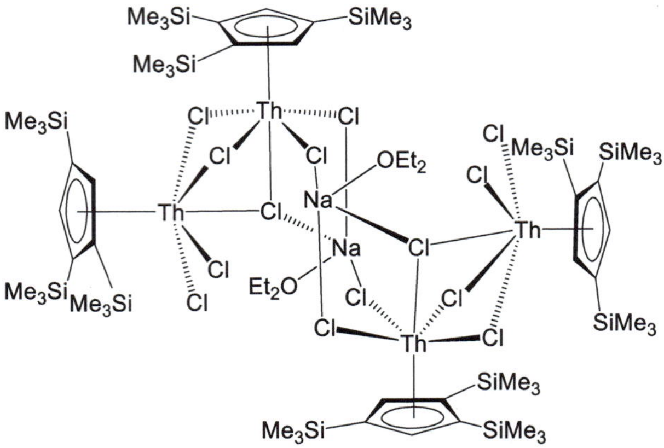

Figure 7.9. Thorium "ate" complex $[\{(Cp^{Si3})ThCl_2(\mu\text{-}Cl)_2Th(Cp^{Si3})(\mu\text{-}Cl)_2(\mu_3\text{-}Cl)Na(OEt_2)\}_2]$.

diethyl ether resulted in the formation of a tetrathorium cluster consisting of two $(Cp^{Si3})_2Th_2Cl_7^-$ units bridged by $Na^+(OEt_2)$ cations (Figure 7.9).

The chemistry of the actinides involving a single pentamethylcyclopentadienyl ligand (Cp*) offers significant advantages relative to chemistries involving the unsubstituted cyclopentadienyl ligand. Base stabilized actinide complexes supported by a single Cp* ancillary, such as $[Cp^*AnX_3(THF)]$,[209,210] may be accessed through reaction of the corresponding metal tetrachloride salt with $(C_5Me_5)MgCl$. Furthermore, these compounds may be alkylated to produce rare examples of stable trialkyl actinide complexes, $[Cp^*AnR_3]$ (An = Th; R = C_3H_5, $CH_2C_6H_5$, *o*-$C_6H_4NMe_2$. An = U; R = C_3H_5, 2-methylallyl, CH_2Ph; Figure 7.10).[17,209,211] In addition, reaction of $[Cp^*ThBr_3(THF)_x]$ with 1 equiv. of KOAr (Ar = 2,6-$C_6H_3{}^tBu_2$) afforded $[Cp^*ThBr_2(OAr)(THF)]$, which was alkylated using Me_3SiCH_2MgCl to form $[Cp^*Th(CH_2SiMe_3)_2(OAr)]$ (Figure 7.10), and subsequent reaction with H_2 provided $[Cp^*ThH_2(OAr)]_3$.[210]

7.6.5 *Thermal stability and bond disruption enthalpies of $[Cp_{(4-x)}AnR_x]$ complexes*

The thermal stability of actinide alkyl complexes is often not reported in the literature. However, bis-cyclopentadienyl alkyl complexes of sufficient steric bulk are generally robust and thermally stable,

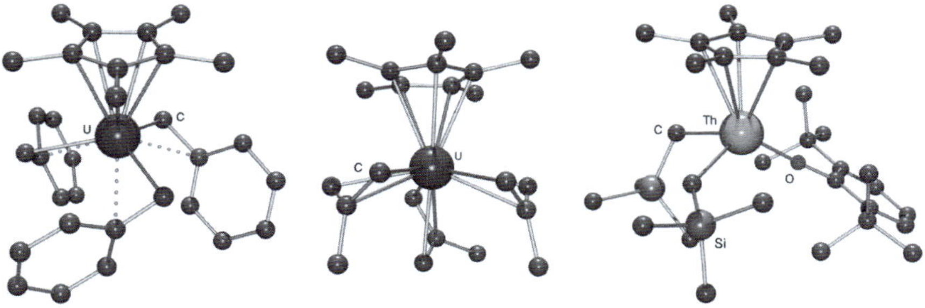

Figure 7.10. X-ray crystal structures of $[Cp^*U(CH_2Ph)_3]$,[16] $[Cp^*U(2\text{-methylallyl})_3]$[211] and $[Cp^*Th(CH_2SiMe_3)_2(OAr)]$ (Ar = 2,6-$C_6H_3{}^tBu_2$).[209]

decomposing only at elevated temperatures (80°–100°C for thorium).[212] Particularly high dialkyl complex thermal stability was reported for $[Cp^*{}_2ThMe_2]$, which was only 50% decomposed after heating for 1 week at 100°C.[113] By contrast, $[Cp^*{}_2UMe_2]$ was 50% decomposed after heating at 100°C for just 16 hours,[113] and more sterically open $[\{Me_2Si(C_5Me_4)_2\}Th(CH_2SiMe_3)_2]$ decomposed at 60 °C.[192] Tris-cyclopentadienyl complexes $[Cp_3ThR]$ {R = ^{n}Bu, ^{i}Pr, allyl, $CH_2{}^tBu$} demonstrate even greater thermal stability than bis-cyclopentadienyl complexes, in some cases showing no sign of decomposition after months at 167 °C, with stability decreasing in the order R = $CH_2{}^tBu$ > allyl > ^{n}Bu > ^{i}Pr.[182] In comparison, $[Cp_3UR]$ complexes decomposed in the 70°–100 °C range, with stability decreasing in the order R = *trans*-CMe=CHMe > Me > ^{n}Bu > allyl > ^{i}Pr > C_6F_5 > ^{t}Bu.[155] This order of stability gives the general trend of I° > II° > III° for alkyl complexes, and it is interesting to note that $[Cp_3Th{}^nBu]$ decomposes to form $[Cp_2Th(\mu\text{-}\eta^5{:}\eta^1\text{-}C_5H_4)_2ThCp_2]$ via σ-bond metathesis to release *n*-butane, rather than via a pathway involving β-hydride elimination.[155,181]

Thermochemical measurements can be used to gain insights into some of the properties of these molecules. Bond disruption enthalpies (BDEs) for complexes of the type $[Cp^*{}_2ThR_2]$ (R = alkyl, aryl) have been acquired by exploiting the propensity of organothorium compounds of this type to undergo rapid, sequential and quantitative protonolysis; a process that is highly exothermic. In this manner, using anhydrous *tert*-butanol, Th–R bond disruption energies, D(Th–R), have been measured, and a summary of the first and second BDE for selected thorium compounds can be found in Table 7.1;[213] BDEs for the process in Eq. (7.1) are defined by Eq. (7.2).

$$L_xAn - R \rightarrow L_xAn + R^{\bullet} \quad (7.1)$$

$$D(An - R) = \Delta H_f^\circ[L_xAn] + \Delta H_f^\circ[R^{\bullet}] - \Delta H_f^\circ[L_xAnR] \quad (7.2)$$

The thorium-hydrocarbyl BDEs in Table 7.1 follow the order Th–Ph > Th–Me ≈ Th–CH_2SiMe_3 > Th–CH_2R (R = Me, ^{n}Pr or ^{t}Bu) > Th–CH_2Ph, consistent with increased s-character in metal–aryl vs. –alkyl bonds, and increased steric hindrance in Th–CH_2R compounds relative to Th–Me compounds. The higher strength of Th–CH_2SiMe_3 vs. Th–$CH_2{}^tBu$ bonds may be due to steric or electronic effects, but the low BDE for Th–CH_2Ph bonds can be attributed to resonance stabilization of the benzyl radicals formed via bond dissociation. Additionally, the second BDE in non-metallacyclic bis-hydrocarbyl complexes is in all cases higher, by ∼3 kcal/mol relative to the first BDE, indicating that the Th–R bonds in the [$Cp^*_2Th(O^tBu)R$] intermediates are slightly stronger than those in [$Cp^*_2ThR_2$]; this is perhaps due to destabilization of the thorium(III) product of bond dissociation as a result of coordination to a hard alkoxide ligand. By comparison, the first BDE in metallacyclic [$Cp^*_2Th\{(CH_2)_2EMe_2\}$] (E = C or Si) complexes is much lower than the second BDE as a consequence of ring strain.

Similar data have been reported for [Cp_3ThR],[214] [$Cp^*_2UR_2$], [Cp^*_2URX][215] and [$(Cp^{Si})_3UR$][216] complexes. However, for [Cp_3ThR], no reaction occurred with ^{t}BuOH, so CF_3CH_2OH was used as the titrant, and for bis-hydrocarbyl uranium complexes, clean stepwise protonolysis by ^{t}BuOH was not observed, so BDE values are an average of the first and second BDEs. Trends in this work are summarized in Figure 7.11, and the key points are: (1) thorium(IV) alkyl BDEs are 5–10 kcal/mol higher than uranium(IV) alkyl BDEs due to the greater accessibility of uranium(III) compared with thorium(III) (i.e. less negative $An^{III/IV}$ redox potentials for U vs. Th compounds), and (2) the thorium–alkyl bonds in Cp_3ThR compounds are 5–7 kcal/mol stronger than those in $Cp^*_2ThR_2$ analogues, illustrating the extent to which ancillary ligation can influence D(M–R) values in organoactinide complexes.

These BDE values suggest that thorium alkyl bonds are particularly strong, especially for Me and CH_2SiMe_3 ligands (81–90

Table 7.1. Solution bond disruption enthalpies for the An–Alkyl bonds in selected bis- and tris-cyclopentadienyl actinide alkyl and phenyl complexes. Parentheses refer to 2σ for 6–12 determinations, and error limits do not include uncertainties that are constant throughout the series.

Th compound	**1st BDE /kcal mol^{-1}**	**2nd BDE / kcal mol^{-1}**	**U compound**	**Ave. BDE /kcal mol^{-1}**
$[Cp^*_2Th(Me)_2]$	81.2(0.8)	83.6(0.9)	$[Cp^*_2U(Me)_2]$	71.8 (3.3)
$[Cp^*_2Th(Et)_2]$	73.5(1.6)	76.3(1.6)	—	—
$[Cp^*_2Th(^nBu)_2]$	71.6(1.0)	73.6(3.4)	—	—
$[Cp^*_2Th(Ph)_2]$	88.9 (2.4)	92.4 (2.1)	—	—
—	—	—	$[Cp^*_2U(CH_2Ph)_2]$	58.3 (2.1)
$[Cp^*_2Th(CH_2^tBu)_2]$	72.3(3.8)	76.9(3.7)	—	—
$[Cp^*_2Th(CH_2SiMe_3)_2]$	80.0(3.1)	82.2(3.1)	$[Cp^*_2U(CH_2SiMe_3)_2]$	73.3 (3.1)
$[Cp^*_2Th\{(CH_2)_2CMe_2\}]$	65.3(2.3)	78.6(2.6)	—	—
$[Cp^*_2Th\{(CH_2)_2SiMe_2\}]$	75.5(3.2)	83.0(3.4)	—	—
$[Cp^*_2ThCl(Et)]$	72.2 (1.8)	n/a	$[Cp^*_2UCl(Me)]$	74.6 (1.6)
$[Cp^*_2ThCl(CH_2Ph)]$	68.2 (1.4)	n/a	$[Cp^*_2UCl(CH_2Ph)]$	63.0 (2.7)
$[Cp^*_2ThCl(Ph)]$	89.2 (2.2)	n/a	$[Cp^*_2UCl(Ph)]$	85.5 (2.6)
$[Cp_3Th(Me)]$	89.6 (1.1)	n/a	$[(Cp^{Si})_3U(Me)]$	44.8 (1.1)
$[Cp_3Th(^iPr)]$	81.8 (2.6)	n/a	—	—
$[Cp_3Th(CH_2SiMe_3)]$	87.9 (3.6)	n/a	$[(Cp^{Si})_3U(CH_2SiMe_3)]$	39.3 (2.3)
$[Cp_3Th(CH_2Ph)]$	75.3 (2.2)	n/a	$[(Cp^{Si})_3U(CH_2Ph)]$	39.3 (2.3)

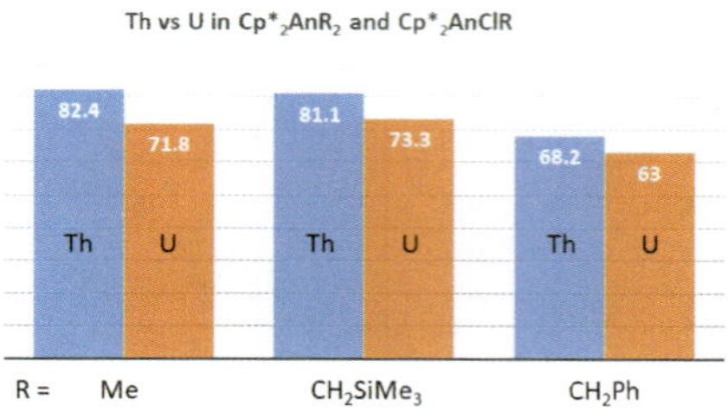

(a)

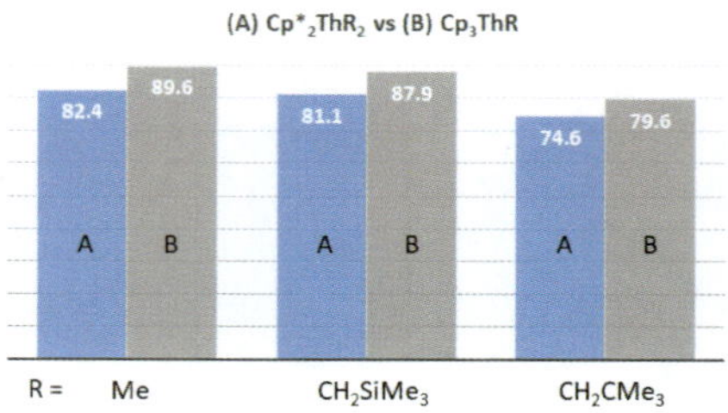

(b)

Figure 7.11. Histograms comparing solution phase actinide-alkyl bond disruption enthalpies (BDEs; kcal/mol; y-axis). Graph (a) compares analogous [$Cp^*_2AnR_2$] (R = Me and CH_2SiMe_3) and [$Cp^*_2UCl(CH_2Ph)$] complexes of thorium (blue bars) and uranium (orange bars). Graph (b) compares $Cp^*_2ThR_2$ (blue bars) and Cp_3Th-R (grey bars) compounds (R = Me, CH_2SiMe_3 and $CH_2{}^tBu$). For dialkyl complexes, values are the average of the first and second BDEs. Additional values (kcal/mol) for comparison are: [$Cp^*_2UCl(Ph)$] (85.5), [$Cp^*_2UCl(Me)$] (74.6),[215] [$Cp_3Th(^iPr)$] (81.8) and [$Cp_3Th(CH_2Ph)$] (75.3).[214]

kcal/mol). By comparison, early to mid-transition metal alkyl complexes have lower M–C bond disruption energies that are more similar to uranium–alkyl BDEs. For example, BDEs of 67, 68 and 73 kcal/mol are reported for [$Cp^*_2MMe_2$] complexes of titanium, zirconium and hafnium, respectively.[217] Actinide–alkyl BDEs are also large compared with those of samarium(III) and mid- and late transition metal complexes. For example, the BDE for the Sm–alkyl bond in [$Cp^*_2Sm\{CH(SiMe_3)_2\}$] is 48.2 kcal/mol, [Cp_2MMe_2] (M = Mo and W) and [$MMe(CO)_5$] (M = Mn and Re) have M–Me BDEs between 30 and 55 kcal/mol,[213,217] and the M–Me BDEs in [$CpPtMe_3$], [$Cp^*IrMe_2(PMe_3)$] and [$PtMe_2(PEt_3)_2$] range from ~40 kcal/mol for the platinum(IV) compound to ~60 kcal/mol in the lower oxidation state iridium(III) and platinum(II) compounds.[217] The U–E (E = Si, Ge and Sn) bond disruption enthalpies in Cp_3U–EPh_3 have also been measured, and in comparison to U–C bond disruption enthalpies they fall in a narrow range between 37(4) and 39(5) kcal/mol, indicative of comparatively weak U–heteroatom bonding, at least in the specific case of EPh_3 compounds.[216]

7.6.6 *Low-valent cyclopentadienyl complexes*

The trivalent oxidation state is increasingly stabilized upon traversing from thorium to plutonium and becomes the dominant oxidation

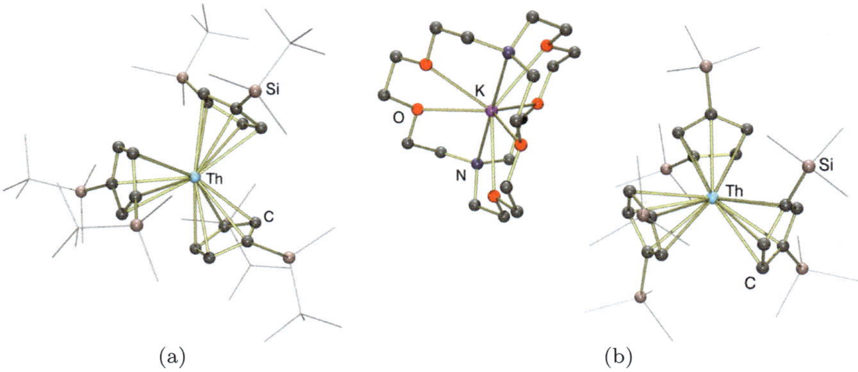

Figure 7.12. X-ray crystal structures of (a) $[Th\{C_5H_3(SiMe_2^tBu)_2\}_3]$[227] and (b) $[K(2,2,2\text{-crypt})][Th(Cp^{Si2})_3]$.[231]

state for the more lanthanide-like elements beyond americium. Illustrating this trend, the E^0 value predicted for aqueous $Th^{III/IV}$ is approx. −3.35 V (vs. the Standard Hydrogen Electrode; SHE),[218] whereas the corresponding values for $U^{III/IV}$, $Np^{III/IV}$ and $Pu^{III/IV}$ couples in acidic aqueous solution are approx. −0.6 V, 0.2 V and 1.0 V (vs. SHE), respectively.[219] This trend is also illustrated in the electrochemistry of $[Cp_3AnCl]$ (An = U and Np) complexes which undergo reversible one-electron reduction at $E_{1/2} = -1.25$ V and −0.74 V vs. SCE in THF.[220]

Unsubstituted tris-cyclopentadienyl actinide(III) complexes, $[AnCp_3]$, have been isolated for U,[221] Np,[222] Pu,[223,224] Am,[6] Cm,[7] Bk and Cf,[8] and perhaps also Th.[221,225] These complexes have been the subject of detailed computational studies, leading to the conclusion that An–Cp orbital overlap, and therefore covalency in the traditional sense, decreases from U to Cm, even though near degeneracy of the 5f and highest-energy filled cyclopentadienyl orbitals for the Am and Cm complexes gives rise to the largest orbital mixings in the series.[226] More sterically hindered $[AnCp^x_3]$ complexes have also been isolated and structurally characterized, including $[UCp^*{}_3]$ which was not predicted to be stable due to steric constraints, $[Np(Cp^{Si})_3]$ which was shown to be a monomer by X-ray crystallography,[222] and dark blue d^1 $[Th(Cp^x)_3]$ ($Cp^x = Cp^{Si2}$ and $C_5H_3(SiMe_2{}^tBu)_2$; Figure 7.12); the latter compound is volatile and thermally robust, which allowed measurement of its first ionization potential by UV photoelectron spectroscopy, yielding a value intermediate between that of Na and K.[227]

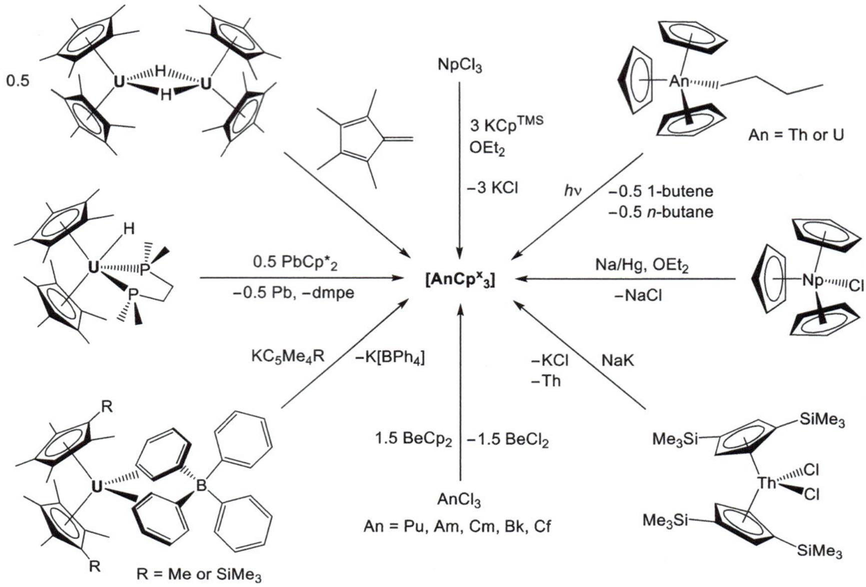

Scheme 7.8. Synthetic routes to [$AnCp^x_3$] complexes.

A variety of synthetic approaches have been employed to prepare [$AnCp^x_3$] complexes (Scheme 7.8). Heavy actinide complexes (An = Pu–Cf) were accessible via reactions of $AnCl_3$ with $BeCp_2$,[223] Am,[6] Cm,[7] Bk and Cf),[8] Additionally, [$PuCp_3$] was prepared via reaction of $MgCp_2$ with $PuCl_3$ (slow) or Cs_2PuCl_6 (rapid),[224] and base-free [$NpCp_3$] was prepared by *in situ* reduction of $NpCl_4$ to $NpCl_3$ in OEt_2, followed by addition of NaCp (3 equiv.).[222] Thorium and uranium [$An(C_5H_3R_2)_3$] (R = H or $SiMe_3$) complexes have typically been prepared by reduction of a [Cp^x_3AnCl] or [$Cp^x_2AnCl_2$] complex. However, tris-cyclopentadienyl thorium(III) and uranium(III) compounds were also accessible by UV photolysis of [Cp_3AnR] (R = Me, ^{i}Pr or ^{n}Bu) precursors, eliminating a 1:1 mixture of 1-butene and n-butane in the case of the n-butyl compounds.[183,225] Extremely hindered [UCp^*_3] was not accessible by the aforementioned routes. However, it was synthesized via the reactions of: (i) [$\{Cp^*_2U(\mu\text{-}H)_2\}$] with tetramethylfulvene (Me_4C_5 =CH_2), (ii) [$Cp^*_2UH(dmpe)$] with 0.5 equiv. of $PbCp^*_2$, eliminating Pb and dmpe, and presumably also H_2, and (iii) [$Cp^*_2U(L)_n$][BPh_4] (L = THF or dmpe; n = 1 or 0) with [$K(18\text{-crown-}6)_n$][Cp^*].[106]

Coordination of $[U(Cp^x)_3]$ compounds to CO, CNR and ECp* ligands is described in Section 7.12, and low-valent cyclopentadienyl hydride complexes including $[\{Cp^*{}_2U(\mu\text{-H})\}_2]$, $[(Cp^x)_3UH]^-$, $[(Cp^x)_3U\text{–H–}U(Cp^x)_3]^-$ and $[(Cp^{Si2})_2Th(\mu\text{-H})_3ThH(Cp^{Si2})_2]^-$ are discussed in Section 7.9. Anionic actinide(III) compounds such as $[(\eta^5\text{-}Cp^x)_3An(\eta^1\text{-}Cp^x)]^-$ (An = U, $Cp^x = Cp^{Si}$;[229] An = Np, Cp^x = Cp)[223] $[Cp_3UX]^-$ (X = Me, ^{n}Bu or BH_4), $[Cp_2U(BH_4)_2]^-$, $[(Cp^{Si2})_2UMe_2]^-$, and $[\{Cp_3U\}_2(\mu\text{-Cl})]^-$ have also been reported,[229] as have neutral and cationic derivatives with fewer than 3 cyclopentadienyl rings; for example, $[Cp^*{}_2U][BPh_4]$,[106] $[Cp^*{}_2U(allyl)(THF)_x]$ (x = 0 or 1),[187] and $[Cp^*{}_2Th\{(N^iPr)_2CMe\}]$.[230] Low-valent Cp/COT derivatives are also known, and the reactivity of $[(COT)(Cp^x)U(THF)_x]$ is discussed in Section 7.9.

Calculated electron affinities for $[An(Cp^{Si})_3]$ compounds are 1.39 (Th), 1.45 (Pa), 1.61 (U), 1.59 (Np), 1.87 (Pu) and 2.49 eV (Am), respectively,[232] which follows the general trend estimated for $An^{II/III}$ aqueous redox potentials (vs. SHE): −4.9 V (Th), −4.7 V (Pa), −4.7 V (U), −4.7 V (Np), −3.5 V (Pu), −2.3 V (Am) and −1.6 V (Cf).[233] These data indicate that divalent actinide complexes will be extremely reducing for Th–Np. Nevertheless, Furche and Evans reported the landmark syntheses of $[K(2,2,2\text{-crypt})][U(Cp^{Si})_3]$[234] and $[K(2,2,2\text{-crypt})][Th(Cp^{Si2})_3]$[231] (Figure 7.12) in 2013 and 2015, respectively, and the isolation of thermally-unstable dark-brown crystals of a complex tentatively assigned as $[K(2,2,2\text{-crypt})][Np(Cp^{Si})_3]$ was reported by Walter and Arnold *et al.* in 2017.[222]

Other cyclopentadienyl compounds with the potential to behave as divalent actinide synthetic equiv. (synthons) include $[Cp^*XU(\mu\text{-}\eta^6\text{:}\eta^6\text{-arene})UCp^*X]$ [X = Cp*, $CH(SiMe_3)_2$, $N(SiMe_3)_2$, $OC_6H_2(^tBu\text{-}o)_2(R\text{-}p)$, or $MeC(N^iPr)_2$], $[(Cp^x)_2U(bipy)]$ (Cp^x = Cp*, Cp^{t2} or Cp^{t3}), $[Cp^*{}_2U\{\eta^2\text{-}C_2(SiMe_3)_2\}]$, and $[(Cp^{t2})_2Th(\eta^4\text{-}C_4H_6)]$, and the reactivity of these complexes is discussed in Sections 7.7.4 and 7.7.8, respectively.

7.6.7 *High-valent cyclopentadienyl complexes*

Uranium(IV) alkyl compounds such as $[Cp^x_2UMe_2]$ and $[Cp^*{}_2UMe(am)]$ (am = amidinate) react with AgOTf,[107,137] CuX (X = Br or I),[137,138] Ph_3CX (X = Cl or Br),[235] Ph_3COTf[236] or $NEt_3(HF)_3$[127] to yield triflate and halide complexes in which

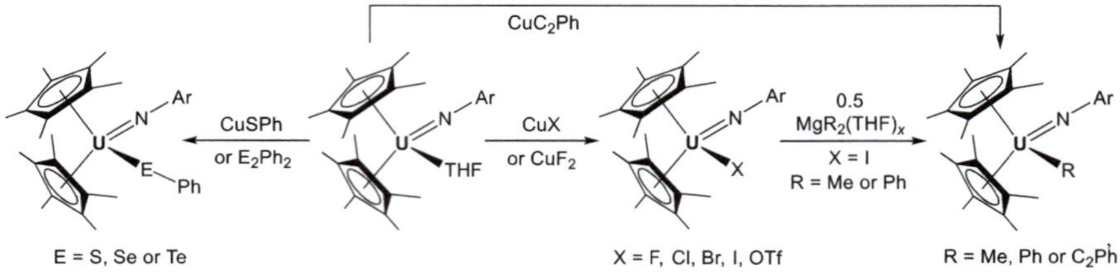

Scheme 7.9. Synthesis of [Cp*$_2$U(=NAr)X] complexes.

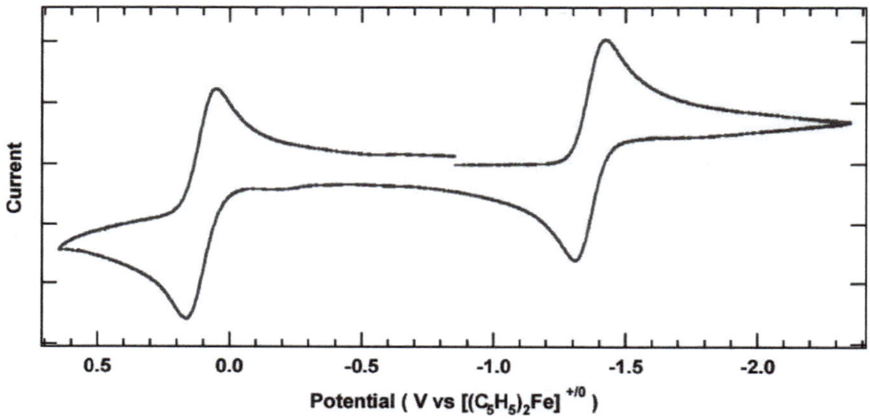

Figure 7.13. Cyclic voltammogram of [Cp*$_2$U(=NAr)I] in THF/[NBu$_4$][B(C$_6$F$_5$)$_4$]. Reproduced from reference 237 with permission from the American Chemical Society.

the oxidation state is maintained. By contrast, uranium(IV) imido compounds such as [Cp*$_2$U(=NAr)(THF)$_x$] (Ar = $C_6H_3^iPr_2$-2,6; Scheme 7.9) react with 1 equiv. of CuI, CuBr, CuCl, CuF$_2$, CuOTf, CuSPh or CuC$_2$Ph to afford uranium(V) iodide,[237,238] bromide,[238] chloride,[239] fluoride,[238,239] triflate,[237] phenylthiolate,[237] or acetylide[240] complexes. Alternatively, [Cp*$_2$U(=NAr)(SPh)] could be prepared by reaction of [Cp*$_2$U(=NAr)(THF)] with S$_2$Ph$_2$ (0.5 equiv.), and analogous reactions with Se$_2$Ph$_2$ and Te$_2$Ph$_2$ afforded the SePh and TePh analogues.[241] Methyl and phenyl uranium(V) imido complexes, [Cp*$_2$U(=NAr)(R)] (R = Me or Ph) were also accessed by reaction of [Cp*$_2$U(=NAr)(I)] with MgMe$_2$ or MgPh$_2$(THF)$_2$.[239] The ability of imido ligands to stabilize high

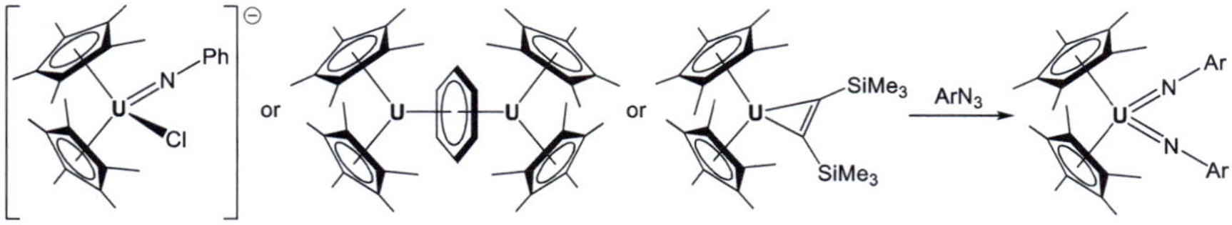

Scheme 7.10. Synthetic routes to $[Cp^*_2U(=NAr)_2]$ complexes.

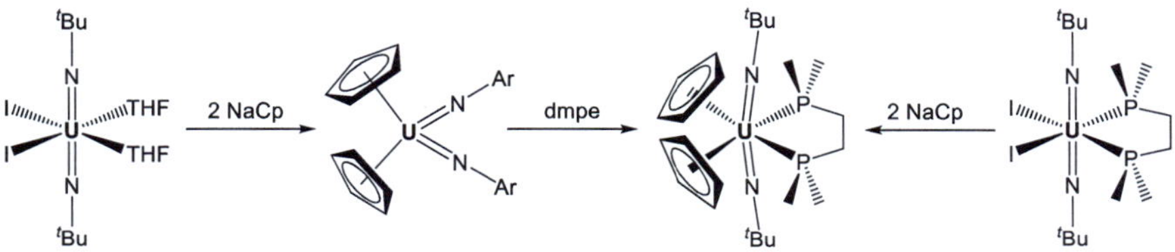

Scheme 7.11. Reactions to prepare $[Cp_2U(=N^tBu)_2]$ and $[Cp_2U(=N^tBu)_2(dmpe)]$.

oxidation states of uranium is also highlighted by the cyclic voltammogram of $[Cp^*_2U(=NAr)I]$ (Figure 7.13), which not only features a reversible $U^{V/IV}$ reduction but also a reversible $U^{V/VI}$ oxidation.

Cyclopentadienyl-supported bis-imido uranium(V) and (VI) complexes have also been prepared by a variety of routes. For example, the uranium(IV) imido complex $[Li(tmeda)][Cp^*_2U(=NPh)Cl]$ reacted with phenyl azide (PhN_3) to form $[Cp^*_2U(=NPh)_2]$ with a *cis*-configuration of the imido groups.[242] Alternatively, the uranium(III) complex, $[Cp^*_2UCl(NaCl)]$, reacted as a source of "Cp^*_2U", generating a 1:1 mixture of $[Cp^*_2U(=NPh)_2]$ and $[Cp^*_2UCl_2]$ upon reaction with PhN_3 or azobenzene (PhNNPh).[243] Analogous bis-imido complexes were also accessed by reaction of the divalent uranium synthons $[(Cp^*_2U)_2(\mu{:}\eta^6{:}\eta^6\text{-}C_6H_6)]$ (0.5 equiv.)[110] and $[Cp^*_2U\{\eta^2\text{-}C_2(SiMe_3)_2\}]$[244] with 2 equiv. of 1-adamantyl- or *p*-tolyl-azide (Scheme 7.10).

Unsubstituted cyclopentadienyl (i.e. Cp) derivatives were also prepared by Boncella *et al.* (Scheme 7.11) by installation of the imido ligands prior to the Cp ligands; reaction of $[U(=N^tBu)_2I_2(THF)_2]$ with 2 equiv. of NaCp afforded $[Cp_2U(=N^tBu)_2]$ with *cis*-imido ligands, which reacted with dmpe to form $[Cp_2U(=N^tBu)_2(dmpe)]$ with

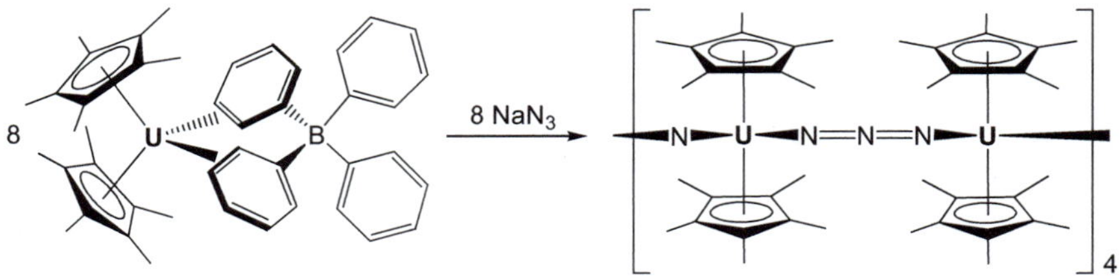

Scheme 7.12. Reaction of $[Cp^*{}_2U][BPh_4]$ with NaN_3 to afford $[\{(\mu\text{-}N)UCp^*{}_2(\mu\text{-}N_3)UCp^*{}_2\}_4]$.

trans-disposed imido groups. The latter compound was also accessible via the reaction of $[U(=N^tBu)_2I_2(dmpe)]$ with 2 equiv. of NaCp, and in the reaction with 1 equiv. of NaCp, a mono-Cp analogue, $[CpU(=N^tBu)_2I(dmpe)]$, was isolated.[245]

High-valent cyclopentadienyl uranium oxo complexes have also been isolated. For example, reaction of $[Cp^*{}_2U(OAr)(THF)]$ or $[Cp^*{}_2U(=NAr)(THF)]$ (Ar = 2,6-diisopropylphenyl) with pyridine-*N*-oxide afforded pentavalent $[Cp^*{}_2U(OAr)(=O)]$ and hexavalent $[Cp^*{}_2U(=NAr)(=O)]$, respectively.[246] In addition, reaction of $[Cp^*{}_2U(CN)_5]^{3-}$ with 2 equiv. of pyridine-*N*-oxide afforded hexavalent $[Cp^*U(=O)_2(CN)_3]^{2-}$, presumably via a pathway involving uranium, cyanide and Cp* oxidation.[247] Furthermore, a range of unique mixed Cp*/2,6-bis(imido)pyridine uranium oxo and imido complexes were recently reported by Bart *et al.*[248]

Other oxygen transfer reagents used to prepare cyclopentadienyl uranium oxo species include N_2O and ^tBuNCO. For example, reaction of $[(^{tips2}COT)Cp^*U(THF)]$ $\{^{tips2}COT$ = 1,4-bis(triisopropylsilyl)cyclooctatetraenide$\}$ with ^tBuNCO yielded pentavalent $[(^{tips2}COT)Cp^*U(=O)(CN^tBu)]$, which extruded CN^tBu under vacuum to afford $[(^{tips2}COT)Cp^*U(=O)]$. By contrast, reaction of $[(^{tips2}COT)Cp^*U(THF)]$ with N_2O afforded the uranium(IV) dimer, $[\{(^{tips2}COT)Cp^*U\}_2(\mu\text{-}O)]$, and this same uranium(III) precursor reacted with NaN_3 (1 equiv.) to provide a rare uranium nitride complex, $[(^{tips2}COT)Cp^*U\{\equiv N\text{–}Na(OEt_2)_2\}]$.[249] This reactivity contrasts that of $[Cp^*{}_2U][BPh_4]$ with NaN_3 (1 equiv.), which afforded octametallic $[\{(\mu\text{-}N)UCp^*{}_2(\mu\text{-}N_3)UCp^*{}_2\}_4]$ containing uranium(IV) centers linked by alternating nitride and azide bridges (Scheme 7.12).[250]

7.7 Organoactinide Complexes Bearing Non-Cyclopentadienyl π-Ligands

7.7.1 *Cyclobutadienyl and related complexes*

Cyclobutadienyl actinide complexes were unknown until 2013, when Liddle *et al.* reported the synthesis of both cyclobutadienyl and diphosphacyclobutadienyl uranium(IV) complexes. These complexes, $[\{(Ts^{Xyl})U\}_2(\mu\text{-}\eta^5\text{:}\eta^5\text{-}C_4Ph_4)]$ and $[\{(Ts^{p-tolyl})U\}_2(\mu\text{-}\eta^4\text{:}\eta^4\text{-}C_2{}^tBu_2P_2\text{-}1,3)]$ (Figure 7.14; Ts^{Ar} = $HC(SiMe_2NAr)_3{}^{3-}$) were prepared by reaction of $[\{(Ts^{Ar})U\}_2(\mu\text{-}\eta^6\text{:}\eta^6\text{-toluene})]$ (considered to be a uranium(V) complex with a bridging $C_7H_8{}^{4-}$ tetraanion; Section 7.7.4), with excess diphenylacetylene or 2 equiv. of *tert*-butyl phosphaalkyne ($^tBuC{\equiv}P$). In both structures the $C_2R_2E_2$ (E = CPh, R = Ph; E = P; R =^{t}Bu) ring is sandwiched between two uranium centers and the structural, magnetic, spectroscopic and computational data are consistent with a central $C_2R_2E_2{}^{2-}$ dianion. However, it is interesting to note that two of the phenyl rings in the butadienyl complex are strongly distorted out of the plane of the butadienyl ring resulting in a unique $\mu\text{-}\eta^5\text{:}\eta^5$-bonding mode in which the *ipso*-carbon atom of each of these phenyl rings interacts with one of the uranium centers.[251]

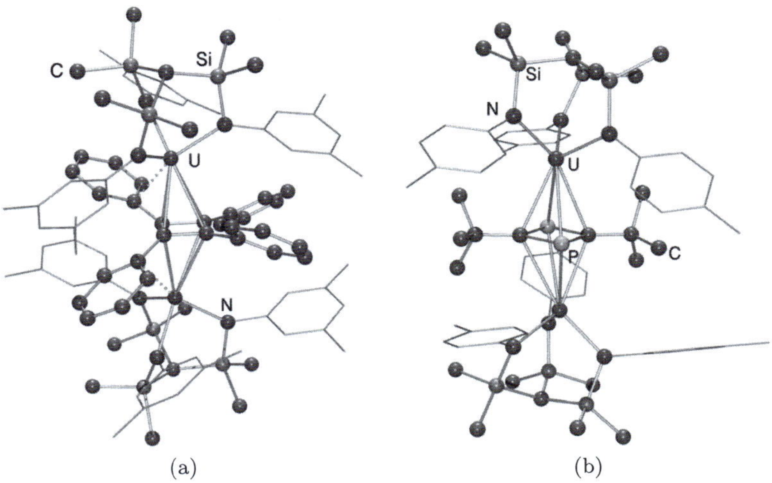

Figure 7.14. X-ray Crystal structures for (a) $[\{(Ts^{Xyl})U\}_2(\mu\text{-}\eta^5\text{:}\eta^5\text{-}C_4Ph_4)]$ and (b) $[\{(Ts^{p-tolyl})U\}_2(\mu\text{-}\eta^4\text{:}\eta^4\text{-}C_2{}^tBu_2P_2\text{-}1,3)]$ (Ts^{Ar} = $HC(SiMe_2NAr)_3^{3-}$).[251]

Further extension of the $C_4R_4{}^{2-}$ and $C_2R_2P_2{}^{2-}$ series would yield P_4^{2-} ligated complexes, and indeed [{(Ar^{t}BuN)$_3$U}$_2$(μ-η^4:η^4-P_4)] (Ar = 3,5-$C_6H_3Me_2$),[252] [{Cp*(^{tips2}COT)U}$_2$(μ-η^2:η^2-P_4)][253] and [{(Cp^{Si2})$_3$Th}$_2$(μ-η^1:η^1-P_4)][254] were reported in 2004, 2011 and 2015 from the Cummins, Cloke and Mills groups. The P_4 ligand in these three complexes adopts very different bridging modes, but in all cases it is considered to be a dianion.

7.7.2 *Phospholyl, arsolyl, pyrrolyl and related complexes*

The tetramethylphospholyl (TMP) anion shares many similarities with the Cp* anion, although it is slightly less bulky, slightly less electron donating ($U^{III/IV}$ redox potentials are shifted to more positive potentials by ∼0.2 V relative to Cp* analogues), and has the potential to engage in κ^1-coordination via phosphorus, in addition to η^5-coordination. Pentahapto-coordinated TMP complexes with Cp* structural analogues include [(η^5-TMP)UR_3] (R = BH_4 or CH_2Ph), [(η^5-TMP)$_2$UX_2] (X = Cl, BH_4 ((a) in Figure 7.15), Me or CH_2SiMe_3), [(η^5-TMP)Cp*UX_2] (X = BH_4, Me or CH_2SiMe_3), [(η^5-TMP)$_3$UX] (X = Cl or Me), and [(COT)U(η^5-TMP)(OR)] (R is Et).[104,255–257] Furthermore, mixed TMP/COT complexes have been prepared, including the uranium(IV) complexes [(COT)(TMP)U(BH_4)(THF)], [(COT)U(η^5-TMP)(hmpa)$_2$][BPh_4] and the uranium(III) complex [(COT)U(TMP)(hmpa)$_2$], and it is notable that reaction of the former complex with [$HNEt_3$][BPh_4] yielded [(COT)U(BH_4)(THF)$_2$][BPh_4], via protonation of the TMP ligand rather than a COT or BH_4 ligand.[258] All of the aforementioned TMP complexes are monometallic, but in less coordinatively saturated bis-TMP uranium(III) complexes, [{(η^5-TMP)(μ-η^5:η^1-TMP)U(BH_4)}$_2$] ((b) in Figure 7.15) and [{Cp*(μ-η^5:η^1-TMP)U(BH_4)}$_2$], dimeric structures are observed with bridging TMP ligands; bridging TMP ligands (η^5-coordinated to uranium, and η^1-coordinated to nickel) were also observed in heterometallic [Ni{(μ-η^5:η^1-TMP)$_2$UCl_2}$_2$] and [{Ni_2(μ-TMP)$_2$}{(μ-η^5:η^1-TMP)^{2}UCl_2}$_2$].[259]

A series of mixed-sandwich C_4Me_4E (E = P, N and As) / ^{tips2}COT {^{tips2}COT = 1,4-C_8H_6(Si^{i}Pr$_3$)$_2$} uranium complexes have also been reported by Cloke *et al.* Reaction of UI_3 with K[C_4Me_4E] followed by K_2[^{tips2}COT] afforded base-free [{(^{tips2}COT)U(C_4Me_4E)}$_n$]

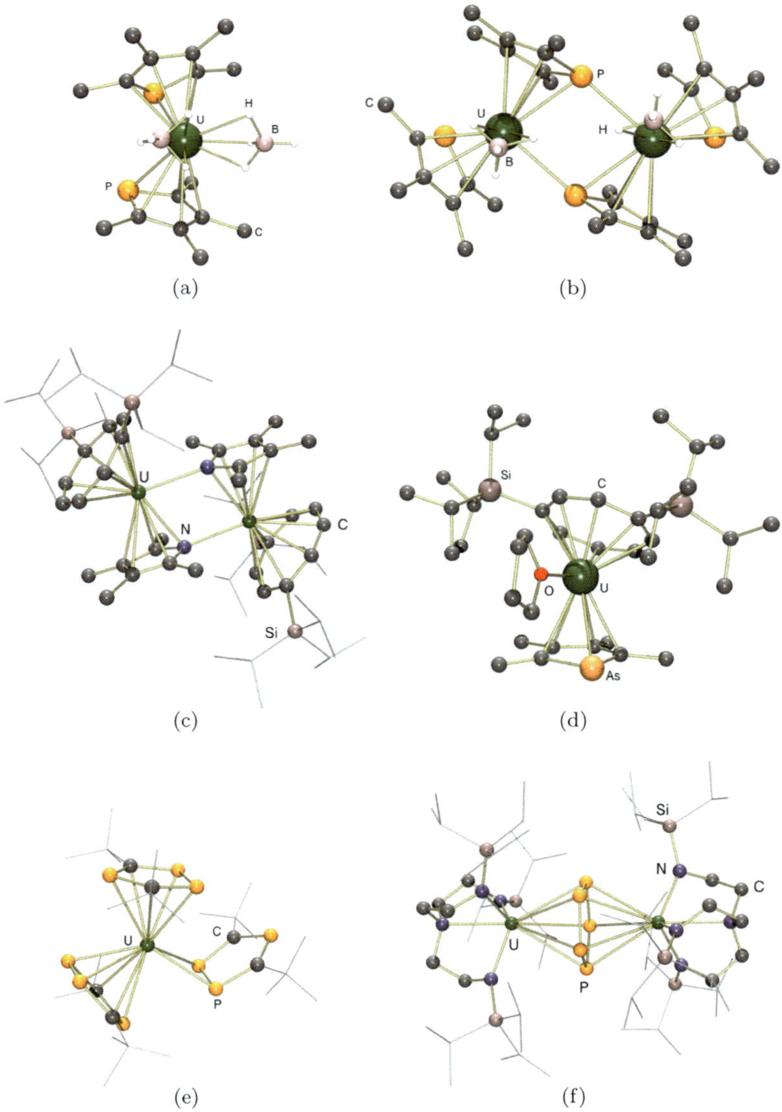

Figure 7.15. X-ray structures of (a) $[(\eta^5\text{-TMP})_2\text{U}(\text{BH}_4)_2]$,[260] (b) $[\{(\eta^5\text{-TMP})(\mu\text{-}\eta^5{:}\eta^1\text{-TMP})\text{U}(\text{BH}_4)\}_2]$,[257] (c) $[\{(^{\text{tips2}}\text{COT})\text{U}(\mu\text{-}\eta^5{:}\eta^1\text{C}_4\text{Me}_4\text{N})\}_2]$ $\{^{\text{tips2}}\text{COT}=\ 1,4\text{-C}_8\text{H}_6(\text{Si}^i\text{Pr}_3)_2\}$, (d) $[(^{\text{tips2}}\text{COT})\text{U}(\eta^5\text{-C}_4\text{Me}_4\text{As})(\text{THF})]$, (e) $[(\eta^5\text{-C}_2{}^t\text{Bu}_2\text{P}_3\text{-1,2,5})_2\text{U}(\eta^2\text{-C}_2{}^t\text{Bu}_2\text{P}_3\text{-1,2,5})]$,[261] and (f) $[\{(\text{tren}^{\text{TIPS}})\text{U}\}_2(\mu\text{-}\eta^5{:}\eta^5\text{-P}_5)]$ $\{\text{tren}^{\text{TIPS}} = \text{N}(\text{CH}_2\text{CH}_2\text{NSi}^i\text{Pr}_3)_3\}$[262]

(n = 1 or 2) complexes ((c) in Figure 7.15, for E = N), and the tetramethylpyrrolyl complex was shown to exist as a dimer with μ-η^5:η^1-bridging heterocyclic ligands. These complexes coordinate THF to generate [(^{tips2}COT)U(η^5-C_4Me_4E)(THF)] monomers ((d) in Figure 7.15; for E = As) and react with CO_2 (for E = N and P) to afford μ-oxo uranium(IV) complexes with bridging carbamate or phosphacarbonate ligands.[263]

Conceptual substitution of the "CR" units in a cyclopentadienyl ligand for phosphorus can be further extended to $C_2R_2P_3^-$ ligands, and [(η^5-$C_2{}^tBu_2P_3$-1,2,5)$_2$U(η^2-$C_2{}^tBu_2P_3$-1,2,5)] ((e) in Figure 7.15), which features both η^5- and η^2-P_2-coordination modes, was prepared from UI_3 and 3 equiv. of K[$C_2{}^tBu_2P_3$-1,2,5].[261] In addition, the first example of a *cyclo*-P_5 actinide complex, [{(tren^{TIPS})U}$_2$(μ-η^5:η^5-P_5)] {Tren^{TIPS} = N($CH_2CH_2NSi^iPr_3$)$_3$} ((f) in Figure 7.15), was prepared via the reaction of [U(tren^{TIPS})] with P_4. However, for the *cyclo*-P_5 complex, computational data are suggestive of charge transfer from uranium to afford a P_5^{2-} dianion, and unlike the bonding in cyclopentadienyl complexes, the major bonding interactions in the U(P_5)U unit appear to involve a pair of polarized δ-bonds. These differences are attributed to the larger size of the P_5 ring relative to C_5R_5, and the greater acceptor ability of P_5 combined with the strongly reducing nature of uranium(III).[262]

A range of bis-,[122,264–266] tris-[267] and macrocyclic tetrakis-pyrrolyl ligands[268–271] have also been utilized to develop the chemistry of thorium and uranium, and even neptunium,[272] with prominent ligands illustrated in Figure 7.16. The coordinative flexibility of pyrrole ligands is illustrated by the solid-state structures of roughly isostructural [(dme)K({(−CH_2−)$_5$}$_4$-calix[4]tetrapyrrole)U(dme)] (Figure 7.17) and [(dme)K(Et_8-calix[4]tetrapyrrole)U(dme)], in which both uranium and potassium are $\kappa^1 N$-coordinated to two of the four pyrrolyl rings and are π-coordinated to the other two. Subsequent reduction of the octaethyl complex using K[naphthalenide] under N_2 afforded mixed-valent [K(dme)$_4$][{(dme)K(Et_8-calix[4]tetrapyrrole)U(μ-NK)}$_2$] in which N_2 has been cleaved to afford bridging nitrido groups (Figure 7.17).[270]

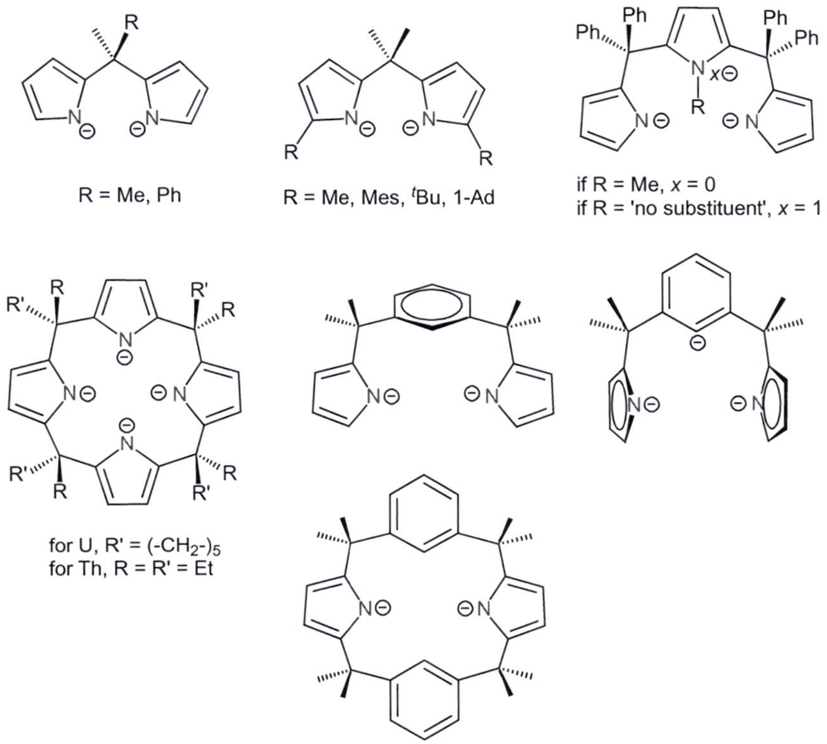

Figure 7.16. Structures of poly-pyrrolyl ligands utilized in thorium and uranium chemistry.

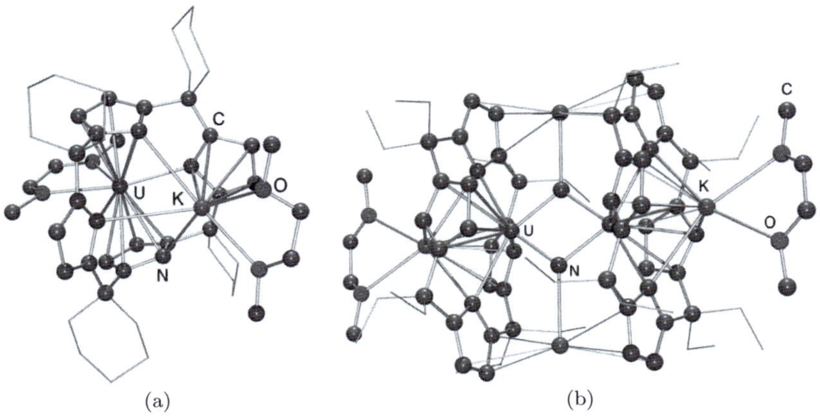

Figure 7.17. X-ray Crystal Structures of (a) [(dme)K({(−CH$_2$−)$_5$}$_4$-calix[4]tetrapyrrole)U(dme)][271] and (b) [K(dme)$_4$][{(dme)K(Et$_8$-calix[4]tetrapyrrole)U(μ-NK)}$_2$].[270]

Changes in the coordination mode of poly-pyrrolyl ligands to actinide centers can also be induced through actinide Lewis base coordination (Scheme 7.13),[265] Lewis acid coordination at the pyrrolyl nitrogen atom (Scheme 7.14; with accompanying ligand backbone deprotonation),[273] or changes in oxidation state (Scheme 7.15; the change in pyrrolyl ring coordination mode, from $\kappa^1 N$ in the uranium(III) complex to η^5 in the uranium(IV) complex, is accompanied by a change in arene hapticity from η^6 to η^1).[122] Furthermore, the aforementioned ligand backbone deprotonation reactivity, which occurred upon reaction of $[Li(dme)_3][\{1,3\text{-}[(2\text{-}NC_4H_3)CMe_2]_2C_6H_4\}UCl_3]$ with $AlEt_3$, highlights the non-innocent behaviour of certain poly-pyrrolyl ancillaries. Demethylation of the central pyrrole ring in 2,5-$[(2\text{-}NC_4H_3)CPh_2]_2C_4H_2NMe$ ligands has also been observed in reactions of thorium complexes carried out under highly reducing conditions, as has ring-opening of the neutral pyrrole ring in a monoanionic $(C_4H_3N)CPh_2(C_4H_3NMe)$ ligand.[267]

7.7.3 *Carborane complexes*

The dicarbollide dianion, *nido*-1,2-$C_2B_9H_{11}{}^{2-}$, has been used extensively in transition metal chemistry as a dianionic analogue of the cyclopentadienyl monoanion, given that it possesses an open pentagonal face and can serve as a 6 π-electron donor. The first actinide dicarbollide complex, $[(\eta^5\text{-}C_2B_9H_{11})_2UCl_2]^{2-}$ was reported by Raymond *et al.* in 1977. This dianionic metallocene analogue was prepared from UCl_4 and $Na_2[C_2B_9H_{11}]$, and while subsequent reaction with $K_2[COT]$ rapidly produced $[U(COT)_2]$, excess (dme)NaCp was not able to completely replace the $C_2B_9H_{11}{}^{2-}$ ligands, indicating that the dicarbollide dianion has ligating properties intermediate between those of COT^{2-} and Cp^-.[274] After this initial report, actinide carborane chemistry remained dormant for almost 20 years before Rabinovich and Abney reported $[(\eta^5\text{-}C_2B_9H_{11})_2AnX_2]^{2-}$ (An = U, X = Br; An = Th, X = Cl, Br or I; (a) in Figure 7.18) analogues, as well as the mono-dicarbollide thorium(IV) anion $[(\eta^5\text{-}C_2B_9H_{11})ThX_3(THF)]^-$. In addition, the uranium(III) complexes $[(\eta^5\text{-}C_2B_9H_{11})UI_2(THF)_2]^-$ and $[(\eta^5\text{-}C_2B_9H_{11})_2UX(THF)]^{2-}$ (X = Br or I) were prepared starting either from $UI_3(THF)_4$ and 1 or 2 equiv. of a $C_2B_9H_{11}{}^{2-}$ salt, or by Na/Hg reduction of $[(\eta^5\text{-}C_2B_9H_{11})_2UBr_2]^{2-}$.[275,276]

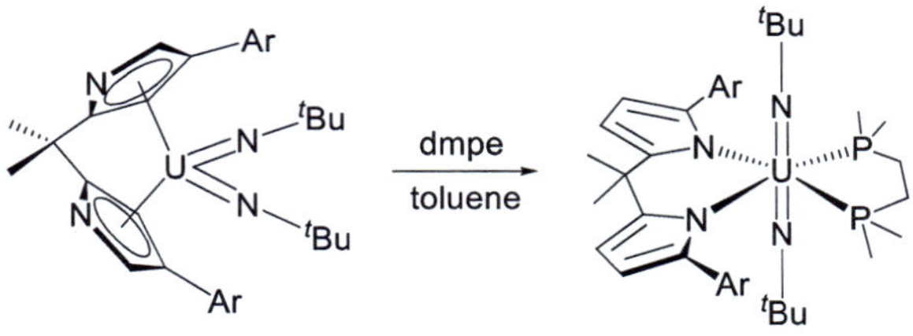

Scheme 7.13. Changes in pyrrolyl ring coordination mode induced by actinide–Lewis base coordination.[265]

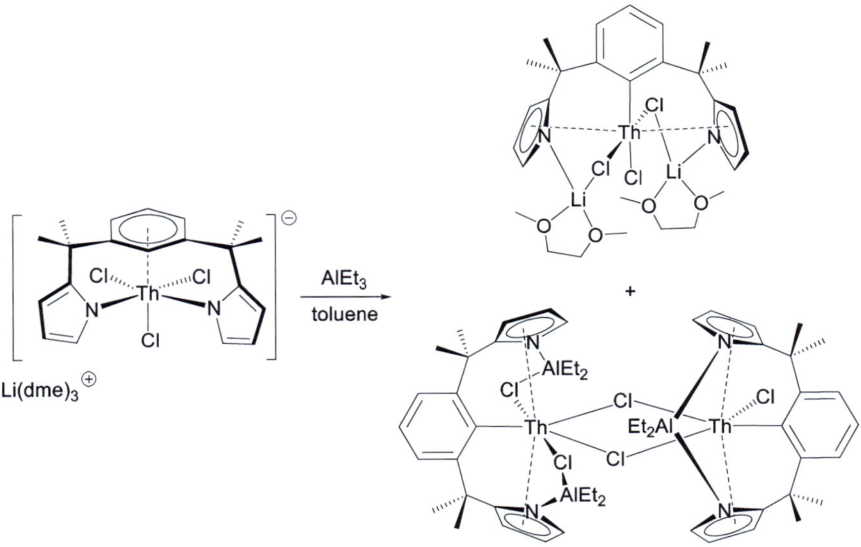

Scheme 7.14. Changes in pyrrolyl ring coordination mode induced by reactivity leading to ligand deprotonation and Lewis acid coordination of the pyrrolyl nitrogen atoms.[273]

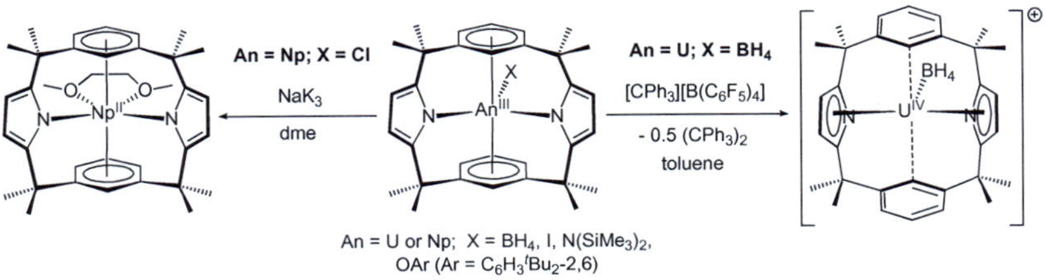

Scheme 7.15. Actinide chemistry with the *trans*-calix[2]benzene[2]pyrrolide (calix) ligand, highlighting access to neptunium(II),[272] and oxidation state-induced changes in the coordination mode of the arene and pyrrolyl rings to uranium (the arene rings are not deprotonated in any of these complexes).[122]

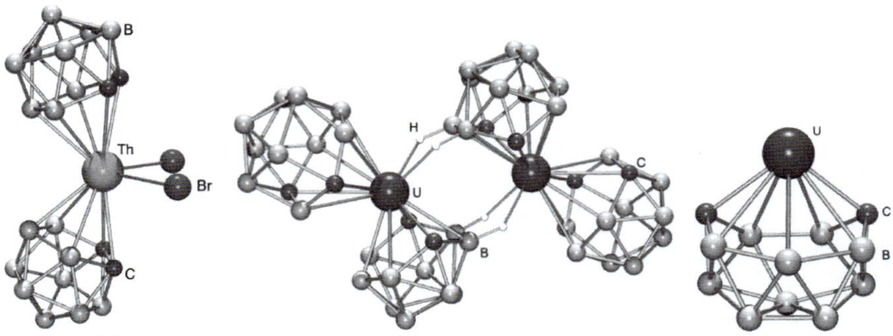

(a) (b) (c)

Figure 7.18. X-ray crystal structure diagrams showing (a) the anionic portion of $[Li(THF)_4]_2[(\eta^5\text{-}C_2B_9H_{11})_2ThBr_2]$,[276] (b) $[K_2(THF)_5]_2[\{[(\eta^7\text{-}C_2B_{10}H_{12})(\eta^6\text{-}C_2B_{10}H_{12})U\}_2]$ with $K(THF)_x^+$ countercations omitted, and (c) the η^7-interaction between uranium and the *arachno*-$C_2B_{10}H_{12}{}^{4-}$ ligand in the structure shown in (b).[277]

A very unusual uranium complex featuring η^6-coordinated *nido*-$C_2B_{10}H_{12}{}^{2-}$ and η^7-coordinated *arachno*-$C_2B_{10}H_{12}{}^{4-}$ ligands, $[\{[K_2(THF)_5]_2][U(\eta^7\text{-}C_2B_{10}H_{12})(\eta^6\text{-}C_2B_{10}H_{12})]\}_2]$ ((b) and (c) in Figure 7.18), was prepared by Xu *et al.* via the reaction of UCl_4 with 2 equiv. of *closo*-$C_2B_{10}H_{12}$ and greater than 6 equiv. of potassium in THF. This is the first example of a metallacarborane bearing an η^7-$C_2B_{10}H_{12}{}^{4-}$ ligand and is the first organoactinide compound featuring an η^6-$C_2B_{10}H_{12}{}^{2-}$ ligand.[277]

7.7.4 *Arene complexes*

7.7.4.1 *Terminal arene complexes*

Between 1985 and 1987, Cotton *et al.* reported a series of uranium(III) and (IV) hexamethylbenzene complexes.[278] The uranium(IV) complexes, trimetallic $[\{(\eta^6\text{-}C_6Me_6)U^{IV}Cl_2(\mu\text{-}Cl)_3\}_2(UCl_2)]$ and dimetallic $[\{(\eta^6\text{-}C_6Me_6)U^{IV}Cl_2\}_2(\mu\text{-}Cl)_3][AlCl_4]$ ((a) in Figure 7.19) were prepared by refluxing UCl_4, C_6Me_6, and differing amounts of $AlCl_3$ in hexanes. The uranium(III) complexes, trimetallic $[\{(\eta^6\text{-}C_6Me_6)U(\kappa^2\text{-}AlCl_4)(\mu\text{-}Cl)\}_3(\mu_3\text{-}Cl)_2][AlCl_4]$ ((b) in Figure 7.19) and monometallic $[(\eta^6\text{-}C_6Me_6)U(\kappa^2\text{-}AlCl_4)_3]$ ((c) in Figure 7.19) were prepared via similar reactions conducted

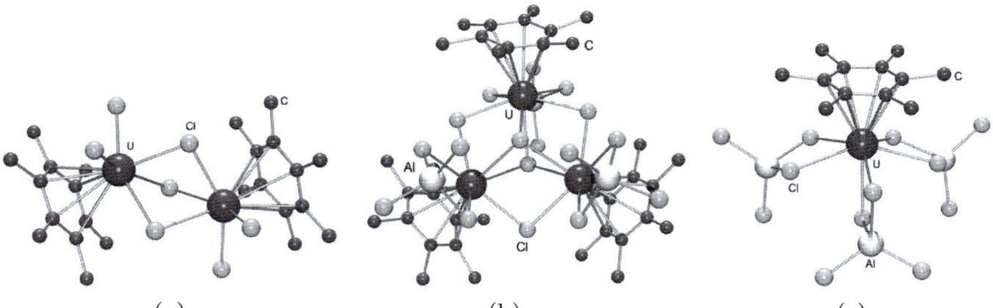

Figure 7.19. X-ray crystal structure diagrams showing the cationic or neutral portions of (a) dimetallic $[\{(\eta^6\text{-C}_6\text{Me}_6)\text{U}^{\text{IV}}\text{Cl}_2\}_2(\mu\text{-Cl})_3][\text{AlCl}_4]$, (b) trimetallic $[\{(\eta^6\text{-C}_6\text{Me}_6)\text{U}^{\text{III}}(\kappa^2\text{-AlCl}_4)(\mu\text{-Cl})\}_3(\mu_3\text{-Cl})_2][\text{AlCl}_4]$, and (c) monometallic $[(\eta^6\text{-C}_6\text{Me}_6)\text{U}^{\text{III}}(\kappa^2\text{-AlCl}_4)_3]$.

in toluene, with addition of aluminum metal after an initial heating period. The borohydride analogue of the $[(\eta^6\text{-C}_6\text{Me}_6)\text{U}(\kappa^2\text{-AlCl}_4)_3]$ complex, $[(\eta^6\text{-mesitylene})\text{U}(\kappa^3\text{-BH}_4)_3]$, was reported in 1988 by Baudry *et al.*, and was prepared simply by refluxing $[\text{U(BH}_4)_4]$ in mesitylene. The mesitylene ligand in this complex was readily displaced by addition of hexamethylbenzene, yielding $[(\eta^6\text{-C}_6\text{Me}_6)\text{U}(\kappa^3\text{-BH}_4)_3]$, whereas in toluene and benzene, only partial and reversible arene exchange was observed. All five hexamethylbenzene complexes described earlier were crystallographically characterized, with very similar average U–C_{arene} distances (2.91–2.94 Å).[60,279]

Other crystallographically characterized actinide complexes featuring intermolecular interactions with a neutral arene are limited to a series of thorium and uranium alkyl complexes reported by Emslie *et al.* Reaction of the neutral dialkyl complexes $[(\text{XA}_2)\text{Th}(\text{CH}_2\text{R})_2]$ (R = SiMe_3 or Ph; XA_2 = a rigid-backbone dianionic NON-donor ligand = 4,5-bis(2,6-diisopropylanilido)-2,7-di-*tert*-butyl-9,9-dimethylxanthene) with $[\text{CPh}_3][\text{B}(\text{C}_6\text{F}_5)_4]$ in benzene or toluene afforded $[(\text{XA}_2)\text{Th}(\text{CH}_2\text{SiMe}_3)(\eta^6\text{-C}_6\text{H}_5\text{R})][\text{B}(\text{C}_6\text{F}_5)_4]$ (R = H or Me) and $[(\text{XA}_2)\text{Th}(\text{CH}_2\text{Ph})(\eta^n\text{-toluene})][\text{B}(\text{C}_6\text{F}_5)_4]$ (Figure 7.20) in which 1 equiv. of arene solvent is π-coordinated to the cationic metal center; in $[(\text{XA}_2)\text{Th}(\text{CH}_2\text{SiMe}_3)(\eta^6\text{-benzene})]^+$, the arene is η^6-coordinated in the solid state (Th–C_{arene} (ave.) = 3.26 Å; Figure 7.20), whereas in $[(\text{XA}_2)\text{Th}(\text{CH}_2\text{Ph})(\eta^n\text{-toluene})]^+$, two Th–$C_{\text{arene}}$ distances are similar to those in the benzene complex (3.21,

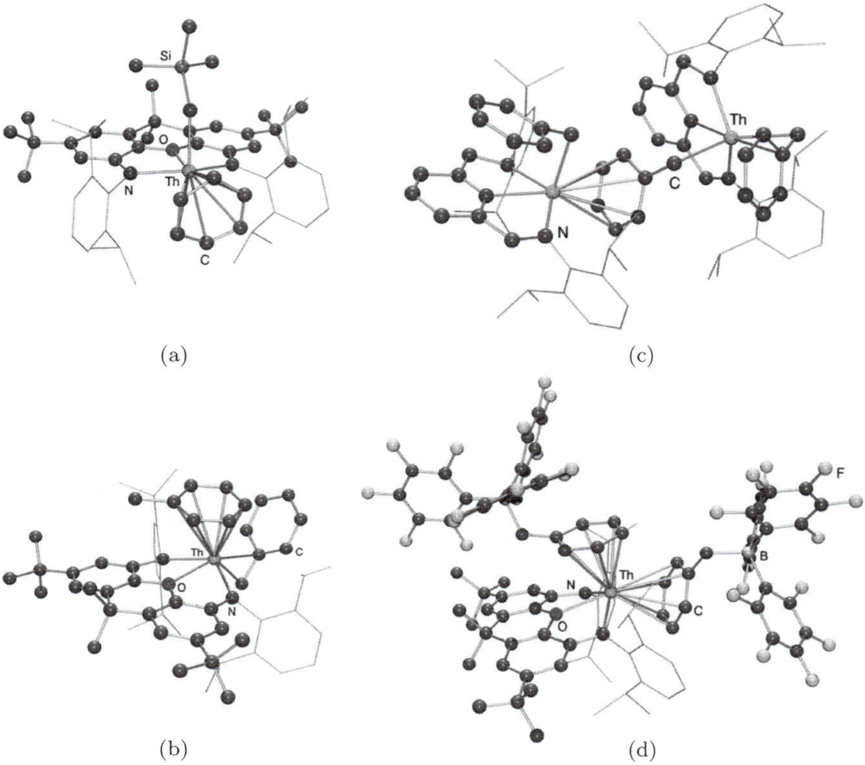

Figure 7.20. X-ray crystal structures of (a) $[(XA_2)Th(CH_2SiMe_3)(\eta^6\text{-}C_6H_6)][B(C_6F_5)_4]$, (b) $[(XA_2)Th(CH_2Ph)(\eta^6\text{-}C_6H_5Me)][B(C_6F_5)_4]$, (c) $[(BDPP)Th(CH_2Ph)(\mu\text{-}\eta^1{:}\eta^6\text{-}CH_2Ph)Th(CH_2Ph)(BDPP)][B(C_6F_5)_4]$, and (d) $[(XA_2)Th][\eta^6\text{-}PhCH_2B(C_6F_5)_3]_2$.

3.28 Å), two are shorter (3.06, 3.09 Å), and two are longer (3.37, 3.44 Å). For $[(XA_2)Th(CH_2SiMe_3)(\eta^n\text{-toluene})]^+$, d_5-bromobenzene does not displace toluene from the metal center to any observable extent, and coordinated and free toluene only undergo slow exchange on the NMR timescale at room temperature.[280]

The uranium analogues, $[(XA_2)U(CH_2SiMe_3)(\eta^x\text{-}C_6H_5R)][B(C_6F_5)_4]$ (R = H or Me) were also recently reported, with η^6-benzene coordination ($U\text{–}C_{arene}$ = 3.099(3)–3.249(3) Å) in the solid state, and slipped toluene coordination with U–C distances ranging from 3.05(2) Å (para) to 3.78(2) Å (*ipso*). In solution distinct ^{1}H NMR signals were again observed for free and coordinated benzene or toluene in d^5-bromobenzene. However, in this case, significant displacement of benzene or toluene by bromobenzene was observed, yielding appreciable amounts of $[(XA_2)U(CH_2SiMe_3)(C_6H_5Br)][B(C_6F_5)_4]$.[281] The An–$C_{arene}$ distances in the arene-coordinated thorium and uranium alkyl cations are consistent with the larger ionic radius of Th vs. U, the reduced donor ability of benzene and toluene vs. hexamethylbenzene in Cotton and Baudry's complexes, and much greater steric encumbrance in the XA_2 complexes relative to the hexamethylbenzene $AlCl_4$ and BH_4 complexes.

In contrast to the reactions of $[CPh_3][B(C_6F_5)_4]$ with $[(XA_2)Th(CH_2Ph)_2]$, the analogous reaction with $[(BDPP)Th(CH_2Ph)_2]$ {BDPP=2,6-bis(2,6-diisopropylanilidomethyl)pyridine} afforded $[(BDPP)Th(CH_2Ph)(\mu\text{-}\eta^1{:}\eta^6\text{-}CH_2Ph)Th(CH_2Ph)(BDPP)]$ $[B(C_6F_5)_4]$ (Figure 7.20) when a sub-stoichiometric amount of $[CPh_3][B(C_6F_5)_4]$ was employed. This compound may be viewed as a "$(BDPP)Th(CH_2Ph)^+$" cation π-coordinated to the phenyl ring of a benzyl group in the starting material [Th–C_{arene} (ave.) = 3.13 Å], and represents the only example of a μ-η^1:η^6-benzyl group.[280] Additionally, reaction of $[(XA_2)Th(CH_2Ph)_2]$ with 1 or 2 equiv. of $B(C_6F_5)_3$ (Section 7.7.8) afforded $[(XA_2)Th(CH_2Ph)][\eta^6\text{-}PhCH_2B(C_6F_5)_3]$ and $[(XA_2)Th][\eta^6\text{-}PhCH_2B(C_6F_5)_3]_2$ (Th–C_{arene} (ave.) = 3.06–3.07 Å; Figure 7.20), respectively, due to benzyl group abstraction and coordination of the resulting benzylborate anion to the cationic or dicationic thorium center.[282] Arene-coordination as part of a tight ion pair has also been observed by Evans *et al.* for $[Cp^x_2U\{(\mu\text{-}Ph)_2BPh_2\}]$ (Cp^x = Cp^{Me4} or Cp*), in which uranium(III) is coordinated to two phenyl rings of a tetraphenylborate anion. In the solid state, the phenyl rings of the BPh_4 anion are η^1-coordinated (U–C = 3.05 Å) and η^6-coordinated (U–C = 2.87–3.07 Å) in the Cp^{Me4} complex,[283] whereas they are both η^2-coordinated in the Cp* complex, with one short {2.857(7) or 2.880(7) Å} and one long {3.138(8) or 3.166(8) Å} U–C contact for each phenyl ring.[106]

Intramolecular actinide–arene interactions have also been observed for bulky aryloxide complexes as well as multidentate ligands with arene units in the ligand backbone. Complexes in the first category are $[\{U(OC_6H_3^iPr_2\text{-}2,6)_3\}_2]$,[32] $[U\{OC_6H_2(CHPh_2\text{-}o)_2$

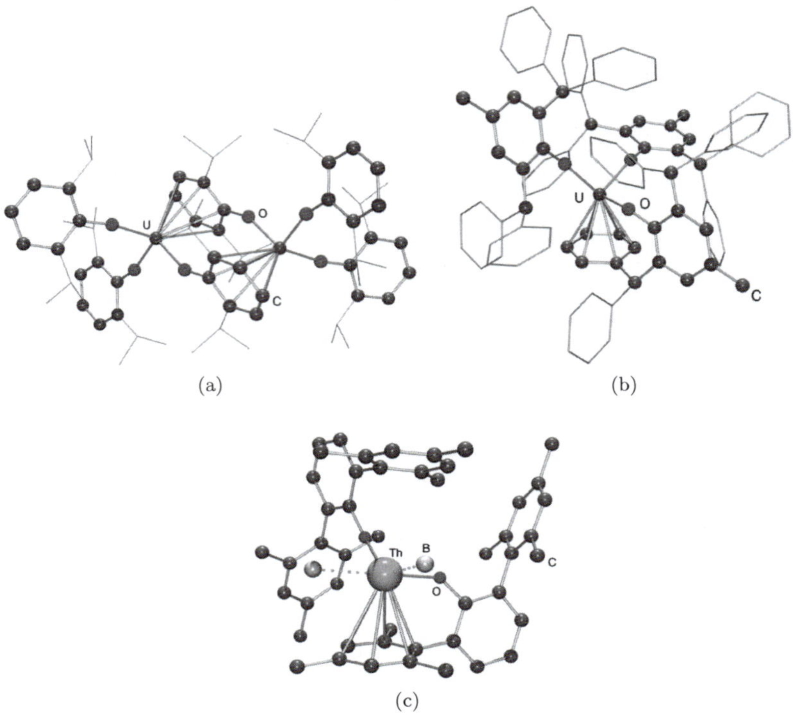

Figure 7.21. X-ray crystal structures of (a) $[\{U(OC_6H_3^iPr_2\text{-}2,6)_3\}_2]$,[32] (b) $[U(OC_6H_2(CHPh_2\text{-}o)_2(Me\text{-}p)\}_3]$,[284] and (c) $[Th(OC_6H_3Mes_2\text{-}2,6)_2(BH_4)_2]$.[285]

$(Me\text{-}p)\}_3]$,[284] and $[Th(OC_6H_3Mes_2\text{-}2,6)_2(BH_4)_2]$[285] (Figure 7.21) with average An–C_{arene} distances of 2.92, 2.85 and 3.14 Å, respectively. The former uranium complex reacted with THF to form $[U(OC_6H_3{}^iPr_2\text{-}2,6)_3(THF)]$ which no longer features a U–arene interaction, whereas $[U\{OC_6H_2(CHPh_2\text{-}o)_2(Me\text{-}p)\}_3]$ reacted with THF to form an adduct in which the U–arene interaction is maintained, albeit with a longer average U–C_{arene} distance of 2.96 Å. However, upon treatment with N_2O in THF, $[U\{OC_6H_2(CHPh_2\text{-}o)_2(Me\text{-}p)\}_3]$ was converted to $[U(=O)\{OC_6H_2(CHPh_2\text{-}o)_2(Me\text{-}p)\}_3(THF)]$ in which the U–arene interaction has been eliminated.

Complexes in the second category include those bearing the following multidentate ligands: $(^{tBu}ArO)_3Mes$ $[(^{tBu}ArO)_3Mes = C_6Me_3\{CH_2C_6H_2(O\text{-}o)(^tBu\text{-}m)_2\}_3]$,[286,287] macrocyclic *trans*-calix[2]benzene[2]pyrrolide (calix),[62,122,272] and 1,3-$C_6H_4(CMe_2Pyr)_2$ (Pyr

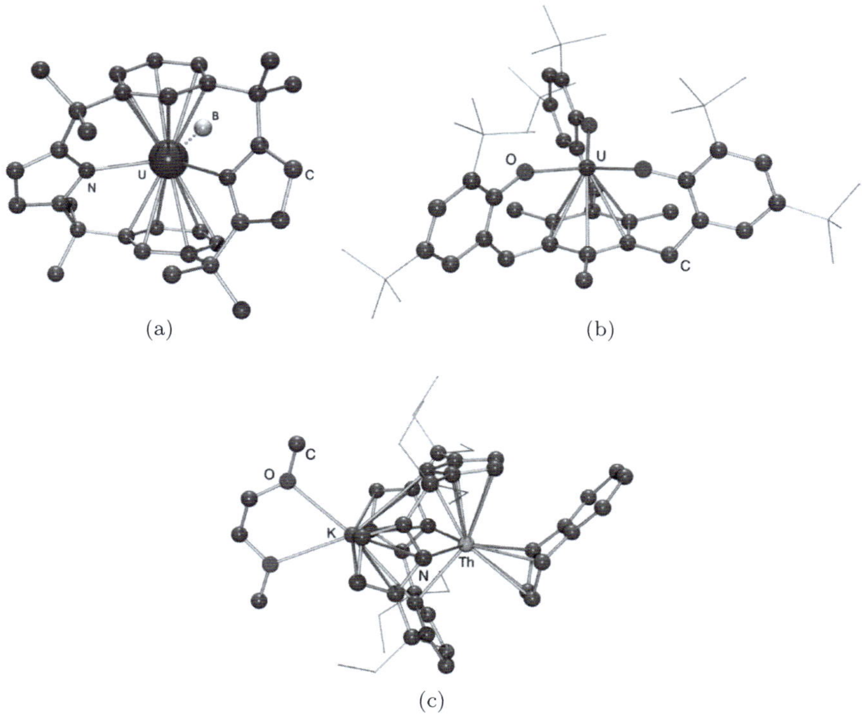

Figure 7.22. X-ray crystal structures for (a) [(calix)U(BH_4)],[62] (b) [{(^{tBu}ArO)$_3$Mes}U],[286] and (c) [(dme)K(Et$_8$-calix[4]tetrapyrrole)Th (η^4-naphthalene)]$^-$.[269]

= 2-pyrrolyl) ligands.[266] In these complexes the An–C distances can be expected to be influenced by the spatial arrangement and coordination mode of the attached heteroatom donors, transmitted through geometric constraints imposed by ligand rigidity. However, this can lead to interesting cooperative effects, as discussed for [(calix)U(BH_4)] ((a) in Figure 7.22).[122] By contrast, in uranium (^{tBu}ArO)$_3$Mes complexes, more modest changes in U–arene coordination are observed in response to changes in oxidation state and coordination number; the average U–C_{arene} distance in the uranium(III) complex [{(^{tBu}ArO)$_3$Mes}U] ((b) in Figure 7.22) is 2.73 Å,[286] compared with 2.92 Å in the uranium(IV) complex [{(^{tBu}ArO)$_3$Mes}U^{IV}F], and 2.99–3.01 Å in THF-coordinated analogues, [{(^{tBu}ArO)$_3$Mes}U^{IV}X(THF)] (X = F, Cl, Br or I).[287]

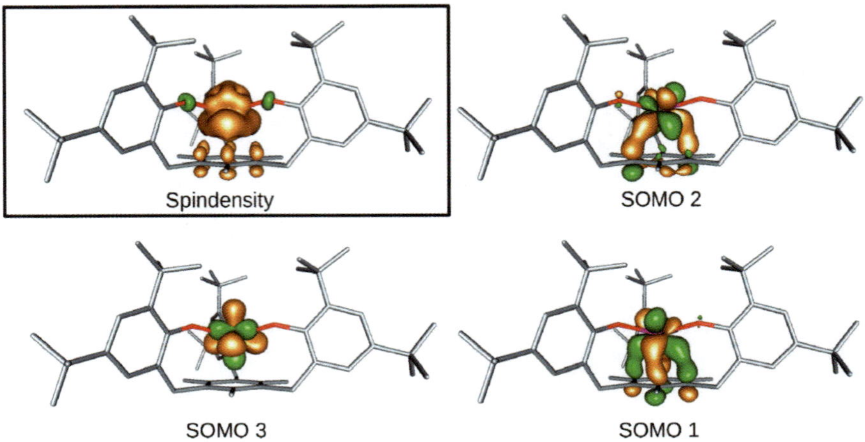

Figure 7.23. Calculated SOMOs and Spin Density for $[\{(^{tBu}ArO)_3Mes\}U]$, not including spin orbit coupling.[286] Reproduced from reference 286 with permission from the American Chemical Society.

The uranium(III) $(^{tBu}ArO)_3Mes$ complex, $[\{(^{tBu}ArO)_3Mes\}U]$, was also investigated computationally (Figure 7.23), revealing that it is indeed best regarded as a uranium(III) complex rather than a uranium(IV) complex of an arene radical anion; the highest energy singly occupied molecular orbital (SOMO 3) is primarily a non-bonding f-orbital while the SOMO 1 and SOMO 2 are involved in δ-backbonding, and are composed of uranium 5f orbitals (70%) and empty π^* orbitals on the mesitylene ring (30%). Uranium has a Mulliken spin density of 2.72, compared with 3.11 for $[\{(ArO)_3tacn\}U]$ $\{(ArO)_3tacn$ = an analogue of $(^{tBu}ArO)_3Mes$ in which the arene has been replaced by 1,4,7-triazacyclononane ring$\}$, with 15% of additional spin density on the mesitylene carbon atoms.[286] By contrast, reduced terminally-bound arenes have been reported by Gambarotta *et al.* in the chemistry of thorium.[269,288] For example, in $[Li(dme)_3][(dme)K(Et_8\text{-calix}[4]tetrapyrrole)Th(\eta^4\text{-}$naphthalene)] ((c) in Figure 7.22), the bond lengths and angles associated with the naphthalene ligand are consistent with an η^4-coordinated but-2-ene-1,4-diyl ligand (i.e. a doubly reduced butadiene fragment).[269]

Furthermore, related $[\{(^{Ad,Me}ArO)_3Mes\}U]$ displays a reversible reduction wave in the cyclic voltammogram at −2.50 V vs. $[FeCp_2]^{0/+}$,[289,290] and low-temperature chemical reduction of deep

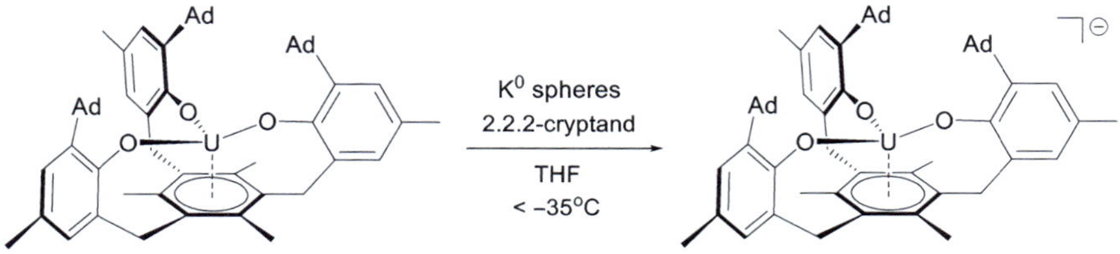

Scheme 7.16. Reduction of [{(^{Ad,Me}ArO)$_3$Mes}U] to [K(2,2,2-crypt)][{(^{Ad,Me}ArO)$_3$Mes}U].

purple [{(^{Ad,Me}ArO)$_3$Mes}U] using potassium spheres in the presence of 2,2,2-cryptand generated red-brown, thermally unstable [K(2,2,2-crypt)][{(^{Ad,Me}ArO)$_3$Mes}U] (Scheme 7.16). This compound was investigated computationally, crystallographically, and by EPR and SQUID magnetometry. These data, including significant changes in U–O distances, support metal- rather than arene-centered reduction, and the geometry optimized structure with four unpaired electrons much more closely matches the solid-state structure (especially the geometry of the arene ligand) than the structure calculated with two unpaired electrons. Calculations without spin-orbit coupling point to an essentially uranium-centered reduction, whereas those with spin-orbit coupling suggest some degree of charge transfer to the arene. Overall, the [{(^{Ad,Me}ArO)$_3$Mes}U]$^-$ anion is best regarded as a unique example of a 5f^4 uranium(II) complex stabilized by δ-bonding to an arene incorporated into a chelating ligand framework (Figure 7.24).[290]

7.7.4.2 *Inverse-sandwich arene complexes*

In 2000, Cummins *et al.* reported that the reduction of [UI(NR′Ar)$_3$] (R′ = ^{t}Bu or Ad (1-adamantyl); Ar = $C_6H_3Me_2$-3,5) with KC_8 (3 equiv.) in benzene or toluene afforded the arene-bridged inverse sandwich complexes, [(ArR′N)$_2$U(μ-η^6:η^6-C_6H_5R)U(NR′Ar)$_2$] (R = H or Me). The crystallographically characterized toluene complex (R′ = Ad) features particularly short U–C$_{arene}$ distances (ave. 2.59 Å) and behaves as a U(II) synthetic equivalent in its reactions with N_2Ph_2 and S_2Ph_2.[291] However, XANES spectroscopy and X-ray crystallography suggest that these inverse sandwich complexes are best considered uranium(III) compounds of a bridging arene dianion, and this assignment is supported

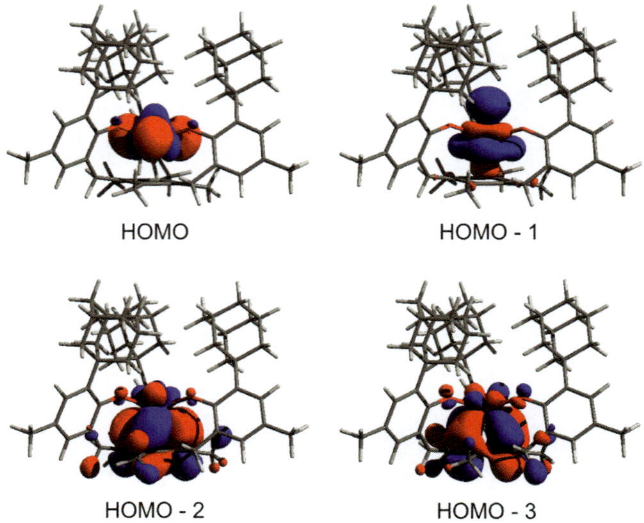

Figure 7.24. The four highest singly occupied orbitals of the $[\{(^{Ad,Me}ArO)_3Mes\}U]^-$ anion ($m_s = 2$), not including spin orbit coupling effects.[290] Reproduced from reference 290 with permission from Elsevier.

by DFT and CASSCF/CASPT2 calculations which reveal that the highest energy occupied orbitals are four SOMOs, below which lie two filled orbitals involving δ-bonding between uranium 5f-orbitals and arene orbitals with two nodal planes (Figure 7.25). Solid-state magnetic measurements on $[\{(Ar^tBuN)_2U\}_2(\mu\text{-}\eta^6\text{:}\eta^6\text{-toluene})]$ show a transition to antiferromagnetic behaviour between 95 and 125 K, and this complex also shows significantly more intense f-f transitions in the NIR region than $[(ArAdN)_3U(THF)]$ and $[(ArRN)_3UX]$ (R = ^{t}Bu or Ad), suggesting a high degree of covalency in the bonding between the bridging arene and the uranium centers.[292]

Since 2000, inverse sandwich complexes featuring uranium(III) centers bridged by an arene dianion have been prepared in combination with a variety of supporting ligands. This yielded neutral $[L_xU(\mu\text{-}\eta^6\text{:}\eta^6\text{-arene})UL_x]$ complexes with the following L_xU fragments: $\{(Me_3Si)_2N\}_2U$, $(3,5\text{-}^tBu_2H_3C_6O)_2U$,[293] $(\kappa^2\text{-}NN^{Fc})U$ $\{NN^{Fc} = Fe(\eta^5\text{-}C_5H_4NSi^tBuMe_2)_2\}$,[294] $Cp^*{}_2U$,[109] Cp^*XU [X = $CH(SiMe_3)_2$, $N(SiMe_3)_2$, $OC_6H_2(^tBu\text{-}o)_2(R\text{-}p)$ (R = H, Me or ^{t}Bu), or $MeC(N^iPr)_2$],[110] $\{\kappa^3\text{-}(Me_3SiN{=}PPh_2)_2CH\}IU$,[295] and $\{\kappa^2\text{-}CH(CMeNAr)_2\}IU$ (Ar = 2,6-diisopropylphenyl),[296] and bridging

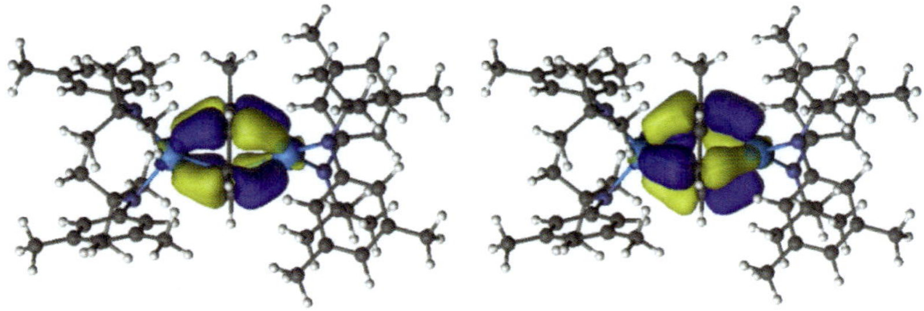

Figure 7.25. The δ-bonding natural orbitals in $[(Ar^tBuN)_2U(\mu\text{-}\eta^6{:}\eta^6\text{-toluene})U(N^tBuAr)_2]$.[292] Reproduced from reference 292 with permission from the American Chemical Society.

arenes including benzene, toluene, *p*-xylene and biphenyl. Furthermore, the dianionic derivatives, $M_2[\{(Mes^tBuC{=}N)_3U\}_2(\mu\text{-}\eta^6{:}\eta^6\text{-arene})]$ {M = Na or K; arene = naphthalene (Figure 7.26), biphenyl, *trans*-stilbene, and *p*-terphenyl} were isolated via reduction of $[(Mes^tBuC{=}N)_3UI(dme)]$ with 4 equiv. of KC_8 or Na metal in dme in the presence of 0.5 equiv. of the desired arene;[297,298] this reactivity contrasts that observed with $[UI(NR'Ar)_3]$ (*vide infra*) where 1 equiv. of KNR′Ar was eliminated to afford neutral derivatives.[291]

Some key features of these complexes are: (a) Benzene displaces toluene and *p*-xylene in $[\{Cp^*{}_2U\}_2(\mu\text{-}\eta^6{:}\eta^6\text{-arene})]$, consistent with preferential coordination to the more readily reduced arene.[109] However, naphthalene and anthracene, which are much more readily reduced, were unable to out-compete benzene in $[(2,6\text{-}{}^tBu_2H_3C_6O)_2U\}_2(\mu\text{-}\eta^6{:}\eta^6\text{-benzene})]$.[293] (b) Naphthalene, biphenyl and *trans*-stilbene in $M_2[\{(Mes^tBuC{=}N)_3U\}_2(\mu\text{-}\eta^6{:}\eta^6\text{-arene})]$ (M = K or Na) bridge via opposite faces of the same ring, rather than opposite faces of different rings.[297,298] (c) The U–C_{arene} distances in crystallographically characterized benzene, toluene and *p*-xylene complexes fall within the fairly narrow range of 2.50 to 2.75 Å, and the arene C–C bonds in these inverse sandwich compounds are typically elongated by ∼0.05 Å relative to the free arenes.[299] However, of the aforementioned inverse sandwich complexes, $[(\{\kappa^3\text{-}(Me_3SiN{=}PPh_2)_2CH\}IU)_2(\mu\text{-}\eta^6{:}\eta^6\text{-toluene})]$ (Figure 7.26) appears to be unique in that it exhibits single molecule magnetism, it does not display antiferromagnetic coupling, in contrast to $[\{(Ar^tBuN)_2U\}_2(\mu\text{-}\eta^6{:}\eta^6\text{-toluene})]$, and it gives rise to remarkably intense transitions in the NIR region ($\varepsilon \approx 4000\,cm^{-1}$).[295]

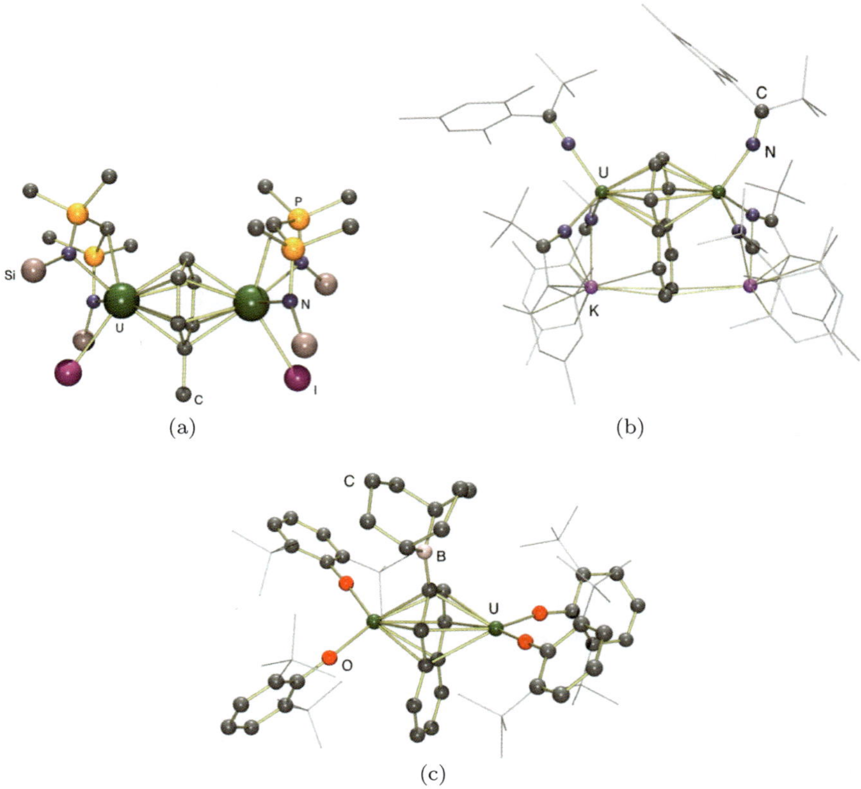

Figure 7.26. X-ray crystal structures of (a) $[(\{\kappa^3\text{-}(Me_3SiN{=}PPh_2)_2CH\}XU)_2(\mu\text{-}\eta^6{:}\eta^6\text{-toluene})]$ (only the *ipso*-C of P-phenyl groups shown, $SiMe_3$ methyl groups removed for clarity; X = 60% Cl, 40% I; only I is shown for clarity),[295] (b) $K_2[\{(Mes^tBuC{=}N)_3U\}_2(\mu\text{-}\eta^6{:}\eta^6\text{-naphthalene})]$,[297] and (c) $[(2,6\text{-}^tBu_2H_3C_6O)_2U\}_2\{\mu\text{-}\eta^6{:}\eta^6\text{-(2-naphthyl-}BC_8H_{14})\}]$.[293]

The majority of the aforementioned complexes were prepared by reduction of a higher valent precursor using a strong reducing agent such as KC_8. However, in 2012, Arnold *et al.* reported that storing solutions of $[UX_3]$ $\{X = N(SiMe_3)_2$ or $OC_6H_3{}^tBu_2\text{-}2,6\}$ at 90°C in neat arene solvents resulted in disproportionation to provide $[X_2U(\mu\text{-}\eta^6{:}\eta^6\text{-arene})UX_2]$ (arene = benzene, toluene, biphenyl and $PhSiH_3$) and $[UX_4]$. Furthermore, heating $[U(OC_6H_3{}^tBu_2\text{-}2,6)_3]$ in benzene with BBN, or addition of BBN to $[(2,6\text{-}^tBu_2H_3C_6O)_2U\}_2(\mu\text{-}\eta^6{:}\eta^6\text{-benzene})]$, afforded $[(2,6\text{-}^tBu_2H_3C_6O)_2U\}_2(\mu\text{-}\eta^6{:}\eta^6\text{-PhBBN})]$ and H_2 via electrophilic aromatic substitution on the reduced arene ligand.

Toluene and biphenyl could also be para-functionalized by BBN at the sandwiched arene ring, and the borylated naphthalene product (Figure 7.26) was isolated when $[U(OC_6H_3{}^tBu_2\text{-}2,6)_3]$ and BBN were reacted in molten naphthalene for 17 hours.[293]

In 2011, Liddle *et al.* reported the reaction of $[\{\kappa^3\text{-}HC(SiMe_2NAr)_3\}UCl(THF)]$ (Ar = $C_6H_3Me_2$-3,5) with KC_8 in toluene to form $[(\{\kappa^3\text{-}HC(SiMe_2NAr)_3\}U)_2(\mu\text{-}\eta^6{:}\eta^6\text{-toluene})]$, which differs from the aforementioned complexes in that each uranium center is coordinated by three anionic donors, but the complex is neutral overall. This complex could be considered to contain uranium(III), (IV) or (V) paired with a neutral, dianionic or tetraanionic bridging arene, respectively, and surprisingly, the U–N bond distances, solid-state magnetic measurements, and DFT calculations indicate that the latter formulation with an aromatic 10 π-electron arene tetraanion is most appropriate.[300] In 2012/2013, Mazzanti *et al.* reported another example of a uranium(V) arene^{4-} inverse sandwich; $[(\{(^tBuO)_3SiO\}_3U)_2(\mu\text{-}\eta^6{:}\eta^6\text{-toluene})]$ (U–C_{arene} (ave.) = 2.69 Å), prepared simply via toluene reduction by $[(U\{OSi(O^tBu)_3\}_2\{\mu\text{-}OSi(O^tBu)_3\})_2]$. Furthermore, this neutral complex could be reduced in consecutive 1-electron steps by KC_8 to afford mono- and di-anionic $K_x[(\{(^tBuO)_3SiO\}_3U)_2(\mu\text{-}\eta^6{:}\eta^6\text{-toluene})]$ (x = 1 and 2, respectively, with U–C_{arene} (ave.) = 2.65 and 2.61 Å), which were also accessed via unprecedented K^+-induced disproportionation reactivity, utilizing KOTf in place of KC_8 (Scheme 7.17).[301,302]

Remarkably, DFT calculations and solid-state magnetic measurements indicate that all three $K_x[(\{(^tBuO)_3SiO\}_3U)_2(\mu\text{-}\eta^6{:}\eta^6\text{-toluene})]$ (x = 0, 1 and 2) complexes are arene^{4-} complexes with U^V/U^V, U^V/U^{IV} and U^{IV}/U^{IV} oxidation states, respectively.[302] The tendency of these siloxide complexes to feature uranium in the highest possible oxidation state (by transfer of four electrons to toluene) is presumably related to the strong σ- and π-donor capabilities of hard siloxide anions, and the bonding situation in $K_2[(\{(^tBuO)_3SiO\}_3U)_2(\mu\text{-}\eta^6{:}\eta^6\text{-toluene})]$ (U^{IV} with an arene tetraanion) contrasts that in the previously discussed $M_2[\{(Mes^tBuC{=}N)_3U\}_2(\mu\text{-}\eta^6{:}\eta^6\text{-arene})]$ {M = Na or K; arene = naphthalene (Figure 7.26), biphenyl, *trans*-stilbene, and *p*-terphenyl} complexes which feature uranium(III) centers bridged by an arene dianion (*vide supra*).[298]

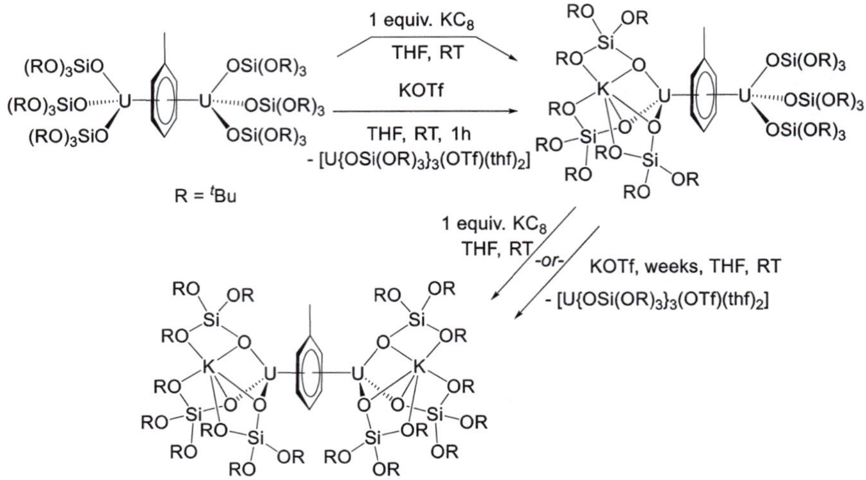

Scheme 7.17. Synthesis of $K_x[(\{(^tBuO)_3SiO\}_3U)_2(\mu\text{-}\eta^6{:}\eta^6\text{-toluene})]$ (x = 0, 1 and 2).[302]

7.7.5 *Cycloheptatrienyl complexes*

The only known actinide cycloheptatrienyl complexes were reported by Ephritikhine *et al.* in 1994 and 1995. These complexes are monoanionic $[U(\eta^7\text{-}C_7H_7)_2]^-$ and $[X_3U(\mu\text{-}\eta^7{:}\eta^7\text{-}C_7H_7)UX_3]^-$ (X = BH_4 and NEt_2), and in all cases, the cycloheptatrienyl ligand is planar and η^7-coordinated to uranium (Figure 7.27). The monometallic complex was prepared via the reaction of UCl_4 with an excess of potassium and cycloheptatriene, while the inverse sandwich complexes were formed from UX_4 with 1.5 equiv. of $K[C_7H_9]$, generating 1 equiv. of C_7H_{10} and KX as by-products.[303–305] The C_7H_7 ligand in these complexes could be regarded as an aromatic 6π-electron cation (a tropylium cation), an antiaromatic monoanion, or an aromatic 10π-electron trianion, and based on EPR and ENDOR studies on $[U(\eta^7\text{-}C_7H_7)_2]^-$,[306] the latter description appears most appropriate, as is typically the case in early transition metal chemistry.[307]

7.7.6 *Cyclooctatetraenide complexes*

Pyrophoric green bis(cyclooctatetraenide)uranium(IV), $[U(\eta^8\text{-}COT)_2]$, was prepared in 1968 from UCl_4 with $K_2[COT]$ by Streitwieser and Müller-Westerhoff,[308] and the D_{8h} solid-state

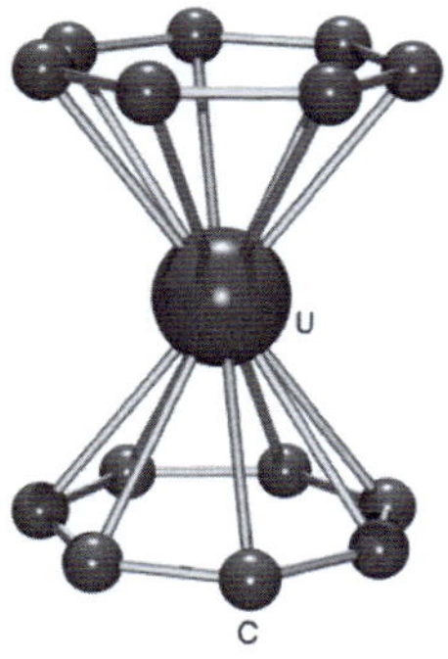

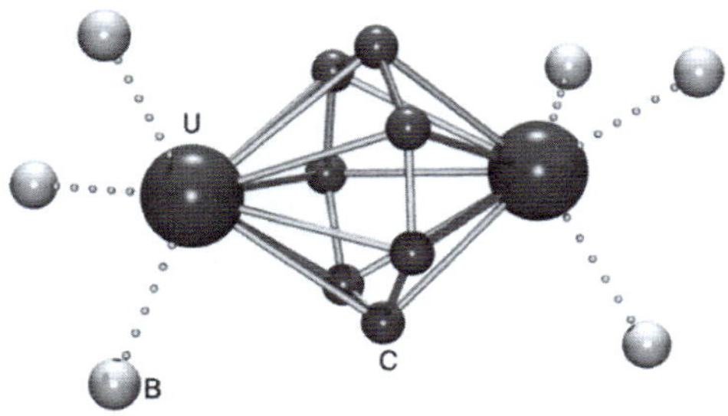

Figure 7.27. X-ray crystal structures of the anionic component in (a) $[K(18\text{-}C\text{-}6)][U(\eta^7\text{-}C_7H_7)_2]$ (18-C-6 = 18-crown-6)[303] and (b) $[U(BH_4)_2(THF)_5][(BH_4)_3U(\mu\text{-}\eta^7{:}\eta^7\text{-}C_7H_7)U(BH_4)_3]$.[304]

structure was published in 1969 by Raymond and Zalkin.[309] This complex is thermally robust, subliming at 180°C (0.03 mm Hg), and hydrolyzes only very slowly in water at neutral pH. It is named uranocene to highlight its similarity to ferrocene, as a sandwich complex involving planar aromatic π-ligands (10π in COT^{2-} vs. 6π in Cp^-), and the bonding in uranocene has been the subject of numerous experimental and theoretical investigations. A qualitative MO diagram for an early actinocene (e.g. uranocene) is shown in Figure 7.28, highlighting the unique ability of the f-orbitals to engage in δ-bonding with e_{2u} symmetric $(COT)_2$ orbitals as well as weak ϕ-backbonding. However, it should be noted that the effects of spin orbit coupling on the bonding picture for uranocene are calculated to be considerable.[310]

Isostructural yellow $[Th(COT)_2]$,[311] yellowish $[Pa(COT)_2]$,[4] and red $[An(COT)_2]$ (An = Np and Pu)[312] were also subsequently prepared, from $AnCl_4$ (An = Th, Pa or Np) or $[NEt_4][PuCl_6]$ with $K_2[COT]$, or by reaction of finely divided pyrophoric thorium or plutonium metal powder (prepared by actinide hydride thermolysis) with cyclooctatetraene. Both plutonocene and hypothetical curocene, $[Cm(COT)_2]$ are calculated to exhibit strong multiconfigurational character, and the multiconfigurational character of the ground state was found to increase across the actinocene series (U < Pu < Cm). Spin-orbit coupling is substantial in the open-shell actinocenes, and has a large effect on their energies and electronic

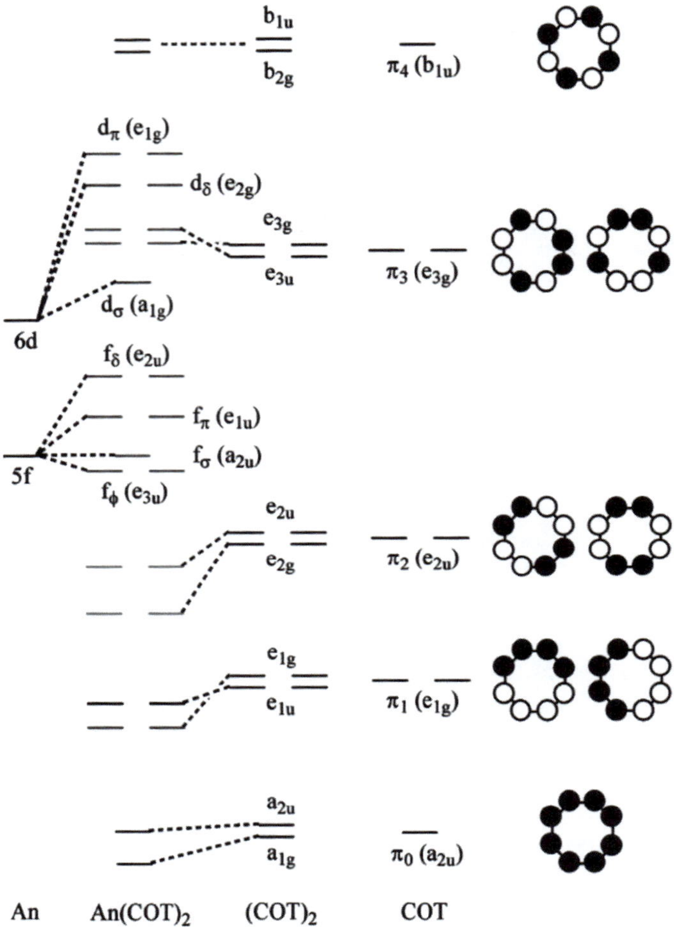

Figure 7.28. Qualitative molecular orbital energy level diagram for an early actinocene.[310]

structures, whereas their geometries are largely unaffected. Additionally, the spin-orbit-free ground states make the largest contribution to the spin-orbit-coupled ground states.[310]

Ansa-actinidocenes have been prepared with an $–SiMe_2(CH_2)_n$ $SiMe_2–$($n = 1$[313] or 2)[314] bridge between the two cyclooctatetraenide rings, and in the structurally characterized $n = 2$ complexes, the An–C bond lengths are analogous to those in unsubstituted $[An(COT)_2]$ complexes and the Cent–An–Cent (Cent = ring centroid) angles of 178°(U) and 177°(Th) are only slightly distorted. Substituted cyclooctatetraenide ligands have also be coordinated to thorium and uranium, with most recent studies focusing on 1,4-$C_8H_6(SiMe_3)_2$ (^{TMS2}COT),[315] 1,3,6-$C_8H_5(SiMe_3)_3$ (^{TMS3}COT),[316] 1,4-$C_8H_6(Si^iPr_3)_2$ (^{tips2}COT),[317] and 1,4-$C_8H_6(Si^tBuMe_2)_2$ (^{TBS2}COT)[9] dianions. The extremely bulky 1,4-$C_8H_6(SiPh_3)_2$ {^{BIG}COT} ligand was also complexed to uranium leading to a unique bent uranocene, [{^{BIG}COT}$_2$U], with a Cent–U–Cent angle of 169°.[318] The vast majority of disubstituted COT ligands are 1,4-substituted due to straightforward synthesis. Nevertheless, [U($^{1,5\text{-}tBu\,2}$COT)$_2$] was prepared from 1,5-di-*tert*-butylcyclooctatetraene ($^{1,5\text{-}tBu\,2}$COT), which was synthesized in 10 steps with an 11 % overall yield.[319] Actinide complexes of mono- and tetrasubstituted cyclooctatetraenide ligands (e.g. ^{tBu}COT and $^{1,3,5,7\text{-}Me4}$COT) have also been reported, as have actinide complexes of fused-ring derivatives such as 1,2-$C_8H_6(CH_2)_3$.[320]

In 2008, unsubstituted uranocene was found to react with [NEt_4]CN to form a mononuclear [NEt_4][(COT)$_2$An(CN)] anion with a Cent–U–Cent angle of 153°.[321] Additionally, both [U(COT)$_2$] and [Th(COT)$_2$] reacted with NaCN/18-crown-6 to form [(COT)$_2$An–CN–Na(18-C-6)].[322] However, thorocene reacted with 1 equiv. of [NEt_4]CN to produce polymeric {[NEt_4][(COT)$_2$Th(μ-CN)]}$_\infty$, due to the larger size of thorium vs. uranium. Furthermore, [Th(COT)$_2$] reacted with 0.5 or 2 equiv. of [NBu_4]CN to form [{(COT)$_2$Th}$_2$(μ-CN)]$^-$ and [(COT)$_2$Th(CN)$_2$]$^{2-}$, respectively, and crystals of trimetallic [{(COT)$_2$Th(μ-CN)}$_2$Th(COT)$_2$] were obtained from the 1:1 reaction of thorocene with [NBu_4]CN.[323] The larger size of thorium also provided access to other Lewis base adducts of thorocene, including 1:1 complexes with pyridine, ^{t}BuNC, bipy and phenanthroline,[324] and 2:1 complexes with H^- and N_3^-.[322]

Actinocenes have also been reduced to yield [An(COT)$_2$]$^-$ anions containing trivalent metal centers. Red-brown [K(solv)$_x$] [U(C_8H_7R)$_2$] (solv = THF or diglyme; R = H or Me),[325] deep-orange [K(dme)$_2$][U(^{TBS2}COT)$_2$], and dark-green [K(dme)$_2$][Th(^{TBS2}COT)$_2$] (Figure 7.29)[9] have been isolated, and the latter has a $6d^1$ ground state, giving rise to a room temperature-observable EPR spectrum and intense peaks in the visible region. Additionally, burgundy [K(THF)$_2$][Np(COT)$_2$] and turquoise-green [K(THF)$_2$][Pu(COT)$_2$] have been prepared directly from actinide(III) halide starting materials.[326]

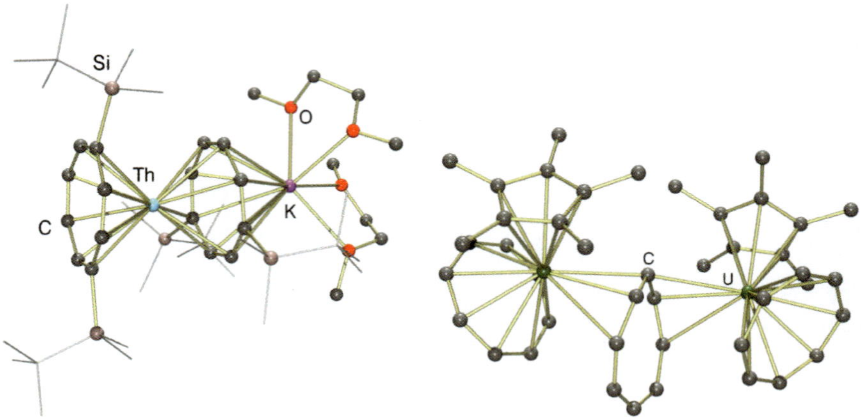

(a) (b)

Figure 7.29. X-ray crystal structures of (a) $[K(dme)_2][Th(^{TBS2}COT)_2]$ and (b) $[\{(COT)(Cp^*)U\}_2(\mu\text{-}\eta^3\text{:}\eta^3\text{-}COT)]$.

Beyond bis-cyclooctatetraenide complexes, a host of mono-cyclooctatetraenide actinide complexes have been reported. These complexes include actinide(III), (IV) and (V) compounds, such as $[(COT)U(hmpa)_3][BPh_4]_n$ (n = 1 and 2),[327] $[(COT)AnCl_2(THF)_2]$ (An = Th or U)[328] and $[(COT)U(NEt_2)_3]^{x-}$ (x = 1 and 0).[329] However, derivatives with organometallic co-ligands are largely confined to the IV oxidation state, and include $[(COT)U(NEt_2)\{CH(SiMe_3)_2\}]$,[330] $[(COT)U(CH_2R)_2(hmpa)_n]$ (R = $SiMe_3$ or Ph), and $[Li(THF)_3][(COT)U(CH_2SiMe_3)_3]$,[331] as well as a range of BH_4 derivatives.[332,333] Mixed ^{x}COT/Cp* thorium[334] and uranium[335] derivatives have also been prepared, as have COT/TMP[258] and ^{tips2}COT/Tp′ complexes {Tp′ = tris(3,5-dimethylpyrazolyl)borate}.[336]

The Evans group has played the major role in the development of COT/Cp* chemistry, including the synthesis of [(COT)(Cp*)UR] (R = Me and Ph) derivatives, the tuck-in complex $[(COT)(C_5Me_4CH_2)U(THF)_x]$ (x = 0 and 1), which undergoes insertion reactions with unsaturated substrates such as ^{t}BuNC and $C(N^iPr)_2$,[337] and bimetallic $[\{(COT)(Cp^*)U\}_2(\mu\text{-}\eta^3\text{:}\eta^3\text{-}COT)]$ (Figure 7.29), which readily eliminates COT and reacts as a source of "(COT)(Cp*)U" in the presence of oxidizing substrates such as phenazine and PhEEPh (E = S,

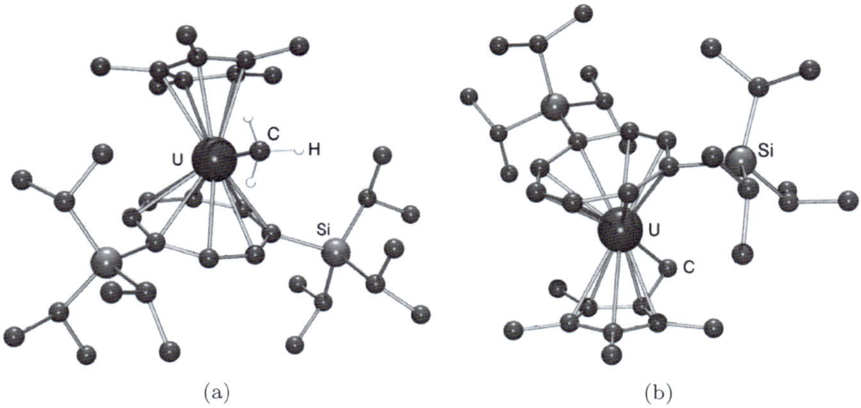

Figure 7.30. X-ray crystal structures of (a) [(^{tips2}COT)(Cp*)UMe] and (b) [(^{tips2}COT)($C_5Me_4CH_2$)U].

Se or Te).[338] More sterically hindered and more crystalline [(^{tips2}COT)(Cp*)UR] {R = H, Me (Figure 7.30), CH_2SiMe_3, CH_2Ph and $CH(SiMe_3)_2$} derivatives, and the tuck-in complexes [(^{tips2}COT)($C_5Me_4CH_2$)U(THF)$_x$] {x = 0 (Figure 7.30) and 1} have also been prepared by Cloke *et al.*,[339] as have the thorium complexes [(^{tips2}COT)(Cp*)Th(CH_2Ph)], [{(^{tips2}COT)(Cp*)ThH}$_n$] (n = 1 or 2), and [{(^{tips2}COT)($C_5Me_4CH_2$)Th}$_2$].[340]

Additionally, Cloke has used the $^{(SiR3)2}$COT/Cp^x ligand set to great advantage in the development of low-valent uranium chemistry and small molecule activation. Reaction of [(^{tips2}COT)(Cp*) U(THF)] with excess CO yielded exclusively the deltate ($C_3O_3^{2-}$) complex, [{(^{tips2}COT)(Cp*)U}$_2$(μ-κ^1:κ^2-C_3O_3)],[317] as a result of reductive CO trimerization. By contrast, the marginally less sterically hindered Cp^{Me4} analogue reacted with excess CO to form only the squarate ($C_4O_4{}^{2-}$) complex, [{(^{tips2}COT)(Cp^{Me4})U}$_2$(μ-κ^2:κ^2-C_4O_4)] (Scheme 7.18; Figure 7.31),[341] and more hindered [(^{TMS2}COT){$C_5Me_4(SiMe_3)$}U] provided only the ynediolate complex, [{(^{TMS2}COT)($C_5Me_4(SiMe_3)$)U}$_2$(μ-κ^1:κ^1-C_2O_2)].[342]

The reaction of [(^{tips2}COT)(Cp*)U(THF)] with excess CO is proposed to occur via (a) initial CO coordination, (b) dimerization by coordination of CO oxygen atoms to a second uranium center, (c) formation of long-lived intermediate containing a bent O–C–C–O linkage which maintains U–C bonding, followed by reaction with another

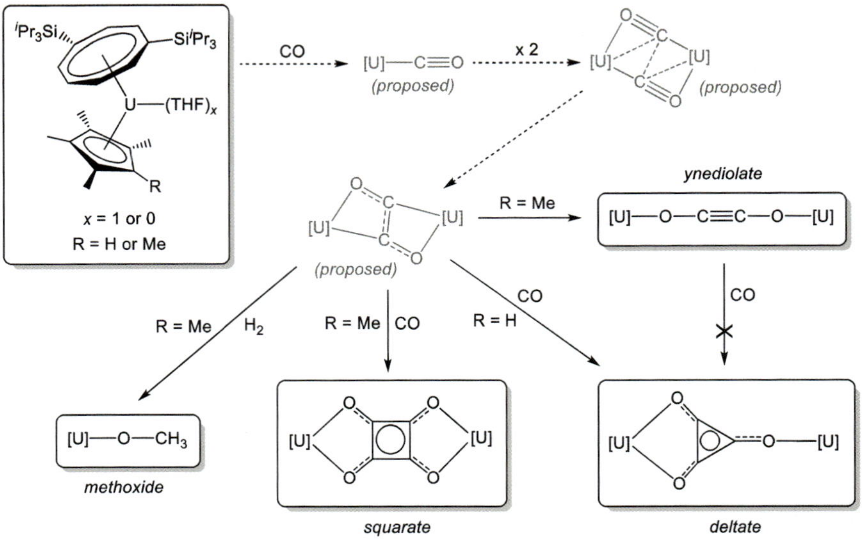

Scheme 7.18. Reactions of $[(^{tips2}COT)(Cp^{x})U(THF)_n]$ ($Cp^{x} = Cp^{*}$ or Cp^{Me4}) with CO. [U] = "$(^{tips2}COT)(Cp^{x})U$". Proposed intermediates in the initial reaction with 1 equiv. of CO per uranium are shown with approximate bond orders.

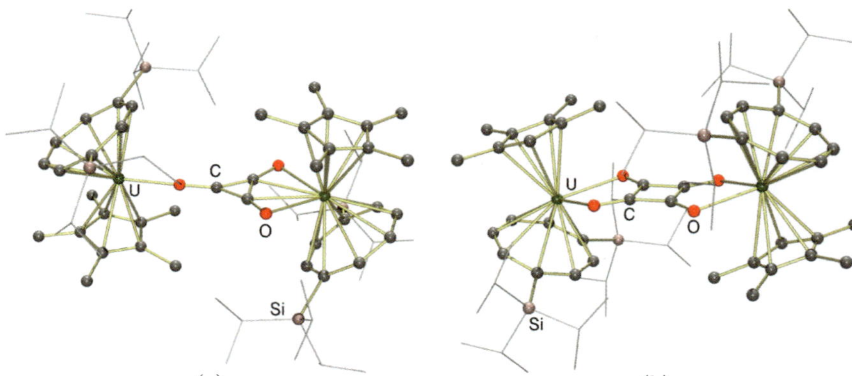

Figure 7.31. X-ray crystal structures of (a) $[\{(^{tips2}COT)(Cp^{*})U\}_2(\mu\text{-}C_3O_3)]$ and (b) $[\{(^{tips2}COT)(Cp^{Me4})U\}_2(\mu\text{-}C_4O_4)]$.

equivalent of CO (Scheme 7.18). However, in the absence of excess CO (i.e. with a 1:1 ratio of CO to uranium), the aforementioned long-lived intermediate was observed to isomerize to form an ynediolate complex, $[\{(^{tips2}COT)(Cp^{*})U\}_2(\mu\text{-}\kappa^1\text{:}\kappa^1\text{-}C_2O_2)]$ which does not react further with CO, even at 60 °C.[343] Furthermore, the ynediolate complex does not react with H_2 at 60 °C, but addition of 1 equiv. of CO to a −78°C solution of [(^{tips2}COT)(Cp*)U] (prepared by heating the THF adduct under high vacuum), followed by addition of 2 equiv. of H_2, afforded [(^{tips2}COT)(Cp*)U(OMe)] upon warming to room temperature; a rare example of CO hydrogenation at ambient or sub-ambient temperatures and pressures (Scheme 7.18).[344]

[(^{tips2}COT)(Cp^x)U(THF)] (Cp^x = Cp* or Cp^{Me4}) also reacted with excess CO_2 to form the carbonate complex [{(^{tips2}COT)(Cp^x)U}$_2$(μ-κ^1:κ^2-CO_3)] and free CO, due to reductive disproportionation of CO_2 (2 CO_2 + 2 e^- → 1 CO_3^{2-} + 1 CO).[345] Furthermore, addition of 1 equiv. of NO to a solution of [(^{tips2}COT)(Cp^x)U] at −78°C, followed by addition of 0.5 equiv. of CO afforded the bimetallic isocyanate complex, [{(^{tips2}COT)(Cp*)U(μ-$\kappa^1 N$:$\kappa^1 O$-NCO)}$_2$], and the oxo complex, [{(^{tips2}COT)(Cp*)U}$_2$(μ-O)], as the two major products; the result of CO/NO reductive co-coupling.[346]

7.7.7 *Pentalene complexes*

Pentalene is effectively composed of two fused cyclopentadienyl rings. However, a pentalene dianion ($C_8R_6{}^{2-}$) features a 10-electron π-system, whereas two cyclopentadienyl rings can provide a total of 12 π-electrons. Furthermore, pentalene is typically η^8-coordinated to f-elements (vs. μ-η^5:η^5-coordination which is common for late transition metals),[347] and the filled orbitals of an η^8-coordinated pentalene dianion can engage in δ-donation, unlike those of a cyclopentadienyl anion. Therefore, pentalene shares more similarities with a cyclooctatraenyl ligand than a pair of cyclopentadienyl ligands. However, the actinide chemistry of pentalene dianions is significantly less developed than that of cyclooctatetraenide dianions, most likely due to less straightforward ligand synthesis. For example, unsubstituted [Li(dme)$_2$]$_2$[Pn] (Pn = C_8H_6) is typically prepared by flash vacuum pyrolysis of cyclooctatetraene vapors at around 600 °C, condensation of the resulting dihydropentalenes at −78 °C, addition of cold hexanes and dme, and dilithiation with ^{n}BuLi at −78 °C. The resulting [Li(dme)$_2$]$_2$[Pn] can serve as a starting material for the synthesis of substituted derivatives, but only 1,4-disubstituted pentalenes are readily prepared. For example, $K_2[Pn^{TIPS2}]$ (Pn^{TIPS2} = 1,4-$C_8H_4(Si^iPr_3)_2$] is accessed by reaction of $[Li(dme)_2]_2[Pn]$ with $^iPr_3SiOTf$ followed by double deprotonation using KNH_2.[348] That said, the permethylpentalene dianion, $[Li(tmeda)_2]_2[Pn^*]$ ($Pn^* = C_8Me_6$), has been prepared in nine steps from tetramethyl-3,7-dihydroxybicyclo[3.3.0]octa-2,6-diene-2,4,6,8-tetracarboxylate, and this route does not involve flash vacuum pyrolysis.[349]

In the chemistry of the actinide elements, both thorium and uranium sandwich complexes, $[An(Pn^{TIPS2})_2]$ (An = Th or U) have been prepared,[348,350] and more recently, $[U(Pn^*)_2]$ has been added to the series ((a) and (b) in Figure 7.32).[351] The fold angle between the two five-membered rings in each pentalene ligand is 24° in $[Th(Pn^{TIPS2})_2]$ and 26° in $[U(Pn^*)_2]$, and a combination of calculations and photoelectron spectroscopy revealed that (a) the top four filled molecular orbitals in $[Th(Pn)_2]$ involve δ-interactions between thorium and the ligands, with greater contributions from the metal 6d-orbitals than the 5f-orbitals, and (b) the donor ability of the carbocyclic ligands in $[UCp_4]$, $[U(COT)_2]$ and $[U(Pn^{TIPS2})_2]$ increases in the order Cp < COT < Pn^{TIPS2}, indicating that low-valent pentalene complexes may be particularly reactive due to very negative $U^{III/IV}$ redox potentials. Mixed ligand uranium complexes have also been prepared, including tetravalent $[(Pn^*)U\{(\mu\text{-}Cl)_2Li(tmeda)\}_2]$, $[(Pn^*)(Cp^x)U\{(\mu\text{-}Cl)_2Li(tmeda)\}]$ (Cp^x = Cp or Cp*), and $[(Pn^*)UCp_2]$,[352] and trivalent $[(Pn^{TIPS2})U(\kappa^3\text{-}Tp')]$[336] and $[(Pn^{TIPS2})UCp^*]$.[24] The latter complex demonstrates appreciable reactivity despite considerable steric encumbrance at the metal center ((c) in Figure 7.32), reacting reversibly with N_2 to form a μ-η^2:η^2-coordinated N_2 dianion,[24] and with $^tBuC{\equiv}P$ to form $[\{(Pn^{TIPS2})Cp^*U\}_2(\mu\text{-}\eta^1 P\text{:}\eta^2 CP\text{-}PC^tBu)]$ ((d) in Figure 7.32) which contains the first example of a doubly reduced phosphaalkyne.[353]

7.7.8 *Acyclic π-ligand complexes*

Actinide acyclic π-ligand complexes include alkene/metallacyclopropane, alkyne/metallacyclopropene, butadiene/metallacyclopentene, metallacyclopentadiene, butadiyne/metallacyclopentatriene, allyl, and pentadienyl complexes. However, allyl complexes are not included in this section since homoleptic allyl derivatives are the focus of Section 7.4.2, and various heteroleptic allyl complexes are included throughout the chapter.

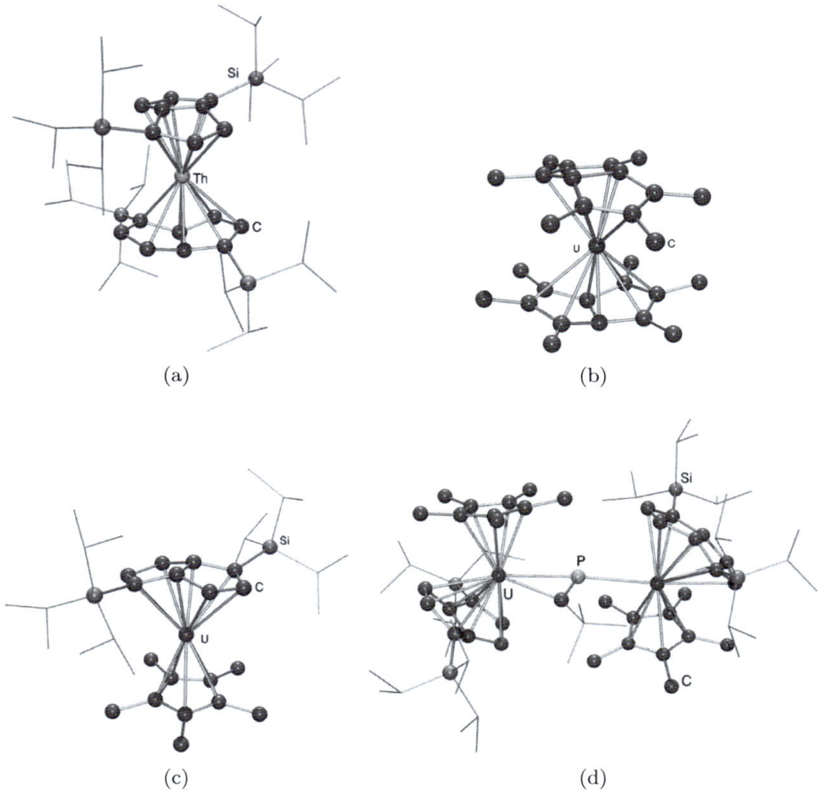

Figure 7.32. X-Ray crystal structures of (a) $[Th(Pn^{TIPS2})_2]$,[350] (b) $[U(Pn^*)_2]$,[351] (c) $[(Pn^{TIPS2})UCp^*]$,[24] and (d) $[\{(Pn^{TIPS2})Cp^*U\}_2(\mu\text{-}\eta^1 P{:}\eta^2 CP\text{-}PC^tBu)]$.[353]

7.7.8.1 *Alkene/metallacyclopropane and alkyne/metallacyclopropene complexes*

A range of actinide complexes have been reported to polymerize ethylene and/or α-olefins, or to be effective for alkene or alkyne hydroelementation, or alkyne oligomerisation. Alkene and alkyne complexes must play a key role in this reactivity, and although intermolecular alkyne complexes, and to a greater extent alkene complexes, are rarely observable/isolable for metal centers that are unable to provide substantial π-backdonation (e.g. d^0 transition metal, trivalent rare earth, tetravalent thorium and to a lesser extent tetravalent uranium complexes), $[(Et_2N)_2U(C{\equiv}C^tBu)(\eta^2\text{-}HC{\equiv}C^tBu)][BPh_4]$ has been identified by IR and NMR spectroscopy.[354]

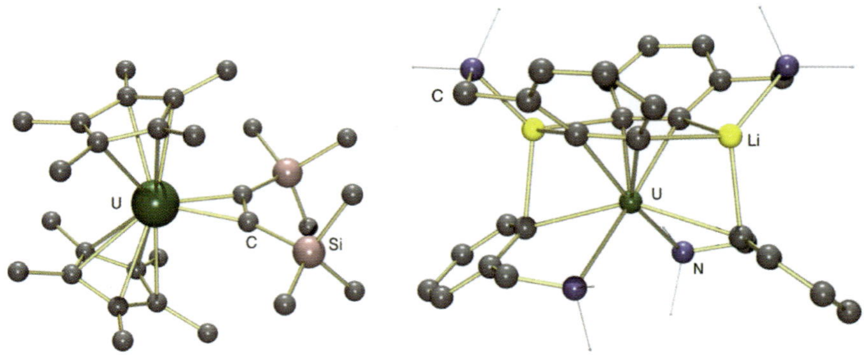

(a) (b)

Figure 7.33. X-ray crystal structures of (a) $[Cp^*{}_2U\{\eta^2\text{-}C_2(SiMe_3)_2\}]$[244] and (b) $Li_2[\{\kappa^2\text{-}C_6H_4(CH_2NMe_2)\text{-}o\}_2U\{\eta^2\text{-}C_6H_3(CH_2NMe_2)\}_2]$.[358]

At the other end of the spectrum, alkynes have been shown to engage in 1:1 reactions with low-valent actinide precursors to generate complexes with considerable metallacyclopropene character. The reversible reaction between $[Cp_3U(THF)]$ and C_2Ph_2 was reported as early as 1987, but the product, $[Cp_3U(\eta^2\text{-}C_2Ph_2)]$, is thermally unstable at room temperature and was not isolated.[355] However, in 2014, reduction of $[Cp^{t3}{}_2ThCl_2]$ with excess KC_8 in the presence of 1 equiv. of C_2Ph_2 was shown to provide $[Cp^{t3}{}_2Th(\eta^2\text{-}C_2Ph_2)]$,[356] and the synthesis of $[Cp^*{}_2U\{\eta^2\text{-}C_2(SiMe_3)_2\}]$ (Figure 7.33) and $[Li(dme)_2][Cp^*{}_2ThCl\{\eta^2\text{-}C_2(SiMe_3)Ph\}]$ soon followed.[244,357] Furthermore, reaction of UCl_4 with 6 equiv. of $LiC_6H_4(CH_2NMe_2)$-*o* afforded the dark blue uranium(IV) metallacyclopropane complex, $Li_2[\{\kappa^2\text{-}C_6H_4(CH_2NMe_2)\text{-}o\}_2U\{\eta^2\text{-}C_6H_3(CH_2NMe_2)\}_2]$ containing two benzyne ligands (Figure 7.33).[358]

The outcome of reactions involving reduction of $[Cp^x{}_2ThCl_2]$ precursors in the presence of an alkyne is highly sensitive to both steric and electronic effects. For example, while reduction of $[Cp^{t3}{}_2ThCl_2]$ in the presence of C_2PhR (R = Me, ^{i}Pr, or Cy) afforded $[Cp^{t3}{}_2Th(\eta^2\text{-}C_2PhR)]$, this product undergoes rapid C–H activation of a cyclopentadienyl *tert*-butyl substituent to generate a cyclometallated vinyl complex, which further converts to an allyl derivative when R is a methyl group (Scheme 7.19). Moreover, while $[Cp^{t3}{}_2Th(\eta^2\text{-}C_2PhR)]$ (R = Ph, ^{i}Pr or Cy) does not react further with additional equiv. of alkyne, reduction $[Cp^{t3}{}_2ThCl_2]$ in the presence of C_2Me_2 afforded

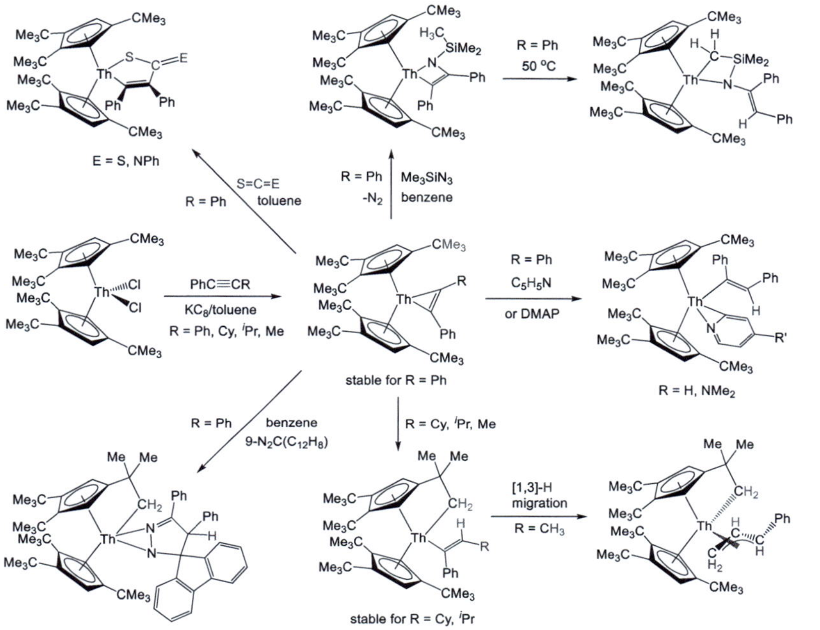

Scheme 7.19. Synthesis of $[Cp^{t3}{}_2Th(\eta^2\text{-}C_2PhR)]$, thermal rearrangement, and reactions with SCE (E = S or NPh), $R'C_5H_4N$ (R′ = H or NMe_2), Diazofluorene $\{9\text{-}N_2C(C_{12}H_8)\}$, and Me_3SiN_3.[356,359]

the metallacyclopentadiene complex, $[Cp^{t3}{}_2Th(\kappa^2\text{-}C_4Me_4)]$.[359] Similarly, while reduction of $[Cp^*{}_2ThCl_2]$ in the presence of $C_2Ph(SiMe_3)$ gave $[Cp^*{}_2ThCl\{\eta^2\text{-}C_2(SiMe_3)Ph\}]^-$, the analogous reactions with C_2Me_2 or C_2PhR (R = Ph, Me, ^{i}Pr, Cy, or $SiHMe_2$) gave only metallacyclopentadiene products (Figure 7.34), and the reaction with $C_2(SiMe_3)_2$ afforded $[Cp^*{}_2ThCl\{\eta^1\text{-}C(SiMe_3){=}CH(SiMe_3)\}]^-$ where the proton originates from toluene solvent.[357]

The thorium metallacyclopropene, $[Cp^{t3}{}_2Th(\eta^2\text{-}C_2Ph_2)]$, underwent insertion reactivity with PhCN, PhNCS, CS_2, (p-ClH_4C_6)CHO and CyNCNCy to form 5-membered metallacycles, reacted with pyridine and pyridine-N-oxide to afford *ortho*-cycometallation products, and afforded more complex products upon reaction with diazofluorene and trimethylsilyl azide (Scheme 7.19).[356,359] However, with the exception of the reaction with Ph_2CO (2 equiv.) to form $[Cp^{t3}{}_2Th(\kappa^2\text{-}OCPh_2CPh_2O)]$, it did not serve as a thorium(II) synthetic equivalent (i.e. as a source of "$Cp^{t3}{}_2Th$").

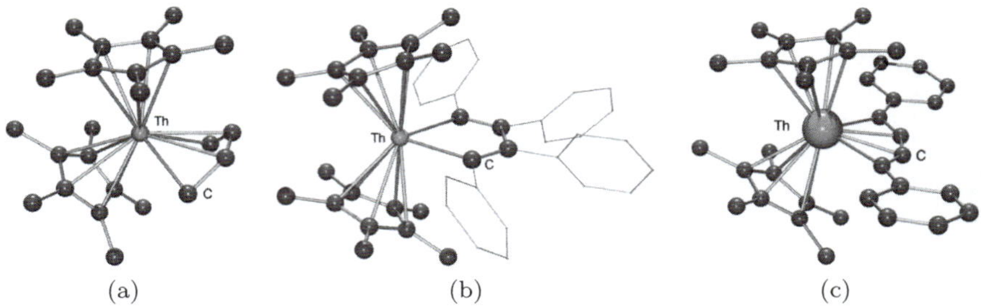

Figure 7.34. X-ray crystal structures of the metallacyclopentene, metallacyclopentadiene, and metallacyclopentatriene complexes (a) [$Cp^*_2Th(\eta^4\text{-}C_4H_6)$],[361] (b) [$Cp^*_2Th(\kappa^2\text{-}C_4Ph_4)$],[363] and (c) [$Cp^*_2Th(\eta^4\text{-}C_4Ph_2)$].[364]

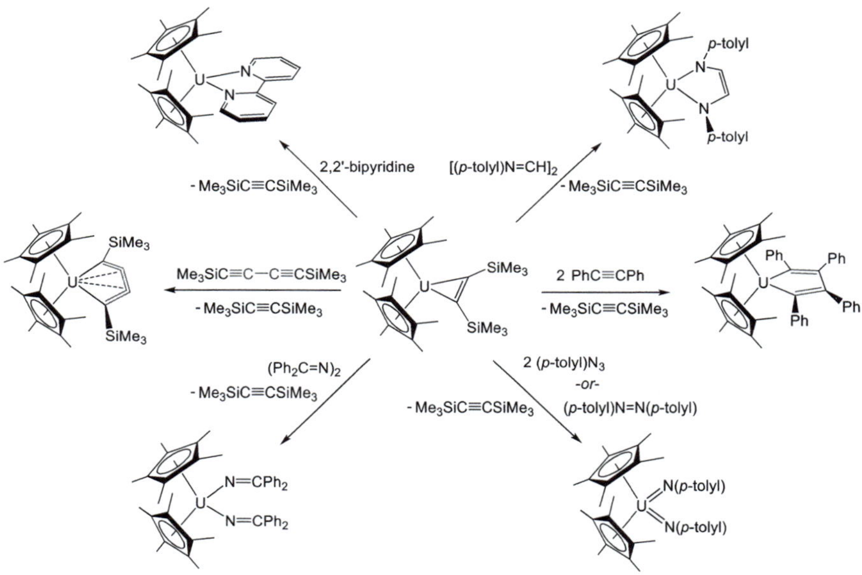

Scheme 7.20. Reactions of [$Cp^*_2U\{\eta^2\text{-}C_2(SiMe_3)_2\}$] with C_2Ph_2 (2 equiv.), $Me_3SiC{\equiv}C\text{–}C{\equiv}CSiMe_3$, 2,2′-bipyridine, ArN=CH–CH=NAr, ArN_3 (2 equiv.) or ArN_2Ar (Ar = *p*-tolyl), and $(Ph_2C{=}N)_2$.[244]

By contrast [$Cp^*_2U\{\eta^2\text{-}C_2(SiMe_3)_2\}$] participates in alkyne exchange reactions with C_2Ph_2 (2 equiv.) and $Me_3SiC{\equiv}C\text{–}C{\equiv}CSiMe_3$ (Scheme 7.20), generating the metallacyclopentadiene and metallacyclopentatriene products [$Cp^*_2U\{\kappa^2\text{-}C_4Ph_4\}$] and [$Cp^*_2U\{\eta^4\text{-}C_4(SiMe_3)_2\text{-}1,4\}$] (Figure 7.34), respectively. Furthermore, [$Cp^*_2U\{\eta^2\text{-}C_2(SiMe_3)_2\}$] reacted with 2,2′-bipyridine to form

$[Cp^*{}_2U(\kappa^2\text{-bipy})]$, with an α-diimine to form the enediamide product, $[Cp^*{}_2U(\kappa^2\text{-ArN–CH=CH–NAr})]$ (Ar = *p*-tolyl), with a 2,3-diazabutadiene $\{(Ph_2C{=}N)_2\}$ to form the diiminato product, $[Cp^*{}_2U(N{=}CPh_2)_2]$, and with either *p*-tolyl azide (2 equiv.) or 4,4′-dimethylazobenzene (1 equiv.) to form the bis-imido product, $[Cp^*{}_2U({=}NAr)_2]$ (Ar = *p*-tolyl) (Scheme 7.20). Based on DFT calculations, the differences in reactivity between the uranium and thorium metallacyclopropene complexes are ascribed to increased covalency and 5f-orbital participation in the U–C bonds of the former, as well as increased π-donation from the remaining π-bond of the metallacyclopropene to uranium, vs. thorium.[244]

7.7.8.2 *Butadiene/metallacyclopentene, M(C_4R_6), complexes*

In 1986, both Erker and Marks reported the synthesis of metallacyclopentene $[Cp^*{}_2An\{\eta^4\text{-}C_4H_6\}]$ (An = Th and U) complexes via the reactions of $[Cp^*{}_2AnCl_2]$ with $Mg(THF)_x(C_4H_6)$. Additionally, Marks *et al.* prepared the uranium complex from the reaction of $[Cp^*{}_2UClMe]$ with 3-butenyl magnesium bromide (Scheme 7.21).[360,361] The latter reaction presumably proceeds with initial formation of $[Cp^*{}_2U(CH_2CH_2CH{=}CH_2)Me]$, followed by β-hydride elimination and reductive elimination of methane. An X-ray structure of $[Cp^*{}_2Th(\eta^4\text{-}C_4H_6)]$ (Figure 7.34) reveals that the but-2-ene-1,4-diyl unit is η^4-coordinated to thorium with average Th–C distances of 2.57 and 2.74 Å to the external and internal carbon atoms, respectively, and C–C bond distances that are equal within error. $[Cp^*{}_2U(\eta^4\text{-}C_4H_6)]$ undergoes insertion reactivity with pyridine and $[Cr(CO)_6]$, whereas $[(Cp^{t2})_2Th(\eta^4\text{-}C_4H_6)]$ reacted with As_4 in refluxing xylene to provide $[(Cp^{t2})_2Th(\eta^4\text{-}As_6)Th(Cp^{t2})_2]$ (Scheme 7.21),[362] suggesting that $[(Cp^{t2})_2Th(\eta^4\text{-}C_4H_6)]$ is able to act as a low valent synthon {i.e. as a source of "$(Cp^{t2})_2Th$"}.

7.7.8.3 *Metallacyclopentadiene, M(C_4R_4), complexes*

The metallacyclopentadiene complexes described in Section 7.7.8.1, $[Cp^x{}_2Th(\kappa^2\text{-}C_4Me_4)]$ ($Cp^x = Cp^{t3}$ or Cp^*), $[Cp^*{}_2Th(\kappa^2\text{-}C_4Ph_2R_2)]$ {R = Ph, Me, ^{i}Pr, Cy, or $SiHMe_2$} and $[Cp^*{}_2U\{\kappa^2\text{-}C_4Ph_4\}]$, were prepared by reduction of $[Cp^x{}_2AnCl_2]$ in the presence of 2 equiv.

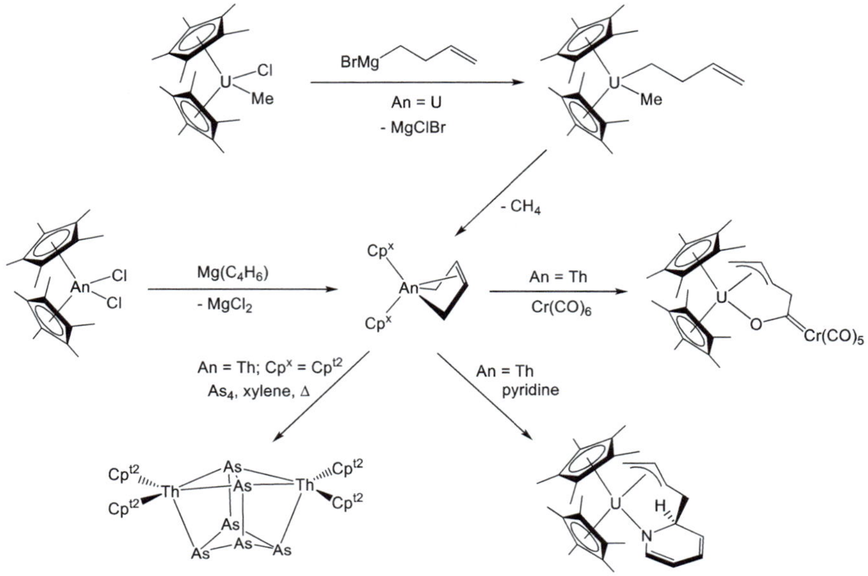

Scheme 7.21. Synthesis of [Cp$^x{}_2$An(η^4-C$_4$H$_6$)], and reactions with As$_4$, pyridine, and [Cr(CO)$_6$] (Cp^x = Cp* unless otherwise specified).

the appropriate alkyne. The butadiene-1,4-diyl unit in these complexes is κ^2-coordinated, as evidenced by the solid-state structure of [Cp*$_2$Th(κ^2-C$_4$Ph$_4$)], contrasting the η^4-coordination mode of the but-2-ene-1,4-diyl ligand in [Cp*$_2$Th(κ^2-C$_4$H$_6$)] ((b) and (a), respectively, in Figure 7.34).

Metallacyclopentadiene complexes were initially prepared in the 1970s by Marks *et al.* by salt metathesis from [Cp*$_2$AnCl$_2$] and Li$_2$[C$_4$Ph$_4$],[112] and via the disproportionation/oxidative coupling reaction of [{Cp*$_2$U(μ-Cl)}$_3$] (0.66 equiv.) with C$_2$Ph$_2$ (2 equiv.) to form a 1:1 ratio of [Cp*$_2$U{κ^2-C$_4$Ph$_4$}] and [Cp*$_2$UCl$_2$].[365] Furthermore, Evans has reported the synthesis of [Cp*$_2$U{κ^2-C$_4$Ph$_4$}] by reaction of C$_2$Ph$_2$ with the uranium(II) synthetic equivalent [Cp*$_2$U(μ-η^6:η^6-C$_6$H$_6$)UCp*$_2$], and via sterically induced reduction (SIR) reactivity of C$_2$Ph$_2$ with [UCp*$_3$] or [Cp*$_2$U{(μ-Ph)$_2$BPh$_2$}], generating (Cp*)$_2$ or biphenyl accompanied by BPh$_3$, respectively, as by-products.[366] Additionally, Kiplinger recently reported the reactions of [Cp*$_2$AnH$_x$] (An = Th, x = 2; An = U, x = 1 or 2; generated from [Cp*$_2$AnMe$_2$] and PhSiH$_3$) with C$_2$R$_2$ (R = Me or Ph) to form [Cp*$_2$An{κ^2-C$_4$R$_4$}].[114,367] In 2012, Meyer *et al.* also

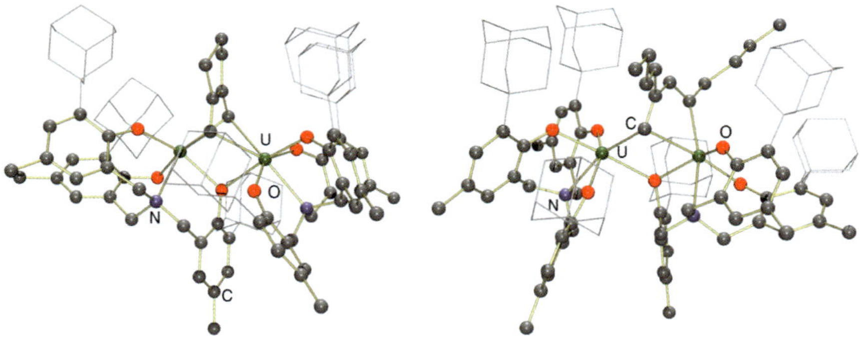

Figure 7.35. X-ray crystal structures of the bridging butadiene-1,4-diyl complexes (a) [{((^{Ad}ArO)$_3$N)U}$_2${μ-1,2-(CH)$_2$(C$_5$H$_8$)}] and (b) [{((^{Ad}ArO)$_3$N)U}$_2$ {μ-1,2-CHC^{n}BuCHC^{n}Bu}].[368]

reported bimetallic complexes (Figure 7.35) containing a bridging C_4R_4 (butadiene-1,4-diyl) unit that is either symmetrically coordinated with κ^2-coordination to both metals, or is κ^2-coordinated to one metal and κ^1-coordinated to the second metal. These complexes are the result of oxidative coupling of two alkynes by two uranium(III) centers.[368]

7.7.8.4 *Butadiyne/metallacyclopentatriene, M(C$_4$R$_2$), complexes*

The synthesis of the uranium metallacyclopentatriene complex, [Cp*$_2$U{η^4-C$_4$(SiMe$_3$)$_2$-1,4}], was described in Section 7.7.8.1, and the phenyl-substituted thorium analogue, [Cp*$_2$Th{η^4-C$_4$Ph$_2$-1,4}], was prepared by reduction of [Cp*$_2$ThCl$_2$] with KC$_8$ in the presence of the butadiyne PhC≡C–C≡CPh. The X-ray structure of this complex is shown in Figure 7.34, highlighting the obtuse (~150°) C–C–C angles at the central positions of the butatriene-1,4-diyl unit, and the essentially identical Th–C distances (1.53–1.54 Å) to all four carbon atoms in this unit. As a significant consequence of increased ring strain, metallacyclopentatriene [Cp*$_2$Th{η^4-C$_4$Ph$_2$-1,4}] proved to be far more reactive than the metallacyclopentadiene analogue, [Cp*$_2$Th{κ^2-C$_4$Ph$_4$}], which failed to react with PhNCS, (^{i}PrN)$_2$C, Ph$_2$CO and PhCN,[364] although it did form a mono-insertion product with (Me$_3$Si)HCN$_2$ and a di-insertion product with ArN$_3$ (Ar = p-tolyl).[363] In contrast, [Cp*$_2$Th{η^4-C$_4$Ph$_2$-1,4}]

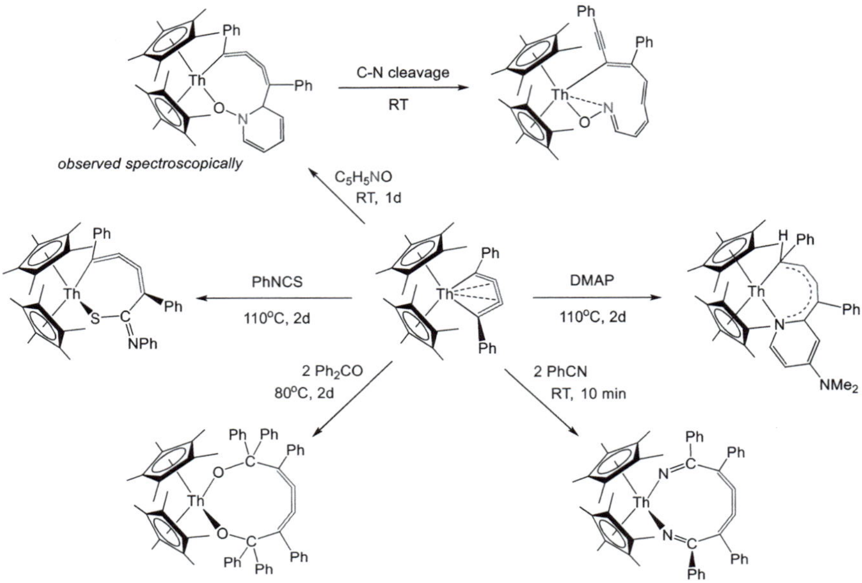

Scheme 7.22. Reactions of the metallacyclopentatriene complex $[Cp^*_2Th\{\eta^4\text{-}C_4Ph_2\text{-}1,4\}]$ with PhNCS, Ph_2CO, PhCN, dmap and pyridine-*N*-oxide.

formed mono-insertion products with PhNCS and $(^i PrN)_2C$ (followed by subsequent product rearrangement in the latter case), and di-insertion products with Ph_2CO and PhCN. Furthermore, $[Cp^*_2Th\{\eta^4\text{-}C_4Ph_2\text{-}1,4\}]$ reacted with dmap to form a 7-membered heterocycle, and with pyridine-*N*-oxide to form a 10-membered metallacycle resulting from pyridine ring opening (Scheme 7.22).[364]

7.7.8.5 *Pentadienyl complexes*

The homoleptic pentadienyl complex, $[U(dmpd)_3]$ (dmpd = 2,4-dimethylpentadienyl) was prepared via the reaction of $UCl_3(THF)_x$ with 3 equiv. of K[dmpd],[369] and subsequent reaction with $[HNEt_3][BPh_4]$ (1 equiv.) or $TlBH_4$ (2 equiv.) provided $[(\eta^5\text{-dmpd})_2U][BPh_4]$ and $[(\eta^5\text{-dmpd})_2U(BH_4)_2]$, respectively. The latter compound could also be prepared by reaction of $[U(BH_4)_4]$ with K(dmpd) to form $[(\eta^5\text{-dmpd})U(BH_4)_3]$ (Figure 7.36), followed by addition of $Mg[dmpd]_2$ (0.5 equiv.). Furthermore, $[(\eta^5\text{-dmch})U(BH_4)_3]$ (dmch = 6,6-dimethylcyclohexadienyl) and several uranium(III) and (IV) bis-dmch chloride and borohydride complexes

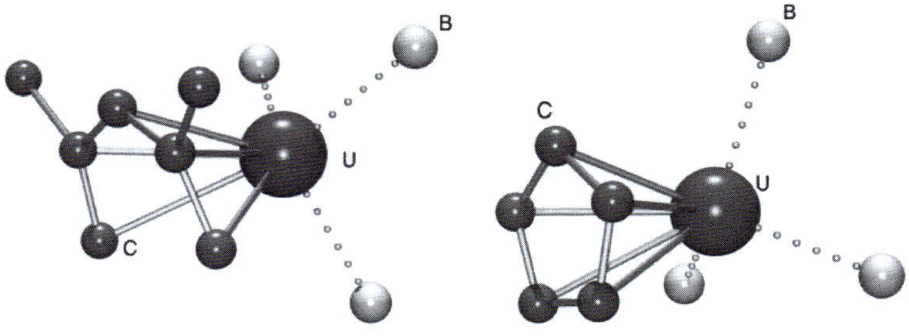

Figure 7.36. Comparison of the X-ray crystal structures of (a) [(η^5-dmpd) U(BH_4)$_3$] and (b) [CpU(BH_4)$_3$].[370]

have been reported.[279,370,371] The larger size of the 2,4-dimethylpentadienyl ligand vs. a cyclopentadienyl ligand is clearly seen in the X-ray structures in Figure 7.36, which also reveal that while the U–C bonds are longer in the dmpd complex vs. the Cp analogue (ave 2.75 vs. 2.67 Å), the distance from uranium to the plane of the ligand is in fact shorter in the pentadienyl complex (2.29 vs. 2.42 Å). In addition, it is notable that in [(η^5-dmpd)U(BH_4)$_3$], the U–C distance to the central carbon atom (2.63 Å) is significantly shorter than the other U–C distances (2.77–2.80 Å), and the outer C–C distances in the pentadienyl unit are significantly shorter than the inner C–C distances (ave 1.35 vs. 1.44 Å), suggesting that the $(CH_2{=}CH)_2CH^-$ resonance structure plays a dominant role in the bonding to uranium. Other pentadienyl complexes have arisen through formal H^- addition to a coordinated arene ligand within a multidentate ligand framework,[266] or deprotonation at a benzylic position in a multidentate ligand (e.g. Scheme 7.8 and Figure 7.34).[289]

7.8 Neutral and Anionic Non-Cyclopentadienyl Hydrocarbyl Complexes

In contrast to actinide alkyl complexes supported by carbocyclic ligands, non-carbocyclic actinide alkyl complexes are less well developed. Prior to 2006, this field was dominated by bulky monodentate amido,[372] alkoxide,[95,373] and aryloxide[374] ligands, as well

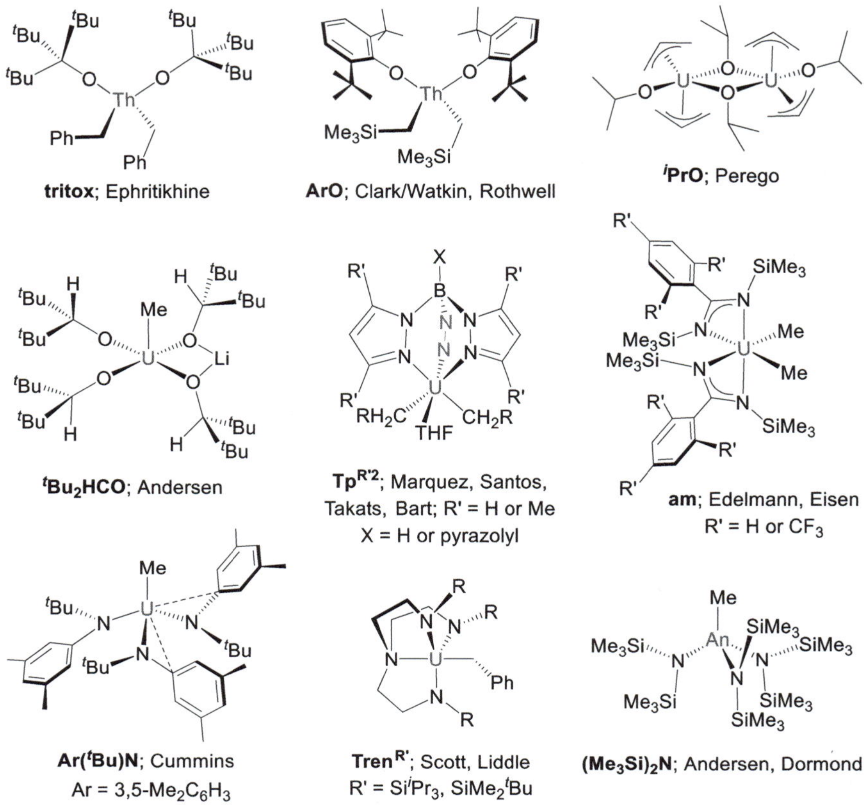

Figure 7.37. Complexes featuring non-cyclopentadienyl supporting ligands applied in actinide alkyl chemistry prior to 2006 (An = Th or U; R is typically H, $SiMe_3$, ${}^{t}Bu$ or Ph). Authors are those who have contributed to organoactinide chemistry, at any time, using each ligand framework.

as amidinate,[375] tris(pyrazolyl)borate (Tp^x)[376] and triamidoamine ($tren^x$; $N(CH_2CH_2NR)_3\}$[377] ligands pioneered by Edelmann, Marquez/Santos/Takats and Scott, respectively (Figure 7.37). Subsequently, the organoactinide chemistry of Tp^x and $tren^x$ ligands has been extended by Bart[99,378,379] and Liddle,[380–382] and new ligand designs have been implemented by the Leznoff,[97,103,383,384] Emslie,[80,100,280,282,385,386] Diaconescu,[88,387–389] Bart,[78,135] Liddle[390] and Maria/Mazzanti[391] groups (Figure 7.38).

Most non-carbocyclic organoactinide complexes were synthesized by salt metathesis using an RLi, RNa, $PhCH_2K$ or RMgBr

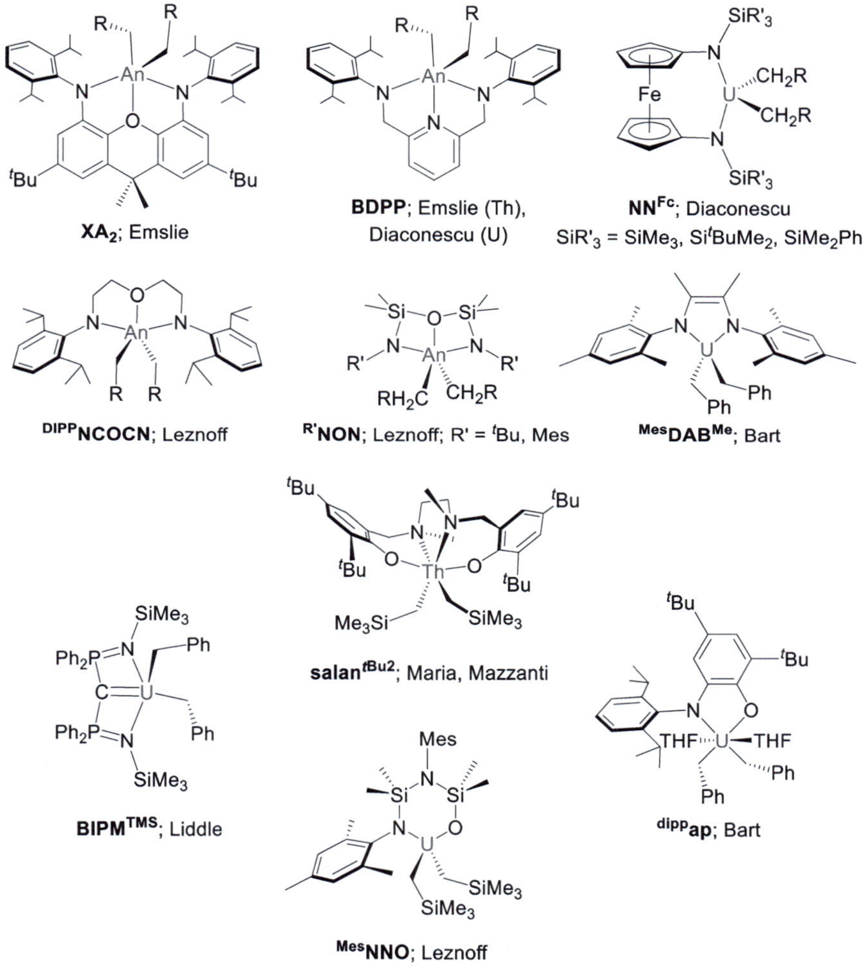

Figure 7.38. Complexes featuring non-cyclopentadienyl ancillary ligands deployed in actinide alkyl chemistry after 2006 (An = Th or U; R is typically H, $SiMe_3$, ^tBu or Ph). Authors are those who have contributed to organoactinide chemistry using each ligand framework.

reagent and an appropriate actinide halide precursor. However, $[\{U(allyl)_2(O^iPr)_2\}_2]$ was prepared by reaction of thermally unstable $[U(allyl)_4]$ with 2 equiv. of ^{i}PrOH, and related reactions with ^{t}BuOH and EtOH were also described.[95] Along similar lines, $[(XA_2)Th(CH_2SiMe_3)_2]$ and $[(BDPP)Th(CH_2SiMe_3)_2]$ could be prepared by initial reaction of $[ThCl_4(dme)_2]$ with 4 equiv.

of $LiCH_2SiMe_3$ to generate "$Th(CH_2SiMe_3)_4(dme)_x$," followed by addition of $H_2(XA_2)$ or $H_2(BDPP)$.[80] In addition, reaction of $[UI_3(THF)_3]$ with 3 equiv. of various MCH_2R (R = Ph, $SiMe_3$ or ^tBu; M = Li or K) reagents, followed by addition of $H_2(BDPP)$ or $H_2(NN^{Fc})$, afforded $[(BDPP)U(CH_2Ph)_2]$ and $[(NN^{Fc})U(CH_2R)_2]$ (R = Ph, $SiMe_3$ or ^tBu), presumably via alkane elimination and disproportionation.[88] Moreover, reactions of $H_2(XA_2)$ and $H_2(BDDP)$ with putative *in situ* generated "$ThCl_2(CH_2SiMe_3)_2(dme)_x$" and "$UI(CH_2Ph)_2(THF)_x$" afforded $[(XA_2)ThCl_2(dme)_2]$[80] and $[(BDPP)UI(CH_2Ph)]$,[88] respectively.

Ancillary ligand installation has also been achieved through redox reactivity. In particular, $[(^{Mes}DAB^{Me})U(CH_2Ph)_2]$ and $[(^{dipp}ap)U(CH_2Ph)_2(THF)_2]$ were prepared by reaction of $[U(CH_2Ph)_4]$ with a neutral redox-active α-diimine ($^{Mes}DAB^{Me}$)[78] or iminoquinone (^{dipp}ap)[135] ligand. In the former case, this reaction occurs via a concerted reductive elimination mechanism, since reaction with a 1:1 mixture of $[U(CH_2C_6H_5)_4]$ and $[U(CD_2C_6D_5)_4]$ yielded only $C_{14}H_{14}$ and $C_{14}D_{14}$. In the latter case, reaction of the iminoquinone with a 1:1 mixture of $[U(CH_2C_6H_5)_4]$ and $[U(CD_2C_6D_5)_4]$ generated 50% of $C_{14}H_7D_7$, supporting a radical mechanism involving homolytic cleavage. This reaction was hypothesised to take place by initial coordination of the iminoquinone ligand with concurrent benzyl radical extrusion to yield a uranium(IV) iminosemiquinone intermediate, $[LU(CH_2Ph)_3]$, followed by ejection of a second benzyl radical to form the 2-amidophenoxide product, $[LU(CH_2Ph)_2(THF)_2]$ (L = ^{dipp}ap; Scheme 7.23). The proposed $[LU(CH_2Ph)_3]$ iminosemiquinone intermediate is considered to be viable based on the accessibility of $[LUI_3(THF)_2]$; an iminosemiquinone complex of uranium(IV) which reacts with 3 equiv. of KCH_2Ph to form the same $[LU(CH_2Ph)_2(THF)_2]$ product (Scheme 7.23).

Alkyl groups have also been installed through unusual alkyl exchange reactivity: Reaction of $[(XA_2)U(CH_2SiMe_3)_2]$ with 2.1 equiv. of $LiCH_2{}^tBu$ in C_6D_6 resulted in complete conversion to $[(XA_2)U(CH_2{}^tBu)_2]$ (Scheme 7.24), and treatment of the bis-neopentyl complex with up to 80 equiv. of $LiCH_2SiMe_3$ did not re-form detectable amounts of the starting bis(trimethylsilyl)methyl complex, implying that the equilibrium in this reaction lies far to the side of di-neopentyl complex. Similar reactivity was observed

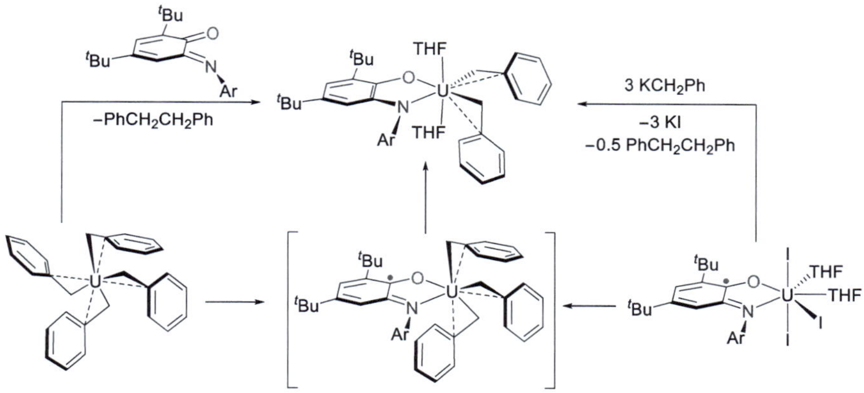

Scheme 7.23. Benzyl radical extrusion reactions to generate [(^{dipp}ap)U(CH_2Ph)$_2$(THF)$_2$].

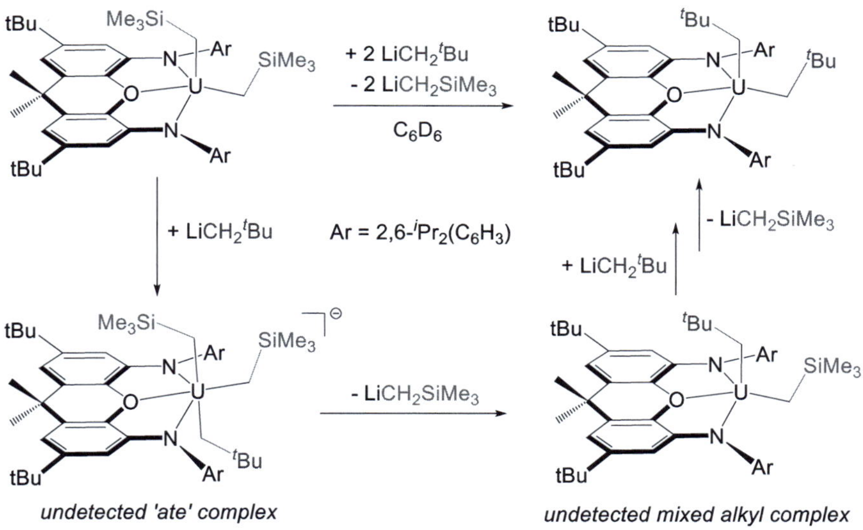

Scheme 7.24. Alkyl exchange reactivity converting [(XA_2)U(CH_2SiMe_3)$_2$] to [(XA_2)U($CH_2{}^tBu$)$_2$].

for thorium, although in this case addition of 2.2 equiv. of $LiCH_2{}^tBu$ to [(XA_2)Th(CH_2SiMe_3)$_2$] yielded a 1:1 mixture of [(XA_2)Th(CH_2SiMe_3)($CH_2{}^tBu$)] and [(XA_2)Th($CH_2{}^tBu$)$_2$], indicative of a significantly smaller equilibrium constant for the thorium reaction relative to the uranium reaction. These reactions

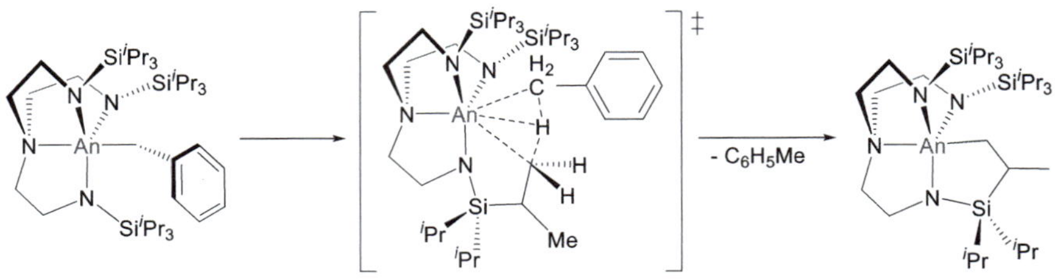

Scheme 7.25. Cyclometallation of the thorium and uranium [(tren^{TIPS})An(CH_2Ph)] complexes.[381]

presumably proceed via undetected trialkyl ate intermediates, such as [(XA_2)An(CH_2SiMe_3)$_2$($CH_2{}^t$Bu)]$^-$, and although no reaction was observed between [(XA_2)U(CH_2SiMe_3)$_2$] and $LiCH_2SiMe_3$ in benzene, this reaction proceeded rapidly in dme to form [Li(dme)$_3$][(XA_2)U(CH_2SiMe_3)$_3$].[385]

Significant differences in the chemistry of thorium and uranium analogues were also reported for the reactions of [(tren^{TIPS})AnI] (TIPS = Si^{i}Pr$_3$) with KCH_2Ph. In the case of thorium, this reaction yielded [(tren^{TIPS})Th(CH_2Ph)], which underwent cyclometallation upon heating to 80°C to afford [(tren^{TIPS-H})Th] {tren^{TIPS-H} = N(CH_2CH_2NSi^{i}Pr$_3$)$_2$$CH_2CH_2$NSi^{i}Pr$_2$CHMe$CH_2$} (Scheme 7.25). By contrast, the reaction of [(tren^{TIPS})UI] with KCH_2Ph proceeded directly to [(tren^{TIPS-H})U] and a benzyl intermediate was not observed, even when the reaction was monitored at –80°C. This reactivity difference was shown computationally to derive from stabilization of the σ-bond metathesis transition state by 5f-orbital participation in the interatom interactions.[381]

Rapid cyclometallation was also observed in the reaction of less sterically encumbered [(tren^{TMS})UI(THF)] (TMS = $SiMe_3$) with KCH_2Ph, but in this case, a dimetallic tuck-in tuck-over complex, [U_2(tren^{TMS-2H})(tren^{TMS})] was formed, containing one doubly-cyclometallated ligand (tren^{TMS-2H}) and one intact tren^{TMS} ligand. Furthermore, subsequent reaction with [$HNEt_3$][BPh_4] in THF did not yield [(tren^{TMS})U(THF)$_x$][BPh_4], but instead resulted in double dearylation of the BPh_4 anion to afford a product containing an NR-$SiMe_2$-CH-BPh_2 linkage (Scheme 7.26).[382]

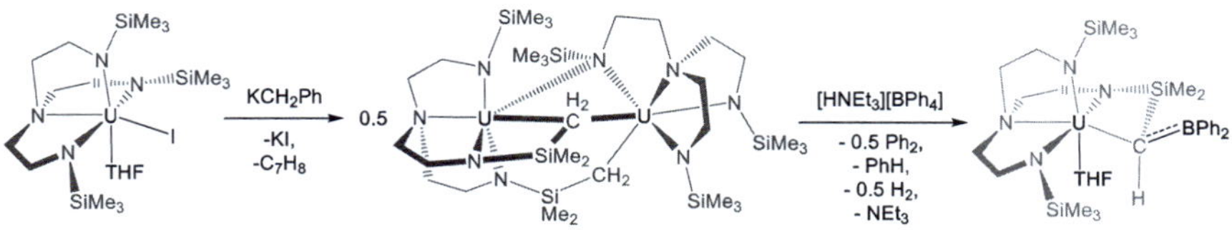

Scheme 7.26. Reaction of $[(\text{tren}^{\text{TMS}})\text{UI(THF)}]$ (TMS = $SiMe_3$) with KCH_2Ph to form dimetallic $[U_2(\text{tren}^{\text{TMS}-2\text{H}})(\text{tren}^{\text{TMS}})]$ containing one doubly-cyclometallated $\text{tren}^{\text{TMS}-2\text{H}}$ ligand and one intact tren^{TMS} ligand, and subsequent reaction with $[HNEt_3][BPh_4]$.

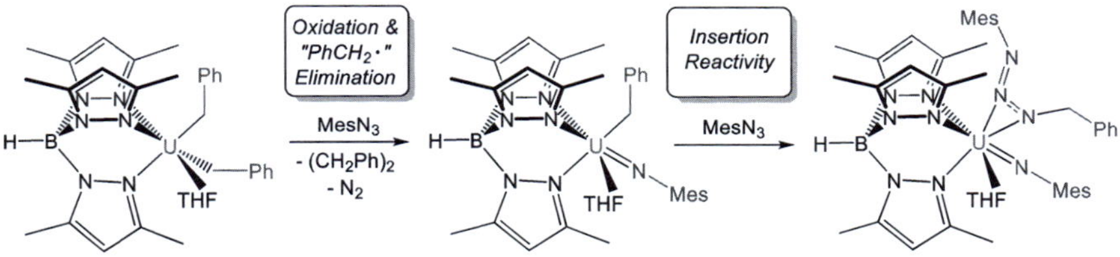

Scheme 7.27. Stepwise reaction of $[\text{Tp}'\text{U}((CH_2Ph)_2(\text{THF})]$ with 2 equiv. of $MesN_3$.

Many other reactions of non-carbocyclic ligand alkyl complexes involve σ-bond metathesis (e.g. with H_2, terminal alkynes, pyridines, acetone, amines, alcohols and thiols) or 1,2-insertion (e.g. with CO_2, ketones or azides). However, in the chemistry of uranium, especially uranium(III), redox reactions with azides and related oxidants must also be considered. For example, reaction of $[\text{Tp}'\text{U}(CH_2Ph)_2(\text{THF})]$ with 1 equiv. of $MesN_3$ afforded $[\text{Tp}'\text{U}^{\text{IV}}(=\text{NMes})(CH_2Ph)(\text{THF})]$ and 0.5 $PhCH_2CH_2Ph$, while reaction with a second equiv. of $MesN_3$ generated the insertion product, $[\text{Tp}'\text{U}^{\text{IV}}(=\text{NMes})(\text{MesN}_3CH_2Ph)(\text{THF})]$ (Scheme 7.27).[379]

A further area of non-carbocyclic ligand organoactinide chemistry which has been explored fairly extensively is the reactivity of $[(\text{NN}^{\text{Fc}})\text{U}(CH_2Ph)_2]$ with heterocycles including pyridine, 2-picoline *N*-methylimidazole, *N*-methylbenzimidazole, benzoxazole, benzothiazole and quinoline. These reactions gave rise to a range of products in good yields, in several cases via multistep mechanisms involving alkyl transfer, C–H bond activation, C–C coupling and/or ring opening (Scheme 7.28).[387,392]

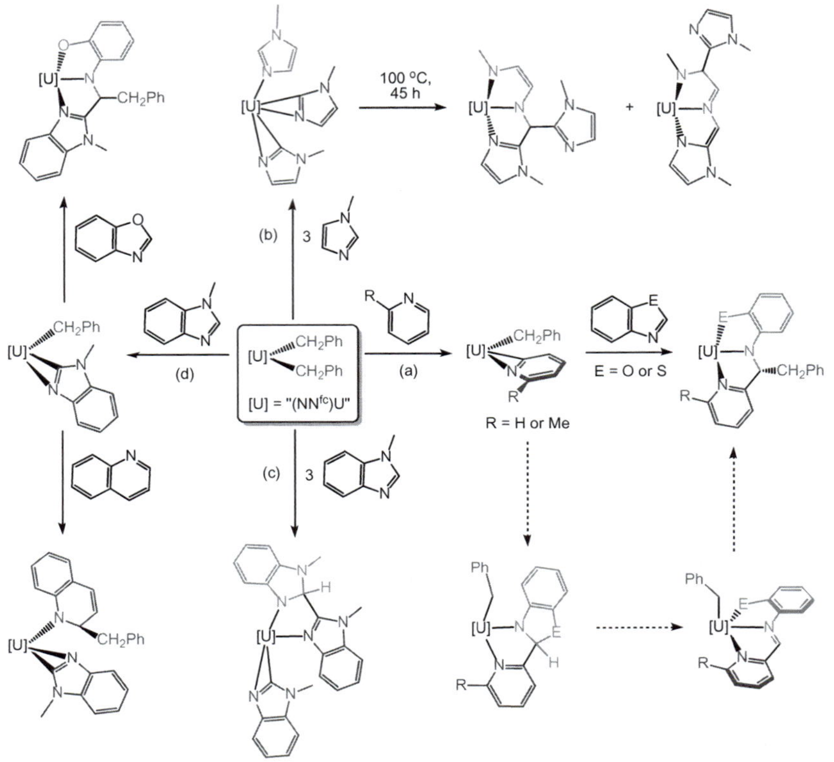

Scheme 7.28. Reactions of $[(NN^{Fc})U(CH_2Ph)_2]$ with: (a) pyridine or 2-picoline followed by benzoxazole or benzothiazole, (b) *N*-methylimidazole (3 equiv.) followed by heating, (c) *N*-methylbenzimidazole (3 equiv.), and (d) *N*-methylbenzimidazole (1 equiv.) followed by benzoxazole or quinoline.

7.9 Neutral and Anionic Hydride Complexes

Exposure of the dialkyl complexes $[Cp^*{}_2AnR_2]$ (An = Th, U; R = CH_3, CH_2SiMe_3) to excess hydrogen gas[115,186] or phenylsilane[114,367] afforded dimeric dihydride complexes, $[\{Cp^*{}_2AnH(\mu\text{-}H)\}_2]$, with two bridging and two terminal hydride ligands (Figure 7.39). Single crystal neutron diffraction studies support the dimeric formulation in the solid state (Scheme 7.29), while variable temperature ^{1}H NMR experiments show that the terminal and bridging hydrogen atoms in the thorium complex undergo rapid exchange down to –85 °C in solution. The thorium complex decomposes slowly at 80 °C, whereas the

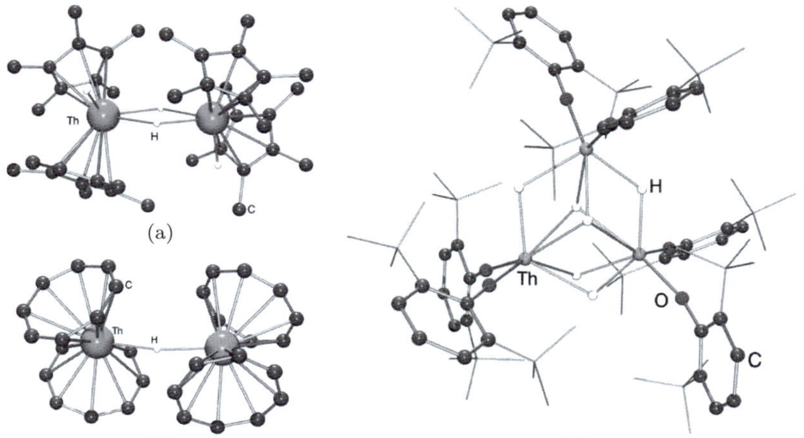

Figure 7.39. Crystal structures of (a) [{Cp*$_2$ThH}$_2$(μ-H)$_2$] (Neutron)**,[395] (b) [{(COT)$_2$Th}$_2$H]$^-$ (X-ray),[322] and (c) [{(ArO)$_2$ThH$_2$}$_3$] (X-ray).[394]
Note: **Atomic coordinates by personal communication from T. J. Marks.

uranium complex quite readily loses H_2 to generate a uranium(III) monohydride (*vide infra*).

Dialkyl complexes of the type [{(C$_5$Me$_4$)$_2$(μ-SiMe$_2$)}ThR$_2$] (R = CH$_2$SiMe$_3$, CH$_2$^{t}Bu, C$_6$H$_5$, ^{n}Bu, CH$_2$Ph) also undergo rapid hydrogenolysis when exposed to an atmosphere of hydrogen gas to yield the light-sensitive dimeric dihydride complex [{(C$_5$Me$_4$)$_2$(μ-SiMe$_2$)}ThH$_2$]$_2$, which was investigated by solution NMR spectroscopy and X-ray diffraction experiments. In the absence of a neutron diffraction structure, determination of the exact nature and position of the hydrogen atoms present in the molecule (bridging or terminal) was not possible.[192] However, the short Th–Th distance of 3.632(2) Å (*cf.* 4.007(8) Å in [{Cp*$_2$ThH(μ-H)}$_2$])[186] and single Th–H stretch in the IR spectrum suggest that the four hydrogen atoms are equivalent, and bridging between the two thorium atoms. Dimetallic [{(^{tips2}COT)Cp*Th(μ-H)}$_2$] was also prepared similarly, by hydrogenolysis of [(^{tips2}COT)Cp*Th(CH$_2$Ph)].[339,340]

In contrast, reaction of [(Cp^{Si})$_2$AnMe$_2$] (Cp^{Si} = C$_5$Me$_4$SiMe$_3$) with H_2 resulted in methane elimination and formation of the double tuck-in complex, [C$_5$Me$_4$-SiMe$_2$-CH$_2$)$_2$U] (Scheme 7.29),[117] presumably through rapid cyclometallation of undetected [(Cp^{Si})$_2$AnH(Me)] or [(Cp^{Si})$_2$AnH$_2$]. Furthermore, the

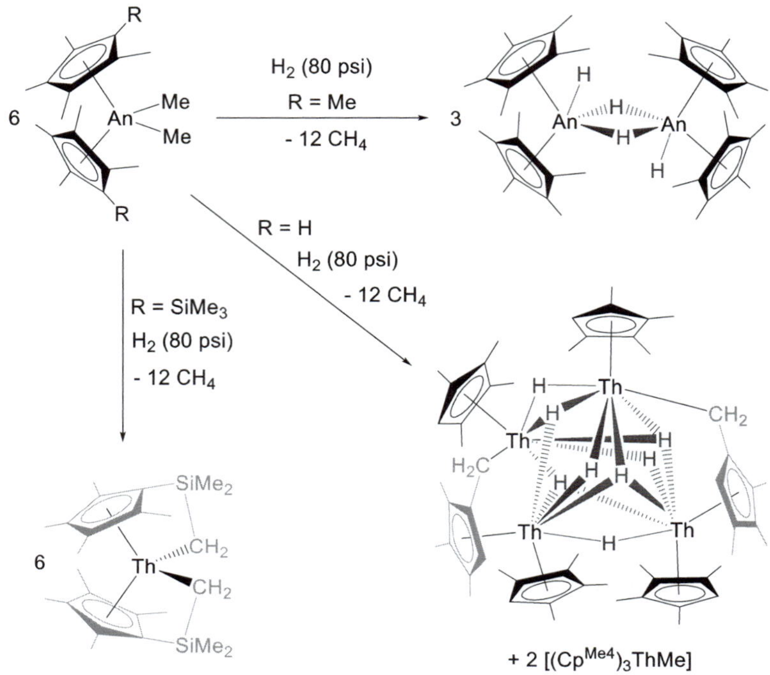

Scheme 7.29. Hydrogenolysis of $[(C_5Me_4R)_2AnMe_2]$ (An = Th or U) where R is Me, H and $SiMe_3$.

analogous reaction of $[(Cp^{Me4})_2ThMe_2]$ with H_2 afforded a 2:1 mixture of the tris-cyclopentadienyl thorium methyl complex, $[(Cp^{Me4})_3ThMe]$, and $[\{(Cp^{Me4})Th(\mu\text{-}\eta^5;\eta^1\text{-}C_5Me_3H\text{-}CH_2)\}_2(\mu\text{-}H)_4(\mu_3\text{-}H)_4\{Th(Cp^{Me4})\}_2]$ (Scheme 7.29).[393] Anionic $[\{(COT)_2Th\}_2H]^-$ (Figure 7.39) was prepared via the reaction of thorocene with NaH or KC_8/18-crown-6,[322,324] and the aryloxide-supported actinide dihydride complexes, $[\{(ArO)_2ThH_2\}_3]$ (Figure 7.39)[394] and $[\{Cp^*(ArO)ThH_2\}_3]$ (Ar = $C_6H_3{}^tBu_2$-2,6)[210] were isolated from the reactions of monometallic dialkyl precursors with H_2. Both complexes are trimers in the solid state, with two triply bridging and four doubly bridging hydride ligands.

By contrast, monomeric actinide(IV) hydride complexes have been isolated in combination with a suitably bulky ligand set; these complexes include $[(Cp^x)_3UH]$ (Cp^x = Cp^{Si}, Cp^t,[154,158,396] or $C_5H_4PPh_2$),[397] $[(TMP)_3UH]$ (TMP = tetramethylphospholide; C_4Me_4P),[255] $[(Cp^x)_3ThH]$ (Cp^x = Cp^*,[179] or Cp^{t2}),[398]

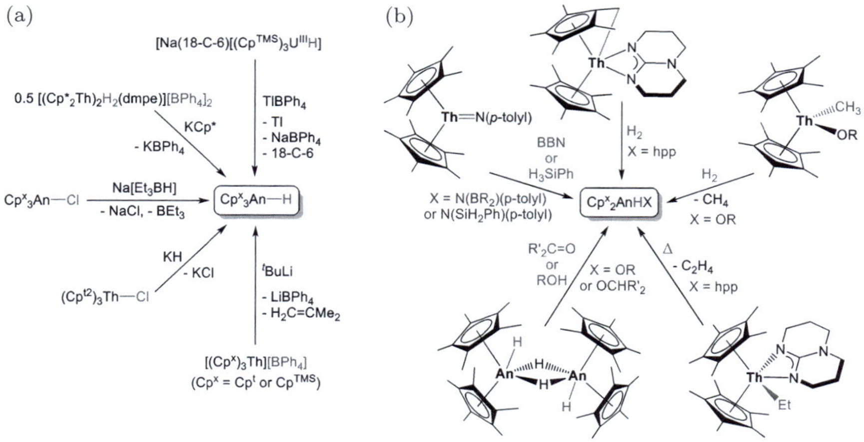

Scheme 7.30. Synthetic pathways to (a) $[(Cp^x)_3AnH]$ and (b) $[(Cp^x)_2AnH(X)]$ compounds.

$[(C_9H_6SiMe_3)_3ThH]$,[399] $[Cp^*{}_2AnH(OR)]$ (R = various, including O^tBu and OSi^tBuMe_2),[400–402] $[(Cp^{t3})_2ThH(NR_2)]$ $\{NR_2 = N(p\text{-tolyl})(BC_8H_{14})\}$,[204] $[Cp^*UH(hpp)]$ (hpp = 1,3,4,6,7,8-hexahydro-2*H*-pyrimido[1,2-*a*]pyrimidinato)[116] and $[\{(Me_3Si)_2N\}_3AnH]$.[403,404] These monomeric actinide(IV) monohydride complexes were prepared via several routes (Scheme 7.30), including (a) reaction of a metal halide or metal BPh_4 salt with KH,[398] $M[Et_3BH]$,[158,396,397,399] or ^tBuLi,[154] (b) reaction of dicationic $[Cp^*{}_4Th_2H_2(dmpe)][BPh_4]_2$ (0.5 equiv) with K(18-C-6)Cp*,[179] (c) oxidation of $[K(18\text{-C-}6)][(Cp^{Si})_3UH]$ with $Tl[BPh_4]$,[396] (d) hydrogenation of an alkyl derivative,[116,215] (e) reaction of $[\{Cp^*{}_2ThH(\mu\text{-}H)\}_2]$ with ROH or $O{=}CR_2$ reagents,[402] (f) β-hydride elimination upon thermolysis of an ethyl complex,[116] (g) reaction of an actinide imido complex, $[(Cp^{t3})_2Th{=}N(p\text{-tolyl})]$ with BBN (9-borabicyclo[3.3.1]nonane),[204] H_3SiPh, or H_2SiPh_2,[203] and (h) the unusual high temperature reaction between $[\{(Me_3Si)_2N\}_3AnCl]$ and $Na[N(SiMe_3)_2]$, where the hydride originates from the THF solvent.[403]

IR stretches for terminal actinide(IV) hydrides are observed in the range 1350–1500 cm^{-1}, while stretches due to bridging hydrides are lower. In the ^1H NMR spectra of thorium hydride compounds, Th*H* signals were typically observed between 12 and 19 ppm, with the exception of $[\{(Me_3Si)_2N\}_3ThH]$ which was reported at 0.90 ppm.

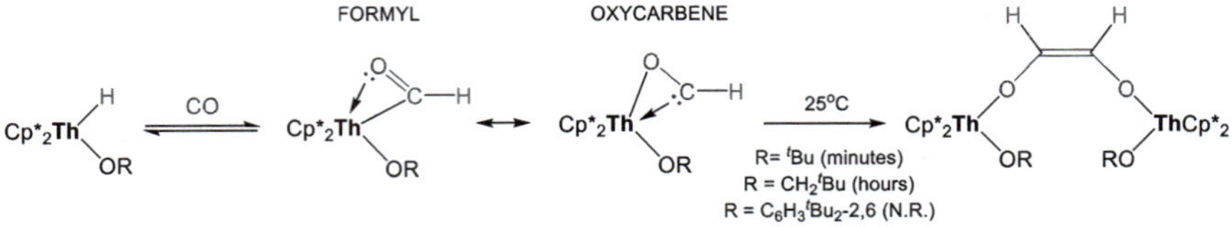

Scheme 7.31. Reactivity of [Cp*$_2$ThH(OR)] with CO (N.R. = No Reaction).

Marks *et al.* have measured metal–ligand bond disruption enthalpies in [Cp*$_2$AnHX] (An = Th and U) complexes and compared these values with alkyl and aryl analogues. In all cases, the thorium–hydride bonds are 5–10 kcal/mol stronger than those of the uranium analogues, and the An–H linkages are approx. 5–20 kcal/mol stronger than An–R (R = CH_2Ph, CH_2SiMe_3 and Me) bonds, but slightly weaker than An–Ph bonds.[213,215]

The more comparable strength of An–C and An–H bonds, compared with transition metal M–C and M–H bonds (the energy difference for middle and late transition metal complexes is typically ~30 kcal/mol)[213,215] leads to various deviations in actinide and transition metal organometallic reactivity. Notable examples are: (1) the pronounced resistance of many β-H containing actinide alkyl complexes towards β-H elimination, and (2) the ability of actinide-hydride complexes to engage in 1,1-insertion reactions with CO (with ensuing enediolate dimer formation in less sterically encumbered complexes; Scheme 7.31). However, it should be noted that the thermodynamic feasibility of An–H/CO insertion reactivity is also increased as a result of a significant interaction between the actinide metal center and the oxygen atom of the formyl/oxycarbene (CHO) ligand in the product. Nevertheless, it is still less exothermic (but more rapid) than analogous alkyl insertion reactivity.

A handful of low-valent actinide hydrides have also been prepared. In fact, the uranium(IV) dihydride dimer, [{Cp*$_2$UH(μ-H)}$_2$] readily loses H_2 in solution to generate the uranium(III) complex, [{Cp*$_2$U(μ-H)}$_2$]. Furthermore, this complex loses a second equiv. of H_2 at high temperature to form a tuck-in tuck-over uranium(IV) hydride complex, [Cp*{$C_5Me_3(CH_2)_2$}U(μ-H)$_2$UCp*$_2$] (Scheme 7.32). This reaction presumably proceeds by (a) cyclometallation of one Cp* methyl group to eliminate H_2, followed by intramolecular oxidative addition of a C–H bond in a second ligand

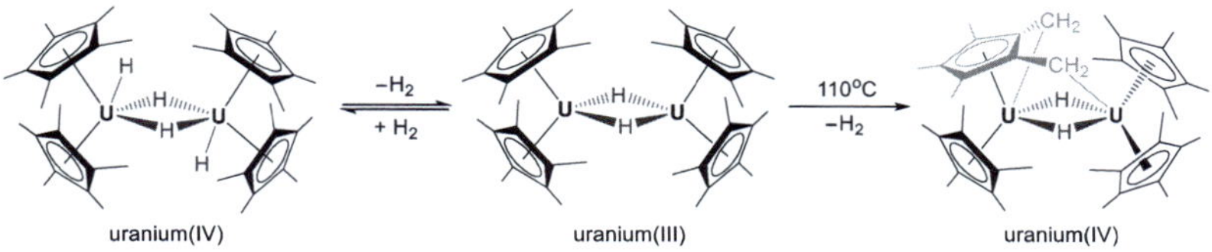

Scheme 7.32. Reductive elimination of H_2 from [(Cp*$_2$UH$_2$)$_2$], and thermolysis of the uranium(III) product, [(Cp*$_2$UH)$_2$], to form a tuck-in tuck-over dimetallic uranium(IV) hydride complex.

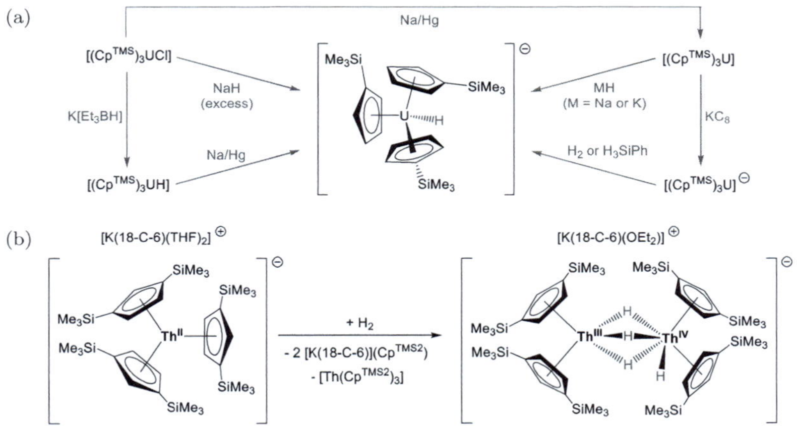

Scheme 7.33. Reactions to form low-valent cyclopentadienyl uranium (a) and thorium (b) hydride complexes.

methyl group, or (b) cyclometallation of two Cp* methyl groups, followed by H_2 oxidative addition.[405]

Other low-valent hydrides are $[(Cp^x)_3UH]^-$ (Cp^x = Cp^t, Cp^{Si}, Cp^{Si2})[234,406] and $[\{(Cp^x)_3U\}_2(\mu\text{-H})]^-$ (Cp^x = Cp or Cp′)[407] paired with THF-, crown-, or cryptand-coordinated K^+ or Na^+ cations.[228,408] These complexes were prepared by reaction of $[(Cp^x)_3U]$ (Cp^x = Cp, Cp′, Cp^{Si} or Cp^t) with NaH or KH, by reduction of $[(Cp^x)_3UH]$ (Cp^x = Cp^{Si} or Cp^t) with Na/Hg, and by reaction of $[(Cp^x)_3UCl]$ (Cp^x = Cp or Cp^{Si}) with excess NaH. Additionally, the unusual uranium(II) species, $[(Cp^{Si})_3U]^-$ reacted with H_2 or H_3SiPh to generate $[(Cp^{Si})_3UH]^-$ ((a) in Scheme 7.33).[228] By contrast, the thorium analogue, $[(Cp^{Si2})_3Th]^-$ reacted with H_2 to form a mixed-valence thorium(III)/(IV) species, $[(Cp^{Si2})_2Th(\mu\text{-H})_3ThH(Cp^{Si2})_2]^-$, accompanied by the formation of

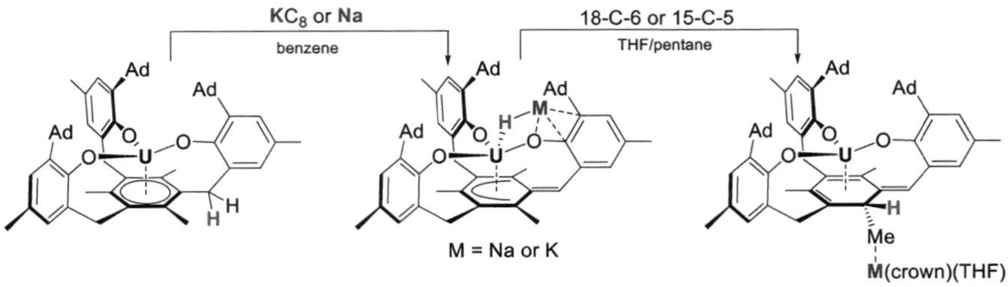

Scheme 7.34. Synthesis of a non-cyclopentadienyl uranium(III) hydride complex, and reaction with 18-crown-6 or 15-crown-5.

1 equiv. of $[(Cp^{Si2})_3Th]$ ((b) in Scheme 7.33), and the Cp* mixed-valence analogue was prepared by reduction of $[(Cp^*{}_2ThH_2)_2]$ with KC_8.[409] The Meyer group[289] also reported a uranium(III) non-cyclopentadienyl hydride complex, which was prepared from the reaction of $[\{(^{Ad,Me}ArO)_3Mes\}U]$ with KC_8 or Na; upon attempted removal of the M^+ cation by addition of crown ethers, this complex decomposed via attack of the hydride on the pentadienyl unit of the ligand backbone (Scheme 7.34).

Actinide hydride complexes engage in insertion reactivity with a range of substrates, including CO (*vide supra*),[402,410] CO_2,[339,411] ketones[402] and alkenes.[401] Additionally, both actinide(III) or (IV) hydride complexes can serve as multielectron reductants where H_2 is eliminated during the course of the reaction. For example, either $[(Cp^*{}_2AnH_2)_2]$ (An = Th and U) or $[(Cp^*{}_2UH)_2]$ reacted with PhS_2Ph and cyclooctatetraene to form $[Cp^*{}_2U(SPh)_2]$ and $[\{Cp^*(COT)U\}_2(\mu\text{-}COT)]$, respectively, where formation of the latter complex also involved $Cp^*{}_2$ elimination.[136] Along similar lines, $[(Cp^*{}_2UH_2)_2]$ or $[(Cp^*{}_2UH)_2]$ reacted with PhN=NPh to form $[Cp^*{}_2U(=NPh)_2]$, and $[(Cp^*{}_2UH)_2]$ reacted with $PhSe_2Ph$,[136] $Ph_2C{=}N{-}N{=}CPh_2$ and RC≡CR (R = Me or Ph)[367] to generate $[Cp^*{}_2U(SePh)_2]$, $[Cp^*{}_2U(N{=}CPh_2)_2]$ and $[Cp^*{}_2U(C_4R_4)_2]$, respectively.

7.10 Cationic Alkyl and Related Complexes

Cationic group 4 transition metal alkyl complexes are widely employed as olefin polymerization catalysts, and f-element alkyl

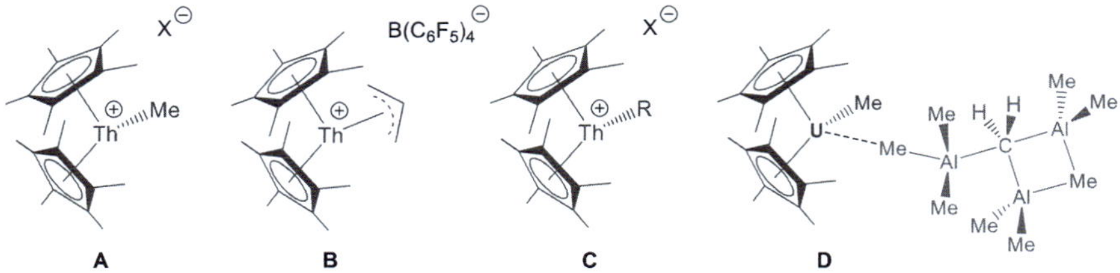

A B C D

Figure 7.40. Base-free cyclopentadienyl actinide alkyl cations [For compound A, X = $MeB(C_{12}F_9)_3$, BPh_4, $B(C_6F_5)_4$, $B(C_6F_4TBS)_4$, $B(C_6F_4TIPS)_4$, $[^tBuCH_2CH\{B(C_6F_5)_2\}_2H]$, $[Co(B_9C_2H_{11})_2]$, or 0.5 $[Fe(B_9C_2H_{11})_2]$. For compound C, R = CH_2SiMe_3 or H, and X = $[Co(B_9C_2H_{11})_2]$, or alternatively, R = CH_2Ph, and X = $B(C_6F_5)_4$].

cations are also of interest for this purpose. However, for the actinide elements, cationic alkyl species are rare. In cyclopentadienyl chemistry, the only base-free and mononuclear examples of actinide alkyl cations were reported by the Marks group, and are of the form $[Cp^*_2ThR][A]$ (R = Me,[412–415,417] CH_2SiMe_3,[415] CH_2Ph,[416] allyl[412] and H[415]) where A is either a weakly-coordinating borate anion, such as BPh_4, $B(C_6F_5)_4$ or $MeB(C_{12}F_9)_3$, or $[M(B_9C_2H_{11})_2]^{x-}$ (M = Co, $x = 1$; M = Fe, $x = 2$) (Figure 7.40); cationic alkyl complexes differing in the nature of the counter anion are of interest due to the ability of the anion to strongly influence polymerization activity and thermal stability through interactions with the cationic metal center, and to modify solubility and crystallinity. Ethylene polymerization activities were reported for methyl complexes paired with borate anions, with $[Cp^*_2ThMe][^tBuCH_2CH\{B(C_6F_5)_2\}_2H]$ achieving an ethylene polymerization activity of 5,800 kg(PE) mol^{-1} h^{-1} atm^{-1} (product $M_w = 2.8 \times 10^5$, $M_n = 1.2 \times 10^5$).[414]

Cationic olefin polymerization catalysts prepared from $[\{Me_2Si(indenyl)_2\}AnMe_2]$ as well as $[Cp^*_2AnMe_2]$ (An = Th or U) are also mentioned in several Marks patents,[418] and the Dow Chemical Company has patented a variety of pentaalkylcyclopentadienyl actinide polymerization catalysts, including those formed from $[Cp^*_2AnX_2]$ and $[Cp^*AnX_3]$ (An = Th and U; X = Cl, Me or CH_2SiMe_3) in combination with activators such as MAO.[419] Moreover, Marks reported highly active heterogeneous olefin polymerization catalysts, which are similar in nature to $[Cp^*_2ThR][A]$, but are formed by reaction of thorium or uranium $[Cp^*_2AnMe_2]$ complexes with dehydroxylated γ-alumina or $MgCl_2$.[420] Anwander and Evans *et al.* also reported the reaction of [$Cp^*{}_2UMe_2$] with excess $AlMe_3$ to form [$Cp^*{}_2UMe$][$Me_3Al(\mu_3-CH_2)(AlMe_2)_2(\mu-CH_3)$] (Figure 7.40) with release of methane. This contact ion pair features an organoaluminum anion, derived from two $AlMe_3$ units bound to a CH_2AlMe_2 group, coordinated to a [$Cp^*{}_2UMe$]$^+$ cation; the U–C distances to the bound methyl group and the interacting methyl group of the anion are 2.395(6) and 2.658(5) Å, respectively.[421]

A number of non-cyclopentadienyl alkyl cations have also been reported. Reaction of neutral [$(XA_2)Th(CH_2SiMe_3)_2$] or [$(XA_2)Th(CH_2Ph)_2$] with [CPh_3][$B(C_6F_5)_4$] in benzene or toluene at room temperature yielded [$(XA_2)Th(CH_2SiMe_3)(\eta^6$-$C_6H_5R)$][$B(C_6F_5)_4$] (R = H or Me) and [$(XA_2)Th(\eta^2$-$CH_2Ph)(\eta^6$-$C_6H_5Me)$][$B(C_6F_5)_4$], respectively; rare examples of arene solvent-separated ion pairs. The reactions of [$(XA_2)Th(CH_2SiMe_3)_2$] and [$(XA_2)Th(CH_2Ph)_2$] with sub-stoichiometric amounts of [CPh_3][$B(C_6F_5)_4$] provided no evidence for dinuclear monocation formation. By contrast, reaction of [$(BDPP)Th(CH_2Ph)_2$] with 0.5 equiv. of [CPh_3][$B(C_6F_5)_4$] precipitated an insoluble oil containing the dinuclear cation, [$(BDPP)Th(\eta^2$-$CH_2Ph)(\mu$-η^1:η^6-$CH_2Ph)Th(\eta^1$-$CH_2Ph)(BDPP)$][$B(C_6F_5)_4$] in which a benzyl group adopts a previously unknown μ-η^1:η^6-bridging mode.[280] Reaction of [$(XA_2)Th(CH_2Ph)_2$] with $B(C_6F_5)_3$ afforded [$(XA_2)Th(\eta^1$-$CH_2Ph)$][η^6-$PhCH_2B(C_6F_5)_3$] in which the benzylborate anion is η^6-coordinated to the metal center, and addition of a 2 equiv. of $B(C_6F_5)_3$ afforded dicationic [$(XA_2)Th$][η^6-$PhCH_2B(C_6F_5)_3$]$_2$ in which both benzylborate anions are η^6-coordinated (Scheme 7.35 and Figure 7.20 in Section 7.7.4.1).[282] The metal center in all of the aforementioned XA_2 and BDPP complexes is π-coordinated, either to neutral arene solvent, a benzyl group in [$(BDPP)Th(CH_2Ph)_2$], or a benzyl group in a $PhCH_2B(C_6F_5)_3$ anion, highlighting a pronounced tendency for these systems to engage in arene π-coordination.

Along similar lines, reaction of [$(XA_2)U(CH_2SiMe_3)_2$] with [CPh_3][$B(C_6F_5)_4$] in benzene, toluene or fluorobenzene afforded [$(XA_2)U(CH_2SiMe_3)(\eta^n$-$C_6H_5R)$][$B(C_6F_5)_4$] (R = H, Me or F) in which uranium is π-coordinated to a molecule of solvent. Under 1 atm of ethylene at room temperature, both the thorium and uranium cations failed to polymerize ethylene when generated in benzene or toluene, whereas cations generated in fluorobenzene or *o*-difluorobenzene displayed reasonable activities {17 (Th) and

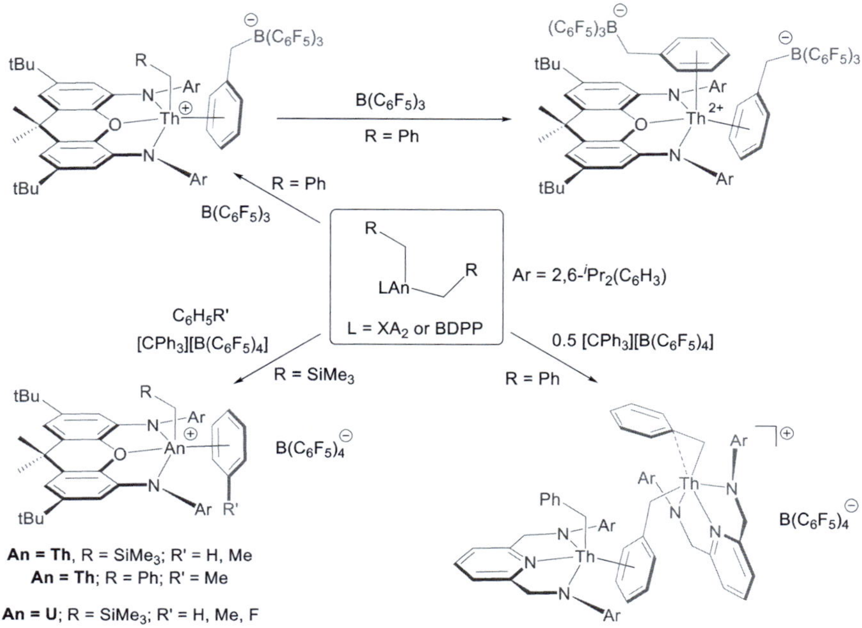

Scheme 7.35. Synthesis of non-cyclopentadienyl actinide alkyl cations free from external ether or amine Lewis base coordination.

11–13 (U) kg(PE) mol^{-1} h^{-1} atm^{-1} at room temperature}, highlighting the strong influence of arene solvents on the polymerization activity of sterically open alkyl cations.[281]

Additionally, Eisen reported bis(amidinate) actinide(IV) chloro complexes of the form [(2-pyridylamidinate)$_2$AnCl(μ − Cl)$_2$ Li(tmeda)] {An = Th or U; 2-pyridylamidinate = $(Me_3SiN)_2C$(2-py)} that serve as precursors for ethylene polymerization catalysis when combined with methylalumoxane (MAO) or a mixture of Al^iBu_3 and $[CPh_3][B(C_6F_5)_4]$, achieving a maximum activity of 10.4 kg(PE) mol^{-1} h^{-1} atm^{-1}.[422] Furthermore, Leznoff has disclosed ethylene polymerization activities of 0.02–0.6 kg(PE) mol^{-1} h^{-1} atm^{-1} for neutral [(^{Dipp}NCOCN)U$(CH_2R)_2$] {R = $SiMe_3$, Ph; ^{Dipp}NCOCN = $O(CH_2CH_2NAr)_2$; Ar = $C_6H_3^iPr_2$-2,6}, [(^{tBu}NON)U$(CH_2SiMe_3)_2$] {^{tBu}NON = $O(SiMe_2N^tBu)_2$}, and dimetallic [(^{tBu}NON)U{CH$(SiMe_3)(SiMe_2CH_2)$}]$_2$, and reduced

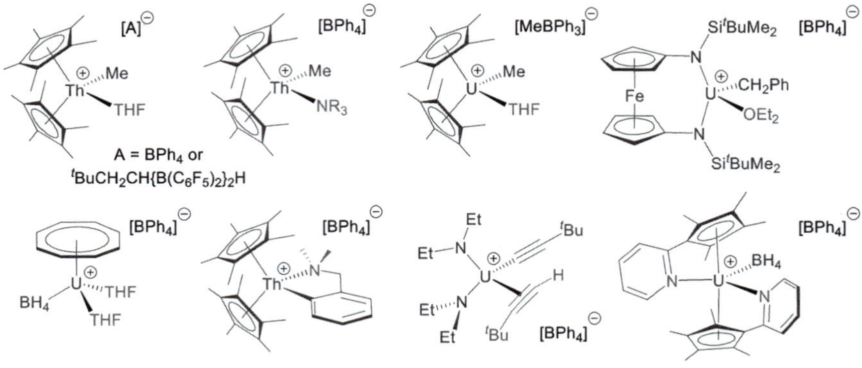

Figure 7.41. Actinide alkyl cations stabilized by Lewis base coordination, and actinide alkynyl or borohydride cations.

activities of less than 0.1 kg(PE) mol^{-1} h^{-1} atm^{-1} after addition of $B(C_6F_5)_3$ as an activator.[383]

Other isolated actinide alkyl cations are Lewis base-stabilized $[Cp^*{}_2ThMe(L)]^+$ (L = THF NMe_3 or NEt_3),[414,417] $[Cp^*{}_2UMe(THF)][MeBPh_3]$,[178] and $[(NN^{Fc})U(CH_2Ph)(OEt_2)][BPh_4]$[389] (Figure 7.41), and dimetallic $[(Cp^*{}_2ThMe)_2(\mu\text{-}Me)][B(C_6F_5)_4]$ which exists in equilibrium with $[Cp^*{}_2ThMe_2]$ and $[Cp^*{}_2ThMe][B(C_6F_5)_4]$ in solution.[412,416] Moreover, cationic aryl and alkynyl complexes (Figure 7.41) have been reported; $[Cp^*{}_2Th(\kappa^2\text{-}C_6H_4CH_2NMe_2\text{-}o)][BPh_4]$ was prepared by reaction of $[Cp^*{}_2Th(\kappa^2\text{-}C_6H_4CH_2NMe_2\text{-}o)Me]$ with $[NEt_3H][BPh_4]$,[417] while $[(Et_2N)_2U(C_2{}^tBu)(HC_2{}^tBu)][BPh_4]$ was observed spectroscopically in the reaction of $[(Et_2N)_3U][BPh_4]$ with 2 equiv. of *tert*-butyl acetylene.[423] Several actinide hydride and borohydride cations have also been prepared. $[(Cp^*{}_2ThH)_2(dmpe)][BPh_4]_2$ was accessed by protonolysis of $[(Cp^*{}_2ThH)_2(\mu\text{-}H)_2]$ with $[HNEt_3][BPh_4]$ in the presence of dmpe,[179] while $[Cp^*{}_2ThH][Co(B_9C_2H_{11})_2]$ was generated in the reaction of $[Cp^*{}_2Th(CH_2SiMe_3)][Co(B_9C_2H_{11})_2]$ with H_2;[415] neither of these complexes was structurally characterized. Borohydride complexes $[(COT)U(BH_4)(THF)_2][BPh_4]$[333] and $[(\eta^5\text{:}\eta^1\text{-}C_5Me_4\text{-pyridyl-}o)_2U(BH_4)][BPh_4]$[424] (Figure 7.41) were isolated from the reactions of $[NEt_3H][BPh_4]$ with $[(COT)U(BH_4)_2(THF)]$ and $[(\eta^5\text{:}\eta^1\text{-}C_5Me_4\text{-pyridyl-}o)U(BH_4)_2]$, respectively. By contrast, $[(calix)U(BH_4)][B(C_6F_5)_4]$ was prepared by oxidation of the

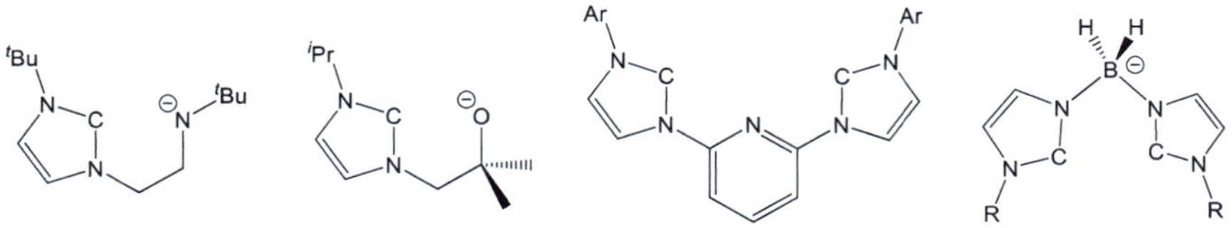

Figure 7.42. Multidentate NHC ligands deployed in actinide chemistry (R = Me or Mes; Ar = $C_6H_3{}^iPr_2$-2,6).

uranium(III) borohydride complex, [(calix)U(BH_4)] (calix = *trans*-calix[2]benzene[2]pyrrolyl), with [CPh_3][B(C_6F_5)$_4$] (this reactivity was accompanied by changes in the coordination mode of the pyrrolyl and arene groups of the calix ligand; Scheme 7.15 in Section 7.7.2).[122]

7.11 Carbene Complexes

A broad range of actinide complexes bearing *N*-Heterocyclic Carbene (NHC) ligands have been prepared and multidentate ligands containing NHC-donors have also been deployed as supporting ligands in actinide chemistry (Figure 7.42).[425] However, NHCs represent a special class of neutral carbene. For example, they have much in common with strongly donating trialkylphosphine ligands, including a demonstrated ability to bind to elements from throughout the periodic table, so they are not discussed further in this section.

Between 1981 and 1990, Cramer and Gilje reported the synthesis and reactivity of tris-cyclopentadienyl uranium [Cp_3U{CH(PR_3)}] (PR_3 = PMe_3, PMe_2Ph, or $PMePh_2$) complexes, in which the CH(PR_3)$_3$ ligand can be described by several resonance structures, including that of a carbene (Scheme 7.5 in Section 7.6.2). These complexes were accessed by reaction of [Cp_3UCl] with 1 equiv. of Li[(CH_2)$_2PR_2$] or 2 equiv. of $H_2C{=}PMeR_2$ ((a) in Scheme 7.36).[169,426] By contrast, a tris(amido) analogue, [{(Me_3Si)$_2$N}$_3$ U{CH(PPh_3)}] (which exists in equilibrium with free $H_2C{=}PPh_3$ and metallacyclic [{(Me_3Si)$_2$N}$_2${(Me_3Si)NSiMe_2CH_2}U]),[427] was prepared by decomposition of the ylide complex [{(Me_3Si)$_2$N}$_3$ U{CH_2(PPh_3)}], forming a 1:1:1 mixture of the target carbene complex alongside [{(Me_3Si)$_2$N}$_3$UMe] and PPh_3 ((b) in Scheme 7.36). Notable structural features of the aforementioned carbene complexes

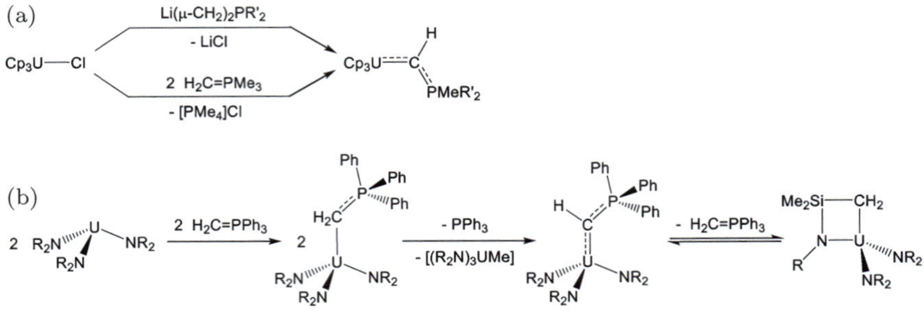

Scheme 7.36. Methods used for $CH(PR_3)$ ligand attachment; (a) from Cp_3UCl and (b) from $U(NR_2)_3$ (R = Me or Ph; R = $SiMe_3$).[169,426]

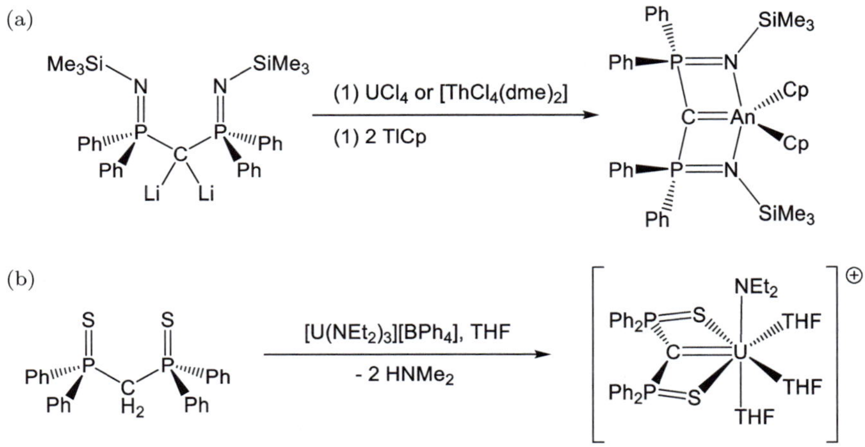

Scheme 7.37. Methods of attachment for (a) NCN[428] and (b) SCS[429] carbene ligands (R = $SiMe_3$).

are obtuse U–C–P angles (142°–152°) and U–C bond distances (2.28–2.29 Å) which are significantly shorter than those for a typical U–C_{alkyl} single bond (2.35–2.50 Å).

Uranium carbene complexes spanning a range of oxidation states have also been isolated through the use of the multidentate NCN[390,428,430,436] and SCS[98,429,437–439] ligands shown in Scheme 7.37, and a handful of thorium(IV) complexes have also been reported.[428,435,439,440] These complexes include mono-, bis- and tris-carbene complexes, and complexes with halide, triflate, borohydride, amido, beznyl, oxo, imido, cyclopentadienyl, cyclooctatetraenide and tris(pyrazolyl)borate co-ligands. Carbene ligand attachment was achieved by reaction of an actinide halide, triflate, or borohydride

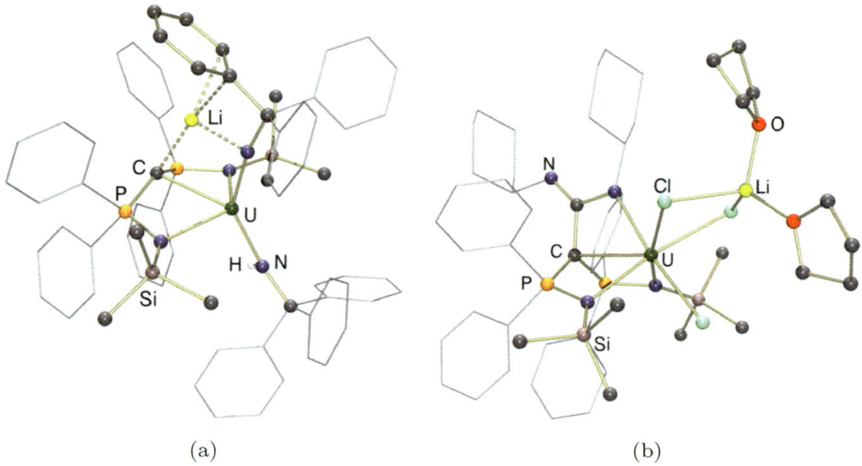

Figure 7.43. X-ray crystal structures of (a) Li[(NCN^{SiMe3})U(=NCPh$_3$)(NHCPh$_3$)][433] and (b) [{C(PPh$_2$ = NSiMe$_3$)$_2$(C(=NCy)NCy)}UCl(μ-Cl)$_2$ Li(THF)$_2$].[432]

precursor with Li$_2$[SCS] or Li$_2$[NCN], or by amine elimination, for example reaction of [U(NEt$_2$)$_3$][BPh$_4$] with H$_2$[SCS] (Scheme 7.37). These carbene complexes provided a unique platform for computational study of actinide–carbon multiple bonding, and key findings include:

- The carbene donors in NCN and SCS actinide complexes are nucleophilic.[437]
- Both thorium and uranium NCN complexes engage in An=C multiple bonding. However, the uranium analogues show higher An=C bond indices with increased actinide participation in bonding.[428] In fact, the level of covalency in [(NCN)M^{IV}(OAr)$_2$] (Ar = C$_6$H$_3{}^i$Pr$_2$-2,6) has been shown to be very similar for U and Ce, but significantly less for Th.[435]
- For the uranium(IV), (V) and (VI) complexes, [(NCN)U^{IV}Cl(μ-Cl)$_2$Li(THF)$_2$] (Figure 7.43), [(NCN)U^{V}Cl$_2$I] and [(NCN)U^{VI}OCl$_2$], the following experimental and computational trends were observed: (a) the U=C bond distances decrease significantly (by ~0.15 Å) across the series, (b) Nalewajski–Mrozek U=C bond orders only increase from 1.43 in the uranium(IV) compound to 1.50–1.54 in the higher oxidation state compounds,[431] (c) the U=C σ- and π-bonds in the uranium(V) and (VI) compounds have

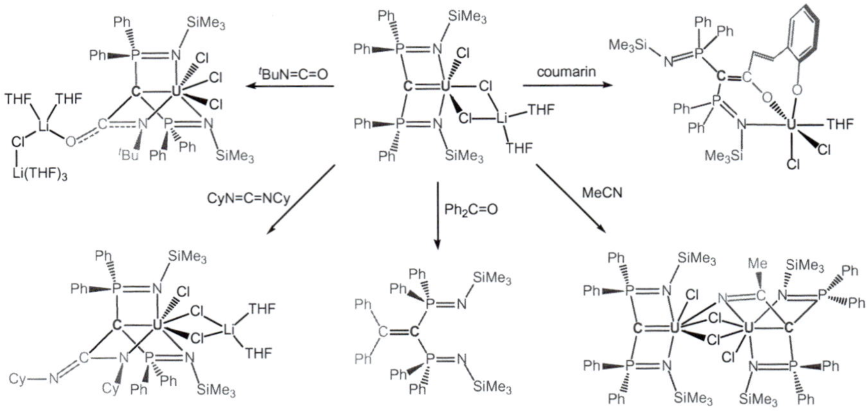

Scheme 7.38. Reactions of [(NCN^{SiMe3})UCl(μ-Cl)$_2$Li(THF)$_2$] with Ph$_2$CO, MeCN, C(NCy)$_2$, ^{t}BuNCO and coumarin (1-benzopyran-2-one).

higher U character than those in the uranium(IV) complex (~25% vs. 18% for the π-component), and (d) for the uranium contribution to the σ- and π-components of the U=C bond, 5f orbital participation ranges from 75% to 95%, with 6d orbitals accounting for the remainder.

- The carbene ligands in [(NCN^{TMS})U(=NCPh$_3$)(bipy)] (bipy = 2,2′-bipyridine) and [(NCN^{TMS})U(=NCPh$_3$)(dmap)$_2$] (dmap = 4-dimethylaminopyridine), which are *cis*- and *trans*- to imido ligands, respectively, exert a *trans*-influence.[434] By contrast, an inverse-*trans*-influence with similarities to that in uranyl (UO_2^{2+}) complexes was observed for the carbene donors in [An(NCN ^{TMS})$_2$] (An = U and Th).[436]

The reactivity of SCS and NCN complexes highlights the nucleophilic character of the carbene donors, consistent with computational predictions. For example, uranium SCS and NCN complexes reacted with ketones and aldehydes (R′R″C=O) to release alkenes {R′R″C=C(PPh$_2$ =E)$_2$; E = S or NR},[431,432,436,437,439] and deprotonation of [(NCN^{SiMe3})U(NHCPh$_3$)$_2$] with (tmeda)LiCH$_2$Ph afforded Li[(NCN^{SiMe3})U(=NCPh$_3$)(NHCPh$_3$)] in which lithium binds to the carbene carbon atom, the imido nitrogen donor and a phenyl ring (Figure 7.43).[433] The reactivity of [(NCN)UCl(μ-Cl)$_2$Li(THF)$_2$] has been particularly well studied, with reagents including Ph$_2$CO, MeCN, C(NCy)$_2$, ^{t}BuNCO and 1-benzopyran-2-one (Scheme 7.38).[428,432]

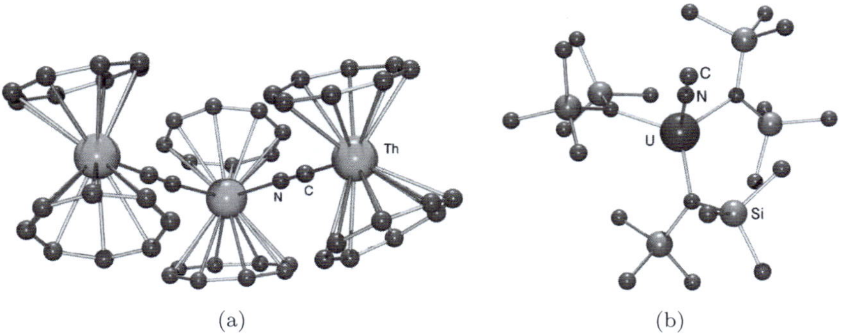

Figure 7.44. X-ray crystal structures of the μ-cyanide complex, (a) $[\{(COT)_2Th\{(\mu\text{-}NC)Th(COT)_2\}_2]^{2-}$ and isocyanide complex (b) $[\{(Me_3Si)_2N\}_3U(NC)]$.

7.12 Cyanide, Carbonyl and Isonitrile Compounds

Actinide complexes bearing terminal cyanide ligands have been prepared for thorium(IV) and uranium(III)-(VI), including $[(COT)_2Th^{IV}(CN)_2]^{2-}$,[323] $[\{(Me_3Si)_2N\}_3U^{III}(CN)]^-$,[441] $[Cp_3U^{IV}(CN)]$,[442] $[Cp^*{}_2U^V(CN)_5]^{2-}$,[443] and $[U^{VI}O_2(CN)_5]^{3-}$.[444] In all of the aforementioned complexes, the cyanide ligand is coordinated via carbon, as is observed in transition metal complexes bearing terminal cyanide ligands. Bridging cyanide ligands in which one actinide binds to carbon and one binds to nitrogen are also well known, and were discussed in Section 7.7.6 on cyclooctatetraenide complexes; a molecular example, $[\{(COT)_2Th\{(\mu\text{-}NC)Th(COT)_2\}_2]^{2-}$,[323] is shown in Figure 7.44. However, whereas the cyanide ligands in $[\{(Me_3Si)_2N\}_3U^{III}(CN)_2]^{2-}$ and $[\{(Me_3Si)_2N\}\{\kappa^2\text{-}(CH_2SiMe_2NSiMe_3)_2\}U^{IV}(CN)_2]^-$ are *C*-coordinated,[441,445] those in $[\{(Me_3Si)_2N\}_3Ce^{III}(NC)_2]^{2-}$,[441] $[\{(Me_3Si)_2N\}_3U^{IV}(NC)_2]^-$, and $[\{(Me_3Si)_2N\}_3U^{IV}(NC)]$ (Figure 7.44)[445] are *N*-coordinated (referred to as isocyanide complexes).

Neutral isonitrile (CNR) and carbonyl (CO) ligands are isoelectronic with cyanide anions. However, while a significant number of cyanide and isonitrile[446,447] compounds have been reported, carbonyl complexes of the actinide elements are rare. The first actinide carbonyl complex, $[(Cp^{Si})_3U(CO)]$, was generated in solution by reaction of $[(Cp^{Si})_3U]$ with CO, and was reported in 1986.[448] Subsequently, this chemistry was extended to include $[Cp^x{}_3U(CO)]$ (Cp^x = Cp^{Si2}, Cp^t, Cp^{Me4} and Cp^*),[447,449,450] and the solid-state structures of the Cp^{Me4} and Cp^* complexes were determined; the U–CO distance is 2.383(6) Å in the former[449] and 2.485(9) Å in the

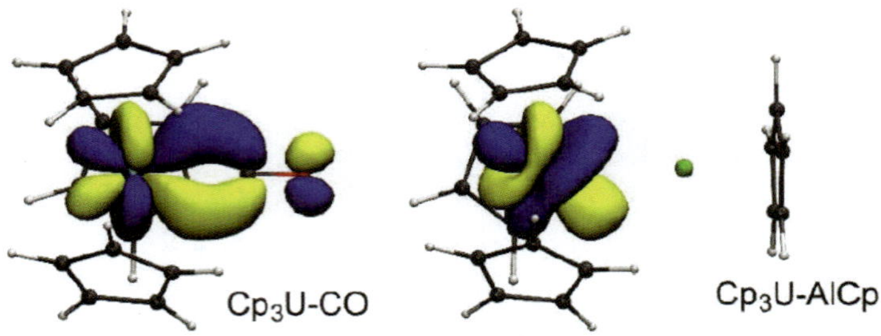

Figure 7.45. Singly occupied molecular orbitals (SOMOs) involved in U–L π-bonding in [$Cp_3U(L)$] (L = CO and AlCp) at identical isovalues. Reproduced from reference 452 with permission from the American Chemical Society.

more sterically hindered Cp* complex.[450] The carbonyl stretching frequencies for the aforementioned uranium(III) CO complexes span an unusually broad range; from 1988 cm^{-1} in [$(Cp^{Si2})_3U(CO)$] to 1880 cm^{-1} in [$(Cp^{Me4})_3U(CO)$]. Such low stretching frequencies, relative to free CO at 2143 cm^{-1}, are typically an indication of substantial π-backdonation and relatively tight CO binding. However, even [$(Cp^{Me4})_3U(CO)$] slowly loses CO in solution, and the solid can only be stored at room temperature for long periods of time under an atmosphere of CO. Uranium–CO bonding in [$(Cp^{x})_3U(CO)$] has been studied computationally by several authors, revealing that both σ-donation and π-backdonation (from orbitals which are primarily metal- or ligand-based) plays an important role.[451,452] Bonding in [$(Cp^{Si2})_3U(ECp^*)$] (E = Al and Ga) has also been studied, and although ECp* ligands bear resemblance to CO due to localization of one σ-donor and two π-acceptor orbitals on the group 13 element, the covalent component of U–ECp* bonding almost exclusively involves σ-donation to uranium, with very little π-backdonation (Figure 7.45).[452]

7.13 Agostic Interactions and Metal–Alkane Coordination

Whereas acute An–C–C angles (often 80°–100°) are typical for actinide benzyl compounds due to multihapto coordination,[282] expanded An–C–Si and An–C–C angles (e.g. 125°–150°) are a common feature of actinide trimethylsilylmethyl complexes.[80,385] These

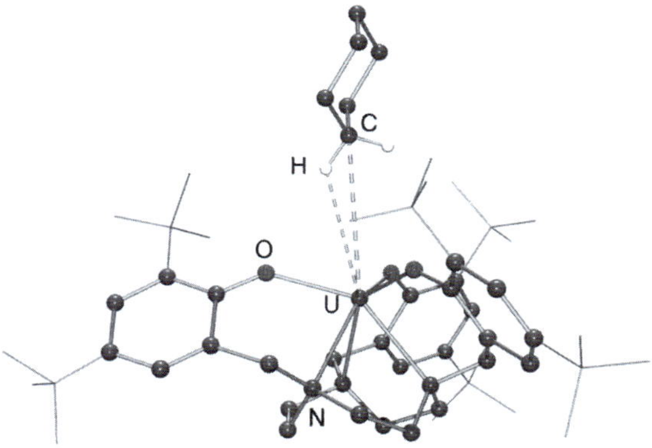

Figure 7.46. X-ray crystal structure of $[\{(ArO)_3tacn\}U^{III}(cyclohexane)]$.[455]

expanded angles are typically considered to be an indication of α-agostic interactions (although steric factors can also play a role), and for diamagnetic complexes, additional evidence for such interactions can be obtained from the $^1J_{C,H}$ coupling constants for the α-hydrogen atoms, which are often below 105 Hz[80] (vs. 118 Hz for $SiMe_4$).[453] Furthermore, the extent to which such interactions take place is in some cases temperature dependent. For example, the average $^1J_{C,H}$ coupling constant for the α-C–H bonds in $[(XA_2)Th(CH_2SiMe_3)_2]$ is 102 Hz at 50 °C, decreasing to 88 and 81 Hz for the inequivalent alkyl groups at −80 °C.[80]

Agostic interactions fall under the umbrella of σ-interactions, which also encompasses intermolecular metal–alkane interactions. Structurally characterized complexes in the latter category are extremely rare.[454,455] However, in 2003, Meyer reported X-ray structures for a series of uranium complexes in which alkanes are located in proximity to uranium(III) (the An–C distances are 0.04–0.17 Å shorter than the sum of the van der Waals radii), sequestered in the hydrophobic binding pocket of the hexadentate $(ArO)_3$tacn $\{(CH_2CH_2N\{CH_2C_6H_2(3,5\text{-}^tBu)(2\text{-}O)\})_3\}$ ligand (Figure 7.46). Bonding in these alkane complexes is proposed to involve σ-donation from the alkane C–H bonds to uranium, and dispersion interactions between the alkane and the surrounding hydrophobic pocket are also likely to make an important contribution to the overall interaction energy between the alkane and the $\{(ArO)_3tacn\}$ U fragment.[455]

References

1. (a) W. B. Jensen, *J. Chem. Educ.*, 1982, *59*, 634; (b) W. B. Jensen, *Found. Chem.*, 2015, *17*, 23.
2. D. Lariviere and N. Guerin, Natural Radioactivity, in *Radionuclides in the Environment. In Encyclopedia of Inorganic Chemistry* (Ed. D. A. Atwood), John Wiley & Sons, Chichester, 2010.
3. D. R. Lide, *CRC Handbook of Chemistry and Physics*, 77 Edn., CRC Press, New York, 1996.
4. D. F. Starks, T. C. Parsons, A. Streitwieser and N. Edelstein, *Inorg. Chem.*, 1974, *13*, 1307.
5. F. Baumgärtner, E. O. Fischer, B. Kanellakopulos and P. Laubereau, *Angew. Chem. Int. Ed. Engl.*, 1969, *8*, 202.
6. F. Baumgartner, E. O. Fischer, B. Kamellakopulos and P. Laubereau, *Angew. Chem. Int. Ed. Engl.*, 1966, *5*, 134.
7. P. G. Laubereau and J. H. Burns, *Inorg. Nucl. Chem. Lett.*, 1970, *6*, 59.
8. P. G. Laubereau and J. H. Burns, *Inorg. Chem.*, 1970, *9*, 1091.
9. J. S. Parry, F. G. N. Cloke, S. J. Coles and M. B. Hursthouse, *J. Am. Chem. Soc.*, 1999, *121*, 6867.
10. H. Geckeis, *Gmelin Handbook of Inorganic and Organometallic Chemistry*, 8th Ed., Thorium Suppl. Vol. C 4, Section 6.1.2: Thorium Tetrachloride, $ThCl_4$, Springer-Verlag, Berlin, 1993, p. 65.
11. O. C. Dean and J. M. Chandler, *Nucl. Sci. Eng.*, 1957, *2*, 57.
12. A. Athimoolam, S. Gambarotta and I. Korobkov, *Organometallics*, 2005, *24*, 1996.
13. T. Cantat, B. L. Scott and J. L. Kiplinger, *Chem. Commun.*, 2010, *46*, 919.
14. N. E. Travia, M. J. Monreal, B. L. Scott and J. L. Kiplinger, *Dalton Trans.*, 2012, *41*, 14514.
15. M. P. Wilkerson, C. J. Burns, R. T. Paine and B. L. Scott, *Inorg. Chem.*, 1999, *38*, 4156.
16. I. A. Khan, H. S. Ahuja, K. W. Bagnall and L. Sinf, *Inorg. Synth.*, 1982, *21*, 187.
17. J. L. Kiplinger, D. E. Morris, B. L. Scott and C. J. Burns, *Organometallics*, 2002, *21*, 5978.
18. D. Patel, A. J. Wooles, E. Hashem, H. Omorodion, R. J. Baker and S. T. Liddle, *New J. Chem.*, 2015, *39*, 7559.
19. D. D. Schnaars, G. Wu and T. W. Hayton, *Dalton Trans.*, 2008, 6121.
20. J.-C. Berthet, P. Thuéry and M. Ephritikhine, *Inorg. Chem.*, 2005, *44*, 1142.

21. A. E. Enriquez, B. L. Scott and M. P. Neu, *Inorg. Chem.*, 2005, *44*, 7403.
22. C. D. Carmichael, N. A. Jones and P. L. Arnold, *Inorg. Chem.*, 2008, *47*, 8577.
23. J.-C. Berthet, M. Lance, M. Nierlich and M. Ephritikhine, *Eur. J. Inorg. Chem.*, 1999, 2005.
24. F. G. N. Cloke and P. B. Hitchcock, *J. Am. Chem. Soc.*, 2002, *124*, 9352.
25. W. J. Evans, S. A. Kozimor, J. W. Ziller, A. A. Fagin and M. N. Bochkarev, *Inorg. Chem.*, 2005, *44*, 3993.
26. D. L. Clark, A. P. Sattelberger, S. G. Bott and R. N. Vrtis, *Inorg. Chem.*, 1989, *28*, 1771.
27. (a) L. R. Avens, S. G. Bott, D. L. Clark, A. P. Sattelberger, J. G. Watkin and B. D. Zwick, *Inorg. Chem.*, 1994, *33*, 2248; (b) D. L. Clark and A. P. Sattelberger, *Inorg. Synth.*, 1997, *31*, 307.
28. L. Natrajan, M. Mazzanti, J. P. Bezombes and J. Pecaut, *Inorg. Chem.*, 2005, *44*, 6115.
29. H. S. La Pierre, F. W. Heinemann and K. Meyer, *Chem. Commun.*, 2014, *50*, 3962.
30. D. C. Moody, A. J. Zozulin and K. V. Salazar, *Inorg. Chem.*, 1982, *21*, 3856.
31. (a) R. G. Jones, G. Karmas, G. A. Martin Jr. and H. Gilman, *J. Am. Chem. Soc.*, 1956, *78*, 4285; (b) D. C. Bradley and M. H. Gitlitz, *J. Chem. Soc. A*, 1969, 980.
32. W. G. Vandersluys, C. J. Burns, J. C. Huffman and A. P. Sattelberger, *J. Am. Chem. Soc.*, 1988, *110*, 5924.
33. J. A. Leary and J. F. Suttle, *Inorg. Synth.*, 1957, *5*, 148.
34. J. D. Hefley, D. M. Mathews and E. S. Amis, *Inorg. Synth.*, 1963, *7*, 146.
35. J.-C. Berthet, M. Lance, M. Nierlich and M. Ephritikhine, *Eur. J. Inorg. Chem.*, 2000, 1969.
36. K. A. Gingerich and D. W. Wilson, *Inorg. Chem.*, 1965, *4*, 987.
37. D. Brown, T. C. Tso and B. Whittaker, *J. Chem. Soc., Dalton Trans.*, 1977, 2291.
38. R. N. R. Mulford and T. A. Wiewandt, *J. Phys. Chem.*, 1966, *69*, 1641.
39. (a) F. Brown, H. M. Ockenden, G. A. Welch, *J. Chem. Soc.* 1955, 3932; (b) R. N. R. Mulford and G. E. Sturdy, *J. Am. Chem. Soc.*, 1956, *78*, 3897.
40. J. W. Roddy, *J. Inorg. Nucl. Chem.*, 1973, *35*, 4141.
41. H. R. Hoekstra and J. J. Katz, *J. Am. Chem. Soc.*, 1949, *71*, 2488.

42. R. H. Banks, N. M. Edelstein, R. R. Rietz, D. H. Templeton and A. Zalkin, *J. Am. Chem. Soc.*, 1978, *100*, 1957.
43. H. I. Schlesinger and H. C. Brown, *J. Am. Chem. Soc.*, 1953, *75*, 219.
44. H. I. Schlesinger and H. C. Brown, *J. Am. Chem. Soc.*, 1953, *75*, 209.
45. A. C. Dunbar, J. E. Gozum and G. S. Girolami, *J. Organomet. Chem.*, 2010, *695*, 2804.
46. V. M. Ehemann and H. Nöth, *Z. Anorg. Allg. Chem.*, 1971, *386*, 87.
47. T. J. Marks and J. R. Kolb, *Chem. Rev.*, 1977, *77*, 263.
48. E. R. Bernstein, W. C. Hamilton, T. A. Keiderling, S. J. LaPlaca, S. J. Lippard and J. J. Mayerle, *Inorg. Chem.*, 1972, *11*, 3009.
49. (a) E. R. Bernstein, T. A. Keiderling, S. J. Lippard and J. J. Mayerle, *J. Am. Chem. Soc.*, 1972, *94*, 2552; (b) P. Charpin, H. Marquetellis and G. Folcher, *J. Inorg. Nucl. Chem.*, 1979, *41*, 1143.
50. R. H. Banks, N. M. Edelstein, B. Spencer, D. H. Templeton and A. Zalkin, *J. Am. Chem. Soc.*, 1980, *102*, 620.
51. H. I. Schlesinger, H. C. Brown, L. Horvitz, A. D. Bond, L. D. Tuck and A. C. Walker, *J. Am. Chem. Soc.*, 1953, *75*, 222.
52. (a) R. Shinomoto, E. Gamp, N. M. Edelstein and D. H. Templeton, *Inorg. Chem.*, 1983, *22*, 2351; (b) W. K. Kot and N. M. Edelstein, *New J. Chem.*, 1995, *19*, 641.
53. (a) A. Zalkin, R. R. Rietz, D. H. Templeton and N. M. Edelstein, *Inorg. Chem.*, 1978, *17*, 661; (b) R. R. Rietz, N. M. Edelstein, H. W. Ruben, D. H. Templeton and A. Zalkin, *Inorg. Chem.*, 1978, *17*, 661; (c) R. R. Rietz, A. Zalkin, D. H. Templeton, N. M. Edelstein and L. K. Templeton, *Inorg. Chem.*, 1978, *17*, 653.
54. D. Baudry and M. Ephritikhine, *J. Organomet. Chem.*, 1988, *349*, 123.
55. (a) D. Baudry, P. Charpin, M. Ephritikhine, M. Lance, M. Nierlich and J. Vigner, *J. Chem. Soc. Chem. Commun.*, 1987, 739; (b) S. R. Daly, M. Ephritikhine and G. S. Girolami, *Polyhedron*, 2012, *33*, 41.
56. (a) R. Shinomoto, A. Zalkin, N. M. Edelstein and D. Zhang, *Inorg. Chem.*, 1987, *26*, 2868; (b) R. Shinomoto, A. Zalkin and N. M. Edelstein, *Inorg. Chim. Acta*, 1987, *139*, 91.
57. J. Brennan, R. Shinomoto, A. Zalkin and N. M. Edelstein, *Inorg. Chem.*, 1984, *23*, 4143.
58. J. E. Gozum, S. R. Wilson and G. S. Girolami, *J. Am. Chem. Soc.*, 1992, *114*, 9483.
59. G. V. Fazakerley, G. Folcher and H. Marquetellis, *Polyhedron*, 1984, *3*, 457.
60. D. Baudry, E. Bulot, P. Charpin, M. Ephritikhine, M. Lance, M. Nierlich and J. Vigner, *J. Organomet. Chem.*, 1989, *371*, 155.

61. T. Arliguie, L. Belkhiri, S. E. Bouaoud, P. Thuery, C. Villiers, A. Boucekkine and M. Ephritikhine, *Inorg. Chem.*, 2009, *48*, 221.
62. P. L. Arnold, C. J. Stevens, J. L. Farnaby, M. G. Gardiner, G. S. Nichol and J. B. Love, *J. Am. Chem. Soc.*, 2014, *136*, 10218.
63. J. F. Le Marechal, M. Ephritikhine and G. Folcher, *J. Organomet. Chem.*, 1986, *309*, C1.
64. (a) H. J. Wasserman, D. C. Moody and R. R. Ryan, *J. Chem. Soc. Chem. Commun.*, 1984, 532; (b) H. J. Wasserman, D. C. Moody, R. T. Paine, R. R. Ryan and K. V. Salazar, *J. Chem. Soc. Chem. Commun.*, 1984, 533.
65. S. R. Daly, P. M. B. Piccoli, A. J. Schultz, T. K. Todorova, L. Gagliardi and G. S. Girolami, *Angew. Chem. Int. Ed.*, 2010, *49*, 3379.
66. S. R. Daly and G. S. Girolami, *Chem. Commun.*, 2010, *46*, 407.
67. S. R. Daly and G. S. Girolami, *Inorg. Chem.*, 2010, *49*, 5157.
68. B. Vlaisavljevich, P. Miro, D. Koballa, T. K. Todorova, S. R. Daly, G. S. Girolami, C. J. Cramer and L. Gagliardi, *J. Phys. Chem. C*, 2012, *116*, 23194.
69. (a) H. Gilman; R. G. Jones; E. Bindschadler; D. Blume; G. Karmas; G. A. Martin; J. F. Nobis Jr, J. R. Thirtle, H. L. Yale and F. A. Yoeman, *J. Am. Chem. Soc.*, 1956, *78*, 2790; (b) T. J. Marks and A. M. Seyam, *J. Organomet. Chem.*, 1974, *67*, 61.
70. E. R. Sigurdson and G. Wilkinson, *Dalton Trans.*, 1977, 812.
71. E. Ciliberto, G. Condorelli, P. J. Fagan, J. M. Manriquez, I. Fragala and T. J. Marks, *J. Am. Chem. Soc.*, 1981, *103*, 4755.
72. H. Lauke, P. J. Swepston and T. J. Marks, *J. Am. Chem. Soc.*, 1984, *106*, 6841.
73. A. Bucaille, T. Le Borgne, M. Ephritikhine and J. C. Daran, *Organometallics*, 2000, *19*, 4912.
74. S. Fortier, B. C. Melot, G. Wu and T. W. Hayton, *J. Am. Chem. Soc.*, 2009, *131*, 15512.
75. L. A. Seaman, J. R. Walensky, G. Wu and T. W. Hayton, *Inorg. Chem.*, 2013, 3556.
76. E. A. Pedrick, P. Hrobarik, L. A. Seaman, G. Wu and T. W. Hayton, *Chem. Commun.*, 2016, *52*, 689.
77. W. G. Vandersluys, C. J. Burns and A. P. Sattelberger, *Organometallics*, 1989, *8*, 855.
78. S. J. Kraft, P. E. Fanwick and S. C. Bart, *J. Am. Chem. Soc.*, 2012, *134*, 6160.
79. S. Fortier, J. R. Walensky, G. Wu and T. W. Hayton, *J. Am. Chem. Soc.*, 2011, *133*, 11732.
80. C. A. Cruz, D. J. H. Emslie, L. E. Harrington, J. F. Britten and C. M. Robertson, *Organometallics*, 2007, *26*, 692.

81. E. Köhler, W. Brüser and K.-H. Thiele, *J. Organomet. Chem.*, 1974, *76*, 235.
82. M. S. Eisen and T. J. Marks, *J. Am. Chem. Soc.*, 1992, *114*, 10358.
83. K.-H. Thiele, R. Opitz and E. Köhler, *Z. Anorg. Allg. Chem.*, 1977, *435*, 45.
84. L. A. Seaman, E. A. Pedrick, T. Tsuchiya, G. Wu, E. Jakubikova and T. W. Hayton, *Angew. Chem. Int. Ed.*, 2013, *52*, 10589.
85. A. C. Behrle, A. J. Myers, P. Rungthanaphatsophon, W. W. Lukens, C. L. Barnesa and J. R. Walensky, *Chem. Commun.*, 2016, *52*, 14373.
86. P. G. Edwards, R. A. Andersen and A. Zalkin, *J. Am. Chem. Soc.*, 1981, *103*, 7792.
87. P. G. Edwards, R. A. Andersen and A. Zalkin, *Organometallics*, 1984, *3*, 293.
88. S. Duhović, S. Khan and P. L. Diaconescu, *Chem. Commun.*, 2010, *46*, 3390.
89. (a) D. L. Clark, J. C. Gordon, P. J. Hay, R. L. Martin and R. Poli, *Organometallics*, 2002, *21*, 5000; (b) L. Perrin, L. Maron, O. Eisenstein and M. F. Lappert, *New J. Chem.*, 2003, *27*, 121.
90. B. D. Zwick, A. P. Sattelberger and L. R. Avens, In *Transuranium Elements: A Half Century*, L. R. Morss, J. Fuger, Eds.; American Chemical Society, Washington, DC: 1992; 239.
91. L. A. Seaman, P. Hrobarik, M. F. Schettini, S. Fortier, M. Kaupp and T. W. Hayton, *Angew. Chem. Int. Ed.*, 2013, *52*, 3259.
92. (a) A. M. Seyam, *Inorg. Chim. Acta*, 1985, *110*, 123; (b) A. M. Seyam, *Inorg. Chim. Acta*, 1982, *58*, 71.
93. G. Wilke, B. Bogdanovic, P. Hardt, P. Heimbach, W. Keim, M. Kroner, W. Oberkirch, K. Tanaka, E. Steinrucke, D. Walter and H. Zimmermann, *Angew. Chem. Int. Ed. Engl.*, 1966, *5*, 151.
94. (a) G. Lugli, W. Marconi, A. Mazzei, N. Paladino and U. Pedretti, *Inorg. Chim. Acta*, 1969, *3*, 253; (b) M. Brunelli, G. Lugli and G. Giacometti, *J. Magn. Reson.*, 1973, *9*, 247.
95. M. Brunelli, G. Perego, G. Lugli and A. Mazzei, *Dalton Trans.*, 1979, 861.
96. C. N. Carlson, T. P. Hanusa and W. W. Brennessel, *J. Am. Chem. Soc.*, 2004, *126*, 10550.
97. K. C. Jantunen, F. Haftbaradaran, M. J. Katz, R. J. Batchelor, G. Schatte and D. B. Leznoff, *Dalton Trans.*, 2005, 3083.
98. J.-C. Tourneux, J.-C. Berthet, P. Thuery, N. Mezailles, P. Le Floch and M. Ephritikhine, *Dalton Trans.*, 2010, *39*, 2494.
99. E. M. Matson, W. P. Forrest, P. E. Fanwick and S. C. Bart, *Organometallics*, 2012, *31*, 4467.

100. C. A. Cruz, T. Chu, D. J. H. Emslie, H. A. Jenkins, L. E. Harrington and J. F. Britten, *J. Organomet. Chem.*, 2010, *695*, 2798.
101. W. S. Ren, G. F. Zi, D. C. Fang and M. D. Walter, *Chem. Eur. J.*, 2011, *17*, 12669.
102. A. A. Danopoulos, D. M. Hankin, S. M. Cafferkey and M. B. Hursthouse, *Dalton Trans.*, 2000, 1613.
103. K. C. Jantunen, R. J. Batchelor and D. B. Leznoff, *Organometallics*, 2004, *23*, 2186.
104. P. Gradoz, D. Baudry, M. Ephritikhine, M. Lance, M. Nierlich and J. Vigner, *J. Organomet. Chem.*, 1994, *466*, 107.
105. C. L. Webster, J. E. Bates, M. Fang, J. W. Ziller, F. Furche and W. J. Evans, *Inorg. Chem.*, 2013, *52*, 3565.
106. W. J. Evans, G. W. Nyce, K. J. Forrestal and J. W. Ziller, *Organometallics*, 2002, *21*, 1050.
107. W. J. Evans, J. R. Walensky, F. Furche, J. W. Ziller, A. G. DiPasquale and A. L. Rheingold, *Inorg. Chem.*, 2008, *47*, 10169.
108. W. G. Van Der Sluys, C. J. Burns and A. P. Sattelberger, *Organometallics*, 1989, *8*, 855.
109. W. J. Evans, S. A. Kozimor, J. W. Ziller and N. Kaltsoyannis, *J. Am. Chem. Soc.*, 2004, *126*, 14533.
110. W. J. Evans, C. A. Traina and J. W. Ziller, *J. Am. Chem. Soc.*, 2009, *131*, 17473.
111. W. E. Piers and D. J. H. Emslie, *Coord. Chem. Rev.*, 2002, *233*, 131.
112. J. M. Manriquez, P. J. Fagan and T. J. Marks, *J. Am. Chem. Soc.*, 1978, *100*, 3939.
113. P. J. Fagan, J. M. Manriquez, E. A. Maatta, A. M. Seyam and T. J. Marks, *J. Am. Chem. Soc.*, 1981, *103*, 6650.
114. J. K. Pagano, J. M. Dorhout, R. Waterman, K. R. Czerwinski and J. L. Kiplinger, *Chem. Commun.*, 2015, *51*, 17379.
115. C. L. Webster, J. W. Ziller and W. J. Evans, *Organometallics*, 2014, *33*, 433.
116. E. Montalvo, K. A. Miller, J. W. Ziller and W. J. Evans, *Organometallics*, 2010, *29*, 4159.
117. W. J. Evans, K. A. Miller, A. G. DiPasquale, A. L. Rheingold, T. J. Stewart and R. Bau, *Angew. Chem. Int. Ed.*, 2008, *47*, 5075.
118. W. W. Lukens Jr., S. M. Beshouri, L. L. Blosch and R. A. Andersen, *J. Am. Chem. Soc.*, 1996, *118*, 901.
119. F. Ossola, N. Brianese, M. Porchia, G. Rossetto and P. Zanella, *J. Chem. Soc. Dalton Trans.*, 1990, 877.
120. P. C. Leverd, T. Arliguie, M. Lance, M. Nierlich, J. Vigner and M. Ephritikhine, *J. Chem. Soc. Dalton Trans.*, 1994, 501.

121. D. Baudry, P. Charpin, M. Ephritikhine, G. Folcher, J. Lambard, M. Lance, M. Nierlich and J. Vigner, *J. Chem. Soc. Chem. Commun.*, 1985, 1553.
122. P. L. Arnold, J. H. Farnaby, M. G. Gardiner and J. B. Love, *Organometallics*, 2015, *34*, 2114.
123. F. Ossola, G. Rossetto, P. Zanella, G. Paolucci and R. D. Fischer, *J. Organomet. Chem.*, 1986, *309*, 55.
124. J.-C. Berthet, C. Boisson, M. Lance, J. Vigner, M. Nierlich and M. Ephritikhine, *J. Chem. Soc. Dalton Trans.*, 1995, 3027.
125. B. D. Stubbert and T. J. Marks, *J. Am. Chem. Soc.*, 2007, *129*, 4253.
126. J. C. Green and O. Watts, *J. Organomet. Chem.*, 1978, *153*, C40.
127. R. K. Thomson, C. R. Graves, B. L. Scott and J. L. Kiplinger, *Dalton Trans.*, 2010, *39*, 6826.
128. (a) I. Castro-Rodriguez, H. Nakai and K. Meyer, *Angew. Chem. Int. Ed.*, 2006, *45*, 2389; (b) K. C. Mullane, A. J. Lewis, H. Yin, P. J. Carroll and E. J. Schelter, *Inorg. Chem.*, 2014, *53*, 9129.
129. (a) S. J. Kraft, P. E. Fanwick and S. C. Bart, *Organometallics*, 2013, *32*, 3279; (b) N. A. Siladke, J. LeDuc, J. W. Ziller and W. J. Evans, *Chem. Eur. J.*, 2012, *18*, 14820.
130. C. Boisson, J.-C. Berthet, M. Lance, M. Nierlich, J. Vigner and M. Ephritikhine, *J. Chem. Soc. Chem. Commun.*, 1995, 543.
131. W. J. Evans, J. R. Walensky and J. W. Ziller, *Inorg. Chem.*, 2010, *49*, 1743.
132. W. J. Evans, G. W. Nyce, M. A. Johnston and J. W. Ziller, *J. Am. Chem. Soc.*, 2000, *122*, 12019.
133. M. Weydert and R. A. Andersen, New Synthetic Routes to Cyclopentadienyluranium(IV) Fluorides — Redox and Atom-Abstraction Reactions, in *ACS Symposium Series: Inorganic Fluorine Chemistry*, Vol. 555, 1994, 383.
134. (a) R. K. Thomson, C. R. Graves, B. L. Scott and J. L. Kiplinger, *Eur. J. Inorg. Chem.*, 2009, 1451; (b) R. K. Thomson, T. Cantat, B. L. Scott, D. E. Morris, E. R. Batista and J. L. Kiplinger, *Nature Chem.*, 2010, *2*, 723.
135. E. M. Matson, S. M. Franke, N. H. Anderson, T. D. Cook, P. E. Fanwick and S. C. Bart, *Organometallics*, 2014, *33*, 1964.
136. W. J. Evans, K. A. Miller, S. A. Kozimor, J. W. Ziller, A. G. DiPasquale and A. L. Rheingold, *Organometallics*, 2007, *26*, 3568.
137. W. J. Evans, J. R. Walensky and J. W. Ziller, *Organometallics*, 2010, *29*, 101.
138. E. Montalvo, J. W. Ziller, A. G. DiPasquale, A. L. Rheingold and W. J. Evans, *Organometallics*, 2010, *29*, 2104.

139. W. J. Evans, S. A. Kozimor and J. W. Ziller, *Inorg. Chem.*, 2005, *44*, 7960.
140. M. K. Takase, J. W. Ziller and W. J. Evans, *Chem. Eur. J.*, 2011, *17*, 4871.
141. A. J. Gaunt, B. L. Scott and M. P. Neu, *Inorg. Chem.*, 2006, *45*, 7401.
142. (a) C. C. Chang, N. K. Sungyu, C. S. Hseu and C. T. Chang, *Inorg. Chem.*, 1979, *18*, 885; (b) N. K. Sungyu, F. F. Hsu, C. C. Chang, G. R. Her and C. T. Chang, *Inorg. Chem.*, 1981, *20*, 2727.
143. (a) E. O. Fischer and A. Treiber, *Z. Naturforschung*, 1962, *17b*, 276; (b) R. Maier, B. Kanellakopulos, C. Apostolidis, D. Meyer and J. Rebizant, *J. Alloy. Compd.*, 1993, *190*, 269.
144. (a) E. O. Fischer and Y. Hristidu, *Z. Naturforschung*, 1962, *17b*, 275; (b) J. H. Burns, *J. Am. Chem. Soc.*, 1973, *95*, 3815.
145. F. Baumgärtner, E. O. Fischer, B. Kanellakopulos and P. Laubereau, *Angew. Chem. Int. Ed. Engl.*, 1968, *7*, 634.
146. (a) J. L. Calderon, F. A. Cotton, B. G. DeBoer and J. J. Takats, *J. Am. Chem. Soc.*, 1971, *93*, 3592; (b) R. D. Rogers, R. V. Bynum and J. L. Atwood, *J. Am. Chem. Soc.*, 1981, *103*, 692.
147. V. I. Kulishov, N. G. Bokii and Y. T. Struchkov, *J. Struct. Chem.*, 1970, *11*, 700.
148. J. Rebizant, M. R. Spirlet, B. Kanellakopulos and E. Dornberger, *J. Less-Common Met.*, 1986, *122*, 211.
149. D. G. Karraker and J. A. Stone, *Inorg. Chem.*, 1979, *11*, 1742.
150. M. J. Tassell and N. Kaltsoyannis, *Dalton Trans.*, 2010, *39*, 6719.
151. M. L. Neidig, D. L. Clark and R. L. Martin, *Coord. Chem. Rev.*, 2013, *257*, 394.
152. J. M. Birmingham and G. Wilkinson, *J. Am. Chem. Soc.*, 1954, *76*, 6210.
153. L. T. Reynolds and G. Wilkinson, *Inorg. Nucl. Chem.*, 1956, *2*, 246.
154. M. Weydert, J. G. Brennan, R. A. Andersen and R. G. Bergman, *Organometallics*, 1995, *14*, 3942.
155. T. J. Marks, A. M. Seyam and J. R. Kolb, *J. Am. Chem. Soc.*, 1973, *95*, 5529.
156. A. E. Gebala and M. Tsutsui, *J. Am. Chem. Soc.*, 1973, *95*, 91.
157. C. Villiers and M. Ephritikhine, *J. Organomet. Chem.*, 1990, *393*, 339.
158. J.-C. Berthet, J. F. Lemarechal, M. Lance, M. Nierlich, J. Vigner and M. Ephritikhine, *J. Chem. Soc. Dalton Trans.*, 1992, 1573.
159. P. C. Leverd, M. Ephritikhine, M. Lance, J. Vigner and M. Nierlich, *J. Organomet. Chem.*, 1996, *507*, 229.
160. (a) P. Zanella, F. Ossola, M. Porchia, G. Rossetto, A. C. Villa and C. Guastini, *J. Organomet. Chem.*, 1987, *323*, 295; (b) M. Porchia,

N. Brianese, F. Ossola, G. Rossetto and P. Zanella, *J. Chem. Soc. Dalton Trans.*, 1987, 691.
161. F. Ossola, N. Brianese, M. Porchia, G. Rossetto and P. Zanella, *J. Organomet. Chem.*, 1986, *310*, C1.
162. (a) M. Porchia, N. Brianese, U. Casellato, F. Ossola, G. Rossetto, P. Zanella and R. Graziani, *J. Chem. Soc. Dalton Trans.*, 1989, 677; (b) M. Porchia, F. Ossola, G. Rossetto, P. Zanella and N. Brianese, *J. Chem. Soc. Chem. Commun.*, 1987, 550.
163. M. Porchia, U. Casellato, F. Ossola, G. Rossetto, P. Zanella and R. Graziani, *J. Chem. Soc. Chem. Commun.*, 1986, 1034.
164. F. Ossola, N. Brianese, M. Porchia, G. Rossetto and P. Zanella, *Inorg. Synth.*, 1992, *29*, 234.
165. G. Paolucci, G. Rossetto, P. Zanella and R. D. Fischer, *J. Organomet. Chem.*, 1985, *284*, 213.
166. (a) P. Zanella, N. Brianese, U. Casellato, F. Ossola, M. Porchia, G. Rossetto and R. Graziani, *Dalton Trans.*, 1987, *8*, 2039; (b) P. Zanella, G. Paolucci, G. Rossetto, F. Benetollo, A. Polo, R. D. Fischer and G. Bombieri, *Chem. Commun.*, 1985, *2*, 96.
167. (a) D. C. Sonnenberger, E. A. Mintz and T. J. Marks, *J. Am. Chem. Soc.*, 1984, *106*, 3484; (b) C. Villiers and M. Ephritikhine, *J. Chem. Soc. Dalton Trans.*, 1994, 3397.
168. M. Weydert, R. A. Andersen and R. G. Bergman, *J. Am. Chem. Soc.*, 1993, *115*, 8837.
169. R. E. Cramer, M. A. Bruck, F. Edelmann, D. Afzal, J. W. Gilje and H. Schmidbaur, *Chem. Ber.*, 1988, *121*, 417.
170. R. E. Cramer, J. Hitt, T. Chung and J. W. Gilje, *New J. Chem.*, 1995, *19*, 509.
171. R. E. Cramer, R. B. Maynard, J. C. Paw and J. W. Gilje, *Organometallics*, 1982, *1*, 869.
172. R. E. Cramer, K. Panchanatheswaran and J. W. Gilje, *Angew. Chem. Int. Ed. Engl.*, 1984, *23*, 912.
173. R. E. Cramer, K. Panchanatheswaran and J. W. Gilje, *J. Am. Chem. Soc.*, 1984, *106*, 1853.
174. R. E. Cramer, H. J. Jong and J. W. Gilje, *Organometallics*, 1987, *6*, 2010.
175. R. E. Cramer, J. H. Jeong and J. W. Gilje, *Organometallics*, 1986, *5*, 2555.
176. (a) R. E. Cramer, K. T. Higa and J. W. Gilje, *J. Am. Chem. Soc.*, 1984, *106*, 7245; (b) R. E. Cramer, K. T. Higa and J. W. Gilje, *Organometallics*, 1985, *4*, 1140.
177. R. E. Cramer, J. H. Jeong, P. N. Richmann, J. W. Gilje, *Organometallics*, 1990, *9*, 1141.

178. W. J. Evans, S. A. Kozimor and J. W. Ziller, *Organometallics*, 2005, *24*, 3407.
179. W. J. Evans, G. W. Nyce and J. W. Ziller, *Organometallics*, 2001, *20*, 5489.
180. W. H. Gordon E. C. Baker and K. N. Raymond, *J. Am. Chem. Soc.*, 1975, *97*, 3049.
181. E. C. Baker, K. N. Raymond, T. J. Marks and W. A. Wachter, *J. Am. Chem. Soc.*, 1974, *96*, 7586.
182. T. J. Marks and W. A. Wachter, *J. Am. Chem. Soc.*, 1976, *98*, 703.
183. J. W. Bruno, D. G. Kalina, E. A. Mintz and T. J. Marks, *J. Am. Chem. Soc.*, 1982, *104*, 1860.
184. R. D. Ernst, W. J. Kennelly, C. S. Day, V. W. Day and T. J. Marks, *J. Am. Chem. Soc.*, 1979, *101*, 2656.
185. (a) A. Zalkin, J. G. Brennan and R. A. Andersen, *Acta Crystallogr. C*, 1987, *43*, 418; (b) A. Zalkin, J. G. Brennan and R. A. Andersen, *Acta Crystallogr. C*, 1987, *43*, 421.
186. R. W. Broach, A. J. Schultz, J. M. Williams, G. M. Brown, J. M. Manriquez, P. J. Fagan and T. J. Marks, *Science*, 1979, *203*, 172.
187. C. L. Webster, J. W. Ziller and W. J. Evans, *Organometallics*, 2012, *31*, 7191.
188. (a) W. W. Lukens, S. M. Beshouri, L. L. Blosch, A. L. Stuart and R. A. Andersen, *Organometallics*, 1999, *18*, 1235; (b) W. Ren, N. Zhao, L. Chen and G. F. Zi, *Inorg. Chem. Commun.*, 2013, *30*, 26.
189. G. F. Zi, L. Jia, E. L. Werkema, M. D. Walter, J. P. Gottfriedsen and R. A. Andersen, *Organometallics*, 2005, *24*, 4251.
190. G. F. Zi, L. L. Blosch, L. Jia and R. A. Andersen, *Organometallics*, 2005, *24*, 4602.
191. A. K. Dash, I. Gourevich, J. Q. Wang, J. X. Wang, M. Kapon and M. S. Eisen, *Organometallics*, 2001, *20*, 5084.
192. C. M. Fendrick, L. D. Schertz, V. W. Day and T. J. Marks, *Organometallics*, 1988, *7*, 1828.
193. B. D. Stubbert, C. L. Stern and T. J. Marks, *Organometallics*, 2003, *22*, 4836.
194. S. D. Wobser and T. J. Marks, *Organometallics*, 2013, *32*, 2517.
195. (a) J. Maynadié, J.-C. Berthet, P. Thuéry and M. Ephritikhine, *J. Am. Chem. Soc.*, 2006, *128*, 1082; (b) J.-C. Berthet, J. Maynadie, P. Thuéry and M. Ephritikhine, *Dalton Trans.*, 2010, *39*, 6801.
196. J. Maynadie, J. C. Berthet, P. Thuéry and M. Ephritikhine, *Organometallics*, 2007, *26*, 4585.

197. K. C. Jantunen, C. J. Burns, I. Castro-Rodriguez, R. E. Da Re, J. T. Golden, D. E. Morris, B. L. Scott, F. L. Taw and J. L. Kiplinger, *Organometallics*, 2004, *23*, 4682.
198. W. J. Evans, N. A. Siladke and J. W. Ziller, *Chem. Eur. J.*, 2010, *16*, 796.
199. J. M. Manriquez, P. J. Fagan, T. J. Marks, C. S. Day and V. W. Day, *J. Am. Chem. Soc.*, 1978, *100*, 7112.
200. J. A. Pool, B. L. Scott and J. L. Kiplinger, *J. Am. Chem. Soc.*, 2005, *127*, 1338.
201. C. M. Fendrick and T. J. Marks, *J. Am. Chem. Soc.*, 1986, *108*, 425.
202. H. Braunschweig, A. Gackstatter, T. Kupfer, K. Radacki, S. Franke, K. Meyer, F. Fucke and M. H. Lemee-Cailleau, *Inorg. Chem.*, 2015, *54*, 8022.
203. W. S. Ren, E. W. Zhou, B. Fang, G. F. Zi, D. C. Fang and M. D. Walter, *Chem. Sci.*, 2014, *5*, 3165.
204. E. W. Zhou, W. S. Ren, G. H. Hou, G. F. Zi, D. C. Fang and M. D. Walter, *Organometallics*, 2015, *34*, 3637.
205. L. Doretti, P. Zanella, G. Faraglia and S. Faleschini, *J. Organomet. Chem.*, 1972, *43*, 339.
206. K. W. Bagnall, M. J. Plews and D. Brown, *J. Less-Common Met.*, 1983, *90*, 29.
207. D. Baudry, P. Dorion and M. Ephritikhine, *J. Organomet. Chem.*, 1988, *356*, 165.
208. (a) M. A. Edelman and M. F. Lappert, *Inorg. Chim. Acta*, 1987, *139*, 185; (b) M. A. Edelman, P. B. Hitchcock, J. Hu and M. F. Lappert, *New J. Chem.*, 1995, *19*, 481.
209. E. A. Mintz, K. G. Moloy, T. J. Marks and V. W. Day, *J. Am. Chem. Soc.*, 1982, *104*, 4692.
210. R. J. Butcher, D. L. Clark, S. K. Grumbine, B. L. Scott and J. G. Watkin, *Organometallics*, 1996, *15*, 1488.
211. T. H. Cymbaluk, R. D. Ernst and V. W. Day, *Organometallics*, 1983, *2*, 963.
212. J. W. Bruno, G. M. Smith, T. J. Marks, C. K. Fair, A. J. Schultz and J. M. Williams, *J. Am. Chem. Soc.*, 1986, *108*, 40.
213. J. W. Bruno, T. J. Marks and L. R. Morss, *J. Am. Chem. Soc.*, 1983, *105*, 6824.
214. D. C. Sonnenberger, L. R. Morss and T. J. Marks, *Organometallics*, 1985, *4*, 352.
215. J. W. Bruno, H. A. Stecher, L. R. Morss, D. C. Sonnenberger and T. J. Marks, *J. Am. Chem. Soc.*, 1986, *108*, 7275.
216. S. P. Nolan, M. Porchia and T. J. Marks, *Organometallics*, 1991, *10*, 1450.
217. J. A. M. Simões and J. L. Beauchamp, *Chem. Rev.*, 1990, *90*, 629.

218. G. Ionova, C. Madic and R. Guillaumont, *Polyhedron*, 1998, *17*, 1991.
219. (a) S. Kihara, Z.Yoshida, H. Aoyagi, K. Maeda, O. Shirai, Y. Kitatsuji and Y. Yoshida, *Pure Appl. Chem.*, 1999, *71*, 1771; (b) S. G. Bratsch, *J. Phys. Chem. Ref. Data*, 1989, *18*, 1.
220. (a) D. C. Sonnenberger and J. G. Gaudiello, *Inorg. Chem.*, 1988, *27*, 2747; (b) J. R. Aranzaes, M.-C. Daniel and D. Astruc, *Can. J. Chem.*, 2006, *84*, 288.
221. B. Kanellakopulos, E. Fischer, E. Dornberger and F. Baumgirmer, *J. Organomet. Chem.*, 1970, *24*, 507.
222. M. S. Dutkiewicz, C. Apostolidis, O. Walter and P. L. Arnold, *Chem. Sci.*, 2017, *8*, 2553.
223. F. Baumgartner, E. Fischer, B. Kanellakopulos and P. Laubereau, *Angew. Chem. Int. Ed. Engl.*, 1965, *4*, 878.
224. L. R. Crisler, W. G. Eggerman and *J. Inorg. Nucl. Chem.*, 1974, *36*, 1424.
225. D. G. Kalina, T. J. Marks and W. A. Wachter, *J. Am. Chem. Soc.*, 1977, *99*, 3877.
226. I. Kirker and N. Kaltsoyannis, *Dalton Trans.*, 2011, *40*, 124.
227. P. C. Blake, N. M. Edelstein, P. B. Hitchcock, W. K. Kot, M. F. Lappert, G. V. Shalimoff and S. Tian, *J. Organomet. Chem.*, 2001, *636*, 124.
228. C. J. Windorff, M. R. MacDonald, K. R. Meihaus, J. W. Ziller, J. R. Long and W. J. Evans, *Chem. Eur. J.*, 2016, *22*, 772.
229. (a) J. F. Lemarechal, E. Bulot, D. Baudry, M. Ephritikhine, D. Hauchard and R. Godard, *J. Organomet. Chem.*, 1988, *354*, C17; (b) J. F. Lemarechal, C. Villiers, P. Charpin, M. Nierlich, M. Lance, J. Vigner and M. Ephritikhine, *J. Organomet. Chem.*, 1989, *379*, 259; (c) N. A. Siladke, J. W. Ziller and W. J. Evans, *Z. Anorg. Allg. Chem.*, 2010, *636*, 2347.
230. J. R. Walensky, R. L. Martin, J. W. Ziller and W. J. Evans, *Inorg. Chem.*, 2010, *49*, 10007.
231. R. R. Langeslay, M. E. Fieser, J. W. Ziller, F. Furche and W. J. Evans, *Chem. Sci.*, 2015, *6*, 517.
232. Q. Y. Wu, J. H. Lan, C. Z. Wang, Z. P. Cheng, Z. F. Chai, J. K. Gibson and W. Q. Shi, *Dalton Trans.*, 2016, *45*, 3102.
233. R. J. M. Konings, L. R. Morss and J. Fuger, Thermodynamic Properties of Actinides and Actinide Compounds, in *The Chemistry of the Transactinide Elements* (Eds. L. R. Morss and N. M. Edelstein, J. Fuger, 4th Edn., Vol. 4, Ch. 19, Springer, Dordrecht, 2006, p. 2113.
234. M. R. MacDonald, M. E. Fieser, J. E. Bates, J. W. Ziller, F. Furche and W. J. Evans, *J. Am. Chem. Soc.*, 2013, *135*, 13310.

235. E. J. Schelter, J. M. Veauthier, C. R. Graves, K. D. John, B. L. Scott, J. D. Thompson, J. A. Pool-Davis-Tournear, D. E. Morris and J. L. Kiplinger, *Chem. Eur. J.*, 2008, *14*, 7782.
236. J. L. Kiplinger, K. D. John, D. E. Morris, B. L. Scott and C. J. Burns, *Organometallics*, 2002, *21*, 4306.
237. C. R. Graves, B. L. Scott, D. E. Morris and J. L. Kiplinger, *J. Am. Chem. Soc.*, 2007, *129*, 11914.
238. C. R. Graves, P. Yang, S. A. Kozimor, A. E. Vaughn, D. L. Clark, S. D. Conradson, E. J. Schelter, B. L. Scott, J. D. Thompson, P. J. Hay, D. E. Morris and J. L. Kiplinger, *J. Am. Chem. Soc.*, 2008, *130*, 5272.
239. C. R. Graves, A. E. Vaughn, E. J. Schelter, B. L. Scott, J. D. Thompson, D. E. Morris and J. L. Kiplinger, *Inorg. Chem.*, 2008, *47*, 11879.
240. C. R. Graves, B. L. Scott, D. E. Morris and J. L. Kiplinger, *Organometallics*, 2008, *27*, 3335.
241. C. R. Graves, B. L. Scott, D. E. Morris and J. L. Kiplinger, *Chem. Commun.*, 2009, 776.
242. D. S. J. Arney, C. J. Burns and D. C. Smith, *J. Am. Chem. Soc.*, 1992, *114*, 10068.
243. B. P. Warner, B. L. Scott and C. J. Burns, *Angew. Chem. Int. Ed.*, 1998, *37*, 959.
244. L. Zhang, G. H. Hou, G. F. Zi, W. J. Ding and M. D. Walter, *J. Am. Chem. Soc.*, 2016, *138*, 5130.
245. L. P. Spencer, R. L. Gdula, T. W. Hayton, B. L. Scott and J. M. Boncella, *Chem. Commun.*, 2008, 4986.
246. D. S. J. Arney and C. J. Burns, *J. Am. Chem. Soc.*, 1993, *115*, 9840.
247. J. Maynadié, J.-C. Berthet, P. Thuéry and M. Ephritikhine, *Chem. Commun.*, 2007, 486.
248. (a) D. P. Cladis, J. J. Kiernicki, P. E. Fanwick and S. C. Bart, *Chem. Commun.*, 2013, *49*, 4169; (b) J. J. Kiernicki, D. P. Cladis, P. E. Fanwick, M. Zeller and S. C. Bart, *J. Am. Chem. Soc.*, 2015, *137*, 11115; (c) J. J. Kiernicki, M. G. Ferrier, J. S. Pacheco, H. S. La Pierre, B. W. Stein, M. Zeller, S. A. Kozimor and S. C. Bart, *J. Am. Chem. Soc.*, 2016, *138*, 13941.
249. N. Tsoureas, A. F. R. Kilpatrick, C. J. Inman and F. G. N. Cloke, *Chem. Sci.*, 2016, *7*, 4624.
250. W. J. Evans, S. A. Kozimor and J. W. Ziller, *Science*, 2005, *309*, 1835.
251. D. Patel, J. McMaster, W. Lewis, A. J. Blake and S. T. Liddle, *Nature Commun.*, 2013, *4*, 2323.
252. F. H. Stephens, PhD Thesis, Massachusetts Institute of Technology, 2004.

253. A. S. P. Frey, F. G. N. Cloke, P. B. Hitchcock and J. C. Green, *New J. Chem.*, 2011, *35*, 2022.
254. A. Formanuik, F. Ortu, R. Beekmeyer, A. Kerridge, R. W. Adams and D. P. Mills, *Dalton Trans.*, 2016, *45*, 2390.
255. P. Gradoz, C. Boisson, D. Baudry, M. Lance, M. Nierlich, J. Vigner and M. Ephritikhine, *Chem. Commun.*, 1992, 1720.
256. P. Gradoz, D. Baudry, M. Ephritikhine, F. Nief and F. Mathey, *Dalton Trans.*, 1992, 3047.
257. P. Gradoz, M. Ephritikhine, M. Lance, J. Vigner and M. Nierlich, *J. Organomet. Chem.*, 1994, *481*, 69.
258. (a) S. M. Cendrowski-Guillaume, M. Nierlich and M. Ephritikhine, *J. Organomet. Chem.*, 2002, *643*, 209; (b) S. M. Cendrowski-Guillaume, G. Le Gland, M. Nierlich and M. Ephritikhine, *Eur. J. Inorg. Chem.*, 2003, 1388.
259. T. Arliguie, M. Ephritikhine, M. Lance and M. Nierlich, *J. Organomet. Chem.*, 1996, *524*, 293.
260. D. Baudry, M. Ephritikhine, F. Nief, L. Ricard and F. Mathey, *Angew. Chem. Int. Ed. Engl.*, 1990, *29*, 1485.
261. G. K. B. Clentsmith, F. G. N. Cloke, M. D. Francis, J. R. Hanks, P. B. Hitchcock and J. F. Nixon, *J. Organomet. Chem.*, 2008, *693*, 2287.
262. B. M. Gardner, F. Tuna, E. J. L. McInnes, J. McMaster, W. Lewis, A. J. Blake and S. T. Liddle, *Angew. Chem. Int. Ed.*, 2015, *54*, 7068.
263. R. J. Kahan, F. G. N. Cloke, S. M. Roe and F. Nief, *New J. Chem.*, 2015, *39*, 7602.
264. (a) A. Athimoolam, S. Gambarotta, I. Korobkov, *Can. J. Chem.*, 2005, *83*, 832; (b) R. J. Batrice, N. Fridman and M. S. Eisen, *Inorg. Chem.*, 2016, *55*, 2998.
265. D. L. Swartz, L. P. Spencer, B. L. Scott, A. L. Odom and J. M. Boncella, *Dalton Trans.*, 2010, *39*, 6841.
266. I. Korobkov, B. Vidjayacoumar, S. I. Gorelsky, P. Billone and S. Gambarotta, *Organometallics*, 2010, *29*, 692.
267. A. Arunachalampillai, P. Crewdson, I. Korobkov and S. Gambarotta, *Organometallics*, 2006, *25*, 3856.
268. (a) I. Korobkov and S. Gambarotta, *Organometallics*, 2004, *23*, 5379; (b) I. Korobkov, S. Gambarotta, G. P. A. Yap, L. Thompson and P. J. Hay, *Organometallics*, 2001, *20*, 5440.
269. I. Korobkov, S. Gambarotta and G. P. A. Yap, *Angew. Chem. Int. Ed.*, 2003, *42*, 814.
270. I. Korobkov, S. Gambarotta and G. P. A. Yap, *Angew. Chem. Int. Ed.*, 2002, *41*, 3433.

271. I. Korobkov, S. Gambarotta and G. P. A. Yap, *Organometallics*, 2001, *20*, 2552.
272. M. S. Dutkiewicz, J. H. Farnaby, C. Apostolidis, E. Colineau, O. Walter, N. Magnani, M. G. Gardiner, J. B. Love, N. Kaltsoyannis, R. Caciuffo and P. L. Arnold, *Nature Chem.*, 2016, *8*, 797.
273. I. Korobkov and S. Gambarotta, *Inorg. Chem.*, 2010, *49*, 3409.
274. F. R. Fronczek, G. W. Halstead and K. N. Raymond, *J. Am. Chem. Soc.*, 1977, *99*, 1769.
275. (a) F. M. de Rege, W. H. Smith, B. L. Scott, J. B. Nielsen and K. D. Abney, *Inorg. Chem.*, 1998, *37*, 3664; (b) D. Rabinovich, C. M. Haswell, B. L. Scott, R. L. Miller, J. B. Nielsen and K. D. Abney, *Inorg. Chem.*, 1996, *35*, 1425.
276. D. Rabinovich, R. M. Chamberlin, B. L. Scott, J. B. Nielsen and K. D. Abney, *Inorg. Chem.*, 1997, *36*, 4216.
277. Z. W. Xie, C. G. Yan, Q. C. Yang and T. C. W. Mak, *Angew. Chem. Int. Ed.*, 1999, *38*, 1761.
278. (a) F. A. Cotton and W. Schwotzer, *Organometallics*, 1987, *6*, 1275; (b) F. A. Cotton, W. Schwotzer and C. Q. Simpson, *Angew. Chem. Int. Ed.*, 1986, *25*, 637; (c) G. C. Campbell, F. A. Cotton, J. F. Haw and W. Schwotzer, *Organometallics*, 1986, *5*, 274; (d) Cotton, F. A. and Schwotzer, W. *Organometallics*, 1985, *4*, 942.
279. D. Baudry, E. Bulot and M. Ephritikhine, *J. Chem. Soc. Chem. Commun.*, 1988, 1369.
280. C. A. Cruz, D. J. H. Emslie, C. M. Robertson, L. E. Harrington, H. A. Jenkins and J. F. Britten, *Organometallics*, 2009, *28*, 1891.
281. N. R. Andreychuk, Ph. D. Thesis, McMaster University, 2017.
282. C. A. Cruz, D. J. H. Emslie, L. E. Harrington and J. F. Britten, *Organometallics*, 2008, *27*, 15.
283. W. J. Evans, S. A. Kozimor, W. R. Hillman and J. W. Ziller, *Organometallics*, 2005, *24*, 4676.
284. S. M. Franke, B. L. Tran, F. W. Heinemann, W. Hieringer, D. J. Mindiola and K. Meyer, *Inorg. Chem.*, 2013, *52*, 10552.
285. J. McKinven, G. S. Nichol and P. L. Arnold, *Dalton Trans.*, 2014, *43*, 17416.
286. S. C. Bart, F. W. Heinemann, C. Anthon, C. Hauser and K. Meyer, *Inorg. Chem.*, 2009, *48*, 9419.
287. D. P. Halter, H. S. La Pierre, F. W. Heinemann and K. Meyer, *Inorg. Chem.*, 2014, *53*, 8418.
288. I. Korobkov, S. Gambarotta and G. P. A. Yap, *Angew. Chem. Int. Ed.*, 2003, *42*, 4958.
289. H. S. La Pierre, H. Kameo, D. P. Halter, F. W. Heinemann and K. Meyer, *Angew. Chem. Int. Ed.*, 2014, *53*, 7154.

290. H. S. La Pierre, A. Scheurer, F. W. Heinemann, W. Hieringer and K. Meyer, *Angew. Chem. Int. Ed.*, 2014, *53*, 7158.
291. P. L. Diaconescu, P. L. Arnold, T. A. Baker, D. J. Mindiola and C. C. Cummins, *J. Am. Chem. Soc.*, 2000, *122*, 6108.
292. B. Vlaisayljevich, P. L. Diaconescu, W. L. Lukens, L. Gagliardi and C. C. Cummins, *Organometallics*, 2013, *32*, 1341.
293. P. L. Arnold, S. M. Mansell, L. Maron and D. McKay, *Nature Chem.*, 2012, *4*, 668.
294. M. J. Monreal, S. I. Khan, J. L. Kiplinger and P. L. Diaconescu, *Chem. Commun.*, 2011, *47*, 9119.
295. D. P. Mills, F. Moro, J. McMaster, J. van Slageren, W. Lewis, A. J. Blake and S. T. Liddle, *Nature Chem.*, 2011, *3*, 454.
296. A. J. Wooles, W. Lewis, A. J. Blake and S. T. Liddle, *Organometallics*, 2013, *32*, 5058.
297. P. L. Diaconescu and C. C. Cummins, *J. Am. Chem. Soc.*, 2002, *124*, 7660.
298. P. L. Diaconescu and C. C. Cummins, *Inorg. Chem.*, 2012, *51*, 2902.
299. S. T. Liddle, *Coord. Chem. Rev.*, 2015, *293*, 211.
300. D. Patel, F. Moro, J. McMaster, W. Lewis, A. J. Blake and S. T. Liddle, *Angew. Chem. Int. Ed.*, 2011, *50*, 10388.
301. V. Mougel, C. Camp, J. Pecaut, C. Coperet, L. Maron, C. E. Kefalidis and M. Mazzanti, *Angew. Chem. Int. Ed.*, 2012, *51*, 12280.
302. C. Camp, V. Mougel, J. Pecaut, L. Maron and M. Mazzanti, *Chem. Eur. J.*, 2013, *19*, 17528.
303. T. Arliguie, M. Lance, M. Nierlich, J. Vigner and M. Ephritikhine, *Chem. Commun.*, 1995, 183.
304. T. Arliguie, M. Lance, M. Nierlich, J. Vigner and M. Ephritikhine, *Chem. Commun.*, 1994, 847.
305. T. Arliguie, M. Lance, M. Nierlich and M. Ephritikhine, *Dalton Trans.*, 1997, 2501.
306. D. Gourier, D. Caurant, T. Arliguie and M. Ephritikhine, *J. Am. Chem. Soc.*, 1998, *120*, 6084.
307. G. Menconi and N. Kaltsoyannis, *Organometallics*, 2005, *24*, 1189.
308. A. Streitwieser Jr. and U. Müller-Westerhoff, *J. Am. Chem. Soc.*, 1968, *90*, 7364.
309. A. Zalkin and K. N. Raymond, *J. Am. Chem. Soc.*, 1969, *91*, 5667.
310. A. Kerridge and N. Kaltsoyannis, *J. Phys. Chem. A*, 2009, *113*, 8737.
311. A. Streitwieser and N. Yoshida, *J. Am. Chem. Soc.*, 1968, *91*, 7528.
312. D. G. Karraker, J. A. Stone, E. R. Jones Jr. and N. Edelstein, *J. Am. Chem. Soc.*, 1970, *92*, 4841.
313. A. Streitwieser, M. T. Barros, H.-K. Wang and T. R. Boussie, *Organometallics*, 1993, *12*, 5023.

314. H. Braunschweig, M. A. Celik, K. Dìck, F. Hupp and I. Krummenacher, *Chem. Eur. J.*, 2015, *21*, 9339.
315. N. C. Burton, F. G. N. Cloke, P. B. Hitchcock, H. C. Delemos and A. A. Sameh, *J. Chem. Soc. Chem. Commun.*, 1989, 1462.
316. (a) U. Kilimann, R. Herbst-Irmer, D. Stalke and F. T. Edelmann, *Angew. Chem. Int. Ed. Engl.*, 1994, *33*, 1618; (b) C. Apostolidis, F. T. Edelmann, B. Kanellakopulos and U. Reißmann, *Z. Naturforsch. B*, 1999, *54*, 960.
317. O. T. Summerscales, F. G. N. Cloke, P. B. Hitchcock, J. C. Green and N. Hazari, *Science*, 2006, *311*, 829.
318. V. Lorenz, B. M. Schmiege, C. G. Hrib, J. W. Ziller, A. Edelmann, S. Blaurock, W. J. Evans and F. T. Edelmann, *J. Am. Chem. Soc.*, 2011, *133*, 1257.
319. M. H. Lyttle, A. Streitwieser and M. J. Miller, *J. Org. Chem.*, 1989, *54*, 2331.
320. D. Seyferth, *Organometallics*, 2004, *23*, 3562.
321. J. C. Berthet, P. Thuéry and M. Ephritikhine, *Organometallics*, 2008, *27*, 1664.
322. A. Hervé, N. Garin, P. Thuéry, M. Ephritikhine and J.-C. Berthet, *Chem. Commun.*, 2013, *49*, 6304.
323. A. Herveí, P. Thueíry, M. Ephritikhine and J.-C. Berthet, *Organometallics*, 2014, *33*, 2088–2098.
324. J.-C. Berthet, P. Thuéry and M. Ephritikhine, *Compt. Rend. Chim.*, 2014, *17*, 526.
325. (a) F. Billiau, G. Folcher, H. Marquetellis, P. Rigny and E. Saito, *J. Am. Chem. Soc.*, 1981, *103*, 5603; (b) D. C. Eisenberg, A. Streitwieser and W. K. Kot, *Inorg. Chem.*, 1990, *29*, 10; (c) T. R. Boussie, D. C. Eisenberg, J. Rigsbee, A. Streitwieser and A. Zalkin, *Inorg. Chem.*, 1991, *10*, 1922.
326. D. G. Karraker and J. A. Stone, *J. Am. Chem. Soc.*, 1974, *96*, 6885.
327. S. M. Cendrowski-Guillaume, M. Nierlich and M. Ephritikhine, *Eur. J. Inorg. Chem.*, 2001, 1495.
328. T. R. Boussie, R. M. Moore Jr., A. Streitwieser, A. Zalkin, J. Brennan and K. A. Smith, *Organometallics*, 1990, *9*, 2010, and references therein.
329. C. Boisson, J.-C. Berthet, M. Lance, J. Vigner, M. Nierlich and M. Ephritikhine, *J. Chem. Soc. Dalton Trans.*, 1996, 947.
330. C. Boisson, J.-C. Berthet, M. Ephritikhine, M. Lance, M. Nierlich, *J. Organomet. Chem.*, 1996, *522*, 249.
331. J.-C. Berthet, J. F. Lemarechal and M. Ephritikhine, *J. Organomet. Chem.*, 1994, *480*, 155.

332. (a) D. Baudry, E. Bulot, M. Ephritikhine, M. Nierlich, M. Lance and J. Vigner, *J. Organomet. Chem.*, 1990, *388*, 279; (b) T. Arliguie, D. Baudry, M. Ephritikhine, M. Nierlich, M. Lance and J. Vigner, *J. Chem. Soc. Dalton Trans.*, 1992, 1019.
333. S. M. Cendrowski-Guillaume, M. Lance, M. Nierlich and M. Ephritikhine, *Organometallics*, 2000, *19*, 3257.
334. T. M. Gilbert, R. R. Ryan and A. P. Sattelberger, *Organometallics*, 1989, *8*, 857.
335. W. J. Evans, S. A. Kozimor and J. W. Ziller, *Polyhedron*, 2006, *25*, 484.
336. J. H. Farnaby, F. G. N. Cloke, M. P. Coles, J. C. Green and G. Aitken, *Compt. Rend. Chim.*, 2010, *13*, 812.
337. M. K. Takase, N. A. Siladke, J. W. Ziller and W. J. Evans, *Organometallics*, 2011, *30*, 458.
338. (a) W. J. Evans, G. W. Nyce and J. W. Ziller, *Angew. Chem. Int. Ed.*, 2000, *39*, 240; (b) W. J. Evans, M. K. Takase, J. W. Ziller, A. G. DiPasquale and A. L. Rheingold, *Organometallics*, 2009, *28*, 236.
339. J. A. Higgins, F. G. N. Cloke and S. M. Roe, *Organometallics*, 2013, *32*, 5244.
340. Z. E. Button, J. A. Higgins, M. Suvova, F. G. N. Cloke and S. M. Roe, *Dalton Trans.*, 2015, *44*, 2588.
341. O. T. Summerscales, F. G. N. Cloke, P. B. Hitchcock, J. C. Green and N. Hazari, *J. Am. Chem. Soc.*, 2006, *126*, 9602.
342. N. Tsoureas, O. T. Summerscales, F. G. N. Cloke and S. M. Roe, *Organometallics*, 2013, *32*, 1353.
343. A. S. Frey, F. G. N. Cloke, P. B. Hitchcock, I. J. Day, J. C. Green and G. Aitken, *J. Am. Chem. Soc.*, 2008, *130*, 13816.
344. A. S. P. Frey, F. G. N. Cloke, M. P. Coles, L. Maron and T. Davin, *Angew. Chem. Int. Ed.*, 2011, *50*, 6881.
345. O. T. Summerscales, A. S. P. Frey, F. G. N. Cloke and P. B. Hitchcock, *Chem. Commun.*, 2009, 198.
346. A. S. P. Frey, F. G. N. Cloke, M. P. Coles and P. B. Hitchcock, *Chem. Eur. J.*, 2010, *16*, 9446.
347. F. G. N. Cloke, *Pure Appl. Chem.*, 2001, *73*, 233.
348. F. G. N. Cloke, J. C. Green and C. N. Jardine, *Organometallics*, 1999, *18*, 1080.
349. (a) A. E. Ashley, A. R. Cowley and D. O'Hare, *Chem. Commun.*, 2007, 1512; (b) A. E. Ashley, A. R. Cowley and D. O. Hare, *Eur. J. Org. Chem.*, 2007, 2239.
350. F. G. N. Cloke and P. B. Hitchcock, *J. Am. Chem. Soc.*, 1997, *119*, 7899.

351. F. M. Chadwick, A. Ashley, G. Wildgoose, J. M. Goicoechea, S. M. Randall and D. O'Hare, *Dalton Trans.*, 2010, *39*, 6789.
352. F. M. Chadwick and D. M. O'Hare, *Organometallics*, 2014, *33*, 3768.
353. N. Tsoureas, A. F. R. Kilpatrick, O. T. Summerscales, J. F. Nixon, F. G. N. Cloke and P. B. Hitchcock, *Eur. J. Inorg. Chem.*, 2013, *2013*, 4085.
354. J. Q. Wang, A. K. Dash, J.-C. Berthet, M. Ephritikhine and M. S. Eisen, *Organometallics*, 1999, *18*, 2407.
355. M. Foyentin, G. Folcher and M. Ephritikhine, *J. Chem. Soc. Chem. Commun.*, 1987, 494.
356. B. Fang, W. S. Ren, G. H. Hou, G. F. Zi, D. C. Fang, L. Maron and M. D. Walter, *J. Am. Chem. Soc.*, 2014, *136*, 17249.
357. B. Fang, G. H. Hou, G. F. Zi, W. J. Ding and M. D. Walter, *Organometallics*, 2016, *35*, 1384.
358. E. A. Pedrick, L. A. Seaman, J. C. Scott, L. Griego, G. Wu and T. W. Hayton, *Organometallics*, 2016, *35*, 494.
359. B. Fang, L. Zhang, G. H. Hou, G. F. Zi, D. C. Fang and M. D. Walter, *Chem. Sci.*, 2015, *6*, 4897.
360. G. Erker, T. Muhlenbernd, R. Benn and A. Rufinska, *Organometallics*, 1986, *5*, 402.
361. G. M. Smith, H. Suzuki, D. C. Sonnenberger, V. W. Day and T. J. Marks, *Organometallics*, 1986, *5*, 549.
362. O. J. Scherer, J. Schulze and G. Wolmershauser, *J. Organomet. Chem.*, 1994, *484*, C5.
363. B. Fang, G. H. Hou, G. F. Zi, D. C. Fang and M. D. Walter, *Dalton Trans.*, 2015, *44*, 7927.
364. B. Fang, L. Zhang, G. H. Hou, G. F. Zi, D. C. Fang and M. D. Walter, *Organometallics*, 2015, *34*, 5669.
365. J. M. Manriquez, P. J. Fagan, T. J. Marks, S. H. Vollmer, C. S. Day and V. W. Day, *J. Am. Chem. Soc.*, 1978, *101*, 5075.
366. W. J. Evans, S. A. Kozimor and J. W. Ziller, *Chem. Commun.*, 2005, 4681.
367. J. K. Pagano, J. M. Dorhout, K. R. Czerwinski, D. E. Morris, B. L. Scott, R. Waterman and J. L. Kiplinger, *Organometallics*, 2016, *35*, 617.
368. B. Kosog, C. E. Kefalidis, F. W. Heinemann, L. Maron and K. Meyer, *J. Am. Chem. Soc.*, 2012, *134*, 12792.
369. T. H. Cymbaluk, J. Z. Liu and R. D. Ernst, *J. Organomet. Chem.*, 1983, *255*, 311.
370. D. Baudry, E. Bulot, P. Charpin, M. Ephritikhine, M. Lance, M. Nierlich and J. Vigner, *J. Organomet. Chem.*, 1989, *371*, 163.

371. D. Baudry, E. Bulot and M. Ephritikhine, *Chem. Commun.*, 1989, 1316.
372. (a) H. W. Turner, R. A. Andersen, A. Zalkin and D. H. Templeton, *Inorg. Chem.*, 1979, *18*, 1221; (b) S. J. Simpson, H. W. Turner and R. A. Andersen, *J. Am. Chem. Soc.*, 1979, *101*, 7728; (c) S. J. Simpson and R. A. Andersen, *J. Am. Chem. Soc.*, 1981, *103*, 4063; (d) D. Baudry, A. Dormond and A. Hafid, *J. Organomet. Chem.*, 1995, *494*, C22; (e) P. L. Diaconescu, A. L. Odom, T. Agapie and C. C. Cummins, *Organometallics*, 2001, *20*, 4993.
373. (a) J. L. Stewart and R. A. Andersen, *J. Chem. Soc. Chem. Commun.*, 1987, 1846; (b) C. Baudin and M. Ephritikhine, *J. Organomet. Chem.*, 1989, *364*, C1; (c) C. Baudin, D. Baudry, M. Ephritikhine, M. Lance, A. Navaza, M. Nierlich and J. Vigner, *J. Organomet. Chem.*, 1991, *415*, 59.
374. (a) D. L. Clark, S. K. Grumbine, B. L. Scott and J. G. Watkin, *Organometallics*, 1996, *15*, 949; (b) S. M. Beshouri, P. E. Fanwick, I. P. Rothwell and J. C. Huffman, *Organometallics*, 1987, *6*, 2498.
375. M. Wedler, F. Knösel, F. T. Edelmann and U. Behrens, *Chem. Ber.*, 1992, *125*, 1313.
376. (a) M. A. Antunes, Â. Domingos,; I. C. dos Santos, N. Marques and J. Takats, *Polyhedron*, 2005, *24*, 3038; (b) M. Silva, Â Domingos, A. P. de Matos, N. Marques and S. Trofimenko, *Dalton Trans.* 2000, 4628; (c) M. Silva, N. Marques and A. P. de Matos, *J. Organomet. Chem.* 1995, *493*, 129; (d) M. P. C. Campello, Â. Domingos, A. Galvao, A. P. de Matos and I. Santos, *J. Organomet. Chem.*, 1999, *579*, 5; (e) M. P. C. Campello, M. J. Calhorda, Â. Domingos, A. Galvão, J. P. Leal, A. P. de Matos and I. Santos, *J. Organomet. Chem.*, 1997, *538*, 223; (f) Â. Domingos, N. Marques, A. P. de Matos, I. Santos and M. Silva, *Organometallics*, 1994, *13*, 654.
377. (a) P. Roussel, R. Boaretto, A. J. Kingsley, N. W. Alcock and P. Scott, *Dalton Trans.*, 2002, 1423; (b) R. Boaretto, P. Roussel, A. J. Kingsley, I. J. Munslow, C. J. Sanders, N. W. Alcock and P. Scott, *Chem. Commun.*, 1999, 1701.
378. (a) E. M. Matson, J. J. Kiernicki, P. E. Fanwick and S. C. Bart, *Eur. J. Inorg. Chem.*, 2016, 2527; (b) E. M. Matson, M. G. Crestani, P. E. Fanwick and S. C. Bart, *Dalton Trans.*, 2012, *41*, 7952; (c) E. M. Matson, W. P. Forrest, P. E. Fanwick and S. C. Bart, *J. Am. Chem. Soc.*, 2011, *133*, 4948.
379. E. M. Matson, W. P. Forrest, P. E. Fanwick and S. C. Bart, *Organometallics*, 2013, *32*, 1484.
380. B. M. Gardner, W. Lewis, A. J. Blake and S. T. Liddle, *Organometallics*, 2015, *34*, 2386.

381. B. M. Gardner, P. A. Cleaves, C. E. Kefalidis, J. Fang, L. Maron, W. Lewis, A. J. Blake and S. T. Liddle, *Chem. Sci.*, 2014, *5*, 2489.
382. B. M. Gardner, J. McMaster, W. Lewis, A. J. Blake and S. T. Liddle, *J. Am. Chem. Soc.*, 2009, *131*, 10388.
383. C. E. Hayes and D. B. Leznoff, *Organometallics*, 2010, *29*, 767.
384. (a) C. E. Hayes, R. H. Platel, L. L. Schafer and D. B. Leznoff, *Organometallics*, 2012, *31*, 6732; (b) C. E. Hayes, D. E. Gill, M. L. Brown and D. B. Leznoff, *Eur. J. Inorg. Chem.*, 2014, 3690; (c) C. E. Hayes and D. B. Leznoff, *Coord. Chem. Rev.*, 2014, *266*, 155.
385. N. R. Andreychuk, S. Ilango, B. Vidjayacoumar, D. J. H. Emslie and H. A. Jenkins, *Organometallics*, 2013, *32*, 1466.
386. C. A. Cruz, D. J. H. Emslie, H. A. Jenkins and J. F. Britten, *Dalton Trans.*, 2010, *39*, 6626.
387. (a) M. J. Monreal and P. L. Diaconescu, *J. Am. Chem. Soc.*, 2010, *132*, 7676; (b) S. Duhovic, M. J. Monreal and P. L. Diaconescu, *J. Organomet. Chem.*, 2010, *695*, 2822; (c) M. J. Monreal, S. Khan and P. L. Diaconescu, *Angew. Chem. Int. Ed.*, 2009, *48*, 8352.
388. P. L. Diaconescu, *Acc. Chem. Res.*, 2010, *43*, 1352.
389. M. J. Monreal and P. L. Diaconescu, *Organometallics*, 2008, *27*, 1702.
390. E. Lu, O. J. Cooper, J. McMaster, F. Tuna, E. J. L. McInnes, W. Lewis, A. J. Blake and S. T. Liddle, *Angew. Chem. Int. Ed.*, 2014, *53*, 6696.
391. E. Mora, L. Maria, B. Biswas, C. Camp, I. C. Santos, J. Pecaut, A. Cruz, J. M. Carretas, J. Marcalo and M. Mazzanti, *Organometallics*, 2013, *32*, 1409.
392. K. L. Miller, B. N. Williams, D. Benitez, C. T. Carver, K. R. Ogilby, E. Tkatchouk, W. A. Goddard, III and P. L. Diaconescu, *J. Am. Chem. Soc.*, 2010, *132*, 342.
393. N. A. Siladke, C. L.Webster, J. R. Walensky, M. K. Takase, J. W. Ziller, D. J. Grant, L. Gagliardi and W. J. Evans, *Organometallics*, 2013, *32*, 6522.
394. D. L. Clark, S. K. Grumbine, B. L. Scott and J. G. Watkin, *J. Am. Chem. Soc.*, 1995, *117*, 9089.
395. R. W. Broach, A. J. Schultz, J. M. Williams, G. M. Brown, J. M. Manriquez, P. J. Fagan and T. J. Marks, *Science*, 1979, *203*, 172.
396. J.-C. Berthet, J.-F. L. Marechal and M. Ephritikhine, *Chem. Commun.*, 1991, 360.
397. D. Baudry, A. Dormond and I. A. Abdallaoui, *J. Organomet. Chem.*, 1994, *476*, C15.
398. W. Ren, N. Zhao, L. Chen, H. Song and G. Zi, *Inorg. Chem. Commun.*, 2011, *14*, 1838.

399. X. Jemine, J. Goffart, M. Ephritikhine and J. Fuger, *J. Organomet. Chem.*, 1993, *448*, 95.
400. Z. Lin and T. J. Marks, *J. Am. Chem. Soc.*, 1987, *109*, 7979.
401. Z. Lin and T. J. Marks, *J. Am. Chem. Soc.*, 1990, *112*, 5515.
402. K. G. Moloy and T. J. Marks, *J. Am. Chem. Soc.*, 1984, *106*, 7051.
403. S. J. Simpson, H. W. Turner and R. A. Andersen, *J. Am. Chem. Soc.*, 1979, *101*, 2782.
404. R. A. Andersen, A. Zalkin and D. H. Templeton, *Inorg. Chem.*, 1981, *20*, 622.
405. W. J. Evans, K. A. Miller, A. G. DiPasquale, A. L. Rheingold, T. J. Stewart and R. Bau, *Angew. Chem. Int. Ed.*, 2008, *47*, 5075.
406. J.-C. Berthet, J. F. Lemarechal and M. Ephritikhine, *J. Chem. Soc. Chem. Commun.*, 1991, 360.
407. J. F. Lemarechal, C. Villiers, P. Charpin, M. Lance, M. Nierlich, J. Vigner and M. Ephritikhine, *J. Chem. Soc. Chem. Commun.*, 1989, 308.
408. J.-C. Berthet, C. Villiers, J. F. Lemarechal, B. Delavauxnicot, M. Lance, M. Nierlich, J. Vigner and M. Ephritikhine, *J. Organomet. Chem.*, 1992, *440*, 53.
409. R. R. Langeslay, M. E. Fieser, J. W. Ziller, F. Furche and W. J. Evans, *J. Am. Chem. Soc.*, 2016, *138*, 4036.
410. P. J. Fagan, K. G. Moloy and T. J. Marks, *J. Am. Chem. Soc.*, 1981, *103*, 6959.
411. K. G. Moloy and T. J. Marks, *Inorg. Chim. Acta*, 1985, *110*, 127.
412. L. Jia, X. Yang, C. L. Stern and T. J. Marks, *Organometallics*, 1997, *16*, 842.
413. (a) Y.-X. Chen, C. L. Stern, S. Yang and T. J. Marks, *J. Am. Chem. Soc.*, 1996, *118*, 12451; (b) Y. X. Chen, M. V. Metz, L. T. Li, C. L. Stern and T. J. Marks, *J. Am. Chem. Soc.*, 1998, *120*, 6287.
414. L. Jia, X. Yang, C. Stern and T. J. Marks, *Organometallics*, 1994, *13*, 3755.
415. X. Yang, W. A. King, M. Sabat and T. J. Marks, *Organometallics*, 1993, *12*, 4254.
416. X. Yang, C. L. Stern and T. J. Marks, *Organometallics*, 1991, *10*, 840.
417. Z. Lin, J. F. Le Marechal, M. Sabat and T. J. Marks, *J. Am. Chem. Soc.*, 1987, *109*, 4127.
418. See for example: (a) T. J. Marks, L. Ja and X. Yang, U.S. Patent 5,477,895, 1995; (b) T. J. Marks and Y.-X. Chen, U.S. Patent 6,229,034 B1, 2001; (c) T. J. Marks and Y.-X. Chen, U.S. Patent 6,274,752 B1, 2001; (d) T. J. Marks and Y.-X. Chen, U.S. Patent 6,403,732 B2, 2002; (e) T. J. Marks and Y.-X. Chen, U.S. Patent 6,388,114 B1, 2002.

419. R. E. Campbell Jr., (The Dow Chemical Company) U.S. Patent 4,665,046, 1987.
420. (a) T. J. Marks, *Acc. Chem. Res.*, 1992, *25*, 57; (b) W. C. Finch, R. D. Gillespie, D. Hedden and T. J. Marks, *J. Am. Chem. Soc.*, 1990, *112*, 6221; (c) D. Hedden and T. J. Marks, *J. Am. Chem. Soc.*, 1988, *110*, 1647; (d) P. J. Toscano and T. J. Marks, *J. Am. Chem. Soc.*, 1985, *107*, 653; (e) M.-Y. He, G. Xiong, P. J. Toscano, R. L. Burwell, Jr. and T. J. Marks, *J. Am. Chem. Soc.*, 1985, *107*, 641.
421. H. M. Dietrich, J. W. Ziller, R. Anwander and W. J. Evans, *Organometallics*, 2009, *28*, 1173.
422. E. Domeshek, R. J. Batrice, S. Aharonovich, B. Tumanskii, M. Botoshansky and M. S. Eisen, *Dalton Trans.*, 2013, *42*, 9069.
423. A. K. Dash, J. X. Wang, J.-C. Berthet, M. Ephritikhine and M. S. Eisen, *J. Organomet. Chem.*, 2000, *604*, 83.
424. L. Moisan, T. Le Borgne, C. Villiers, P. Thuéry and M. Ephritikhine, *Compt. Rend. Chim.*, 2007, *10*, 883.
425. (a) S. A. Mungur, S. T. Liddle, C. Wilson, M. J. Sarsfield and P. L. Arnold, *Chem. Commun.*, 2004, 2738; (b) P. L. Arnold, A. L. Blake and C. Wilson, *Chem. Eur. J.*, 2005, *11*, 6095; (c) D. Pugh, J. A. Wright, S. Freeman and A. A. Danopoulos, *Dalton Trans.*, 2006, 775; (d) K. R. Meihaus, S. G. Minasian, W. W. Lukens, S. A. Kozimor, D. K. Shuh, T. Tyliszczak and J. R. Long, *J. Am. Chem. Soc.*, 2014, *136*, 6056; (e) M. E. Garner, S. Hohloch, L. Maron and J. Arnold, *Organometallics* 2016, *35*, 2915.
426. (a) R. E. Cramer, R. B. Maynard and J. W. Gilje, *Inorg. Chem.*, 1981, *20*, 2466; (b) R. E. Cramer, R. B. Maynard, J. C. Paw and J. W. Gilje, *J. Am. Chem. Soc.*, 1981, *103*, 3589.
427. S. Fortier, J. R. Walensky, G. Wu and T. W. Hayton, *J. Am. Chem. Soc.*, 2011, *133*, 6894.
428. G. B. Ma, M. J. Ferguson, R. McDonald and R. G. Cavell, *Inorg. Chem.*, 2011, *50*, 6500.
429. J.-C. Tourneux, J.-C. Berthet, T. Cantat, P. Thuéry, N. Mezailles, P. Le Floch and M. Ephritikhine, *Organometallics*, 2011, *30*, 2957.
430. (a) O. J. Cooper, J. McMaster, W. Lewis, A. J. Blake and S. T. Liddle, *Dalton Trans.*, 2010, *39*, 5074; (b) O. J. Cooper, D. P. Mills, J. McMaster, F. Moro, E. S. Davies, W. Lewis, A. J. Blake and S. T. Liddle, *Angew. Chem. Int. Ed.*, 2011, *50*, 2383.
431. D. P. Mills, O. J. Cooper, F. Tuna, E. J. L. McInnes, E. S. Davies, J. McMaster, F. Moro, W. Lewis, A. J. Blake and S. T. Liddle, *J. Am. Chem. Soc.*, 2012, *134*, 10047.
432. O. J. Cooper, D. P. Mills, W. Lewis, A. J. Blake and S. T. Liddle, *Dalton Trans.*, 2014, *43*, 14275.

433. E. Lu, F. Tuna, W. Lewis, N. Kaltsoyannis and S. T. Liddle, *Chem. Eur. J.*, 2016, *22*, 11554.
434. E. Lu, O. J. Cooper, F. Tuna, A. J. Wooles, N. Kaltsoyannis and S. T. Liddle, *Chem. Eur. J.*, 2016, *22*, 11559.
435. M. Gregson, E. Lu, F. Tuna, E. J. L. McInnes, C. Hennig, A. C. Scheinost, J. McMaster, W. Lewis, A. J. Blake, A. Kerridge and S. T. Liddle, *Chem. Sci.*, 2016, *7*, 3286.
436. M. Gregson, E. Lu, D. P. Mills, F. Tuna, E. J. L. McInnes, C. Hennig, A. C. Scheinost, J. McMaster, W. Lewis, A. J. Blake, A. Kerridge and S. T. Liddle, *Nature Commun.*, 2017, *8*, 14137.
437. T. Cantat, T. Arliguie, A. Noel, P. Thuéry, M. Ephritikhine, P. Le Floch and N. Mezailles, *J. Am. Chem. Soc.*, 2009, *131*, 963.
438. J.-C. Tourneux, J.-C. Berthet, T. Cantat, P. Thuéry, N. Mezailles and M. Ephritikhine, *J. Am. Chem. Soc.*, 2011, *133*, 6162.
439. W. S. Ren, X. B. Deng, G. F. Zi and D. C. Fang, *Dalton Trans.*, 2011, *40*, 9662.
440. E. Lu, W. Lewis, A. J. Blake and S. T. Liddle, *Angew. Chem. Int. Ed.*, 2014, *53*, 9356.
441. A. Herve, Y. Bouzidi, J.-C. Berthet, L. Belkhiri, P. Thuéry, A. Boucekkine and M. Ephritikhine, *Inorg. Chem.*, 2014, *53*, 6995.
442. B. Kanellakopulos, E. Dornberger and H. Billich, *J. Organomet. Chem.*, 1974, *76*, C42.
443. J. Maynadié, N. Barros, J.-C. Berthet, P. Thuéry, L. Maron and M. Ephritikhine, *Angew. Chem. Int. Ed.*, 2007, *46*, 2010.
444. J.-C. Berthet, P. Thuéry and M. Ephritikhine, *Chem. Commun.*, 2007, 604.
445. A. Herve, Y. Bouzidi, J.-C. Berthet, L. Belkhiri, P. Thuéry, A. Boucekkine and M. Ephritikhine, *Inorg. Chem.*, 2015, *54*, 2474.
446. (a) S. M. Mansell and P. L. Arnold, *Polyhedron*, 2016, *116*, 82; (b) W. J. Evans, T. J. Mueller and J. W. Ziller, *Chem. Eur. J.*, 2010, *16*, 964; (c) M. Schultz, C. J. Burns, D. J. Schwartz and R. A. Andersen, *Organometallics*, 2001, *20*, 5690; (d) W. G. Vandersluys and A. P. Sattelberger, *Inorg. Chem.*, 1989, *28*, 2496.
447. M. D. Conejo, J. S. Parry, E. Carmona, M. Schultz, J. G. Brennan, S. M. Beshouri, R. A. Andersen, R. D. Rogers, S. Coles and M. Hursthouse, *Chem. Eur. J.*, 1999, *5*, 3000.
448. J. G. Brennan, R. A. Andersen and J. L. Robbins, *J. Am. Chem. Soc.*, 1986, *108*, 335.
449. J. Parry, E. Carmona, S. Coles and M. Hursthouse, *J. Am. Chem. Soc.*, 1995, *117*, 2649.

450. W. J. Evans, S. A. Kozimor, G. W. Nyce and J. W. Ziller, *J. Am. Chem. Soc.*, 2003, *125*, 13831.
451. (a) L. Maron, O. Eisenstein and R. A. Andersen, *Organometallics*, 2009, *28*, 3629; (b) W. W. Lukens, M. Speldrich, P. Yang, T. J. Duignan, J. Autschbach and P. Kogerler, *Dalton Trans.*, 2016, *45*, 11508.
452. S. G. Minasian, J. L. Krinsky, J. D. Rinehart, R. Copping, T. Tyliszczak, M. Janousch, D. K. Shuh and J. Arnold, *J. Am. Chem. Soc.*, 2009, *131*, 13767.
453. R. B. Nazarski and W. Makulski, *Phys. Chem. Chem. Phys.*, 2014, *16*, 15699.
454. (a) D. R. Evans, T. Drovetskaya, R. Bau, C. A. Reed and P. D. W. Boyd, *J. Am. Chem. Soc.*, 1997, *119*, 3633; (b) S. D. Pike, A. L. Thompson, A. G. Algarra, D. C. Apperley, S. A. Macgregor and A. S. Weller, *Science*, 2012, *337*, 1648; (c) N. R. Andreychuk and D. J. H. Emslie, *Angew. Chem. Int. Ed.*, 2013, *52*, 1696; (d) F. M. Chadwick, N. H. Rees, A. S. Weller, T. Kramer, M. Iannuzzi and S. A. Macgregor, *Angew. Chem. Int. Ed.*, 2016, *55*, 3677.
455. I. Castro-Rodriguez, H. Nakai, P. Gantzel, L. N. Zakharov, A. L. Rheingold and K. Meyer, *J. Am. Chem. Soc.*, 2003, *125*, 15734.

https://doi.org/10.1142/9781800610163_0008

Chapter 8

Small Molecule Activation by Lanthanide Complexes

Conrad A. P. Goodwin and David P. Mills†*
Department of Chemistry, School of Natural Sciences, The University of Manchester, Oxford Road, Manchester, M13 9PL, UK
**conrad.goodwin@manchester.ac.uk*
†david.mills@manchester.ac.uk

8.1 Introduction and Scope of the Review

The small molecule activation chemistry of lanthanide (Ce–Lu, Ln) complexes is described herein, along with analogous reactivity studies of group 3 (Sc, Y, La) metal complexes. The group 3 and Ln metals, collectively the rare earths, are frequently discussed together due to their chemical similarity, where electrostatics typically dictate bonding regimes and the +3 oxidation state predominates.[1] Further to this, the eight-coordinate ionic radii of La(III) (1.16 Å) and Ce(III) (1.143 Å) are very similar, as are Y(III) (1.019 Å) and Ho(III) (1.015 Å),[2] giving these ions comparable charge densities and reactivity profiles. Although the term "rare earth" is often used to encompass group 3 and Ln metals, we will use "Ln" to cover both groups throughout for brevity. Examples of Ln small molecule activation chemistry in this chapter are limited to molecular complexes that react to give clearly defined structurally characterised products, rather than catalytic processes mediated by LnO_x mixtures. The "small molecules" covered include the environmentally relevant gases CO, CO_2, CH_4, NO_x and N_2, together with selected analogues that warrant inclusion (e.g. COS and CS_2), and organoazides, RN_3, for completeness.

Ln small molecule activation chemistry is dictated by the predominantly electrostatic bonding between Ln cations and coordinated ligands, as polarised Ln–ligand bonds can be easily cleaved and reformed, and undergo insertion chemistry readily.[1] This can be attributed to the relatively limited radial extension of valence 4f orbitals, which have a spatial and energetic mismatch with ligand orbitals, reducing the influence of ligands on bonding regimes.[2] Despite the dominance of the +3 oxidation state in Ln chemistry and the fact that Ln(II) starting materials are only commercially available for Eu, Yb, Sm, Tm, Dy and Nd,[3] Ln small molecule activation chemistry is well represented by Ln(II) complexes, which are typically strong reducing agents. The ease of preparation and reactivity of Ln(II) complexes can be rationalised by reference to Ln(III)→Ln(II) reduction potentials, which are largely dictated by ground state electron configurations. Within the last 20 years isolated examples of Ln(II) complexes have been expanded to include every Ln save for radioactive Pm, opening up this area further.[4]

Ln small molecule activation chemistry is somewhat underdeveloped, so where there are few reported reactivity studies it can be assumed that this is more symptomatic of the paucity of previous research rather than this being the limit of the field. Hence, comprehensive future studies are vital for new and important discoveries to be made. Since the first landmark results in Ln small molecule activation chemistry were reported 35–40 years ago, this area has blossomed, and in this chapter, we will highlight the most representative examples. Important concepts for the reader to consider while reading this chapter are summarised below:

- Applications — As Ln chemistry is immature in comparison to d-transition metal chemistry, most transformations herein provide proof of concept rather than current industrial applicability. Selected examples are limited to those that form well-defined products and these are used to rationalise reactivity profiles and reaction pathways. This knowledge can also be transferred to heterogeneous systems for potential future exploitation.
- HSAB mismatches — As all $Ln^{2+/3+}$ cations are hard, interactions with substrates such as CO (a soft π-acceptor ligand) are relatively

unfavourable when compared with donor solvent interactions and oligomerisation decomposition pathways. This explains the paucity of small molecule activation chemistry reported to date.

- 1 e^- processes — Due to the inability of Ln^{n+} cations to undergo facile 2 e^- oxidation [with the exception of rare Ce(II) complexes, which could conceivably be oxidised to Ce(IV) in one step] their small molecule activation chemistry is often limited to 1 e^- processes, such as single electron transfer by Ln(II) complexes or σ-bond metathesis by Ln(III) complexes.
- Ligand effects — Previous chapters have highlighted the influence of ligands on the physicochemical properties of f-element complexes. Some transformations described herein are not general to a wide range of complexes; therefore, the ligand environment is key.
- Choice of solvent — Non-aqueous systems dominate lanthanide small molecule activation chemistry. This is rationalised by aqueous solutions providing stabilising hard O-donor ligands and large coordination numbers to complexes, tempering their reactivity.

8.2 CO Reactivity

CO activation by metal complexes has important implications for industrial processes as it can be a C_1 feedstock in the synthesis of organic molecules; however, Ln complex-mediated CO activation is a hugely under-developed area. Homoleptic metal carbonyl complexes of many d-transition metals can be readily synthesised from metal powders and CO gas.[5] These complexes exhibit synergic π-backbonding and can gain additional stabilisation by adhering to the 18-electron rule. In contrast, Ln complexes do not obey any electron count rules and their "core-like" valence 4f-orbitals typically preclude any significant π-backbonding. Further to this, carbonyl ligands are soft Lewis bases and the orbital energy mismatch with hard Ln cations is also in opposition to the formation of simple binary Ln–CO complexes. There is evidence that some Ln(II) complexes can form adducts with CO, which are in equilibrium with the unbound complex, however the starting materials predominate the equilibrium mixture and as such no products have been isolated to date.[6] It is noteworthy that actinide carbonyl complexes have been reported (see Chapter 9).

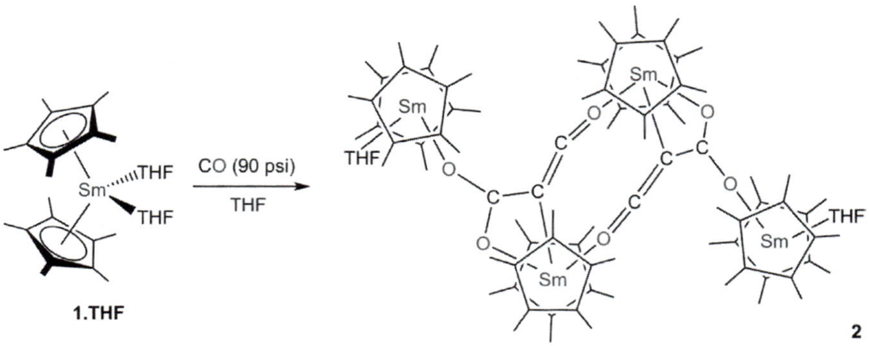

1.THF

2

Scheme 8.1. Synthesis of **2** from **1.THF** and CO at 90 psi; aromaticity not shown on **2** for clarity.[7]

Ln complexes typically exhibit two types of reactivity with CO; reductive homologation and 1,1-migratory insertion, and the reactivity mode is often not simple to predict in advance. One of the earliest examples of well-defined Ln-mediated CO activation was by $[Sm(Cp^*)_2(THF)_2]$ (**1.THF**, $Cp^* = \{C_5Me_5\}^-$), which does not react with CO at 1 atm, but at 90 psi of CO the tetranuclear ketene-carboxylate Sm(III) complex, $[\{Sm(Cp^*)_2\}_2(\mu^3\text{-}\eta^2{:}\eta^1{:}\eta^1\text{-}O_2CCCO)(THF)]_2$ (**2**), was isolated (Scheme 8.1).[7] Complex **2** forms via four 1 e^- reductions and coupling of organic fragments deriving from six molecules of CO. It is noteworthy that the reductive homologation of CO is unique for f-element complexes to date, as simple adducts typically form in d-transition metal carbonyl chemistry.

More recently, the reductive homologation of CO by the direduced N_2-bridged La(III) complex $[\{La(Cp^*)_2(THF)\}_2(\mu\text{-}\eta^2{:}\eta^2\text{-}N_2)]$ (**3**) was observed to give the structurally analogous La(III) product, $[\{La(Cp^*)_2\}_2(\mu^3\text{-}\eta^2{:}\eta^1{:}\eta^1\text{-}O_2CCCO)(THF)]_2$ (**4**), at 90 psi CO (Scheme 8.2) (Ln dinitrogen complexes are covered in more detail in Section 8.5).[8] Formally, complex **4** is produced from six molecules of CO via two 2 e^- reductions and coupling of organic fragments in a similar fashion to the formation of complex **2**, but on this occasion the N_2^{2-} ligand provides the electrons instead of Ln(II) centres.

Reduction of the Y(III) silylamide complex $[Y(N'')_3]$ (**5a**, $\{N'' = \{N(SiMe_3)_2\}^-)$ with KC_8 in the presence of 18-crown-6 (18-C-6) at low temperature gives a Y(II) species *in situ*, which

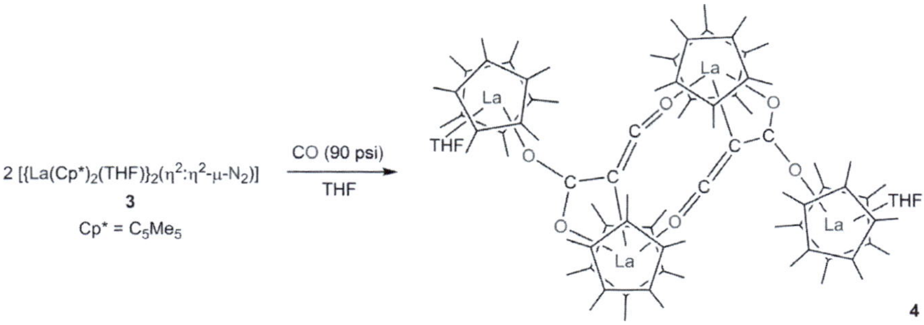

Scheme 8.2. Synthesis of **4** from **3** and CO at 90 psi; aromaticity not shown on **4** for clarity.[8]

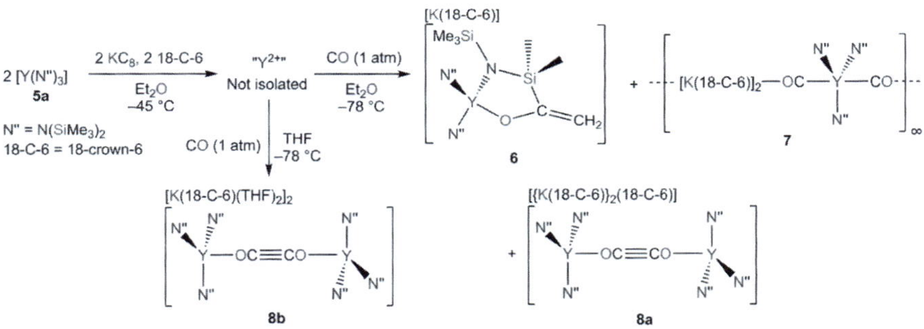

Scheme 8.3. Synthesis of **6–8** from **5a**, KC_8 and CO at 1 atmosphere.[9]

exhibits divergent reactivity with CO depending on the conditions employed to furnish a range of crystalline products; $[K(18\text{-}C\text{-}6)][Y(N'')_2\{[N(SiMe_3)\{SiMe_2C(O){=}CH_2\}\text{-}\kappa^2\text{-}N,O]\}]$ (**6**), $[\{Y(N'')_3\}(\mu\text{-}CO)\{K(18\text{-}C\text{-}6)_2\}]_\infty$ (**7**), $[\{K(18\text{-}C\text{-}6)\}_2(18\text{-}C\text{-}6)][\{Y(N'')_3\}_2(\mu\text{-}OCCO)]$ (**8a**), and $[K(18\text{-}C\text{-}6)(THF)_2]_2[\{Y(N'')_3\}_2(\mu\text{-}OCCO)]$ (**8b**) (Scheme 8.3).[9] The multiple reaction pathways stem from the elevated reactivity of the Y(II) intermediate, which could not be isolated. Complex **6** likely results from the 1,2-migratory insertion of CO into the Y–C bond of a potassium salt of the yttrium cyclometallate, $[Y\{N''\}_2\{N(SiMe_3)(SiMe_2CH_2)\text{-}\kappa^2\text{-}N,C\}]^-$, which is a common by-product in the reduction of **5a**. Complex **7** forms via a 1 e^- reduction of a CO molecule by the Y(II) intermediate, followed by coordination and reduction of a second CO, presumably consuming a sacrificial equivalent of **5a**. Complexes **8a–b** are formed by the radical homologation of two CO molecules to form an ynediolate unit.

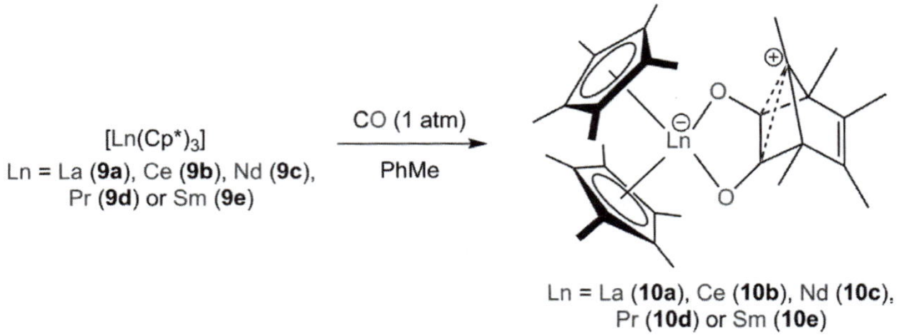

Scheme 8.4. Synthesis of **10a–e** from **9a–e** and CO.[11]

If THF is present when the Y(II) intermediate is exposed to CO then the reaction proceeds with fewer detectable side-products, giving **8b** almost exclusively.

The remaining examples of well-defined CO activation by Ln complexes in the literature are mediated by reactive M–C bonds. Ln complexes containing Cp^R (Cp^R = substituted Cp) ligands can adopt an η^1-Cp^R binding mode in solution, forming transient M–C bonds that are susceptible to migratory insertion reactions, thus giving them a Ln alkyl-like reactivity profile. For example, $[Sc(Cp^{Me4})_3]$ (Cp^{Me4} = $\{C_5Me_4H\}^-$) and $[Lu(Cp^*)_2(Cp^{Me4})]$ were found to exhibit one η^1- and one η^3-Cp^{Me4} binding mode, respectively, in the solid state by X-ray crystallography as the metal centres are sterically saturated, but it is noteworthy that at −80 °C in solution ^{1}H NMR spectroscopy indicates that all the Cp^{Me4} rings display an η^5-binding mode.[10] η^5-Cp* is also in equilibrium with η^1-Cp* in $[Ln(Cp^*)_3]$ [Ln = La (**9a**), Ce (**9b**), Nd (**9c**), Pr (**9d**), Sm (**9e**)]. The Ln alkyl-like reactivity of **9a–e** is exemplified by their reactions with CO, where two 1,2-migratory insertions and subsequent C–C bond formation gives the norbornadienyl complexes $[Ln(Cp^*)_2\{C(Me)[OCC(Me)C(Me)]_2\text{-}\kappa^2\text{-}O,O'\}]$ [Ln = La (**10a**), Ce (**10b**), Nd (**10c**), Pr (**10d**), Sm (**10e**)] (Scheme 8.4).[11] X-ray crystallography data showed that **10a–e** formally contain carbocations, and these are stabilised by the $\{Ln(Cp^*)_2(OR)_2\}^-$ fragments.

A range of Ln alkyl complexes have been shown to undergo 1,1-migratory insertion reactions with CO; for example, $[Lu(Cp)_2(^tBu)(THF)]$ (**11**) reacts with 1 equiv. of CO to form the acyl complex $[Lu(Cp)_2(OC^tBu)]$ (**12**) (Scheme 8.5).[12] However,

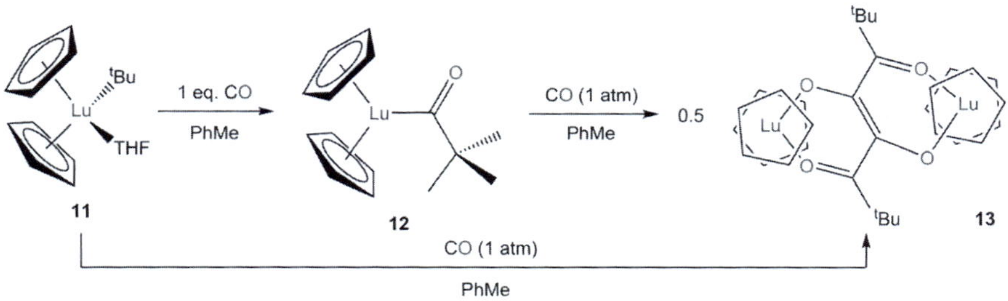

Scheme 8.5. Synthesis of **12** and **13** from **11** and CO.[12]

when an excess of CO is added, the enedionediolate complex [{Lu(Cp)$_2$}$_2${[μ-OC(^{t}Bu)C(O)]$_2$}] (**13**) was isolated via initial formation of **12**, followed by insertion of 2 equiv. of CO and subsequent coupling to give a C=C bond (Scheme 8.5). A 1,1-migratory insertion reaction was also observed when the La(III) complex [La{CH(SiMe$_3$)$_2$}{1,1′-(2-OC$_{10}$H$_5$-3-SiPh$_3$)$_2$}(Et$_2$O)] was treated with CO, giving a transient oxycarbene intermediate [La{OCCH(SiMe$_3$)$_2$}{1,1′-(2-OC$_{10}$H$_5$-3-SiPh$_3$)$_2$}], which subsequently undergoes a 1,2-SiMe$_3$ migration to give [La{OC(SiMe$_3$) = CH(SiMe$_3$)}{1,1′-(2-OC$_{10}$H$_5$-3-SiPh$_3$)$_2$}].[13] Migratory insertion of unsaturated substrates into Ln–C bonds is observed extensively due to the high polarity of these linkages.[14]

Migratory insertion of CO into other polarised Ln–E bonds, and the subsequent reactivity of these complexes, has also been reported. The reaction of the Sc(III) complex [Sc(NacNac^{NiPr2})(μ-PAr)]$_2$ [NacNac^{NiPr2} = {MeC(NDipp)CHC(Me)(NCH$_2$CH$_2$N^{i}Pr$_2$)}$^-$, Ar = (C$_6$H$_3$Me$_2$-2,6)] (**14**) with CO proceeds via two 1,2-migratory insertions into the Sc–P bonds, leading to Sc–P bond cleavage, C=O bond scission to form a Sc–O–Sc moiety and coupling reactions to form C=C and P–P bonds, giving [{Sc(NacNac^{NiPr2})}$_2$(μ-O){μ-OC=C[P(Ar)]$_2$-κ^2-*O*,*C*}] (**15**) (Scheme 8.6).[15]

8.3 CO$_2$/CS$_2$/COS Reactivity

CO$_2$ is an environmentally relevant small molecule, it is abundant and could potentially be a useful C$_1$ feedstock gas. However, the strong C=O bonds in CO$_2$ are notoriously unreactive (532 kJ mol^{-1}).[16] CS$_2$ and COS are analogues of CO$_2$ in that they

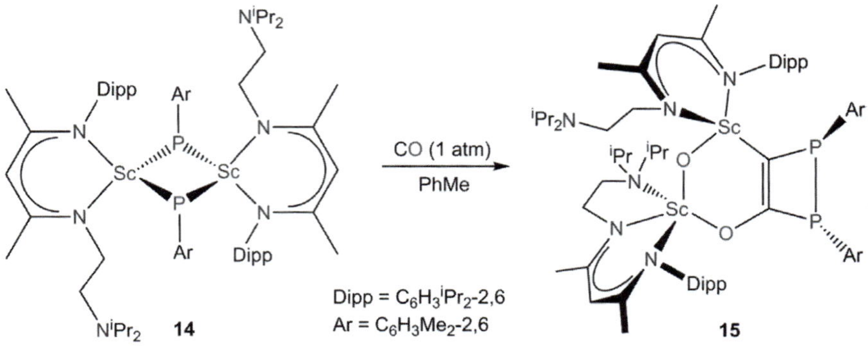

Scheme 8.6. Synthesis of **15** from **14** and CO.[15]

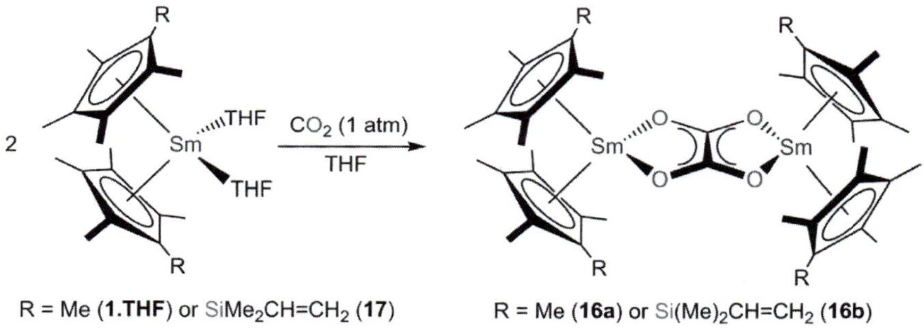

Scheme 8.7. Synthesis of **16a–b** from **1.THF** or **17** and excess CO_2.[17]

contain an unsaturated sp-hybridised carbon atom solely bound to two group 16 elements, but their chemistry differs. Reduction of CO_2 can often lead to the release of CO, whereas CS is not a common by-product from COS or CS_2 reduction due to its latent instability.[17] COS reactivity studies additionally provide complementary results by virtue of O vs. S binding in isolated products. As with CO chemistry, Ln(II) and Ln(III) reactivity studies differ and Ln(II) chemistry is presented first.

Complex **1.THF** provided the first example of Ln(II)-mediated CO_2 activation, reacting with an excess of CO_2 to give the dimeric Sm(III) oxalate complex $[\{Sm(Cp^*)_2\}_2(\mu,\eta^2{:}\eta^2\text{-}C_2O_4)]$ (**16a**) (Scheme 8.7).[17] An analogous complex, $[\{Sm[C_5Me_4\{SiMe_2(CH_2CH{=}CH_2)\}]_2\}_2(\mu,\eta^2{:}\eta^2\text{-}C_2O_4)]$ (**16b**), was obtained with the tethered alkene complex $[Sm\{C_5Me_4[SiMe_2CH_2CH{=}$

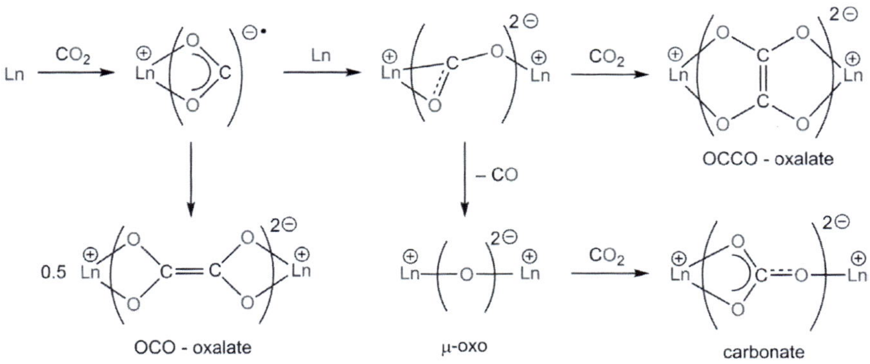

Scheme 8.8. Carbon dioxide activation products by reducing Ln systems (CS_2 and COS also go via similar routes).[18]

$CH_2]\}_2(THF)_2]$ (**17**). The $\{LnCp^*_2\}$ units in **16a–b** provide a protective shell around the oxalate dianion, with the four Cp* centroids forming an approximate square plane.

The OCCO-oxalate in **16a–b** is postulated to form via the direduction of CO_2 by two Ln(II) complexes to give a $\{Ln_2(\mu\text{-}\eta^1\text{:}\eta^2\text{-}CO_2)\}$ intermediate, followed by insertion of a second CO_2 molecule (Scheme 8.8).[18] Once a $\{Ln_2(\mu\text{-}\eta^1\text{:}\eta^2\text{-}CE_2)\}(E = O, S)$ intermediate has formed the reaction can proceed by two different pathways, which are not mutually exclusive, depending on both the ancillary ligands and whether CO_2, CS_2 or COS is used. An OCO-bound oxalate would result from an alternative mechanism, whereby two molecules of CO_2 are each singly reduced to give two Ln–$CO_2^{\cdot}$ radical intermediates, which subsequently couple.

The Sm(II) complex [Sm(OEPG-Me_2)$(THF)_2$] (**18**, 5,10,15,20-Octaethyl-21,23-di-*N*-methylporphyrinogen, $\{cyclo\text{-}[(C_4H_2N)_2(C_4H_2NMe)_2\text{-}\mu\text{-}C(C_2H_5)_2]_4\}^{2-}$) reacts with an excess of CO_2 to give the dimeric carbonate complex [{Sm(OEPG-Me_2)}$_2$($\mu\text{-}\eta^2\text{:}\eta^2\text{-}CO_3\text{-}\kappa$-O,O′)] (**19**), with concomitant release of CO (Scheme 8.9).[19] In contrast, the reaction of an excess of CO_2 and KC_8 with the *tetra*-siloxide Yb(II) potassium-ate complex $[K_2Yb\{OSi(O^tBu)_3\}_4]$ (**20**) gave a mixture of potassium carbonate and oxalate reduction products, as well as extruding CO (Scheme 8.10).[20] The Yb(III) carbonate and oxalate intermediates in this reaction can be reduced by KC_8 to regenerate **20**, and several cycles of this process were observed.

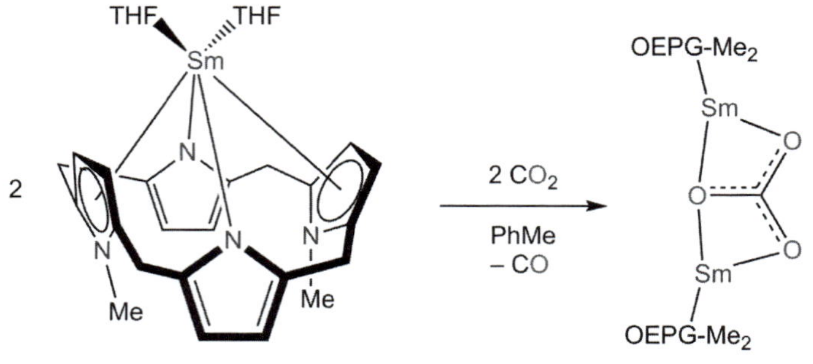

18
(*meso*-ethyl groups omitted for clarity)

19

OEPG-Me$_2$ = {*cyclo*-[(C$_4$H$_2$N)$_2$(C$_4$H$_2$NMe)$_2$-μ-C(C$_2$H$_5$)$_2$]$_4$}

Scheme 8.9. Synthesis of **19** from **18** and 2 equiv. of CO_2.[19]

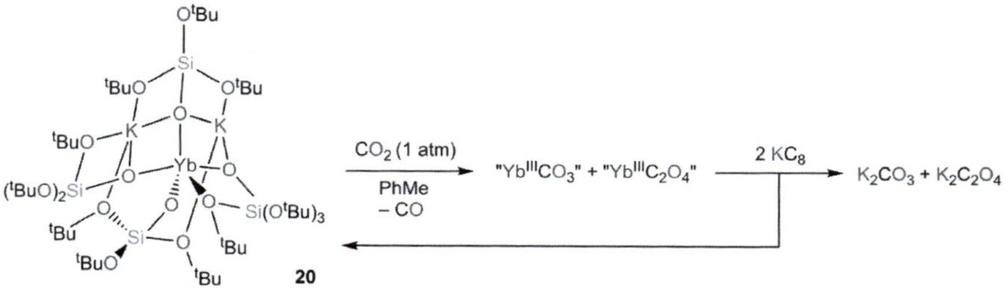

20

Scheme 8.10. Synthesis of potassium carbon-oxide species from **20** and CO_2 followed by KC_8.[20]

The separate reactions of **20** and its Eu(II) analogue (**20-Eu**) with CS_2 were also performed, and in each case similar products were obtained; namely thiocarbonate, thiooxalate and traces of thiomalonate.[20] The product distribution was dependent on the size of the Ln(II) ion, where thiocarbonate was the main product for **20-Eu** and thiooxalate and thiomalonate were in the majority for **20**. It is striking that **20-Eu** is capable of reducing CS_2 as the reduction potential of Eu(III) → Eu(II) ($E^0 = -0.35$ V) is relatively small (*c.f.* Sm(III) → Sm(II) $E^0 = -1.55$ V),[21] showing that the potassium siloxide ligands control the Ln(II) reduction potential and are active participants in these transformations. Similarly, a homoleptic Eu(II) complex with strongly donating ligands, [Eu(BPA)$_2$] (BPA = *bis*-(2-pyridylmethyl)(2-hydroxybenzyl)amine),

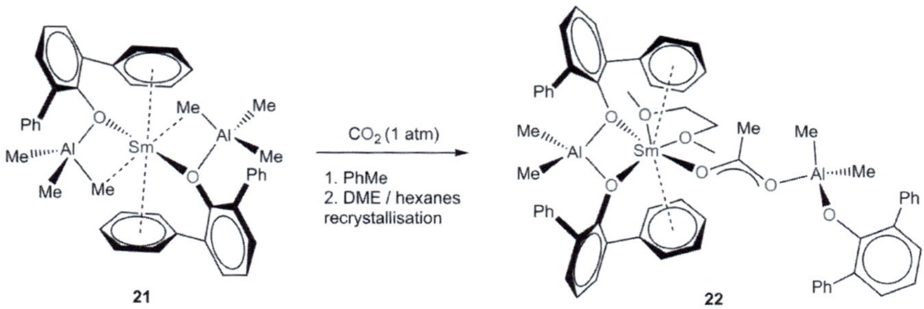

Scheme 8.11. Synthesis of **22** from **21** and excess CO_2.[23]

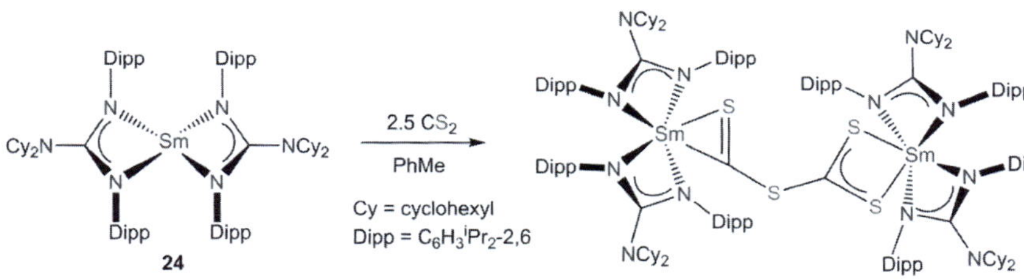

Scheme 8.12. Synthesis of **24** from **23** and CS_2.[24]

was found to reduce CS_2 with subsequent extrusion of "CS" to form $[Eu(BPA)_2(\eta^2{:}CS_3)(DMSO)_2]$ after treatment with DMSO.[22]

Ligand effects may also outweigh metal-based reactivity, as seen in the reaction of CO_2 with the Sm(II) *bis*-alkoxide aluminate complex $[Sm\{OPh(2,6\text{-}Ph)\}_2(AlMe_3)_2]$ (**21**). Complex **21** has the capacity to reduce CO_2, but CO_2 insertion into an Al–C bond is facile and $[Sm\{Me_2Al(OPh\{2,6\text{-}Ph\})\}(DME)\{OC(Me)OAl(Me_2)OPh(2,6\text{-}Ph)\}]$ (**22**) is the resultant product from CO_2 insertion and subsequent rearrangement of a phenoxide moiety (Scheme 8.11).[23] This illustrates that the choice of supporting ligand is of vital importance when designing complexes with targeted reactivity profiles.

The Sm(II) complex $[Sm(Giso)_2]$ (**23**) (Giso = $\{[N(Dipp)]_2CNCy_2\}^-$) reacted with CS_2 to give the dinuclear complex $[\{Sm(Giso)_2\}_2\{\mu\text{-}\eta^2{:}\eta^2\text{-}SCSCS_2\text{-}\kappa^4\text{-}C,S,S',S''\}]$ (**24**) (Scheme 8.12).[24] The formation of a C–S bond to give the $\{SCSCS_2\}^{2-}$ fragment in **24** is thought to be favoured over the generation of a thiooxalate or thiocarbonate for two reasons: (i) Thiooxalate or thiocarbonate formation is thought to proceed via a Sm-SCS-Sm intermediate, which

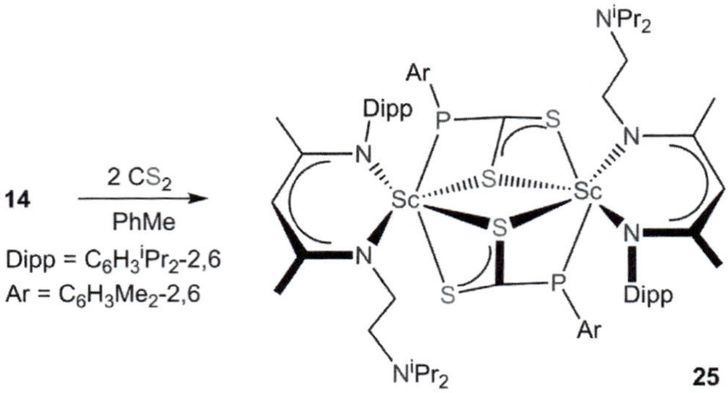

Scheme 8.13. Synthesis of **25** from **14** and CS_2.[15]

may be too hindered to form with such bulky ligands; (ii) "CS" extrusion is energetically unfavourable and thus thiocarbonate formation is challenging.[25] Finally, insertion of CS_2 into polar Ln–E bonds has been observed, such as in the reaction of CS_2 with the scandium phosphinidine complex **14** to give the dimeric complex $[\{Sc(NacNac^{NiPr2})\}_2(\mu\text{-}\eta_2{:}\eta^2\text{-}S_2CPAr\text{-}\kappa^3)_2]$ (**25**, Ar = $C_6H_3Me_2$-2,6) by Sc–S and P–C bond formation (Scheme 8.13).[15]

8.4 CH_4 Reactivity

Selective C–H bond activation is a synthetically challenging but desirable process for the production of fine chemicals and pharmaceuticals.[26] Direct metal-mediated C–H transformations are hampered by several factors, particularly so for methane. Firstly, C–H bonds are notoriously hard to activate, with bond dissociation energies (*ca.* 435 $kJ\,mol^{-1}$) that are much greater than the enthalpy of formation of the resultant C–M bond. Secondly, the pK_a values of alkanes are extremely high, e.g. pK_a CH_4 ~56, and finally, there are no low-lying vacant orbitals in alkanes that can interact with filled metal orbitals. As a result, CH_4 activation by Ln complexes is essentially limited to methyl exchange reactions. Some examples are presented here, as well as related examples of unsaturated hydrocarbons for comparison.

In general, the reactions of Ln hydride or methyl complexes with H_2 or CH_4 is dictated by the energetics of the possible 4-centred

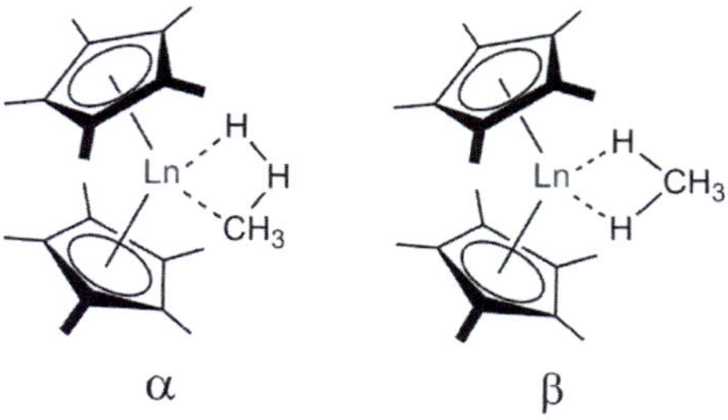

Figure 8.1. The α- and β-transition states for Ln–CH_4/H reacting with H_2/CH_4.[27]

transition states (α- or β-), and the β-transition state is energetically unfavourable (Figure 8.1).[27] Ln hydride complexes can react with CH_4 to give either a Ln hydride complex via C–H bond scission and proton exchange, or a Ln methyl complex with elimination of H_2. Computational analysis suggests that proton exchange is unfavourable as it would proceed via the β-transition state, whereas there is a lower activation barrier to form Ln methyl complexes.[27] However, in practice Ln hydride complexes are routinely synthesised by hydrogenolysis of Ln methyl complexes,[28] and it is noteworthy that complexes of the general formula $[Ln(Cp^*)_2(R)]_n$ (R = H or alkyl) can metallate a wide range of substrates including benzene, pyridine and furan.[29,30] As Ln(II) complexes typically react with substrates via 1 e^- transfer, it is not surprising that CH_4 is not readily activated by these species.

The earliest examples of molecular Ln complexes activating alkane C–H bonds were in the exchange reactions of $[Ln(Cp^*)_2(CH_3)]_n$ [Ln = Sc (**26a**), Y (**26b**) or Lu (**26c**)] with $^{13}CH_4$ to give $[Ln(Cp^*)_2(^{13}CH_3)]_n$ [Ln = Sc (**27a**), Y (**27b**) or Lu (**27c**)] (Scheme 8.14).[28] The same exchange reaction was found to proceed with CD_4. In solution, complexes **26a–c** exist as equilibrium mixtures of dimeric and monomeric forms (**26a–c.mono**), with **26a–c.mono** being the active species. The postulated mechanism requires a β-transition state and while the proposed Ln $\cdots$ CH_4 interaction is relatively weak, metalation of the solvent followed by proton transfer, or extrusion of CH_4 to form a "tuck-in" complex followed by insertion and proton transfer, were both found to not occur. Interestingly, complexes **26a–c** do not activate the d_{12}-cyclohexane solvent at all, and only limited reactivity with higher alkanes such as ethane

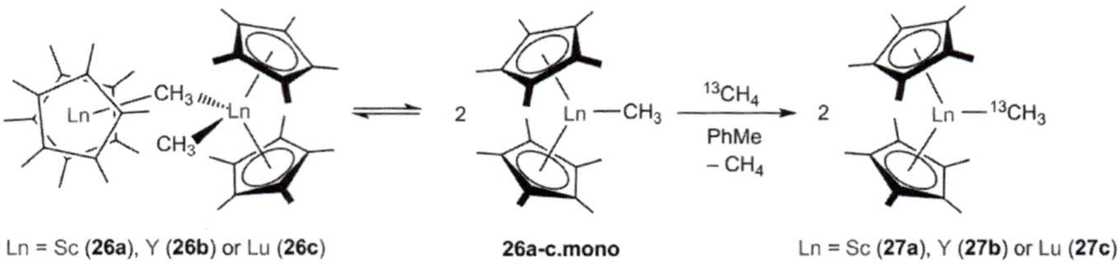

Scheme 8.14. Exchange reactions of **26a–c** with $^{13}CH_4$ to give **27a–c**.[28]

was observed; this was attributed to the transition state being much higher in energy for these substrates.[28] It is germane to note that alkenes which have low-lying π-orbitals can react with Ln(II) complexes; for example, **1** reacts with a series of alkenes to yield the allyl complexes [Sm(Cp*)$_2$(η^3-CH_2CHCHR)].[31]

8.5 N$_2$ Reactivity

The functionalisation of N_2 by molecular complexes is of great interest as it can be a convenient source of atomic "N" if the relatively inert N≡N triple bond (bond dissociation energy = 945 kJ mol^{-1})[16] can be cleaved in a controlled fashion. A small crop of crystals of the first Ln dinitrogen complex, [{Sm(Cp*)$_2$}$_2$(μ-η^2:η^2-N_2)] (**3.Sm**), were discovered during a purification of **1** under an N_2 atmosphere.[32] The electron-rich Sm centre in **1** can form an adduct with N_2 in non-coordinating solvents with a temperature-dependent equilibrium that heavily favours **1** (Scheme 8.15). However, **1** and **3.Sm** have differing solubilities (bimetallic complexes are typically less soluble than parent monomers), thus **3.Sm** can be preferentially crystallised and separated. In the solid state, the N≡N bond distances in **3.Sm** are similar to those in gaseous N_2 [**3.Sm**: 1.088(12) Å; $N_{2(g)}$: 1.0975 Å],[16,32] which suggests that the N_2 ligand in **3.Sm** is neutral and binds as a simple adduct. However, the rest of the analytical data obtained suggests that Sm(III) ions are present, indicating that **3.Sm** has a multiconfigurational ground state that is a complex admixture of Sm(II)$_2$/N_2 and Sm(III)$_2$/N_2^{2-} formulations. A key feature of **3.Sm** (and other Ln dinitrogen complexes, see below) is the orientation of the N_2 moiety, which is bound side-on, and the SmCp*$_2$ units are perpendicular with respect to each other. This observation contrasts with d-transition metal dinitrogen complexes,

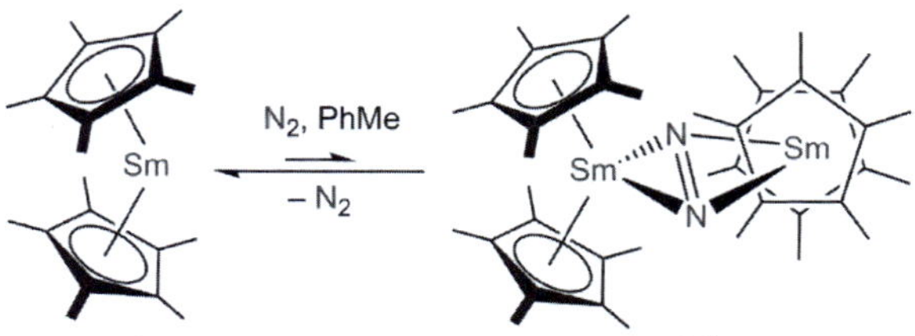

Scheme 8.15. Equilibrium of **1** and **3.Sm** in N_2 and toluene.[30]

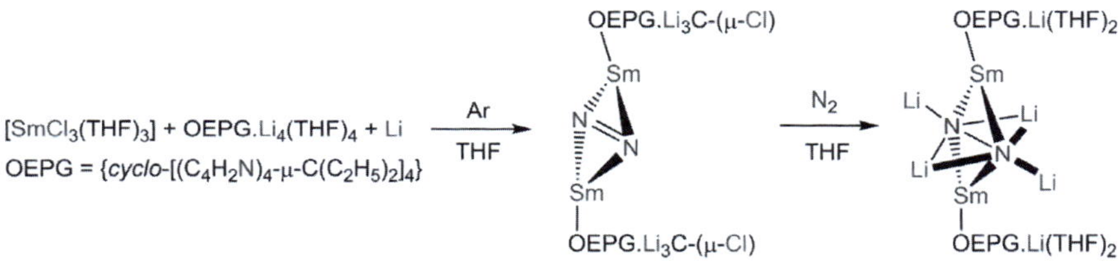

Scheme 8.16. Synthesis of **28–29** from $[SmCl_3(THF)_3]$, $OEPG.Li_4(THF)_4$ and Li.[34]

which typically exhibit terminal N_2 binding, although several group 4 complexes with bridging dinitrogen ligands have been reported.[33]

Following on from the report of **3.Sm**, a second Sm-N_2 complex, $[\{Sm[OEPG.Li(THF)_2]\}_2(Li)_4(\mu^6\text{-}\eta^2{:}\eta^2{:}\eta^2{:}\eta^2{:}\eta^1{:}\eta^1\text{-}N_2)]$ OEPG = 5,10,15,20-octaethyl-porphyrinogen, $\{cyclo\text{-}[(C_4H_2N)_4\text{-}\mu\text{-}C(C_2H_5)_2]_4\}^{4-}$ (**28**), was prepared by reduction of a Sm(III) precursor with Li metal followed by exposure to dinitrogen (Scheme 8.16).[34] Complex **28** is comprised of two Sm(III) centres bridged by an N_2^{4-} fragment bound by four lithium cations. The N–N bond distance in **28** [1.525(4) Å] is consistent with a hydrazido tetraanionic assignment. It has been established that further functionalisation of the N_2^{n-} (n = 1–4) fragment is usually facilitated by the tetraanionic rather than the dianionic form,[34b] thus the N_2^{4-} moiety in **28** should demonstrate further reactivity. During the initial synthesis of **28**, before N_2 is introduced, a green intermediate was observed. This was later isolated and identified as $[\{Sm[OEPG.Li_3(\mu^3\text{-}Cl)]\}_2(\mu\text{-}\eta^2{:}\eta^2\text{-}N_2)]$ (**29**), which contains a side-on bound formally neutral N_2 unit with a characteristic N≡N triple bond distance [1.08(3) Å].[36b] The reaction of this intermediate with Li_4L yields **28**.[34b]

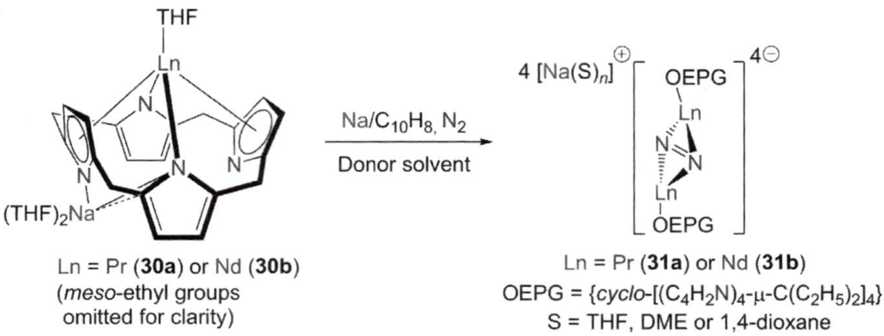

Scheme 8.17. Synthesis of **31a–b** from **30a–b**, N_2 and Na/naphthene.[35]

This work was extended by the reduction of the Ln(III) "ate" complexes [Ln{OEPG.Na(THF)$_2$}(THF)] [Ln = Pr (**30a**) or Nd (**30b**)] with Na and a catalytic amount of naphthalene under a dinitrogen atmosphere to afford $[Na(S)_n]_4[\{Ln(OEPG)\}_2(\mu\text{-}\eta^2\text{:}\eta^2\text{-}N_2)]$ [Ln = Pr (**31a**) or Nd (**31b**), S = THF, DME or 1,4-dioxane] (Scheme 8.17).[35] Complexes **31a–b** again exhibit side-on N_2 binding, but the N=N distances [1.254(7) Å (**31a**); 1.234(8) Å (**31b**)] are more consistent with a N_2^{2-} formulation, regardless of the proximity of Na centres. Given that the measured Sm(III)/(II) reduction potential ($E^0 = -1.55$ V) is smaller than the theoretical values for Pr(III)/(II) ($E^0 = -2.7$ V) and Nd(III)/(II) ($E^0 = -2.6$ V),[21] and that **28** contains a more reduced N_2 fragment than **31a–b**, it is probable that Li or another highly reduced species generated *in situ* during the synthesis of **28** plays a role in reducing the N_2 moiety further.[33b,34]

Further examples of Ln dinitrogen activation include the reduction of [Ln(N″)$_3$] (**5a–k**) by KC_8 in the presence of N_2 to give $[\{Ln(N'')_2(THF)\}_2(\mu\text{-}\eta^2\text{:}\eta^2\text{-}N_2)]$ [Ln = Sc (**32a**), Y (**32b**), La (**32c**), Nd (**32d**), Gd (**32e**), Tb (**32f**), Dy (**32g**), Ho (**32h**), Er (**32i**), Tm (**32j**) and Lu (**32k**)] as well as other species, including various potassium salts of the yttrium cyclometallate $[Y\{N''\}_2\{N(SiMe_3)(SiMe_2CH_2)\text{-}\kappa^2\text{-}N,C\}]^-$, as mentioned in Section 8.2 (Scheme 8.18).[36] A variety of side products such as $[K(18\text{-}C\text{-}6)(THF)_2][\{Ln(N'')_2(THF)\}_2(\mu\text{-}\eta^2\text{:}\eta^2\text{-}N_2)]$ [Ln = Y (**33a**), Gd (**33b**), Tb (**33c**), Dy (**33d**), Ho (**33e**), Er (**33f**) and Lu (**33g**)] were also isolated from these reaction mixtures.[37] Complexes **33a–g** contain N_2^{3-} radical anions that can promote magnetic coupling between the two Ln centres, leading to interesting single molecule

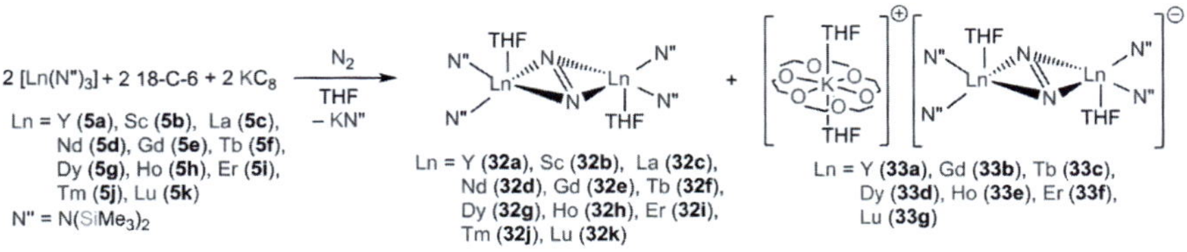

Scheme 8.18. Synthesis of **32a–k** and **33a–g** from **5a–k** and KC_8.[36,37]

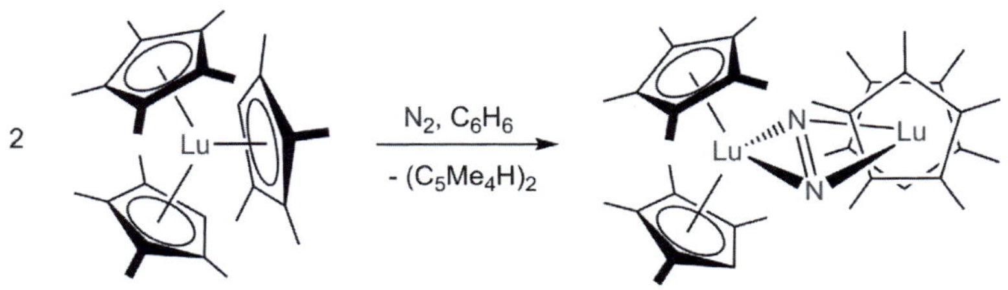

34 **35**

Scheme 8.19. Synthesis of **35** from **34**.[39]

magnet behaviour in some cases.[37] This work has been extended to include the reduction of $[Y(Cp^*)_2H]_2$, or $[Ln(Cp^R)_2][\mu\text{-}Ph_2(BPh_2)]$ with KC_8 under N_2 to produce a series of complexes with the same Ln_2N_2 core.[38,39] Further work showed that isolated Ln(II) species, or transiently produced Ln(II) species (as for the LnL_3/K systems) can reduce N_2 rather than the alkali metal present.[38]

Some Ln dinitrogen complexes have been prepared by more diverse synthetic routes. For example, benzene solutions of $[Lu(Cp^*)(Cp^{Me4})_2]$ (**34**) were found to convert to $[\{Lu(Cp^*)(Cp^{Me4})\}_2(\mu\text{-}\eta^2{:}\eta^2\text{-}N_2)]$ (**35**) under a dinitrogen atmosphere, but in the absence of KC_8 (Scheme 8.19).[39] This observation was later attributed to a light-induced radical activation mechanism, arising from one of the Cp^{Me4} rings in **34** being relatively weakly bound in an η^3- fashion and forming $(Cp^{Me4})_2$ by radical coupling.[40] DFT studies of the yttrium analogue **34.Y** showed that the LUMO in this complex exhibits charge transfer from the η^3-Cp^{Me4} ring to the metal centre, hence $[Y(Cp^*)(Cp^{Me4})]^{\cdot}$ and $Cp^{Me4\cdot}$ radicals can form, with the Ln radical reducing N_2 analogously to a discreet Ln(II) complex. Further examples of photoactive Ln complexes containing an η^3-carbanion include $[Ln(Cp^*)_2(L)]$ [Ln = Y, Lu; L = CH_2CHCH_2, $CH_2C(Me)CH_2$], which also reduce dinitrogen under

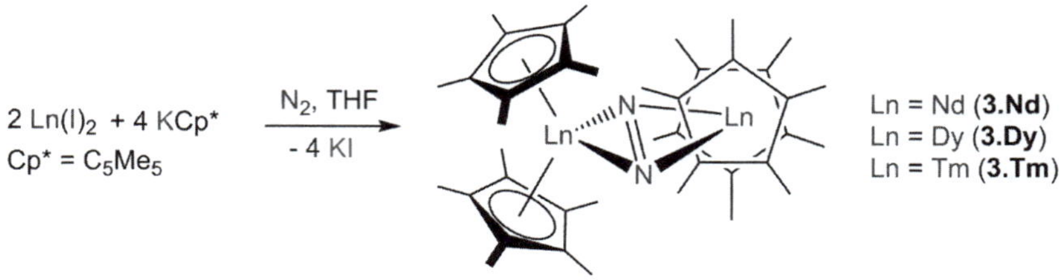

Scheme 8.20. Synthesis of **3.Nd**, **3.Dy** and **3.Tm** from LnI_2, N_2 and KCp*.[43]

photolytic conditions to give $[\{Ln(Cp^*)_2\}_2(\mu\text{-}\eta^2{:}\eta^2\text{-}N_2)]$ (Ln = Y, Lu).[41] It is noteworthy that the addition of 2,5-di-*tert*-butyl-aniline to $[Tm(Tp^{Me,tBu})_2\{CH(SiMe_3)_2\}]$ ($Tp^{Me,tBu}$ = $\{HB(C_3HN_2Me\text{-}3\text{-}{}^tBu\text{-}5)_3\}^-$) under a dinitrogen atmosphere gave $[Tm(Tp^{Me,tBu})_2(\mu\text{-}\eta^2{:}\eta^2\text{-}N_2)]$, but this proceeded via an unspecified route.[42]

The development of synthetic routes to NdI_2, DyI_2 and TmI_2[3] allowed the direct preparation of the dinuclear complexes $[\{Ln(Cp^*)_2\}_2(\mu\text{-}\eta^2{:}\eta^2\text{-}N_2)]$ [Ln = Nd (**3.Nd**), Dy (**3.Dy**), Tm (**3.Tm**)] using KCp* under an atmosphere of dinitrogen (Scheme 8.20).[43] This combined salt metathesis and oxidation methodology was extended to allow the preparation of a series of N″, $OC_6H_3{}^tBu\text{-}2,6$ and Cp″ ($\{C_5H_3(SiMe_3)_2\text{-}1,3\}^-$) analogues.[43] In these syntheses N_2 reduction occurs directly due to the highly reducing Ln(II) ions and supporting ligands. NdI_2 and DyI_2 are unstable as THF or DME solutions at room temperature, but they do not reduce N_2, thus the electron donating ancillary ligands facilitate this process. Finally, the Nd(II)-ate complex, $[Nd(Cp^{ttt})_2\{(\mu\text{-}I)K(18\text{-}C\text{-}6)\}]$ (**36**), which is not stable at room temperature under an Ar atmosphere, reacts with N_2 to yield the Nd(III) complex $[\{Nd(Cp^{ttt})_2\}_2(\mu\text{-}\eta^2{:}\eta^2\text{-}N_2)]$ (**37**) (Scheme 8.21).[44] This is the first reported example of dinitrogen reduction by a discreet, isolated Ln(II) organometallic complex without the aid of an external reductant.[46]

The geometrical arrangement of the $L_4Ln_2N_2$ cores in N_2-bridged Ln complexes indicates that there may be some degree of orbital overlap, and perhaps covalency, between the Ln centres and the N_2^{2-} fragment.[36b,45] Using models of $[\{Sc(Cp^*)_2\}_2(\mu\text{-}\eta^2{:}\eta^2\text{-}N_2)]$ (**3.Sc**) and $[\{Y(N'')_2(THF)\}_2(\mu\text{-}\eta^2{:}\eta^2\text{-}N_2)]$ (**32a**) for theoretical studies, it was found that occupied *n*d orbitals in the L_2Ln fragments are high enough in energy to engage in π-backbonding with the N_2-π^* orbitals. The result is a polarised covalent interaction that can

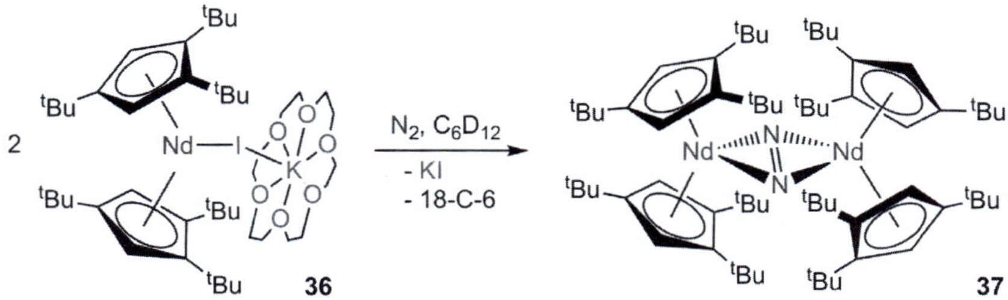

36 37

Scheme 8.21. Synthesis of **37** from **36** and N_2.[44]

be assigned as a 4-centre, 2 e^- bond in the Ln_2N_2 fragment, reducing the formal bond order between the N atoms to 2.[45] Formal N_2 adducts of Ln complexes are limited to a single characterised example (**3.Sm**), but theoretical studies suggest that an adduct between N_2 and [Ln(Cp)$_2$] (Ln = Eu, Yb) or [Ln(Cp*)$_2$] (Ln = Eu, Yb) is energetically favourable.[46] However, the calculated energy gain is exceptionally small ($\Delta E < 5$ kcal mol^{-1}) and the predicted stretching frequency of the N_2 fragment was barely perturbed for any of the complexes ($\Delta\nu < 4\,cm^{-1}$), thus these adducts are likely to be very weakly bound.

8.6 NO_x Reactivity

The reactivity of Ln complexes with nitrogen oxides is currently not well developed. NO is an odd-electron species which can undergo a facile 1 e^- reduction, and N_2O eliminates $N_{2(g)}$ following O-atom transfer. Therefore, NO_x species are typically strong oxidants, especially with highly reduced metal complexes.[47] In common with the CO reactions described previously, the reactivity profiles of Ln complexes with NO contrast with d-block chemistry, as simple adducts are not readily isolated and the highly polarised Ln–ligand bonds are a source of rich further chemistry. The first examples of NO_x reactivity with a Ln complex were when **1.THF** was treated with both NO and N_2O to give [{Sm(Cp*)$_2$}$_2$(μ-O)] (**38**) (Scheme 8.22).[48] The solid-state structure of **38** is different to that of previous examples in this chapter that contain two LnCp*$_2$ moieties, as the four Cp* rings form a protective tetrahedron about the Sm–O–Sm fragment.

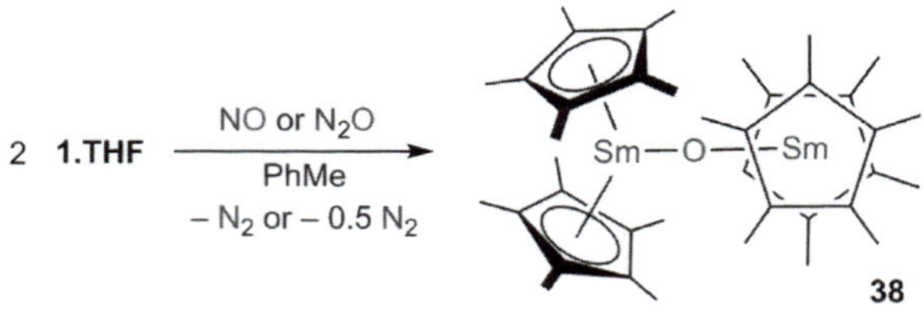

Scheme 8.22. Synthesis of **38** from **1.THF** and NO_x.[48]

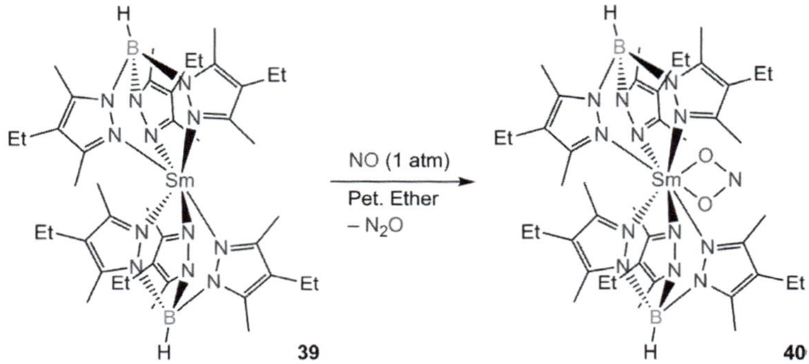

Scheme 8.23. Synthesis of **40** from **39** and excess NO.[49]

The Ln(II) complex $[Sm(Tp^{Me2,4\text{-}Et})_2]$ (**39**, $Tp^{Me2\text{-}4Et}$ = $\{HB(C_3N_2Me_2\text{-}3,5\text{-}Et\text{-}4)_3\}^-$) reacts with an excess of NO to give $[Sm(Tp^{Me2,4\text{-}Et})_2(ONO\text{-}\kappa\text{-}O,O')]$ (**40**) (Scheme 8.23).[49] Formally 1 equiv. of "N" has been lost in this process, so in the proposed mechanism a terminal nitrosyl intermediate reacts with 2 further equiv. of NO to give **40** and 1 equiv. of N_2O. An analogous NO reactivity profile has been previously observed in d-transition metal chemistry.[50] The reactions of $[Ln(Tp^{Me2})_2]$ (Ln = Sm, Eu or Yb; Tp^{Me2} = $\{HB(C_3HN_2Me_2\text{-}3,5)_3\}^-$) and $[Ln(Tp^{Me2,4\text{-}Et})_2]$ (Ln = Eu or Yb) with NO were also investigated, but the products could not be structurally characterised.[49] However, as the FTIR spectra of the products were essentially identical to that of **40**, the reactions were presumed to proceed by the same mechanism.

18-C-6-free analogues of the previously described Ln complexes, **33a–g**, exhibit a close contact between the $\mu\text{-}N_2{}^{3-}$ moiety and the K^+ counter ion and can act as potent 2 or $3e^-$ reducing agents.[36b] This was exploited in the reaction of $[\{Y(N'')_2(THF)\}_2(\mu\text{-}\eta^2\text{:}\eta^2\text{-}N_2)K]$ (**41**) with 1.5 eq. of NO to give $[\{Y(N'')_2(THF)\}_2(\mu\text{-}\eta^2\text{:}\eta^2\text{-}NO)]$ (**42**), by elimination of $N_{2(g)}$ and concomitant loss of K^+ with unidentified

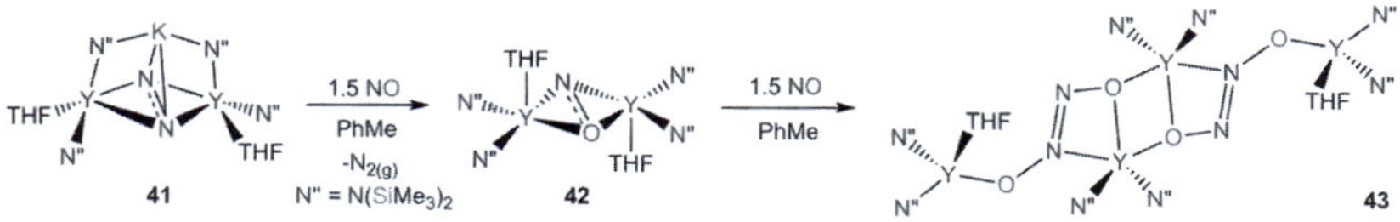

Scheme 8.24. Synthesis of **43** and **42** from **41** and different amounts of NO.[51]

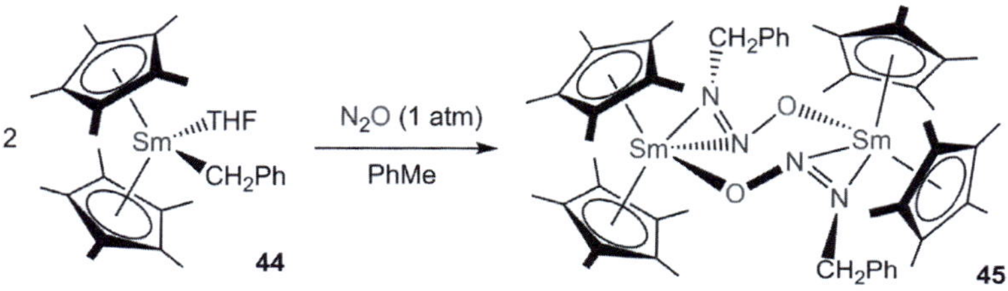

Scheme 8.25. Reaction of **44** with N_2O to give **45**.[53]

by-products (Scheme 8.24).[51] Complex **42** is the first example of a μ-NO^{2-} species, with this formulation verified by EPR spectroscopy. The reaction of **42** with NO gave [{Y(N″)$_2$(THF)}$_2${Y(N″)$_2$}$_2$(μ^3-η^1:η^1:η^2-ON=NO-κ^3-N,O,O')$_2$] (**43**) by NO homologation.[51]

In common with CO, nitrogen oxides can readily insert into Ln–C$_{\text{alkyl}}$ bonds. For example, N_2O reacts with the Sm(III) complex, [Sm(Cp*)$_2$(CH$_2$Ph)(THF)] (**44**), to give the dimeric complex [{Sm(Cp*)$_2$}{μ-η^1:η^2-N(CH$_2$Ph)=NO-κ^3-N,N',O}]$_2$ (**45**) (Scheme 8.25).[52] Analogous N_2O insertion chemistry was observed for the Ln-allyl complexes [Ln(Cp^R)$_2$(η^3-C$_3$H$_5$)] (Ln = Sc, Y; Cp^R = Cp*, Cp^{Me4}), which yielded complexes with bridging {μ-η^1:η^2-N(C$_3$H$_5$) =NO-κ^3-N,N',O} moieties.[53] In contrast, 2 equiv. of NO insert into one of the Ln–C$_{\text{allyl}}$ bonds in [Ln(Cp*)$_2$(η^3-C$_3$H$_5$)] [Ln = Y (**46a**), La (**46b**), Sm (**46c**)], with subsequent dimerisation to give [Ln(Cp*)$_2${μ-η^1:η^2-ONN(C$_3$H$_5$)O-κ^3-N,O,O'}]$_2$ [Ln = Y (**47a**), La (**47b**), Sm (**47c**)] (Scheme 8.26).[54] In the case of **47a**, thermolysis in toluene yielded monomeric [Y(Cp*)$_2${μ-η^1:η^2-ONN(C$_3$H$_5$)O-κ^3-N,O,O'}]$_2$ (**48**), where the biologically relevant {ONN(C$_3$H$_5$)O} ligand has changed its binding mode.

8.7 RN$_3$ Reactivity

Organoazides are readily activated by Ln complexes and exhibit a reactivity profile that is unlike most other heteroallenes, as they can

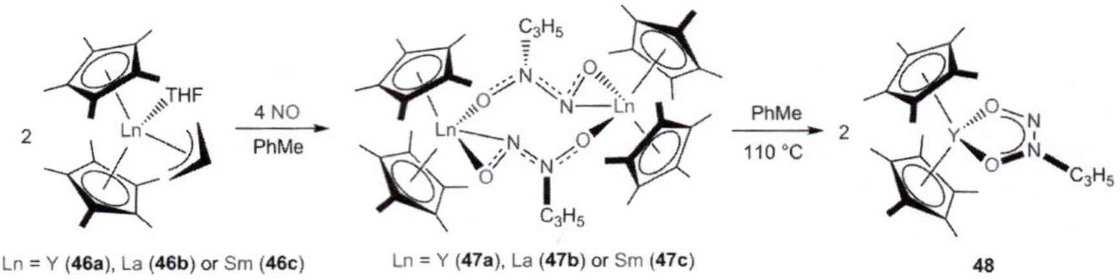

Scheme 8.26. Synthesis of **48** and **47a–c** from **46a–c** and NO.[54]

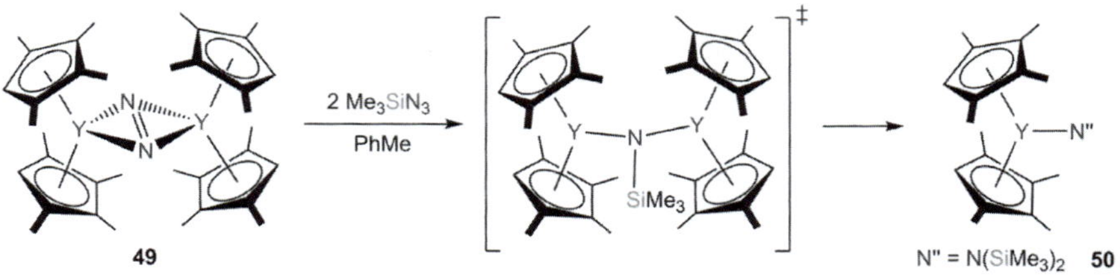

Scheme 8.27. Synthesis of **50** from **49** and trimethylsilylazide.[56]

act as 2 e^- oxidants to yield a $\{RN{=}\}^{2-}$ moiety and $N_{2(g)}$ as a by-product. However, most Ln complexes are not able to undergo a 2 e^- oxidation process and insertion into Ln–ligand bonds is more common, though a Ce(II) complex could theoretically be oxidised to Ce(IV).[55] An alternative to a discrete Ln 2 e^- process is to use a highly reduced ligand which can serve as a leaving group upon oxidation. This was demonstrated when the yttrium complex $[\{Y(Cp^{Me4})_2\}_2(\mu\text{-}\eta^2{:}\eta^2\text{-}N_2)]$ (**49**) was reacted with 2 equiv. of Me_3SiN_3 to give $[Y(Cp^{Me4})_2(N'')]$ (**50**) (Scheme 8.27).[56] Presumably **50** forms via a 2 e^- oxidation of **49** to give an imido bridged intermediate, which undergoes further reactivity to generate the N″ group in **50** along with other unidentified products.

The reaction of the alkyl alkylidene complex $[Y(BIPM^{TMS})(CH_2Ph)(THF)]$ (**51**, $BIPM^{TMS} = \{C(PPh_2NSiMe_3)_2\}^{2-}$) with adamantyl azide gave the triazenido alkylidene $[Y(BIPM^{TMS})(\eta^3\text{-}N_3Ad\text{-}1,CH_2Ph\text{-}3\text{-}\kappa^2\text{-}N,N')(THF)]$ (**52**) via a 1,2-migratory insertion (Scheme 8.28).[57] Interestingly, in this reaction the Y=C bond does not react with AdN_3 while other heteroallenes such as isocyanates (Bu^tNCO) and carbodiimides $[(CyN)_2C]$ additionally

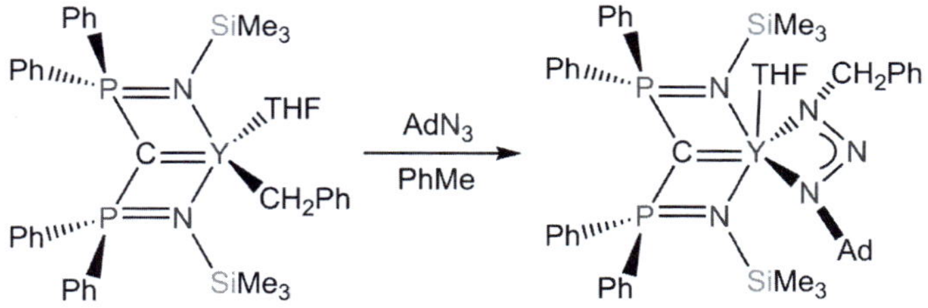

51 **52**

Scheme 8.28. Synthesis of **52** from **51** and adamantyl azide.[57]

undergo a [2 + 2] cycloaddition reaction with **51**.[57] A 1,2-migratory insertion of adamantyl azide into a Ln–C bond was also observed with the Ln complexes [Ln{$OC(Me)_2CH_2[NCH_2CH_2N(Dipp)C]$}$(N'')_2$] (Ln = Y or Ce) to yield [Ln{η^3-$OC(Me)_2CH_2[NCH_2CH_2N(Dipp)C$-1-$N_3$Ad-3]-$\kappa^3$-$N$,$N'$,$O$}$(N'')_2$] (Ln = Y or Ce).[58]

8.8 Ce(IV) Reactivity

The small molecule activation chemistry of Ce(IV) complexes is described separately, as Ce is unique among the Ln in having a readily accessible +4 oxidation state.[1] However, organometallic Ce(IV) complexes are rare,[59] and as a result this field is currently poorly developed. As the oxidation state of a Ln increases, the valence orbital energies decrease, reducing the metal-ligand orbital energy mismatch and thus increasing the extent of covalency.[60] Therefore Ce(IV) complexes should form multiple bond interactions to p-block elements more readily than other Ln(III) complexes, as in the Ce(IV) complex [Ce{CoCp[$P(O)(OEt)_2]_3$}$_2(OH_2)(O)${MeC(O)NH_2}]$_2$ (**53**).[61] The bonding in the formal Ce=O moiety in **53** is best described as a highly polarised double bond, and treatment with CO_2 yields the carbonate species [Ce{CoCp[$P(O)(OEt)_2]_3$}$_2$(η^2-CO_3)] (**54**) by insertion into the Ce=O bond (Scheme 8.29). Furthermore, **53** reacts with CO to give a dinuclear carbonate species, [{Ce[CoCp {$P(O)(OEt)_2$}$_3]_2$}(μ-η^2:η^1-CO_3){Ce[CoCp{$P(O)(OEt)_2$}$_3]_2(OH_2)$}] (**55**), where each Ce=O group has provided one of the necessary O atoms for $CO_3{}^{2-}$ formation.[61]

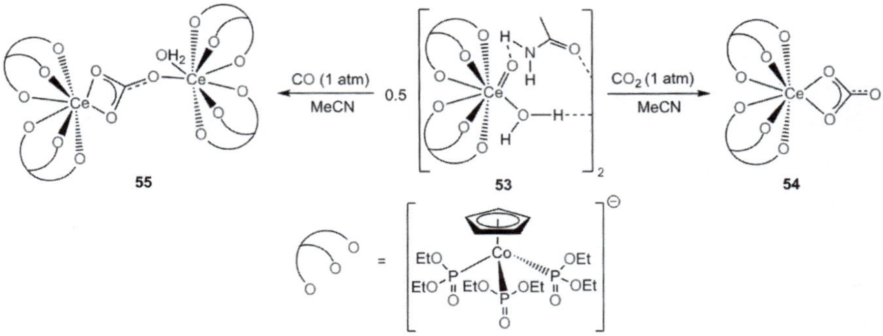

Scheme 8.29. Synthesis of **54** and **55** from **53** and CO_2 or CO, respectively.[61]

8.9 Summary

As molecular Ln chemistry is in its infancy the small molecule activation chemistry of these complexes is underdeveloped, hence the relatively limited number of examples presented in this chapter.[62] However, these complexes promote diverse and interesting reactivity profiles that are in stark contrast with analogous d-block chemistry, and therefore fundamental investigations in this field have been fully justified. Further to this, the work presented in this chapter is neither exhaustive for the substrates shown nor is Ln small molecule activity limited to these substrates. For example, Ln complexes react with molecular element precursors such as S_8 and P_4 to provide clusters that could potentially be used for stoichiometric delivery of E^{n-}. Some of the CO and N_2 activation chemistry by Ln complexes summarised in this chapter have provided the first examples for such reactivity in the periodic table, affording unseen pathways, molecular fragments and complex geometries. These were only subsequently seen in the d-block after further research, with the side-on binding of N_2 discovered for Zr after the Ln's.

In the future we expect Ln(III) complexes that exhibit unique reactivity will be targeted, such as $[Ln(Cp^*)_3]$, which degrades ethereal solvents. In Ln(II) chemistry, complexes of the non-traditional Ln(II) metals such as $[K(donor)][Ln(Cp')_3]$ ($Cp'' = \{C_5H_4(SiMe_3)\}^-$) are potent 1 e^- reducing agents, and their reactivity has yet to be fully explored. In general, more Ln-ligand combinations and novel complexes will generate new and unprecedented reactivity, but there are many synthetic challenges to overcome to develop molecular Ln chemistry for future exploitation. For example, many d-block catalysts contain terminal unsupported M=E multiple bonds that undergo insertion as part of the catalytic cycle. To date there is a paucity of complexes that exhibit Ln=E multiple bonding, and there are only a handful of examples where these bonds are not protected or supported by co-ligands.[63] The synthesis of more Ln=E complexes could open up this area and allow d-transition metal-type reactivity and catalytic processes to be mimicked, though these Ln-mediated transformations would inevitably contrast to the d-block due to the prevalence of 1 e^- processes for Ln complexes. As many Ln are significantly more abundant than industrially relevant platinum group metals, it is likely that economic factors could also drive this area of research in future.

References

1. (a) H. C. Aspinall, *Chemistry of the f-Block Elements*, Gordon and Breach, London, 2001, pp. 61–156; (b) S. A. Cotton, *Lanthanide and Actinide Chemistry*, Wiley, Chichester, 2006, pp. 1–144; (c) S. T. Liddle, *Lanthanides: Organometallic Chemistry, Encyclopaedia of Inorganic and Bioorganic Chemistry*, Wiley, Chichester, 2012; (d) D. A. Atwood, *The Rare Earth Elements: Fundamentals and Applications*, Wiley, Chichester, 2012.
2. R. D. Shannon, *Acta Crystallogr., Sect. A*, 1976, *A32*, 751–767.
3. (a) P. Girard, J. L. Namy and H. B. Kagan, *J. Am. Chem. Soc.*, 1980, *102*, 2693–2698; (b) M. N. Bochkarev, I. L. Fedushkin, A. A. Fagin, T. V. Petrovskaya, J. W. Ziller, R. N. R. Broomhall-Dillard and W. J. Evans, *Angew. Chem. Int. Ed. Engl.*, 1997, *36*, 133–135; (c) W. J. Evans, N. T. Allen and J. W. Ziller, *J. Am. Chem. Soc.*, 2000, *122*, 11749–11750; (d) M. N. Bochkarev, I. L. Fedushkin, S. Dechert, A. A. Fagin and H. Schumann, *Angew. Chem. Int. Ed.*, 2001, *40*, 3176–3178; (e) F. Nief, *Dalton. Trans.*, 2010, *39*, 6589–6598.
4. (a) G. Meyer, *Angew. Chem. Int. Ed.*, 2014, *53*, 3550–3551; (b) F. T. Edelmann, *Coord. Chem. Rev.*, 2014, *261*, 73–155, and references cited therein.
5. H. H. Storch, N. Golumbic and R. B. Anderson, *The Fischer–Tropsch Reaction and Related Synthesis*, Wiley, New York, 1951.

6. (a) M. Schultz, C. J. Burns, D. J. Schwartz and R. A. Andersen, *Organometallics*, 2001, *20*, 5690–5699; (b) P. Selg, H. H. Brintzinger, M. Schultz and R. A. Andersen, *Organometallics*, 2002, *21*, 3100–3107.
7. W. J. Evans, J. W. Grate, L. A. Hughes, H. Zhang and J. L. Atwood, *J. Am. Chem. Soc.*, 1985, *107*, 3728–3730.
8. W. J. Evans, D. S. Lee, J. W. Ziller and N. Kaltsoyannis, *J. Am. Chem. Soc.*, 2006, *128*, 14176–14184.
9. M. Fang, J. H. Farnaby, J. W. Ziller, J. E. Bates, F. Furche and W. J. Evans, *J. Am. Chem. Soc.*, 2012, *134*, 6064–6067.
10. (a) J. M. O'Connor and C. P. Casey, *Chem. Rev.*, 1987, *87*, 307–318; (b) S. Demir, T. J. Mueller, J. W. Ziller and W. J. Evans, *Angew. Chem. Int. Ed.*, 2011, *50*, 515–518.
11. (a) W. J. Evans, K. J. Forrestal and J. W. Ziller, *J. Am. Chem. Soc.*, 1995, *117*, 12635–12636; (b) W. J. Evans, S. A. Kozimor, G. W. Nyce and J. W. Ziller, *J. Am. Chem. Soc.*, 2003, *125*, 13831–13835.
12. W. J. Evans, A. L. Wayda, W. E. Hunter and J. L. Atwood, *J. Chem. Soc., Chem. Commun.*, 1981, 706–708.
13. C. J. Schaverien, N. Meijboom and A. G. Orpen, *J. Chem. Soc. Chem. Commun.*, 1992, 124–126.
14. See X. Zhou and M. Zhu, *J. Organomet. Chem.*, 2002, *647*, 28–49, and references cited therein.
15. Y. Lu, C. E. Kefalidis, J. Zhou, L. Maron, X. Leng and Y. Chen, *J. Am. Chem. Soc.*, 2013, *135*, 14784–14796.
16. (a) D. R. Lide (ed.), *CRC Handbook of Chemistry and Physics, 73rd edn.*, CRC Press, Boca Raton, 1992; (b) E. K. Moltzen, K. J. Klabunde and S. Senning, *A. Chem. Rev.*, 1988, *88*, 391–406.
17. (a) W. J. Evans, C. A. Seibel and J. W. Ziller, *Inorg. Chem.*, 1998, *37*, 770–776; (b) W. J. Evans, J. M. Perotti, J. C. Brady and J. W. Ziller, *J. Am. Chem. Soc.*, 2003, *125*, 5204–5212.
18. L. Castro, S. Labouille, D. R. Kindra, J. W. Ziller, F. Nief, W. J. Evans and L. Maron, *Chem. Eur. J.*, 2012, *18*, 7886–7895.
19. N. W. Davies, A. S. P. Frey, M. G. Gardiner and J. Wang, *Chem. Commun.*, 2006, 4853–4855.
20. J. Andrez, J. Pécaut, P.-A. Bayle and M. Mazzanti, *Angew. Chem. Int. Ed.*, 2014, *53*, 10448–10452.
21. L. J. Nugent, R. D. Baybarz, J. L. Burnett and J. L. Ryan, *J. Phys. Chem.*, 1973, *77*, 1528–1539.
22. J. Andrez, G. Bozoklu, G. Nocton, J. Pécaut, R. Scopelliti, L. Dubois and M. Mazzanti, *Chem. Eur. J.*, 2015, *21*, 15188–15200.
23. I. Korobkov and S. Gambarotta, *Organometallics*, 2009, *28*, 4009–4019.
24. D. Heitmann, C. Jones, D. P. Mills and A. Stasch, *Dalton. Trans.*, 2010, *39*, 1877–1882.

25. O. P. Lam, L. Castro, B. Kosog, F. W. Heinemann, L. Maron and K. Meyer, *Inorg. Chem.*, 2012, *51*, 781–783.
26. See P. L. Arnold, M. W. McMullon, J. Rieb and F. E. Kühn, *Angew. Chem. Int. Ed.*, 2015, *54*, 82–100, and references cited therein.
27. L. Maron, L. Perrin and O. Eisenstein, *J. Chem. Soc., Dalton Trans.*, 2002, 534–539, and references cited therein.
28. (a) P. L. Watson, *J. Am. Chem. Soc.*, 1983, *105*, 6491–6493; (b) P. L. Watson and G. W. Parshall, *Acc. Chem. Res.*, 1985, *18*, 51–56; (c) M. E. Thompson, S. M. Baxter, A. Ray. Bulls, B. J. Burger, M. C. Nolan, B. D. Santarsiero, W. P. Schaefer and J. E. Bercaw, *J. Am. Chem. Soc.*, 1987, *109*, 203–219; (d) D. S. Levine, T. D. Tilley and R. A. Andersen, *Organometallics*, 2015, *34*, 4647–4655.
29. P. L. Watson, *J. Chem. Soc., Chem. Commun.*, 1983, 276–277.
30. (a) W. J. Evans, J. M. Perotti and J. W. Ziller, *J. Am. Chem. Soc.*, 2005, *127*, 3894–3909; (b) K. H. den Haan, Y. Wielstra and J. Teuben, *Organometallics*, 1987, *6*, 2053–2060; S. N. Ringelberg, A. Meetsma, S. I. Troyanov, B. Hessen and J. H. Teuben, *Organometallics*, 2002, *21*, 1759–1765.
31. W. J. Evans, T. A. Ulibarri and J. W. Ziller, *J. Am. Chem. Soc.*, 1990, *112*, 2314–2324.
32. W. J. Evans, T. A. Ulibarri and J. W. Ziller, *J. Am. Chem. Soc.*, 1988, *110*, 6877–6879.
33. (a) P. J. Chirik, *Dalton Trans.*, 2007, *25*, 16–25; (b) S. Gambarotta and J. Scott, *Angew. Chem. Int. Ed.*, 2004, *43*, 5298–5308.
34. (a) J. Jubb and S. Gambarotta, *J. Am. Chem. Soc.*, 1994, *116*, 4477–4468; (b) J. Guan, T. Dubé, S. Gambarotta and G. P. A. Yap, *Organometallics*, 2000, *19*, 4820–4827.
35. E. Campazzi, E. Solari, C. Floriani and R. Scopelliti, *Chem. Commun.*, 1998, 2603–2604.
36. (a) W. J. Evans, D. S. Lee, D. B. Rego, J. M. Perotti, S. A. Kozimor, E. K. Moore and J. W. Ziller, *J. Am. Chem. Soc.*, 2004, *126*, 14574–14582; (b) W. J. Evans, M. Fang, G. Zucchi, F. Furche, J. W. Ziller, R. M. Hoekstra and J. I. Zink, *J. Am. Chem. Soc.*, 2009, *131*, 11195–11202; (c) W. J. Evans, J. E. Bates, S. E. Lorenz, D. S. Lee, D. B. Rego, J. W. Ziller, F. Furche and W. J. Evans, *Inorg. Chem.*, 2011, *50*, 1459–1469.
37. (a) J. D. Rinehart, M. Fang, W. J. Evans and J. R. Long, *Nature Chem.*, 2011, *3*, 538–542; (b) J. D. Rinehart, M. Fang, W. J. Evans and J. R. Long, *J. Am. Chem. Soc.*, 2011, *133*, 14236–14239.
38. (a) B. M. Schmiege, J. W. Ziller and W. J. Evans, *Inorg. Chem.*, 2010, *49*, 10506–10511; (b) M. R. MacDonald, J. W. Ziller and W. J. Evans, *J. Am. Chem. Soc.*, 2011, *133*, 15914–15917.

39. T. J. Mueller, M. E. Fieser, J. W. Ziller and W. J. Evans, *Chem. Sci.*, 2011, *2*, 1992–1996.
40. M. E. Fieser, J. E. Bates, J. W. Ziller, F. Furche and W. J. Evans, *J. Am. Chem. Soc.*, 2013, *135*, 3804–3807.
41. M. E. Fieser, C. W. Johnson, J. E. Bates, J. W. Ziller, F. Furche and W. J. Evans, *Organometallics*, 2015, *34*, 4387–4393.
42. J. Cheng, J. Takats, M. J. Ferguson and R. McDonald, *J. Am. Chem. Soc.*, 2008, *130*, 1544–1545.
43. (a) W. J. Evans, N. T. Allen and J. W. Ziller, *J. Am. Chem. Soc.*, 2001, *123*, 7927–7928; (b) W. J. Evans, N. T. Allen and J. W. Ziller, *Angew. Chem. Int. Ed.*, 2002, *41*, 359–361; (c) W. J. Evans, G. Zucchi and J. W. Ziller, *J. Am. Chem. Soc.*, 2003, *125*, 10–11.
44. F. Jaroschik, A. Momin, F. Nief, X.-F. Le Goff, G. B. Deacon and P. C. Junk, *Angew. Chem. Int. Ed.*, 2009, *48*, 1117–1121.
45. S. Demir, S. E. Lorenz, M. Fang, F. Furche, G. Meyer, J. W. Ziller and W. J. Evans, *J. Am. Chem. Soc.*, 2010, *132*, 11151–11158.
46. L. Perrin, L. Maron, O. Eisenstein, D. J. Schwartz, C. J. Burns and R. A. Andersen, *Organometallics*, 2003, *22*, 5447–5453.
47. H. Yu, G. Jia and Z. Lin, *Organometallics*, 2009, *28*, 1158–1164.
48. W. J. Evans, J. W. Grate, I. Bloom, W. E. Hunter and J. L. Atwood, *J. Am. Chem. Soc.*, 1985, *107*, 405–409.
49. G. H. Maunder, M. R. Russo and A. Sella, *Polyhedron*, 2004, *23*, 2709–2714.
50. C. E. Ruggiero, S. M. Carrier and W. B. Tolman, *Angew. Chem. Int. Ed.*, 1994, *33*, 895–897.
51. W. J. Evans, M. Fang, J. E. Bates, F. Furche, J. W. Ziller, M. D. Kiesz and J. I. Zink, *Nature Chem.*, 2010, *2*, 644–647.
52. T. Labahn, A. Mandel and J. Magull, *Z. Anorg. Allg. Chem.*, 1999, *625*, 1273–1277.
53. S. Demir, E. Montalvo, J. W. Ziller, G. Meyer and W. J. Evans, *Organometallics*, 2010, *29*, 6608–6611.
54. I. J. Casely, Y. Suh, J. W. Ziller and W. J. Evans, *Organometallics*, 2010, *29*, 5209–5214.
55. (a) P. B. Hitchcock, M. F. Lappert, L. Maron and A. V. Protchenko, *Angew. Chem. Int. Ed.*, 2008, *47*, 1488–1491; (b) M. E. Fieser, M. R. MacDonald, B. T. Krull, J. E. Bates, J. W. Ziller, F. Furche and W. J. Evans, *J. Am. Chem. Soc.*, 2015, *137*, 369–382.
56. S. E. Lorenz, B. M. Schmiege, D. S. Lee, J. W. Ziller and W. J. Evans, *Inorg. Chem.* 2010, *49*, 6655–6663.
57. D. P. Mills, L. Soutar, O. J. Cooper, W. Lewis, A. J. Blake and S. T. Liddle, *Organometallics*, 2013, *32*, 1251–1264.

58. Z. R. Turner, R. Bellabarba, R. P. Tooze and P. L. Arnold, *J. Am. Chem. Soc.*, 2010, *132*, 4050–4051.
59. (a) O. Eisenstein, P. B. Hitchcock, A. G. Hulkes, M. F. Lappert and L. Maron, *Chem. Commun.*, 2001, *17*, 1560–1561; (b) P. B. Hitchcock, A. G. Hulkes and M. F. Lappert, *Inorg. Chem.*, 2004, *43*, 1031–1038; (c) U. J. Williams, P. J. Carroll and E. J. Schelter, *Inorg. Chem.*, 2014, *53*, 6338–6345; (d) M. P. Coles, P. B. Hitchcock, A. V. Khvostov, M. F. Lappert, Z. Li and A. V. Protchenko, *Dalton Trans.*, 2010, *39*, 6780–6788; (e) M. Gregson, E. Lu, J. McMaster, W. Lewis, A. J. Blake and S. T. Liddle, *Angew. Chem. Int. Ed.*, 2013, *52*, 13016–13019; (f) E. J. Schelter, *Nature Chem.*, 2013, *5*, 348.
60. D. L. Clark, J. C. Gordon, P. J. Hay and R. Poli, *Organometallics*, 2005, *24*, 5747–5758.
61. Y.-M. So, G.-C. Wang, Y. Li, H. H.-Y. Sing, I. D. Williams, Z. Lin and W.-H. Leung, *Angew. Chem. Int. Ed.*, 2014, *53*, 1626–1629.
62. W. J. Evans, *J. Alloys Compd.*, 2009, *488*, 493–510.
63. (a) G. R. Giesbrecht and J. C. Gordon, *Dalton Trans.*, 2004, 2387–2393; (b) O. T. Summerscales and J. C. Gordon, *RSC Adv.*, 2013, *3*, 6682–6693; (c) S. T. Liddle, D. P. Mills and A. J. Wooles, *Organomet. Chem.*, 2010, *36*, 29–55; (d) S. T. Liddle, D. P. Mills and A. J. Wooles, *Chem. Soc. Rev.*, 2011, *40*, 2164–2176.

https://doi.org/10.1142/9781800610163_0009

Chapter 9

Small Molecule Activation by Actinide Complexes

Christopher Hoerger and Karsten Meyer†*

Department of Chemistry and Pharmacy,
Friedrich-Alexander-Universität Erlangen-Nürnberg,
Egerlandstr. 1, Erlangen, 91058, Germany
**chris.hoerger@gmx.de*
†karsten.meyer@fau.de

9.1 Introduction and Scope of the Review

Since the emergence of organometallic uranium and thorium chemistry in the mid-1950s,[1] advances in classic Werner-type coordination chemistry of the actinides have lagged behind significantly from those of their transition metal counterparts. This was likely due to the lack of well-defined and soluble uranium(III) starting materials[2] until they were first introduced by Alfred Sattelberger and David Clark,[3,4] and further expanded upon by more recent contributions.[5–8] With the introduction of these suitable precursors, the coordination chemistry of uranium began to thrive and the occurrence of transition-metal-like actinide coordination chemistry led to a boom of reactivity studies as well as studies of molecular and electronic structure properties. However, low valent thorium(III) starting complexes still remain scarce.[9] Over the last two decades, the uranium-mediated activation of small molecules of industrial and biological importance received particular attention and remains an active field of current research, transcending into the field of catalysis and molecular magnetism, spectroscopic and computational analyses of the complexes' electronic structures.[10]

The advantages of utilizing actinides for small molecule activation and transformation chemistry include the large ionic radii, the range of accessible oxidation states and the relatively low but significant degree of covalent bonding. This is especially true for complexes with uranium and thorium metal centers, where coordination numbers range from low-coordinate 3 to a record-high 15 for thorium borohydride complexes.[11] Oxidation states in uranium chemistry range from +II to +VI, with the oxidation state of +II being accessed in the early 2010s.[12,13] The redox-chemistry of thorium is noticeably less rich and is primarily dominated by thorium(IV) and to a lesser extent, thorium(III). A complex with the unusual thorium oxidation state of +II was postulated back in the 1980s,[14] and finally realized approximately 30 years later.[15] Most importantly, and in contrast to the lanthanide 4f orbitals, the 5f orbitals of the early actinides may overlap with ligand orbitals to participate to a larger degree of covalent bonding.[10,16] Thus, the ability of f-orbitals to participate in bonding leads to covalent but highly polarized M–L bonds. This results in higher ligand (substrate) binding on/off rates and noticeably different reactivity in actinides compared to the predominantly covalent d-block metal and ionic lanthanide complexes.

This chapter covers selected examples of general importance up to the year 2016 but is not conceived to give a complete overview of the rapidly growing field of small molecule activation with actinides. In particular, thorium and uranium present the most examples due to their high natural occurrence in comparison to the other actinides as well as their radioactive stability. This review highlights examples of CO, CO_2, N_2O, NO, N_2, H_2O and N_3^- activation; for more complete coverage of actinide reactivity, other published reviews are recommended,[9,10,17–29] including an extensive review concerning small molecule activation by uranium.[30]

9.2 Carbon Monoxide

Carbon monoxide has significant industrial relevance despite it being a rather inert molecule, featuring the strongest bond ($1076.5 \pm 0.4\,\text{kJ mol}^{-1}$) between two elements in the periodic table.[31] For the production of liquid fuels, analogous to the Fischer–Tropsch process, reactive metal complexes are sought with uranium compounds emerging as potential candidates.[32,33] Compared to the limited amount of uranium(III) and uranium(V) complexes known to react with carbon monoxide, the CO chemistry of uranium(IV) is more extensive with insertion chemistry involving U–C as well as U–N and U–H bonds.[30] For thorium, theoretical investigations as well as reactivity studies have been published, focusing mostly on CO insertion chemistry with the binuclear insertion products $[(Cp)_2Th_2(CO)_n]$(n = 2 to 5) and $[(C_8H_8)_2Th_2(CO)_n]$(n = 1 to 5).[34–36]

Reversible CO activation by $[(TMSC_5H_4)_3U]$ ($TMSC_5H_4$ = trimethylsilylcyclopentadienyl anion) forming the complex $[(TMSC_5H_4)_3U(CO)]$ was initially observed in 1986.[37] The successful coordination of the CO ligand was confirmed by IR spectroscopic labeling studies. The first crystallographic analysis of a monomeric actinide carbonyl complex was reported by Carmona and co-workers in 1995.[38] They obtained the carbonyl complex $[(C_5Me_4H)_3U(CO)]$ (Figure 9.1) by treatment of $[(C_5Me_4H)_3U]$ (C_5Me_4H = tetramethylcyclopentadienyl anion) with CO. Both CO complexes, $[(TMSC_5H_4)_3U(CO)]$ and $[(C_5Me_4H)_3U(CO)]$, revert to their respective precursors when exposed to vacuum.

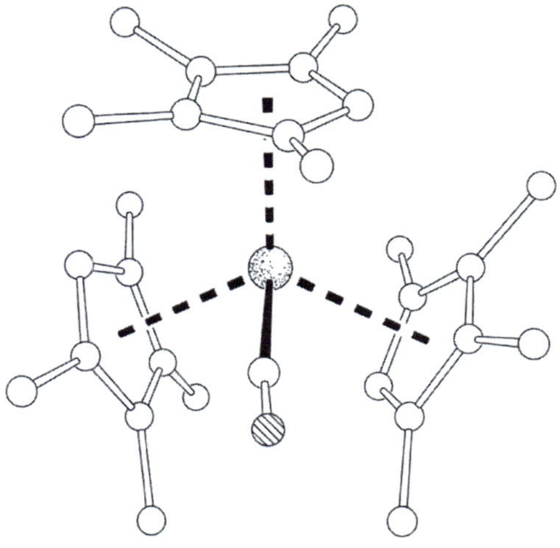

Figure 9.1. Molecular structure of $[(C_5Me_4H)_3U(CO)]$. Reproduced from Ref. [39].

In 2003, Evans *et al.* highlighted another example of a terminally bound carbonyl unit by employing the Cp* ligand (Cp* = pentamethylcyclopentadienyl anion).[39] The complex $[(Cp^*)_3U(CO)]$ as well as the aforementioned $[(C_5Me_4H)_3U(CO)]$ feature the CO bound to the uranium center via the carbon atom, as suggested by theoretical investigations several years prior.[40–43] Evans' publication included a direct comparison to the corresponding lanthanide CO reactivity of the complex $[(Cp^*)_3Nd]$. This study revealed that the 5f system uranium complex binds CO terminally, whereas its 4f element counterpart with neodymium undergoes CO insertion chemistry. The complex $[(Cp^*)_3Nd]$ inserts 2 equiv. of CO to form $[(Cp^*)_2Nd(O_2C_7Me_5)]$; thus, featuring a 7-norbornadienyl carbocation ligand. This insertion was similarly observed for the samarium complex $[(Cp^*)_2Sm(O_2C_7Me_5)]$.[44] These examples highlight the differences in reactivity between the 4f and 5f systems studied.

Aside from terminal coordination, a unique example of a bridging CO unit is also known. This binding mode is observed for the CO ligand coordinating to $[((^{tBu,tBu}ArO)_3tacn)U]$ $((^{tBu,tBu}ArOH)_3$ tacn = 1,4,7-tris(3,5-di-*tert*-butyl-2-hydroxybenzylate)-1,4,7-triazacyclononane),[45] resulting in $[\{((^{tBu,tBu}ArO)_3tacn)U\}_2(\mu,\kappa^1{:}\kappa^1\text{-CO})]$ with a radical anionic CO bridging the dinuclear uranium(III/IV) complex.

In 2006, Cloke *et al.* reported the homologation of CO by $[(\eta^8\text{-}C_8H_6(1,4\text{-}Si^iPr_3)_2)(\eta^5\text{-}Cp^*)U(THF)]$ (Scheme 9.1) to selectively form the deltate dianion, $C_3O_3^{2-}$.[32] The compound $[\{(\eta^8\text{-}C_8H_6(1,4\text{-}Si^iPr_3)_2)(\eta^5\text{-}Cp^*)U\}_2(\mu\text{-}\kappa^1{:}\kappa^1\text{-}C_3O_3)]$ (Scheme 9.1, middle) features agostic interactions between the f-orbital of one U(IV) center and the C_3 core of the deltate anion. The unusual agostic interaction and resulting distortion of the $C_3O_3^{2-}$ fragment was verified by DFT calculations. Following this discovery, a number of publications provided computational and structural insight into the outcome of the CO homologation process subject to varying steric and electronic properties of the chelating Cp* ligand and varying CO stoichiometry.[46–48] Replacing Cp* by C_5Me_4H yields the squarate dianion $C_4O_4^{2-}$ (Scheme 9.1, bottom), which is not distorted and in line with the standard bonding picture attributed to smaller average O–C–C bond angles.[33] Addition of 1 equiv. of CO yields a dimeric ethyne diolate complex $[\{(\eta^8\text{-}C_8H_6(1,4\text{-}Si^iPr_3)_2)(\eta^5\text{-}Cp^*)U\}_2(\mu\text{-}\kappa^1{:}\kappa^1\text{-}C_2O_2)]$ (Scheme 9.1, top).[49] Interestingly, when

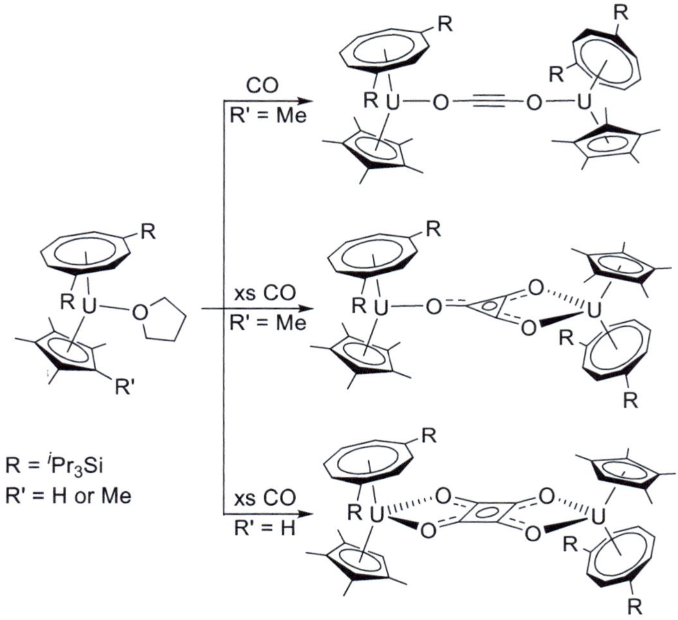

Scheme 9.1. Reductive homologation of CO by $[(\eta^8\text{-}C_8H_6(1,4\text{-}Si^iPr_3)_2)(L)U(thf)]$ ($L = C_5Me_4H$ or Cp^*).

$[(\eta^8\text{-}C_8H_6(1,4\text{-}Si^iPr_3)_2)(\eta^5\text{-}Cp^*)U]$ is exposed to a 1:2 mixture of ^{13}CO and H_2, the uranium(IV) methoxide complex $[(\eta^8\text{-}C_8H_6(1,4\text{-}Si^iPr_3)_2)(\eta^5\text{-}Cp^*)U(O^{13}CH_3)]$ is formed,[50] a reaction reminiscent of Fischer–Tropsch catalysis.[20]

The ynediolate moiety, $C_2O_2^{2-}$, is prevalent in recent literature. Arnold *et al.* were able to isolate ynediolate complexes starting from $[U(N(TMS)_2)_3]$ and $[U(ODtbp)_3]$ (ODtbp = (O-2,6-^{t}BuC$_6$H$_3$ anion) as well as $[U(OTtbp)_3]$ (OTtbp = O-2,4,6-^{t}BuC$_6$H$_2$ anion).[51,52] Liddle *et al.* conducted research on the triamidoamine complex $[U(Tren^{DMSB})]$ ($Tren^{DMSB}$ = $N(CH_2CH_2NSiMe_2^tBu)_3$ trianion), which reductively couples excess CO to form $C_2O_2^{2-}$ (Scheme 9.2).[53] The resulting complex $[\{(Tren^{DMSB})U\}_2(\mu,\kappa^1{:}\kappa^1\text{-}C_2O_2)]$ further reacts with organosilyl halides to form the uranium(IV) halide complex with concomitant elimination of bis(organosiloxy)acetylene. Subsequent reduction of the uranium(IV) halide with potassium restores the starting complex, closing the synthetic cycle by recovering the precursor complex.

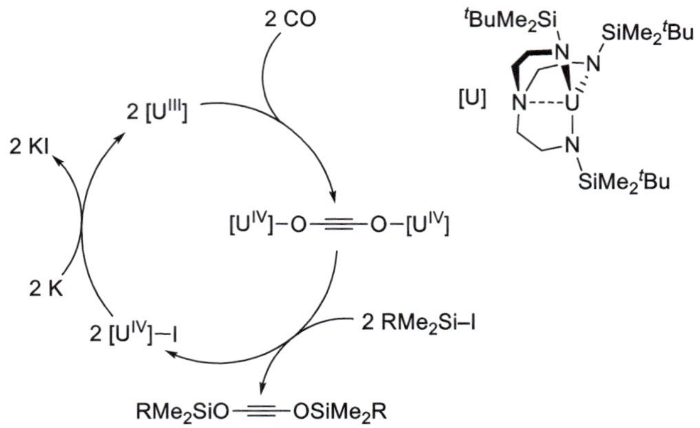

Scheme 9.2. Reductive coupling of CO by [U(Tren^{DMSB})] complex (upper right corner).

Examples of carbon monoxide coordination to uranium(V) and uranium(VI) complexes are rare. Multiple-bond metathesis of CO mediated by a sterically pressured uranium(V) imido complex was first presented by Meyer *et al.* by treating[(($^{Ad,{}^tBu}$ArO)$_3$tacn) U(NTMS)] with CO to yield the corresponding isocyanate [(($^{Ad,{}^tBu}$ArO)$_3$tacn)U(η^1-NCO)].[54] In this reaction, the uranium imido entity acts as a nucleophile on the π-accepting CO ligand and the trimethylsilyl radical becomes a leaving group. Nucleophilic attack of alkyl halide substrates on the newly formed heterocumulene NCO ligand of [(($^{Ad,{}^tBu}$ArO)$_3$tacn)U(κ^1-NCO)] yields products with newly formed C–N bonds. Reduction of the resulting uranium(IV) halide complex recovers the uranium(III) starting material, thereby closing the synthetic cycle of CN/CO multiple-bond metathesis. In concluding, a nitrogen atom is transferred from a uranium(V) imido complex to the π-acid CO. This forms a uranium(IV) isocyanate that can be reduced to regenerate the uranium(III) precursor. Similarly, reaction of the imide with methyl isocyanide forms cyanimides and functionalized carbodiimides. This chemistry occurs in successive one-electron steps. In 2014, Liddle *et al.* showed that the unique uranium(VI) nitride complex [(Tren^{TIPS})U(N)] (Tren^{TIPS} = N(CH$_2$CH$_2$NSi(^{i}Pr)$_3$)$_3$) (*vide infra*) also forms an isocyanate complex. Treatment of [(Tren^{TIPS})U(N)] with CO yields the uranium(IV) isocyanate that was further reduced with KC$_8$ to the corresponding uranium(III) complex [K(B-15-C-5)$_2$][UN(Tren^{TIPS})(η^1-NCO)] (B-15-C-5 = benzo-15-crown-5 ether).[55]

9.3 Carbon Dioxide

The activation of the thermodynamically stable greenhouse gas CO_2 continues to be a desirable research goal, as reductive CO_2 functionalisation is an attractive route to the production of renewable C1 feedstocks.[56,57] Reports on reactions of uranium complexes with CO_2 range in its activation with a variety of binding modes to U–C, U–H, U–N and U–S bond insertion chemistry.

CO_2 can also be utilized as an oxygen transfer agent to form uranium-oxo complexes with the concomitant release of CO. Early accounts cover the reductive bond cleavage of CO_2 with $[(TMSC_5H_4)_3U]$ to form $[\{(TMSC_5H_4)_3U\}_2(\mu\text{-}O)]$ (Scheme 9.3) with a suggested bridging CO_2 transition state followed by expulsion of CO.[58] This claim of a bridged transition state is based on the observation that the complex also reacts with CS_2 to yield a CS_2 bridged dinuclear species and was later supported by theoretical investigations that revealed the formation of CS to be less favourable than CO.[59]

Mazzanti *et al.* showcased the influence of multimetallic cooperativity on CO_2 reactivity. Differences in reactivity were observed between the uranium(III) complex $[K(OSi(O^tBu)_3)_4U]$ with the potassium cation in close proximity to the uranium center vs. the complex with the potassium secluded by a 18-crown-6 ether crown ether moiety.[60] The close proximity of the potassium to the uranium(III) ion facilitates the two electron reduction of CO_2, resulting in a complex with a terminal uranium(V) monooxo entity. Encapsulation of the potassium ion by a crown ether facilitates the formation of a mononuclear uranium(IV) carbonate complex. The authors' mechanistic proposals are supported by DFT calculations.

Formation of oxalate ligands remains rare in actinide CO_2 activation chemistry. Cloke *et al.* reported the formation of the

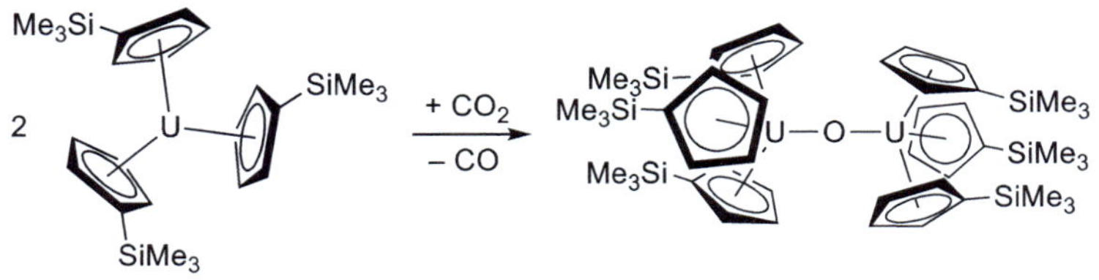

Scheme 9.3. Reactivity of $[(TMSC_5H_4)_3U]$ with CO_2.

first uranium oxalate complex.[61] When $[(\eta^8\text{-}C_8H_6(1,4\text{-}SiMe_3)_2)U(\eta^5\text{-}Cp^{Me4R})]$ (R = Me, Et, ^{i}Pr) reacts with CO_2, reductive coupling to the dianionic oxalate ligand readily occurs. Another example of a uranium oxalate complex $[\{((^{Neop,Me}ArO)_3tacn)U\}_2(\mu,\kappa^2{:}\kappa^2\text{-}C_2O_4)]$ (Neop = neopentyl) emerged shortly thereafter and is one of the two products resulting from the reaction of $[((^{Neop,Me}ArO)_3tacn)U]$ with CO_2, the other product being the bridged carbonate complex $[\{((^{Neop,Me}ArO)_3tacn)U\}_2(\mu,\kappa^2{:}\kappa^2\text{-}CO_3)]$.[62] The formation of both products occurs in analogy to the dithionate and thiocarbonate complexes obtained from the uranium-mediated activation of SO_2, described in the same report.

Along with uranyl carbonates, bridging carbonates are among the more prominent products of uranium CO_2 reactivity. Clokes' complex $[\{\eta^8\text{-}C_8H_6(1,4\text{-}Si^iPr_3)_2\}U(\eta^5\text{-}Cp^{Me4H})(THF)]$ triggers reductive CO_2 disproportionation to form CO gas and a $CO_3{}^{2-}$ ligand, resulting in a dimer with the carbonate unit bound in a $\mu,\kappa^1{:}\kappa^2$ coordination mode.[63] In another example by Meyer *et al.*, the reaction of excess CO_2 with tris(aryloxide) uranium(III) complexes featuring different anchoring units A (A = N or mesityl) yields the carbonate bridged dimeric products $[\{((^{Ad,Me}ArO)_3N)U\}_2(\mu,\kappa^1{:}\kappa^2\text{-}CO_3)]$ and $[\{((^{tBu,tBu}ArO)_3mes)U\}_2(\mu,\kappa^2{:}\kappa^2\text{-}CO_3)]$ with different carbonate coordination modes (Scheme 9.4).[64] The authors propose a μ-O

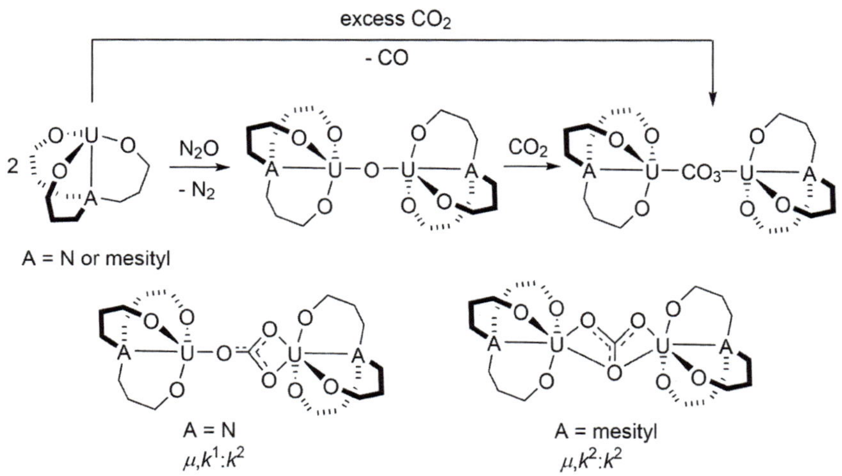

Scheme 9.4. Reactivity of $[((^{Ad,Me}ArO)_3N)U]$ and $[((^{tBu,tBu}ArO)_3mes)U]$ with CO_2 and N_2O.

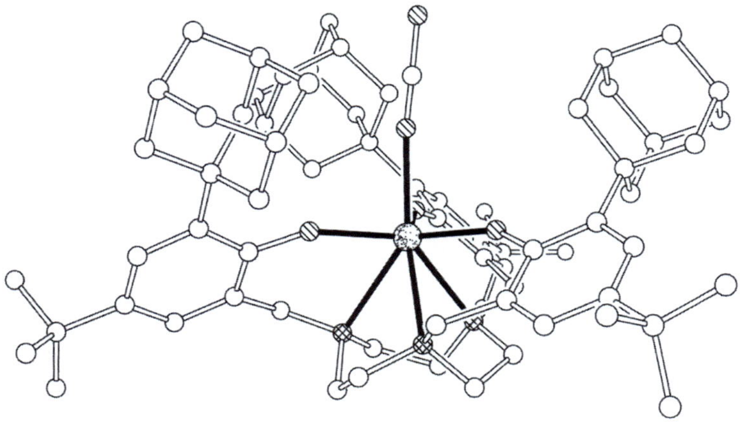

Figure 9.2. Molecular structure of $[((^{Ad,^tBu}ArO)_3tacn)U(\kappa^1\text{-}CO_2)]$. Reproduced from Ref. [71].

intermediate complex in the formation of both complexes. This proposed intermediate was synthesized via an alternative route employing N_2O and yields the identical carbonate complexes when exposed to CO_2.

A unique, linear end-on bound CO_2 ligand was successfully isolated by sterically constraining the reactive cavity of the trivalent $[((^{Ad,^tBu}ArO)_3tacn)U]$ complex with a custom-tailored ligand environment.[65] The complex undergoes a one-electron oxidation with CO_2 to give a monomeric, charge separated uranium(IV)–η^1–$CO_2^{\bullet -}$ species, namely $[((^{Ad,^tBu}ArO)_3tacn)U(\eta^1\text{-}CO_2)]$ (Figure 9.2) that is stable at room temperature. The radical anionic nature of the activated CO_2 ligand was confirmed by spectroscopic and X-ray crystallography studies.

Reports on CO_2 insertion into the U–C bond started to emerge in 1985.[66] More recently, Bart *et al.* reported CO_2 as well as CS_2 insertion into the U–C bond of a uranium (III) benzyl complex.[67] In their experiments, exposure of the complex $[Tp^*_2U(CH_2Ph)]$ (Tp^* = hydro-tris(3,5-dimethylpyrazolyl)borate) to 1 atm of CO_2 resulted in the generation of the carboxylate complex $[Tp^*_2U(\kappa^2\text{-}O_2CCH_2Ph)]$. The carboxylate unit can be released by treatment with TMS chloride or iodide, forming the silyl ester and the uranium monohalide species. When the monohalide compound reacts with benzyl potassium, the precursor $[Tp^*_2U(CH_2Ph)]$ is recovered, and the synthetic

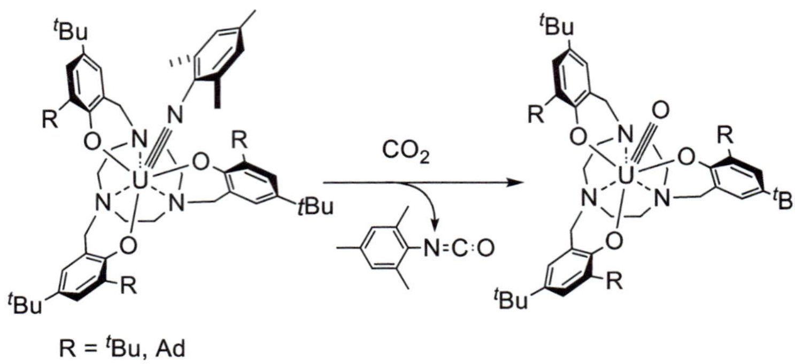

Scheme 9.5. Reaction of $[((^{R,^{t}Bu}ArO)_3tacn)U(Nmes)]$ with CO_2.

cycle is completed. In a similar fashion, this insertion chemistry was extended to U–S and U–N bonds also known from the earlier works of Ephritikine.[68–70]

Meyer *et al.* reported the first high valent uranium oxo complex synthesized from CO_2 by utilizing the mesityl imido complex $[((^{R,^{t}Bu}ArO)_3tacn)U(Nmes)]$ (mes = mesityl, R = ^{t}Bu, Ad). In a multiple-bond metathesis reaction, treatment of both complex derivatives (R = ^{t}Bu or Ad) with CO_2 yields one molecule of mesityl isocyanate and the corresponding uranium terminal oxo complex, $[((^{R,^{t}Bu}ArO)_3tacn)U(O)]$(Scheme 9.5).[71]

Although only few examples of small molecule activation with thorium complexes are known, there are several reports on CO_2 activation. Initial accounts by Marks *et al.* report the formation of thorium acetate and formate complexes by CO_2 insertion into thorium methyl and hydride bonds.[66] Later, reports of CO_2 addition to thorium–arene π bonds[72] as well as the formation of dimeric carboxylate and oxalate thorium complexes have also emerged.[73]

9.4 Nitric and Nitrous Oxide

The chemistry of actinides with nitric oxide (NO) and nitrous oxide (N_2O) is rather underdeveloped. In comparison to the abundance of metal-nitrosyl complexes in transition metal chemistry and despite its prediction as early as 1989,[42] only one crystallographically verified example of NO coordination to an actinide is known to date. The uranium(III) complex $[(C_5Me_4H)_3U]$ reacts with 1 equiv. of NO

Figure 9.3. Molecular structure of [(C_5Me_4H)$_3$U(NO)]. Reproduced from Ref. [74].

at −78 °C to form the nitrosyl complex [(C_5Me_4H)$_3$U(NO)] that was successfully isolated and characterized.[74] Although the linear U–N–O bond angle of 180.0(4)° hints at a formal NO^{1+} unit, the bond lengths, magnetic susceptibility as well as IR and NIR spectroscopy suggest the presence of a NO^{1-} unit and a corresponding oxidized uranium(IV) metal center (Figure 9.3). In analogy to the reactivity of transition metal NO^{1-} complexes, the basic oxygen atom in [(C_5Me_4H)$_3$U(NO)] reacts with the Lewis acid Al_2Me_6 to form the adduct [(C_5Me_4H)$_3$U(N-OAlMe_3)]; thus, providing further evidence for the presence of a NO^{1-} ligand.

Aside from direct coordination, NO can also be utilized as an oxygen transfer reagent. Upon addition of 0.5 equiv. of NO to [U(ODtbp)$_3$], Burns *et al.* found that the bridging species [{(ODtbp)$_3$U}$_2$(μ-O)] is formed, although the fate of the nitrogen atom in that reaction was not reported.[75] In addition, [{(ODtbp)$_3$U}$_2$(μ-O)] can be isolated when [U(ODtbp)$_3$] is oxidized with N_2O with concurrent release of N_2, suggesting related O-transfer reactivity between the two nitrous oxides in this case.

Cloke *et al.* reported that the reaction of [(η^8-C_8H_6(1,4-Si^{i}Pr$_3$)$_2$)(η^5-Cp*)U] with 1 equiv. of NO in toluene led to a colour change of

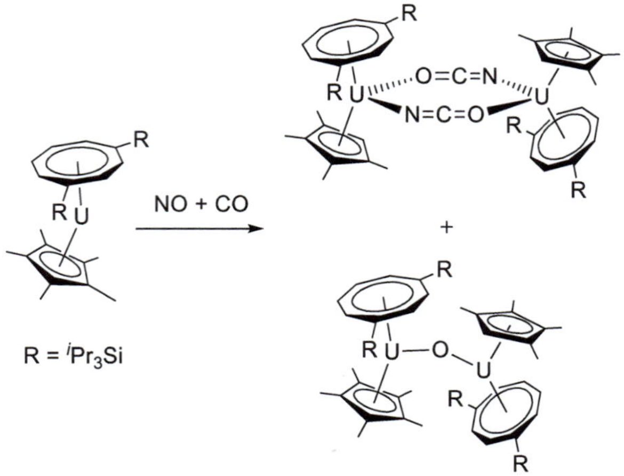

Scheme 9.6. Reaction of $[(\eta^8\text{-}C_8H_6(1,4\text{-}Si^iPr_3)_2)(\eta^5\text{-}Cp^*)U]$ with CO and NO.

the solution, but no product could be isolated.[76] In another attempt, 0.5 equiv. of CO was added, following the addition of 1 equiv. NO, and a mixture of the unique bridging bis(cyanate) complex $[\{(\eta^8\text{-}C_8H_6(1,4\text{-}Si^iPr_3)_2)(\eta^5\text{-}Cp^*)U\}_2(\mu,\kappa^1{:}\kappa^1\text{-}NCO)_2]$ (Scheme 9.6, top) and μ-oxo complex $[\{(\eta^8\text{-}C_8H_6(1,4\text{-}Si^iPr_3)_2)(\eta^5\text{-}Cp^*)U\}_2(\mu\text{-}O)]$ (Scheme 9.6, bottom) was isolated. Meyer *et al.* reported the reductive disproportionation of NO with a trivalent uranium tris(aryloxide) complex, $[U(OAr^{Ad,Ad,Me})_3]$.[77] Employing the sterically encumbered 2,6-di-adamantyl-4-methyl-derivatized phenol, $(^{Me,Ad,Ad}ArO^-)$, the reaction of $[U(OAr^{Ad,Ad,Me})_3]$ with NO yields the oxidized complex $[(^{Me,Ad,Ad}ArO)_3U(O)]$ and N_2O; in the solid-state and in solution. Mechanistic and computational studies suggest that the nitrous oxide N–N bond is formed by the coupling of a κ^1-O coordinated NO ligand with gaseous NO, to form a $\kappa^1\text{-}(N_2O_2)^{1-}$ fleeting intermediate, prior to N_2O elimination and formation of the uranium(V) oxo complex.

Several examples of uranium complexes with bridging μ-O formed from nitrous oxide, N_2O, have been reported.[64,75,78] Predating the aforementioned oxo-bridged complex $[\{(ODtbp)_3\}_2U(\mu\text{-}O)]$ (Chapter 1), Ephritikine *et al.* reported crystallographic evidence of the μ-O bridged complex $[\{(TMSC_5H_4)_3U\}_2(\mu\text{-}O)]$ synthesized from $[(TMSC_5H_4)_3U]$ and either CO_2 or N_2O.[58]

Terminal oxo species generated from N_2O are exceptionally rare. Early reports by Burns *et al.* focus on the first crystallographically characterized uranium(VI) terminal oxo complexes [(Cp*)$_2$U(N-2,6-^{i}Pr$_2$C$_6$H$_3$)(O)] and [(Cp*)$_2$U(N-2,4,6-^{t}Bu$_3$C$_6$H$_3$)(O)] that were synthesized from the uranium(IV) metallocene imido precursors and N_2O.[79,80] The only other known example of a terminal oxo species generated from N_2O is the pentavalent [(Ar*O)$_3$U(O)(THF)] (Ar* = 2,6-Ph$_2$-C$_6$H$_4$-Me), which is formed upon addition of N_2O gas to the bulky uranium(III) tris(aryloxide) precursor complex [(Ar*O)$_3$U(THF)].[81]

9.5 Dinitrogen

Compared to the abundance of known transition metal N_2 complexes, N_2 activation chemistry is rarely observed for the actinides. In fact, it was not until 1998 that Scott *et al.* crystallographically characterized the first diuranium dinitrogen complex. When a solution of [U(Tren^{DMSB})] in benzene is exposed to an atmosphere of N_2, a colour change of the solution from purple to red can be observed. The isolated product, namely [{(Tren^{DMSB})U}$_2$(μ,η^2:η^2-N$_2$)], features the N_2 ligand in a side-on bridging coordination mode (Figure 9.4).[82–84] An end-on coordination mode can be observed for the monometallic complex [(Cp*)$_3$U(η^1-N$_2$)], which coordinates N_2 only under pressure (80 psi) and loses the dinitrogen ligand upon reducing the pressure.[85] However, X-ray crystallographic analysis of the [(Cp*)$_3$U(η^1-N$_2$)] crystals did not reveal significant activation of the N_2 from its bond length, and the N_2 stretching frequency in the vibrational spectrum is comparable to that of free dinitrogen.

Stable N_2 compounds have been synthesized using precursor complexes with monodentate aryloxide ligands, such as [U(OTtbp)$_3$] as well as [U(ODtbp)$_3$]. These complexes, first reported by Sattelberger *et al.*, are exclusively monomeric if synthesized under an argon atmosphere.[86] If, instead, the complexes are synthesized in an atmosphere of N_2, the dimeric complexes [{(OAr)$_3$U}$_2$(μ,η^2:η^2-N$_2$)] (Ar = Ttbp or Dtbp) form (Scheme 9.7).[51] This complex, first isolated by Arnold *et al.*, has a side-on bound dinitrogen bridging two uranium centers. An elongated N–N bond suggests the presence of a reduced diazenido

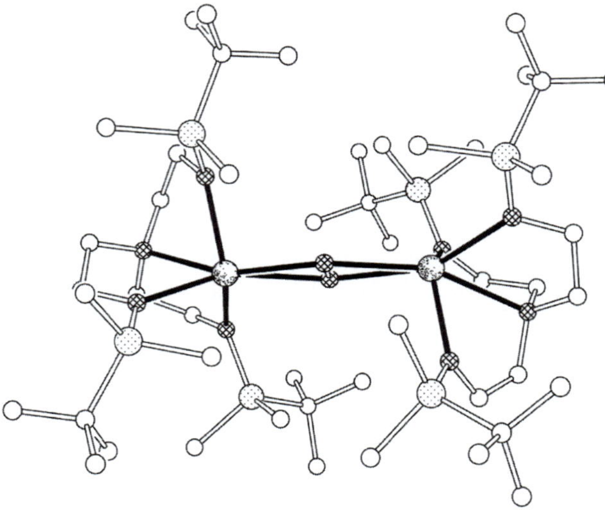

Figure 9.4. Molecular structure of [{(Tren^{DMSB})U}$_2$(μ,η^2:η^2-N_2)]. Reproduced from Ref. [83].

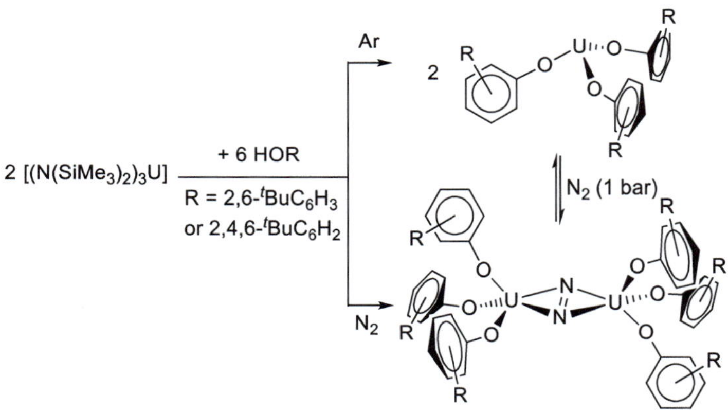

Scheme 9.7. Synthesis of [U(OTtbp)$_3$] and [U(ODtbp)$_3$] under an atmosphere of N_2 or Ar.

unit, N_2^{2-}. This reduced entity is also known for the bimetallic molybdenum/uranium complex [(Ar(^{t}Bu)N)$_3$U(μ,η^1:η^1-N_2)Mo(N(^{t}Bu)Ph)$_3$], reported by Cummins *et al.*, as well as Cloke's [{(Cp*)(η^8-C_8H_4(1,4-Si^{i}Pr$_3$)$_2$)U}$_2$(μ,η^2:η^2-N_2)].[87,88] When precursor complexes with siloxide ligands are utilized, the corresponding dimeric complex

$[\{(OSi(mes)_3)_3U\}_2(\mu,\eta^2{:}\eta^2\text{-}N_2)]$ forms, in which the $N_2{}^{2-}$ dianion was confirmed by Raman spectroscopy. The diazenido ligand in the latter complex exhibits heightened stability in vacuum as well as increased thermal stability.[89]

9.6 Water

Uranium(III) complexes are known to react violently with H_2O to form bridging oxo complexes or uranium(VI) uranyl species featuring the omnipresent $[O{=}U(VI){=}O]^{2+}$ entity.[27] There are only a few exceptional examples of controlled uranium-mediated water activation.[90–94]

The only uranium(III) hydroxo complexes known to date were synthesized by Andersen *et al.* in 1996.[90] Hydrolysis of $[(Cp'')_3U]$ ($Cp'' = 1,3\text{-}(TMS)_2C_5H_3$) or $[(Cp^{\ddagger})_3U]$ ($Cp^{\ddagger} = 1,3\text{-}(^tBu)_2C_5H_3$) gives the dimeric hydroxide complexes $[\{(Cp'')_2U\}_2(\mu\text{-}OH)_2]$ and $[\{(Cp^{\ddagger})_2U\}_2(\mu\text{-}OH)_2]$. At elevated temperatures, H_2 is released to form the uranium(IV) bridging oxo complexes $[(Cp'')_3U(\mu\text{-}O)_2]$ and $[(Cp^{\ddagger})_3U(\mu\text{-}O)_2]$, respectively. The mononuclear uranium(IV) hydroxo complex $[(OTf)_3(py)_4U(OH)]$ (OTf = triflate) forms in roughly 20% yield when UH_3 is reacted with triflic acid containing traces of water.[91]

The first homogenous uranium complex for electrocatalytic reduction of H_2 from water was reported by Meyer *et al.*[94] observed with the uranium(III) complex, $[(^{Ad,Me}ArO)_3mes)U]$ (Scheme 9.8, (a)).[95,96] In a controlled reaction in THF solution, the trivalent uranium coordination complex cleanly and quantitatively reacts with water to yield 0.5 equiv. of dihydrogen and the corresponding U(IV) hydroxide, $[(^{Ad,Me}ArO)_3mes)U(OH)]$ (Scheme 9.8, (b)). *In situ* EPR experiments were employed to study the mechanism of this reaction, suggesting an initial 2 e^- oxidative addition of H_2O to U(III) to yield the fleeting hydroxo/hydride intermediate $[(^{Ad,Me}ArO)_3mes)U(OH)(H)]$, followed by elimination of H_2 and formation of $[(^{Ad,Me}ArO)_3mes)U(O)]$. The latter, formally pentavalent, oxo complex subsequently comproportionates with U(III) and 1 equiv. of H_2O to yield the isolable tetravalent $[(^{Ad,Me}ArO)_3mes)U(OH)]$.[116]

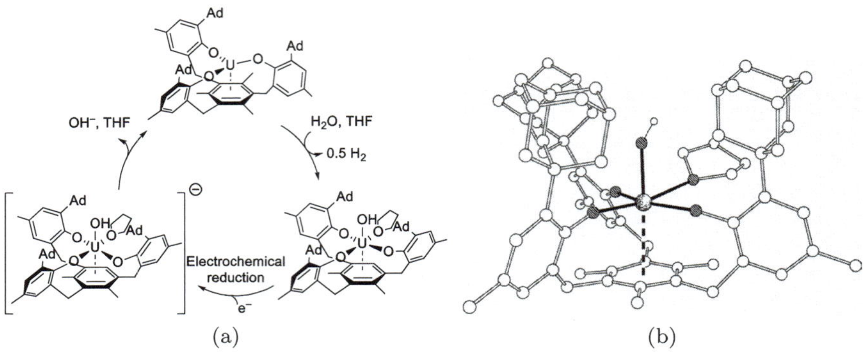

Scheme 9.8. (a) Electrocatalytic cycle for the production of H_2 from H_2O and (b) molecular structure of $[(^{Ad,Me}ArO)_3mes)U(OH)]$. Reproduced from Ref. [94].

9.7 Azides

Activation of organic and inorganic azides by low-valent uranium complexes often leads to uranium azides as well as imide formation driven by the release of N_2. Prominent examples of organic azide activation by uranium include the reaction of $[(C_5MeH_4)_3U]$ with phenyl azide to yield the first uranium imido species, $[(C_5MeH_4)_3U(NPh)]$ (Scheme 9.9).[97] The U–N linkage of this uranium imido complex is best described as a formal triple bond. The uranium(VI) bis-imido complex, $[(Cp^*)_2U(NPh)_2]$, was isolated soon after.[98]

Azide activation at a uranium(III) center forming a uranium azide complex was published in 2003 by Meyer and co-workers.[99] Treatment of $[((^{R,{}^tBu}ArO)_3tacn)U]$ (R= ^{t}Bu, Ad) with trimethylsilyl azide yielded hexamethyl disilane (Me_6Si_2) and the corresponding uranium(IV) azide $[((^{R,{}^tBu}ArO)_3tacn)U(N_3)]$ in rather low yields. Improved yields were obtained employing trityl azide as the $N_3^{\bullet}$ radical source. Both complex derivatives (R = ^{t}Bu and Ad) feature η^1-bound, near-linear azide units, but the steric demand of the derivatized aryloxide chelating ligand is reflected in different U–N–N bond angles of 145.9(9)° for the ^{t}Bu derivatized and 175.6(3)° for the bulkier adamantyl derivatized chelate. Utilizing these complexes, uranium nitrides could not be isolated. One of the earlier successful nitride syntheses, reported by Evans *et al.*, is the reaction of $[(Cp^*)_2U((\mu,\eta^2{:}\eta^1\text{-Ph})_2BPh_2)]$ with NaN_3 to form 24-membered uranium nitrogen rings.[100] These rings feature alternating nitrido and

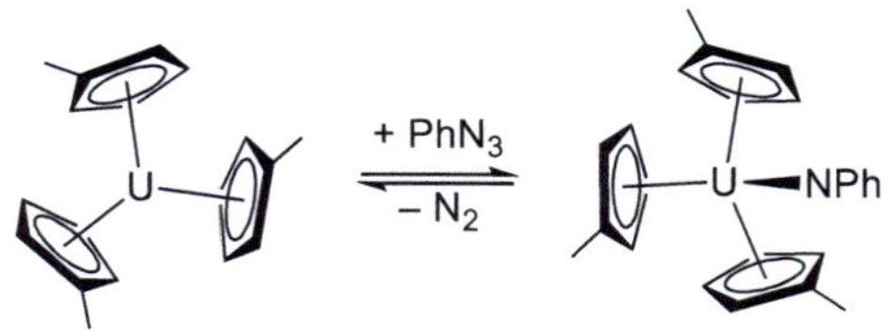

Scheme 9.9. Reaction of $[(C_5MeH_4)_3U]$ with phenyl azide.

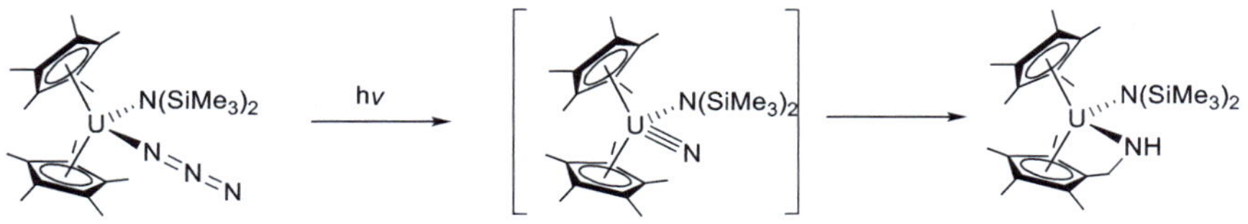

Scheme 9.10. Photolysis of $[(Cp^*)_2U(N_3)(N(TMS)_2)]$ to give $[(Cp^*)U(\eta^5{:}\kappa^1\text{-}C_5Me_4CH_2NH)(N(TMS)_2)]$.

azido linkages $(UNUN_3)_4$, with the two Cp* rings being coordinated trans to each other. Another mixed azido/nitride cluster was generated from $[UI_3(THF)_4]$ with $Cs_3[U(N_3)_7]$ as the nitrogen source.[101] This complex features a nitrido ligand bridging four uranium centers.

Following the isolation of bridging nitrides by Gambarotta *et al.*,[102] Cummins *et al.* synthesized a series of μ-nitrido bridged uranium complexes. The trivalent precursor $[(N(^tBu)Ar)_3U(THF)]$ ($Ar = 3,5\text{-}Me_2C_6H_3$) reacts with NaN_3 to give $Na[\{(N(^tBu)Ar)_3)U\}_2(\mu\text{-}N)]$, featuring a bridging nitride between two uranium(IV) centers,[103] which can be further oxidized to give the uranium(IV/V) and (V/V) couples, respectively. A terminal nitride could not be isolated utilizing this route, as well as from boron-capped nitrides.[104]

Following the concept of azide photolysis that is routinely and successfully applied for the generation of transition metal nitride complexes, Kiplinger *et al.* were able to observe intramolecular C–H activation by a proposed intermediate uranium nitride species.[105] The photolysis of $[(Cp^*)_2U(N_3)(N(TMS)_2)]$, a starting material prepared by employing the versatile azide transfer agent $[(Ph_3P)Au(N_3)]$,[106] triggers the formation of a nucleophilic terminal uranium(VI) nitride intermediate that, ultimately, results in the formation of the insertion product $[(Cp^*)U(\eta^5{:}\kappa^1\text{-}C_5Me_4CH_2NH)(N(TMS)_2)]$ (Scheme 9.10).

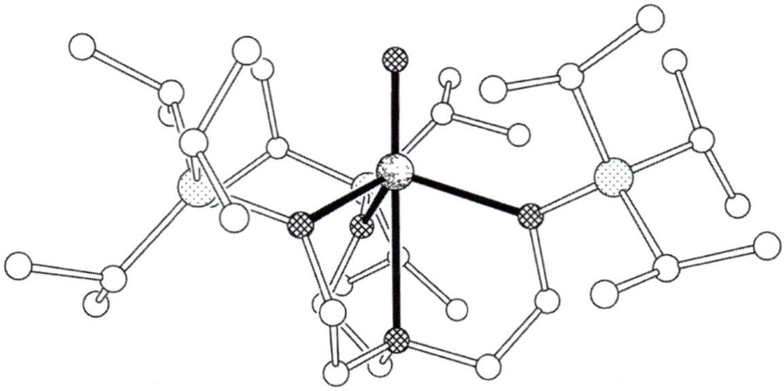

Figure 9.5. Molecular structure of [(Tren^{TIPS})U(N)] in crystals of [Na(12-C-4)$_2$][(Tren^{TIPS})U(N)]. Reproduced from Ref. [109].

Isolation and crystallographic characterization of the first monomeric uranium nitride complex was achieved by Liddle and co-workers.[107] This class of compounds remain rare, with one additional example being accessed only recently.[108] The complex [U(Tren^{TIPS})] reacts with sodium azide to give the sodium bridged dinuclear μ-nitrido species [{(Tren^{TIPS})U(μ-N)(Na)}$_2$] with release of N_2. Addition of 2 equiv. of 12-crown-4 allows for the isolation of the terminal uranium(V) nitrido complex, namely [Na(12-C-4)$_2$] [(Tren^{TIPS})U(N)] (Figure 9.5). DFT calculations suggest a significant degree of covalency within the formal U–N triple bond. Oxidation to the uranium(VI) nitride complex was also achieved.[109]

Several thorium azide compounds as well as one neptunium azide complex have been structurally characterized.[110–114] The metallocene complexes [(Cp*)$_2$ThCl$_2$] and [(C$_5$H$_4$Et)$_2$ThCl$_2$] react with NaN_3 to form 1D infinite coordination polymers[109] instead of ring structures, as observed for the comparable uranium metallocene counterparts mentioned earlier.[100,115]

References

1. L. T. Reynolds and G. Wilkinson, *J. Inorg. Nucl. Chem.*, 1956, *2*, 246.
2. P. J. Fagan, J. M. Manriquez, T. J. Marks, C. S. Day, S. H. Vollmer and V. W. Day, *Organometallics*, 1982, *1*, 170.

3. L. R. Avens, S. G. Bott, D. L. Clark, A. P. Sattelberger, J. G. Watkin and B. D. Zwick, *Inorg. Chem.*, 1994, *33*, 2248.
4. D. L. Clark, A. P. Sattelberger, S. G. Bott and R. N. Vrtis, *Inorg. Chem.*, 1989, *28*, 1771.
5. M. J. Monreal, R. K. Thomson, T. Cantat, N. E. Travia, B. L. Scott and J. L. Kiplinger, *Organometallics*, 2011, *30*, 2031.
6. C. D. Carmichael, N. A. Jones and P. L. Arnold, *Inorg. Chem.*, 2008, *47*, 8577.
7. R. J. Baker, *Coord. Chem. Rev.*, 2012, *256*, 2843.
8. W. J. Evans, S. A. Kozimor, J. W. Ziller, A. A. Fagin and M. N. Bochkarev, *Inorg. Chem.*, 2005, *44*, 3993.
9. F. Ortu, A. Formanuik, J. R. Innes and D. P. Mills, *Dalton Trans.*, 2016, *45*, 7537.
10. S. T. Liddle, *Angew. Chem. Int. Ed.*, 2015, *54*, 8604.
11. S. R. Daly, P. M. B. Piccoli, A. J. Schultz, T. K. Todorova, L. Gagliardi and G. S. Girolami, *Angew. Chem. Int. Ed.*, 2010, *49*, 3379.
12. M. R. MacDonald, M. E. Fieser, J. E. Bates, J. W. Ziller, F. Furche and W. J. Evans, *J. Am. Chem. Soc.*, 2013, *135*, 13310.
13. H. S. La Pierre, A. Scheurer, F. W. Heinemann, W. Hieringer and K. Meyer, *Angew. Chem. Int. Ed.*, 2014, *53*, 7158.
14. P. C. Blake, M. F. Lappert, R. G. Taylor, J. L. Atwood, W. E. Hunter and H. Zhang, *J. Chem. Soc. Chem. Commun.*, 1986, 1394.
15. R. R. Langeslay, M. E. Fieser, J. W. Ziller, F. Furche and W. J. Evans, *Chem. Sci.*, 2015, *6*, 517.
16. M. Mazzanti, R. Wietzke, J. Pecaut, J.-M. Latour, P. Maldivi and M. Remy, *Inorg. Chem.*, 2002, *41*, 2389.
17. S. C. Bart and K. Meyer, *Struct. Bond.*, 2008, *127*, 119.
18. A. R. Fox, S. C. Bart, K. Meyer and C. C. Cummins, *Nature*, 2008, *455*, 341.
19. P. L. Arnold, *Chem. Commun.*, 2011, *47*, 9005.
20. O. P. Lam and K. Meyer, *Angew. Chem. Int. Ed.*, 2011, *50*, 9542.
21. O. P. Lam and K. Meyer, *Polyhedron*, 2012, *32*, 1.
22. B. M. Gardner and S. T. Liddle, *Eur. J. Inorg. Chem.*, 2013, *2013*, 3753.
23. T. W. Hayton, *Chem. Commun.*, 2013, *49*, 2956.
24. B. M. Gardner and S. T. Liddle, *Chem. Commun.*, 2015, *51*, 10589.
25. W. J. Evans and S. A. Kozimor, *Coord. Chem. Rev.*, 2006, *250*, 911.
26. M. Ephritikhine, *Dalton Trans.*, 2006, 2501.
27. I. Korobkov and S. Gambarotta, *Prog. Inorg. Chem.*, 2005, *54*, 321.
28. T. W. Hayton, *Dalton Trans.*, 2010, *39*, 1145.
29. O. P. Lam, C. Anthon and K. Meyer, *Dalton Trans.*, 2009, 9677.
30. H. S. La Pierre and K. Meyer, *Prog. Inorg. Chem.*, 2014, *58*, 303.

31. W. B. Tolman, *Activation of Small Molecules: Organometallic and Bioinorganic Perspectives*, 2006.
32. O. T. Summerscales, F. G. N. Cloke, P. B. Hitchcock, J. C. Green and N. Hazari, *Science*, 2006, *311*, 829.
33. O. T. Summerscales, F. G. N. Cloke, P. B. Hitchcock, J. C. Green and N. Hazari, *J. Am. Chem. Soc.*, 2006, *128*, 9602.
34. H. Li, H. Feng, W. Sun, R. B. King and H. F. Schaefer III, *Inorg. Chem.*, 2013, *52*, 6893.
35. H. Li, H. Feng, W. Sun, Q. Fan, R. B. King and H. F. Schaefer III, *New J. Chem.*, 2014, *38*, 6031.
36. D. C. Sonnenberger, E. A. Mintz and T. J. Marks, *J. Am. Chem. Soc.*, 1984, *106*, 3484.
37. J. G. Brennan, R. A. Andersen and J. L. Robbins, *J. Am. Chem. Soc.*, 1986, *108*, 335.
38. J. Parry, E. Carmona, S. Coles and M. Hursthouse, *J. Am. Chem. Soc.*, 1995, *117*, 2649.
39. W. J. Evans, S. A. Kozimor, G. W. Nyce and J. W. Ziller, *J. Am. Chem. Soc.*, 2003, *125*, 13831.
40. L. Maron, O. Eisenstein and R. A. Andersen, *Organometallics*, 2009, *28*, 3629.
41. B. E. Bursten and R. J. Strittmatter, *J. Am. Chem. Soc.*, 1987, *109*, 6606.
42. B. E. Bursten, L. F. Rhodes and R. J. Strittmatter, *J. Am. Chem. Soc.*, 1989, *111*, 2758.
43. B. E. Bursten, L. F. Rhodes and R. J. Strittmatter, *J. Am. Chem. Soc.*, 1989, *111*, 2756.
44. W. J. Evans, K. J. Forrestal and J. W. Ziller, *J. Am. Chem. Soc.*, 1995, *117*, 12635.
45. I. Castro-Rodriguez and K. Meyer, *J. Am. Chem. Soc.*, 2005, *127*, 11242.
46. D. McKay, A. S. P. Frey, J. C. Green, F. G. N. Cloke and L. Maron, *Chem. Commun.*, 2012, *48*, 4118.
47. G. Aitken, N. Hazari, A. S. P. Frey, F. G. N. Cloke, O. Summerscales and J. C. Green, *Dalton Trans.*, 2011, *40*, 11080.
48. N. Tsoureas, O. T. Summerscales, F. G. N. Cloke and S. M. Roe, *Organometallics*, 2013, *32*, 1353.
49. A. S. P. Frey, F. G. N. Cloke, P. B. Hitchcock, I. J. Day, J. C. Green and G. Aitken, *J. Am. Chem. Soc.*, 2008, *130*, 13816.
50. A. S. P. Frey, F. G. N. Cloke, M. P. Coles, L. Maron and T. Davin, *Angew. Chem. Int. Ed.*, 2011, *50*, 6881.
51. S. M. Mansell, N. Kaltsoyannis and P. L. Arnold, *J. Am. Chem. Soc.*, 2011, *133*, 9036.

52. P. L. Arnold, Z. R. Turner, R. M. Bellabarba and R. P. Tooze, *Chem. Sci.*, 2011, *2*, 77.
53. B. M. Gardner, J. C. Stewart, A. L. Davis, J. McMaster, W. Lewis, A. J. Blake and S. T. Liddle, *Proc. Natl. Acad. Sci. USA*, 2012, *109*, 9265.
54. I. Castro-Rodriguez, H. Nakai and K. Meyer, *Angew. Chem. Int. Ed.*, 2006, *45*, 2389.
55. P. A. Cleaves, D. M. King, C. E. Kefalidis, L. Maron, F. Tuna, E. J. L. McInnes, J. McMaster, W. Lewis, A. J. Blake and S. T. Liddle, *Angew. Chem. Int. Ed.*, 2014, *53*, 10412.
56. E. A. Quadrelli, G. Centi, J.-L. Duplan and S. Perathoner, *ChemSusChem*, 2011, *4*, 1194.
57. E. E. Benson, C. P. Kubiak, A. J. Sathrum and J. M. Smieja, *Chem. Soc. Rev.*, 2009, *38*, 89.
58. J.-C. Berthet, J. F. Le Marechal, M. Nierlich, M. Lance, J. Vigner and M. Ephritikhine, *J. Organomet. Chem.*, 1991, *408*, 335.
59. O. P. Lam, L. Castro, B. Kosog, F. W. Heinemann, L. Maron and K. Meyer, *Inorg. Chem.*, 2012, *51*, 781.
60. O. J. Cooper, C. Camp, J. Pecaut, C. E. Kefalidis, L. Maron, S. Gambarelli and M. Mazzanti, *J. Am. Chem. Soc.*, 2014, *136*, 6716.
61. N. Tsoureas, L. Castro, A. F. R. Kilpatrick, F. G. N. Cloke and L. Maron, *Chem. Sci.*, 2014, *5*, 3777.
62. A.-C. Schmidt, F. W. Heinemann, C. E. Kefalidis, L. Maron, P. W. Roesky and K. Meyer, *Chem. Eur. J.*, 2014, *20*, 13501.
63. O. T. Summerscales, A. S. P. Frey, F. G. N. Cloke and P. B. Hitchcock, *Chem. Commun.*, 2009, 198.
64. O. P. Lam, S. C. Bart, H. Kameo, F. W. Heinemann and K. Meyer, *Chem. Commun.*, 2010, *46*, 3137.
65. I. Castro-Rodriguez, H. Nakai, L. N. Zakharov, A. L. Rheingold and K. Meyer, *Science*, 2004, *305*, 1757.
66. K. G. Moloy and T. J. Marks, *Inorg. Chim. Acta*, 1985, *110*, 127.
67. E. M. Matson, W. P. Forrest, P. E. Fanwick and S. C. Bart, *J. Am. Chem. Soc.*, 2011, *133*, 4948.
68. E. M. Matson, P. E. Fanwick and S. C. Bart, *Organometallics*, 2011, *30*, 5753.
69. C. Lescop, T. Arliguie, M. Lance, M. Nierlich and M. Ephritikhine, *J. Organomet. Chem.*, 1999, *580*, 137.
70. P. C. Leverd, M. Ephritikhine, M. Lance, J. Vigner and M. Nierlich, *J. Organomet. Chem.*, 1996, *507*, 229.
71. S. C. Bart, C. Anthon, F. W. Heinemann, E. Bill, N. M. Edelstein and K. Meyer, *J. Am. Chem. Soc.*, 2008, *130*, 12536.
72. I. Korobkov and S. Gambarotta, *Organometallics*, 2004, *23*, 5379.

73. Z. E. Button, J. A. Higgins, M. Suvova, F. G. N. Cloke and S. M. Roe, *Dalton Trans.*, 2015, *44*, 2588.
74. N. A. Siladke, K. R. Meihaus, J. W. Ziller, M. Fang, F. Furche, J. R. Long and W. J. Evans, *J. Am. Chem. Soc.*, 2012, *134*, 1243.
75. L. R. Avens, D. M. Barnhart, C. J. Burns, S. D. McKee and W. H. Smith, *Inorg. Chem.*, 1994, *33*, 4245.
76. A. S. P. Frey, F. G. N. Cloke, M. P. Coles and P. B. Hitchcock, *Chem. Eur. J.*, 2010, *16*, 9446.
77. C. J. Hoerger, H. S. La Pierre, L. Maron, A. Scheurer, F. W. Heinemann and K. Meyer, *Chem. Commun.*, 2016, *52*, 10854.
78. A.-C. Schmidt, A. V. Nizovtsev, A. Scheurer, F. W. Heinemann and K. Meyer, *Chem. Commun.*, 2012, *48*, 8634.
79. D. S. J. Arney and C. J. Burns, *J. Am. Chem. Soc.*, 1995, *117*, 9448.
80. D. S. J. Arney and C. J. Burns, *J. Am. Chem. Soc.*, 1993, *115*, 9840.
81. S. M. Franke, B. L. Tran, F. W. Heinemann, W. Hieringer, D. J. Mindiola and K. Meyer, *Inorg. Chem.*, 2013, *52*, 10552.
82. P. Roussel, P. Scott and D. N. Tinker, *J. Alloys Compd.*, 1998, *271–273*, 150.
83. P. Roussel and P. Scott, *J. Am. Chem. Soc.*, 1998, *120*, 1070.
84. P. Roussel, W. Errington, N. Kaltsoyannis and P. Scott, *J. Organomet. Chem.*, 2001, *635*, 69.
85. W. J. Evans, S. A. Kozimor and J. W. Ziller, *J. Am. Chem. Soc.*, 2003, *125*, 14264.
86. W. G. Van der Sluys, C. J. Burns, J. C. Huffman and A. P. Sattelberger, *J. Am. Chem. Soc.*, 1988, *110*, 5924.
87. A. L. Odom, P. L. Arnold and C. C. Cummins, *J. Am. Chem. Soc.*, 1998, *120*, 5836.
88. F. G. N. Cloke and P. B. Hitchcock, *J. Am. Chem. Soc.*, 2002, *124*, 9352.
89. S. M. Mansell, J. H. Farnaby, A. I. Germeroth and P. L. Arnold, *Organometallics*, 2013, *32*, 4214.
90. W. W. Lukens Jr., S. M. Beshouri, L. L. Blosch and R. A. Andersen, *J. Am. Chem. Soc.*, 1996, *118*, 901.
91. L. Natrajan, M. Mazzanti, J.-P. Bezombes and J. Pecaut, *Inorg. Chem.*, 2005, *44*, 6115.
92. K. A. N. S. Ariyaratne, R. E. Cramer, G. B. Jameson and J. W. Gilje, *J. Organomet. Chem.*, 2004, *689*, 2029.
93. G. H. John, I. May, C. A. Sharrad, A. D. Sutton, D. Collison, M. Helliwell and M. J. Sarsfield, *Inorg. Chem.*, 2005, *44*, 7606.
94. D. P. Halter, F. W. Heinemann, J. Bachmann and K. Meyer, *Nature*, 2016, *530*, 317.

95. H. I. Karunadasa, C. J. Chang and J. R. Long, *Nature*, 2010, *464*, 1329.
96. V. S. Thoi, Y. Sun, J. R. Long and C. J. Chang, *Chem. Soc. Rev.*, 2013, *42*, 2388.
97. J. G. Brennan and R. A. Andersen, *J. Am. Chem. Soc.*, 1985, *107*, 514.
98. D. S. J. Arney, C. J. Burns and D. C. Smith, *J. Am. Chem. Soc.*, 1992, *114*, 10068.
99. I. Castro-Rodriguez, K. Olsen, P. Gantzel and K. Meyer, *J. Am. Chem. Soc.*, 2003, *125*, 4565.
100. W. J. Evans, S. A. Kozimor and J. W. Ziller, *Science*, 2005, *309*, 1835.
101. G. Nocton, J. Pecaut and M. Mazzanti, *Angew. Chem. Int. Ed.*, 2008, *47*, 3040.
102. I. Korobkov, S. Gambarotta and G. P. A. Yap, *Angew. Chem. Int. Ed.*, 2002, *41*, 3433.
103. A. R. Fox, P. L. Arnold and C. C. Cummins, *J. Am. Chem. Soc.*, 2010, *132*, 3250.
104. A. R. Fox and C. C. Cummins, *J. Am. Chem. Soc.*, 2009, *131*, 5716.
105. R. K. Thomson, T. Cantat, B. L. Scott, D. E. Morris, E. R. Batista and J. L. Kiplinger, *Nature Chem.*, 2010, *2*, 723.
106. R. K. Thomson, C. R. Graves, B. L. Scott and J. L. Kiplinger, *Eur. J. Inorg. Chem.*, 2009, 1451.
107. D. M. King, F. Tuna, E. J. L. McInnes, J. McMaster, W. Lewis, A. J. Blake and S. T. Liddle, *Science*, 2012, *337*, 717.
108. N. Tsoureas, A. F. R. Kilpatrick, C. J. Inman and F. G. N. Cloke, *Chem. Sci.*, 2016, *7*, 4624.
109. D. M. King, F. Tuna, E. J. L. McInnes, J. McMaster, W. Lewis, A. J. Blake and S. T. Liddle, *Nature Chem.*, 2013, *5*, 482.
110. M. J. Monreal, L. A. Seaman, G. S. Goff, R. Michalczyk, D. E. Morris, B. L. Scott and J. L. Kiplinger, *Angew. Chem. Int. Ed.*, 2016, *55*, 3631.
111. W. Ren, E. Zhou, B. Fang, G. Hou, G. Zi, D.-C. Fang and M. D. Walter, *Angew. Chem. Int. Ed.*, 2014, *53*, 11310.
112. A. Herve, N. Garin, P. Thuéry, M. Ephritikhine and J.-C. Berthet, *Chem. Commun.*, 2013, *49*, 6304.
113. A. N. Dame, M. S. Bharara, C. L. Barnes and J. R. Walensky, *Eur. J. Inorg. Chem.*, 2015, 2996.
114. N. A. Budantseva, G. B. Andreev, A. M. Fedoseev and M. Y. Antipin, *Koordinat. Khimiya*, 2003, *29*, 265.
115. W. J. Evans, K. A. Miller, J. W. Ziller and J. Greaves, *Inorg. Chem.*, 2007, *46*, 8008.
116. D. P. Halter, F. W. Heinemann, L. Maron and K. Meyer, *Nature Chem.* 2018, *10*, 259.

https://doi.org/10.1142/9781800610163_0010

Chapter 10

Modern Applications of the Actinides in Catalysis

Rami J. Batrice[*,‡], *Isabell S. R. Karmel*[*,§], *Guy Yardeni*[†,¶], *and Moris S. Eisen*[*,||]

[*]*Schulich Faculty of Chemistry, Technion — Israel Institute of Technology, Technion City, Haifa 3200008, Israel*
[†]*Chemistry Department, Nuclear Research Centre Negev, Beer Sheva, Israel*
[‡]*batricer@gmail.com*
[§]*sikarmel@gmail.com*
[¶]*yardeniguy@gmail.com*
[||]*chmoris@technion.ac.il*

10.1 General

The chemistry of the early actinide elements and the compounds they form has reached a high level of sophistication and complexity in recent years. While the coordination chemistry of these metals has long been an area of intense investigation, their reactivity, especially in catalytic reactions, has only recently undergone closer scrutiny. While some of their catalytic activity has been found to provide analogous reactivity to some transition metal and lanthanide complexes, studies in recent decades have revealed chemical transformations of organic molecules unique to the 5f block. Although the factors which govern the observed transformations remain an area of great interest in modern research, it is generally understood that the structure and bonding of the early actinides are strongly influenced by their unique steric and electronic properties which stand in contrast to other metal groups. These features have led to a breadth of actinide-catalysed transformations such as small molecule activations, hydroelementations, coupling reactions, polymerizations and more. This chapter presents select examples from recent years which exemplify the advances of organic molecule transformations mediated by early actinide complexes.

10.2 Introduction

The inception of organoactinide chemistry is marked by the preparation of uranocene [(η^8-$C_8H_8)_2U$] by Streitwieser and Mueller-Westerhoff 50 years ago,[1] and the characterization of this compound one year later by Zalkin and Raymond.[2] Although this early report presents a simple example of an organoactinide complex, the coordination chemistry of the early actinides has rapidly developed to a high level of complexity owing to the diverse array of organic ligands used to stabilize the metal center.[3–16] Such complexes have been shown to perform several impressive stoichiometric transformations of organic molecules,[17–22] however, the catalytic activity of the organoactinides has only received attention in recent years.

In comparing the structure of d-block complexes to their actinide analogs, the diffuse 5f orbitals give rise to compounds with high coordination numbers. In the transition block, achieving such high coordination numbers is precluded due to electronic saturation of the metal center; as such, coordinative saturation exists as the principal component of the metal complex stability in the early actinides. This stability has been described more thoroughly by the sum of the cone angles of the ligands, wherein "over saturation" results in destabilization of the complex in question.[23–25] Throughout such investigations of the actinides, the vast majority has been devoted to the two naturally occurring elements of this series, thorium and uranium. Especially in the realm of catalytic studies, the tetravalent ions of these two metals have been the most widely used, however, recent works have also found considerable utility in the ability of the uranyl cation (UO_2^{2+}) to catalysed organic transformations.

In order to better understand the thermodynamic factors governing the observed catalytic behaviour, several factors are considered. One such factor is calculations using bond disruption enthalpies

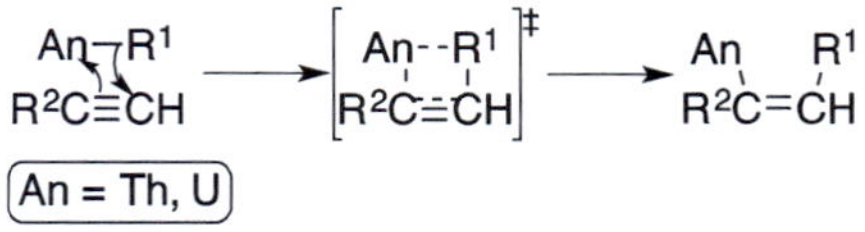

Scheme 10.1. Migratory insertion mechanism operative in actinide-mediated catalysis.

to determine the thermodynamically favoured intermediates and products of a catalytic cycle.[26–32] Moreover, thermodynamic calculations and experimental observations inform that an oxidative addition/reductive elimination mechanism is energetically disfavored; rather, a migratory insertion mechanism through a four-centered transition state which conserves the metal oxidation state is found to be operative in actinide-mediated catalysis (Scheme 10.1).

This chapter presents recent advances of actinide catalysis in homogeneous systems and the diverse applications of these systems.

10.3 Catalytic Reactions of Alkynes

Of the catalytic studies performed with actinide ions, the transformations of terminal alkynes are by far the most prevalent. Use of various substituents on the internal *sp*-hybridized carbon of the alkyne permits the investigation of various steric, electron and chelating effects, and facilitates the formation of numerous organic products through atom-economical methods.

10.3.1 *Oligomerization and cyclotrimerization of terminal alkynes*

One of the earliest examples of actinide-mediated catalysis was demonstrated in the oligomerisation of terminal alkynes to produce organic enynes by tetravalent uranium and thorium complexes.[33,34] The first examples of this reaction utilized metallocene dialkyl complexes of the formula $Cp^*_2AnMe_2$, (An = Th (**1**), U (**2**)), and showed impressive chemoselectivity using sterically encumbered alkynes such as *tert*-butyl acetylene, albeit with limited regioselectivity arising from head-to-head or head-to-tail insertion of the alkyne (Scheme 10.2).

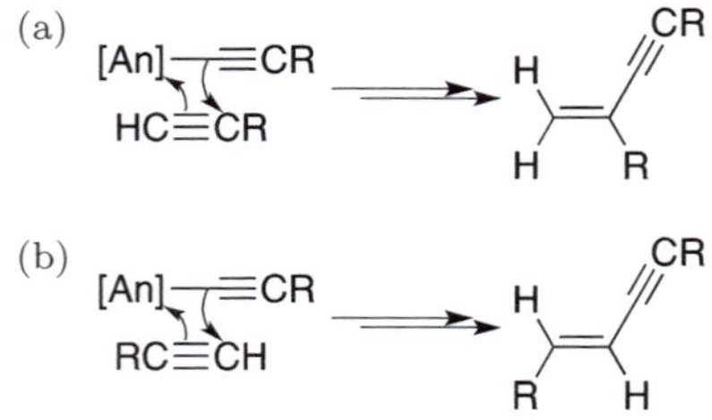

Scheme 10.2. Head-to-tail (a) and head-to-head (b) insertion of alkynes generating *cis* and *trans* organic enynes, respectively.

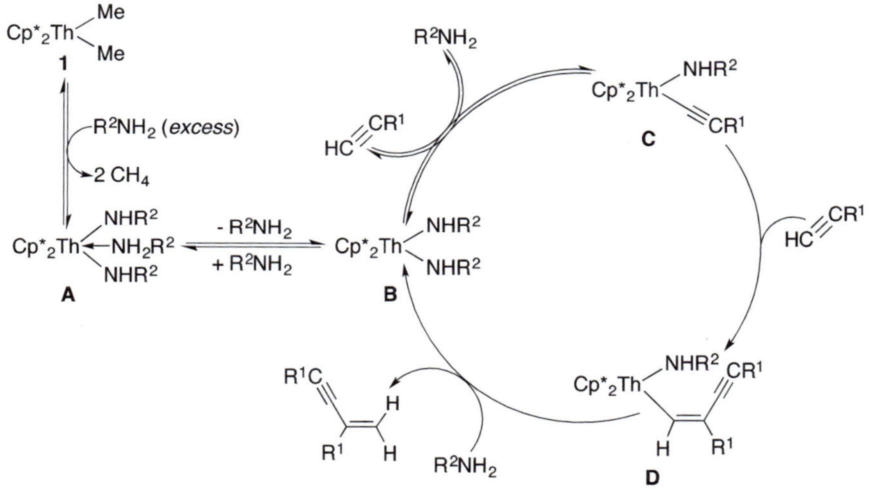

Scheme 10.3. A plausible mechanism of thorium-catalysed oligomerization of terminal alkynes assisted by an external primary amine.

When smaller substituents were placed on the alkyne (*iso*propyl, phenyl or trimethylsilyl) and studied in similar reactions, oligomersation was similarly achieved, however, the selectivity seen with larger substituents was lost, evident by the formation of larger oligomers up to decamers. Interestingly, a later study revealed a convenient method to control the extent of oligomerisation for these substrates.[35] By the addition of an external primary amine to serve as a protolytic agent, dimers were observed as the dominant product for all substrates studied. Investigation of this process suggested the catalytic mechanism shown in Scheme 10.3.

In the proposed mechanism, the precatalyst **1** is first activated through protonolysis by the external amine to generate the saturated amide complex **A** with concomitant formation of 2 equiv. of methane. Dative coordination of the neutral amine exists in a rapid equilibrium, and decoordination of the amine generates the active catalyst **B**. From the active catalyst, an additional rapid equilibrium exists from protolytic cleavage by the terminal alkyne to generate intermediate **C**; head-to-tail insertion of an additional equivalent of alkyne generates the metallaenyne **D** which undergoes rapid protolytic cleavage by the primary amine to form the *cis*-dimer product and regenerate the active catalyst **B**.

In a recent study by Eisen *et al.*, trivalent and tetravalent uranium amides of the formula $U[N(SiMe_3)_2]_3$ (**3**) and $[(Me_3Si)_2N]U[\kappa^2\text{-}(N,C)\text{-}CH_2Si(CH_3)_2N(SiMe_3)]$ (**4**), respectively, were similarly studied in the catalytic transformation of terminal alkynes; interestingly, low catalyst loadings (1 mol%) generated oligomerisation products analogous to those discussed earlier, however, increasing the catalyst loading to 10 mol% was found to favour a [2+2+2] cycloaddition product which generated 1,2,4- and 1,3,5-trisubstituted benzenes (Scheme 10.4).[36] These reactions were also studied using the analogous tetravalent thorium metallacycle, $[(Me_3Si)_2N]Th[\kappa^2\text{-}(N,C)CH_2Si(CH_3)_2N(SiMe_3)]$ (**5**), yet no cyclotrimerization product was formed, suggesting strongly that the 5f orbital electrons of the actinides are active in the observed structure and bonding.

Using either of the uranium amide precatalysts **3** and **4**, protolytic cleavage by terminal alkyne provides the active catalyst **A**. Head-to-tail or head-to-head migratory insertion of an additional equivalent of alkyne generates the metallaenyne intermediates **B** and **C**, respectively, and a subsequent insertion of the substrate in a head-to-tail or head-to-head manner yields the metalladienyne intermediates **D** and **E**, respectively. Activation of the internal alkyne of these intermediates by an additional equivalent of uranium catalyst results in distortion of the linear *sp* carbon–carbon bond and cyclization by the bound uranium center through transition states **F** and **G** to yield intermediates **H** and **I**; protolytic cleavage by an additional equivalent of terminal alkyne respectively yields the 1,3,5- and 1,2,4-trisubstituted benzene products accompanied by regeneration of the active catalyst **A**.

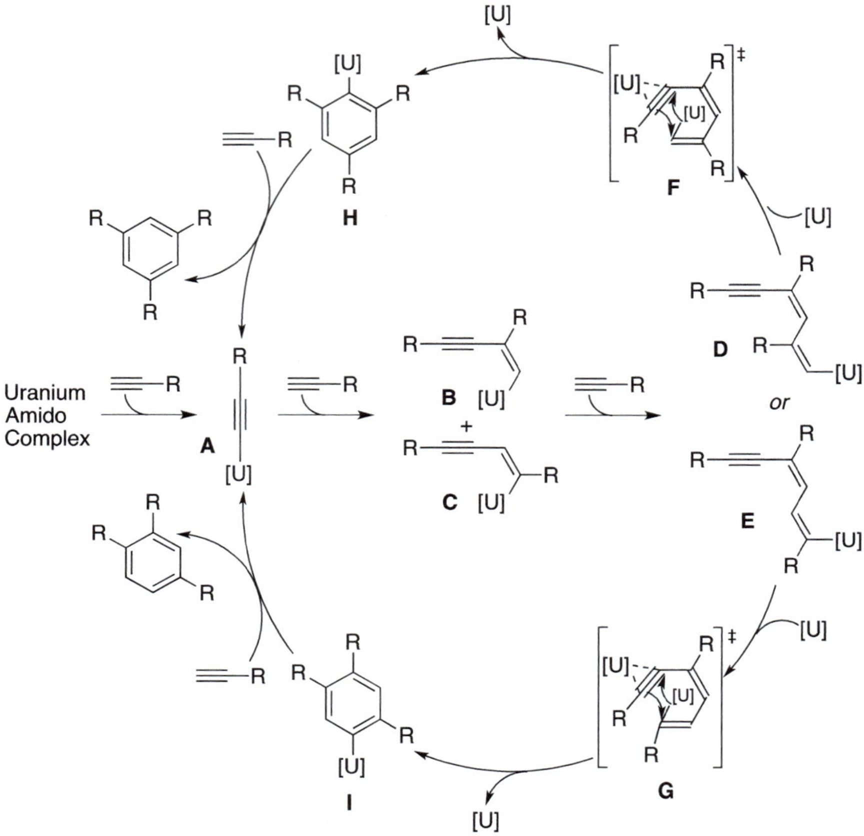

Scheme 10.4. Proposed mechanism for the uranium-mediated cyclotrimerization of terminal alkynes.

10.3.2 *Isonitrile coupling to terminal alkynes*

In addition to the oligimerisation experiments described, the coupling of isonitriles to terminal alkynes to provide α,β-unsaturated aldimines has also been studied using the metallocene dialkyl complexes **1** and **2** as well as the cationic uranium amide $[(Et_2N)_3U][BPh_4]$ (**6**).[37] Systematic studies of the reaction described showed that the choice of catalyst and the substrate ratios were of considerable consequence to the selectivity of the reaction and the products yielded. This is particularly evident through the formation of the isonitrile mono-addition product **E** achieved by the thorium catalyst. This product can also insert into the thorium catalyst

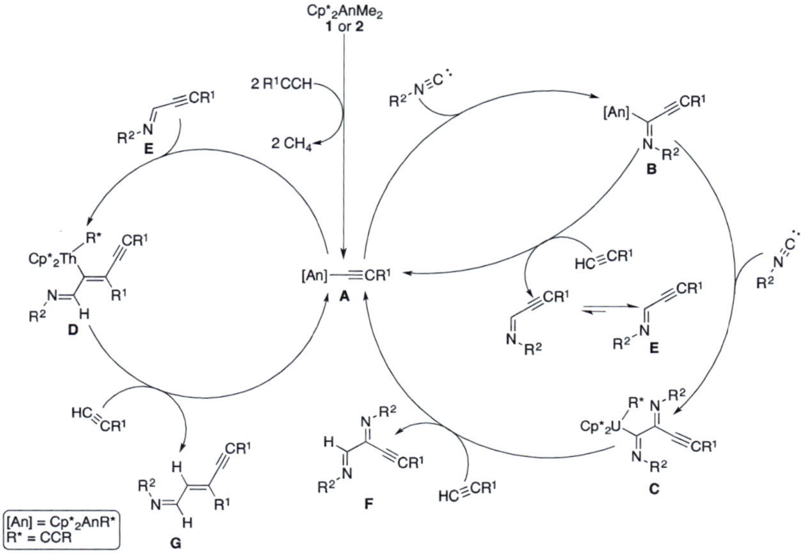

Scheme 10.5. The proposed mechanism of catalytic isonitrile insertion into terminal alkynes mediated by actinide complexes **1**, **2** and **6**.

A to ultimately form the organic product **G**. Conversely when the uranium complexes are used in this reaction, both mono- **E** and bis-addition products **F** are formed (Scheme 10.5).

Consideration of the characteristics of the metal catalysts used in the aforementioned reactions alludes to the selectivity of the products formed. It was proposed in this study that the steric encumbrance around the catalyst center created by the pentamethylcyclopentadienyl ligands forces the R^2 substituent of the isonitrile *syn* to the alkyne; subsequent protolytic cleavage generates the organic product which rapidly isomerizes to yield the *trans* product **E**. When a slight excess of isonitrile is used relative to alkyne, protonolysis of the intermediate **B** occurs more slowly, permitting 1,1-insertion of an additional equivalent of isonitrile to provide intermediate **C** which generates organic product **F** and active catalyst **A** after protolytic cleavage.

10.3.3 *Hydroelementation of terminal alkynes*

Of the catalytic reactions occurring on carbon–carbon unsaturated bonds, few have received as much attention as hydroelementation

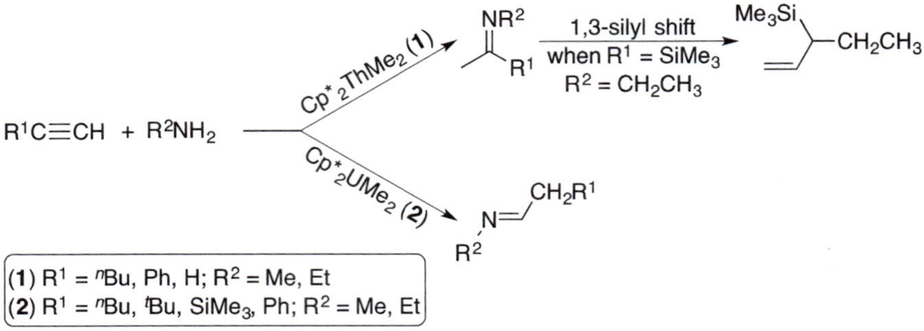

Scheme 10.6. Intermolecular hydroamination products generated from the catalytic reaction with **1** or **2**.

reactions. While hydroelementations have long been known to be mediated by numerous transition metal and rare earth metal catalysts,[38–48] the actinides have only recently emerged as suitable complexes to achieve these transformations. This section presents the array of inter- and intramolecular hydroelementation reactions mediated by actinide catalysts.

10.3.3.1 *Inter- and intramolecular hydroamination*

In the library of catalytic hydroelementations performed using actinide complexes, intermolecular hydroamination of terminal alkynes was one of the first studied. Using the actinide metallocene dialkyl complexes **1** and **2**, or the cationic uranium amide **6**, hydroaminations were achieved using a variety of alkyne and amine substrates.[49–51] From these studies, it was found that the identity of the metal center and the amine used strongly influenced the observed chemo- and regioselectivity of the products, however, the choice of alkyne proved inconsequential in these reactions.

For example, use of the thorium complex **1** afforded the *syn*-imine product after isomerization, as well as some degree of enyne or dienyne product from alkyne oligomerisation. In contrast, the analogous uranium complex **2** furnished only the hydroamination product with no evidence of oligomer formation (Scheme 10.6).

Kinetic studies of the catalytic cycle revealed the first-order behaviour of the precatalyst, zero-order behaviour in alkyne, and inverse first-order behaviour in amine; these findings allowed for the proposal of the catalytic cycle shown in Scheme 10.7.

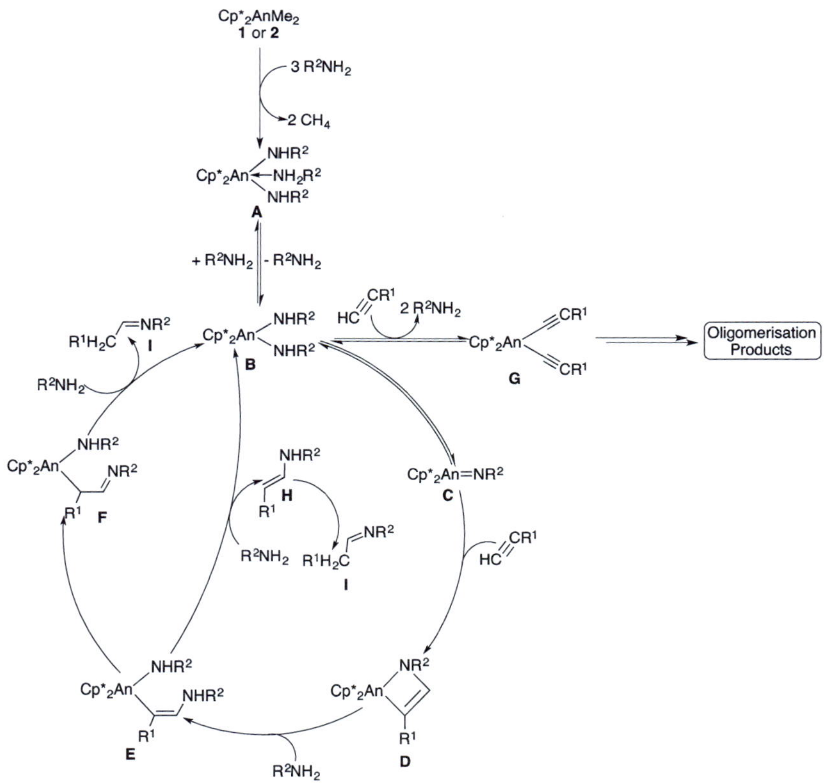

Scheme 10.7. The proposed mechanism of intermolecular hydroamination of terminal alkynes mediated by actinide catalysts.

According to the proposed mechanism, activation of the precatalyst occurs via protolytic cleavage by a primary amine to generate the coordinatively saturated complex **A** with concomitant formation of 2 equiv. of methane. Decoordination of the dative bound amine generates the active catalyst **B**, which after elimination of one bound amide generates the actinide-imido intermediate **C**. Migratory insertion of the terminal alkyne into the resulting An=N bond forms the metallacyclobutane intermediate **D** which is rapidly cleaved by an additional equivalent of amine to yield intermediate **E**. From this intermediate, two potential pathways are possible; the first of these may occur by tautomerization of the organoactinide to form intermediate **F** which undergoes protolytic cleavage by additional amine to form the organic product **I** and regenerate the active catalyst **B**.

Conversely, intermediate **E** can directly undergo protolytic cleavage to regenerate the active catalyst and form product **H** which rapidly tautomerizes to form the thermodynamically favoured product **I**.

A later study prepared variations of the classical actinide metallocene dialkyl complexes in the constrained-geometry complexes (CGC) of the formula (CGC)An$(NR_2)_2$ (CGC = $Me_2Si(Me_4C_5)$ (^{t}BuN), An = Th (**7**), U (**8**)) and applied these complexes to the intramolecular hydroamination/cyclization of alkynyl amines.[52,53] By virtue of the ancillary ligands being tethered together while bound to the metal center, the coordination sphere about the actinide ion is more open, facilitating higher reaction rates as compared to complexes **1** and **2**. By this logic, the larger metal ion should also display higher reactivity, and in fact, the thorium analogue **7** showed higher activity as compared to **8**.[54] In the reaction of 5-aminopentyne with either of the actinide-CGC complexes, the desired hydroamination/cyclization occurred to generate 2-methylenepyrrolidine according to the proposed mechanism (Scheme 10.8).

Activation of the precatalyst occurs by protolytic cleavage from 5-aminopentyne to form the active catalyst **A**. Migratory insertion of the tethered alkyne through a 4-centered transition state, **A'**, results in cyclization to give intermediate **B**. Protolytic cleavage by an additional equivalent of substrate yields product **C** and regenerates the active catalyst **A**.

In addition to the classical metallocene-based actinide complexes, ligand sets bearing anionic *N*-donors have also been prepared and show considerable utility in facilitating hydroaminations of alkynes (Figure 10.1). One such example was a uranium(IV) complex stabilised by a ferrocene-diamide ligand of the formula $(NN^{fc})U(Bn)_2$ (**9**) which was found to facilitate intra- and intermolecular hydroamination reactions.[55] Another study by Leznoff and co-workers presented the synthesis of two uranium(IV) diamido-ether complexes (**10** and **11**) and their successful application in hydroamination/cyclization reactions.[56]

10.3.3.2 *Intramolecular hydroalkoxylation/cyclization of alkynols*

Similar to hydroamination, hydroalkoxylation occurs by the formal addition of an O–H bond across an unsaturated C–C bond and

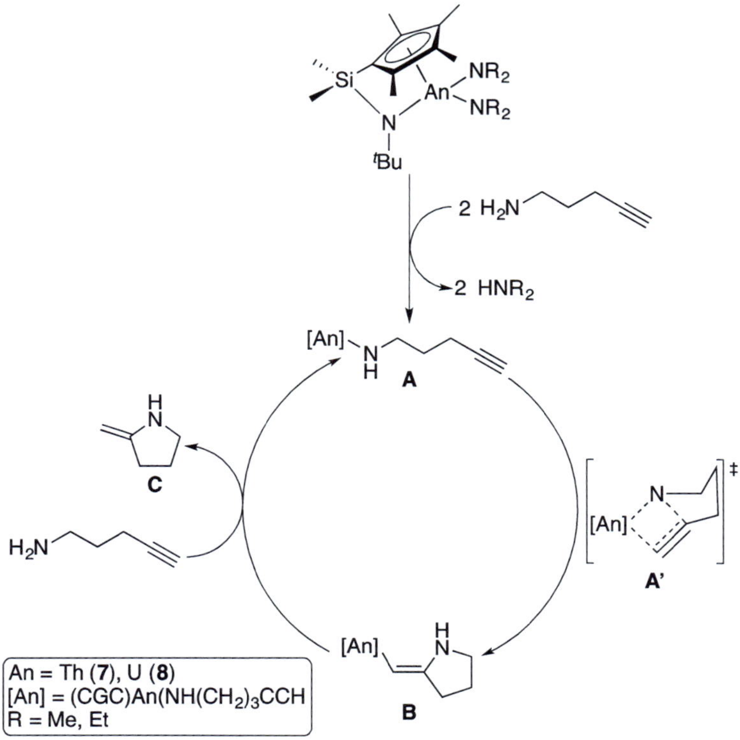

Scheme 10.8. The proposed mechanism of intramolecular hydroamination/cyclization of 5-aminopentyne catalysed by complexes **7** and **8**.

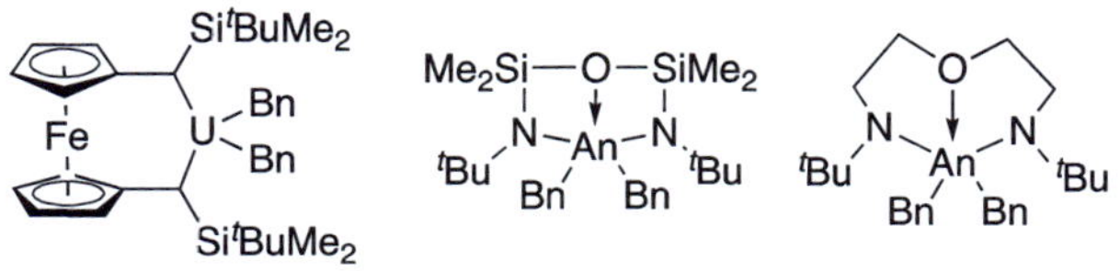

9 **10** **11**

Figure 10.1. Precatalysts **9**–**11** used in hydroamination reactions.

represents a highly atom economical method for the formation of C–O bonds.[41,57] In the realm of actinide catalysis, reactions using alcoholic substrates such as hydroalkoxylations were disregarded for years owing to the high bond disruption enthalpy of the actinide–oxygen bond formed (Th–O = 208.2 kcal mol^{-1}, U–O = 181.2 kcal mol^{-1}),[58] suggesting that the thermodynamic barrier for overcoming this interaction and achieving turnover would be far too

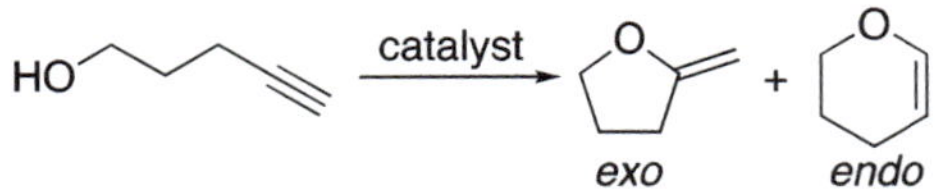

Figure 10.2. Possible *exo*- and *endo*-products of intramolecular hydroalkoxylation/cyclization.

high for practical applications. This prevailing theory was disproven by Marks *et al.* in a study wherein the thorium complexes **1**, **5** and **7** were employed in the catalytic intramolecular hydroalkoxylation/cyclization of numerous organic substrates bearing alcohols (1° and 2°) and unsaturated carbon–carbon moieties (alkynes and allenes).[59] From the kinetic analysis of the catalytic cycle, it was found that the reaction is first-order in catalyst and zero-order in the substrate; in addition, substrates bearing primary alcohols and terminal alkynes were found to convert most rapidly, suggesting that steric factors dominate the turnover of the catalytic cycle. This steric control explains the highest activity being observed for the CGC-thorium complex **7** owing to the open coordination sphere about the metal. The products formed in this study indicate that the transition state of the reaction occurs through a Markovnikov *exo*-methylene insertion (Figure 10.2).

According to kinetic and thermodynamic studies performed in this research, the mechanism was proposed to occur in an analogous manner to that seen in the intramolecular hydroamination/cyclization (Scheme 10.9).

The precatalyst is activated by protolytic cleavage by the alcohol moiety to generate the active catalyst **A**. As described, a Markovnikov *exo*-insertion of the terminal alkyne through transition state **A'** gives intermediate **B** as the turnover-limiting step. A subsequent rapid protolytic cleavage by an additional equivalent of substrate yields the furan product **C** and regenerates the active catalyst **A**.

10.3.3.3 *Hydrothiolation of terminal alkynes*

Hydrothiolation reactions by actinide complexes have provided a convenient route to provide vinyl sulfides. Contrary to transition metal systems which favour anti-Markovnikov addition,[60–63] the actinides have been found to achieve both Markovnikov and

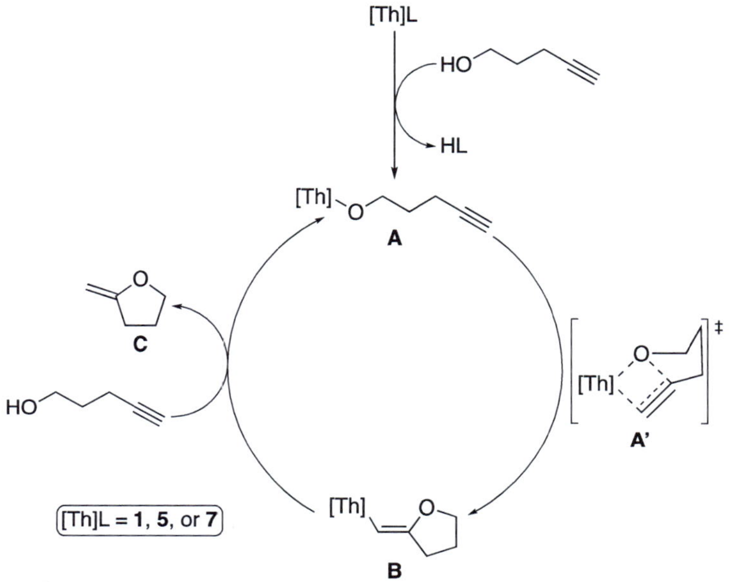

Scheme 10.9. The proposed mechanism of intramolecular hydroalkoxylation/ cyclization mediated by thorium complexes **1**, **5** and **7**.

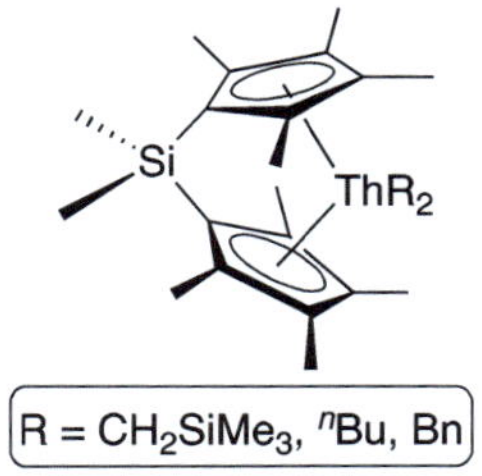

Figure 10.3. *Ansa*-bridged thorium metallocene **12**.

anti-Markovnikov additions.[64–66] In addition to the bis(pentameth ylcyclopentadienyl)actinide dialkyl complexes **1** and **2**, Marks and co-workers have prepared the *ansa*-bridged metallocene (Figure 10.3) of the formula ($[Me_2SiCp^R{}_2]ThR_2$ (**12**), ($Cp^R = Me_4C_5$, R = alkyl, benzyl), and studied the hydrothiolation of terminal alkynes mediated by these complexes.[67] It was proposed that the efficient catalysis observed in these reactions is facilitated by the highly exothermic nature of the alkyne insertion into the An–S bond.[68]

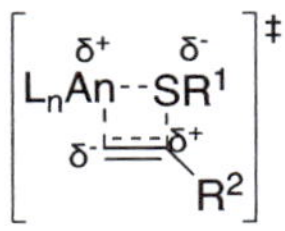

Figure 10.4. Markovnikov addition seen in the hydrothiolation of terminal alkynes catalyzed by actinide complexes.

The results showed that complex **12** provided the greatest catalytic activity and substrate scope, achieving insertion of aliphatic, aromatic and benzylic thiols across various terminal alkynes through Markovnikov addition. The regioselectivity of this insertion is stabilized by both electronic and steric effects (Figure 10.4).

By this mode of insertion, the partial negative charge situated on the terminal carbon of the alkyne is suitably disposed in proximity to the positive charge on the metal center. Additionally, the sterically encumbering R^2 group of the alkyne is favourably situated away from the sterically encumbered metal center.

The mechanism of catalysis was proposed to follow a similar mechanism to the previously discussed hydroelementation reactions (Scheme 10.10). Protonolysis of the thorium–alkyl bond by thiol liberates 2 equiv. of tetramethylsilane and generates the active thorium–thiolate catalyst **A**. Subsequent to activation, head-to-tail insertion of terminal alkyne generates the thorium-vinylidene sulfide intermediate **B** which then undergoes protonolysis by additional thiol to liberate the product **C** and regenerate the catalyst.

10.3.3.4 *Catalytic hydrosilylation of terminal alkynes*

The considerable utility of organosilicon compounds in industrial and research applications has earned this class of compounds a great deal of attention and fostered the development of versatile synthetic techniques; to this end, hydrosilylation stands as one of the methods of choice for the production of bulk and fine organosilanes.[69–71] Contrary to the previously discussed catalytic hydroelementations mediated by the actinides, hydrosilylations can produce geminal products in addition to the *cis*- and *trans*-conformer, but the product yielded is found to be highly dependent on the catalyst, substrate and reaction conditions.[72–75]

In an early study of this process using the actinide metallocenes **1** and **2**, it was found that hydrosilylation was achieved in a non-polar

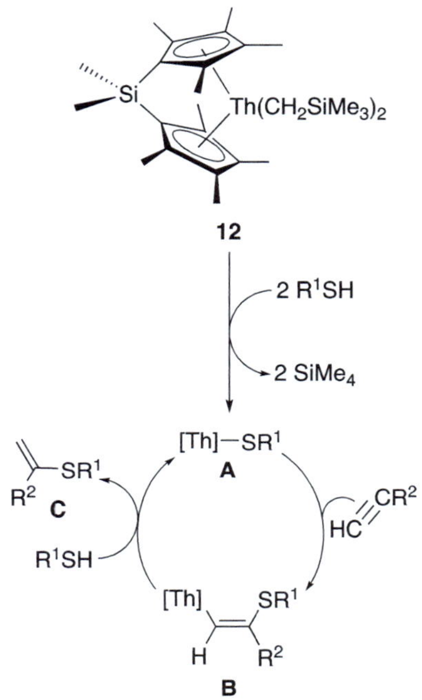

Scheme 10.10. A plausible mechanism of hydrothiolation of terminal alkynes catalysed by thorium metallocene **12**.

solvent such as benzene upon heating, however, the identity of the products obtained relied heavily on the reaction temperature.[76] As a model system, the insertion of phenysilane into *iso*-propylacetylene catalysed by the thorium complex **1** was studied and proposed to occur by the mechanism shown in Scheme 10.11.

At ambient temperature, the alkyne protolytically cleaves the methyl groups of the precatalyst to form the active bis-acetylide catalyst with concomitant formation of 2 equiv. of methane. Following this step, σ-bond metathesis of the Si–H bond generates the hydride intermediate **B** and forms the internal silyl-substituted alkyne **D**. From the thorium–hydride, migratory insertion of additional alkyne occurs in equilibrium to yield the vinyl–thorium **C**. From this intermediate, several potential reaction pathways may be operative, although it is worth note that the "high temperature" pathway is dominant when the uranium precatalyst **2** is used.

The first of these pathways involves a second σ-bond metathesis of phenylsilane by **C** to regenerate the hydride intermediate **B** and

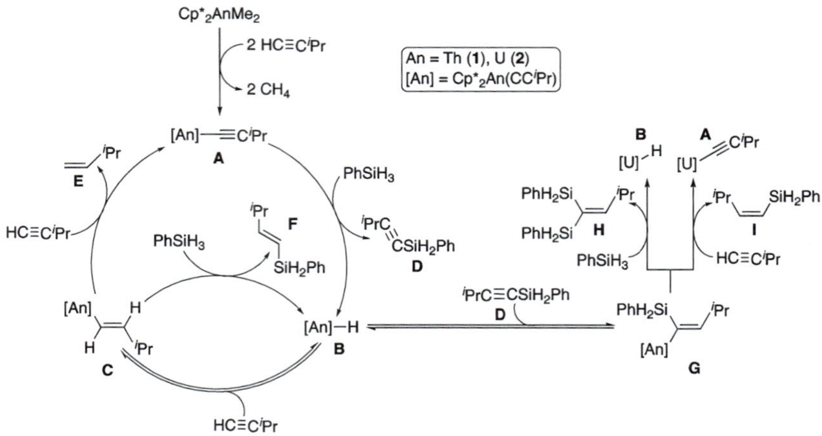

Scheme 10.11. The proposed mechanism of phenysilane insertion into *iso*-propylacetylene catalyzed by actinide metallocenes **1** and **2**.

liberate the *trans*-vinylsilane **F**. From the same vinyl–thorium species **C**, protolytic cleavage by the alkyne substrate is also possible, forming 3-methylbutene **E** and regenerating the catalyst **A**. Conversely, the organic product **D** can insert into the thorium–hydride **B** to form the metallovinylsilane **G**, but upon elevating the temperature of the reaction (50°–80°C), two additional reaction pathways are possible. The first of these involves the reaction of **G** with *iso*-propylacetylene via a protolytic cleavage to regenerate catalyst **A** along with the formation of *cis*-vinylsilane **I**. Alternatively, σ-bond metathesis with phenylsilane produces the geminal disilylalkene product **H** and the intermediate actinide–hydride **B**.

Hydrosilylation stands apart from the aforementioned hydroelementations further in that changing solvent previously showed little to no effect in the products formed. Conversely, performing the actinide-catalysed hydrosilylation in a coordinating solvent such as tetrahydrofuran accelerates protonolysis of intermediate **C** by *iso*-propylacetylene and increases the amount of product **E** formed. This effect is proposed to arise from solvent-silane interactions that inhibit the other reactive pathways seen in Scheme 10.11.[70] From kinetic and thermodynamic studies of this reaction, several aspects of the catalytic cycle were identified. Namely, the reaction order was found to be first-order in the catalyst, alkyne and silane, and the reaction of silane with **C** is identified as the rate-limiting step of the reaction.

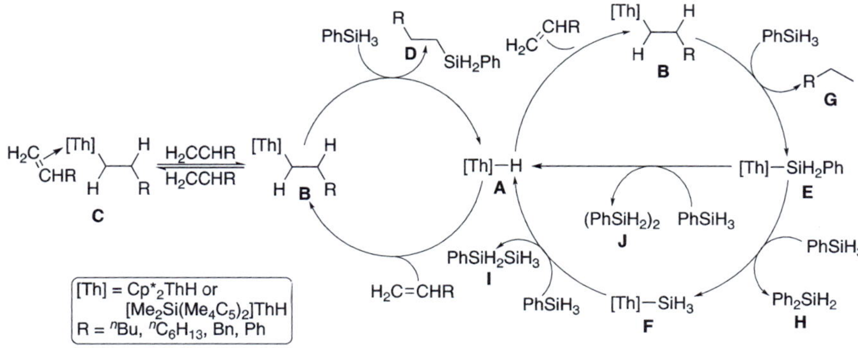

Scheme 10.12. Proposed mechanism of hydrosilylation of alkenes catalyzed by complexes **1** and **12**.

Note: *Precatalyst activation occurs by σ-bond metathesis of $PhSiH_3$ and the accompanying formation of $PhSiH_2Me$ or $PhSiH_2^nBu$ when using catalyst **1** or **2**, respectively.

Substitution of precatalyst **1** with the *ansa*-bridged metallocene **12** provided enhanced reaction rates, regioselectivities and chemoselectivities, typified by the greater than 92% conversion to the *trans*-vinylsilane **F**.[77] Similar to the greater reactivity seen for **12** in other reactions, this effect is proposed to arise from the more open coordination sphere about the metal center. The enhanced selectivity seen when using this catalyst is explained by the hindered equatorial girdle created around the thorium center by the bridged ligand which restricts the approaching substrate to a single, discriminating orientation.[78]

Up to now, the hydroelementations performed have been restricted to *sp*-hybridized carbon moieties, however, the thorium metallocenes **1** and **12** were found to be suitable catalysts in the hydrosilylation of alkenes, although the selectivity was found to diminish with the change in hybridization.[79] This reaction was investigated using aliphatic, benzylic and aromatic alkenes, and the proposed catalytic cycle diverged significantly from the operative mechanism in the hydrosilylation of alkynes (Scheme 10.12).

The aforementioned catalytic cycle does not show activation of the precatalyst, however, this has been experimentally determined to occur by a σ-bond metathesis with phenylsilane to form the active catalyst **A**. After formation of the active catalyst, two potential reactive pathways are proposed to be possible. The first of these pathways occurs by migratory insertion of the alkene substrate into the thorium–hydride **A**, forming the alkyl–thorium intermediate **B**. This intermediate exists in a rapid equilibrium with product **C** wherein side-on coordination of alkene removes **B** from the catalytic cycle and slows the progression of the reaction. This observed inhibitory effect of the alkene justifies the inverse first-order behaviour determined from kinetic experiments on the reaction. After decoordination of alkene to again form intermediate **B**, σ-bond metathesis with phenylsilane occurs to form alkylsilane **D** and regenerate the active catalyst **A**.

The competitive reaction in this cycle similarly begins with the migratory insertion of the alkene substrate into **A** to form intermediate **B**, however, this alternative pathway varies beyond this step. Protonolysis by phenylsilane liberates alkane **G** and forms the thorium-silane intermediate **E**. This intermediate also is capable of undergoing two different reactions, the first of which being σ-bond metathesis of another equivalent of phenylsilane to regenerate the active catalyst and form 1,2-diphenyldisilane **J**. Alternatively, from intermediate **E**, insertion of the silicon-carbon bond of phenylsilane liberates diphenylsilane **H** and furnishes the thorium-silane **F**. This intermediate then may undergo an addition σ-bond metathesis with phenylsilane to regenerate catalyst **A** and produce phenyldisilane **I**.

Beyond the metallocenes used, the cationic uranium(IV)-amide (**6**) was also investigated as a catalyst to facilitate these reactions. Systematic studies using similar reaction and substrate conditions revealed reaction selectivity comparable to the uranium metallocene **2**.[80]

10.3.3.5 *σ-bond metathesis of silylalkynes catalyzed by organouranium*

Inspired by the structure and reactivity of the *ansa*-bridged thorium complex **12**, similar conditions were employed to form the analogous uranium complex. Serendipitously, a thermally stable oxo-bridged uranium *ansa*-metallocene (**13**) was formed (Figure 10.5).[81]

Optimization of the reaction to form **13** was performed which allowed for subsequent catalytic studies; hydrosilylation reactions were executed with this complex; however, an unexpected σ-bond metathesis of the Si–C bond in silylalkynes was found, and the

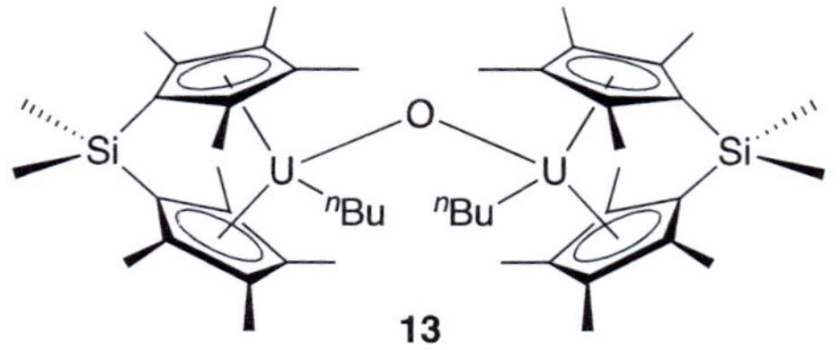

Figure 10.5. Oxo-bridged uranium *ansa*-metallocene dialkyl complex **13**.

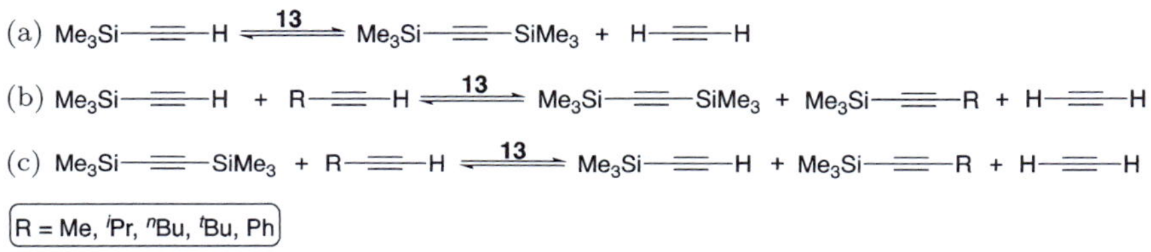

Scheme 10.13. Potential products of σ-bond metathesis of silylalkynes catalyzed by complex **13**. (a) trimethylsilylacetylene, (b) trimethylsilylacetylene in the presence of terminal alkynes and (c) bistrimethylsilyl acetylene in the presence of terminal alkynes.

products formed were found to vary based on the nature of other alkyne substrates present in the reaction solution (Scheme 10.13).

Of the reactions described, autometathesis of trimethylsilylacetylene (Scheme 10.13(a)) occurred readily at room temperature, however, the cross-metathesis in the presence of additional alkynes (Scheme 10.13(a) and (b)) required elevated temperatures (66°C). A plausible mechanism of this metathesis process was presented in this study, and the cycle was hypothesized to occur through an identical mechanism despite the alkynes used (Scheme 10.14).

Protonolysis of the *n*-butyl groups on **13** occurs by trimethylsilylacetylene to liberate 2 equiv. of butane and generate the uranium-acetylide **A**. The subsequent σ-bond metathesis of the Si–C bond from an additional equivalent of trimethylsilylacetylene exists in equilibrium and occurs through a four-centered transition state **A'**, ultimately forming the uranium-acetylide **B** and bistrimethylsilylacetylene. From intermediate **B**, another equivalent of trimethylsilylacetylene undergoes a similar σ-bond metathesis through an analogous reversible four-centered transition state **B'**, thus regenerating the active catalyst **A** with concomitant formation of acetylene gas.[77]

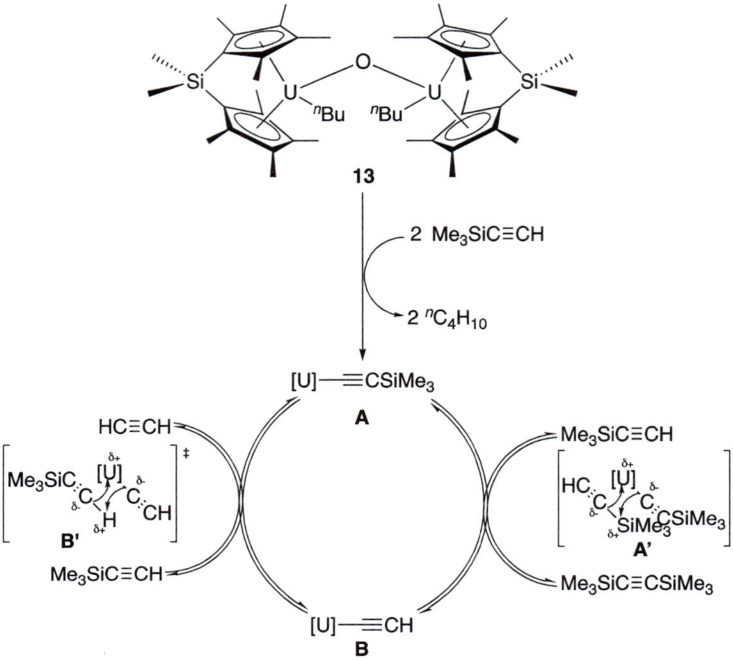

Scheme 10.14. Proposed mechanism for the autometathesis of trimethylsilylacetylene catalyzed by complex **13**.

10.4 Catalytic Synthesis of Esters from Aldehydes

As aldehydes lack the Brønsted–Lowry acidity of alcohols given the absence of a labile proton, they remained absent from actinide catalysis studies for decades due to the presumed deactivation of the catalyst center which would occur upon coordination of the oxygen atom.[82] However, in recent years, this assumption was disproven through the catalytic formation of esters beginning from aldehydes using two distinct synthetic techniques. This section describes recent works toward these goals using various actinide complexes.

10.4.1 *Catalytic Tishchenko reaction mediated by actinide complexes*

The earliest method toward the catalytic formation of esters using actinide complexes was demonstrated through the coupling of

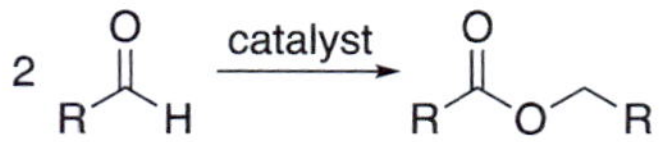

Figure 10.6. Archetypical Tishchenko reaction forming an ester from aldehydes.

aromatic aldehydes mediated, also known as the Tishchenko reaction (Figure 10.6).[83,84]

The first example of this reaction was demonstrated using various thorium catalysts. While the formation of an intermediate thorium–oxygen bond would require considerable energy to disrupt, the driving force of the reaction was determined to be the continuous, thermoneutral formation of Th–O bonds in each step of the catalytic cycle, also informing that entropic contributions exist as the main energetic barrier in achieving a turnover.

The thorium complexes $Cp^*_2ThMe_2$ (**1**), $[Me_2Si(Cp")_2]Th(^nBu)_2$ (**12**) and $Th(NEtMe)_4$ (**14**) were employed in this study and proved to be useful catalysts to achieve the desired transformation. As the coordination sphere was subject to lower levels of steric encumbrance, the reactivity was seen to increase, resulting in the highest reactivity being seen for the thorium *ansa*-metallocene complex, and the lowest reactivity being observed for the bis(pentamethylcyclopentadienyl)thorium dimethyl complex (1).[84] The reaction rates were found to be fastest for electron-deficient aldehydes, and homocoupling was efficiently achieved using these catalysts, however, attempts to prepare asymmetric esters by using two different aldehyde substrates proved ineffective and lacked any discernible selectivity. The catalytic behavior of complex **1** in the coupling of benzaldehyde was investigated in greater detail, and a plausible mechanism proposed for the thorium-catalysed Tishchenko reaction (Scheme 10.15).

The reaction is initiated by migratory insertion of the C=O bond of benzaldehyde into the thorium–methyl bond of **1** to produce the thorium-alkoxide **A**. A second equivalent of aldehyde inserts similarly into the resulting Th–O bond, generating **B**, which in the presence of an additional benzaldehyde molecule undergoes a Zimmerman–Traxler transition state **B'** with concomitant hydride transfer to yield **E** and the active catalyst **C**. A similar insertion of two benzaldehyde moieties occurs to furnish **D**, form the chair-like transition

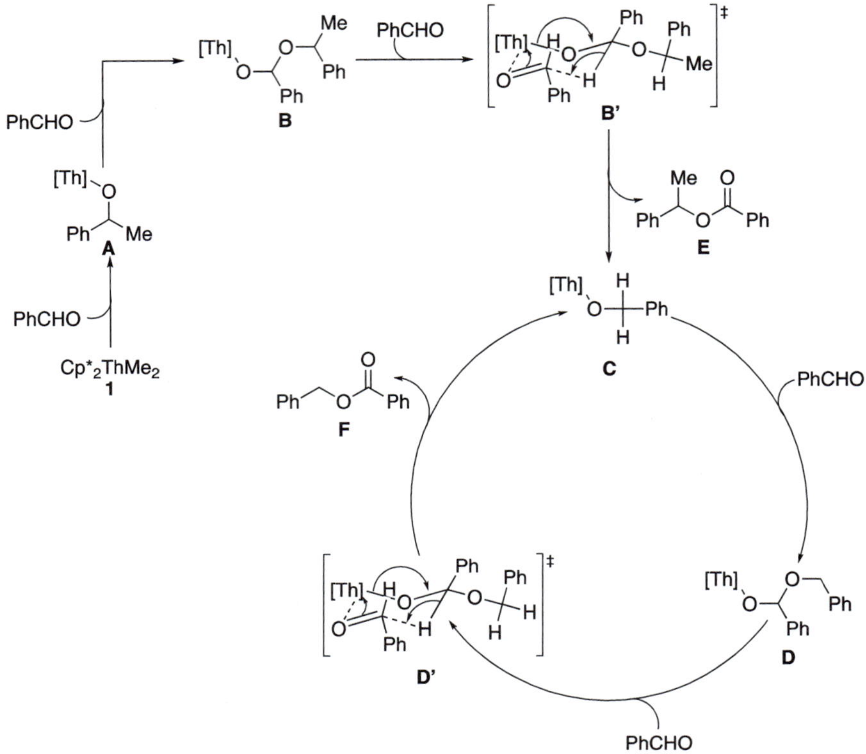

Scheme 10.15. Proposed mechanism for the catalytic Tishchenko reaction by complex **1**.

state **D'**, and after hydride transfer, regenerate the catalyst (**C**) with concomitant formation of the desired ester product.

After these early reports, the catalytic Tishchenko reaction was expanded further and performed with novel imidazolin-2-iminato thorium complexes of the formula ($Cp^*{}_2Th(Im^{Dipp}N)(Me)$ (**15**) and $(Im^{Dipp}N)Th(N(SiMe_3)_2)_3$ (**16**) (Figure 10.7).[81]

Due to the increased electron density residing on the metal center afforded by the imidazolin-2-iminato, (thus reducing the metal's oxophilicity), the reactivity of the complexes was expected to be enhanced as compared to the pure metallocene complexes. Performing the reactions with **15** in the presence of aromatic, heteroaromatic, cyclic and branched aliphatic aldehydes mediated the desired coupling and displayed considerable tolerance for a breadth of substrates

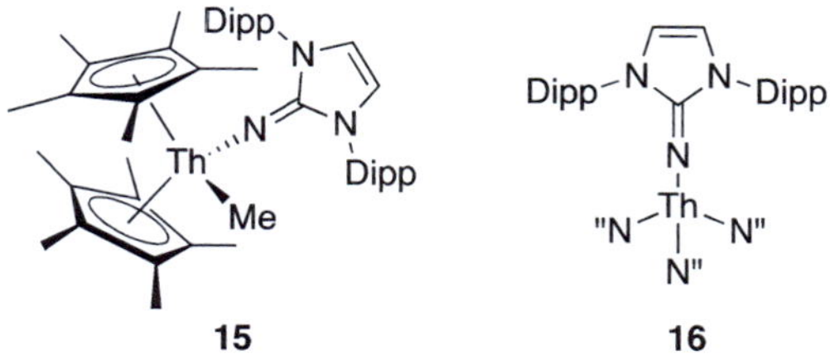

Figure 10.7. Imidazolin-2-iminato thorium complexes **15** and **16** used in the catalytic Tishchenko reaction.

with various steric and electronic influences.[85] Mechanistic studies using this complex revealed a similar catalytic cycle to that seen for the thorium metallocenes.

When similar studies were employed using complex **16**, impressive reactivity was also observed, however, the tris-amido thorium complex additionally succeeded in generating asymmetric esters from the coupling of an aromatic and aliphatic aldehyde. Due to the lower reactivity of aromatic aldehydes as compared to aliphatic aldehydes using this complex, a 4:1 mixture of aromatic to aliphatic aldehyde was introduced to the reaction mixture to efficiently and selectively form the desired asymmetric product.[86]

10.4.2 *Tandem proton-transfer esterification of alcohols with aldehydes*

In addition to the formation of esters from the actinide-mediated Tishchenko reaction, proton-transfer esterification reactions utilizing aldehydes and alcohols as substrates has similarly been achieved. Unlike the Tishchenko reaction discussed previously which is shown to produce a mixture of products when using more than one aldehyde, the latter catalytic process is particularly advantageous in the selective synthesis of asymmetric esters (Scheme 10.16).

The actinide metallacycles **4** and **5** and uranium tris(amide) (**3**) were found to be suitable precatalysts to achieve this transformation in the presence of α,α,α-trifluoromethylacetophenone, a sacrificial ketone which competitively coordinates to the actinide center and undergoes preferential hydride transfer.[87] Moreover, the yields of the

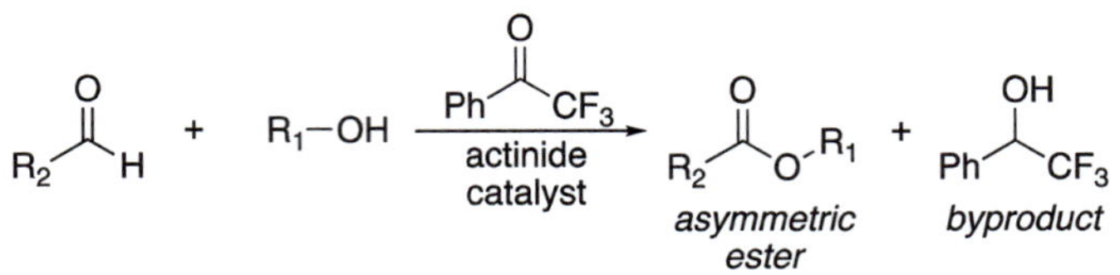

Scheme 10.16. Tandem proton-transfer esterification between aldehydes and alcohols mediated by actinide catalysts.

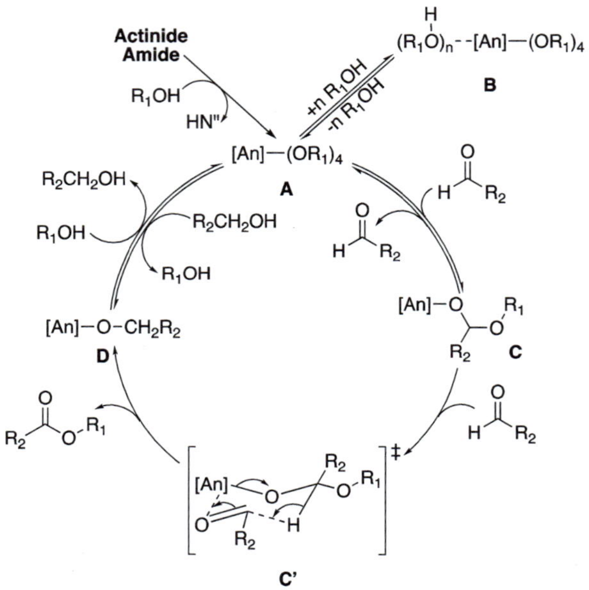

Scheme 10.17. Proposed mechanism for actinide-mediated tandem proton-transfer esterification.

products obtained in these reactions were found to be highly dependent on the steric and electronic characteristics of the substrates used. Generally, electron withdrawing substituents on the aldehyde, as well as sterically unencumbered substrates, were found to accelerate the rate of the reaction observed.

The catalytic mechanism for this transformation was investigated and proposed to initiate through activation of the actinide-amide precatalyst through protonolysis by an excess of alcohol to generate the active catalyst **A** (Scheme 10.17).

The actinide tetra(alkoxide) **A** exists in rapid equilibrium with the saturated complex **B**, though upon decoordination of the neutral alcohol, the catalyst undergoes insertion into the aldehyde to furnish **C**. Subsequent hydride transfer from the hemiacetalate to an additional equivalent of aldehyde through a Zimmerman–Traxler transition state **C′** yields the desired ester with subsequent formation of an intermediate actinide-alkoxide **D**. Further protonolysis by alcohol present in solution regenerates the active catalyst **A** and turns over the catalytic cycle. From isotopic studies using deuterium labeled alcohol, it was seen that hydride transfer as seen in the chair-like transition state is rate determining in the reaction.

10.5 Catalytic Addition of Protic Nucleophiles to Heterocumulenes

The insertion of heterocumulenes such as carbodiimides or isocyanates into protic nucleophiles provides a convenient means to produce synthetically and biologically useful molecules such as guanidines or ureas, yet the majority of catalytic studies which facilitate this transformation have focused on lanthanide-based systems.[88–94] Research by Evans *et al.* reported the stoichiometric insertion of carbodiimides into actinide–carbon bonds,[95,96] however, attempts to achieve a catalytic turnover in these reactions was not shown until recently. A recent report utilized the thorium imidazolin-2-iminato complex **16** to catalyse the insertion of carbodiimides, isocyanates and isothiocyanates into several aromatic and aliphatic amines, phosphines and thiols and achieved moderate to high conversions under mild conditions with excellent functional group tolerance.[97] Studies of the operative mechanism of the catalytic insertion provided the reaction shown in Scheme 10.18.

The reaction begins with the insertion of carbodiimide into the Th–N bond of the bis(trimethylsilyl) moiety, generating the thorium-amidinate **A**. Protolytic cleavage by diphenylphosphine liberates the amidine byproduct **D** and creates the thorium–phosphide active catalyst **B**. Another equivalent of carbodiimide inserts into the thorium–phosphorus bond and forms the thorium-phosphaguanidinate **C**, which after protonolysis by diphenylphosphine forms the phosphaguanidine product **E** and regenerates the

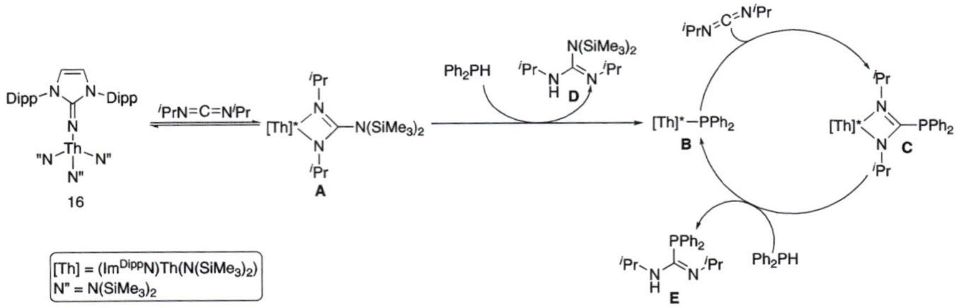

Scheme 10.18. The proposed mechanism of the catalytic addition of diphenylphosphine to di-isopropyl carbodiimide mediated by thorium complex **16**. *Note:* *Only one active catalyst site is shown for clarity.

active thorium–phosphide catalyst **B**. Isotopic labeling studies of this reaction revealed that the protolytic cleavage step was rate-determining for the catalytic cycle.

Another study which investigated actinide-catalysed nucleophilic addition to heterocumulenes utilized the previously discussed uranium tris-amide (**3**), and the thorium and uranium metallacycles **4** and **5**, respectively.[98] In comparing the reactivity of the actinide-amides to the thorium complex **16**, the former showed accelerated reactions rates for the addition of aromatic amines to carbodiimides, and near quantitative conversion in short reaction times for the addition of thiols to carbodiimides. The study further elaborated this reaction by the actinide-catalysed addition of terminal alkynes (RCCH; R = ^{t}Bu, $SiMe_3$, Ph) to di-*iso*-propylcarbodiimide, producing alkynylamidines and representing the sole example of an actinide complex facilitating the addition of a C–H bond across a carbon–heteroatom unsaturated bond. In contrast to the activation of complex **16**, reaction of the actinide-amides **3**–**5** begins with protolytic cleavage to directly form the active catalyst. After thorough characterization of these reactions and the products formed, the addition of primary, secondary and tertiary alcohols to carbodiimides was successfully attempted.[99] Kinetic studies using complex **3** showed first-order behaviour in catalyst and carbodiimide, and inverse first-order in alcohol, suggesting the alternative mechanism proposed in this study (Scheme 10.19).

According to the mechanism, precatalyst **3** undergoes protolytic cleavage by the alcohol substrate accompanied by the liberation of

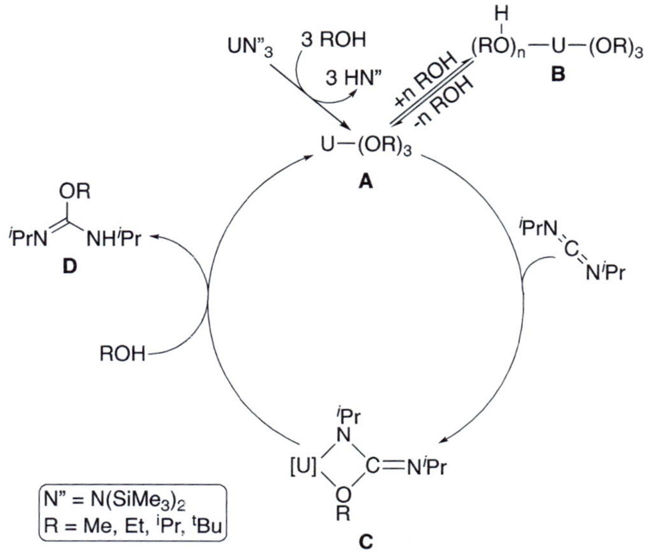

Scheme 10.19. The proposed mechanism of actinide-catalyzed addition of alcohols to carbodiimide.

3 equiv. of bis(trimethylsilyl)amine to give active catalyst **A** which exists in rapid equilibrium with **B** in the presence of excess alcohol. From **A**, the C=N moiety of the carbodiimide inserts across the U–O bond through a four-centered transition state to provide intermediate **C**. From this intermediate, protolytic cleavage by additional alcohol liberates the isourea product **D** with concomitant regeneration of the catalyst **A**.

Since these studies, other actinide complexes bearing benzimidazolin- and perimidin-2-iminato have been prepared conveniently by protonolysis of the actinide metallacycles **4** and **5** with the neutral ligand (Scheme 10.20), and the resulting complexes have been applied to similar catalytic reactions with great success.[100]

While the steric influence of the substrates did not appear to affect reaction rates or conversions when utilizing complexes **3**–**5**, these factors proved to be central to the selectivity of the products formed when using complexes **17**–**24**. The robust precatalysts prepared were also suitable for transformation of more elaborate substrates such as diols and triols which yielded isourea products bearing multiple nitrogen and oxygen donor atoms in the organic scaffold. Interestingly,

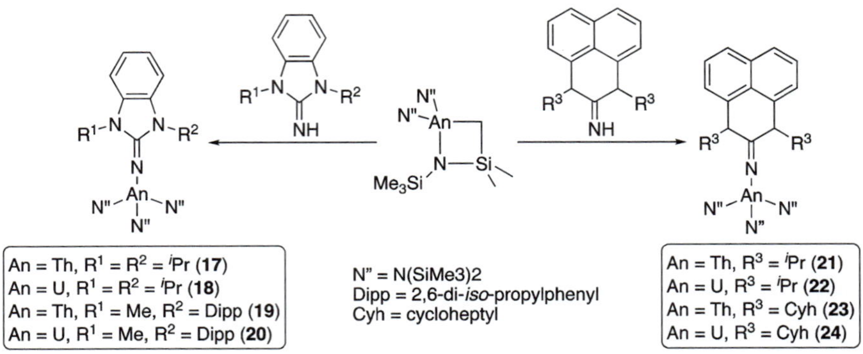

Scheme 10.20. Synthesis of benzimidazolin-2-iminato and perimidin-2-iminato actinide complexes **17**–**24**.

kinetic studies performed with complexes **19** and **21** were both found to be first-order in catalyst and carbodiimide, however, the former of these compounds showed inverse first-order behaviour in alcohol, while compound **21** exhibited first-order kinetics for alcohol.

10.6 Dehydrocoupling of Silanes and Amines

Silicon nitrides are rigid, chemically inert solids[101] derived from silazanes that have gained widespread attention in both industry and research owing to their vast utility in electronics,[102–104] automobiles[105,106] and medical devices.[107–111] The ubiquity of these materials has inspired extensive research into the preparation of silazanes as precursors to silicon nitrides and catalysts capable of effecting these transformations. To date, the actinides have not been studied thoroughly toward this goal, however, the cationic uranium(IV) amide **6** was investigated as a precatalyst for the coupling of primary and secondary amines with phenylsilane to yield aminosilanes of the general formula $PhSiH_{3-n}(NHR)_n$ (n = 1–3).[112] The reaction conditions were shown to affect the product selectivity to a limited degree, however, the choice of amine was found to be most consequential, wherein primary amines reacted much more rapidly than secondary amines. In addition to this, the steric encumbrance of the amine substituent was found to directly influence the observed selectivity. For example, use of *n*-propylamine was found to produce $PhSiH(NH^nPr)_2$ and $PhSi(NH^nPr)_3$, yet the use of the bulkier *tert*-butylamine exclusively

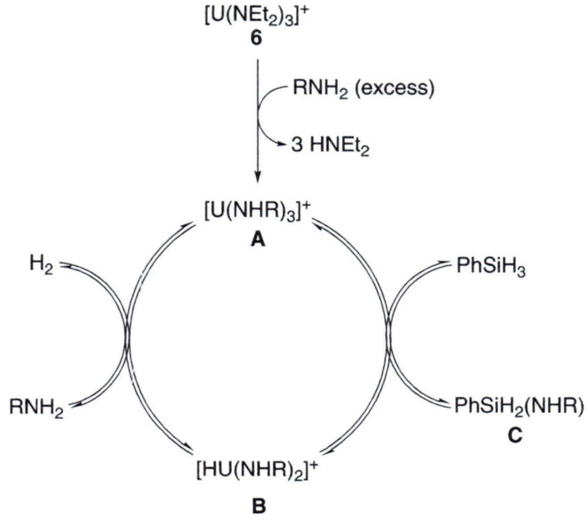

Scheme 10.21. Proposed mechanism for the dehydrocoupling of silanes and amines catalyzed by compound **6**.

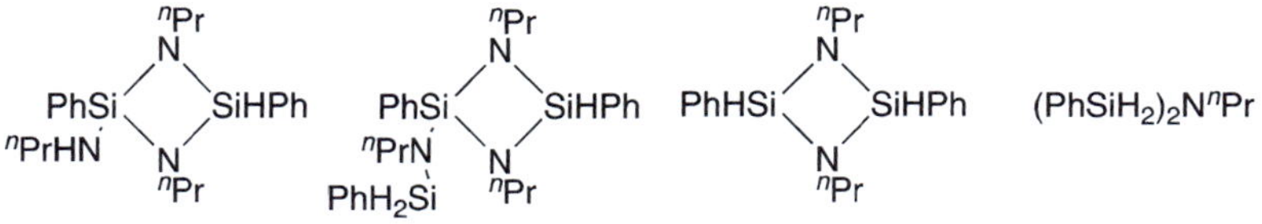

Figure 10.8. Products of the uranium-catalysed dehydrocoupling of silanes and primary amines.

formed $PhSiH_2NH^tBu$. Within this study, the plausible mechanism was also proposed using experimental data (Scheme 10.21).

The cationic uranium(IV)-amide **6** reacts rapidly with excess of amine to liberate 3 equiv. of diethylamine and form catalyst **A**. From the active catalyst, σ-bond metathesis of the silane Si–H bond exists in equilibrium, ultimately generating silazane **C** and the uranium-hydride intermediate **B**. An additional equilibrium process exists with intermediate **B** wherein the amine substrate reacts with the uranium-hydride to regenerate the active catalyst accompanied by the production of dihydrogen. In this reaction, the ability for product **C** to enter the catalytic cycle twice more has been observed, forming the products seen in Figure 10.8; however, permutations of the reaction in the proposed mechanism have yielded the other products shown.

10.7 Uranium-Catalyzed Reduction of Azides and Hydrazines

While the studies discussed thus far have focused on the tetravalent oxidation state of uranium, the easily accessible hexavalent oxidation state has given rise to interesting redox processes. One such example is the two-electron reduction of organic molecules including azides and hydrazines by Burns and co-workers.[113] The report detailed the preparation of a bis(imido) uranium(VI)metallocene (**25**) and its reduction by dihydrogen to the bis(amido) analog (**26**) (Scheme 10.22).

Although this reaction occurred only stoichiometrically, it was turned catalytic in a later study by the addition of an organic azide in benzene solution (Scheme 10.23).[114]

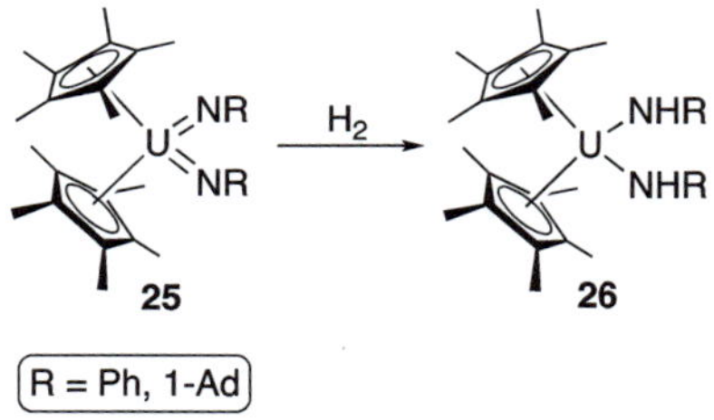

Scheme 10.22. Reduction of uranium bis(imide) **25** to uranium bis(amide) **26** by dihydrogen.

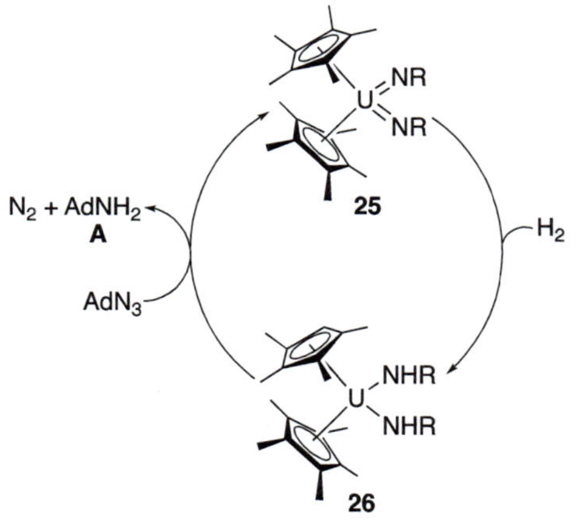

Scheme 10.23. Proposed mechanism for the catalytic reduction of azide by complex **25**.

Upon heating to 55°C in a dihydrogen atmosphere, the bis(imide) **25** is rapidly reduced to the tetravalent bis(amide) complex **26**. Azidoadamantane subsequently reacts with the reduced species and oxidizes the uranium center back to a hexavalent state with concomitant formation of dinitrogen and adamantylamine **A**. The reduction of *N*,*N*′-diphenylhydrazine Doccurs via an analogous mechanism; however, the reduction is found to ensue even in the absence of dihydrogen, informing that hydrazine is both an oxidizing and reducing within this catalytic cycle.

10.8 Other Catalytic Reactions

10.8.1 *Asymmetric Diels–Alder reaction*

A class of tetravalent uranium binolate complexes (**27**–**29**) (Figure 10.9) has been developed by Walsh *et al.* and studied in asymmetric Diels–Alder reactions (Scheme 10.24).[115]

Of the complexes studied, **29** provided the highest enantioselectivity (63% ee) and a 78% yield of Diels–Alder product. Further investigation was dedicated to elucidating the influence of the lithium ion on the observed catalytic activity. In contrast to previously reported

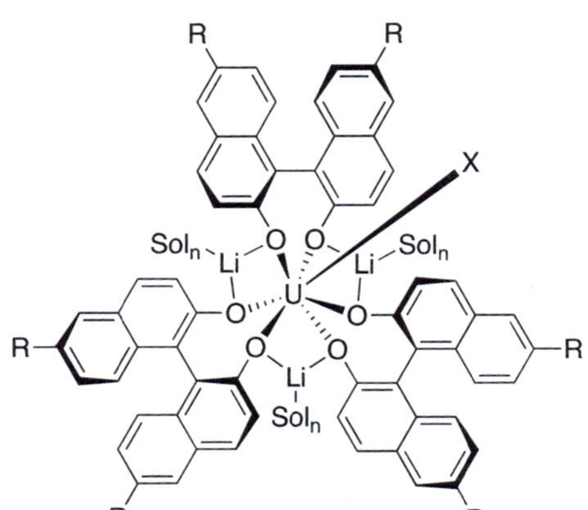

R = H, Sol = THF, n = 2, X = Cl (**27**)
R = H, Sol = DMEDA, n = 1, X = Cl (**28**)
R = Br, Sol = THF, n =2, X = Cl,
X = 6,6′-Br_2-NINOLate(**29**)

Figure 10.9. Structure of uranium/lithium binolates **27**–**29**.

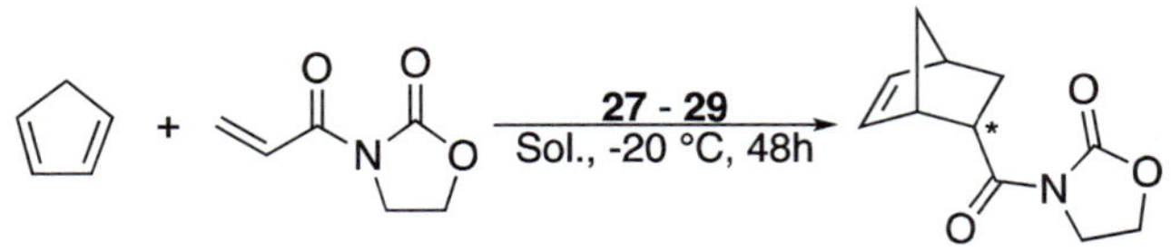

Scheme 10.24. Diels–Alder reaction mediated by uranium binolates **27–29**.

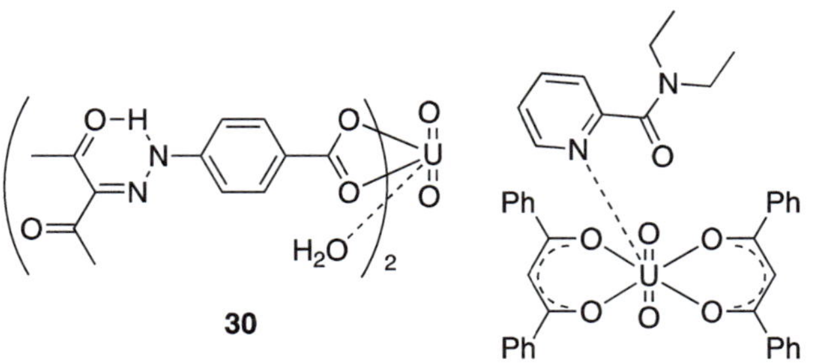

30

31

Figure 10.10. Uranyl complexes **30** and **31** used in the catalyzed Henry reaction.

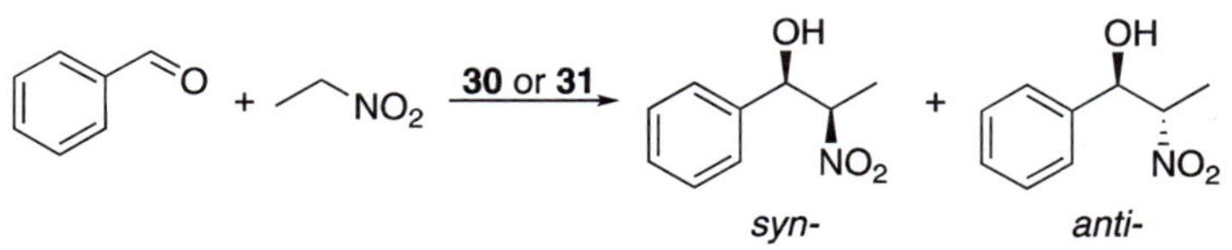

Scheme 10.25. Nitroaldol reaction of benzaldehyde with nitroethane catalysed by **30** or **31**.

lanthanide analogs,[116] saturation of the metal center was found to enhance the enantioselectivity observed.

10.8.2 *Uranyl-catalyzed Henry addition*

While the majority of catalytic studies performed with uranium have focused on the tetravalent oxidation state, some reports have shown the ability of the hexavalent uranyl ion to catalyse select reactions. One such example was presented by Mahmudov *et al.* wherein the reactivity of two aqueous uranyl complexes was investigated (Figure 10.10).[117]

The aforementioned complexes were found to be effective precatalysts for the Henry addition of nitroethane to benzaldehyde

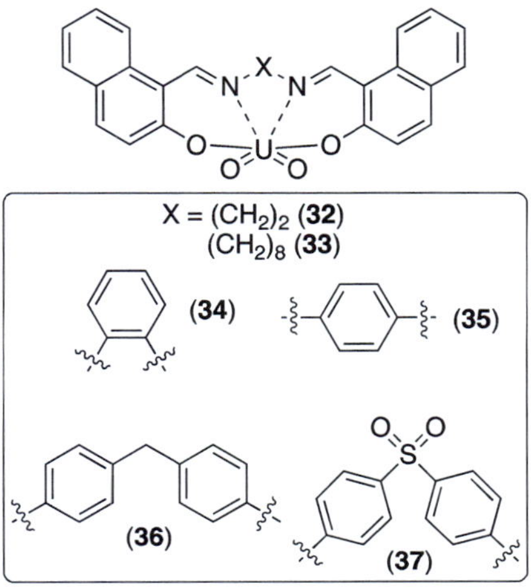

Figure 10.11. Uranyl Schiff base complexes **32**–**37**.

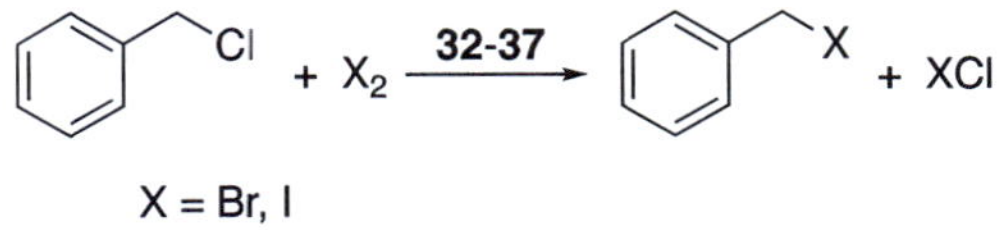

Scheme 10.26. Catalytic halogen exchange mediated by **32**–**37**.

(Scheme 10.25). Studies on the effects of solvent on the efficacy of the reaction showed a strong dependence on the solvent used, where methanol and acetonitrile provided moderate diastereoselectivities and low yield. In contrast, running the catalytic reactions in water provided the highest yields, yet unimpressive diastereoselectivity.

10.8.3 *Halogen exchange reactions*

A series of uranyl Schiff base compounds were prepared by Naeimi and co-workers (Figure 10.11) and applied to a mild catalytic halogen exchange between benzyl chloride and elemental bromine or iodine (Scheme 10.26).[118]

The precatalysts in this study were shown to be highly efficient in mediating the described transformation, achieving the desired halogen exchange in high yields and short reaction times.

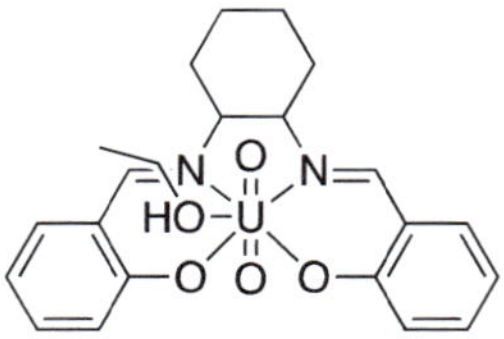

Figure 10.12. Uranyl salen complex **38**.

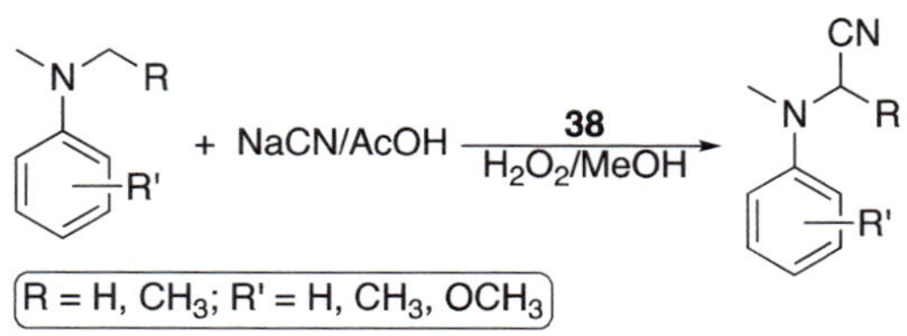

Scheme 10.27. Visible light induced C–H bond functionalisation by complex **38**.

10.8.4 *C–H bond functionalization*

In addition to the catalysis achieved by uranyl Schiff base complexes described earlier, a chiral uranyl compound bearing a salen type ligand has been developed by Jain *et al.* and its catalytic behaviour studied (Figure 10.12).[119]

A reaction mixture of dialkylanalines, sodium cyanide and acetic acid, in the presence of **38** and H_2O_2 as an oxidant was prepared and irradiated with visible light. Upon irradiation, the catalyst was activated and achieved the desired bond activation characterized by the formation of α-aminonitriles (Scheme 10.27).

10.8.5 *Uranyl-mediated acylation of alcohols*

A series of uranyl complexes (**39**–**42**) (Figure 10.13) were prepared by Takao *et al.* and studied in the acylation of alcohols in the presence of acetic anhydride (Scheme 10.28).[120]

It was found that complex **42** was most efficient in achieving this transformation as compared to the other three uranyl compounds. In addition, pivalic anhydride was also studied as the nucleophile present. Kinetic and spectroscopic investigation alluded to the interaction of $[UO_2(OPPh_3)_4]^{2+}$ with Ac_2O, forming

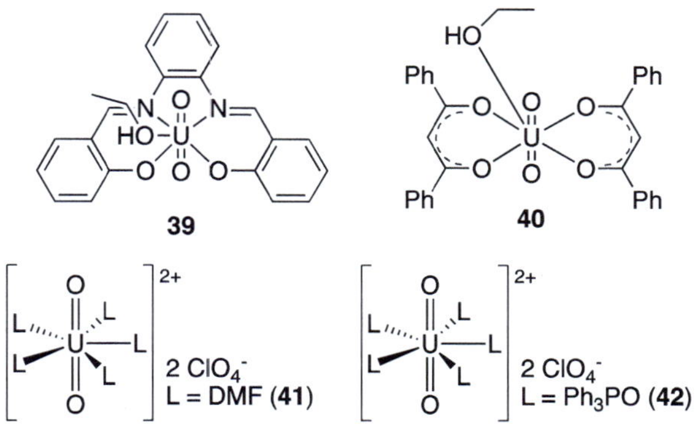

Figure 10.13. Uranyl compounds **39–42** studied in the acylation of alcohols.

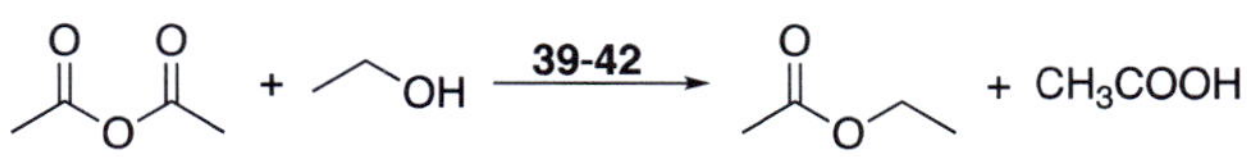

Scheme 10.28. Uranyl-catalyzed acylation in the presence of acetic anhydride.

$[UO_2(Ac_2O)(OPPh_3)_3]^{2+}$. This interaction was found to be rate limiting in the catalytic turnover.

10.9 Actinide-Catalyzed Polymerization Reactions

The utility of the actinides beyond the transformation of organic molecules has been known for decades since the earliest reports of butadiene polymerization by $[(\eta^3\text{-}C_3H_5)_3UCl]$ in the presence of $[EtAlCl_2]$ as co-catalyst; the reaction produced 1,4-*cis*-polybutadiene with mechanical properties superior to polybutadiene produced from transition metal-catalysed reactions.[121–124] Since this report, the actinides have enjoyed considerable attention in the polymerization of various α-olefins and unsaturated substrates.

10.9.1 *Polymerization of α-olefins*

The first studies of α-olefin polymerization mediated by actinide catalysts was performed by the cationic thorium metallocenes $[Cp^*_2ThMe][BPh_4]$ and $[Cp^*_2ThMe][B(C_6F_5)_4]$.[125,126] These initial studies proved informative to the influence of the counterion used on the efficacy of polymerization and spurred an extensive study

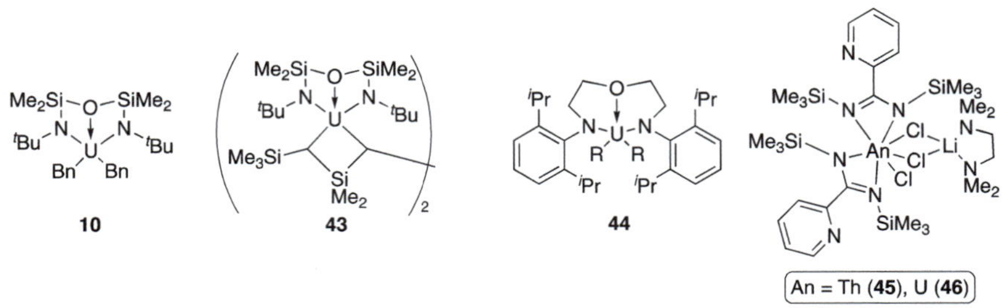

Figure 10.14. Diamido ether uranium(IV) complexes and actinide-amidinates used in the polymerization of ethylene.

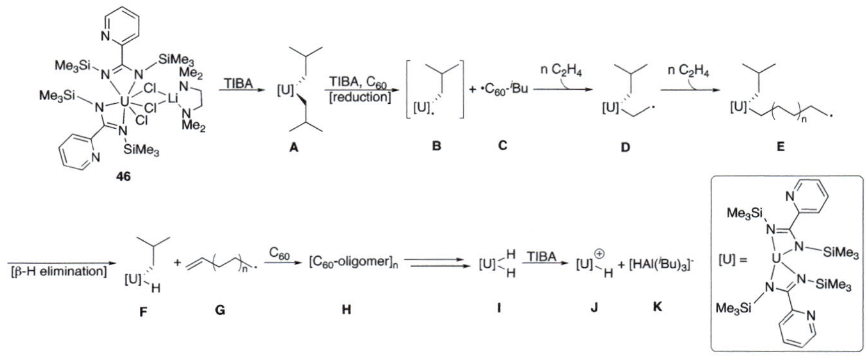

Scheme 10.29. A plausible mechanism of ethylene polymerization catalyzed by uranium bis(amidinate) **46**.

preparing a library of cationic actinide metallocenes bearing different anions as well as reactivity studies of the isolated complexes. Even the most reactive of these complexes fell short of the observed activity of analogous group IV metallocenes,[127] however, changing the ancillary ligand to organic molecules containing *N*-donors, such as diamido ethers or amidinates, was found to drastically improve the catalytic activity in the polymerization of ethylene (Figure 10.14).[128]

Of the catalysts described, the bis(amidinate) actinides **45** and **46** revealed the best reactivity in the polymerization of ethylene when tri-*iso*-butylaluminum (TIBA) was used as the co-catalyst.[129] Experimental studies including mass spectrometry and fullerene radical trapping experiments informed of the operative mechanism proposed in Scheme 10.29.

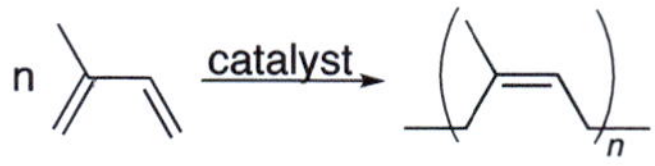

Scheme 10.30. General reaction scheme for the polymerization of isoprene.

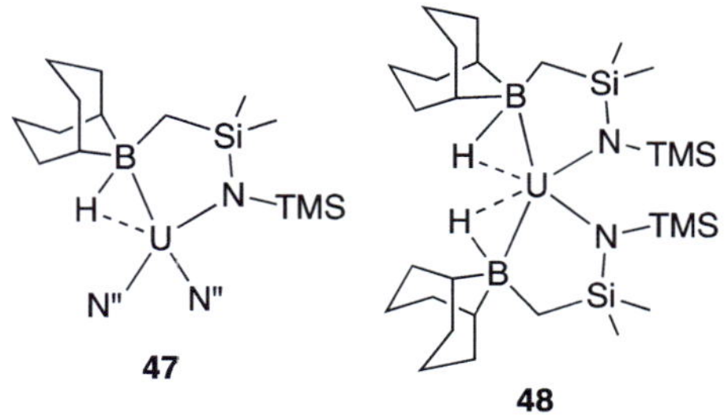

Figure 10.15. Uranium amido-borohydrides (**47** and **48**) utilized in isoprene polymerization.

Alkylation of the precatalyst with TIBA yields intermediate **A** that undergoes reduction by another equivalent of TIBA to give intermediate **B** which is trapped by fullerene, forming product **C** which is detected by MALDI-TOF spectrometry. The radical intermediate is oxidized by ethylene to furnish **D** which undergoes additional radical insertions by the olefin to generate the catalyst-bound polymer **E**. β-hydride elimination of this intermediate releases the radical polymer chain **G** and produces the uranium-hydride **F**. Reaction of the radical polymer chain **G** with fullerene in solution forms **H** which is similarly detected by mass spectrometric methods. The uranium–dihydride species **I** reacts with TIBA to produce the active cationic uranium catalyst **J**. When the reaction was performed in the absence of fullerene, no difference in reactivity or product formation was discernible, indicating that radical polymerization is not the dominant pathway in the reaction.

In addition to the aforementioned ethylene polymerization, the polymerization of other olefins such as isoprene have also been reported (Scheme 10.30), however, the ability to catalyse these reactions using actinide complexes remain scarce.

One prominent example of this was recently reported by Arnold *et al.*, wherein two uranium(IV) amido-borohydrides (**47** and **48**) were prepared and applied to this reaction (Figure 10.15).[130]

Similar to the actinide-amidinates, different cocatalysts or activating agents were found to impart a variety properties in the final product. When treated with $B(C_6F_5)_3$ and TIBA, minimal polymerization occurred. Conversely, use of $[HNMePh][B(C_6F_5)_4]$ as the cocatalyst resulted in 66% conversion after heating for 17 hours at 50°C. Of the soluble product isolated, 95.5% was determined to assume *cis* conformation of the unsaturated product.

10.9.2 *Ring-opening polymerization of cyclic esters*

In the advent of the "green revolution", the focus of polymer chemistry has shifted toward the production of biodegradable polymers as biocompatible materials,[131] packing materials,[132] microelectronics[133] and adhesives.[134] Of the potential polymers suitable to fill this niche, polycaprolactone (PCL) and polylactide (PLA) emerge as ideal candidates. Several metal catalysts have been employed to produce these materials, however, only in the past decade were the actinides applied to ring-opening polymerizations (ROP). The first of these reports was presented by Ephritikhine and co-workers wherein trivalent rare earth complexes and an isostructural uranium complex was used to catalyse the ROP of ε-caprolactone; the results showed far lower activity for the uranium complex as compared to the lanthanide analogs, likely due to the significantly higher oxophilicity of the uranium center.[135] A later study utilized the actinide metallocenes **1** and **2** as well as the cationic uranium amide **6** in the ROP of L-lactide.[136] In this study, complex **6** exhibited the highest catalytic activity, likely due to the open coordination sphere. Each catalyst generated PLA with high molecular weights and extremely low molecular weight distributions (PDI = 1.01), informing that the polymerization proceeds in a living manner. When these complexes were used in the ROP of ε-caprolactone, similar results were obtained. Further analysis of the reaction allowed for the proposal of a plausible mechanism for the ring-opening polymerization (Scheme 10.31).

Upon introduction of ε-caprolactone to a toluene or THF solution of the metallocene precatalyst, α-hydrogen abstraction of the monomer by the precatalyst forms the active catalyst **A**. This species is proposed to exist in a rapid equilibrium favouring the tautomerized enolate form **B**. Recurrent migratory insertion of the monomer into

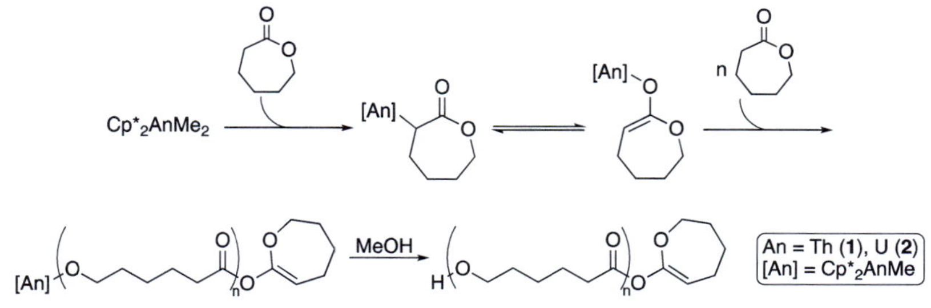

Scheme 10.31. A plausible mechanism of the ROP of ε-caprolactone catalysed by actinide metallocenes **1** and **2**.

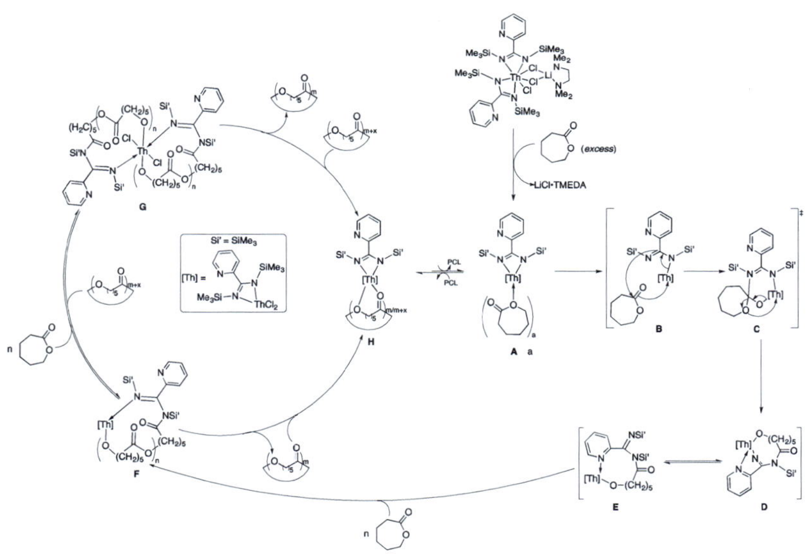

Scheme 10.32. A plausible mechanism of the cyclooligomerization of ε-caprolactone catalysed by bis(amidinate) thorium **45**.

the An–O bond forms the metal-bound polymer **C** which is liberated after quenching the reaction with methanol, yielding PCL **D**. The imidazolin-2-iminato thorium complex **15** was also studied in this reaction and found to occur through a similar reaction mechanism.[85]

In developing a reaction portfolio containing a breadth of actinide coordination compounds, several bis(amidinate) actinide complexes have been studied in the ROP of ε-caprolactone; a thorough mechanistic study was performed using complex **45** and revealed a divergent mechanistic pathway from that observed when using the actinide metallocenes, yielding a cyclo-PCL (Scheme 10.32).[137]

From the aforementioned mechanism, two operative pathways are possible, and from the experimental data, the two different modes of reactivity either form cyclic pentamers or cyclic undecamers up to tridecamers. Coordination of monomer to the precatalyst liberates a bridging chloride moiety in the form of a lithium chloride:TMEDA complex, ultimately furnishing intermediate **A**. The reaction pathway to the right of the scheme occurs due to nucleophilic attack by the amidinate ligand of the metal complex on the carbonyl-carbon of the monomer through transition states **B** and **C**, subsequently forming the dative bound thorium intermediate **D** which exists in equilibrium with intermediate **E**. A successive insertion of ε-caprolactone into these intermediates produces complex **F** which may undergo elimination of the cyclic polymer to form PCL which then may recoordinate to the metal center, furnishing intermediate **H**. This intermediate reacts with ε-caprolactone to ultimately liberate the PCL product and regenerate the active catalyst **A**. The aforementioned second reaction pathway deviates from the mechanism described earlier after formation of intermediate **F**. This pathway occurs by the reaction of the second metal-bound amidinate moiety on an additional equivalent of ε-caprolactone, generating intermediate **G**; from this intermediate, reversible elimination can occur to again form **F**, or rather, both coordinated oligomers on **G** may eliminate to form the intermediate **H** accompanied by the formation to two separate cyclo-PCL moelcules.

Stemming from these early studies with a simple amidinate ligand, other actinide-amidinates have been prepared wherein the ancillary ligands is elaborated further to create vastly different steric and electronic environments surrounding the active center. An impressive example of this type of coordination chemistry showed the preparation and reactivity of a series of actinide amidinates bearing a tertiary amine as a coordinating side-arm (Figure 10.16).[138,139]

Utilization of these complexes in the ROP of ε-caprolactone was found to produce linear polycaprolactone polymers with narrow polydispersities. The activity of these catalysts was found to increase in the order of **50** $<$ **49** $<$ **53** $<$ **52** $<$ **51**, clearly displaying the higher activity of the thorium complexes as compared to the uranium analogs.

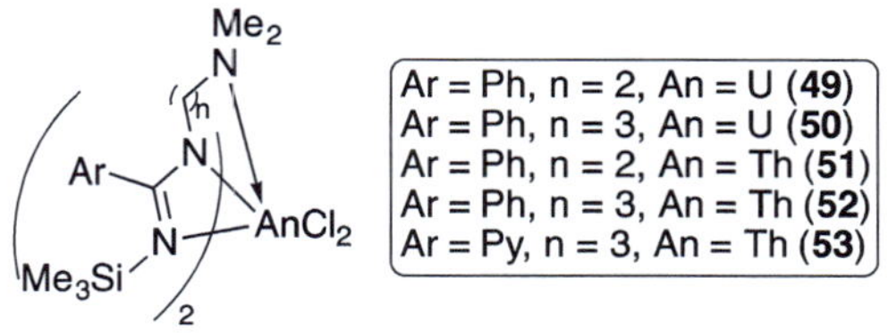

Figure 10.16. Actinide bis(amidinate) complexes bearing a coordinating side-arm.

A uranyl aryloxide of the formula $(ArO)_2U(O)_2(THF)_2$ (**54**) was recently prepared by Baker *et al.* and studied using experimental and computational methods in the ROP of cyclic esters.[140] Introduction of cyclic esters such as ε-caprolactone or δ-valerolactone to a solution containing the uranyl precatalyst **54** resulted in efficient cyclooligomerisation of these cyclic esters, however, when smaller esters were employed such as β- or γ-butyrolactone, no catalytic reaction was observed. From these studies, a proposed bimetallic mechanism of catalysis is proposed which is initiated by displacement of coordinated THF through nucleophilic attack by the incoming monomer to produce transition state **A** (Scheme 10.33).

A ring-opening transition state **B** occurs next, resulting in formation of the uranyl-alkoxide intermediate **C**. Propagation of the polymerization proceeds through an intermolecular attack on the oxygen anti-bonding orbital of the monomer by a second catalytic center **D**, and subsequent transfer to the second catalytic site **E**, thus forming the binuclear intermediate proposed **F**. Successive insertions of monomer into this intermediate further facilitate polymer propagation, and elimination of intermediate **G** liberates the cyclic polycaprolactone product and yields transition **A**.

The ROP of other esters such as *rac*-lactide has also been studied, and a series of thorium metallocenes bearing *tert*-butyl-substituted cyclopentadienyl ligands (**55**–**57**) were found to affect the desired polymerization of this monomer, producing atactic PLA with low polydispersities (Figure 10.17).[141,142]

The ROP of both rac- and L-lactide by actinide catalysts has additionally been demonstrated by a series of uranium and thorium complexes stabilized by diamido ether ligands;[143] this study showed the considerable influence of both the metal center employed and the ligand architecture on the observed catalytic activity.

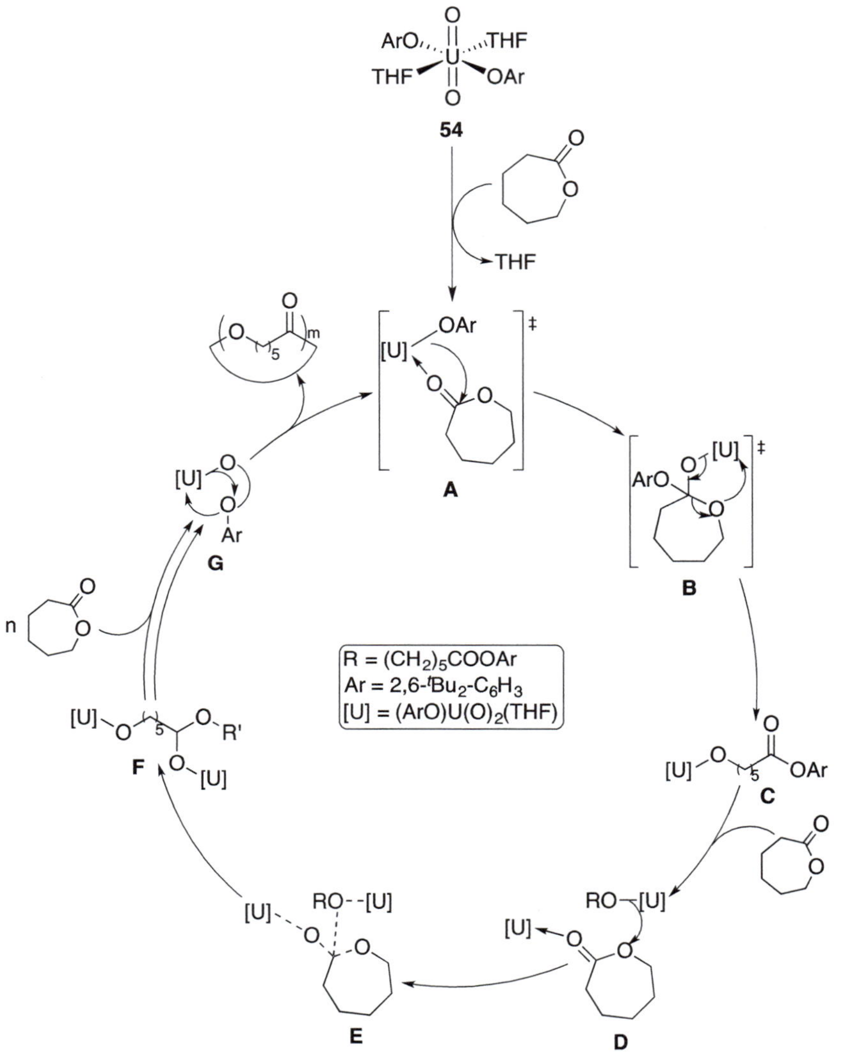

Scheme 10.33. Proposed mechanism for the cyclic oligomerisation of ε-caprolactone catalysed by uranyl complex **54**.

10.9.3 *Ring-opening polymerization of epoxides*

Although much less studied than the ROP of cyclic esters, the ring-opening polymerization of epoxides such as propylene or cyclohexene oxide has recently been demonstrated by Baker and co-workers

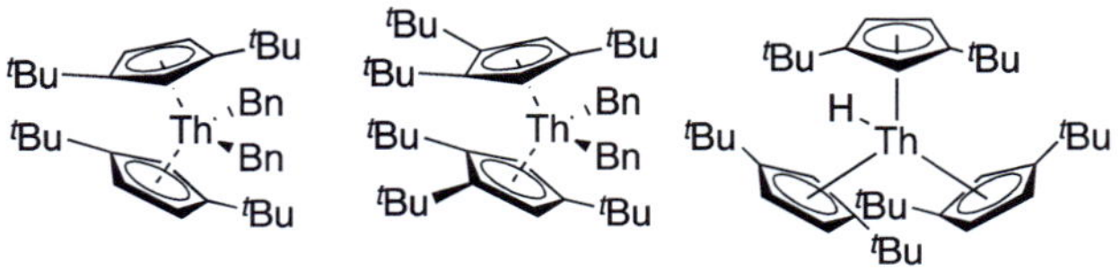

Figure 10.17. Modified thorium metallocenes used in the ROP of *rac*-lactide.

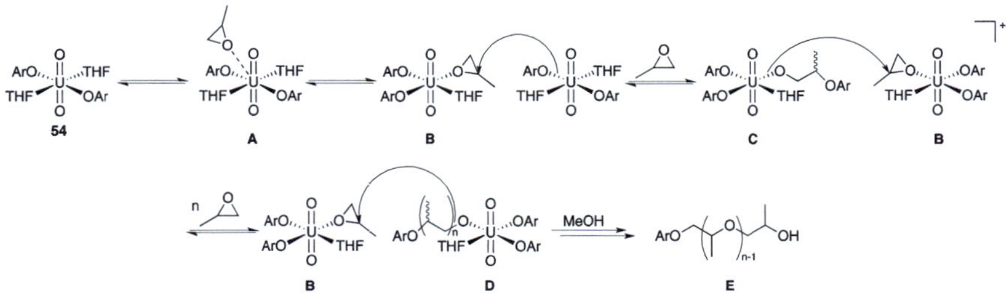

Scheme 10.34. The proposed mechanism of ROP of propylene oxide catalysed by complex **54**.

by a series of uranyl chlorides and aryloxides.[144,145] Three complexes of the formula $(ArO)_2U(O)_2(THF)_2$ (**54**), $Cl_2U(O)_2(THF)_3$ (**58**) and $[Cl_2U(O)_2(THF)_2]_2$ (**59**) were employed in these reactions, and similar to the energetic factors driving the actinide-catalysed Tishchenko reaction (*vide supra*), the recurring formation of uranium–oxygen bonds serves as the thermodynamic impetus for the achieved catalysis. The proposed mechanism begins by coordination of propylene oxide **A** followed by the displacement of a coordinated THF by the monomer to form intermediate **B** (Scheme 10.34).

Nucleophilic attack on the bound propylene oxide from the aryloxide of a second equivalent of **54** yields the uranyl alkoxide intermediate **C**; propagation of the polymerization occurs in this step by nucleophilic attack of the newly formed alkoxide on **C** on the coordinated epoxide of another equivalent of **B**, ultimately providing the uranyl-bound polymeric intermediate **D**. Quenching of the reaction deactivates the catalyst and consequently liberates the polymer product **E**.

10.10 Uranium-Catalyzed Electrocatalytic Production of Dihydrogen from Water and Coupling of Carbon Monoxide

The reactivity of uranium complexes in the stoichiometric activation of small molecules such as H_2O, CO_2, CO or N_2 is well recognized.[146–152] However, the catalytic involvements of those small molecules with actinides is still a challenge. Low-valent uranium is able to react with water producing H_2, but the reaction is uncontrollable.[153] The reductive recovery of the catalyst, via a cleavage of the U–O bond, makes this transformation extremely difficult, however, Meyer and co-workers have reported a uranium tris-alkoxide complex able to produce H_2 from H_2O.[154] The complex (**60**) contains a bulky adamantyl group for the steric protection of the U(III) core from water and serves directly as the active catalyst in this process (Scheme 10.35).

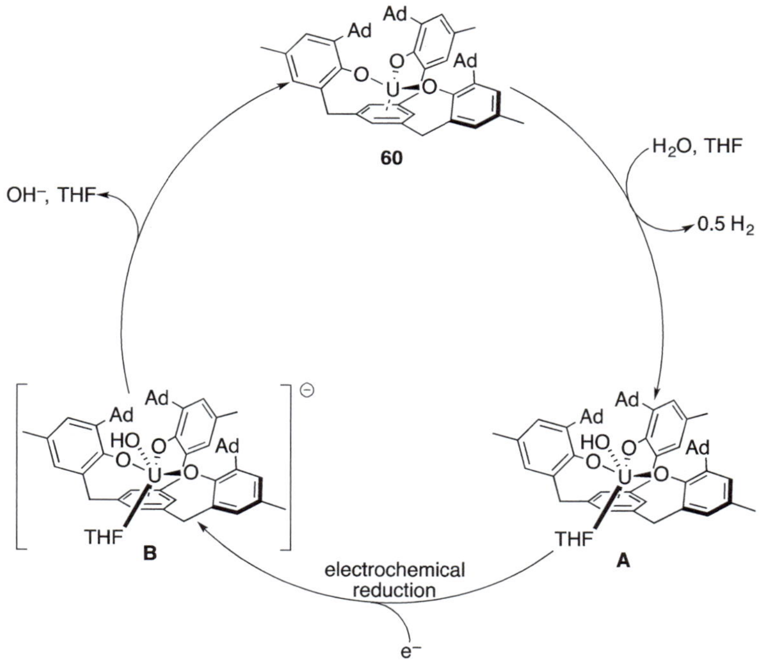

Scheme 10.35. Proposed mechanism for the electrocatalytic formation of H_2 from water mediated by the U(III) alkoxide complex.

The conditions resulted in a six-coordinated terminal hydroxo species **A** with the concomitant release of dihydrogen. The U(IV) undergoes an electrochemical reduction to produce the unstable anionic compound **B** which rapidly releases the bound hydroxide ion and a coordinated THF molecule to complete the catalytic cycle.

10.11 Reductive Homologation and Functionalization of Carbon Monoxide

Carbon monoxide (CO) is a primordial industrial feedstock and the direct one-carbon homologation is of great scientific and industrial significance. Liddle *et al.* have shown that the triamidoamine

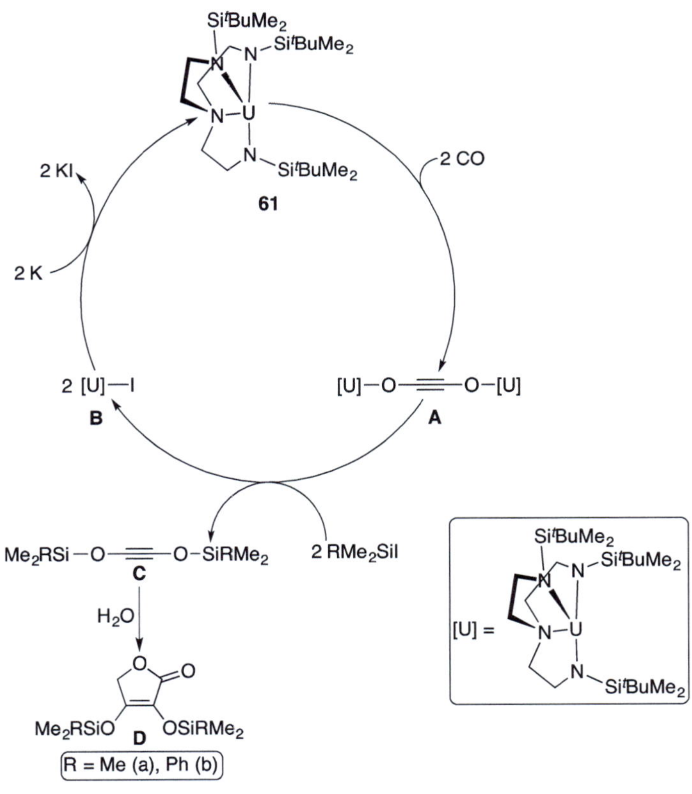

Scheme 10.36. Catalytic cycles for the U(III)-catalyzed reductive homologation and funtionalisation of carbon monoxide.

uranium (III) complex will catalytically reductively dimerize CO under mild conditions as shown in the mechanism in Scheme 10.36.[155]

The reaction of $[U(Tren^{DMSB})]$ ($Tren^{DMSB} = N(CH_2CH_2NSiMe_2{}^tBu)_3$) **61** with an excess of CO produces the dimeric complex **A**; while this complex continues to react in the catalytic cycle, it is also possible to isolate **A** in high yields as a highly crystalline product. In the presence of RMe_2SiI (R = Ph, Me), **A** reacts to form bis(organosiloxy)acetylenes **C** of the general formula $Me_2RSiOC{\equiv}COSiRMe_2$ (R = Me (**a**), Ph (**b**)) and the iodo-uranium complex **B**. In the presence of water, **C** cyclizes to ultimately produce the bis(siloxy)furanone **D**. Rather, if rigorous anaerobic conditions are maintained in the presence of elemental potassium, the uranium intermediate **B** is subsequently reduced to the active catalyst **61** with concomitant formation of potassium iodide as a byproduct of the reaction.

10.12 Conclusions

The application of the actinides to catalyse challenging chemical transformations has advanced immensely in recent years. Although the inception of organoactinide chemistry was marked by the introduction of rather simple metallocene complexes, studies of the coordination chemistry of the 5f block have flourished and produced highly elaborate complexes with unique structural features displaying impressive activity in both stoichiometric and catalytic reactions. The breadth of reactions mediated by the actinides provides invaluable information about structure and bonding of the 5f metals but also alludes to the possibility for this class of metals to effect chemical transformations not easily achieved by transition metal or rare earth complexes. While the growth in this field of research is impressive, the level of knowledge and understanding of the actinides lags far behind that of the d– and 4f-block, emphasizing the need for further studies of actinide coordination chemistry and reactivity.

Acknowledgments

This work was supported by the Israel Science Foundation, administered by the Israel Academy of Science and Humanities under Contract No. 184/18, and by the PAZY Foundation Fund under

Contract Number 128-19 (2020) administered by the Israel Atomic Energy Commission.

References

1. A. Streitwieser and U. Mueller-Westerhoff, *J. Am. Chem. Soc.*, **1968**, *90*, 7364–7364.
2. A. Zalkin and K. N. Raymond, *J. Am. Chem. Soc.*, 1969, *91*, 5667–5668.
3. P. L. Arnold, G. M. Jones, S. O. Odoh, G. Schreckenbach, N. Magnani and J. B. Love, *Nature Chem.*, 2012, *4*, 221–227.
4. P. L. Arnold, A. J. Blake, C. Wilson and J. B. Love, *Inorg. Chem.*, 2004, *43*, 8206–8208.
5. W. J. Evans, S. A. Kozimor and J. W. Ziller, *Polyhedron*, 2004, *23*, 2689–2694.
6. B. Kosog, C. E. Kefalidis, F. W. Heinemann, L. Maron and K. Meyer, *J. Am. Chem. Soc.*, 2012, *134*, 12792–12797.
7. I. Castro-Rodriguez, H. Nakai, P. Gantzel, L. N. Zakharov, A. L. Rheingold and K. Meyer, *J. Am. Chem. Soc.*, 2003, *125*, 15734–15735.
8. I. Castro-Rodriguez, H. Nakai, L. N. Zakharov, A. L. Rheingold and K. Meyer, *Science*, 2004, *305*, 1757–1759.
9. I. Castro-Rodriguez, K. Olsen, P. Gantzel and K. Meyer, *J. Am. Chem. Soc.*, 2003, *125*, 4565–4571.
10. P. L. Diaconescu, P. L. Arnold, T. A. Baker, D. J. Mindiola and C. C. Cummins, *J. Am. Chem. Soc.*, 2000, *122*, 6108–6109.
11. P. L. Arnold, N. A. Potter, N. Magnani, C. Apostolidis, J.-C. Griveau, E. Colineau, A. Morgenstern, R. Caciuffo and J. B. Love, *Inorg. Chem.*, 2010, *49*, 5341–5343.
12. P. L. Arnold, A-F. Pécharman and J. B. Love, *Angew. Chem. Int. Ed.*, 2011, *123*, 9628–9630.
13. W. J. Evans, K. A. Miller, A. G. DiPasquale, A. L. Rheingold, T. J. Stewart and R. Bau, *Angew. Chem. Int. Ed.*, 2008, *120*, 5153–5156.
14. H. Nakai, X. Hu, L. N. Zakharov, A. L. Rheingold and K. Meyer, *Inorg. Chem.*, 2004, *43*, 855–857.
15. W. J. Evans, K. A. Miller and J. W. Ziller, *Angew. Chem. Int. Ed.*, 2008, *47*, 589–592.
16. W. J. Evans, N. A. Siladke and J. W. Ziller, *Comptes Rend. Chim.*, 2010, *13*, 775–780.
17. W. Ren, E. Zhou, B. Fang, G. Hou, G. Zi, D.-C. Fang and M. D. Walter, *Angew. Chem. Int. Ed.*, 2014, *53*, 11310–11314.

18. E. Zhou, W. Ren, G. Hou, G. Zi, D.-C. Fang and M. D. Walter, *Organometallics*, 2015, *34*, 3637–3647.
19. A. R. Fox, S. C. Bart, K. Meyer and C. C. Cummins, *Nature*, 2008, *455*, 341–349.
20. S. M. Mansell, N. Kaltsoyannis and P. L. Arnold, *J. Am. Chem. Soc.*, 2011, *133*, 9036–9051.
21. E. M. Matson, A. T. Breshears, J. J. Kiernicki, B. S. Newell, P. E. Fanwick, M. P. Shores, J. R. Walensky and S. C. Bart, *Inorg. Chem.*, 2014, *53*, 12977–12985.
22. B. Fang, W. Ren, G. Hou, G. Zi, D.-C. Fang, L. Maron and M. D. Walter, *J. Am. Chem. Soc.*, 2014, *136*, 17249–17261.
23. X.-F. Li, X.-Z. Feng, Y.-T. Xu, H.-T. Wang, J. Shi, L. Liu and P.-N. Sun, *Inorg. Chim. Acta*, 1986, *116*, 85–93.
24. X.-F. Li and A.-L. Guo, *Inorg. Chim. Acta*, 1987, *134*, 143–153.
25. X.-F. Li, Y.-T. Xu, X.-Z. Feng and P.-N. Sun, *Inorg. Chim. Acta*, 1986, *116*, 75–83.
26. J. P. Leal, N. Marques, A. Pires de Matos, M. J. Calhorda, A. M. Galvao and J. A. M. Simoes, *Organometallics*, 1992, *11*, 1632–1637.
27. J. P. Leal and J. A. M. Simoes, *J. Chem. Soc., Dalton Trans.*, 1994, 2687–2691.
28. X. Jemine, J. Goffart, J.-C. Berthet and M. Ephritikhine, *J. Chem. Soc. Dalton Trans.*, 1992, 2439.
29. X. Jemine, J. Goffart, M. Ephritikhine and J. Fuger, *J. Organomet. Chem.*, 1993, *448*, 95–98.
30. T. J. Marks, M. R. Gagne, S. P. Nolan, L. E. Schock, A. M. Seyam and D. Stern, *Pure Appl. Chem.*, 1989, 61.
31. W. A. King and T. J. Marks, *Inorg. Chim. Acta*, 1995, *229*, 343–354.
32. W. A. King, T. J. Marks, D. M. Anderson, D. J. Duncalf and F. G. N. Cloke, *J. Am. Chem. Soc.*, 1992, *114*, 9221–9223.
33. T. Straub, A. Haskel and M. S. Eisen, *J. Am. Chem. Soc.*, 1995, *117*, 6364–6365.
34. A. Haskel, T. Straub, A. K. Dash and M. S. Eisen, *J. Am. Chem. Soc.*, 1999, *121*, 3014–3024.
35. A. Haskel, J. Q. Wang, T. Straub, T. G. Neyroud and M .S. Eisen, *J. Am. Chem. Soc.*, 1999, *121*, 3025–3034.
36. R. J. Batrice, J. McKinven, P. L. Arnold and M. S. Eisen, *Organometallics*, 2015, *34*, 4039–4050.
37. E. Barnea, T. Andrea, J.-C. Berthet, M. Ephritikhine and M. S. Eisen, *Organometallics*, 2008, *27*, 3103–3112.
38. M. Beller, J. Seayad, A. Tillack, H. Jiao and *Angew. Chem. Int. Ed.*, 2004, *43*, 3368–3398.

39. T. E. Müller, K. C. Hultzsch, M. Yus, F. Foubelo and M. Tada, *Chem. Rev.*, 2008, *108*, 3795–3892.
40. L. Huang, M. Arndt, K. Gooßen, H. Heydt and L. J. Gooßen, *Chem. Rev.*, 2015, *115*, 2596–2697.
41. F. Alonso, I. P. Beletskaya and M. Yus, *Chem. Rev.*, 2004, *104*, 3079–3160.
42. A. Z. Voskoboynikov, A. K. Shestakova and I. P. Beletskaya, *Organometallics*, 2001, *20*, 2794–2801.
43. S. Seo, X. Yu and T. J. Marks, *J. Am. Chem. Soc.*, 2009, *131*, 263–276.
44. Y. Nakajima and S. Shimada, *RSC Adv.*, 2015, *5*, 20603–20616.
45. X. Zeng, *Chem. Rev.*, 2013, *113*, 6864–6900.
46. A. Dondoni and A. Marra, *Eur. J. Org. Chem.*, 2014, *2014*, 3955–3969.
47. G. A. Molander and J. A. C. Romero, *Chem. Rev.*, 2002, *102*, 2161–2186.
48. D. Riegert, J. Collin, A. Meddour, E. Schulz and A. Trifonov, *J. Org. Chem.*, 2006, *71*, 2514–2517.
49. A. Haskel, T. Straub and M. S. Eisen, *Organometallics*, 1996, *15*, 3773–3775.
50. T. Straub, A. Haskel, T. G. Neyroud, M. Kapon, M. Botoshansky and M. S. Eisen, *Organometallics*, 2001, *20*, 5017–5035.
51. J. Wang, A. K. Dash, M. Kapon, J.-C. Berthet, M. Ephritikhine and M. S. Eisen, *Chem. Eur. J.*, 2002, *8*, 5384–5396.
52. B. D. Stubbert and T. J. Marks, *J. Am. Chem. Soc.*, 2007, *129*, 6149–6167.
53. B. D. Stubbert, C. L. Stern and T. J. Marks, *Organometallics*, 2003, *22*, 4836–4838.
54. R. Shannon, *Acta Crystallogr. Sect. A*, 1976, *32*, 751–767.
55. E. M. Broderick, N. P. Gutzwiller and P. L. Diaconescu, *Organometallics*, 2010, *29*, 3242–3251.
56. C. E. Hayes, R. H. Platel, L. L. Schafer and D. B. Leznoff, *Organometallics*, 2012, *31*, 6732–6740.
57. A. S. K. Hashmi, *Chem. Rev.*, 2007, *107*, 3180–3211.
58. W. M. Haynes, *CRC Handbook of Chemistry and Physics*. 97th ed.; CRC Press: Boca Raton, FL, 2016–2017.
59. S. D. Wobser and T. J. Marks, *Organometallics*, 2013, *32*, 2517–2528.
60. K. Kuciński, P. Pawluć and G. Hreczycho, *Adv. Synth. Catal.*, 2015, *357*, 3936–3942.
61. K. Griesbaum, *Angew. Chem. Int. Ed. Engl.*, 1970, *9*, 273–287.
62. L. Benati, P. C. Montevecchi and P. Spagnolo, *J. Chem. Soc. Perkin Trans. I*, 1991, 2103–2109.
63. Y. Yang and R. M. Rioux, *Green Chem.*, 2014, *16*, 3916–3925.
64. C. J. Weiss and T. J. Marks, *Dalton Trans.*, 2010, *39*, 6576.

65. C. J. Weiss, S. D. Wobser and T. J. Marks, *J. Am. Chem. Soc.*, 2009, *131*, 2062–2063.
66. C. J. Weiss, S. D. Wobser and T. J. Marks, *Organometallics*, 2010, *29*, 6308–6320.
67. C. M. Fendrick, E. A. Mintz, L. D. Schertz and T. J. Marks, *Organometallics*, 1984, *3*, 819–821.
68. M. Sharma and M. S. Eisen, Metallocene Organoactinide Complexes, in *Structure and Bonding*, Springer-Verlag Berlin Heidelberg, Berlin, 2008, pp. 1–85.
69. J. A. Reichl and D. H. Berry, Recent Progress in Transition Metal-Catalyzed Reactions of Silicon, Germanium, and Tin, in *Advances in Organometallic Chemistry* (Eds. W. Robert, F. H. Anthony), Vol. 43, Academic Press, Cambridge, 1999, pp. 197–265.
70. S. Patai, Z. Rappoport and Y. Apeloig, *The Chemistry of Organic Silicon Compounds*. Wiley, New York, 1989.
71. J. L. Speier, Homogeneous Catalysis of Hydrosilation by Transition Metals, in *Advances in Organometallic Chemistry* (Eds. F. G. A. Stone and W. Robert), Vol. 17, Academic Press, Cambridge, 1979, pp. 407–447.
72. B. Fengyu, K.-I. Kanno and T. Takahashi, *Trends Org. Chem.*, 2008, *12*, 1–17.
73. D. A. Rooke, Z. A. Menard and E. M. Ferreira, *Tetrahedron*, 2014, *70*, 4232–4244.
74. M. Iglesias, F. J. Fernández-Alvarez and L. A. Oro, *ChemCatChem*, 2014, *6*, 2486–2489.
75. M. D. Greenhalgh, A. S. Jones and S. P. Thomas, *ChemCatChem*, 2014, *7*, 190–222.
76. A. K. Dash, J. Q. Wang and M. S. Eisen, *Organometallics*, 1999, *18*, 4724–4741.
77. A. K. Dash, Y. Gurevizt, J. Q. Wang, J. Wang, M. Kapon and M. S. Eisen, *J. Alloys Compd.*, 2002, *344*, 65–69.
78. P. J. Fagan, J. M. Manriquez, E. A. Maatta, A. M. Seyam and T. J. Marks, *J. Am. Chem. Soc.*, 1981, *103*, 6650–6667.
79. A. K. Dash, I. Gourevich, J. Q. Wang, J. Wang, M. Kapon and M. S. Eisen, *Organometallics*, 2001, *20*, 5084–5104.
80. A. K. Dash, J. X. Wang, J.-C. Berthet, M. Ephritikhine and M. S. Eisen, *J. Organomet. Chem.*, 2000, *604*, 83–98.
81. J. Wang, Y. Gurevich, M. Botoshansky and M. S. Eisen, *J. Am. Chem. Soc.*, 2006, *128*, 9350–9351.
82. Z. Lin and T. J. Marks, *J. Am. Chem. Soc.*, 1987, *109*, 7979–7985.
83. T. Andrea, E. Barnea and M. S. Eisen, *J. Am. Chem. Soc.*, 2008, *130*, 2454–2455.

84. M. Sharma, T. Andrea, N. J. Brookes, B. F. Yates and M. S. Eisen, *J. Am. Chem. Soc.*, 2011, *133*, 1341–1356.
85. I. S. R. Karmel, N. Fridman, M. Tamm and M. S. Eisen, *Organometallics*, 2015, *34*, 2933–2942.
86. I. S. R. Karmel, N. Fridman, M. Tamm and M. S. Eisen, *J. Am. Chem. Soc.*, 2014, *136*, 17180–17192.
87. H. Liu and M. S. Eisen, *Organometallics*, 2017, *36*, 1461–1464.
88. W. Yi, J. Zhang, L. Hong, Z. Chen and X. Zhou, *Organometallics*, 2011, *30*, 5809–5814.
89. Z. Du, H. Zhou, H. Yao, Y. Zhang, Y. Yao and Q. Shen, *Chem. Commun.*, 2011, *47*, 3595.
90. A. C. Behrle and J. A. R. Schmidt, *Organometallics*, 2013, *32*, 1141–1149.
91. J. Tu, W. Li, M. Xue, Y. Zhang and Q. Shen, *Dalton Trans.*, 2013, *42*, 5890.
92. Y. Cao, Z. Du, W. Li, J. Li, Y. Zhang, F. Xu and Q. Shen, *Inorg. Chem.*, 2011, *50*, 3729–3737.
93. X. Zhang, C. Wang, C. Qian, F. Han, F. Xu and Q. Shen, *Tetrahedron*, 2011, *67*, 8790–8799.
94. Z. Li, M. Xue, H. Yao, H. Sun, Y. Zhang and Q. Shen, *J. Organomet. Chem.*, 2012, *713*, 27–34.
95. W. J. Evans, J. R. Walensky, J. W. Ziller and A. L. Rheingold, *Organometallics*, 2009, *28*, 3350–3357.
96. W. J. Evans, M. K. Takase, J. W. Ziller and A. L. Rheingold, *Organometallics*, 2009, *28*, 5802–5808.
97. I. S. R. Karmel, M. Tamm and M. S. Eisen, *Angew. Chem. Int. Ed.*, 2015, *54*, 12422–12425.
98. R. J. Batrice and M. S. Eisen, *Chem. Sci.*, 2016, *7*, 939–944.
99. R. J. Batrice, C. E. Kefalidis, L. Maron and M. S. Eisen, *J. Am. Chem. Soc.*, 2016, *138*, 2114–2117.
100. T. Ghatak, N. Fridman and M. S. Eisen, *Organometallics*, 2017, *36*, 1296–1302.
101. C. E. Boardwine, B. Ceccaroli, H. M. Rong, A. Schei and G. Schüssler, *Progress in Organosilicon Chemistry* (Eds. B. Marciniec and J. Chojnowski), Gordon and Breach, Basel, 1995, pp. 555–570.
102. M. Ohring, *Materials science of thin films deposition and structure*. Academic Press: San Diego, 2002.
103. L. B. Schein, *Electrophotography and Development Physics*, Springer, Berlin, 1988.
104. H. O. Pierson, *Handbook of Chemical Vapor Deposition (CVD): Principles, Technology, and Applications*, Noyes Publications, Park Ridge, 1992.

105. D. W. Freitag and D. W. Richerson, Opportunities for Advanced Ceramics to Meet the Needs of the Industries of the Future, in *USDOE* (Ed. U.S. Advanced Ceramics Association and Oak Ridge National Laboratory), Oak Ridge, Tennessee, 1998.
106. F. L. Riley, *J. Am. Ceram. Soc.*, 2000, *83*, 245–265.
107. A. Arafat, K. Schroën, L. C. P. M. de Smet, E. J. R. Sudhölter and H. Zuilhof, *J. Am. Chem. Soc.*, 2004, *126*, 8600–8601.
108. M. Mazzocchi and A. Bellosi, *J. Mater. Sci.: Mater. Med.*, 2008, *19*, 2881–2887.
109. J. Olofsson, T. M. Grehk, T. Berlind, C. Persson, S. Jacobson and H. Engqvist, *Biomatter*, 2012, *2*, 94–102.
110. M. C. Anderson and R. Olsen, *J. Biomed. Mater. Res. A*, 2010, *92A*, 1598–1605.
111. T. J. Webster, A. A. Patel, M. N. Rahaman and B. S. Bal, *Acta Biomater.*, 2012, *8*, 4447–4454.
112. J. X. Wang, A. K. Dash, J. C. Berthet, M. Ephritikhine and M. S. Eisen, *J. Organomet. Chem.*, 2000, *610*, 49–57.
113. D. S. J. Arney and C. J. Burns, *J. Am. Chem. Soc.*, 1995, *117*, 9448–9460.
114. R. G. Peters, B. P. Warner and C. J. Burns, *J. Am. Chem. Soc.*, 1999, *121*, 5585–5586.
115. J. R. Robinson, P. J. Carroll, P. J. Walsh and E. J. Schelter, *Organometallics*, 2013, *32*, 1493–1499.
116. A. J. Wooten, P. J. Carroll and P. J. Walsh, *J. Am. Chem. Soc.*, 2008, *130*, 7407–7419.
117. Z. Ma, M. Sutradhar, A. V. Gurbanov, A. M. Maharramov, R. A. Aliyeva, F. S. Aliyeva, F. N. Bahmanova, V. I. Mardanova, F. M. Chyragov and K. T. Mahmudov, *Polyhedron*, 2015, *101*, 14–22.
118. H. Naeimi and Z. S. Nazifi, *Russ. Chem. Bull.*, 2015, *64*, 1814–1818.
119. M. Azam, S. I. Al-Resayes, A. Trzesowska-Kruszynska, R. Kruszynski, P. Kumar and S. L. Jain, *Polyhedron*, 2017, *124*, 177–183.
120. K. Takao and S. Akashi, *RSC Adv.*, 2017, *7*, 12201–12207.
121. G. P. Guiliani, E. Sorta and M. Bruzzone, *Angew. Makromol. Chem.*, 1976, *50*, 87–99.
122. A. De Chirico, P. C. Lanzani, E. Raggi and M. Bruzzone, *Die Makromol. Chem.*, 1974, *175*, 2029–2038.
123. L. Gargani, G. P. Giuliani, F. Mistrali and M. Bruzzone, *Angew. Makromol. Chem.*, 1976, *50*, 101–113.
124. G. Lugli, A. Mazzei and S. Poggio, *Die Makromol. Chem.*, 1974, *175*, 2021–2027.
125. L. Jia, X. Yang, C. Stern and T. J. Marks, *Organometallics*, 1994, *13*, 3755–3757.

126. X. Yang, C. Stern and T. J. Marks, *Organometallics*, 1991, *10*, 840–842.
127. L. Jia, X. Yang, C. L. Stern and T. J. Marks, *Organometallics*, 1997, *16*, 842–857.
128. C. E. Hayes and D. B. Leznoff, *Organometallics*, 2010, *29*, 767–774.
129. E. Domeshek, R. J. Batrice, S. Aharonovich, B. Tumanskii, M. Botoshansky and M. S. Eisen, *Dalton Trans.*, 2013, *42*, 9069.
130. S. M. Mansell, F. Bonnet, M. Visseaux and P. L. Arnold, *Dalton Trans.*, 2013, *42*, 9033–9039.
131. M. Bhavsar and M. Amiji, *AAPS PharmSciTech*, 2008, *9*, 288–294.
132. Y. Ikada and H. Tsuji, *Macromol. Rapid Commun.*, 2000, *21*, 117–132.
133. J. L. Hedrick, T. Magbitang, E. F. Connor, T. Glauser, W. Volksen, C. J. Hawker, V. Y. Lee and R. D. Miller, *Chem. Eur. J.*, 2002, *8*, 3308.
134. P. Joshi and G. Madras, *Polym. Degrad. Stab.*, 2008, *93*, 1901–1908.
135. C. Villiers, P. Thuéry and M. Ephritikhine, *Eur. J. Inorg. Chem.*, 2004, 4624–4632.
136. E. Barnea, D. Moradove, J.-C. Berthet, M. Ephritikhine and M. S. Eisen, *Organometallics*, 2006, *25*, 320–322.
137. E. Rabinovich, S. Aharonovich, M. Botoshansky and M. S. Eisen, *Dalton Trans.*, 2010, *39*, 6667–6676.
138. I. S. R. Karmel, T. Elkin, N. Fridman and M. S. Eisen, *Dalton Trans.*, 2014, *43*, 11376–11387.
139. I. S. R. Karmel, N. Fridman and M. S. Eisen, *Organometallics*, 2015, *34*, 636–643.
140. A. Walshe, J. Fang, L. Maron and R. J. Baker, *Inorg. Chem.*, 2013, *52*, 9077–9086.
141. W. Ren, N. Zhao, L. Chen, H. Song and G. Zi, *Inorg. Chem. Commun.*, 2011, *14*, 1838–1841.
142. W. Ren, N. Zhao, L. Chen and G. Zi, *Inorg. Chem. Commun.*, 2013, *30*, 26–28.
143. C. E. Hayes, Y. Sarazin, M. J. Katz, J.-F. Carpentier and D. B. Leznoff, *Organometallics*, 2013, *32*, 1183–1192.
144. R. J. Baker and A. Walshe, *Chem. Commun.*, 2012, *48*, 985–987.
145. J. Fang, A. Walshe, L. Maron and R. J. Baker, *Inorg. Chem.*, 2012, *51*, 9132–9140.
146. S. T. Liddle, *Angew. Chem. Int. Ed.*, 2015, *54*, 8604–8641.
147. W. W. Lukens, S. M. Beshouri, L. L. Blosch and R. A. Andersen, *J. Am. Chem. Soc.*, 1996, *118*, 901–902.
148. F. G. N. Cloke and P. B. Hitchcock, *J. Am. Chem. Soc.*, 2002, *124*, 9352–9353.

149. W. J. Evans, S. A. Kozimor, G. W. Nyce and J. W. Ziller, *J. Am. Chem. Soc.*, 2003, *125*, 13831–13835.
150. O. T. Summerscales, F. G. N. Cloke, P. B. Hitchcock, J. C. Green and N. Hazari, *Science*, 2006, *311*, 829–831.
151. O. T. Summerscales, F. G. N. Cloke, P. B. Hitchcock, J. C. Green and N. Hazari, *J. Am. Chem. Soc.*, 2006, *128*, 9602–9603.
152. N. A. Siladke, K .R. Meihaus, J. W. Ziller, M. Fang, F. Furche, J. R. Long and W. J. Evans, *J. Am. Chem. Soc.*, 2012, *134*, 1243–1249.
153. I. Korobkov and S. Gambarotta, *Prog. Inorg. Chem.*, 2005, *54*, 321–348.
154. D. P. Halter, F. W. Heinemann, J. Bachmann and K. Meyer, *Nature*, 2016, *530*, 317–321.
155. B. M. Gardner, J. C. Stewart, A. L. Davis, J. McMaster, W. Lewis, A. J. Blake and S. T. Liddle, *Proc. Natl. Acad. Sci. USA*, 2012, *109*, 9265–9270.

https://doi.org/10.1142/9781800610163_0011

Chapter 11

Computational Aspects of f-Element Chemistry

Andrew Kerridge
Department of Chemistry, Lancaster University, Lancaster, LA1 4YB, UK
a.kerridge@lancaster.ac.uk

11.1 Introduction

The simulation of f-element compounds presents a unique and formidable challenge to the computational chemist. Broadly speaking, the origin of this complexity is dominated by two phenomena: special relativity and electron correlation. These two phenomena combine to produce compounds with highly complex electronic structures whose simulation requires the application of sophisticated and technically challenging theoretical techniques. Only by careful consideration of these phenomena is it possible to produce quantitatively accurate simulations from which experimental observations can be both explained and predicted. While both relativity and electron correlation will be considered in detail later in this chapter, it is of benefit here to outline the relevant concepts and effects in brief.

Relativistic effects manifest themselves in situations where electrons are moving at an appreciable fraction of the speed of light. The vicinities of f-element nuclei provide such an environment and such high velocities lead to a relativistic mass increase and subsequent contraction of core orbitals. Conversely, this contraction results in enhanced shielding of the nuclear charge, leading to expansion of valence orbitals with little core penetration. In addition to such *scalar* relativistic effects, replacing the non-relativistic Schrödinger equation with the relativistic Dirac equivalent results in a coupling between the orbital and spin components of angular momentum. Such *spin–orbit coupling* leads to subtle but chemically relevant variation in electronic structure.

Electron correlation is an entirely anthropic concept, since it is not possible for the motion of electrons, or indeed any interacting bodies, to be entirely uncorrelated. Nevertheless, the difficulties in solving the many-body equations of motion (whether classical or quantum mechanical) make it necessary, or at least theoretically convenient, to assume uncorrelated motion in the first instance. As we shall see, electron correlation manifests itself in two conceptually different ways. The first, *dynamical correlation*, is due to the instantaneous Coulomb interaction between electrons, whereas the second, *static correlation*, is due to the energetic near-degeneracy of different electronic configurations.

These complexities differentiate f-element compounds from the vast remainder of the periodic table, yet there are noticeable differences in the f-elements themselves, particularly between the lanthanides, characterised by a partially occupied 4f shell, and the actinides, characterised by a partially occupied 5f shell. The 4f orbitals of the lanthanides have no radial nodes and therefore penetrate the atomic core region, resulting in a compact, core-like radial distribution. In contrast, the 5f orbitals of the actinides, possessing a single radial node, penetrate the core to a lesser degree and therefore experience greater shielding, leading to more diffuse, valence-like character. The core-like nature of the 4f orbitals results in the characterisation of lanthanides as being hard Lewis acids with dominant ionic bonding character, experiencing only weak crystal fields. The valence-like nature of the 5f orbitals results in the actinides being characterised as slightly softer ions, capable of covalent bonding interactions and experiencing crystal fields of comparable magnitude to spin–orbit interactions. These different characteristics should be used to inform the choice of simulation methodology.

11.2 Simulation Methodologies

When considering the simulation of an f-element complex there are three important factors to consider: the degree to which electron

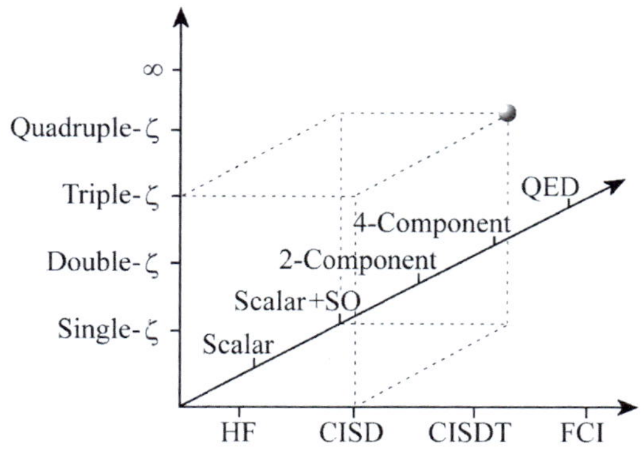

Figure 11.1. Pople diagram representing the quality of basis set (x-axis), treatment of relativistic effects (y-axis) and incorporation of electron correlation (z-axis). Reproduced from Ref. [103], with permission from John Wiley and Sons.

correlation will be incorporated, the sophistication of the relativistic Hamiltonian to be employed, and the quality of the electronic basis set (see Figure 11.1). Each of these factors will be discussed in this section.

In terms of incorporating electron correlation there are two commonly applied approaches, based on either wavefunction theory (WFT) or density functional theory (DFT). WFT-based approaches provide a basis for a systematically improvable model of electronic structure, but suffer from high computational cost and poor scaling with system size, whereas DFT-based methodologies are computationally efficient and exhibit attractive scaling properties, but do not offer a systematic approach to improved simulation accuracy. For this reason, only WFT methods are considered in Figure 11.1.

11.2.1 *Density functional theory*

DFT has its conceptual origins in the Thomas-Fermi model of electronic structure,[1,2] further developed by Dirac to incorporate the exchange interaction.[3] Modern implementations, however, are based on the seminal work of Kohn and co-workers.[4,5] There exist two theorems, proven by Hohenburg and Kohn,[4] which demonstrate that (i) the ground state electron density uniquely determines the potential and thus all properties of the system and (ii) the electron density is subject to a variational principal and so the ground state energy of a system is uniquely determined by the ground state density.

The Hohenburg–Kohn theorems provide a simplified description of the system of interest, allowing it to be described in just three spatial variables as opposed to the $3N$-dimensional spatial wavefunction (with N being the number of electrons in the system). However, computational tractability is only achieved by approximating the real system of interacting electrons as that of a set of non-interacting electrons (experiencing an effective potential) with a density exactly equivalent to that of the real system.[5] This approximation reduces the N-electron problem to N coupled one-electron problems, for which a computational methodology exists. This approach is known as Kohn–Sham density functional theory (KS-DFT), and is by far the most popular approach to DFT in current research. In KS-DFT a set of Kohn–Sham equations must be solved. In atomic units:

$$\left(-\frac{1}{2}\nabla^2 + v_{\text{eff}}(\mathbf{r})\right)\psi_i(\mathbf{r}) = \varepsilon_i\psi_i(\mathbf{r}) \tag{11.1}$$

where v_{eff} is the effective potential, defined as follows:

$$v_{\text{eff}}(\mathbf{r}) = v_{\text{ext}}(\mathbf{r}) + \int \frac{\rho(\mathbf{r}) - \rho(\mathbf{r}')}{\mathbf{r} - \mathbf{r}'} d\mathbf{r}' + \frac{\delta E_{\text{XC}}[\rho\mathbf{r})]}{\delta\rho(\mathbf{r})} \tag{11.2}$$

Here the first term is an external potential which, at a minimum, represents electron–nuclear interactions, the second term represents the Coloumb repulsion between electrons (known as the Hartree potential) and the final term is the *exchange-correlation potential*, detailed below. The solutions to the Kohn–Sham equations, $\psi_i(\mathbf{r})$ are known as Kohn–Sham (molecular) orbitals, each of which has a corresponding energy, ε_i. The total electron density, $\rho(\mathbf{r})$, may be constructed from the Kohn–Sham orbitals:

$$\rho(\mathbf{r}) = \sum_i^N |\psi_i(\mathbf{r})|^2 \tag{11.3}$$

The total energy of the system is then given as a functional (a function which itself takes a function as an argument, returning a number) of the density:

$$E[\rho] = T_{\text{s}}[\rho] + E_{\text{ext}}[\rho] + E_{\text{H}}[\rho] + E_{\text{XC}}[\rho] \tag{11.4}$$

$T_{\text{s}}[\rho]$ gives the kinetic energy of the non-interacting system, $E_{\text{ext}}[\rho]$ gives the energy due to the external potential, $E_{\text{H}}[\rho]$ gives the energy due to Coulomb repulsion between electrons and $E_{\mathrm{XC}}[\rho]$ gives the exchange-correlation energy, which accounts for all energy terms not defined elsewhere. This expression is, in principle, exact for a system which can be described by a single electronic configuration. However, the exact functional form of $E_{\mathrm{XC}}[\rho]$ is not known and so a hierarchy of approximations have been developed.

11.2.1.1 *Approximations to the exchange-correlation energy*

In 2001, Perdew introduced "Jacob's ladder", which summarises the hierarchy of commonly available approximations to the exact exchange-correlation functional[6] (see Figure 11.2). The first rung of this ladder corresponds to the Local Density Approximation (LDA), which assumes that the exchange-correlation energy for a given value of the density, ρ, can be obtained from the equivalent energy of an homogeneous electron gas (HEG) of the same density. $E_{\mathrm{XC}}[\rho]$ is then defined as follows:

$$E_{\mathrm{XC}}^{\mathrm{LDA}}[\rho(\mathbf{r})] = \int \epsilon_{\mathrm{XC}}[\rho(\mathbf{r})]\rho(\mathbf{r})d\mathbf{r} \tag{11.5}$$

Since $\epsilon_{\mathrm{XC}}[\rho]$, the exchange-correlation energy density, can be evaluated with a high degree of accuracy for the HEG, the LDA depends only on the parametrisation, with the most popular being that due

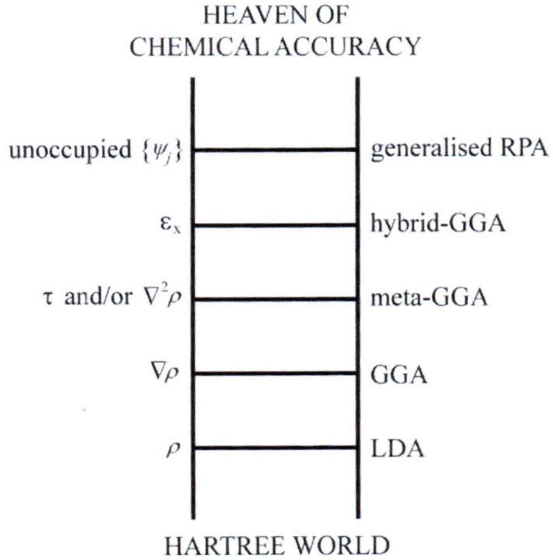

Figure 11.2. Jacob's ladder of approximations to the exchange correlation functional. Reproduced from Ref. [6].

to Vosko, Wilk and Nusair[7] (VWN). While the LDA performs well for simple metallic systems with loosely bound valence electrons, it is known to strongly overbind in molecular environments and is therefore of little practical use when considering the f-elements.

The second rung on the ladder corresponds to the generalised gradient approximation (GGA). In contrast to the HEG, a molecular electron density is strongly varying, particularly in nuclear regions, and including a contribution to $E_{\mathrm{XC}}[\rho]$ dependent on the gradient of the density, $\nabla\rho$, has been shown to significantly improve the performance of DFT when applied to molecular systems. Under this approximation, $E_{\mathrm{XC}}[\rho]$ is defined as follows:

$$E_{\mathrm{XC}}^{\mathrm{GGA}}[\rho(\mathbf{r})] = \int \epsilon_{\mathrm{XC}}[\rho(\mathbf{r}), \nabla\rho(\mathbf{r})]\rho(\mathbf{r})d\mathbf{r} \tag{11.6}$$

There are numerous GGA functionals available, with some of the most popular including PW91,[8] PBE,[9] BLYP[10,11] and BP86.[10,12] Of these, PW91 and PBE are non-empirical, being constructed so as to satisfy certain conditions that the exact exchange-correlation functional would itself satisfy. The construction of the BLYP and BP86 functionals are instead based on empirical data, with the aim of minimising the error in the properties of a set of test molecules. It should be noted that *none* of the commonly used test sets include any f-element compounds: for example, the commonly used G2 set[13] includes no elements beyond the second row of the periodic table. GGA functionals typically perform well in the prediction of molecular structures, but are less accurate with respect to other molecular properties.

The next rung on the ladder corresponds to functionals constructed within the meta-generalised gradient approximation (meta-GGA). Analogous to the improvement of GGA over LDA via inclusion of gradient dependent terms, meta-GGA functionals incorporate dependence on second derivatives of the density, giving a general form for $E_{\mathrm{XC}}[\rho]$ of

$$E_{\mathrm{XC}}^{\mathrm{mGGA}}[\rho(\mathbf{r})] = \int \epsilon_{\mathrm{XC}}[\rho(\mathbf{r}), \nabla\rho(\mathbf{r}), \nabla^2\rho(\mathbf{r})]\rho(\mathbf{r})d\mathbf{r} \tag{11.7}$$

Meta-GGA functionals offer only a modest improvement over those based on the GGA, however, this improvement still outweighs the near-negligible increase in computational cost. There are currently few meta-GGAs commonly available. Among the most popular is the non-empirical TPSS functional,[14] which can be thought of as meta-GGA extension of the PBE functional.

The fourth rung on the ladder, corresponding to the hybrid-GGA approach, offers a departure from the previous rungs. The exchange contribution to E_{XC}, which must be approximated in "pure" exchange-correlation functionals, can be evaluated exactly within the confines of the Hartree–Fock approximation. The exact exchange can be combined with approximate exchange functionals to produce "hybrid" exchange-correlation functionals. The most popular hybrid-GGA functionals are B3LYP,[15,16] containing a 20% contribution of exact exchange and parametrised with respect to the G2 test set, and the non-empirical PBE0 functional,[17] containing a perturbatively derived 25% contribution of exact exchange. It is worth mentioning here that, since the introduction of Jacob's ladder, a new hybrid variant has been developed: the Coulomb-attenuated, or range-separated, hybrid functional. The percentage contribution of exact exchange in these functionals is dependent on the interelectronic separation, leading to better long-range behaviour, particularly noticeable in excited state simulations. The most popular Coulomb-attenuated functional is known as CAM-B3LYP.[18] As a final point, it should be noted that hybrid functionals, due to the incorporation of exact exchange, are significantly more computationally expensive than pure LDA, GGA or meta-GGA functionals.

11.2.2 *Correlated wavefunction approaches*

In contrast to DFT methodologies, correlated wavefunction approaches explicitly model the full N-electron wavefunction. Almost invariably, such approaches are based on an initial wavefunction generated from a Hartree–Fock calculation, and are part of a set of *post* Hartree–Fock methods.

The basis of correlated wavefunction approaches is that the exact, correlated, N-electron wavefunction can be constructed to arbitrary accuracy from a complete set of N-electron functions. In this section we will focus on configuration interaction (CI) and coupled cluster (CC) based approaches, for which the set of functions from which the correlated wavefunction is constructed are typically Slater determinants (antisymmetrised products of one-electron molecular orbitals).

11.2.2.1 *Configuration interaction (CI) theory*

Configuration interaction is based on the fact that the exact wavefunction can be constructed as a linear combination of electronic configurations, with each configuration differing from the reference (Hartree–Fock) configuration through the degree of occupation of higher energy molecular orbitals.

$$\left|\Psi^{\mathrm{CI}}\right\rangle = c_0 \left|\Psi_0\right\rangle + \sum c_i^a \left|\Psi_i^a\right\rangle + \sum c_{ij}^{ab} \left|\Psi_{ib}^{ab}\right\rangle + \sum c_{ijk}^{abc} \left|\Psi_{ijk}^{abc}\right\rangle + \cdots \tag{11.8}$$

Here $|\Psi_i^a\rangle$ represents a singly-excited configuration formed by deoccupying molecular orbital (MO) ψ_a and occupying ψ_i. Configurations with higher degrees of excitation can be constructed in an analogous manner. This expansion makes it clear that the CI wavefunction can be thought of as correcting the Hartree–Fock reference configuration $|\Psi_0\rangle$, via sets of singly, doubly, triply, etc., excited configurations. The CI coefficients, along with the total CI energy, are obtained by constructing and diagonalising the CI Hamiltonian matrix, $\mathbf{H}^{\mathrm{CI}}$.

Expanding the exact wavefunction in this manner, and allowing configurations in which up to N electrons are excited, is known as full CI. While full CI, in principle, allows for the recovery of the exact wavefunction and energy, in practice it is so computationally expensive as to only be practical for very small systems. Instead, computationally less intensive truncated CI methods are employed. These methods approximate the correlated wavefunction by including only those configurations in which the number of excited electrons is equal to or less than a certain threshold. A popular approach is to allow only singly and doubly excited configurations to contribute to the CI expansion, defining the CISD method.

11.2.2.2 *Coupled cluster (CC) theory*

In contrast to CI, CC theory expands the exact wavefunction using an exponential *ansatz*:

$$\left|\Psi^{\mathrm{CC}}\right\rangle = e^{\hat{T}} \left|\Psi_0\right\rangle \tag{11.9}$$

where the cluster operator, $\hat{T}$, is defined as $\hat{T} = \hat{T}_1 + \hat{T}_2 + \hat{T}_3 + \cdots + \hat{T}_n$, and $\hat{T}_i$ generates all determinants having i excitations from the HF reference, along with corresponding expansion coefficients. In the limit that $n = N$, the CC approach recovers the exact wavefunction and so is equivalent to full CI. For computational efficiency however, truncated forms of CC are again employed. A popular approximation is CCSD, where $\hat{T} = \hat{T}_1 + \hat{T}_2$, while CCSD(T), in which the parentheses indicate that triple excitations are treated perturbatively, is considered the "gold-standard" of quantum chemistry for systems in which multiconfigurational character is not pronounced.

CC theory, though not variational, is often preferred to CI, and this is due in part to its size-consistency, which has its origins in the fact that, for a given truncation, CC incorporates more configurations than CI. This is easily seen by comparing CID (CI with only doubly excitations) to CCD. Here the CI wavefunction is given by

$$\left|\Psi^{\mathrm{CID}}\right\rangle = \left(1 + \hat{C}_2\right)\left|\Psi_0\right\rangle \tag{11.10}$$

whereas the CC wavefunction is defined as

$$\left|\Psi^{\mathrm{CCD}}\right\rangle = e^{\hat{T}_2}\left|\Psi_0\right\rangle = \left(1 + \hat{T}_2 + \frac{\hat{T}_2^2}{2!} + \frac{\hat{T}_2^3}{3!} + \cdots\right)\left|\Psi_0\right\rangle \tag{11.11}$$

while CI and CC based approaches give, in principle, a method for evaluating the *exact* electronic wavefunction, the computational expense of these methods mean that in practise only truncated expansions are practical (see Figure 11.3). For systems exhibiting no

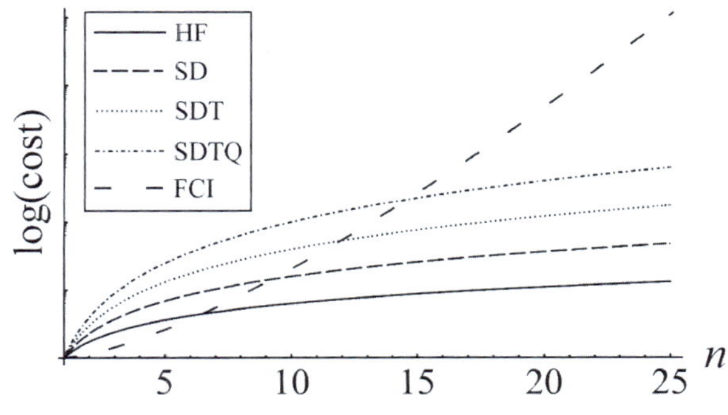

Figure 11.3. Comparison of computational cost of truncated CI/CC methods in comparison to Hartree–Fock theory and full CI. n represents the number of MOs in the complex. Reproduced from Ref. [103], with permission from John Wiley and Sons.

pronounced multireference character the CCSD(T) approach, often achieves chemical accuracy of 1 kcal/mol for some molecular properties.

11.2.2.3 *Complete-active-space self-consistent-field (CASSCF) theory*

Many f-elements compounds exhibit multiconfigurational character, the origin of which is often in the weak crystal fields experienced by the ion. These weak fields result in the 4f/5f-manifold being near-degenerate and consequently there may be several electronic configurations with almost identical energies. In situations where one configuration is lower in energy than a set of other (near-degenerate) configurations, truncated CI or CC approaches are suitable. In situations where the stability of this one configuration is pronounced, DFT may also suitable. However, in situations where there is no single dominant configuration, *multireference* approaches are required. Multireference approaches approximate the exact wavefunction without expansion from a single Hartree–Fock reference, instead using a number of configurations as reference points for the expansion. The most popular of these is the CASSCF approach.[19] Here a chemically meaningful orbital subspace is defined as being "active" (see

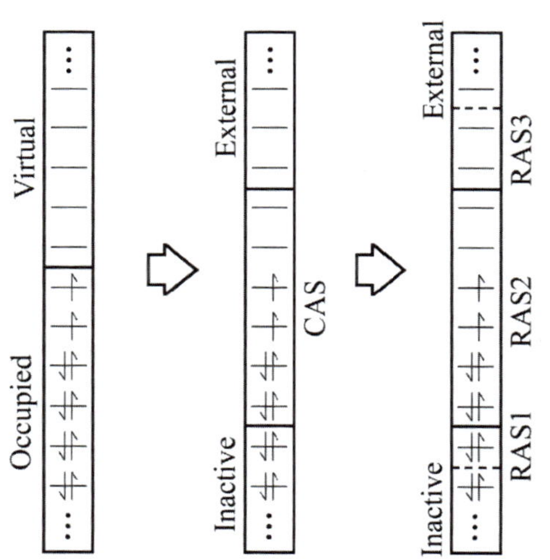

Figure 11.4. Partitioning of orbitals subspaces in monodeterminantal, CASSCF and RASSCF approaches.

Figure 11.4). Within this active subspace, a full CI expansion is performed, while no other configurations are allowed to contribute. A judicious choice of active space allows the correlation effects associated with the weak crystal fields experienced by the f-elements to be simulated with high accuracy. An extension of the CASSCF methodology is known as restricted-active-space self-consistent-field (RASSCF) theory.[20] In the RASSCF approach, three orbitals subspaces are defined: while a full CI expansion is performed in the second (RAS2) subspace, a truncated CI expansion associated with excitations between the first (RAS1) and third (RAS3) subspaces is also included, allowing for contributions from a greater number of orbitals, albeit with reduced accuracy. In CAS/RASSCF approaches, dynamical correlation effects can be included perturbatively.[21–23]

11.2.2.4 *Relativity*

The large nuclear charges of the f-elements generate forces that result in core electrons experiencing pronounced relativistic effects. The magnitude of such effects is governed by the Lorentz factor:

$$\gamma = \frac{1}{\sqrt{1 - \frac{v^2}{c^2}}} \tag{11.12}$$

where v is the velocity and c is the speed of light. The (classical) electronic velocity can be estimated by the following argument. Via the virial theorem, the average kinetic energy of a Coulombic system, T, can be defined in terms of the average potential energy, V, by $2T = -V$. For a hydrogenic system, the total energy of the nth level is given by:

$$\begin{aligned} E_n &= -\frac{1}{2}\frac{m_e e^4}{(4\pi\varepsilon_0)^2\hbar^2}\frac{Z}{n^2} \\ &= -\frac{Z^2}{2n^2}\,\text{a.u.} \end{aligned} \tag{11.13}$$

where the final expression is given in atomic units (a.u.), with $e = m_e = \hbar = 4\pi\varepsilon_0 = 1$. The (classical) kinetic energy of an electron is given in a.u. by $T = v^2/2$ and the virial theorem allows us to write $E = T+V = -T$, giving for the $n = 1$ ground state of the hydrogenic system:

$$\frac{v^2}{2} = \frac{Z^2}{2} \longrightarrow v = Z \tag{11.14}$$

i.e. the (classical) velocity of a 1s electron in a hydrogenic atom, when expressed in atomic units, is approximately equal to the nuclear charge of the atom. In a.u., the speed of light can be defined in terms of the fine structure constant, $c = 1/\alpha \approx 137$. The Lorentz factor can therefore be easily determined as a function of nuclear charge.

The electron experiences a relativistic mass increase given by $m_e{}' = \gamma m_e$ and so the Bohr radius contracts accordingly, $a'_0 = a_0/\gamma$ (see Figure 11.5): Taking uranium as an example, the 1s electrons are

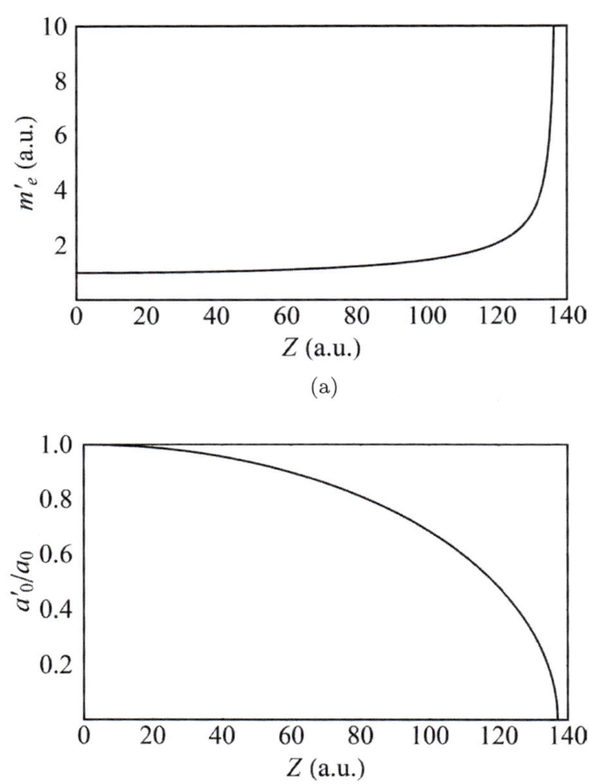

(b)

Figure 11.5. Effect of the Lorentz factor on (a) the electron mass and (b) the Bohr radius of a 1s electron in a hydrogenic atom of charge Z.

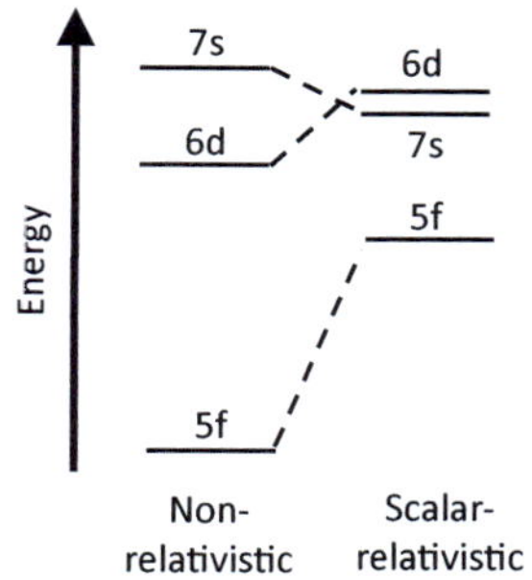

Figure 11.6. Schematic of scalar-relativistic effects on the energies of the 5f, 6d and 7s shells of the uranium atom.

moving with radial velocity of $\approx 92/137 = 0.67c$, a substantial fraction. Overall, relativistic effects result in a contraction and stabilisation of core orbitals. These contracted core orbitals shield the nucleus more effectively, leading to destabilisation of valence orbitals. These effects can be pronounced and, in part, explain the complexities of actinide bonding where the 5f, 6d and 7s shells can be considered as valence, with the 6p shell not fully core-like (see Figure 11.6).

The effects discussed above are termed scalar relativistic effects, and can be included via approximate Hamiltonians[24–26] or relativistic effective core potentials (RECPs),[27,28] however complete relativistic treatments of actinide complexes begin from the Dirac equation:

$$(c(\boldsymbol{\alpha} \cdot \mathbf{p}) + \boldsymbol{\beta} mc^2)\Psi = i\hbar \frac{\partial}{\partial t}\Psi \tag{11.15}$$

$\boldsymbol{\alpha}$, $\boldsymbol{\beta}$ and $\mathbf{p}$ are 4×4 matrices and so solutions to the Dirac equation have four components: two associated with electronic and two with positronic degrees of freedom. While "4-component" computational methods exist, it is more common to attempt to eliminate or decouple the positronic degrees of freedom via unitary transformations, leading to the more popular "2-component" methodologies which maintain the spin–orbit coupling manifested in the $\boldsymbol{\alpha} \cdot \mathbf{p}$ term of the Dirac equation. There are numerous (in fact, infinitely many) transformations which allow this, at least approximately. The Foldy–Wouthuysen transformation[29] involves an approximate decoupling and leads to the Pauli Hamiltonian, which is only appropriate for use in perturbative calculations, whereas an alternative approximate decoupling leads to the regular Hamiltonians:[26] the zeroth order regular approximation (ZORA) is perhaps the most popular two-component method employed today and been implemented so as to include or neglect spin–orbit coupling. Attempts have also been made to implement exact decoupling approaches, leading to the Douglas–Kroll–Hess[24,25] (DKH) and Barysz–Sadlej–Snijders[30] (BSS) Hamiltonians while, more recently, exact 2-component (X2C) methods[31,32] have been developed and implemented.

It is important to recognise that while these transformations lead to practical two-component Hamiltonians, the quantum mechanical operators corresponding to observable properties must also be appropriately transformed so as to avoid "picture change" errors.

11.2.2.5 *Basis sets*

Integral to any quantum chemical calculation is the choice of basis set. The basis set is the set of functions used to construct the molecular orbitals which, in turn, are used to construct the many electron wavefunction/density. Here, we focus on basis sets used in molecular, as opposed to periodic, systems.

The vast majority of commonly available codes use either Slater-type orbitals (STOs), with an $e^{-\zeta r}$ dependence, or Gaussian-type orbitals (GTOs), with an $e^{-\zeta r^2}$ dependence. STOs offer greater accuracy both in the core region and at long range but are computationally unwieldy, whereas GTOs lack this accuracy but are computationally efficient. The relative efficiency of GTOs over STOs is such that it is still computationally expedient to approximate STOs with linear combination of GTO "primitives".

A given atomic basis set is usually divided into core and valence subsets. Core orbitals of an atom are largely unaffected by the chemical environment and as such each can be represented by a single basis function. In contrast, valence orbitals vary substantially, depending on the nature of coordinating species, charge, and oxidation state. To simulate this variation, a number of functions are used to represent each valence orbital. If only a single function is used to represent each orbital, the basis sets would be described as minimal; two functions representing each orbital constitutes a valence double-ζ (or split-ζ) basis set: three functions constitute a valence triple-ζ basis set, etc. Further flexibility can be incorporated by using polarisation and diffuse functions. As a rule of thumb, polarised valence double-ζ

(DZVP) basis sets are typically of sufficient quality to accurately reproduce structural data, whereas polarised valence triple-ζ (TZVP) basis sets are a minimum requirement for accurate electronic properties. If diffuse electronic structures such as those associated with anionic complexes, electronically excited states or Rydberg states are to be considered, larger and more flexible basis sets including diffuse basis functions are recommended.

It should also be borne in mind that an appropriate choice of basis set is dependent on the methodology employed: CI/CC based methodologies, which represent the many-electron wavefunction via an expansion in electronic configurations, require a large basis set (TZVP or larger) to produce meaningful results, whereas this requirement is largely absent in DFT-based approaches. Four-component relativistic calculations require basis functions for both the electronic and positronic components of the wavefunction, provided by, for example, the Dyall basis sets.[33,34] These basis sets may also be employed in the X2C methods described in the previous section. If only scalar-relativistic Hamiltonians are to be considered, then appropriately contracted standard basis sets, e.g. the SARC basis sets of Pantazis and Neese,[35,36] may be employed, although relativistic effective core potentials (RECPs)[27,28] provide a computationally inexpensive alternative. Replacing the core electrons of a heavy element with an appropriately constructed effective potential implicitly accounts for scalar relativistic effects. This allows calculations to be performed using standard, non-relativistic methods, although the property of interest must be considered beforehand. In particular, if the property is dependent on knowledge of the wavefunction in the vicinity of the nucleus, as is the case for NMR parameters, Mössbauer shifts and hyperfine couplings, then all-electron basis sets and explicit relativistic Hamiltonians are a necessity.

11.2.3 *Analysis methods*

Although the MOs of a complex have historically been used in order to analyse simulation data, they do not provide an unambiguous description of electronic structure since unitary transformation of the occupied orbitals leave all physical observables unaltered (see Figure 11.7). This, coupled with the fact that integer orbital occupation numbers are not maintained in state of the

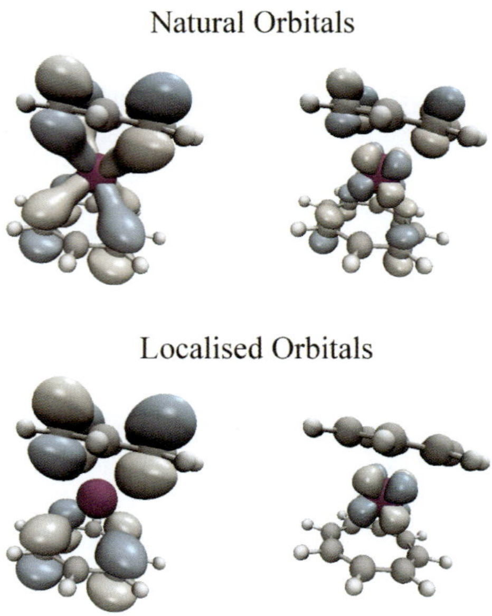

Figure 11.7. Examples of natural and localised molecular orbitals of cerocene, $CeCOT_2$, involving the interaction of Ce $4f_\delta$ with ligand π-orbitals. Appropriate configurational admixture allows the electronic structure of cerocene to be equivalently represented using either orbital representation. Reproduced from Ref. [85].

art multiconfigurational simulation methods mean that analytical methods based on the MOs, such as Mulliken analysis,[37] are best avoided. The electron density of a system is invariant to unitary transformations of the MOs from which it is constructed and, as such, modern analysis methods are usually based on some partitioning and decomposition of the electron density. Of these, the most popular are Hirshfeld,[38] natural bond orbital (NBO),[39] and the quantum theory of atoms in molecules (QTAIM)[40] analyses. Clarke *et al.* have compared several methods when analysing a series of uranyl complexes,[41] concluding that QTAIM performed best in terms of predicting charges, although it lacked sensitivity when considering the electron donating abilities of ligands, whereas NBO performed well in evaluating bond orders. Vallet *et al.* have probed chemical bonding in uranyl complexes via NBO and QTAIM methods,[42] finding that axial bonding is strongly affected by equatorial coordination, purely electrostatic interactions play a significant role in perturbing the axial bond, and that equatorial covalency is strongly dependent on the nature of the ligating species. Tassell and Kaltsoyannis have compared orbital- and density-based analysis techniques in order to assess covalency in $An(C_5H_5)_4$, finding that QTAIM predicts the early actinides to exhibit greater covalent character than heavier members of the series.[43] Kirker and Kaltsoyannis have performed an analogous study of $An(C_5H_5)_3$, concluding that although orbital based methods show greater metal–ligand hybridisation as the actinide series is traversed, QTAIM data reveal that there is no associated accumulation of charge density in the metal–ligand bonding region.[44] Kerridge has applied density-based analysis methods to investigate covalency and oxidation state in f-element metallocenes $M(C_8H_8)_2$, finding that variation in covalency as the actinide series is traversed is due to a commensurate reduction in f-electron contribution to bonding.[45,46] Foroutan-Nejad *et al.* have used QTAIM extensively to characterise the U–U bond in the endohedral $U_2@C_{80}$ and related complexes.[47]

11.2.4 *Applications*

In this section, we briefly present some highlights from the computational literature regarding the more commonly encountered f-element complexes. By far the largest proportion of simulations have been performed using DFT, due in part to its affordable scaling with system size, but also because of the large number of properties that can be efficiently evaluated within the methodological framework. This section will, however, place emphasis on studies where more sophisticated methodologies were applied where possible. For more comprehensive summaries of the current state of the art, the interested reader is referred to several reviews which expound on the successful applications of computational f-element chemistry in far greater detail.[48–56]

11.2.4.1 *Actinyls, $AnO_2{}^{n+}$*

The linear AnO_2 structural motif is very common among the higher oxidations states of the early actinides, and the size of these systems make them amenable to advanced quantum chemical simulation techniques. Matsika *et al.* have extensively studied the ground and excited states electronic structure of $AnO_2{}^{n+}$ with An = U – Pu using spin–orbit coupled configuration interaction methods.[57–59] Pierloot and Van Besien have applied the CASPT2 methodology in the investigation of the structural, vibrational and optical excitation properties of both free uranyl and its chloro complex,[60] evaluating ground and excited state bond lengths as well clarifying the dense excited state electronic structure. Calculated properties were shown to be in excellent agreement with experiment. The plutonyl cation has also be extensively investigated[61] and the ground states of the actinyls, AnO_2^{2+} (An = Th – Cm) have been elucidated with the same methodology.[62] The CCSD(T) approach has also been used to investigate the ground state and vibrational properties of uranyl,[63] providing the most accurate structural characterisation to date, and equation-of-motion (EOM) coupled cluster theory has been applied to investigate its excited state electronic structure.[64]

The electronic structure of the uranyl unit is highly sensitive to its equatorial coordination environmental and the larger size of these coordination complexes necessitates the use of lower level DFT simulations, which nonetheless have been demonstrated to provide accurate structural and vibrational data.[42,65–68] These studies also highlight the applicability of a variety of analytical methods in the elucidation of the effect of equatorial complexation on U–O_{yl} bond strength.

The ^{235}U chemical shifts of a series of uranyl complexes have been computationally investigated using DFT[69] and large chemical shifts of >21, 000 ppm were been predicted, with spin-orbit coupling required to obtain acceptable agreement with experimental studies. While f-element chemical shifts are rarely directly observed, those of species in the first coordination shell are more accessible. Gendron *et al.* have evaluated dipolar ^{13}C chemical shifts of actinyl carbonates using CAS-based methods,[70] identifying a pronounced negative shift for plutonyl, in line with experimental observations, whereas Hrobárik *et al.* have investigated ^{1}H shifts in UO_2H_2 using two-component relativistic DFT, predicting a "giant" shift of 220.9 ppm. This shift was identified as being almost entirely due to spin-orbit coupling effects.[71]

11.2.4.2 *Hexahalide complexes of the f-elements*

The hexahalides of the f-elements, MX_6^{n-}, have been the subject of numerous studies due to their size, high symmetry and, in the case of UF_6, relevance to the nuclear power industry, with the majority of these studies having been performed using density functional theory. Straka *et al.* have used hybrid-DFT to demonstrate[72] that the 5f orbitals have a well-defined effect on structural parameters in fluorides of An = Th – Np, while it has been shown that both HF and hybrid-DFT simulations are capable of accurately reproducing the vibrational properties of AnF_6 with An = Th – Pu when suitably scaled, although simulations were most accurate for the earlier actinides with fewer unpaired 5f electrons.[73,74] Schreckenbach and Shamov have shown that hybrid DFT-functionals predict UF_6 (and UF_5) defluorination reaction enthalpies in good agreement with both CCSD(T) and experimental data,[50] a result also found by Batista *et al.*[75]

Recently, surprising experimental evidence for a covalent contribution to Ce–Cl bonding in $CeCl_6^{2-}$ has been provided via X-ray absorption spectroscopy, with covalent contributions being as large as that found in analogous uranium complexes.[76,77] Subsequent RASSCF simulations support this finding, with Ce–Cl covalent character almost indistinguishable from that of the U–Cl bond.[78]

Schreckenbach has investigated the ^{19}F chemical shifts in UF_nL_{6-n} (L=Cl, OCH_3) using spin-orbit coupled DFT,[69,79] obtaining "moderate to excellent" agreement with experimentally available data. Similar calculations have been performed by Straka and Kaupp on uranium chlorofluorides[80] with pronounced dependence on the choice of exchange-correlation functional identified and attributed to variations in the description of σ- and π-bonding interactions. Hrobárik, *et al.* have investigated ^1H shifts in HUF_5 using two-component relativistic DFT, again predicting an enormous shift of 254.8 ppm arising almost exclusively from spin-orbit coupling effects.[71]

11.2.4.3 *Organometallic complexes of the f-elements*

The organometallic chemistry of the f-elements is well developed and is dominated by the cyclopentadienyl (Cp, C_5H_5) and, uniquely, the cyclooctatetraene (COT, C_8H_8) ligands. The former results in complexes primarily of the form MCp_4, MCp_3, MCp_3X and MCp_2X_2, whereas the latter gives rise to the widely-studied f-element metallocene sandwich complexes, $MCOT_2$. Mixed sandwich complexes of the form MCpCOT are also synthetically accessible.

The f-element metallocenes have been extensively investigated from a theoretical perspective, partly due to their high (D_{8h}) symmetry but also due to their capacity for delta-bonding interactions. Furthermore, it was suggested in 1989 that cerocene, $CeCOT_2$, may behave in a manner analogous to condensed phase Kondo systems,[81] with subsequent experimental studies supporting this assertion.[82] Dolg *et al.* applied multiconfigurational methods in the study cerocene and its actinide analogue, thorocene.[83,84] While the ground state of thorocene was found to be well described by a simple $5f^0$ Th(IV) configuration, cerocene exhibited far greater complexity, being best described by a combination $4f^1$ Ce(III) and $4f^0$ Ce(IV) configurations. While Dolg *et al.* found the Ce(III) configuration to contribute ~80% to the ground state, subsequent CASSCF studies showed this interpretation to be dependent on the degree of orbital mixing present in the simulation,[85,86] see Figure 11.7.

While cerium is the only lanthanide with an easily accessible +4 oxidation state, this state is accessible for all of the early actinides and simulations of the actinocenes reflect this. Uranocene, $UCOT_2$, has been extensively studied with both monodeterminantal[87,88] and multiconfigurational[89,90] methods, the latter of which yielded a predicted magnetic moment of $2.30\mu_B$, in excellent agreement with the experimental value of $2.36\mu_B$. A series of actinocenes with An = Th – Cm have been studied using the CASSCF methodology[46,91] where the increasing stability of the +3 oxidation was identified.

Organometallic complexes of the lanthanides have emerged as excellent single molecule magnets (SMMs), for which there are numerous potential technological applications: the wealth of research in this area has recently been reviewed.[92] Here, multiconfigurational (typically CASSCF) methodologies are essential for the accurate representation of magnetic properties. Gendron *et al.* have investigated the magnetization of $Ln(COT_2)^-$ with Ln = Ce – Yb,[93] Boulon *et al.* have elucidated[94] the magnetic anisotropy in ErCp*COT, while LeRoy *et al.* have investigated the origin of low magnetic relaxation in f-element metalocenes.[95] These examples highlight the fact that advanced simulation methods are now suitably accurate to allow the rational design of synthetic targets for technological relevant applications.

11.2.4.4 *An–An bonding*

While unsupported An–An bonds have yet to be found experimentally, the endohedral actinide fullerene system U_2@C80, in which the U_2 system is encapsulated within the fullerene, has been detected via mass spectroscopy[96] Foroutan-Nejad *et al.* have investigated the electronic structure of this complex using DFT, identifying an "unwilling" U–U bond with an interaction energy of −18 kcal/mol and a bond order close to unity.[47] The authors also investigated the unsupported U_2 molecule at the same level of theory, identifying a bond order of approximately five. This is in excellent agreement with previous CASSCF studies of U_2 which characterised the U–U interaction in terms of an (albeit unconventional) quintuple bond composed of three traditional two-electrons bonds and four "one electron bonds", giving a total of ten electrons in bonding orbitals.[97]

A series of An dimers (An = Ac – U) have been studied using the CASSCF methodology.[97,98] Unlike the transition metal analogues Mo_2 and W_2, which both exhibit sextuple bonds, these studies revealed an increase in bond order from two to five as the actinide series is traversed: Pa_2 is predicted to have an equivalent bond order to U_2, but a bond strength some four times greater. This was rationalised in terms of stronger occupation of bonding orbitals in Pa_2 as well as a greater contribution of the more strongly interacting 6d orbitals to the Pa–Pa bond.

11.3 Summary and Outlook

Computational f-element chemistry is a rapidly developing field, where ever-increasing computational resources, coupled with methodological developments and advances, allow physically and chemically realistic compounds to be simulated with increasing accuracy. However, the complex electronic structures which characterises these compounds means that great care must be taken in both the choice of simulation methods applied and the interpretation of results obtained. While there is no single methodological approach efficient and accurate enough to be applied to all systems, prior understanding of the relative importance of competing phenomena in the system of interest allows the researcher to make informed decisions as to the approach best suited to the problem at hand. Recent and continuing developments in methodology, allowing, e.g. the application of advanced techniques to larger systems[99] and/or the explicit consideration of environmental effects[100–102] will serve to further strengthen the link between simulation and experiment but, wherever possible, systems should be investigated with multiple simulation and analysis methods in order to develop a robust, dependable and trustworthy theoretical description. Where this is not possible, careful consideration of the system, informed by chemical intuition and empirical data, is required for the results of simulations to have meaning.

References

1. L. H. Thomas, *Math. Proc. Cambridge Philos. Soc.*, 1927, *23*, 542.
2. E. Fermi, *Rend. Accad. Naz. Lincei*, 1927, *6*, 602–607.
3. P. A. M. Dirac, *Math. Proc. Cambridge Philos. Soc.*, 1930, *26*, 376.
4. P. Hohenberg and W. Kohn, *Phys. Rev.*, 1964, *155*, B864–B871.
5. W. Kohn and L. J. Sham, *Phys. Rev.*, 1965, *140*, A1133–A1138.
6. J. P. Perdew and K. Schmidt, *AIP Conf. Proc.*, 2001, *577*, 1–20.
7. S. H. Vosko, L. Wilk and M. Nusair, *Can. J. Phys.*, 1980, *58*, 1200–1211.
8. J. P. Perdew and Y. Wang, *Phys. Rev. B*, 1992, *45*, 13244–13249.
9. J. Perdew, K. Burke and M. Ernzerhof, *Phys. Rev. Lett.*, 1996, *77*, 3865–3868.
10. A. D. Becke, *Phys. Rev. A*, 1988, *38*, 3098–3100.
11. C. Lee, W. Yang and R. G. Parr, *Phys. Rev. B*, 1988, *37*, 785–789.
12. J. P. Perdew, *Phys. Rev. B*, 1986, *33*, 8822–8824.
13. L. Curtiss and K. Raghavachari, *J. Chem. Phys.*, 1991, *94*, 7221–7230.
14. J. Tao, J. Perdew, V. Staroverov and G. Scuseria, *Phys. Rev. Lett.*, 2003, *91*, 146401.
15. A. D. Becke, *J. Chem. Phys.*, 1993, *98*, 5648.
16. P. Stephens, F. Devlin, C. Chabalowski and M. Frisch, *J. Phys. Chem.*, 1994, *98*, 11623–11627.
17. C. Adamo and V. Barone, *J. Chem. Phys.*, 1999, *110*, 6158–6170.
18. T. Yanai, D. P. Tew and N. C. Handy, *Chem. Phys. Lett.*, 2004, *393*, 51–57.
19. B. Roos, P. R. Taylor and P. E. M. Siegbahn, *Chem. Phys.*, 1980, *48*, 157–173.
20. P.-Å. Malmqvist, A. Rendell and B. O. Roos, *J. Phys. Chem.*, 1990, *94*, 5478.

21. K. Anderson, P.-Å. Malmqvist, B. O. Roos, A. J. Sadlej and K. Wolinskit, *J. Phys. Chem.*, 1990, 5483–5488.
22. P.-Å. Malmqvist, K. Pierloot, A. R. M. Shahi, C. J. Cramer and L. Gagliardi, *J. Chem. Phys.*, 2008, *128*, 204109.
23. C. Angeli, R. Cimiraglia, S. Evangelisti, T. Leininger and J.-P. Malrieu, *J. Chem. Phys.*, 2001, *114*, 10252.
24. M. Douglas and N. Kroll, *Ann. Phys. (N. Y).*, 1974, *155*, 89–155.
25. B. Hess, *Phys. Rev. A*, 1986, *33*, 3742–3748.
26. E. van Lenthe, E. J. Baerends and J. G. Snijders, *J. Chem. Phys.*, 1993, *99*, 4597.
27. M. Dolg, H. Stoll and H. Preuss, *J. Chem. Phys.*, 1989, *90,* 1730–1734.
28. W. Küchle, M. Dolg, H. Stoll and H. Preuss, *J. Chem. Phys.*, 1994, *100*, 7535.
29. L. L. Foldy and S. A. Wouthuysen, *Phys. Rev.*, 1950, *78*, 29–36.
30. M. Barysz, A. J. Sadlej and J. G. Snijders, *Int. J. Quantum Chem.*, 1997, *65*, 225–239.
31. W. Kutzelnigg and W. Liu, *J. Chem. Phys.*, 2005, *123*, 241102.
32. M. Iliaš and T. Saue, *J. Chem. Phys.*, 2007, *126*, 64102.
33. A. S. P. Gomes, K. G. Dyall and L. Visscher, *Theor. Chem. Acc.*, 2010, *127*, 369–381.
34. K. G. Dyall, *Theor. Chem. Acc.*, 2007, *117*, 491–500.
35. D. A. Pantazis and F. Neese, *J. Chem. Theory Comput.*, 2009, *5*, 2229.
36. D. A. Pantazis and F. Neese, *J. Chem. Theory Comput.*, 2011, *7*, 677–684.
37. R. S. Mulliken, *J. Chem. Phys.*, 1955, *23*, 1833.
38. F. L. Hirshfeld, *Theor. Chim. Acta*, 1977, *44*, 129–138.
39. A. E. Reed, R. B. Weinstock and F. Weinhold, *J. Chem. Phys.*, 1985, *83*, 735–746.
40. R. F. W. Bader, *Atoms in Molecules: A Quantum Theory*, Oxford University Press, Oxford, 1990.
41. A. E. Clark, J. L. Sonnenberg, P. J. Hay and R. L. Martin, *J. Chem. Phys.*, 2004, *121*, 2563–2570.
42. V. Vallet, U. Wahlgren and I. Grenthe, *J. Phys. Chem. A*, 2012, *116*, 12373–12380.
43. M. J. Tassell and N. Kaltsoyannis, *Dalton Trans.*, 2010, *39*, 6576–6588.
44. I. Kirker and N. Kaltsoyannis, *Dalton Trans.*, 2011, *40*, 124–131.
45. A. Kerridge, *Dalton Trans.*, 2013, *42*, 16428–16436.
46. A. Kerridge, *RSC Adv.*, 2014, *4*, 12078–12086.
47. C. Foroutan-Nejad, J. Vícha, R. Marek, M. Patzschke and M. Straka, *Phys. Chem. Chem. Phys.*, 2015, *17*, 24182–24192.
48. A. Kovács, R. J. M. Konings, J. K. Gibson, I. Infante and L. Gagliardi, *Chem. Rev.*, 2015, 150213131248008.

49. D. Wang, W. F. van Gunsteren and Z. Chai, *Chem. Soc. Rev.*, 2012, *41*, 5836–5865.
50. G. Schreckenbach and G. A. Shamov, *Acc. Chem. Res.*, 2010, *43*, 19–29.
51. V. Vallet, P. Macak, U. Wahlgren and I. Grenthe, *Theor. Chem. Acc.*, 2006, *115,* 145–160.
52. N. Kaltsoyannis, *Chem. Soc. Rev.*, 2003, *32*, 9–16.
53. G. Schreckenbach, P. J. Hay and R. L. Martin, *J. Comput. Chem.*, 1999, *20*, 70–90.
54. L. Gagliardi and B. O. Roos, *Chem. Soc. Rev.*, 2007, *36*, 893–903.
55. N. Kaltsoyannis, P. J. Hay, J. Li, J.-P. Blaudeau and B. E. Bursten, in *The Chemistry of the Actinide and Transactinide Elements* (Eds. L. R. Morss, N. M. Edelstein, J. Fuger and J. J. Katz), 3rd edn., Springer, Dordrecht, 2006, pp. 1893–2012.
56. A. Kerridge, *Chem. Commun.*, 2017, *53*, 6685–6695.
57. S. Matsika and R. M. Pitzer, *J. Phys. Chem. A*, 2000, *104*, 4064–4068.
58. S. Matsika, R. M. Pitzer and D. T. Reed, *J. Phys. Chem. A*, 2000, *104*, 11983–11992.
59. S. Matsika, Z. Zhang, S. R. Brozell, J.-P. Blaudeau, Q. Wang and R. M. Pitzer, *J. Phys. Chem. A*, 2001, *105*, 3825–3828.
60. K. Pierloot and E. van Besien, *J. Chem. Phys.*, 2005, *123*, 204309.
61. G. La Macchia, I. Infante, J. Raab, J. K. Gibson and L. Gagliardi, *Phys. Chem. Chem. Phys.*, 2008, *10*, 7278–7283.
62. I. Infante, A. Kovacs, G. La Macchia, A. R. M. Shahi, J. K. Gibson and L. Gagliardi, *J. Phys. Chem. A*, 2010, *114*, 6007–6015.
63. V. E. Jackson, R. Craciun, D. A. Dixon, K. A. Peterson and W. A. De Jong, *J. Phys. Chem. A*, 2008, *112*, 4095–4099.
64. P. Tecmer, N. Govind, K. Kowalski, W. a de Jong and L. Visscher, *J. Chem. Phys.*, 2013, *139*, 34301.
65. J. L. Sonnenberg, P. J. Hay, R. L. Martin and B. E. Bursten, *Inorg. Chem.*, 2005, *44*, 2255–2262.
66. K. I. M. Ingram, L. J. L. Häller and N. Kaltsoyannis, *Dalton Trans.*, 2006, *2*, 2403–2414.
67. S. Tsushima, *Dalton Trans.*, 2011, *40*, 6732–6737.
68. P. Di Pietro and A. Kerridge, *Inorg. Chem.*, 2016, *55*, 573–583.
69. G. Schreckenbach, *Inorg. Chem.*, 2002, *41*, 6560–6572.
70. F. Gendron, B. Pritchard and J. Autschbach, *Inorg. Chem.*, 2014, *53*, 8577–8592.
71. P. Hrobarik, V. Hrobarikova, A. H. Greif and M. Kaupp, *Angew. Chemie. Int. Ed.*, 2012, *51*, 10884–10888.
72. M. Straka, P. Hrobárik and M. Kaupp, *J. Am. Chem. Soc.*, 2005, *127*, 2591–2599.

73. T. Privalov, B. Schimmelpfennig, U. Wahlgren and I. Grenthe, *J. Phys. Chem. A*, 2002, *106*, 11277–11282.
74. B. Schimmelpfennig, T. Privalov, U. Wahlgren and I. Grenthe, *J. Phys. Chem. A*, 2003, *107*, 9705–9711.
75. E. R. Batista, R. L. Martin, P. J. Hay, J. E. Peralta and G. E. Scuseria, *J. Chem. Phys.*, 2004, *121*, 2144–2150.
76. S. G. Minasian, J. M. Keith, E. R. Batista, K. S. Boland, D. L. Clark, S. D. Conradson, S. A. Kozimor, R. L. Martin, D. E. Schwarz, D. K. Shuh, G. L. Wagner, M. P. Wilkerson, L. E. Wolfsberg and P. Yang, *J. Am. Chem. Soc.*, 2012, *134*, 5586–5597.
77. M. W. Löble, J. M. Keith, A. B. Altman, S. C. E. Stieber, E. R. Batista, K. S. Boland, S. D. Conradson, D. L. Clark, J. Lezama Pacheco, S. A. Kozimor, R. L. Martin, S. G. Minasian, A. C. Olson, B. L. Scott, D. K. Shuh, T. Tyliszczak, M. P. Wilkerson and R. A. Zehnder, *J. Am. Chem. Soc.*, 2015, *137*, 2506–2523.
78. R. Beekmeyer and A. Kerridge, *Inorganics*, 2015, *3*, 482–499.
79. G. Schreckenbach, S. K. Wolff and T. Ziegler, *J. Phys. Chem. A*, 2002, *104*, 8244–8255.
80. M. Straka and M. Kaupp, *Chem. Phys.*, 2005, *311*, 45–56.
81. C. Neumann and P. Fulde, 1989, *278*, 277–278.
82. C. Booth, M. Walter, M. Daniel, W. Lukens and R. Andersen, *Phys. Rev. Lett.*, 2005, *95*, 267202.
83. M. Dolg, P. Fulde, W. Küchle, C.-S. Neumann and H. Stoll, *J. Chem. Phys.*, 1991, *94*, 3011.
84. M. Dolg, P. Fulde, H. Stoll, H. Preuss, A. Chang and R. M. Pitzer, *Chem. Phys.*, 1995, *195*, 71–82.
85. A. Kerridge, R. Coates and N. Kaltsoyannis, *J. Phys. Chem. A*, 2009, *113*, 2896–2905.
86. O. Mooßen and M. Dolg, *Chem. Phys. Lett.*, 2014, *594*, 47–50.
87. N. Rosch and A. Streitwieser, *J. Am. Chem. Soc.*, 1983, *105*, 7237–7240.
88. P. M. Boerrigter, E. J. Baerends and J. G. Snijders, *Chem. Phys.*, 1988, *122*, 357–374.
89. A. H. H. Chang and R. M. Pitzer, *J. Am. Chem. Soc.*, 1989, *111*, 2500–2507.
90. W. Liu, M. Dolg and P. Fulde, *J. Chem. Phys.*, 1997, *107*, 3584.
91. A. Kerridge and N. Kaltsoyannis, *J. Phys. Chem. A*, 2009, *113*, 8737–8745.
92. D. N. Woodruff, R. E. P. Winpenny and R. A. Layfield, *Chem. Rev.*, 2013, *113*, 5110–5148.
93. F. Gendron, B. Pritchard, H. Bolvin and J. Autschbach, *Dalton Trans.*, 2015, 19886–19900.

94. M. E. Boulon, G. Cucinotta, S. S. Liu, S. Da Jiang, L. Ungur, L. F. Chibotaru, S. Gao and R. Sessoli, *Chem. A Eur. J.*, 2013, *19*, 13726–13731.
95. J. J. Le Roy, S. I. Gorelsky, I. Korobkov and M. Murugesu, *Organometallics*, 2015, *34*, 1415–1418.
96. K. Akiyama, Y. Zhao, K. Sueki, K. Tsukada, H. Haba, Y. Nagame, T. Kodama, S. Suzuki, T. Ohtsuki, M. Sakaguchi, K. Kikuchi, M. Katada and H. Nakahara, *J. Am. Chem. Soc.*, 2001, *123*, 181–182.
97. L. Gagliardi and B. O. Roos, *Nature*, 2005, *433*, 848–851.
98. B. O. Roos, P.-A. Malmqvist and L. Gagliardi, *J. Am. Chem. Soc.*, 2006, *128*, 17000–17006.
99. D. Ghosh, J. Hachmann, T. Yanai and G. K.-L. Chan, *J. Chem. Phys.*, 2008, *128*, 144117.
100. T. A. Wesolowski and A. Warshel, *J. Phys. Chem. A*, 1993, *97*, 8050–8053.
101. A. Severo Pereira Gomes and C. R. Jacob, *Annu. Reports Sect. "C" (Physical Chem.)*, 2012, *108*, 222–277.
102. T. Dresselhaus, J. Neugebauer, S. Knecht, S. Keller, Y. Ma and M. Reiher, *J. Chem. Phys.*, 2015, *142*, 044111.
103. A. Kerridge (2015), The Complete-Active-Space Self-Consistent-Field Approach and Its Application to Molecular Complexes of the f-Elements, in *Computational Methods in Lanthanide and Actinide Chemistry* (Ed. M. Doig), Wiley, Chichester, 2015, pp. 121–144.

https://doi.org/10.1142/9781800610163_0012

Chapter 12

Spectroscopy of the Actinides

*Yvonne Rechkemmer and Joris van Slageren**
Institut für Physikalische Chemie, Universität Stuttgart, Pfaffenwaldring 55, 70569 Stuttgart, Germany
**slageren@ipc.uni-stuttgart.de*

12.1 General

X-rays are ionising radiation, i.e. the absorption of an X-ray photon leads to emission of an electron due to the well-known photoelectric effect. A great many X-ray and other high-energy techniques have been employed in actinide science.[1] X-ray absorption spectra display so-called absorption edges corresponding to the ionisation energies of electrons in a given shell, where K, L and M edges correspond to ejection of electrons with principal quantum number $n = 1, 2, 3$. While the K edge is a single feature, the L edge consists of three components, namely the L_1, L_2 and L_3 edges. The L_1 edge corresponds to excitation of a 2s electron and the L_2 and L_3 edges to excitation of a 2p electron. The L_2–L_3 energy difference is due to spin–orbit coupling of the spin and orbital angular momenta of hole in the 2p shell created by ionisation, leading to $J = 1/2$ (L_2) and $J = 3/2$ (L_3 edge) states. The latter edge is more intense due to the larger degeneracy of the $J = 3/2$ state. Similar observations can be made for the M edges. The absorption intensity can be measured directly or by measuring the intensity of the fluorescence or the emission of secondary emitted electrons resulting from the hole being filled by

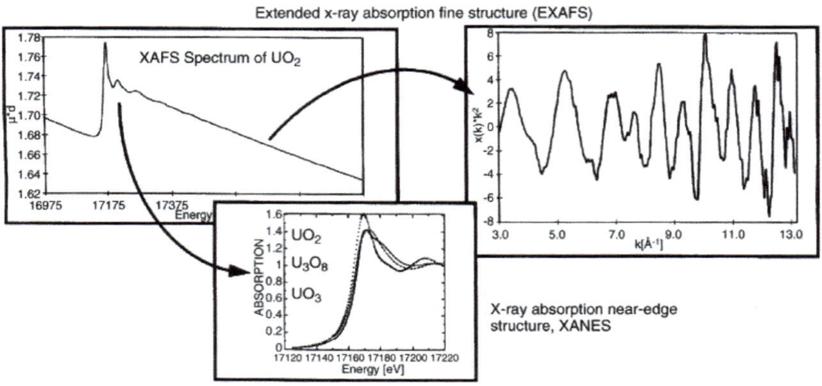

Figure 12.1. (Top left) Raw X-ray absorption spectrum recorded on UO_2 at the uranium L_3 edge; (top right) k^3 weighted extended X-ray absorption fine structure (EXAFS) data derived from the raw data; (bottom) X-ray absorption near-edge structure (XANES) of UO_2 compared to other uranium oxides. Reprinted from Ref. [2]. Copyright (2006) with permission from Elsevier.

an electron from a higher-lying state.[2] The X-ray absorption spectrum displays pronounced structure both around the edge energy as well as at energies up to some hundreds of eV beyond the edge (Figure 12.1). Collectively this structure is known as X-ray absorption fine structure (XAFS). The structure beyond the absorption edge is called extended X-ray absorption fine structure (EXAFS) and is due to the fact that the photoelectron is scattered by the neighbouring electrons. Because the De Broglie wavelength is comparable to interatomic distances, interference between the outgoing and scattered waves occurs, leading to a modulation of the absorption intensity, normally expressed as a function of photoelectron wavevector. Fourier transformation gives the radial distribution from which distances to neighbouring atoms. Comparison with simulated EXAFS curves for model species allows extraction of relevant structural information. The structure around the edge (X-ray absorption near-edge structure, XANES, also known as near edge X-ray absorption fine structure, NEXAFS) is partly due to excitation of core electrons to bound states (pre-edge structure) but mainly due to multiple scattering processes that are the consequence of the long wavelength of the low-energy photoelectrons resulting in long mean free path lengths. These multiple scattering events make data analysis according to

EXAFS procedures complicated. XANES is therefore mainly used for chemical characterisation, due to its sensitivity to metal oxidation state and fingerprint characterisation of specific coordination geometries. The distinction between XANES and EXAFS energy ranges is somewhat arbitrary. In a magnetic field, left- and right-circularly polarised X-rays are absorbed to different extents. This phenomenon is called X-ray magnetic circular dichroism (XMCD) and can be used to study the magnetic properties of (ferromagnetic) materials including the determination of spin and orbital angular momenta.[3] The linear dichroism analogous technique is mainly used for studying antiferromagnets.[4] X-rays can also be scattered rather than absorbed by the sample. The advantage of X-ray absorption methods compared to other spectroscopic methods is that they are element specific and not limited to particular (e.g. paramagnetic) species. On the other hand, if the atom under study is present in several surroundings in the sample, the XAFS spectra are due to the sum of these and separation of the contributions is challenging. Although elastic scattering (including high-energy X-ray scattering, Ref. [5]) is outside of the scope of this chapter, *inelastic* X-ray scattering contains spectroscopic information on the material under study. Especially when the incident X-ray frequency approaches that of an absorption edge, resonant enhancement of the scattering cross-section can occur (resonant inelastic X-ray scattering, RIXS). This technique yields information on excitations in the sample, especially electronic excitations. The advantage compared to optical Raman techniques is that the full dispersion of the excitation can be investigated. Finally, rather than monitoring the X-ray absorption or scattering, the primary photoelectron can be detected (X-ray photoelectron spectroscopy, XPS). Here the kinetic energy of the photoelectrons is measured, which can be related to the energy of the (core) orbital from which the electron originated. This technique is often used for the chemical characterisation especially of surface species, e.g. the determination of oxidation states.

In the following, the applications of X-ray spectroscopic techniques in actinide science are briefly outlined, where the focus is on techniques that have been used to study compounds of chemical significance.

12.2 EXAFS

A number of excellent recent reviews deal with the subject of EXAFS studies of actinides.[2,6,7] A main topic of research is actinide speciation in environmentally relevant surroundings, i.e. aqueous solutions at varying pHs. A large number of EXAFS studies have resulted in a consensus that the predominant uranium species in water at neutral to acidic pH is the uranyl(VI) ion $[UO_2(H_2O)_5]]^{3-}$,[6] with a pentagonal bipyramidal structure. Similar results were found for other actinyl ions. A large amount of work deals with the coordination of additional ions to the actinyl unit, such as halide, sulfate, perchlorate and oxalate (Figure 12.2).[6,8] There is some controversy about the hydroxide coordination number of uranyl(VI) in basic solution, where

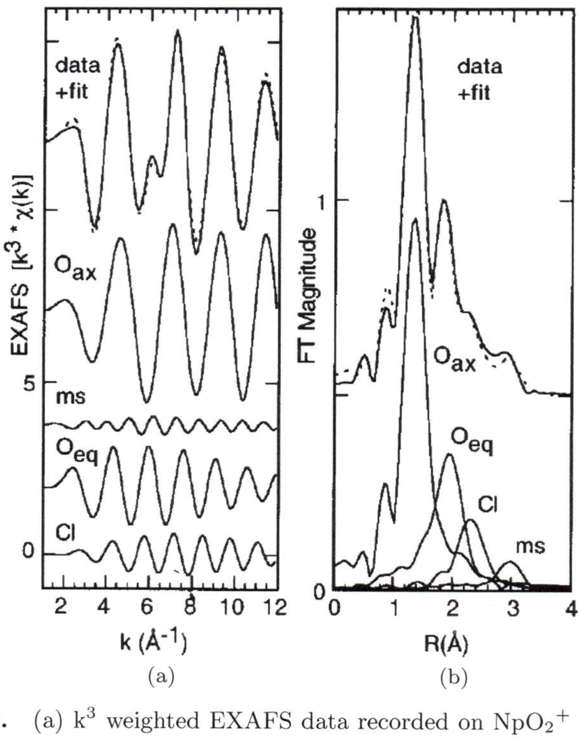

Figure 12.2. (a) k^3 weighted EXAFS data recorded on $NpO_2{}^+$ in 10 M LiCl. (b) Real-space data obtained after Fourier transformation. The figure shows the contributions of the various coordinating atoms to the spectra. MS indicates the contribution of multiple scattering processes. Reprinted from Ref. [8]. Copyright (1997) with permission from American Chemical Society.

reports of coordination numbers of both 4 and 5 have appeared.[9–11] Recent EXAFS data suggest that the experimental evidence for the pentahydroxo species may actually indicate a species with three uranium-oxo bonds, namely $[UO_3(OH)_3]^{3-}$.[12] The dimeric species $\{[UO_2(OH)_4]^{2-}[UO_3(H_2O)_3]^{3-}\}^-$ was proposed to play a crucial role in oxo ligand exchange.[12] Multinuclear species are also very important in aqueous uranyl chemistry in the presence of carbonate, with a number of di- and trinuclear compounds, e.g. $[(UO_2)_3(CO_3)_6]^{6-}$, observed in EXAFS experiments.[6] These investigations are also relevant because of the predominance of carbonate minerals in many soils, and many studies deal with the influence of carbonates and other minerals on actinyl speciation in the environment.[2,6] EXAFS has also contributed to the investigation of the structures of uranium minerals themselves.[13] Space is lacking here to discuss the many investigations of aqueous interactions with natural substances, of different oxidation states, of non-uranium and non-actinyl actinide species.[2,6,7] An interesting time-resolved study was that of the formation of the poly-oxo cluster U_{38} in non-aqueous solution, where EXAFS was combined with XANES, powder X-ray diffraction, as well as UV/Vis and NMR spectroscopies.[14]

12.3 XANES

Although scattering of the photoelectron is just as much the cause for the fine structure in XANES as it is in EXAFS, the analysis of the near-edge structure is much more challenging and no simple theoretical framework exists.[2] The sensitivity of the absorption edge position to the oxidation state can be used to assign oxidation states to actinide ions (Figure 12.1).[2,15] A combined X-ray diffraction and XANES study on sodium actinates relevant to nuclear fuel (An = U, Np, Pu) in oxidation states IV to VII investigated the correlation between oxidation state and actinide L_3 absorption energy threshold.[16] This correlation was revealed out to be linear for neptunium but not for uranium. In contrast, for molecular compounds, a linear correlation was found.[17] The short actinide-oxo bonds in actinyl ions give rise to characteristic XANES features,[2] but similar features were also found in non-actinyl species.[16] Careful comparison of a large series of plutonium species revealed that influences of the composition and order in the matrix can have influences on the XANES spectrum that are of the same order as those due to the oxidation state.[15] Most studies are carried out at the actinide L_3 edge (at 16–19 keV for the light actinides).[2] Higher (e.g. N−) edges and transitions to bound states can be probed by means of soft X-rays (< 1 keV for uranium).[18] In addition to actinide absorption edges, 1s XANES/NEXAFS studies of light elements, where transitions to bound states (e.g. 1s$\rightarrow \pi^*, \sigma^*$) can also provide useful information. Thus, oxygen edge XANES was employed to investigate the effect of equatorial coordination on the U=O bond in uranyl complexes.[19] Scanning transmission X-ray microscopy combines XANES spectroscopy with raster scanning giving morphological and microchemical information.[2] In the uranium-oxo U_{38} cluster, XANES revealed the gradual transition from uranium(IV) to uranium(VI).[14]

12.4 XMD

Of the X-ray magnetic dichroism (XMD) techniques, XMCD has been used to determine the static and dynamic magnetic properties of paramagnetic molecular species on surfaces, including spin and orbital angular momenta, magnetic couplings, magnetisation relaxation times and quantum tunnelling of the magnetisation,[20–22] in addition to more traditional applications in magnetically ordered materials.[3] Applications in actinide science appear to be scarce and limited to heavy fermions, Laves phase alloys and uranium monochalcogenides US, USe and UTe. The last of these dealt with the investigation of the localisation, and orbital and spin moments of 5f and 6d electrons.[23] In addition paramagnetic β-US_2 was investigated.[24]

12.5 High Resolution X-ray Absorption, High-Energy-Resolution Fluorescence-Detection X-ray Absorption Spectroscopy and Resonant Inelastic X-ray Scattering Spectroscopy

All these (relatively modern) techniques have in common that a specific X-ray emission line is monitored rather than the X-ray absorption or total fluorescence, which leads to improved resolution. X-ray emission spectroscopy itself has been mainly used for the metals and simple compounds.[1] Currently a wide range of related techniques with a corresponding range of terminology is being developed and applied. In the technique of high-resolution XANES, the monitored transition is one where an electron fills the primary hole (e.g. at the L_3 edge, i.e. $2p_{3/2}$) from a higher lying state, e.g. as in the $L\alpha1$ ($3d_{5/2}$–$2p_{3/2}$) fluorescence line.[25] This has the advantage that the resolution is now limited by the longer $3d_{5/2}$ core-hole lifetime rather than the $2p_{3/2}$ core-hole lifetime, increasing the resolution by a factor of two.[26] This technique has been especially useful in investigating pre-edge features, corresponding to transitions to bound states. From polarisation-dependent investigations on single-crystal $Cs_2UO_2Cl_4$ together with quantum chemical calculations conclusions regarding the covalency of bonds involving the uranium ion were drawn.[27] In ThO_2, it proved possible to resolve a shoulder at 0.8 eV above the main maximum at the M_4 edge, which was attributed to crystal field splitting in the excited state of the thorium(IV) ion.[28] Emission methods also allowed measurements with very soft X-rays, up to the $N_{6,7}$ edges of uranium.[29]

A further method in this area is resonant inelastic X-ray scattering (RIXS).[30–32] Because the scattering cross-section is very small compared to the absorption intensity, very bright X-ray sources, i.e. synchrotrons are required for RIXS experiments. A second-order perturbation theory approach to the scattering yields a differential scattering cross-section proportional to the scattering intensity $F(\omega,\omega')$:

$$F(\omega, \omega') = \sum_f \left| \sum_n \frac{\langle f|T_2|n\rangle\langle n|T_1|g\rangle}{E_g - E_n + \omega + i\Gamma} \right|^2 \qquad (12.1)$$

where ω and ω' are the energies of the incoming and scattered X-ray photons, respectively, and g, n and f denote ground, intermediate and final states, respectively, with E_g and E_n the corresponding energies. T_1 and T_2 are the operators (e.g. electric dipole, electric quadrupole) connecting the ground and final states to the intermediate states. Γ represents the lifetime broadening of the core-hole state. Equation (12.1) shows that when the incoming photon energy is close to an electronic excitation energy $E_g - E_n$, resonant enhancement of the scattering occurs. Importantly, as in HR-XANES, the linewidth is determined by the hole lifetime in the final state, where this hole may be located in the valence shell, leading to much higher resolutions. In contrast to HR-XANES, in RIXS the X-ray fluorescence intensity as a function of energy transfer at fixed incident energy is recorded. However, the physical principles are the same and the distinction between the two techniques is somewhat arbitrary. Actinide materials that were studied include UO_2, where the 6d t_{2g}-e_g crystal field splitting was determined to be 3.5 eV, a value that is larger than that obtained from optical spectroscopy.[33] It was also shown by RIXS that mobile uranium(VI) and neptunium(V) are reduced to the UO_2 and NpO_2 solids in the presence of elemental iron.

12.6 XPS

One of the main aims of XPS is the quantitative analysis of the elements in a sample. As such it is an established technique and will not be discussed further here. Because of the limited mean free path length depth of the photoelectrons, XPS is especially useful in thin film and surface investigations. The binding energies of core electrons, derived from XPS are somewhat sensitive to the chemical surroundings of the atom being ionised, leading to a chemical shift of the photoemission peak. In actinides, this can be used to determine oxidation states, similar to the XANES lineshape.[34] The ejection of the photoelectron can be accompanied by excitation of a second electron to a bound state (i.e. a partially filled or empty orbital) leading to so-called shake-up features to the high-energy side of the main peak.[1] Often these features correspond to the transfer of a ligand electron to an unoccupied metal orbital. Angular momentum (exchange) coupling between the core-hole and the open f shell electrons gives rise to a multiplet structure and splitting of the XPS peaks.[35] This has been little studied in actinides, because the most intense XPS is obtained for the 4f levels of actinides, which have very small multiplet splittings. One example is given by the $5p_{3/2}$ XPS line of UO_2.[35] XPS investigations appear to have been limited to the metals and intermetallics as well as simple compounds such as oxides and nitrides. Valence electrons can also be studied by photoelectron spectroscopy, usually in the vacuum UV, but also soft X-rays can be used. In single crystals, the emission of photoelectrons is angle-dependent, which allows obtaining information on the band structure of the (thin film) material.[36]

References

1. C. Bonnelle, N. Spector, *Rare-Earths and Actinides in High Energy Spectroscopy*, Springer Netherlands, Dordrecht, 2015, pp. 327–373.
2. M. A. Denecke, *Coord. Chem. Rev.*, 2006, *250*, 730–754.
3. H. Wende, C. Antoniak, in *Magnetism and Synchrotron Radiation: New Trends* (Eds. E. Beaurepaire, H. Bulou, F. Scheurer, K. Jean-Paul), Springer Berlin Heidelberg, Berlin, Heidelberg, 2010, pp. 145–167.
4. G. van der Laan, in *Magnetism and Synchrotron Radiation: Towards the Fourth Generation Light Sources: Proceedings of the 6th International School "Synchrotron Radiation and Magnetism", Mittelwihr (France), 2012* (Eds. E. Beaurepaire, H. Bulou, L. Joly, F. Scheurer), Springer International Publishing, Cham, 2013, pp. 239–256.
5. K. E. Knope, L. Soderholm, *Chem. Rev.*, 2013, *113*, 944–994.
6. M. R. Antonio, L. Soderholm, *The Chemistry of the Actinide and Transactinide Elements* (Eds. L. R. Morss, N. M. Edelstein, J. Fuger), Springer Netherlands, Dordrecht, 2006, pp. 3086–3198.
7. C. Den Auwer, E. Simoni, S. Conradson, C. Madic, *Eur. J. Inorg. Chem.*, 2003, 3843–3859.
8. P. G. Allen, J. J. Bucher, D. K. Shuh, N. M. Edelstein, T. Reich, *Inorg. Chem.*, 1997, *36*, 4676–4683.
9. D. L. Clark, S. D. Conradson, R. J. Donohoe, D. W. Keogh, D. E. Morris, P. D. Palmer, R. D. Rogers, C. D. Tait, *Inorg. Chem.*, 1999, *38*, 1456–1466.
10. U. Wahlgren, H. Moll, I. Grenthe, B. Schimmelpfennig, L. Maron, V. Vallet, O. Gropen, *J. Phys. Chem. A*, 1999, *103*, 8257–8264.
11. H. Moll, T. Reich, Z. Szabó, *Radiochim. Acta*, 2000, *88*, 411.
12. H. Moll, A. Rossberg, R. Steudtner, B. Drobot, K. Müller, S. Tsushima, *Inorg. Chem.*, 2014, *53*, 1585–1593.
13. R. J. Baker, *Coord. Chem. Rev.*, 2014, *266–267*, 123–136.
14. C. Falaise, C. Volkringer, C. Hennig, T. Loiseau, *Chem. Eur. J.*, 2015, *21*, 16654–16664.
15. S. D. Conradson, K. D. Abney, B. D. Begg, E. D. Brady, D. L. Clark, C. den Auwer, M. Ding, P. K. Dorhout, F. J. Espinosa-Faller, P. L. Gordon, R. G. Haire, N. J. Hess, R. F. Hess, D. W. Keogh, G. H. Lander, A. J. Lupinetti, L. A. Morales, M. P. Neu, P. D. Palmer,

P. Paviet-Hartmann, S. D. Reilly, W. H. Runde, C. D. Tait, D. K. Veirs, F. Wastin, *Inorg. Chem.*, 2004, *43*, 116–131.
16. A. L. Smith, P. Martin, D. Prieur, A. C. Scheinost, P. E. Raison, A. K. Cheetham, R. J. M. Konings, *Inorg. Chem.*, 2016, *55*, 1569–1579.
17. B. Kosog, H. S. La Pierre, M. A. Denecke, F. W. Heinemann, K. Meyer, *Inorg. Chem.*, 2012, *51*, 7940–7944.
18. J. G. Tobin, S. W. Yu, C. H. Booth, T. Tyliszczak, D. K. Shuh, G. van der Laan, D. Sokaras, D. Nordlund, T. C. Weng, P. S. Bagus, *Phys. Rev. B*, 2015, *92*, 035111.
19. C. Fillaux, D. Guillaumont, J.-C. Berthet, R. Copping, D. K. Shuh, T. Tyliszczak, C. D. Auwer, *Phys. Chem. Chem. Phys.*, 2010, *12*, 14253–14262.
20. A. Candini, D. Klar, S. Marocchi, V. Corradini, R. Biagi, V. De Renzi, U. del Pennino, F. Troiani, V. Bellini, S. Klyatskaya, M. Ruben, K. Kummer, N. B. Brookes, H. Huang, A. Soncini, H. Wende, M. Affronte, *Sci. Rep.*, 2016, *6*, 21740.
21. C. Wäckerlin, F. Donati, A. Singha, R. Baltic, S. Rusponi, K. Diller, F. Patthey, M. Pivetta, Y. Lan, S. Klyatskaya, M. Ruben, H. Brune, J. Dreiser, *Adv. Mater.*, 2016, *28*, 5195–5199.
22. M. Mannini, F. Pineider, C. Danieli, F. Totti, L. Sorace, P. Sainctavit, M. A. Arrio, E. Otero, L. Joly, J. C. Cezar, A. Cornia, R. Sessoli, *Nature*, 2010, *468*, 417–421.
23. N. Kernavanois, P. D. d. Réotier, A. Yaouanc, J. P. Sanchez, V. Honkimäki, T. Tschentscher, J. McCarthy, O. Vogt, *J. Phys. Cond. Mat.*, 2001, *13*, 9677.
24. Y. Saitoh, Y. Fukuda, Y. Takeda, H. Yamagami, S. Takahashi, Y. Asano, T. Hara, K. Shirasawa, M. Takeuchi, T. Tanaka, H. Kitamura, *J. Synchrotron Rad.*, 2012, *19*, 388–393.
25. K. O. Kvashnina, Y. O. Kvashnin, S. M. Butorin, *J. Electron. Spectrosc. Relat. Phenom.*, 2014, *194*, 27–36.
26. T. Vitova, K. O. Kvashnina, G. Nocton, G. Sukharina, M. A. Denecke, S. M. Butorin, M. Mazzanti, R. Caciuffo, A. Soldatov, T. Behrends, H. Geckeis, *Phys. Rev. B*, 2010, *82*.
27. T. Vitova, J. C. Green, R. G. Denning, M. Loble, K. Kvashnina, J. J. Kas, K. Jorissen, J. J. Rehr, T. Malcherek, M. A. Denecke, *Inorg. Chem.*, 2015, *54*, 174–182.
28. S. M. Butorin, K. O. Kvashnina, J. R. Vegelius, D. Meyer, D. K. Shuh, *Proc. Natl. Acad. Sci.*, 2016, *113*, 8093–8097.
29. S. M. Butorin, A. Modin, J. R. Vegelius, M.-T. Suzuki, P. M. Oppeneer, D. A. Andersson, D. K. Shuh, *Anal. Chem.*, 2016, *88*, 4169–4173.
30. P. Glatzel, U. Bergmann, *Coord. Chem. Rev.*, 2005, *249*, 65–95.
31. A. Kotani, S. Shin, *Rev. Mod. Phys.*, 2001, *73*, 203–246.

32. S. M. Butorin, *Actinide Nanoparticle Research* (Eds. S. N. Kalmykov, M. A. Denecke), Springer Berlin Heidelberg, Berlin, Heidelberg, 2011, pp. 63–103.
33. K. O. Kvashnina, Y. O. Kvashnin, J. R. Vegelius, A. Bosak, P. M. Martin, S. M. Butorin, *Anal. Chem.*, 2015, *87*, 8772–8780.
34. E. S. Ilton, P. S. Bagus, *Surf. Interface Anal.*, 2011, *43*, 1549–1560.
35. P. S. Bagus, C. J. Nelin, Y. Al-Salik, E. S. Ilton, H. Idriss, *Surf. Sci.*, 2016, *643*, 142–149.
36. B. L. Scott, J. J. Joyce, T. D. Durakiewicz, R. L. Martin, T. M. McCleskey, E. Bauer, H. Luo, Q. Jia, *Coord. Chem. Rev.*, 2014, *266–267*, 137–154.

https://doi.org/10.1142/9781800610163_0013

Chapter 13

Electronic Spectroscopy on Non-Metallic Actinide Systems

*Yvonne Rechkemmer and Joris van Slageren**
Institut für Physikalische Chemie, Universität Stuttgart, Pfaffenwaldring 55, 70569 Stuttgart, Germany
**slageren@ipc.uni-stuttgart.de*

13.1 Methods

Electronic spectroscopy is one of the most useful experimental tools for probing the electronic structures of actinide compounds.[1] The most commonly applied methods are electronic absorption, magnetic circular dichroism (MCD) and luminescence spectroscopy. While absorption spectra, especially those recorded at low temperatures, provide information about excited state energies, luminescence spectroscopy is especially useful for the determination of energy levels arising from the ground multiplet.[1] In an MCD experiment, the differential absorption of left and right circularly polarised light in the presence of a longitudinal magnetic field is recorded.[2] In comparison to unpolarised light, the well-defined z-component of angular momentum of circularly polarised light leads to more specific selection rules, facilitating the assignment of experimentally observed transitions.[3]

13.2 Types of Transitions and Comparison to Lanthanide Systems

Electronic transitions occurring in the UV–Vis-NIR range include intra-configurational f-f-transitions, inter-configurational f-d-transitions as well as charge-transfer (CT) transitions.[1,4–6] The f-f-transitions correspond to electron rearrangement within the 5f-shell and are the greatest source of information concerning the electronic level structure of actinide compounds. Experimentally determined f-f-transition energies can be used for empirically modelling the free-ion and crystal-field (CF) interactions. In this parametric approach, the matrix elements of the corresponding Hamiltonians are written as products of angular and radial integrals, where the radial dependencies are described by adjustable parameters:[1,4,5,7–9]

$$\mathcal{H}_{\text{free ion}} = E_{\text{ave}} + \sum_{k=2,4,6} F^k f_k + \zeta_{5\text{f}} A_{\text{so}} + H_{\text{corr}} \tag{13.1}$$

and

$$V_{\text{cf}} = \sum_{k,q} B_{kq} C_q^{(k)} \tag{13.2}$$

The free-ion Hamiltonian $\mathcal{H}_{\text{free ion}}$ takes into account the spherically symmetric perturbations, described by E_{ave}, electrostatic repulsion, described by the product of the radial integrals F^k and the angular parts f_k, as well as spin–orbit interaction, given by the spin–orbit coupling constant $\zeta_{5\text{f}}$ and the angular part of the spin–orbit interaction A_{so}. The correction term H_{corr} accounts for configuration interaction, which is more important in actinide compounds compared to lanthanide-based systems. V_{cf} denotes the crystal-field potential being described as a linear combination of spherical tensor operators $C_q^{(k)}$ of rank k with the coefficients B_{kq}. The relevant values of k and q are restricted by the site symmetry of the actinide ion in a given compound.

In comparison to lanthanides, the larger extension of the 5f-shell and the stronger contribution of 5f-electrons to the bonding in actinide compounds result in reduced intra-configurational electrostatic interactions, leading to lower values of F^k. For analogous systems, the radial electrostatic integrals for actinides are about two thirds of those of lanthanides.[1,4] For a given actinide, the F^k values are a measure of the nephelauxetic effect, i.e. the expansion of the f-shell due to hybridisation and covalency.[4] The strength of the spin–orbit interaction for actinides is about twice that for analogous lanthanide systems, leading to a larger splitting of the ^{2S+1}L parental terms and thus in turn to a more pronounced mixing of states with the same J-values.[1,4] Since the spin–orbit coupling and the Coulomb interaction are of the same order of magnitude, neither the Russell–Saunders coupling scheme nor j-j-coupling correctly describe the electronic structure of actinide compounds. As recently demonstrated in a comparative study by Karbowiak and Rudowicz, calculations perform best in the intermediate coupling scheme, taking into account the mixing of LS terms.[10] Strictly speaking, labelling a multiplet as $^{2S+1}L_J$ becomes incomplete; however, this notation is often used for depicting the major component.[1,7,11] Due to the stronger interaction of 5f-electrons with the environment, also the crystal-field splittings in actinide systems are larger than those in comparable 4f-systems. The CF parameters for actinide compounds are approximately twice those of the corresponding lanthanide systems and can be of comparable size to the spin–orbit coupling constants.[4,11] The increased crystal-field splittings often lead to significant overlap between CF levels of different J-multiplets and thus to J-mixing. In CF calculations, J-mixing is accounted for by simultaneous diagonalisation of the free-ion and CF Hamiltonians.[1,4,7] As illustrated in Figure 13.1, the combined effects of reduced electrostatic interactions, increased spin–orbit coupling as well as increased CF splittings may lead to rather large densities of nested $5f^N$ states, which complicates the interpretation of actinide spectra in comparison to those of lanthanides. The small energy gaps between different J-multiplets not only lead to complex spectral patterns but also to reduced luminescence lifetimes due to phonon-induced relaxation.[1]

Most f-f-transitions are electric dipole transitions. This means they are Laporte-forbidden, but in non-centrosymmetric compounds they can gain intensity by admixing states of opposite parity into the $5f^N$-configuration, induced by the odd terms of the crystal-field potential.[1,4,5,7] f-f-transition intensities can be analysed within the framework of Judd–Ofelt theory, which has been subject of an extensive review by Görller-Walrand and Binnemans.[12] Due to the larger

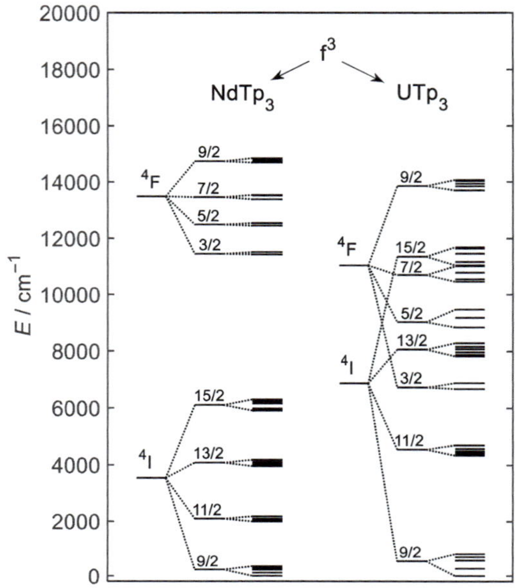

Figure 13.1. Comparison of energy levels arising from the $4f^3$ and $5f^3$ configurations in $NdTp_3$ and UTp_3 (both in D_{3h} symmetry), respectively. For reasons of clarity, only the states arising from the ^4F and ^4I terms are shown. Reproduced from Refs. [13] and [14].

interaction with environment compared to lanthanide systems, the induced intensity of actinide f-f-transitions is higher, e.g. the oscillator strengths in spectra of aquated An^{3+} ions are about 10–100 times greater than in the corresponding Ln^{3+} spectra.[1] The admixture of opposite parity states into the $5f^N$ configuration is of the order of 0.1%, which is large enough to partially allow the transitions, but sufficiently weak to preserve the f-f-character.[1,7] In centrosymmetric compounds, no intensity-inducing odd CF-terms appear; however, non-zero intensity is obtained by coupling to ungerade vibrational modes.[1,6–8] Such vibrational side-bands may also be observed in non-centrosymmetric systems and care has to be taken in distinguishing them from pure electronic transitions. Vibronic features are usually broader and weaker than the pure electronic f-f-transitions and show a different temperature-dependence.[7]

In addition to f-f-transitions, electronic spectra of actinide systems are governed by inter-configurational $5f^N$-$5f^{N-1}6d$ transitions.[1,4,5] Since the states involved are of opposite parity, these transitions are orbitally allowed and thus more intense than f-f-transitions. Furthermore, enhanced coupling to lattice vibrations leads to broad vibronic bands, even in the absence of inversion symmetry.[1,4,5] In comparison to the excited $4f^{N-1}5d$ configurations in lanthanides, the relative energies of $5f^{N-1}6d$ states in actinides are significantly lower. The corresponding transitions may be already observed in the visible range (above approx. 20,000 cm^{-1}) and overshadow the f-f-transitions, thus complicating spectral assignments.[6] Another complication arises from the occurrence of broad charge-transfer bands, where electron density is either transferred from the ligands to the unoccupied states of the actinide ions (LMCT) or *vice versa* (MLCT).[4–6] The maxima are usually observed in the UV region; however, the tails may very well reach far into the visible region. Generally, the energies of CT transitions depend on the electronegativity of the ligands, the electron affinity of the actinide ion and the metal ligand distances. The higher the covalency, the higher CT transition energies are expected.[4,5]

13.3 Actinyl Compounds and Related Systems

Due to their spectral similarities, actinyl compounds can be treated as a separate class in actinide electronic spectroscopy. Strictly speaking, actinyl compounds are only those containing an AnO_2^{2+} unit with the actinide adopting the oxidation state VI.[15] However, related systems are the AnO_2^{+} and AnO_2^{3+} building blocks, often referred to as actinyl(V) and actinyl(VII) ions.[15] They are all characterised by a nearly linear O–An–O symmetry where the axial crystal field plays a dominant role compared to electronic repulsion, spin–orbit coupling and the crystal field of equatorial ligands.[3,15–17]

The most thoroughly studied systems are those based on the UO_2^{2+} moiety, which is, due to the lack of 5f electrons, a typical LMCT system. Its electronic structure can be described in the framework of molecular orbital (MO) theory applying a molecular Russell–Saunders coupling scheme (electronic repulsion > spin–orbit coupling).[1,3,15–19] A typical simplified MO scheme based on the uranium 6d, 5f and oxygen 2p orbitals only is depicted in Figure 13.2. The highest occupied MOs are the mainly oxygen-based orbitals π_u, π_g, σ_g^+ and σ_u^+, which are rather close in energy. The ground state is

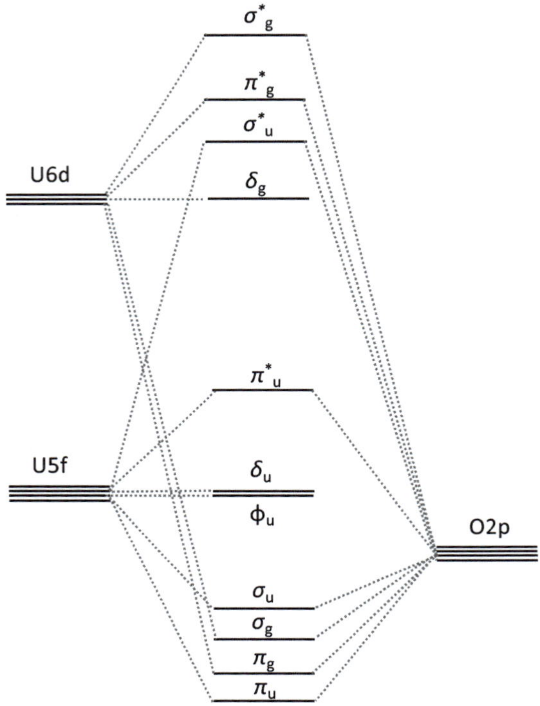

Figure 13.2. Schematic MO scheme for the uranyl ion UO_2^{2+}. Reproduced from Ref. [19].

thus given by $^1\Sigma_g^+$. The lowest unoccupied MOs are the non-bonding 5f-based $5f\delta_u$ and $5f\phi_u$ orbitals and excitations to these orbitals lead to absorption spectra with characteristic band structures between ca. 20,000 and 30,000 cm^{-1}.[3,15–20] Above ca. 27,000–30,000 cm^{-1} usually a structureless continuum is observed.[17] The observed lower-energy bands are attributed to the transitions $\Sigma_g^+ \Delta_g$ $(\sigma_u^+ \delta_u)$ and Σ_g^+ Φ_g $(\sigma_u^+ \phi_u)$, which are Laporte-forbidden, but can gain intensity by the removal of inversion symmetry by the presence of equatorial ligands.[3,15–20] Furthermore, electronic absorption in actinyl compounds is accompanied by pronounced vibrational progression, which is mainly attributed to the symmetrical O–U–O stretching mode with typical stretching frequencies of 750–900 cm^{-1}.[1,3,6,20–22] In high-resolution spectra of well-defined systems, e.g. single crystals, other vibronic lines due to coupling with other ligand vibrational modes including the bending deformations (in the range of 200 cm^{-1}) may be observed.[1,23]

Due to the large energy gap between the ground state and the first excited CT state the excited states of uranyl compounds are relatively long lived and in contrast to many other actinide systems, strong luminescence may be observed even in solution.[1] The corresponding emission peaks are usually observed between 16,000 and 21,000 cm^{-1} and similarly to absorption, they often show vibrational fine structure as well. However, the distance between the individual vibronic lines is higher because of the lower force constant of the excited states compared to the ground state.[1,21,23]

The MCD spectra of uranyl compounds are governed by progressions of temperature-independent derivative-shaped Faraday A-terms, as expected for non-degenerate ground states and degenerate excited states being involved in the transitions. It has been shown that the intensities and signs of these A-terms strongly depend on the coordination symmetry of the equatorial ligands and sign prediction is possible by applying Judd–Ofelt theory, facilitating spectral assignments.[3,20] Accordingly, the A-term sign can be predicted with the help of Equation (13.3), as follows:

$$\Delta\Omega = -(\rho + q) \tag{13.3}$$

where Ω is the projection of the total angular momentum along the O–U–O axis, q is the intensity-inducing component of the odd CF potential and $\rho = -1$ represents left circularly polarised light.[3,20]

In contrast to the UO_2^{2+} ion, the other actinyl ions AnO_2^{2+} (with An = Np, Pu, Am) as well as the related systems AnO_2^{+} and AnO_2^{3+} (with An = U, Np, Pu, Am) have partially filled $5f\delta_u/5f\phi_u$ orbitals and f-f-transitions may be observed.[1,15,23–26] According to Figure 13.2, the lowest-energy f-f-transitions are those between these partially filled orbitals.[15] Higher-energy f-f-transitions correspond to excitations to $5f\pi_u$; however, they lie in the same energy range as the LMCT transitions and thus may be overshadowed.[15,24–26]

13.3.1 *f^0: U(VI), Np(VII)*

In addition to U(VI) in uranyl compounds, spectra of $5f^0$ systems have been spectroscopically studied, e.g. in the halides UX_6 (with $X^- = F^-, Cl^-$). Typical LMCT transitions are observed above ca.

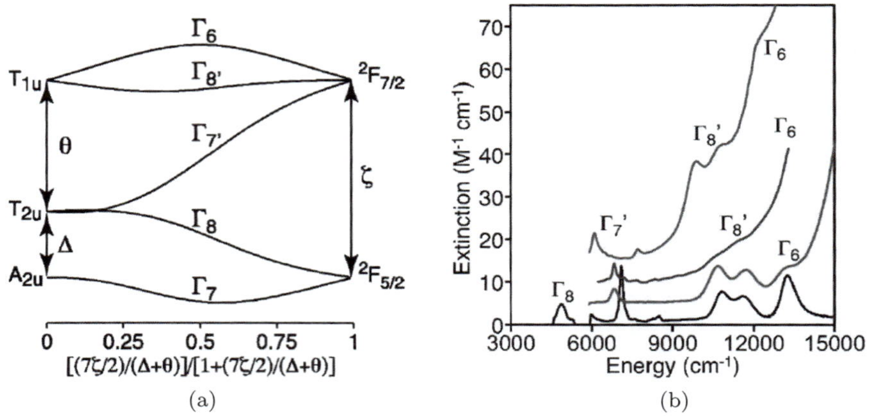

Figure 13.3. (a) Energy levels in octahedral f^1 systems depending on the relative size of the crystal-field splitting and the spin–orbit interaction. (b) NIR absorption spectra of octahedral UX_6^- complexes with ketimide, alkyl, amide and alkoxide ligands (from top to bottom). Reprinted from Ref. [28]. Copyright (2013) with permission from American Chemical Society.[28]

24,000 cm^{-1} and may show vibrational fine structure due to the symmetrical stretching mode.[1,27]

13.3.2 *f^1: Pa(IV), U(V), Np(VI)*

The $5f^1$ configuration is the simplest one showing actinide-based f-f-transitions. Since there is no electron-electron repulsion to be considered, the energy level structure is governed by spin–orbit coupling and crystal-field effects, which have the same order of magnitude. The ^2F ground term is split by spin–orbit coupling into the two multiplets $^2F_{7/2}$ and $^2F_{5/2}$ with the latter being lower in energy (Figure 13.3(a)).[1,6,28,29] Depending on the point group symmetry of the actinide ion, these multiplets are then further split by the crystal-field into maximally four and three Kramers doublets, respectively.[29]

However, the simplicity of the $5f^1$ configuration may turn out to be a disadvantage regarding CF parametrisation because the restricted number of spectroscopically accessible energy levels might be insufficient for the reliable determination of all the required parameters. This is one of the reasons why the majority of cf analyses for f^1 systems has been performed on highly symmetrical compounds, e.g. the halides UX_6^- exhibiting a nearly octahedral point symmetry[1] or more recently, on a series of octahedral UX_6^- complexes with alkyl, alkoxide, amide and ketimide ligands.[28] In octahedral symmetry, only two CF parameters appear, commonly denoted as θ and Δ.[28] As shown in Figure 13.3(a), an octahedral crystal field leads to a splitting of the 5f orbitals into an orbital singlet A_{2u} ground state and the excited triplets T_{2u} and T_{1u}. Switching on spin–orbit coupling then results in further splitting and mixing, finally producing the states Γ_7, Γ_8, $\Gamma_{7'}$, $\Gamma_{8'}$ and Γ_6, with Γ_8 and $\Gamma_{8'}$ being orbitally two-fold degenerate.[28]

The expected f-f-transitions generally appear in the IR and NIR region;[28,29] however, the observation of the transition to the first excited state Γ_8 is often precluded by technical limitations or solvent absorption while the transitions to the highest-lying states $\Gamma_{8'}$ and Γ_6 may be hidden by CT bands.[29] As an example, Figure 13.3(b) shows the NIR spectra of the aforementioned UX_6^- complexes.[28]

13.3.3 *f^2: U(IV), Np(V), Pu(VI)*

$5f^2$ systems possess much more complicated electronic structures compared to $5f^1$ systems and a wealth of f-f-transitions may be observed in experimental spectra. Tetravalent uranium has been studied in various crystal lattices and clear f-f-transitions are usually observed in the NIR absorption spectra. Above ca. 30,000 cm^{-1} the more intense f-d transitions in combination with CT transitions may lead to a UV cut-off.[1,30,31] U^{4+} solid-state luminescence is only rarely observed;[7] however, well-known exceptions are, e.g. U^{4+} doped into ThX_4 (X = Cl, Br), $ThSiO_4$ or $LiYF_4$, where emission from the excited $5f^1 6d^1$ configuration to the ground state $^3H_4$ of the $5f^2$ configuration occurs in the visible and near-UV range.[7,23] The f^2 configuration represents a non-Kramers system and the ground CF state is often characterised by a magnetic total angular momentum quantum number of $m_J = 0$. In this case, the low-temperature MCD spectra are dominated by temperature-independent Faraday A-terms, evidencing the non-degenerate ground state. Similarly to the A-terms of uranyl compounds, the signs and intensities of the corresponding signals are useful for assigning the states involved, as shown, e.g. for the organometallic U(IV) complexes $[Cp_3UX_2]^q$ (with $q = -1, 0, +1$ and $X = NCS^-$, $NCBH_3^-$, H_2O and D_2O).[32]

13.3.4 f^3: U(III), Np(IV), Pu(V)

The f-f-transitions arising from the $5f^3$ configuration have been subjected to numerous crystal field analyses, most of them being performed on ionic systems, e.g. on U^{3+} doped into $LaCl_3$ or on Np^{4+}: $ThSiO_4$.[1,7,33–38] The first CF analysis for a molecular U(III) compound, namely for the single-ion magnet $U(Tp)_3$ (with Tp^- = trispyrazolylborate), has been reported in 2010.[13] Absorption spectra of U(III) compounds are usually divided into two main regions: Below approximately 15,000 cm^{-1} (ca. 13,400 cm^{-1} for UI_3, ca. 23,000 cm^{-1} for UF_3) only relatively sharp peaks due to f-f-transitions appear, while at higher energies these peaks overlap with those arising from f-d-transitions (Figure 13.4).[11,13,35,37,39] In the initial cf fitting step, only the clearly assignable peaks should be included.[34,40] Since U(III) easily oxidises to U(IV) the spectra may show additional peaks arising from the $5f^2$ configuration, which have to be carefully separated.[34,35,39,41]

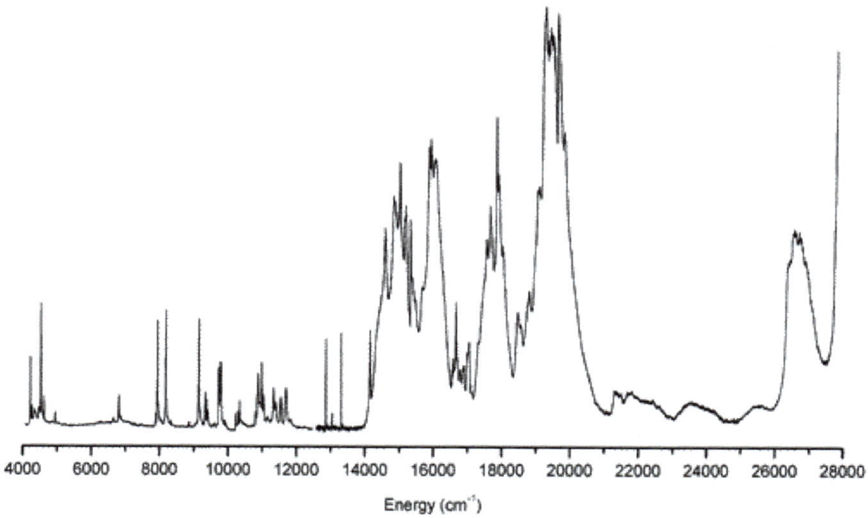

Figure 13.4. Absorption spectrum of U^{3+}:$CsCdBr_3$ at 7 K. Reprinted from Ref. [35]. Copyright (2004) with permission from American Chemical Society.[35]

Many U(III) and Np(IV) compounds show intense luminescence and the corresponding f-f-emission spectra are observed either by excitation of the f-d-transitions or direct f-f-excitation.[7,11,34,37,39,40,42] Luminescence spectroscopy proved to be an extraordinarily useful method for the electronic structure determination of U(III) systems since it not only allows for the ground state energy level determination but also for distinguishing lines belonging to different U(III) sites or those of U(III) and U(IV).[35,39] Line separation is for example possible by recording site-selective excitation spectra or by comparing luminescence lifetimes. Frequently, upconversion emission is observed, e.g. in $K_2LaI_5{:}U^{3+}$ green emission was observed after excitation of the $^2H_{9/2}$ multiplet either by red laser light or by using IR radiation.[11,37] Upconversion in U(III) may be caused either by cross-relaxation, i.e. the energy exchange between two nearby U(III) centres, or by excited state absorption (ESA).[37] It was pointed out that ESA is more effective in actinides compared to lanthanides due to the participation of parity-allowed f-d-transitions in this process.[35,39] The proximity of the $5f^26d^1$ levels was also made responsible for the pronounced thermal luminescence quenching.[35,39] Further quenching mechanisms include multi-phonon relaxation, cross-relaxation or reabsorption by coexistent U(IV).[11,39]

13.3.5 *f^4: Np(III), Pu(IV)*

Optical spectra of Np(III) and Pu(IV) have been reported for the aquated ions and for several solid-state systems such as the halides or Pu^{4+} doped into $ThSiO_4$.[1,7,33,38] The spectra of $Np^{3+}{:}LaCl_3$ and $NpCl_3$ were part of a comprehensive study of the trivalent actinide halides by Carnall.[33] However, apart from those, the reliable determination of free-ion and CF parameters for the $5f^4$ configuration based on optical spectroscopy is very rare due to the high density of states above 10,000 cm^{-1}.[7]

13.3.6 *f^5: Pu(III), Am(IV)*

The optical spectra obtained for the $5f^5$ configuration are extremely complex due to the high density of states and similarly to the $5f^4$ configuration, thorough analyses are essentially limited to $PuCl_3$ and Pu^{3+}: $LaCl_3$.[33,38] In spite of the numerous data available for these compounds, a lack of sensitivity was found for some of the fit parameters because the large energy range of the $5f^5$ configuration (up to ca. 90,000 cm^{-1}) precludes the spectroscopic determination of all the transitions.[33,38] Recently, the electronic absorption spectrum of the new plutonium molybdate $PuMoO_4Br(H_2O)$ was published and the characteristic f-f-transitions around 580, 620 and 900 nm were used for confirming the +IV oxidation state.[43]

13.3.7 *f^6: Am(III)*

Am(III) is the actinide analogue of Eu(III) and is characterised by a non-degenerate $^7F_0$ ground state. The first excited state arising from the $^7F_1$ multiplet lies above 2,000 cm^{-1} (e.g. 2,100 cm^{-1} for Am^{3+}: $LaCl_3$) and no temperature-dependent transitions from thermally populated excited states occur.[33] However, extensive vibronic structure may be present in the spectra and line sharpening as well as intensity increasing of electronic transitions can be observed at low temperature.[33] Am(III) is especially known for its luminescence properties, e.g. for Am^{3+} in aqueous solution excitation at 503 nm leads to four emission bands between 570 and 1,100 nm.[23] These bands are attributed to transitions from the metastable excited $^5D_1$ multiplet to the lower-lying ^7F_J terms (with $J = 0, 1, 2, 3$).[1,23] Based on the results of absorption and luminescence spectroscopy, crystal-field analyses have been published for Am^{3+} doped into different host lattices, e.g. Am^{3+}: $LaCl_3$ or more recently, Am^{3+}: $LiYF_4$.[33,44]

13.3.8 *f^7: Cm(III), Bk(IV)*

The electronic ground state for the $5f^7$ systems Cm(III) and Bk(IV) is given by $^8S_{7/2}$, which is an orbitally non-degenerate state. However, compared to the lanthanide analogue Gd^{3+}, the larger spin–orbit interaction leads to significant admixing of higher lying $J = 7/2$ states, resulting in a comparatively large crystal-field splitting.[1] For instance, an overall ground state cf splitting of ca. 40 cm^{-1} was found in Cm^{3+}: ThO_2 and an even larger one of ca. 60 cm^{-1} in Bk^{4+}: CeF_4.[1,45] The first excited multiplet is well separated from the ground state (ca. 16,000 cm^{-1}), which makes Cm^{3+} the most studied actinide ion in luminescence spectroscopy.[23,45,46] Furthermore, the metastability of the excited state allows for excited state absorption studies and thus in turn for the experimental determination of the energy levels above 40,000 cm^{-1}, which are usually not accessible by conventional absorption spectroscopy.[45] The combined application of absorption, laser-selective excitation and luminescence spectroscopy may thus lead to a rather complete description of the energy level structure and electronic structure determination has been performed, e.g. for Cm^{3+} doped into $LaCl_3$, $LuPO_4$, $CsCdBr_3$, ThO_2 and Cs_2NaYCl_6.[1,45]

Time-resolved laser-luminescence measurements proved to be a useful tool for the speciation of Cm(III), i.e. the characterisation and quantification of chemically different Cm(III) species in solutions or mineral phases.[45] If the respective Cm(III) species show different luminescence lifetimes, multi-exponential decay is observed. Sensitised luminescence, i.e. luminescence after the energy transfer from an excited ligand leads to increased intensities and provides additional information about the coordination of the Cm(III) centre.[45] It is also worth mentioning that Cm^{3+} complexed by chiral ligands was the first actinide ion, for which recently sensitised circularly polarised 5f-5f-emission was reported.[46]

References

1. G. Liu, J. V. Beitz, *The Chemistry of the Actinide and Transactinide Elements* (Ed. L. R. Morss), Vol. 3, Ch. 18, Springer, Dordrecht, 2006, pp. 2013–2111.
2. S. B. Piepho, P. Schatz, *Group Theory in Spectroscopy with Applications to Magnetic Circular Dichroism*, Wiley-Interscience, New York, 1983.
3. C. Görller-Walrand, W. Colen, *Inorg. Chim. Acta*, 1984, *94*, 183–188.
4. J. C. Krupa, *J. Alloys Compd.*, 1995, *225*, 1–10.
5. J. C. Krupa, *J. Solid State Chem.*, 2005, *178*, 483–488.
6. S. A. Cotton, *Lanthanides & Actinides*, Macmillan Education Ltd, Houndmills, 1991.
7. J. C. Krupa, *Inorg. Chim. Acta*, 1987, *139*, 223–241.
8. C. Görller-Walrand, K. Binnemans, *Handbook on the Physics and Chemistry of Rare Earths* (Eds. A. Gschneider, L. Eyring), Vol. 23, Elsevier, Amsterdam, 1996.
9. B. G. Wybourne, *Spectroscopic Properties of Rare Earths*, Interscience Publ., New York, 1965.

10. M. Karbowiak, C. Rudowicz, *Polyhedron*, 2015, *93*, 91–98.
11. H. P. Andres, K. Kramer, H. U. Gudel, *Phys. Rev. B*, 1996, *54*, 3830–3840.
12. C. Görller-Walrand, K. Binnemans, in *Handbook on the Physics and Chemistry of Rare Earths* (Eds. A. Gschneider, L. Eyring) Vol. 25, Elsevier, Amsterdam, 1998.
13. C. Apostolidis, A. Morgenstern, J. Rebizant, B. Kanellakopulos, O. Walter, B. Powietzka, M. Karbowiak, H. Reddmann, H. D. Amberger, *Z. Anorg. Allg. Chem.*, 2010, *636*, 201–208.
14. H. Reddmann, C. Apostolidis, O. Walter, D. H. Amberger, *Z. Anorg. Allg. Chem.*, 2006, *632*, 1405–1408.
15. S. Matsika, Z. Zhang, S. R. Brozell, J.-P. Blaudeau, Q. Wang, R. M. Pitzer, *J. Phys. Chem. A*, 2001, *105*, 3825–3828.
16. C. Görller-Walrand, L. G. Vanquickenborne, *J. Chem. Phys.*, 1972, *57*, 1436–1440.
17. C. Görller-Walrand, L. G. Vanquickenborne, *J. Chem. Phys.*, 1971, *54*, 4178–4186.
18. R. G. Denning, *J. Phys. Chem. A*, 2007, *111*, 4125–4143.
19. E. van Besien, K. Pierloot, C. Görller-Walrand, *Phys. Chem. Chem. Phys.*, 2006, *8*, 4311–4319.
20. C. Görller-Walrand, W. Colen, N. Q. Dao, *J. Chem. Phys.*, 1982, *76*, 13–19.
21. K. Binnemans, H. De Leebeeck, C. Görller-Walrand, J. L. Adam, *J. Phys. Cond. Mat.*, 1999, *11*, 4283–4287.
22. C. Görller-Walrand, S. De Jaegere, *Spectrochim. Acta*, 1972, *A28*, 257–268.
23. L. S. Natrajan, *Coord. Chem. Rev.*, 2012, *256*, 1583–1603.
24. R. Copping, C. Talbot-Eeckelaers, D. Collison, M. Helliwell, A. J. Gaunt, I. May, S. D. Reilly, B. L. Scott, R. D. McDonald, O. A. Valenzula, C. J. Jones, M. J. Sarsfield, *Dalton Trans.*, 2009, 5609–5611.
25. G. K. Liu, S. A. Wang, *J. Phys. Chem. A*, 2012, *116*, 8297–8302.
26. N. M. Edelstein, *J. Phys. Chem. A*, 2015, *119*, 11146–11153.
27. W. B. Lewis, L. B. Asprey, L. H. Jones, R. S. McDowell, S. W. Rabideau, A. H. Zeltmann, *J. Chem. Phys.*, 1976, *65*, 2707–2714.
28. W. W. Lukens, N. M. Edelstein, N. Magnani, T. W. Hayton, S. Fortier, L. A. Seaman, *J. Am. Chem. Soc.*, 2013, *135*, 10742–10754.
29. C. R. Graves, P. Yang, S. A. Kozimor, A. E. Vaughn, D. L. Clark, S. D. Conradson, E. J. Schelter, B. L. Scott, J. D. Thompson, P. J. Hay, D. E. Morris, J. L. Kiplinger, *J. Am. Chem. Soc.*, 2008, *130*, 5272–5285.

30. E. J. Schelter, P. Yang, B. L. Scott, J. D. Thompson, R. L. Martin, P. J. Hay, D. E. Morris, J. L. Kiplinger, *Inorg. Chem.*, 2007, *46*, 7477–7488.
31. K. Binnemans, I. Couwenberg, H. De Leebeeck, C. Görller-Walrand, J. L. Adam, *J. Alloys Compd.*, 1999, *285*, 105–111.
32. H. D. Amberger, R. D. Fischer, K. Yunlu, *Organometallics*, 1986, *5*, 2109–2114.
33. W. T. Carnall, *J. Chem. Phys.*, 1992, *96*, 8713–8726.
34. N. Karbowiak, A. Mech, J. Drożdżyński, Z. Gajek, N. M. Edelstein, *New J. Chem.*, 2002, *26*, 1651–1657.
35. M. Karbowiak, J. Drożdżyński, N. M. Edelstein, S. Hubert, *J. Phys. Chem. B*, 2004, *108*, 160–170.
36. M. Karbowiak, J. Drożdżyński, *J. Alloys Compd.*, 2000, *300*, 329–333.
37. J. Drożdżyński, *Coord. Chem. Rev.*, 2005, *249*, 2351–2373.
38. W. T. Carnall, B. G. Wybourne, *J. Chem. Phys.*, 1964, *40*, 3428–3433.
39. M. Karbowiak, A. Mech, J. Drożdżyński, N. M. Edelstein, *Phys. Rev. B*, 2003, *67*, 195108.
40. P. Deren, M. Karbowiak, J. C. Krupa, J. Drożdżyński, *J. Alloys Compd.*, 1998, *275*, 393–397.
41. T. Yamamura, K. Shirasaki, D. X. Li, Y. Shiokawa, *J. Alloys Compd.*, 2006, *418*, 139–144.
42. W. A. Hargreaves, *Phys. Rev. B*, 1970, *2*, 2273–2284.
43. J. N. Cross, S. K. Cary, J. T. Stritzinger, M. J. Polinski, T. E. Albrecht-Schmitt, *Inorg. Chem.*, 2014, *53*, 3148–3152.
44. R. Cavellec, S. Hubert, E. Simoni, *J. Solid State Chem.*, 1997, *129*, 189–195.
45. N. M. Edelstein, R. Klenze, T. Fanghanel, S. Hubert, *Coord. Chem. Rev.*, 2006, *250*, 948–973.
46. G.-L. Law,C. M. Andolina, J. Xu, V. Luu, P. X. Rutkowski, G. Muller, D. K. Shuh, J. K. Gibson, K. N. Raymond, *J. Am. Chem. Soc.*, 2012, *134*, 15545–15549.

https://doi.org/10.1142/9781800610163_0014

Chapter 14

Vibrational Spectroscopy of Non-Metallic Actinide Compounds

*Yvonne Rechkemmer and Joris van Slageren**
Institut für Physikalische Chemie, Universität Stuttgart, Pfaffenwaldring 55, 70569 Stuttgart, Germany
**slageren@ipc.uni-stuttgart.de*

14.1 General

The vast majority of vibrational spectroscopic investigations in actinide complexes and compounds has been concerned with compounds of the uranyl fragment, especially in the hexavalent f^0 state of uranium. These studies fall into two groups. The first group is that which deals with the investigation of the nature of the bonding in uranyl complexes, both from a fundamental viewpoint, as well as for designing novel compounds and understanding reactivity. The second group is focused on the study of uranyl speciation in aqueous solutions, especially from an environmental viewpoint. We will discuss first uranyl and other actinyl vibrational studies and then briefly address work on non-actinyl actinide compounds.

Uranyl ions are unique moieties that have no counterpart in lanthanide chemistry. Their electronic structure is dealt with elsewhere in this volume. They are almost always very close to linear with very short U–O bonds of ca. 1.78 Å. The nature of the ligands in the equatorial plane has a strong influence on the uranyl vibrational spectrum, and, thus, vibrational spectroscopy is a very sensitive tool to investigate the electronic structure and bonding of uranyl complexes. A free uranyl ion would possess three vibrational modes, namely a symmetric stretching mode (ν_1), an asymmetric stretching mode (ν_3) and a doubly degenerate bending mode (ν_2), just like CO_2. In rigorous inversion symmetry conditions, the symmetric stretching mode would be only Raman-active, the asymmetric stretching and the bending modes only IR-active. In real molecules, the symmetry is reduced and the mutual exclusion rule does not strictly hold. In acidic aqueous uranyl solutions, the main species is $[UO_2(H_2O)_5]^{2+}$, for which vibrational frequencies of $\nu_1 = 870\,\text{cm}^{-1}$, $\nu_2 = 254\,\text{cm}^{-1}$ and $\nu_3 = 962\,\text{cm}^{-1}$ have been determined.[1,2] The introduction of electron donating ligands in the equatorial plane causes substantial decreases in the frequencies of the stretching motions. Thus the uranyl symmetric stretching mode was reported at $825\,\text{cm}^{-1}$ for $[UO_2(C_2O_4)_3]^{4-}$,[3] at $808\,\text{cm}^{-1}$ for $[UO_2(THF)_2(O\text{-}2,6\text{-}Ph_2C_6H_3)_2]$[4] and at $786\ \text{cm}^{-1}$ for aqueous $[UO_2(OH)_5]^{3-}$.[5] A number of correlations involving especially ν_1 have been established. Thus, there is a correlation between ν_1 and the reduction potential of the uranyl(VI) species (Figure 14.1).[6] No correlation with ν_3 was found. Correlations were also established with the number of equal equatorial ligands for a number of different ligands, ranging from halides to carboxylates. These correlations followed the general form $\nu_1 = -A\langle n\rangle + 870\,\text{cm}^{-1}$, where $\langle n\rangle$ is the average number of ligands in aqueous solution.[3] The proportionality constant A is positive except for perchlorate. Linear correlations were also found between ν_1 and the overall stability constant of the complex log β_n for $n = 1 - 3$ equatorial ligands.[3] The nature of the equatorial ligands has a dramatic influence on the oxo-ligand exchange half-life decreasing from 40,000 h for the penta-aquo complex to 0.04 s for the penta-hydroxy compound.[5] On the other hand, it was found that there is no simple correlation between bond length and vibrational frequency, as would be expected from Badger's rule which states that the stretching force constant is inversely proportional to the third power of the distance.[7,8] This deviation was elegantly demonstrated by pressure experiments on $Cs_2UO_2Cl_4$[9] In fact, the physical origin of these correlations is not quite clear, and may be covalent or electrostatic in nature.[8] In the former case, the equatorial ligands interact covalently with the 6d metal orbitals, resulting in a change of the uranyl electronic structure. In the latter, the presence of electric charges in the neighbourhood of the

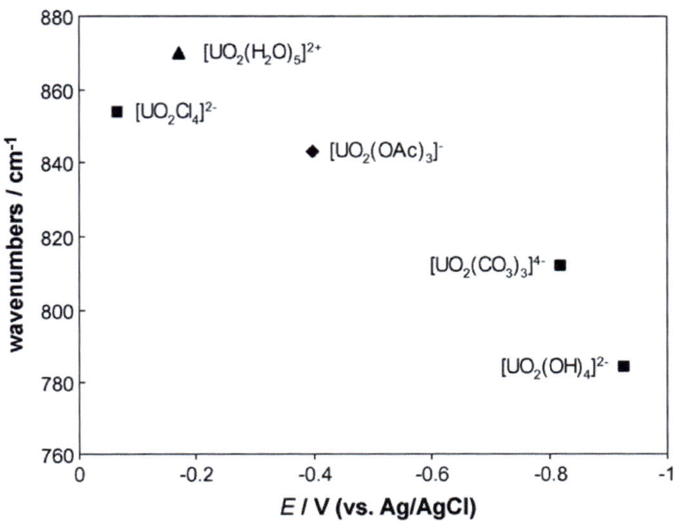

Figure 14.1. Correlation between reduction potential and ν_1 frequency for different uranyl (VI) complexes, taken from 10.1016/j.ccr.2009.06.003.

uranyl combined with charge transfer from the ligand to the uranyl moiety influence the stretching frequency. In this context, work on $[UO_2(L)_n]^{2+}$ where L are neutral ligands is interesting. In the gas-phase technique of infrared multiphoton dissociation (IRMPD), the influence of the equatorial ligands on the *asymmetric* stretching frequency was investigated. There appears to be a correlation between gas phase basicity (GB) of the ligand and a red-shift of the ν_3 frequency. Thus for L = CH_3CN (GB = 748 kJ mol^{-1}, ν_3 is 1019, 1007 and 995 cm^{-1} (for $n = 1, 2, 3$), while for L = acetone (GB = 782 kJ mol^{-1}), ν_3 frequencies of 1017, 1000, and 988 cm^{-1} were found.[10] A stronger red-shift was found for L = tetramethyl-3-oxa-glutaramide ($\nu_3 = 965$ for $n = 2$).[11] The change in IR frequency was attributed to charge donation from the ligand to the uranyl with concomitant bond weakening.[11] A strong correlation of both ν_2 and ν_3 with ligand basicity was found for the complexes $[UO_2(hfac)_2(L)]$ where hfac = hexafluoroacetylacetonate and L are various Lewis basic ligands, both in the vapour phase and in solution.[12]

For hexavalent actinide ions, the actinyl symmetric stretching frequencies strongly decrease with increasing atomic number. Thus, in aqueous solutions at low pHs, ν_1 frequencies of 863, 835 and

796 cm^{-1} for neptunyl(VI), plutonyl(VI) and americyl(VI) species were found.[13] In contrast, the asymmetric stretch is largely metal-atom-independent.[14] Similar findings were reported more recently for the complexes $M_2[AnO_2Cl_4]$ ($M^+ = Rb^+$, Cs^+, Me_4N^+; An = U, Pu).[8] Reduction of uranyl(VI) to uranyl(V) leads to a large red-shift of the ν_3 frequency from 895 to 770 cm^{-1} in the case of the complexes $[UO_2(saloph)DMSO]^-$ (saloph = *N*,*N*′-disalicylidene-*o*-phenylenediaminate, DMSO = dimethyl sulfoxide).[15]

The uranyl(VI) cation is water-soluble and thus of great importance to environmental studies and uranyl speciation under various condition is an important topic of research. While at low pH, the penta-aquo-uranyl(VI) species is predominant, at higher pH $[UO_2(OH)_n]^{2-n}$ ($n = 4, 5$) species form.[5] Uranyl speciation in the presence of a variety of environmentally relevant potentially ligating ions, such as carbonate, nitrate but also citrate, was studied by Raman spectroscopy.[3,16–18] The dimeric species $[(UO_2)_2CO_3(OH)_3]^-$ was identified by means of IR spectroscopy.[19] Such studies also bear relevance to the extraction of uranium, where a potential method involves functionalised ionic liquids. The extraction of uranyl from the aqueous phase by such an ionic liquid was confirmed by the appearance of a new band at 934 cm^{-1}, corresponding to the asymmetric stretching vibration of uranyl.[20]

Literature on vibrational studies of non-actinyl actinide complexes is far less abundant. Vibrational spectroscopy is most often used as chemical characterisation technique for novel molecules and the studies do not directly involve the actinide ion. Most in-depth studies involve species of importance in nuclear fuel processing, including halides and oxides. The gas phase structure of UCl_4 was after a period of controversy shown to be regularly tetrahedral partly by means of gas phase IR spectroscopy.[21] Here, two infrared-active modes, namely ν_3 at 337.4 and ν_4 at 71.7 cm^{-1} were found. Isotope effects, relevant for enrichment procedures were studied for the hexafluoride UF_6, where a 0.6 cm^{-1} shift for the Q branch of the $\nu3$ band in the gas phase was observed going from U-238 to U-235.[22] Uranium oxides are of relevance to storage and disposal of spent nuclear fuel. The species relevant to the gradual oxidation process $UO_4 \rightarrow U_4O_9 \rightarrow U_3O_8 \rightarrow UO_3$ were studied by IR spectroscopy already some years ago.[23]

References

1. F. Quilès, C. Nguyen-Trung, C. Carteret, B. Humbert, *Inorg. Chem.*, 2011, *50*, 2811–2823.
2. M. Gál, P. L. Goggin, J. Mink, *J. Mol. Struct.*, 1984, *114*, 459–462.
3. C. Nguyen-Trung, G. M. Begun, D. A. Palmer, *Inorg. Chem.*, 1992, *31*, 5280–5287.
4. M. P. Wilkerson, C. J. Burns, D. E. Morris, R. T. Paine, B. L. Scott, *Inorg. Chem.*, 2002, *41*, 3110–3120.
5. D. L. Clark, S. D. Conradson, R. J. Donohoe, D. W. Keogh, D. E. Morris, P. D. Palmer, R. D. Rogers, C. D. Tait, *Inorg. Chem.*, 1999, *38*, 1456–1466.
6. S. Fortier, T. W. Hayton, *Coord. Chem. Rev.*, 2010, *254*, 197–214.
7. V. Vallet, U. Wahlgren, I. Grenthe, *J. Phys. Chem. A*, 2012, *116*, 12373–12380.
8. D. D. Schnaars, R. E. Wilson, *Inorg. Chem.*, 2013, *52*, 14138–14147.
9. H. H. Osman, P. Pertierra, M. A. Salvado, F. Izquierdo-Ruiz, J. M. Recio, *Phys. Chem. Chem. Phys.*, 2016, *18*, 18398–18405.
10. G. S. Groenewold, A. K. Gianotto, K. C. Cossel, M. J. Van Stipdonk, D. T. Moore, N. Polfer, J. Oomens, W. A. de Jong, L. Visscher, *J. Am. Chem. Soc.*, 2006, *128*, 4802–4813.
11. J. K. Gibson, H.-S. Hu, M. J. Van Stipdonk, G. Berden, J. Oomens, J. Li, *J. Phys. Chem. A*, 2015, *119*, 3366–3374.
12. R. G. Bray, G. M. Kramer, *Inorg. Chem.*, 1983, *22*, 1843–1848.
13. L. J. Basile, J. C. Sullivan, J. R. Ferraro, P. LaBonville, *Appl. Spectrosc.*, 1974, *28*, 142–145.
14. K. Müller, H. Foerstendorf, S. Tsushima, V. Brendler, G. Bernhard, *J. Phys. Chem. A*, 2009, *113*, 6626–6632.
15. K. Mizuoka, Y. Ikeda, *Inorg. Chem.*, 2003, *42*, 3396–3398.
16. C. Madic, D. E. Hobart, G. M. Begun, *Inorg. Chem.*, 1983, *22*, 1494–1503.
17. G. Lu, T. Z. Forbes, A. J. Haes, *Anal. Chem.*, 2016, *88*, 773–780.
18. M. Basile, D. K. Unruh, K. Gojdas, E. Flores, L. Streicher, T. Z. Forbes, *Chem. Commun.*, 2015, *51*, 5306–5309.
19. K. Guckel, S. Tsushima, H. Foerstendorf, *Dalton Trans.*, 2013, *42*, 10172–10178.
20. P. K. Mohapatra, P. Kandwal, M. Iqbal, J. Huskens, M. S. Murali, W. Verboom, *Dalton Trans.*, 2013, *42*, 4343–4347.
21. A. Haaland, K.-G. Martinsen, O. Swang, H. V. Volden, A. S. Booij, R. J. M. Konings, *J. Chem. Soc. Dalton Trans.*, 1995, 185–190.

22. J. P. Aldridge, E. G. Brock, H. Filip, H. Flicker, K. Fox, H. W. Galbraith, R. F. Holland, K. C. Kim, B. J. Krohn, D. W. Magnuson, W. B. M. II, R. S. McDowell, C. W. Patterson, W. B. Person, D. F. Smith, G. K. Werner, *J. Chem. Phys.*, 1985, *83*, 34–48.
23. G. C. Allen, N. R. Holmes, *Appl. Spectrosc.*, 1994, *48*, 525–530.

https://doi.org/10.1142/9781800610163_0015

Chapter 15

Electron Paramagnetic Resonance of Non-Metallic Actinide Systems

*Yvonne Rechkemmer and Joris van Slageren**
Institut für Physikalische Chemie, Universität Stuttgart, Pfaffenwaldring 55, 70569 Stuttgart, Germany
** slageren@ipc.uni-stuttgart.de*

15.1 General

By applying an external magnetic field, the electronic microstates of a paramagnetic ion experience a splitting due to the Zeeman effect.[1–3] The Zeeman splitting, in some cases accompanied by hyperfine coupling due to the interaction with nuclear spins, contains additional information about the electronic structure and the bonding situation in actinide compounds. However, since the Zeeman splitting is much smaller compared to the main interactions of electrostatic repulsion, spin–orbit coupling (SOC) and the crystal field (CF), it usually cannot be studied by conventional optical methods such as luminescence spectroscopy where the bandwidths are often too large.[3] Instead, electron paramagnetic resonance (EPR) spectroscopy is applied, where transitions between the Zeeman split states are induced by oscillating magnetic fields at microwave frequencies.[1–3] Since the population difference between the states involved in the transitions is rather small and spin-lattice relaxation often increases rapidly with temperature, EPR spectra of actinide systems are usually observed only at low, e.g. liquid helium, temperatures.[2]

The direct consequence is that only the lowest-lying microstates contribute to the spectra and an effective spin approach can be used for their interpretation.[1–3] For instance, in the case of a well-separated ground Kramers doublet, the system may be characterised by an effective spin $\tilde{S} = 1/2$, although the states are actually a complex mixture of wave functions.[1–4] The corresponding simplified Hamiltonian, including only the effects of electronic Zeeman interaction and hyperfine coupling, is given by[1–3]

$$\mathcal{H} = \mu_{\mathrm{B}}(g_{\mathrm{x}}B_{\mathrm{x}}\tilde{S}_{\mathrm{x}} + g_{\mathrm{y}}B_{\mathrm{y}}\tilde{S}_{\mathrm{y}} + g_{\mathrm{z}}B_{\mathrm{z}}\tilde{S}_{\mathrm{z}}) + A_{\mathrm{x}}I_{\mathrm{x}}\tilde{S}_{\mathrm{x}} + A_{\mathrm{y}}I_{\mathrm{y}}\tilde{S}_{\mathrm{y}} + A_{\mathrm{z}}I_{\mathrm{z}}\tilde{S}_{\mathrm{z}} \tag{15.1}$$

where the first three terms the Zeeman splitting with μ_B the Bohr magneton, g_i (with $i = x, y, z$) the components of the effective g-factor along the principal axes and B_i the magnetic field components. The last three terms describe the hyperfine interaction with nuclear spins I, where A_i are the components of the hyperfine coupling constant describing the strength of the interaction. The g-values and the hyperfine splittings are rather sensitive to the state compositions. For 5f-ions, the increased SOC and reduced electronic repulsion compared to 4f-ions lead to a break-down of the Russell–Saunders coupling scheme and an intermediate coupling scheme accounting for the mixing of terms with the same total angular momentum quantum number J has to be applied for calculating the state compositions.[1–3] Another important aspect concerns the strength of the crystal field acting on the actinide ion. If the crystal field is only a weak perturbation, no significant J-J-mixing occurs and J remains a good quantum number. The two components of a Kramers doublet are then described by[1,3,4]

$$|\alpha\rangle = \sum\nolimits_{M_J} a_{M_J}\,|JM_J\rangle \text{ and } |\beta\rangle = \sum\nolimits_{M_J} (-1)^{J-M_J} a^*_{M_J}\,|JM_J\rangle \tag{15.2}$$

In stronger crystal fields J-J mixing has to be accounted for. Experimental evidence for J-J-mixing is, e.g. given by deviations from the relation[3]

$$\mathrm{A_x/g_x = A_y/g_y = A_z/g_z = A}_J/\mathrm{g}_J \tag{15.3}$$

where g_J is the Landé factor given by[1]

$$g_J = \frac{3}{2}\frac{L(L+1) - S(S+1)}{2J(J+1)} \tag{15.4}$$

For pronounced covalent bonding, CF theory is not valid and a more complicated treatment is necessary. Experimental indication of covalency in EPR spectra is, e.g. given by reduced g-values due to reduced spin–orbit coupling, decreased hyperfine coupling constants or the observation of superhyperfine pattern due to the interaction between the actinide ions and its neighbouring nuclei.[2]

15.1.1 f^1: Pa(IV), U(V), Np(VI)

$5f^1$ ions typically show rather small effective g-values with negative sign,[3] as can be shown for example by using circularly polarised radiation.[1,5] However, the individual spectral features strongly depend on the site symmetry of the studied actinide ion and the corresponding crystal field. In octahedral symmetry, the crystal field acting on a $5f^1$ ion is a rather strong perturbation and J is usually no longer a good quantum number. The ground state is therefore commonly labelled by the corresponding irreducible representation of the point group, which in octahedral symmetry is the doublet Γ_7.[1,3] For a pure $^2F_{5/2}$ multiplet in octahedral symmetry, the g-value of this doublet is expected to be isotropic and adopt a value of $g = -5/3\,g_J = -10/7$.[1,2] Experimentally, clearly lower g-values are found due to J-J-mixing, e.g. the EPR spectrum of Pa^{4+} in Cs_2ZrCl_6 was fitted with a g-value of -1.14 while that of Np^{6+} in UF_6 showed an even lower g-value of just -0.604.[1,2,5,6] Hyperfine structure due to the interaction with F^--ligands evidenced covalent contributions to the bonding. Deviations from octahedral symmetry leads to anisotropic g-values, as for example recently found in a series of U(V) complexes with halide, alkyl, alkoxide, amide and ketimide ligands, with mean g-values in the range of $|g| = 0.73-1.43$.[7] Generally, g-values lower than 1 indicate significant covalent bonding character.[3,4]

In distorted tetrahedral symmetry the situation is rather different. Here, the crystal field interaction is much weaker and the states may be described by Equation (15.2), neglecting J-J-mixing. Within the

$^2F_{5/2}$ ground multiplet and assuming C_{2v} symmetry, this leads to state compositions given by[4]

$$a\left|\frac{5}{2},\pm\frac{1}{2}\right\rangle + b\left|\frac{5}{2},\pm\frac{5}{2}\right\rangle + c\left|\frac{5}{2},\mp\frac{3}{2}\right\rangle \tag{15.5}$$

Using Equation (15.5) and the relations[4,8]

$$\begin{aligned} g_x &= \pm 6/7(2\sqrt{5}bc + 4\sqrt{2}ac + 3a^2) \\ g_y &= \pm 6/7(2\sqrt{5}bc - 4\sqrt{2}ac + 3a^2) \\ g_z &= \pm 6/7(5b^2 + a^2 - 3c^2) \end{aligned} \tag{15.6}$$

experimentally observed g-values can be used for the determination of ground state compositions, as has been, e.g. realised for a series of organouranium(V) amide compounds with distorted tetrahedral or bipyramidal symmetry.[4] However, the determined coefficients a, b, c showed slight deviation from the normalisation condition $a^2 + b^2 + c^2 = 1$, indicating small ligand admixtures.[4]

15.1.2 *f^3: U(III), Np(IV), Pu(V)*

EPR properties of trivalent uranium have been predominantly studied in halide host lattices such as $LaCl_3$, CaF_2 or SrF_2.[1,2,9,10] The similarity to the spectra of the $4f^3$ analogue Nd^{3+} confirmed that the lowest doublet in U^{3+} arises from a crystal-field split ground multiplet with the major component $^4I_{9/2}$. For instance, fitting the EPR spectra of U^{3+}: $LaCl_3$ using a common spin Hamiltonian and an effective spin of $\tilde{S} = 1/2$ yielded g-values of $g_{\|} = 4.15$ and $g_{\perp} = 1.52$, while the spectra of Nd^{3+}: $LaCl_3$ were fitted with similar values of $g_{\|} = 4.00$ and $g_{\perp} = 1.76$.[1,9] However, U^{3+} is a typical intermediate-coupled system and the ground multiplet contains also a significant contribution from $^2H_{9/2}$.[3] Furthermore, J-J mixing has to be taken into account. In the spectra of U^{3+} doped into fluoride host lattices, superhyperfine structure due to the fluorine nuclei was observed, which is a difference to most of the spectra of Nd^{3+} and indicates the participation of 5f-orbitals in the bonding.[2,10]

Very recently, EPR spectroscopy in combination with magnetometry has been employed for the cf analysis of the complex $(Cp'')_3U$

and its base adducts $(Cp'')_3U \cdot t$BuNC and $(Cp'')_3U \cdot$ CyNC as well as the corresponding neodymium analogues (with Cp″ = 1,3-bis-(trimethylsily)cyclopentadienyl, *t*BuNC = *tert*-butylisocyanide and CyCN = cyclohexylisocyanide).[11] Although full CF analyses are usually based on optical spectroscopic data, EPR spectroscopy and magnetometry proved to be a well-suited alternative, since the 5f-6d transitions in Cp_3U complexes lie in the visible range and tail into the near infrared, thus complicating the analysis of optical spectra.[11] The CF analyses based on EPR and magnetic susceptibility data were performed using a full crystal-field model, including the simultaneous diagonalisation of the free-ion, crystal-field and Zeeman Hamiltonians. The free-ion parameters were fixed to those determined spectroscopically for the related compound $U(Tp)_3$ (with Tp = hydrotris(1-pyrazolyl)borato)[12] and only the cf parameters, assuming D_{3h} or C_{3v} symmetry, were varied in the fitting procedure. The calculated susceptibility curves as well as the effective *g*-values agreed well with experiments and the results hinted at a significant participation of 5f-orbitals in the bonding, not only to the Cp-ligands but also to the isocyanide ligands.[11]

15.1.3 f^5: *Pu(III), Am(IV)*

EPR spectra of $5f^5$ ions have been mainly recorded on Pu^{3+} and Am^{4+} doped into host lattices providing cubic site symmetries, such as $SrCl_2$, CaF_2, SrF_2, BaF_2, ThO_2 and CeO_2.[2,13–17] Isotropic effective *g*-values were found and the comparison of different host lattices showed that these *g*-values increase with increasing crystal-field strength from $SrCl_2$ to CeO_2. For instance, a *g*-value of 1.297 was found for Pu^{3+} in CaF_2, while $g = 1.333$ was observed for Pu^{3+} in CeO_2. This trend is attributed to increased *J*-*J*-mixing induced by the stronger crystal field. Consistently, the hyperfine coupling constants decrease with increasing cf interaction. In $^{239}Pu^{3+}$: CaF_2 ($I = 1/2$), the hyperfine pattern is accompanied by a clear superhyperfine pattern due to the interaction with adjacent fluoride ions (Figure 15.1).[13,14] Examples of EPR studies performed on Pu^{3+} in non-cubic symmetry are, e.g. those on Pu^{3+}: $LaCl_3$, which was the first Pu^{3+} system studied by EPR, and Pu^{3+}: $LuPO_4$, where axial spectra were obtained.[18,19]

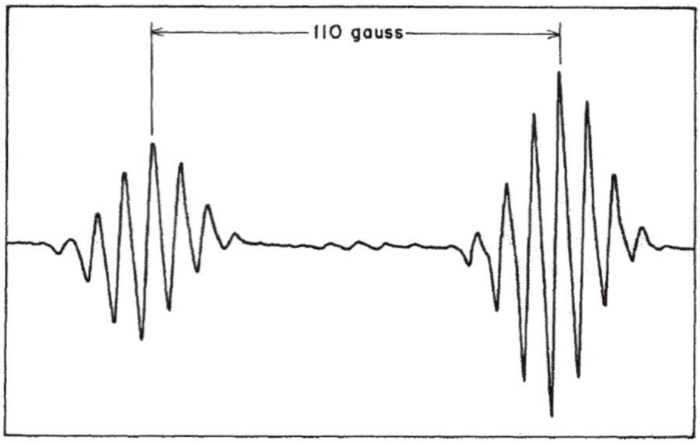

Figure 15.1. Isotropic EPR spectrum of $^{239}Pu^{3+}$ doped into CaF_2. The two hyperfine lines of $^{239}Pu^{3+}$ ($I = 1/2$) are further split by superhyperfine interaction with fluoride ions. Reprinted from Ref. [14]. Copyright (1971) with permission from the American Physical Society.

15.1.4 *f⁷: Am(II), Cm(III), Bk(IV)*

In a pure Russell–Saunders coupling scheme, the ground state of the f^7 configuration is given by $^8S_{7/2}$ without any angular orbital momentum; i.e. no crystal field splitting is expected. Experimentally observed CF splittings are therefore attributed to the admixing of higher-lying excited states into the ground state.[2] For the lanthanide analogue Gd^{3+}, such higher-order effects are rather small and according to the EPR selection rule $\Delta M_J = \pm 1$, its EPR spectra may show a characteristic seven-line pattern due to allowed intra- and inter-Kramers doublet transitions.[2] However, the situation looks different for the $5f^7$ ions Am^{2+}, Cm^{3+}, Bk^{4+}, where the combination of reduced electronic repulsion, increased spin–orbit coupling and increased CF strength leads to much more pronounced state mixing. As a consequence, the ground state CF splitting is larger compared to Gd^{3+} and usually not accessible by the frequencies applied in conventional EPR spectrometers (9.5 GHz). In other words, at low temperatures, only one resonance line due to transitions within the ground doublet will be observed.[2] A typical example is $^{244}Cm^{3+}$ in the cubic sites of CaF_2, showing a single resonance with an isotropic g-value of $|g| = 4.492$ at low fields.[1,20] This transition is attributed to a Γ_6 ground doublet, for which in the absence of state mixing a g-value of $|g| = 7g_J/3 = 4.667$ is expected.[1,2] At higher fields, the EPR spectra show slight anisotropy due to the field-induced admixing of the Γ_8 excited state and the angular dependence of the observed lines can serve to determine the zero-field energy gap between the ground doublet and this excited state.[1,2,20]

15.1.5 f^2, f^4, f^6

The configurations f^2, f^4 and f^6 represent non-Kramers systems, i.e. depending on the symmetry, the degeneracy of the electronic microstates may be completely lifted even in the absence of an external magnetic field. Electron paramagnetic resonance at conventional frequencies is therefore only rarely observed; however, some examples have been reported.[2,3,21–23] One of them is the observation of resonance lines due to U^{4+} (f^2) in trigonal sites of CaF_2.[21–23] Those resonances show an asymmetric shape typical for non-Kramers systems and larger effective g-values than possible for U^{2+}, confirming the tetravalent state.[2]

Another example is the EPR spectrum of the plutonyl ion PuO_2^{2+} in $(UO_2)Rb(NO_3)_3$.[1,2,24] According to Eisenstein and Pryce,[25] its ground doublet can be described by $L_z = \pm5$ and $S_z = \mp1$, for which a parallel g-component of $g_{\|} = 2(5 - 2) = 6$ is expected. The experimentally determined g-value was $g_{\|} = 5.32$ and the deviation was attributed to covalent effects and slight admixture of excited states into the ground state. An orbital reduction factor k accounting for these effects was therefore introduced, modifying the g-value to $g_{\|} = 2(5k–2)$. The experimentally determined g-value thus corresponds to an orbital reduction factor of $k = 0.93$.[1,2,25]

References

1. A. Abragam, B. Bleaney, *Electron Paramagnetic Resonance of Transition Ions*, Oxford University Press, Oxford, 2012.
2. L. A. Boatner, M. M. Abraham, *Rep. Prog. Phys.*, 1978, *41*, 87–155.
3. I. Ursu, A. Lupei, *Bull. Magn. Reson.*, 1984, *6*, 162–224.
4. D. Gourier, D. Caurant, J. C. Berthet, C. Boisson, M. Ephritikhine, *Inorg. Chem.*, 1997, *36*, 5931–5936.
5. C. A. Hutchison, B. Weinstock, *J. Chem. Phys.*, 1960, *32*, 56–61.
6. J. D. Axe, H. J. Stapleton, C. D. Jeffries, *Phys. Rev.*, 1961, *121*, 1630.

7. W. W. Lukens, N. M. Edelstein, N. Magnani, T. W. Hayton, S. Fortier, L. A. Seaman, *J. Am. Chem. Soc.*, 2013, *135*, 10742–10754.
8. S. D. McLaughlan, P. A. Forrester, *Phys. Rev.*, 1966, *151*, 311–314.
9. C. A. Hutchison, P. M. Llewellyn, E. Wong, P. Dorain, *Phys. Rev.*, 1956, *102*, 291–292.
10. B. Bleaney, P. M. Llewellyn, D. A. Jones, *Proc. Phys. Soc. Lond. B*, 1956, *69*, 858–860.
11. W. W. Lukens, M. Speldrich, P. Yang, T. J. Duignan, J. Autschbach, P. Kögerler, *Dalton Trans.*, 2016, *45*, 11508–11521.
12. C. Apostolidis, A. Morgenstern, J. Rebizant, B. Kanellakopulos, O. Walter, B. Powietzka, M. Karbowiak, H. Reddmann, H. D. Amberger, *Z. Anorg. Allg. Chem.*, 2010, *636*, 201–208.
13. N. Edelstein, H. F. Mollet, W. C. Easley, R. J. Mehlhorn, *J. Chem. Phys.*, 1969, *51*, 3281–3285.
14. W. Kolbe, N. Edelstein, *Phys. Rev. B*, 1971, *4*, 2869–2875.
15. L. A. Boatner, M. M. Abraham, C. B. Finch, R. W. Reynolds, *Bull. Am. Phys. Soc.*, 1971, *16*, 361.
16. M. M. Abraham, L. A. Boatner, C. B. Finch, R. W. Reynolds, *Phys. Rev. B*, 1971, *3*, 2864–2868.
17. W. Kolbe, N. Edelstein, C. B. Finch, M. M. Abraham, *J. Chem. Phys.*, 1974, *60*, 607–609.
18. H. Lammermann, H. J. Stapleton, *J. Chem. Phys.*, 1961, *35*, 1514–1516.
19. W. K. Kot, N. M. Edelstein, M. M. Abraham, L. A. Boatner, *Phys. Rev. B*, 1993, *47*, 3412–3414.
20. N. Edelstein, W. Easley, *J. Chem. Phys.*, 1968, *48*, 2110.
21. R. S. Title, P. P. Sorokin, M. J. Stevenson, G. D. Pettit, J. E. Scardefield, J. R. Lankard, *Phys. Rev.*, 1962, *128*, 62–66.
22. A. Yariv, *Phys. Rev.*, 1962, *128*, 1588–1592.
23. S. D. McLaughlan, *Phys. Rev.*, 1966, *150*, 118–120.
24. B. Bleaney, P. M. Llewellyn, M. H. L. Pryce, G. R. Hall, *Phil. Mag.*, 1954, *45*, 773–774.
25. J. C. Eisenstein, M. H. L. Pryce, *Proc. R. Soc. Lon. Ser. A*, 1956, *238*, 31–45.

https://doi.org/10.1142/9781800610163_0016

Chapter 16

Nuclear Magnetic Resonance of Actinides

Peter Kaden
Helmholtz-Zentrum Dresden-Rossendorf, Institute of Resource Ecology, Bautzner Landstraße 400, 01328 Dresden, Germany
p.kaden@hzdr.de

16.1 Introduction

NMR spectroscopy is a bulk method and as such rather insensitive. On the other hand NMR spectroscopy is one of the few methods available to study the electronic interaction of actinides with complexing ligands in a non-destructive and non-invasive manner. Rather large concentrations, at least millimolar scale, are necessary to conduct hyphenated multidimensional heteronuclear NMR spectroscopy. In the actinide series only few isotopes are available that can be studied in normal NMR labs without considering further protection or radiation protection precautions. This is why mainly thorium and uranium isotopes are commonly used to investigate complexes of the actinide series. The common oxidation state of the later actinides is 3+ so that most studies for the understanding of their complexes are performed by using analogous complexes of the stable lanthanides. Especially from research in the area of nuclear fuel reprocessing it is known that there are distinct differences in actinide and lanthanide complexation chemistry that can even be used to separate trivalent actinides from lanthanides.[1] These fundamental differences in the coordination chemistry of the trivalent actinides compared to the lanthanides and the much richer redox chemistry of the early actinides give reason to the necessity to study actinide compounds directly by NMR spectroscopy. Unfortunately, only few institutions worldwide are able to handle the rather active actinides and even fewer institutions are equipped to perform NMR experiments in their radiochemistry labs.

16.2 NMR Spectroscopy of Actinide Containing Compounds

The complexity of NMR spectroscopic investigations increases largely by the number of compounds present in a sample. This is true for rather slow exchange processes leading to different sets of resonances as well as for systems with a set of different complexes present at the same time. In diamagnetic systems diffusion ordered spectroscopy (DOSY)[2] can be used to disentangle resonances of complexes that differ in size and shape. This spectroscopy uses a pair of symmetric gradient pulses separated by a delay for allowing diffusion. Signals of substances that changed their position in respect to the gradient axis will thus experience a loss in signal intensity which can then be correlated to the size and shape of the molecule. This can be seen as a superimposed artificial relaxation. When dealing with paramagnetic complexes of different sizes, however, paramagnetic relaxation enhancement (PRE), inducing faster relaxation on surrounding nuclei, may lead to unreliable results as PRE and diffusion induced signal attenuation cannot easily be distinguished. A thorough knowledge of the solution speciation in the desired concentration range is necessary and available by other techniques like TRLFS (time resolved laser fluorescence spectroscopy) or UV–vis spectroscopy. Effects of the paramagnetically induced relaxation like paramagnetically induced contact relaxation, dipolar relaxation or Cure-spin relaxation are out of the scope of this review. A comprehensive description can be found elsewhere.[3]

NMR spectroscopic investigations may take several hours to several days. During this time, care must be taken to have as little changes in the system as possible. This however is not always true for systems involving actinides due to decay processes. Heat production and associated processes like convection or even boiling of a solvent may be overcome by using long-lived isotopes of the actinides or by reducing the concentration. Decay processes may deposit energy in ligands or surrounding matrices and may introduce degradation by direct interaction or by production of radical species. Consequently, after fission new nuclides are left in solution undergoing redox chemistry related to the matrix and the properties of the element. This might lead to additional paramagnetic species in the sample and thus changes the system's susceptibility and initiates additional chemical shifts or relaxation processes.[4] At a certain point or by using non-purified actinides these systems reach equilibrium with a stable ratio of mother and daughter nuclides in the system. Only the stable end-member of the decay chain builds up in the sample. Influences of solvated paramagnetic daughters and in special cases of stable radicals must be taken into account.[4] In practice the concentration of these interfering species are so low that only minor effects are expected when working with long-lived isotopes.

Comprehensive reviews of NMR related physics of actinide compounds[5] and actinide oxides,[6] mainly in the solid state, were published recently. Thus, solid-state physics of the actinides is out of the scope of this review.

16.2.1 *Direct observation of NMR active actinides*

The NMR properties of the actinides in general vary with the selected nuclide. Most of the earlier actinides possess isotopes with magnetic moments and are thus NMR active. A collection of the most stable isotopes and their NMR properties can be found in a recent review by Farnan and Berthon.[7] However, only two nuclides have favourable nuclear quadrupolar moments enabling their detection under standard spectrometer conditions.

Uranium Uranium-235 is an $I = 7/2$ nucleus and thus quadrupolar in nature. Direct observations of the magnetic resonance of ^{235}U are reported for UF_6 where quadrupole effects are minimised by the cubic symmetry.[8] The signal observed in enriched $^{235}UF_6$ however is still 20 kHz broad and thus ^{235}U studies are restricted to highly symmetric environments. Beyond a study at ultra-low fields[9] and a study of ^{19}F-^{235}U-coupling interactions via ^{19}F linewidths combined with theoretical calculations,[10] no other reports of chemical shifts of ^{235}U or of influenced nuclei *via* spin–spin coupling are published.

Plutonium Plutonium-239 is an $I = 1/2$ nucleus and thus should be detectable by NMR. Only recently, researchers at PNNL discovered the resonance of Pu-239 in a wide scan of the external magnetic field at a temperature of only 4 K and were able to determine the gyromagnetic ratio.[11] Due to the broad lines and not-detectable spin coupling, plus the restriction to very deep temperature and to solid state, the application of Pu-239 NMR has remained limited.[12]

16.2.2 *Effects of actinides on NMR signals of surrounding nuclei*

NMR spectroscopy of f−element containing samples is governed mostly by paramagnetic interactions. While in the (mostly) trivalent lanthanide series two diamagnetic metal ions namely La^{3+} and Lu^{3+} are available, there is no diamagnetic analogue available in NMR relevant quantities for the trivalent actinides. For the tetravalent actinides Th^{4+} and for the hexavalent actinides U(VI) are proper diamagnetic references. All other available actinides and respective oxidation states possess a varying number of unpaired electrons in their valence shell and are thus paramagnetic in nature.

As a discussion of diamagnetic spectra of actinide compounds compared to reference spectra of matrix or ligands alone is straight forward, paramagnetic interactions are more challenging to interpret. It is these paramagnetic contributions, however, which allow for assessing the electronic interaction of metal ion and surrounding nuclei by a number of different influences, like additional chemical shifts, broadening of signals or even bleaching by PRE (paramagnetic relaxation enhancement). These effects are widely used in the lanthanide series to identify metal binding areas in metallo-proteins or by introducing artificial binding sides to probe the tertiary structure of proteins.[3,13–15] Prior to the development of heteronuclear multidimensional spectroscopy paramagnetically induced shifts were used to overcome spectral crowding in 1D spectra by the use of so called shift reagents.[16] From thorough mathematical treatment[3,17,18] of these chemical shifts combined with knowledge from quantum chemical calculations[19,20] more information about the nature of the interaction between metal and ligands is accessible.

16.2.3 *Paramagnetic chemical shifts*

NMR shifts in paramagnetic systems are the most commonly interpreted effect of paramagnetic centres. Paramagnetic shifts may be treated as an additional chemical shift (perturbation theory) to the shifts of a diamagnetic reference compound (σ^{orb}, orbital shift). The major contributors to the observed chemical shift in a paramagnetic system (omitting spin–orbit effects) can be written as follows:

$$\delta^{\text{obs}} = \delta^{\text{orb}} + \delta^{\text{HF}} \cong \delta^{\text{orb}} + (\delta^{\text{FC}} + \delta^{\text{PC}}) \qquad (16.1)$$

The hyperfine contribution (δ^{HF}) consists of two major contributors, the so called Fermi contact shift (δ^{FC}) and the pseudo contact shift (δ^{PC}). All possible contributors can be derived theoretically.[21] The hyperfine shift contribution is isolated from the observed chemical shift (δ^{obs}) by subtracting the chemical shift of a diamagnetic reference which should equal δ^{orb}. Historically, temperature-dependent measurements of the paramagnetic shift and disentangling of the two major contributions, by numerous mathematical methods led to a rather good agreement with theoretical predictions in certain systems (mostly containing light paramagnetic ions),[22] while already lanthanide containing systems cannot be described sufficiently with these methods.[15,23–26] This holds especially true for the actinides as spin–orbit effects, which are omitted in all available mathematical models, are expected to increase dramatically.

Practically, hyperfine effects to the chemical shift of a nucleus are usually observed on light nuclei (^{1}H, ^{13}C, ^{15}N, etc.) in direct proximity to the paramagnetic centre where electron spin-polarisation occurs.[27,28] Electron spin-polarisation, mainly in the valence shell of heavy atoms, is introduced by spin–orbit coupling near the nucleus of the heavy atom, mixing triplet character into the ground state wavefunction in the presence of an external magnetic field.[29] This spin polarisation, mainly found at heavy atoms with non-spherical electron distribution (e.g. unpaired electrons), induces spin polarisation in paired electrons nearby. The induction step utilises either a propagation of polarisation in delocalised electron density in occupied molecular orbitals of the actinide and its ligand (Fermi contact shift) or induces polarisation via a dipolar interaction of the polarised electrons at the heavy atom with electrons of the ligand in close proximity (pseudo-contact shift).

In practice, the major contributor to the paramagnetic chemical shift can be identified by the magnitude and sign of the individual shift. While Fermi-contact shifts may be shielding or deshielding with a magnitude from very small to a multitude of the shielding range of the observed nucleus, pseudo-contact shifts show a symmetrical vectorial dependence (distance and angle dependence from general axis) getting much smaller when getting more remote from the paramagnetic centre.[28,30,31]

In the case of organic ligands, this induced spin polarisation at the ligand may then propagate along bonding orbitals analogous to the well-known spin–spin coupling interactions and thus, for atoms further removed from the paramagnetic centre, follow a Karplus-type dependence on dihedral angles or a damped oscillatory behaviour in aromatic systems.[28,29] Following this transfer of electron spin polarisation an additional contribution to the chemical shifts of the nucleus under investigation is caused by hyperfine interactions, mainly via a Fermi contact. A Fermi contact occurs directly at the nucleus, thus a sizeable s-character in the bond from the observed NMR-active nucleus to its neighbours is essential[28,29] as only those possess a non-vanishing probability of an electron being located at the nucleus. Thus, a periodical dependence seen on ligands' nuclei may point to a Fermi-contact initiated chemical shift.

The simplified form of the equation for paramagnetically induced chemical shifts in the principal magnetic axis system, omitting spin–orbit effects, can be written as:[3]

$$\begin{aligned}\delta_{ij}^{\text{para}} &= \delta_{ij}^{FC} + \delta_{ij}^{PC} = \langle S_Z \rangle_j \cdot F_i \\ &\quad + \frac{1}{2N_A}\left[\left(\chi_{zz}^j - \frac{Tr\chi^j}{3}\right)\cdot\left(\frac{3\cos^2\theta_i - 1}{r_i^3}\right)\right. \\ &\quad \left. + \left(\chi_{xx}^j - \chi_{yy}^j\right)\left(\frac{\sin^2\theta_i\cos^2 2\phi_i}{r_i^3}\right)\right] \qquad (16.2)\end{aligned}$$

Both parts, the Fermi contact (δ_{ij}^{FC}) and the pseudo-contact contribution (δ_{ij}^{PC}), are influenced by terms depending on properties of the paramagnetic centre (j) and the involved nucleus of the ligand (i). The Fermi contact term consists of the contact term F_i, which includes the Fermi hyperfine constant A_i and the influence of the field strength, and the spin expectation value for the paramagnetic centre $\langle S_Z \rangle_j$, which reflects the time-averaged excess of spin along the z-axis and is typically assumed to be due to a pure spin contribution from the paramagnetic centre (j). The pseudo contact term consists of components of the anisotropic magnetic susceptibility tensor ($\chi_{xx}^{j} \neq \chi_{yy}^{j} \neq \chi_{zz}^{j}$) of the paramagnetic centre and reflects the dependence of the dipolar interaction through space by introducing the polar coordinates of the nucleus (i) of the ligand (r_i, θ_i, ϕ_i) in a coordinate systems with the paramagnetic centre at the origin and z being the principal magnetic axis of the system. An in-depth discussion of all components and their theoretical derivation can be found in the review by Piguet and Geraldes[3] and the respective included references. This simplified form gives an estimation of the underlying influences on the observed shift in paramagnetic systems. For a full description and interpretation of the observed chemical shifts, however, this equation is not sufficient.[15,19,23,25,32–34]

A thorough interpretation of underlying mechanisms of the observed paramagnetic shift in the presence of sizeable spin–orbit effects, however, is yet only possible by the use of sophisticated theoretical calculations and subsequent calculation of magnetic properties. As these are very time consuming and very challenging to perform on systems with more than ten atoms or even more than one paramagnetic centre, development of dedicated theoretical methods is necessary and research focusing on advancing quantum chemical approaches is done in several groups at the moment.[19,20] A review of the progress on theoretical calculations is out of the scope of this chapter and can be found elsewhere.[19,20,28]

16.3 New Studies of Actinide Complexes

Preparation and measurement of NMR samples of actinide-containing materials necessitates specific protective equipment. Safe handling, especially of transuranics, requires facilities operating under a nuclear license providing both special safety precautions and monitoring for the scientists. Specific laboratories with special fume hoods, negative pressure gloveboxes and Schlenk-equipment to ensure safe enclosure of the samples on the one hand and controlling the atmosphere to stabilise certain oxidation states on the other hand are crucial for successful and safe sample preparation. In recent years, a number of spectrometers have been commissioned to safely acquire spectra of actinide containing samples.[7] A comprehensive overview of NMR spectroscopy applied to the field of nuclear chemistry in liquid and solid state was published recently.[7] This review is going to highlight recent studies and essential steps towards the full description of paramagnetic NMR spectra of actinide containing samples.

For the simulation of a spectrum of a paramagnetic complex just from a known or postulated structure not only structural parameters need to be known, but all physical constants of the underlying equations have to be determined either by experiment or by theoretical derivation. Those parameters of the lanthanide series have been known for quite some time, in the actinide series, however, most of the parameters still need to be determined.

16.3.1 *Susceptibility*

A step towards the simulation of the behaviour of paramagnetic systems of the actinide series is the thorough determination of the respective susceptibility in solution. Magnetic susceptibilities of actinide containing solutions have been measured by application of the Evans method.[35] The study examines all readily accessible oxidation states of actinides in solution and offers an explanation for the deviations found from predictions based on lanthanide studies. It provides a good starting point for work on direct magnetism effects in actinide sciences.[36,37] Effects of the radioactivity on susceptibility measurements and thus on results of NMR studies of actinides has been shown for ^{243}Am containing solutions and showcase the challenges of NMR studies in actinide containing systems.[4]

16.3.2 *Effects on nuclei directly bound to actinides*

As described earlier, most effects of the actinides' paramagnetism are detected on ligand molecules directly interacting with the respective heavy metal ion. Effects on the nuclei under investigation vary depending on the kind of binding interaction and on the distance to the paramagnetic centre. Only a huge number of studies with thorough interpretation of the measured data might enable scientists in the future to prove theoretically established concepts or to base predictions upon empirically determined knowledge. The following section highlights studies that try to enlighten certain binding schemes and their effects on paramagnetic chemical shifts.

Following the HSAB principle actinides are considered hard to medium with a contracted electronic 7s core but rather diffuse 5f electrons due to spin–orbit coupling. Thus, interaction with hard donor atoms like oxygen most likely results in predominantly ionic interactions.[38] Soft donors however are expected to have a higher probability of overlapping electronic orbitals and are thus more prone to form shared molecular orbitals with a certain *f*-orbital contribution which might be seen as a certain form of covalency.

To probe the bonding situation at a paramagnetic centre, preferably a direct interaction with this centre should be investigated. However, only few NMR-active nuclei (with preferably high receptivity in natural abundance) are known to bind directly to actinides. The following paragraphs will highlight studies of hydrogen, carbon group, chalcogen and pnictogen-series compounds of the actinides.

Hydrogen A comprehensive experimental and theoretical study of transition metal hydrides including the diamagnetic f^0 systems thorium(IV) and uranium(VI) shows that exceptionally large chemical shifts can be found for directly bound hydrogen atoms. As no paramagnetic contribution is expected for f^0 systems, only spin–orbit contributions can cause these shifts. Quantum chemical calculations of related systems predict shifts of up to 150 ppm for selected uranium(VI) hydride complexes, a value NMR spectroscopists would hardly expect to see for hydrogen shifts in diamagnetic systems. Calculations in this study and the comparison to experimental data highlight relativistic effects in NMR spectra of the heavy and super-heavy elements and underline the necessity to include relativity in quantum chemical calculations of actinide containing systems.[39]

Carbon group Only few systems are recently reported with a direct bond between an actinide and carbon. Shifts of carbons coordinated to an U(VI) centre are shown to be exceptionally large by quantum chemical calculations. Up to 810 ppm are calculated and are solely due to spin–orbit effects in these compounds. Experimentally these carbons are found in the predicted range only when following the suggestions of the quantum chemical calculations.[40] Same behaviour has been shown in pentalene complexes, where two pentalene molecules are coordinating one U(IV) centre in an umbrella-shaped η^8-coordination motif yielding high carbon shifts of up to 970 ppm. 2D HMBC data allows for the identification of ^{1}J coupled proton carbon pairs, but could not clarify the position of the quartenary carbons. Variable temperature methods applied to this uranocene are indicative of a simple Curie paramagnetism. SQUID measurements in solid state additionally allowed for determining a magnetic moment of $\mu_{\text{eff}} = 2.90\,\mu_B$.[41]

Pnictogen series Studies of actinide complexes with nitrogen containing ligands are largely stimulated by research centred on actinide-lanthanide separation. Soft donor systems such as nitrogen containing ligands, like bis-triazinyl pyridine (BTP), bis-pyrazole pyridine (BPP) and bis-triazinyl bipyridine (BTBP), are thought to enable a better overlap between the ligands' orbitals at the pnictogens with the rather diffuse *f*-electrons of the actinides. These ligand systems are known for decades to form complexes with both trivalent actinides and lanthanides but with a huge preference for actinides over lanthanides (separation factors > 100).[1] Direct measurement of ^{15}N in isotope enriched complexes of BTP[32,33] and BPP[34] ligands coordinated to Am(III) as well as 2D heteronuclear experiments show exceptionally high nitrogen shifts of nuclei directly bound to the Am^{3+} centre (Figure 16.1). Interpretation of these shifts relative to shifts of not metal-bound adjacent nitrogen nuclei suggests that these large shifts originate largely from Fermi-contact contributions and thus from a sizeable s-character in shared molecular orbitals comprising of both ligand and actinide electron density. These NMR studies were the first substantial contributions proving the existence of a certain covalent character of the actinide–ligand bond in such systems.[33] Complexes of pentalene featuring a η^8-coordination of the

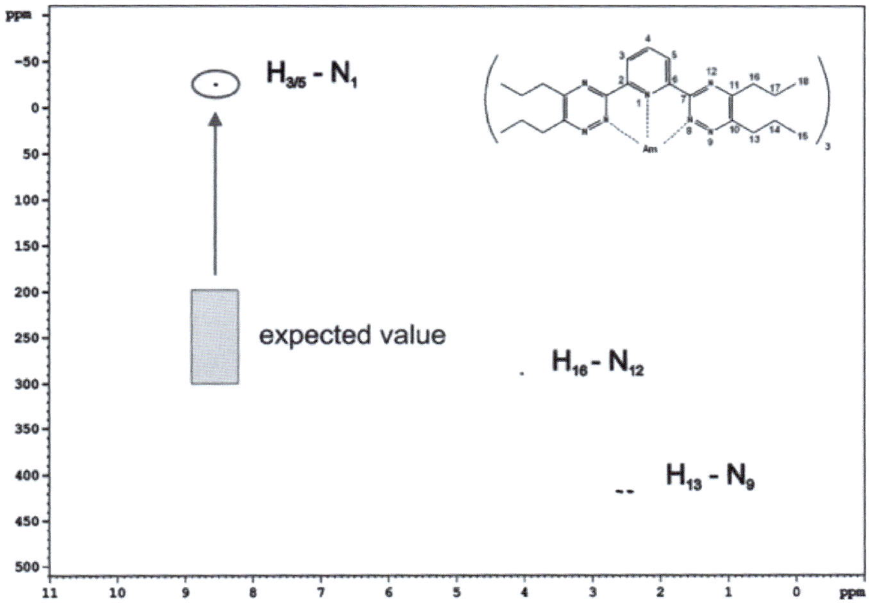

Figure 16.1. Evidence for covalence in a N-donor complex of Am(III) by comparison on the ^{1}H-^{15}N-HMQC spectra of ^{243}Am complexed with unlabelled and 10% labelled ^{n}PrBTP where the expected values range is taken from the uncomplexed ligand and the respective Lu(III) and Sm(III) complexes. Adapted from Ref. [33] with permission from the Royal Society of Chemistry.

metal centre are reported for U(III) forming a umbrella-like structural motif. With a Cp* in a mixed sandwich complex the uranium compound forms a bimetallic dinitrogen complex with a side-on bridging coordination of the N_2. However no data for either carbon shifts of the pentalene, nor for the nitrogens is reported.[42] As this molecule displays an interesting chemistry, a reaction with a phosphaalkyne (^{t}BuCP) resulted in a two electron reduction generating an unusual unsymmetrically (μ-η^2:η^1-^{t}BuCP) bridging complex. ^{31}P NMR shows unreacted ^{t}BuCP at −88 ppm and the product at +1028 ppm.[43]

Chalcogen series Rather exotic ground states of the cubic actinide oxides AnO_2 were identified by ^{17}O NMR studies in the solid state. Temperature dependent measurements at low temperature as well as scans of the polar angles in these ordered systems give insights into the physics of low-temperature states. A number of studies in this field in combination with theoretical calculations is presented in a recent review.[6] High resolution MAS ^{17}O measurements at high MAS rates yield ^{17}O shifts for all early actinides (Th to Am). The observed sign change in shifts through the series is attributed to Fermi-contact interactions occurring already in systems of Am ($5f^6$) which is in contrast to the lanthanide series where it occurs at gadolinium ($4f^7$). The resulting shifts are shown to correlate with the spin expectation value $\langle S_Z \rangle$ which is calculated in this study for all actinide dioxides for the first time.[44] In a study of coordination compounds of the chalcogen-series complexes involving Se- and Te-coordination to actinides Th(IV), Pa(V), and U(VI), all in f^0 state, were analysed by ^{77}Se and ^{125}Te NMR spectroscopy.[45] Using a combination of structural data derived from XRD measurements and analysis of the bonding interactions by DFT calculations, the authors could identify the chemical shifts of Se and Te, respectively, as a quantitative measure of the covalency in the actinide–chalcogen bond. Moreover, the study reports the highest chemical shifts ever observed in diamagnetic substances for Se-containing compounds and presents the shortest known Th–E distances, which is indicative for multi bonding character in the observed complexes. This finding is further substantiated by natural bond analysis and QTAIM metrics. In a similar study a Th(IV) chalcogenolate series is reported including more of these rarely reported ^{77}Se and ^{125}Te resonances.[46]

16.4 Conclusion and Outlook

NMR spectroscopy is a widely used tool to characterise compounds. In metal-organic chemistry, however, little data is available due to paramagnetic effects which are less understood and lead to unexpected shifts. Only few studies try to interpret the found chemical shifts of actinide systems, which is of course still hampered by the fact, that yet no conclusive interpretation of all contributors to the chemical shifts in these paramagnetic molecules of the actinides is available, besides the not sufficient theories derived from Bleaney-type temperature dependent methods. However, sometimes only little effort is taken to even assign all peaks, nor are all accessible

NMR active nuclei reported. From a general NMR point of view these systems are lost for the community and can thus not serve to inspire quantum chemistry, for instance, to calculate and explain the observed behaviour, or to at least be used as a reference for similar systems. Admittedly, paramagnetic systems often suffer from accelerated relaxation hampering the measurement of 2D heteronuclear experiments and thus making a full assignment very difficult. By using different 1D techniques a lot of information can still be generated that often enables a (partial) assignment in combination with suitable references. Only with a broad basis of available, well assigned data, further development of the understanding of underlying mechanisms and paramagnetic influences can be generated in the future. The development of suitable quantum chemical methods which include relativistic effects, methods which are able to calculate open shell and multi reference systems and the combination with response theory to obtain magnetic properties needs suitable well-investigated model systems. Only if chemists working on metal-organic complexes of the actinides take the effort to measure and publish complete NMR data preferably on the whole accessible series of actinides of the same f^n-state or the same oxidation state, quantum chemists will be able in the future to help explain all the paramagnetic contributions to those systems.

References

1. P. J. Panak, A. Geist, *Chem. Rev.*, 2013, *113*, 1199–1236.
2. C. S. Johnson, *Prog. Nucl. Mag. Reson. Spectrosc.*, 1999, *34*, 203–256.
3. C. Piguet, C. F. G. C. Geraldes, *Handbook on the Physics and Chemistry of Rare Earths* (Eds. K. A. Gschneidner, J.-C. G. Bünzli, V. K. Pecharsky), Vol. 33, Elsevier, Amsterdam, 2003, pp. 353–463.
4. M. Autillo, P. Kaden, A. Geist, L. Guerin, P. Moisy, C. Berthon, *Phys. Chem. Chem. Phys.*, 2014, *16*, 8608–8614.
5. R. E. Walstedt, Y. Tokunaga, H. Kato, H. Sakai, T. Fujimoto, S. Kambe, H. Yasuoka, *J. Phys. Soc. Jpn.*, 2006, *75*, 77–81.
6. R. E. Walstedt, Y. Tokunaga, S. Kambe, *Comptes Rend. Physique*, 2014, *15*, 563–572.
7. I. Farnan, C. Berthon, *R. Soc. Chem.*, 2016, *45*, 96–141.
8. H. Le Bail, C. Chachaty, P. Rigny, R. Bougon, *J. Physique Lett.* 1983, *44*, 1017–1019.

9. P. E. Magnelind, A. N. Matlashov, P. L. Volegov, M. A. Espy, *IEEE Trans. Appl. Superconduct.*, 2009, *19*, 816–818.
10. I. Ursu, M. Bogdan, P. Fitori, F. Balibanu, D. E. Demco, *Mol. Phys.*, 1987, *62*, 793–796.
11. H. Yasuoka, G. Koutroulakis, H. Chudo, S. Richmond, D. K. Veirs, A. I. Smith, E. D. Bauer, J. D. Thompson, G. D. Jarvinen, D. L. Clark, *Science*, 2012, *336*, 901–904.
12. A. M. Mounce, H. Yasuoka, G. Koutroulakis, J. A. Lee, H. Cho, F. Gendron, E. Zurek, B. L. Scott, J. A. Trujillo, A. K. Slemmons, J. N. Cross, J. D. Thompson, S. A. Kozimor, E. D. Bauer, J. Autschbach, D. L. Clark, *Inorg. Chem.*, 2016.
13. G. Pintacuda, M. John, X.-C. Su, G. Otting, *Acc. Chem. Res.*, 2007, *40*, 206–212.
14. I. Bertini, C. Luchinat, G. Parigi, *Prog. Nucl. Mag. Res. Spectrosc.*, 2002, *40*, 249–273.
15. O. A. Blackburn, R. M. Edkins, S. Faulkner, A. M. Kenwright, D. Parker, N. J. Rogers, S. Shuvaev, *Dalton Trans.*, 2016, *45*, 6782–6800.
16. A. F. Cockerill, G. L. O. Davies, R. C. Harden, D. M. Rackham, *Chem. Rev.*, 1973, *73*, 553–588.
17. C. F. G. C. Geraldes, S. Zhang, A. D. Sherry, *Inorg Chim Acta*, 2004, *357*, 381–395.
18. A. G. Martynov, Y. G. Gorbunova, A. Y. Tsivadze, *Dalton Trans.*, 2011, *40*, 7165–7171.
19. M. Kaupp, M. Bühl, V. G. Malkin, in *Calculation of NMR and EPR Parameters*, Wiley-VCH Verlag GmbH & Co. KGaA, 2004.
20. J. Autschbach, *Annual Reports in Computational Chemistry* (Ed. A. D. David), Vol. 11, Elsevier, Amsterdam, 2015, pp. 3–36.
21. J. Vaara, P. Manninen, P. Lantto, *Calculation of NMR and EPR Parameters*, Wiley-VCH Verlag GmbH & Co. KGaA, 2004, pp. 209–226.
22. B. Martin, J. Autschbach, *J. Chem. Phys.*, 2015, *142*, 054108.
23. A. M. Funk, K.-L. N. A. Finney, P. Harvey, A. M. Kenwright, E. R. Neil, N. J. Rogers, P. Kanthi Senanayake, D. Parker, *Chem. Sci.*, 2015, *6*, 1655–1662.
24. M. Hiller, M. Maier, H. Wadepohl, M. Enders, *Organometallics*, 2016, *35*, 1916–1922.
25. G. Castro, M. Regueiro-Figueroa, D. Esteban-Gómez, P. Pérez-Lourido, C. Platas-Iglesias, L. Valencia, *Inorg. Chem.*, 2016, *55*, 3490–3497.
26. E. A. Suturina, K. Mason, C. F. G. C. Geraldes, I. Kuprov, D. Parker, *Angew. Chem. Int. Ed.*, 2017, *56*, 12215–12218.

27. M. Kaupp, O. L. Malkina, V. G. Malkin, P. Pyykkö, *Chem. Eur. J.*, 1998, *4*, 118–126.
28. M. Kaupp, *Theoretical and Computational Chemistry* (Ed. S. Peter), Vol. 14, Elsevier, Amsterdam, 2004, pp. 552–597.
29. M. Kaupp, *Calculation of NMR and EPR Parameters*, Wiley-VCH Verlag GmbH & Co. KGaA, 2004, pp. 293–306.
30. M. Enders, *Modeling of Molecular Properties*, Wiley-VCH Verlag GmbH & Co. KGaA, 2011, pp. 49–63.
31. H. M. McConnell, *J. Chem. Phys.*, 1957, *27*, 226–229.
32. C. Adam, V. Rohde, U. Müllich, P. Kaden, A. Geist, P. J. Panak, H. Geckeis, *Proc. Chem.*, 2016, *21*, 38–45.
33. C. Adam, P. Kaden, B. B. Beele, U. Mullich, S. Trumm, A. Geist, P. J. Panak, M. A. Denecke, *Dalton Trans.*, 2013, *42*, 14068–14074.
34. C. Adam, B. B. Beele, A. Geist, U. Mullich, P. Kaden, P. J. Panak, *Chem. Sci.*, 2015, *6*, 1548–1561.
35. D. F. Evans, *J. Chem. Soc.*, 1959, 2003–2005.
36. T. F. Wall, S. Jan, M. Autillo, K. L. Nash, L. Guerin, C. L. Naour, P. Moisy, C. Berthon, *Inorg. Chem.*, 2014, *53*, 2450–2459.
37. M. Autillo, L. Guerin, D. Guillaumont, P. Moisy, H. Bolvin, C. Berthon, *Inorg. Chem.*, 2016, *55*, 12149–12157.
38. N. Kaltsoyannis, *Inorg. Chem.*, 2012, *52*, 3407–3413.
39. P. Hrobárik, V. Hrobáriková, A. H. Greif, M. Kaupp, *Angew. Chem. Int. Ed.*, 2012, *51*, 10884–10888.
40. (a) L. A. Seaman, P. Hrobárik, M. F. Schettini, S. Fortier, M. Kaupp, T. W. Hayton, *Angew. Chem. Int. Ed.*, 2013, *52*, 3259–3263; (b) K. C. Mullane, P. Hrobárik, T. Cheisson, B. C. Manor, P. J. Carroll, E. J. Schelter, *Inorg. Chem.*, 2019, *58*, 4152–4163.
41. F. M. Chadwick, A. Ashley, G. Wildgoose, J. M. Goicoechea, S. Randall, D. O'Hare, *Dalton Trans.*, 2010, *39*, 6789–6793.
42. F. G. N. Cloke, P. B. Hitchcock, *J. Am. Chem. Soc.*, 2002, *124*, 9352–9353.
43. N. Tsoureas, A. F. R. Kilpatrick, O. T. Summerscales, J. F. Nixon, F. G. N. Cloke, P. B. Hitchcock, *Eur. J. Inorg. Chem.*, 2013, *2013*, 4085–4089.
44. L. Martel, N. Magnani, J.-F. Vigier, J. Boshoven, C. Selfslag, I. Farnan, J.-C. Griveau, J. Somers, T. Fanghänel, *Inorg. Chem.*, 2014, *53*, 6928–6933.
45. D. E. Smiles, G. Wu, P. Hrobárik, T. W. Hayton, *J. Am. Chem. Soc.*, 2016, *138*, 814–825.
46. N. S. Settineri, M. E. Garner, J. Arnold, *J. Am. Chem. Soc.*, 2017, *139*, 6261–6269.

https://doi.org/10.1142/9781800610163_0017

Chapter 17

Applications of Rare Earths

Jean-Claude G. Bünzli
Swiss Federal Institute of Technology, Lausanne (EPFL), Institute of Chemical Sciences and Engineering, Lausanne, Switzerland and Hong Kong Baptist University, Department of Chemistry, Kowloon Tong, Hong Kong SAR, P. R. China
jean-claude.bunzli@epfl.ch

17.1 Rare Earths: The Vitamins of High Technology

The group of 17 elements including Sc (21), Y (39) and the lanthanides (Ln = La–Lu, 57–71) is commonly referred to as "rare-earth elements" (REEs)[a,b] and features stunning chemical, magnetic and optical properties conferring it a special status not only in physics, chemistry and materials sciences but, and most importantly, in our contemporary highly technical world.[1] Indeed, a growing number of critical technologies are vitally dependent on these elements. A single example should suffice to make this point clear: magnets. Neodymium-iron-boron magnets are used in a range of items, from wind turbines to computer hard disk drives (HDD), audio devices

[a]Radioactive promethium (Pm, 61), with isotopes having very short half-lives, is absent from rare-earth ores and will therefore be ignored in this chapter. In addition, scandium (Sc, 21) has very different geochemical and chemical properties so that it will only be briefly mentioned.

[b]Although corresponding to specific definitions, rare earths, lanthanides (or lanthanoids, La–Lu) and 4f-elements (Ce–Lu, 58–71) are often loosely used as synonyms.

and all kind of electric motors. Their performances are so much better than those of ferrite magnets that replacing them with the latter would have far reaching consequences. For instance the loudspeakers in the doors of a car would be bulkier, calling for thicker doors, themselves calling for reinforced structures of the vehicle and in the end it is estimated that about 2–300 pounds would be added to the car, with concomitant loss of performances (e.g. mileage!). As a matter of fact, rare-earth elements are presently providing irreplaceable functions in a wealth of items we use daily and are therefore essential elements without which severe loss of performances would have to be accepted, at variance with the current innovation drive!

Because rare earths are at the heart of all electronic, optical and magnetic applications, no military offensive or defensive platform can be deployed and maintained without a dependable supply chain of these elements and their technology. Guidance and control systems relying on rare-earth lasers and optics, and rare-earth containing electronic components are at the heart of unmanned aircrafts (drones), missiles, guided artillery projectiles and smart bombs. Communications, detection, surveillance missions, night vision, jamming devices need fibre optics, displays, radars, sonars, infrared and radiation detectors, and microwave generators, all of them featuring rare earths in their vital components. Lasers are not only used for targeting and guiding but also as weapons or mine-clearing devices. To this list can be added computers and all other devices using permanent magnets, electric motors and generators or air-conditioners for instance. All this explains the crucial strategic importance of rare earths.

17.1.1 *Industrial periods*

The first rare-earth element, yttrium (in fact its sesquioxide Y_2O_3, yttria), has been isolated by the Finnish chemist Johan Gadolin in 1794 from a black mineral now named gadolinite and discovered in 1787 near Ytterby (Sweden) by a Swedish army lieutenant, Carl Axel Arrhenius. It took more than 100 years (1803–1907) to separate and identify the remaining naturally occurring RE elements from minerals in which they appear as intertwined mixtures, while radioactive Pm was synthesised in 1947. The exciting story of the isolation of REEs is full of incorrect claims and heated disputes among would-be

(a) (b) (c)

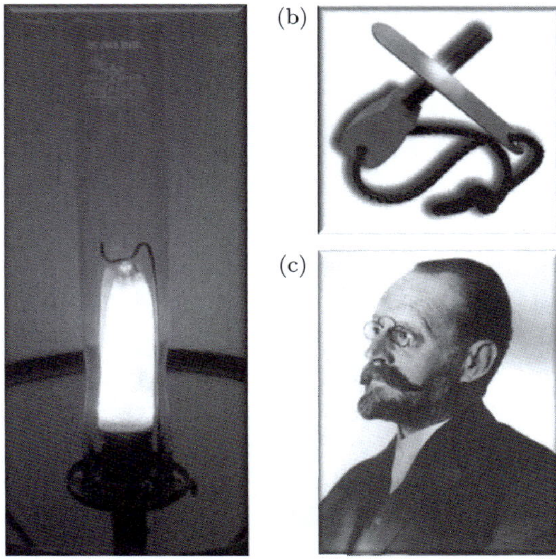

Figure 17.1. Two key inventions of Auer von Welsbach: an Auer gas lamp with the incandescent mantle (a) and Mischmetall in a firesteel lighter (b) (http://firesteel.com, accessed on May 31, 2021). (c) Photograph of Auer von Welsbach and incandescent mantle taken from The Editors of Encyclopaedia Britannica, *Carl Auer, Freiherr von Welsbach*, Encyclopædia Britannica, August 28, 2020, https://www.britannica.com/biography/Carl-Auer-Freiherr-von-Welsbach, accessed on May 31, 2021.

discoverers while reflecting the developments in separation, analytical and spectroscopic techniques that took place during this span of time.

Industrial uses started in 1891 with Carl Auer von Welsbach (1858–1929) being granted a patent for producing incandescent mantles for gas lighting made up of thorium (99%) and cerium (1%) oxides. Twelve years later he invented the "mischmetal — from German Mischmetall", a mixture of La, Ce, Pr and Nd to which iron is usually added and that is used in the manufacturing of flint stones and firesteel or as additive in metallurgy. Both products are still in use today (Figure 17.1). In addition to discovering praseodymium and neodymium (1885, separated from a didymium mixture), Auer von Welsbach added rare-earth fluorides in the electrodes of powerful electric arc lamps used in cinemas or as search lights and also developed tungsten/osmium filaments for electric lighting. His work resulted in the creation of two companies still alive today: Treibacher

Industrie A.G. (1898), producing RE compounds — including mischmetal — and Osram (1906), focusing on lighting devices. This early industrial period of rare earths, dominated by the production of cerium oxide and Mischmetall, ended around 1930 when electricity replaced gas lighting.

The second industrial era, between 1930 and 1965, saw expansion of the REE market governed by several applications of cerium oxide (sometimes containing some La and Nd oxides), as polishing powder replacing iron oxide, opacifier in ceramic glazes, additive in optical glasses — including sunglasses, and, at the beginning of the 1960s, mixed with lanthanide oxide, in zeolite oil-cracking catalysts. Small quantities of cerium oxalate were also used to fight sea sickness and nausea during pregnancy. All of these applications however resulted in relatively small amounts of rare earths being consumed, compared to the large quantities generated by the production of thorium oxide from monazite as catalyst in the Fischer–Tropsch process and feed material in breeder nuclear reactors. The situation started to change in the 1940s: atomic programmes in the US (Manhattan project), England, and USSR generated abundant rare earths as fission products, encouraging scientists to become interested in their properties. Efficient separation processes were developed yielding highly pure rare earth oxides and salts which were incorporated into more sophisticated products. Metallurgical procedures were established to get high-purity RE metals. Sensitisation of lanthanide luminescence by organic molecules was discovered in 1942 and was a great help in developing photonic applications since it overcomes the intrinsically faint molar absorption coefficients of forbidden f-f transitions (antenna effect). Optical applications were also boosted by the advent of lasers, successful operation of the first $Y_3A1_5O_{12}$:Nd (YAG:Nd) laser being demonstrated in 1964. Another prominent example is the red phosphor YVO_4:Eu that was introduced in television screens Y_2O_2S:Eu and/or Y_2O_3:Eu. The latter phosphor is still in use today.

Diversification of the use of rare earths in high-technology products accelerated in the third industrial period, between 1965 and 1990, with the discovery of (i) optical upconversion (1966) later used in security inks and bioanalysis, (ii) three-way catalysts for controlling exhaust emission from automobiles (Pt,Rh/Al_2O_3:CeO_2(1%–5%), 1982), (iii) new fluid catalytic cracking catalysts (CeO_2/$Mg_2Al_2O_5$, 1982), (iv) $Nd_2Fe_{14}B$ permanent magnets (1983), (v) buckeyball, under the form of $C_{60}La$ (1985),

(vi) high-temperature superconductivity in the La–Ba–Cu–O system (30 K, 1986), (vii) erbium-doped amplifiers for optical-fibre telecommunications (1986), (viii) consumer-grade nickel hydride rechargeable batteries using mischmetal (1987–1989), among others. The advent of biomedical applications such as time-resolved immunoassays and contrast agents for magnetic resonance imaging (MRI) in the 1980s can be added to this success list.

The present industrial era of rare earths started around 1990 and has been dominated by growing influence of China in both extraction/purification of rare earths and manufacturing of raw and finished products that led to amazing monopoly 15 years later. Despite mainly building on scientific breakthroughs from the previous period, improving materials, this fourth era has seen its share of novelties. For instance, Er_3Ni spheres have started to be commercially used as component of Gifford–McMahon cryogenic systems in 1992, allowing steady development of superconducting magnets, e.g. for nuclear magnetic resonance spectroscopy (NMR) and associated MRI scanners, or for magnetic levitation high-speed trains that are now operating in several Asian countries (China, Japan, South Korea). Along similar lines, a major impact for refrigeration science was the discovery in 1997 of the giant magnetocaloric effect in $Gd_5(Si_xGe_{1-x})_4$ compounds along with the demonstration that magnetic refrigeration is a viable technique, able to compete with vapour-cycle refrigeration. The mid 1990s have also seen the discovery of switchable metal hydride films LnH_x ($2 < x < 3$), promising materials for hydrogen sensing, the advent of lanthanide-based light-emitting diodes (LEDs), as well as a thorough search for single-molecule magnets containing rare earths — many of them were found and are still being discovered today. Optical refrigeration into the cryogenic regime (below 123 K) has been achieved with an ytterbium-containing crystal in 2013.

In the 21st century, most facets of technology, including defence aspects, depend critically on rare earths. This includes cars (catalysts for refining gasoline and for exhaust gas converters, magnets for motorisation), magnets for wind turbines, television (the screen phosphors as well as electronic components contain REEs), telecommunications, phones and computers (screens, hard-disk drives, speakers, buzzers), lasers and guiding systems, lighting (both compact fluorescent lamps and LEDs), biomedical analyses and imaging, to name but a few.

17.1.2 *Supplies and end uses*

The total amount of rare earths produced annually is difficult to assess precisely because of illegal mining (especially in China, estimated to about 30,000–40,000 tons of equivalent rare-earth oxides, REO, per year at the time of writing). From various reports though one can guess that the **official** worldwide production of rare earths (expressed in terms of REO) was around 5,000 tons per year in 1955–1965 and then grew steadily until it reached 110,000–120,000 tons in 2008. It stayed about constant until 2015, but then increased again, reaching 240,000 tons in 2020. With respect to the world's metal production, rare earths represent only 0.01% in weight (steel: 90%; technology metals: 1%). Given the large specific mass of rare-earth oxides (6.5–9.5 $g \cdot cm^{-3}$) all rare earths mined in one year can easily fit into a space equivalent to 200–250 40-feet maritime containers (i.e. in a quite small cargo ship). As for their prices, 99.9% pure Y, La and Ce oxides are cheap (July 2019): \$2–3 per kg, while "critical" elements[c] such as neodymium, europium, terbium and dysprosium are in the range \$50–600 per kg oxide (>99% purity). Altogether the present commercial value of rare earths produced worldwide is around \$1.5–2 billion (in terms of REO).

Historically, production of rare earths in the US by Molycorp, at Mountain Pass in California, started in 1964 and the company supplied about 50% of the world needs until the mid-1980s when China bluntly entered into the market to become the world's largest supplier by 1991; the share of China's production grew to 97% in 2010 and then slowly decreased to about 85% in 2015 and 62% in 2019 (Figure 17.2), following the short revival of Molycorp (2012–2015) and the opening of a large facility in Australia (Mount Weld, Lynas company). This share has now further decreased to 58% in 2020. But more importantly, China has created a complete industrial chain, from mining to end-products, complemented by strong research and development efforts, resulting in this country dominating the world market for most of the important rare earth containing products.

[c]A critical element is an element that one absolutely needs but the supply of which is not granted at short to medium term (5–10 years). Five RE elements presently enter into this category, Y, Nd, Eu, Tb, Dy.

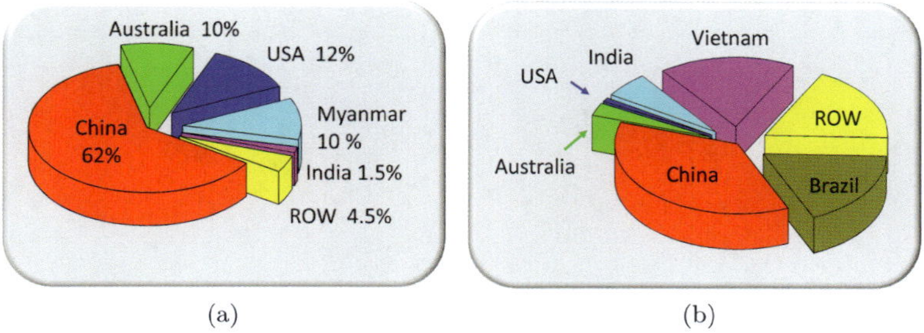

Figure 17.2. (a) Shares of rare earth production by country in 2019 (total production 210,000 tons (REO equivalent). (b) Shares of rare-earth reserves by country in 2019 (total 120 million tons REO). Data from US Geological Survey Mineral Commodity Summaries 2020. ROW: rest of the world.

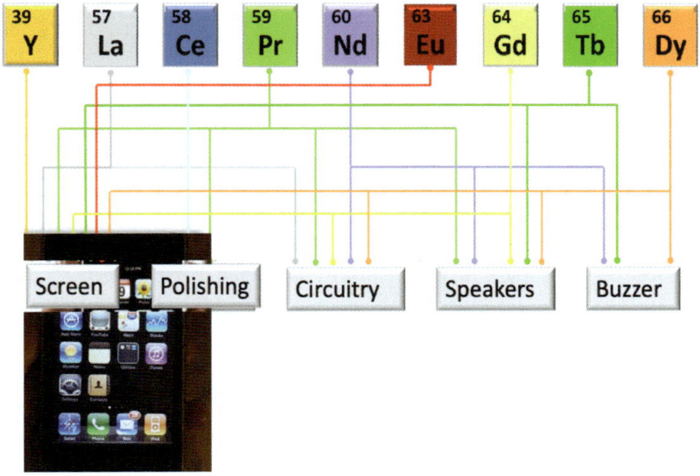

Figure 17.3. Rare earths in a smart phone (approximately 150–250 mg).

End uses of specific lanthanides depend on the countries, e.g. there are large differences between the United States, Europe and Asia. The market is not easy to follow in view of the small amounts of REO used per year and in view of the multitude of products into which rare earths are provided. In addition, some products contain only a very tiny amount of rare earths, e.g. there are nine different rare earths in a smart phone for a total amount of about 0.2 g, representing only a fraction of its price (Figure 17.3)! The same is true for many other appliances and end-products, except magnets, polishing

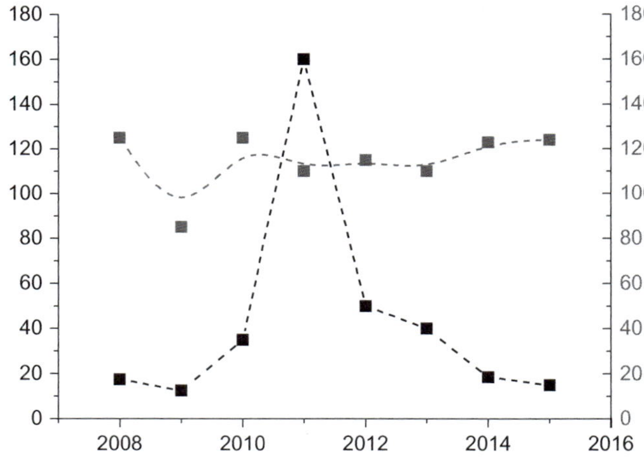

Figure 17.4. Left: Average REO prices in US \$/kg. Right: official annual production in terms in ktons REO/year. Sources: IMCOA, Curtins University (2008–2013), Association of Chinese Rare Earth Industries (2014–2015), and United States Geological Survey Mineral Commodity Summaries January 2016.

powders and some catalysts or alloys and compounds. This feature remained almost un-noticed until 2010 when following a geopolitical incident between China and Japan the stunned world realised that China had a large monopoly on the production and technology of these elements. Military and political staff panicked, prices were multiplied up to 50-fold for some elements, peaking in 2011, and mining operations were reopened or newly opened in the western world while dozens of potential mines started to be prospected. In 2015, the crisis was over with prices comparable to or even lower than those of 2009 (Figure 17.4), but the hefty interest stirred for rare earths remains, several of them being now classified by many governments, including the European Commission, as strategic elements.

Estimated data for rare-earth applications are displayed in Figure 17.5. Catalysts, glass industry (additives and polishing powders), as well as magnets represent almost two-third of the tonnage used. However, magnets and phosphors are the leaders with more than two-third of the commercial value.

Usages with respect to individual elements reflect a large dissymmetry in that La, Ce, Nd, Pr and Y represent 97 wt% of all rare earths used in end products. This is the reason for the "imbalance problem". Indeed if there are plenty of minerals containing rare earths around the world (>200, some say >400), one has to consider that (i) ores

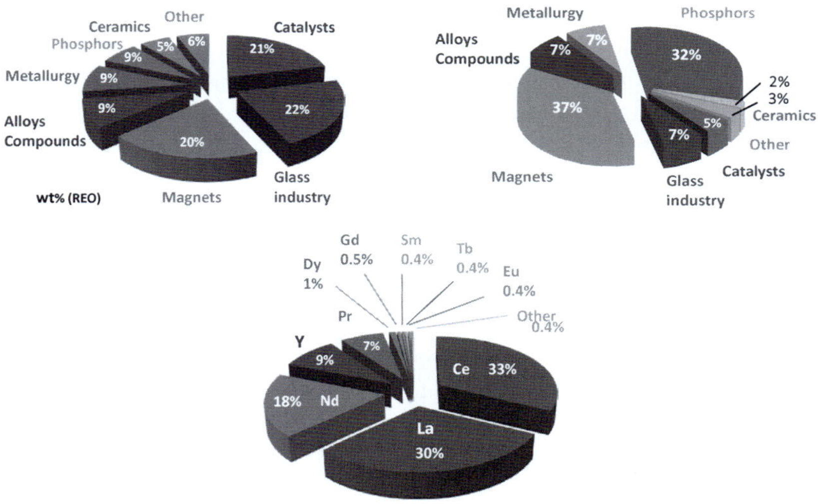

Figure 17.5. Top: Uses of rare earths with respect to main applications; data from US Geological Survey Report 2011–5094 and Mineral Commodity Summaries 2012; percentages expressed with respect to REO (top left) and value in US$ (top right, data from Ref. [21]). Bottom: Uses of rare earths with respect to individual elements (same sources as for top left data).

have usually low concentrations of these elements (1%–10%) and, (ii) minerals contain a range of REEs, typically La–Eu for one type and Y, Eu–Lu for the other one, meaning that once extracted and beneficiated, individual elements have to be painstakingly separated, a difficult and costly task in view of their similar chemical properties. Therefore companies are often left with surpluses of uneconomical rare earths (cerium, lanthanum) while aiming at more technological elements, such as europium and terbium for optical applications for instance. This is the reason why Auer von Welsbach developed the mischmetal since gas mantle production consumed only thorium and a tiny quantity of cerium! As an example, one metric ton of Bastnaesite ore extracted at Mountain Pass (California) yields approximately 40 kg of Ce and 40 kg of other REEs among which 27 kg of La, 9 of Nd, 3 of Pr, 0.9 of Sm and 80 g of Eu and Y each (all data in terms of REO). This, allied with the fact that when prices go up companies are either digging into their stockpiles, looking for substitutes and/or favouring recycling, makes the rare-earth market quite volatile and difficult to predict. Presently, known economically viable reserves of rare earths are large and sufficient for about 100 years of exploitation at current level and recently discovered, unassessed new deposits are numerous so that altogether, concerns about availability can be minimised, especially that recycling is gaining momentum (magnets, lamp phosphors and rechargeable batteries) and contributing solving the imbalance problem.

17.2 Catalysts

Rare earths are active elements in a large number of catalytic reactions and several areas have long reached industrial level, in particular catalysts based on ceria.[2]

17.2.1 *Fluidified cracking catalysts*

The oldest application (1962) is the structural and chemical stabilisation of zeolites for oil cracking by fluidified cracking catalysts (FCCs). These catalysts contain typically 20%–40% zeolite and addition of several per cent of lanthanides (10%–20%) in the latter, mainly lanthanum and/or cerium, allows the catalyst to remain acidic, which is essential for the conversion of high molecular weight molecules into lighter species, especially in the very aggressive conditions of the petroleum industry. Sodium ions of zeolite are exchanged for hydrated $[Ln(H_2O)_x]^{3+}$ ions, followed by thermal dehydration. Dehydrated Ln^{3+} ions then generate strong electrostatic fields leading to dissociation of the remaining water molecules and to the production of acidic Brönsted sites. First, zeolite-X was treated in this way, but soon zeolite-Y with higher Si/Al ratio (2.5–3 vs. 1.2 for zeolite-X) appeared and was also activated and stabilised either with Ln^{3+} or with ammonium ions. FCC catalysts are used for the production of liquid petroleum gas (LPG), gasoline (50% of the world consumption is produced with FCC), light oils used in diesel and jet fuel (kerosene), and heavy fuel oil, as well as for many other applications, including synthetic gas (syn gas) and palm oil production.

17.2.2 *Automotive catalysts*

Due to it redox properties, cerium dioxide (CeO_2, ceria) enters into the composition of automotive post-combustion catalysts. It is a major component of three-way catalysts (TWCs) used in all modern gasoline cars. TWCs lower the level of pollutant emissions from the engine through selective reduction of nitrogen oxides (NO_x) and simultaneous oxidation of carbon monoxide and hydrocarbons. As cerium can be either tri- or tetravalent, redox properties of CeO_2 make it an oxygen buffer stabilizing the composition of exhaust gases and allowing oxidation of CO and hydrocarbons even when the medium is globally reducing. Initial compositions were low in ceria (1–5 wt%), but subsequent ones developed starting in 1992 were made of 100–3,000 ppm of precious metal (Pd, Rh or Pt) dispersed on a mixture of alumina and ceria (≈20 wt%). In addition to its major role in the control of redox conditions of the medium, ceria is highly refractory and allows alumina to keep its surface thermally stable at the elevated temperatures sustained by the catalytic muffler (>1000°C). Finally, it also permits a good, thermally stable, dispersion of the metallic particles, the actual catalyst, preventing them from sintering, which would make them less active. The development of cerium–zirconium mixed oxides (1995) has led to improved performances in terms of oxygen buffering capacity and thermal stability. Based on industrial development of highly efficient chemical methods, special grades of cerium–zirconium oxides, remaining active at high temperatures, have been developed for TWC catalysts.[3]

Diesel engines enjoy growing popularity because of their fuel-efficiency (30% more mileage than a gasoline engine) and drivability. However, they face major challenges in controlling particulate matter (PM) and NO_x emissions. The most efficient way to reduce particulate matter emissions is to filter the exhaust gases. However, because the filter becomes blocked as the soot particles accumulate, the technological challenge has been to find a way to eliminate these particles. The best solution is to burn them in order to regenerate the filter. Rhodia (now Solvay) has developed an efficient filter/catalyst system in 2002 based on cerium oxide (Eolys™). The fuel-borne catalyst facilitates filter regeneration by significantly lowering the temperature at which the soot burns, from >600°C to around 450°C, and by increasing the kinetics of combustion, typically 2–3 minutes instead of 20–25 minutes for some competing technologies. The performance of Eolys™ technology is independent of the NO_x/PM ratio; furthermore, the process does not increase NO_2 emissions and eliminates >99.9% of the emitted particles. In its initial formulation, regeneration was performed every 500–1,000 km and the catalyst was added to the fuel from a 3–5 litre tank at a concentration of about 50 ppm, allowing for a service interval of 120,000–180,000 km for an average passenger car and a similar filter service life. Toxicological studies have demonstrated the non-hazardous nature of Eolys™ and the product is compatible with the majority of plastics and elastomers used in the manufacture of fuel tanks, fuel pumps and fuel lines. Despite some initial problems that led drivers to remove these filters, but are now overcome, more than 10 million vehicles were equipped with this system in 2015; in addition, retrofitting to light-, medium- and heavy-duty vehicles is easy. The newest version of the catalyst, Eolys Powerflex®, launched in 2010, is now available. It compels with the latest European regulations on diesel engine emission (Euro 6) despite requiring a much smaller concentration of the cerium oxide particles (5 ppm). This leads to longer service intervals (250,000 km) and smaller additive tanks (<2 litres).

17.2.3 *Production of artificial rubber and other organic compounds*

Another application of rare-earth catalysts is developing steadily: the use of neodymium ternary salts as polymerisation catalysts for butadiene, isoprene and styrene. For example, with respect to polybutadiene, the use of neodymium carboxylates allows getting high cis-polybutadiene content and a very good control of the molecular weight distribution of the polymer. Advantages of rare-earth salts compared to transition elements are high polymerisation temperatures, reducing the cooling stage of the polymer, and substituting toxic aromatic solvents by aliphatic ones. This led to a significant industrial development of the use of neodymium catalysts for rubber manufacturing. Recent replacement of nickel catalysts for the production of butadiene for automotive tires by neodymium catalysts (containing about 8% Nd) resulted in higher activity and conversion efficiency reaching 99% cis-butadiene. The resulting synthetic rubber features better elasticity, which lowers the rolling resistance, as well as better endurance and durability. As a consequence, fuel efficiency is improved, which helps manufacturers meeting stringent environmental legislation. In 2014, an estimate put the market growth for these "green tires" to 9% per year. In fact neodymium can be replaced by a mixture of Nd and Pr (didymium), which avoids having to separate pure elements and reduces the cost of the catalyst.[4]

Among other tested catalysts are lanthanide triflates for enantioselective synthesis of pharmaceuticals as well as amidinates, guanidinates and borohydride complexes for ring-opening polymerisation. Although not a catalyst, samarium di-iodide ought to be mentioned here as a mild reducing agent that is widely used in organic chemistry.

Furthermore, ceria is added to the dominant catalyst in several chemical processes, e.g. in the production of styrene from methylbenzene. Mixtures of tungsten-doped cerium oxide and ammonia are presently being tested in the selective reduction of nitrogen monoxide. Cerium dioxide also enters into the composition of some coatings in self-cleaning kitchen ovens.

17.3 Pigments and Additives for Glass, Ceramic and Leather Industries

Lanthanides, particularly cerium, are extensively used in glass industry for both their chemical, optical and abrasive properties. For instance, discolouration of glass is obtained through oxidation of impurities of deep-blue divalent iron to pale-yellow trivalent iron by tetravalent cerium. Cerium oxide is a UV-blocking material which is used in sunglasses and could potentially be fitted onto dye-sensitised solar cells to prevent photodegradation of the dye. Ce^{IV} is often added to glasses and polymers (e.g. television screens) as anti-browning agent. Other optical applications involve lanthanum as a component (40 wt%) in high refractive index borate glasses for microscope, telescope, camera lenses and automotive headlight glasses.

Praseodymium (green), neodymium (purple) and erbium (pink) are used to colour glasses and ceramics. Zircon ($ZrSiO_4$) doped by tetravalent praseodymium is the strongest and most stable yellow pigment for ceramics with high firing temperatures. Pigments for the coloration of plastics and paints are based on the optical properties of cerium sulphide derivatives and their nontoxicity. These are welcome alternatives to heavy-metal-based pigments like cadmium sulpho-selenides. Owing to co-doping with alkali or alkaline earths, it has been possible to stabilise a range of colours from orange to bright red and maroon in the form of Ce_2S_3. These pigments feature high brightness and tinting strength, a good stability in the application media (polymers such as polypropylene, acrylonitrile butadiene styrene, polycarbonate), and a strong resistance to weathering and UV.

Another intriguing application of rare-earth pigments is the colouring of leather with yellow-brownish NIR-reflective pigments, one of which is molybdenum-doped gadolinium cerium oxide, $GdCe_{1-x}Mo_xO_{2.5+x}$. The advantage of such dyestuffs is their very large reflectivity in the NIR, rendering leather garments more comfortable to wear, with a feeling of coolness, even under the sun.

Last but not least since it represents sizeable tonnage, cerium oxide (sometimes containing some La and Nd oxides) is the best polishing agent for glass, owing to its natural hardness, a consequence of its compact fluorite structure and of a chemical reaction at the silica–cerium oxide interface. Purity and morphology of the polishing powders made from cerium oxide can be adapted to the polishing quality required and are used to polish any type of glass, from computer screens to house windows and optical components (e.g. the Hubble telescope).

17.4 Magnets

Rare-earth metals have exceptional magnetic properties. At low temperature, magneto-crystalline anisotropy constants are 10–100 times higher than for other elements and absolute saturation magnetisation is much higher than in iron, for example. However, magnetic ordering occurs only at low temperatures, as the internal character of the 4f orbitals induces weak coupling for direct interactions between neighbouring atoms as well as for long-range exchange via conduction electrons. Thus, rare-earth metals have usually very low Curie temperature T_{C}, the temperature at which materials become ferromagnetic — i.e. permanent magnets, except for gadolinium for which $T_C = 292.7$ K. However, Curie temperatures can be increased to commercially interesting levels by alloying lanthanide metals with elements having higher ordering temperatures, such as iron, cobalt or nickel.

The first high-performing magnetic materials to be industrialised were samarium–cobalt magnets, mainly $SmCo_5$, in 1970. These magnets displayed intrinsic coercivity[d] >20 kOe. This enabled the miniaturisation of many electronic devices, such as stepping motors and headphones, for instance for the celebrated Walkman® available since 1979 and the fabrication of which was discontinued in 2010. The growth of applications was however limited by the relative scarcity and high cost of samarium and cobalt. In particular, the district of Kolwezi in Zaire from where most cobalt originated went through a period of civil unrest which, in turn, resulted in a 5-fold increase in the price of cobalt between 1976 and 1978. This bottleneck was removed by the discovery of the 2.5 times more powerful neodymium–iron–boron magnets in 1984. Most magnets with $Nd_2Fe_{14}B$ composition (95%) are produced as dense sintered products using classical powdered metallurgy processes; the remaining are bonded materials in epoxy or nylon (4%) while sintered Sm-Co magnets account for less than 1%. Applications range from computer hard-disk drives (12 g/item, 430 million produced in 2015) to audio speakers and headphones, bicycle dynamos, permanent motor magnets (e.g. driving motor for magnetic levitation high-speed trains), wind turbine generators (1 ton magnet/1.5 MW turbine), diesel engines for large cargo ships and magnetic resonance imaging. Present and future demand for Nd permanent magnets is primarily determined by four major applications: electric motors for hybrid and electric vehicles (cars and bicycles), wind turbines, hard disks, the market of which is plummeting, as well as superconducting magnets. Coercivity of these magnets can be enhanced (particularly at high temperature) by adding a small amount of dysprosium, initially up to about 8%, but the recent RE crisis led engineers to find solutions to reduce this amount to almost nil.

17.4.1 *Electric and hybrid vehicles*

Growth of this market is steady. For instance, the legendary first hybrid vehicle, the Toyota Prius, requires one kilogram of neodymium

[d]Coercivity is the intensity of the magnetic field needed to cancel the magnetisation of a material after it has been magnetised to saturation.

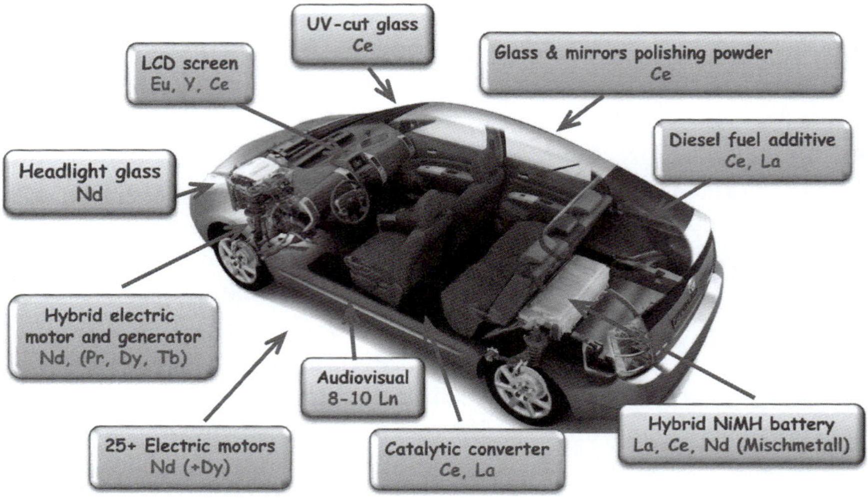

Figure 17.6. Rare earths in a hybrid vehicle (here the Toyota Prius). Adapted from https://www.ultimatecarblog.com/toyota-turns-attention-to-turbos-and-direct-injection/, accessed November 29, 2020.

for its drive electric motors, 5–6 kg La/Ce for its nickel-lanthanide hydride rechargeable battery, that are warrantied 10 years, in addition to probably another kg for all other applications (Figure 17.6). The amount of magnetic, electronic and optical materials containing rare earth in these vehicles will probably not decrease in the future, but replacement of the NiMH batteries with Li batteries is foreseeable at middle term, which might decrease Ce and La usage although, in 2015, Ni–MH battery demand was still increasing; their main asset is their long-term stability without loss of performance.

17.4.2 *Wind turbines*

In 2019, installed wind power capacities were estimated to 651 GW generated by 341,000 turbines and representing 5.3% of worldwide electricity production. Asian share amounts to 45% of the total (of which 36% for China and to 49% of the new installations. Europe ranks second, with 30% of the total installed power but only 25% of the new turbines; nevertheless, about 15% of European electricity needs are covered by wind turbines.

There are several technologies in use for wind turbines, but the direct-drive one taking advantage of Nd-magnets appears to be the best because it does not require a gearbox so that it is the technology of choice for off-shore turbines since maintenance is easier. Indeed the gearboxes is modern turbines are quite complex, with up to 13 gears and pinions and 22 bearings. There is nonetheless some debate as to determine if Nd magnets are the best "green" technology since one standard turbine needs about 500–600 kg of Nd, with associated environmental nuisances for its production. On the other hand, it takes only 3–6 months of electricity production to recoup the energy needed to fabricate, operate and recycle the turbine. Finally, since Nd is a critical element some manufacturers are ready to switch back to iron-only magnets if Nd price reaches an uneconomical level. Altogether though, one can be reasonably optimistic that Nd-magnets will continue to be used in a good share of the future wind turbines.

17.4.3 *Hard disk drives, data storage and electronic components*

In 2015, the Japanese company Shin-Etsu estimated that 30% of the Nd-magnets were used in hard disk drives for personal (PCs) and laptop computers, and 10% in optical and acoustic applications (including buzzers for smart phones and hi-fi loudspeakers). Presently, the number of PCs and laptops produced is decreasing to the benefit of tablets and smart phones and substitution of HDDs with solid-state drive (SSD) technology is expanding, which will probably reduce the use of Nd with respect to these consumer products.

Rare-earth transition-metal alloys are also applied in magneto-optic recording, in which the magnetic and optical (high value of the Kerr rotation angle) properties of (Gd, Tb)(Fe, Co) amorphous powdered alloys are used to obtain high recording densities (300 Mbits/cm^2) in laser erasable–rewritable compact disks (CDs). But again, technology changes with the advent of powerful SDD-based USB keys and hard disks, which will cause a decline in these applications.

Supercomputers contain Gd-based bubble memory substrates consisting in a thin film of a magnetic material including small magnetised areas, called bubbles or domains, each storing one bit of data. The material forms a series of parallel tracks along which the bubbles can move under the action of an external magnetic field. The bubbles are read magnetically at the edge of the material and then rewritten on the far edge to keep the memory cycling through the material.

17.5 Metallurgy, Alloys and Compounds

17.5.1 *Metals and alloys*

The large electropositive nature of rare-earth elements precludes obtaining metals by reduction of oxides or salts with carbon. There are two preparative methods for metals. The first one is the electrolysis of fused oxides or salts in presence of an LnF_3 flux; this can only be done for elements La through Nd because the melting points of the medium and heavier elements are too high. Mischmetal is commonly produced in this way either from a fused chloride or fluoride bath. Alternatively, rare-earth metals can be electrodeposited with a solvent metal (Zn or Mg) and subsequently distilled from the mixture. The second method relies on metallothermal reduction and can be applied to all lanthanide metals and yttrium. Suitable reducing agents are alkali metals, alkaline-earth metals, aluminium or light lanthanide metals prepared by electrolysis. Applications of rare-earth metals and alloys rely on their large affinity for oxygen and sulphur. Small amounts of mischmetal act as a trap for these elements, usually detrimental to the properties of steel or cast iron. This results in better resistance to high temperature oxidation and thermomechanical properties of several metals and alloys, for instance by hardening stainless steel or producing a malleable iron. As said above, pyrophoric Mischmetall is used in flint ignition devices for lighters and torches.

Addition of 3–4% cerium to magnesium alloys in addition to 0.2–0.6% zirconium results in a grain refinement and allows perfect casting of complex shapes. Resistance of aluminium to corrosion can also be improved by addition of cerium. Addition of Y, La or Ce to heat-resistant alloys likewise increases their performances.

A highly magnetostrictive alloy, $Tb_{0.3}Dy_{0.7}Fe_{1.92}$, has magnetic deformation reaching $1.9\,\mu m/cm$ at room temperature. This is 2–5 times greater than piezoelectric ceramics. The alloy was discovered in 1950 and enters into the composition of electroacoustic transducers in sonars, hence its acronym Terfenol-D meaning

Terbium **Fe** **n**aval **o**rdnance **l**aboratory **D**ysprosium. Similarly, ytterbium metal sees it resistivity increasing when subjected to stress, a property on which stress gauges monitoring ground deformations caused by explosions or earthquakes are based.

Scandium is the most potent strengthener of aluminium in Sc–Al alloys (0.1–1 wt% Sc); in addition, these alloys are extremely easy to weld because scandium inhibits grain regrowth within the alloy during welding. Scandium initially found application in Soviet military aircraft and weaponry during the cold war. Presently, aircraft designers believe that Al–Sc alloys can reduce aircraft weight by as much as 15%–20%. Extension to ship building is foreseen. Non-military applications that were proposed in 1990 by a Canadian company are in high-performance materials for high-level sport competition such as bicycle frames, gold clubs, ski poles and baseball bats. Nevertheless, the scandium market remains small, estimated to 5–15 tons/year (including the oxide, see below), not only because the scarcity of its availability but, more importantly, because of its price, $4500–5,000 per kg of oxide (99.99% purity). However, the recent use of scandium in the electrolyte of solid oxide fuel cells will probably boost its market.

17.5.2 *Magnetic cooling*

Several rare-earth metals and alloys have large heat capacities and are used as low-temperature regenerators in cryo-coolers. The initial Er_3Ni alloy that replaced Pb in the low-temperature stage (10–4 K) of Gifford-McMahon cryo-coolers in 1992 was subsequently substituted by Nd, $HoCu_2$ or $GdAlO_3$ while Pb in the intermediate stage (60–10 K) is presently replaced with $Er_{1-x}Pr_x$ ($0 \leq x \leq 0.5$) alloys. New compounds LnM (Ln = Er, Tm; M = transition metal) displaying still larger heat capacities are presently being investigated for improving the performances of the cryo-coolers below 15 K.[5]

17.5.3 *Fuel cells and batteries*

Solid oxide fuel cells represent the most promising technology for fuel cells. They contain several rare-earth compounds. The anode is made of nickel mixed with yttria-stabilised zirconia (8 mol%, Y8SZ). Alternatively, since scandium oxide (scandia) exhibits exceptional electrical conductivity and heat stabilisation qualities it can advantageously replace yttria; a doping concentration of 9 mol% (Sc9SZ) enables lower operating temperature resulting in longer lived equipment. As a matter of fact this application represents the largest use of scandium. Widespread materials for the electrolyte are Y8SZ, Sc9SZ or gadolinium-doped ceria. Finally, lanthanum strontium manganate is the material of choice for the cathode because it is compatible with zirconia electrolytes.

Nickel-lanthanide hydride (Ni–LnH) rechargeable batteries are seen in electric (EV) and hybrid electric (HEV) vehicles and in earlier portable appliances (tablets, smart phones, cameras, for instance). The first really stable design was proposed in 1987 and had formula $La_{0.8}Nd_{0.2}Ni_{2.5}Co_{2.4}Si_{0.1}H_x$ but a more economically viable design called for the lanthanides replaced by mischmetal and was marketed in 1989. Almost 60% of these batteries are built for EVs and HEVs. A typical assembly for HEV contains about 4–6 kg of rare earths. The demand for them will be dominated by the growth of the EV+HEV market on one hand (2.1 million vehicles sold in 2019) and by the feasibility of substituting the metal hydride batteries with lighter and more powerful lithium batteries on the other hand. Overheating problems and the fact that lithium is also a critical element have somewhat slowed large dissemination of these batteries but they have nevertheless now replaced NiMH devices in most electronic devices. Their price has decreased 10 times over the past 12 years and some reports estimate that NiMH battery technology will decrease significantly after 2020.

17.5.4 *Water treatment*

A mixed rare-earth chloride solution marketed under the brand name SorbX-100™ by Molycorp leads to rapid and stable precipitation of phosphates in municipal and industrial wastewater facilities.

17.6 Ceramics

Chemical and structural properties of the rare earths are in line with the needs of the ceramics industry. For instance, minute additions of rare-earth oxides, mainly cerium or yttrium, stabilise tetragonal or cubic forms of zirconia. Among them yttrium oxide (1–10 mol%)

is the best compromise. Stabilised forms of zirconia are used in sensors (for their high ionic conductivity), cutting tools (for their good thermomechanical properties), refractory materials, or imitation jewellery (when the cubic form is fully stabilised, for Y_2O_3 content above 7 mol%).

The Superamic® ceramics from Rhodia have been modelled as dielectric, piezoelectric and ferroelectric materials. In particular, barium titanate doped with Nd^{III} has a dielectric constant that does not change over a large temperature range. These ceramics have very fine particle size (10–500 nm) allowing miniaturisation. They enter into the composition of electronic equipment such as microwave filters, high-performance capacitors, semi-conductors, resonators or oxygen sensors (in particular those used in automotive TWC systems). The RE elements most used are La, Ce, Pr and Nd.

The high-temperature (HTS) superconductors discovered in the late 1980s involve yttrium, e.g. $YBa_2Cu_3O_7$. These ceramics conduct electricity without resistance but still have a limited use since they do not work at room temperature; they can be made into power cables, although of limited length, for example for switching stations.

Glass ceramics containing cerium oxide enter in the composition of some zirconia dental implants. It is noteworthy that in 1966, a radioactive sodium uranium oxide was introduced as fluorescent substance in aluminous veneer porcelains for dentistry; this salts emits a strong greenish-yellow colour but addition of small amounts of cerium oxide results in a bluish-white emission. This makes artificial teeth very much alike natural ones, but since then radioactive fluorescent substances have been banned and this porcelain is no more used.

17.7 Photonics (Phosphors)

This field is burgeoning because all trivalent lanthanide ions, barring lanthanum and lutetium, behave as ideal wavelength converting devices with narrow emission bands covering the entire spectral range from UV (Gd) to visible and near-infrared (up to 3 μm). In addition, trivalent cerium and some divalent ions like europium, display broad but widely tuneable d-f emission. Selected emission spectra of trivalent lanthanides are shown in Figure 17.7. The drawback is that intraconfigurational transitions are forbidden and therefore the

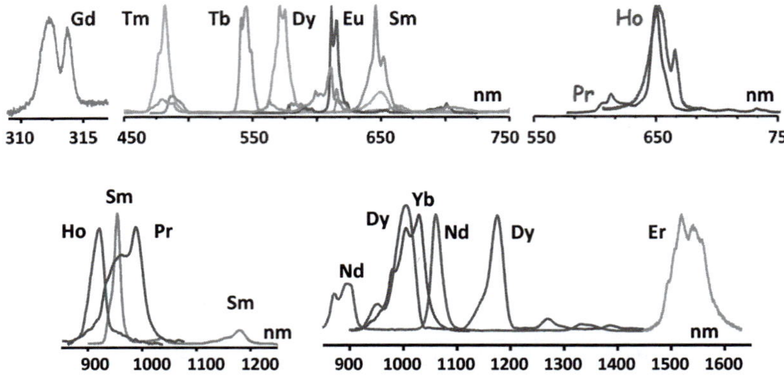

Figure 17.7. Emission spectra of selected trivalent lanthanide ions. Vertical scales are not comparable.

corresponding absorptions are faint ($\varepsilon < 15\ \mathrm{M}^{-1}\,\mathrm{cm}^{-1}$). This can, however, be overcome by the antenna effect discovered in 1942.[6]

17.7.1 *Lasers*

In view of their numerous and well defined electronic levels, lanthanide ions are ideally suited as active materials for solid-state lasers emitting in the UV, visible or NIR. The only limitation is the transparency of the matrix. Lasing materials are often single crystals carefully grown to avoid defects. A good thermal conductivity is also required in order to evacuate heat: yields of lasing are low, so that a great amount of energy is transformed into heat. One of the most widely used lasers is YAG:Nd^{III} with its mythic line at 1.06 μm, the frequency of which can easily be doubled (532 nm, green; e.g. laser pointers for slide presentations), tripled (355 nm, blue) or quadrupled (266 nm, UV), leading to multi-line lasers for excitation of luminescence spectra. To enhance the yield of YAG:Nd lasers, Cr^{III} or Ce^{III} are often introduced in the garnet as sensitisers. The YAG garnet can be doped by several other lanthanide ions, giving rise to devices emitting at 1.03 μm (Yb), 1.93 – 2.04 μm (Tm), 2.1 μm (Ho, Tm) and 2.94 μm (Er). Low-power lasers, particularly those emitting at long wavelengths are used in several medical applications, eye surgery, skin treatment and dentistry or for monitoring the sugar content in blood. NIR-emitting lasers are one of the major components in telecommunication systems. High-power YAG:Nd lasers are used in manufacturing while arrays of high-power Nd-doped phosphate glass lasers provide up to 500 TW power in nuclear fusion experimental facilities, e.g. at National Ignition Facility, Lawrence Livermore Laboratory, USA.

17.7.2 *Telecommunications*

Modern telecommunications and internet-providing systems mostly rely on silica optical fibres but despite their excellent transparency the signals get attenuated after 50 or 100 km and need amplification. The latter is achieved by erbium-doped fibre amplifiers (EDFAs) that were proposed in 1987. Since Er^{III} is emitting in the main telecommunication window (C band, 1.5 μm), silica glasses doped with this ion are ideal waveguide amplifiers.[7] The Er^{III} ion has indeed a rich and adequate electronic configuration, which is partly shown in Figure 17.8(a) along with the one of its best sensitizing ion, Yb^{III}.

In EDFAs, erbium is excited either at 980 nm, on the $^4I_{11/2}$ level, or at 1,480 nm, on a metastable ligand-field sub-level of $^4I_{13/2}$. This results in a two-level laser emitting in the range 1530–1,550 nm depending on the fine-tuning of the materials into which it is doped.

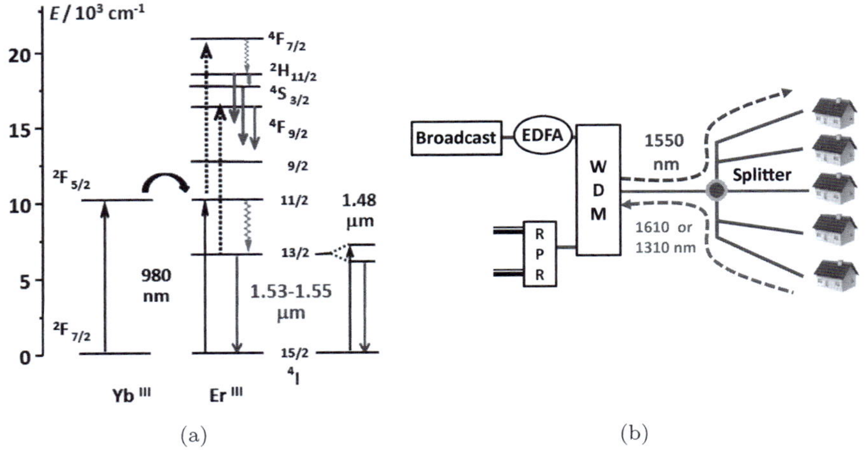

Figure 17.8. (a) Partial electronic level diagram of Er^{III}. (b) Scheme of a fibre-optic network; RPR = resilient packet ring. Reproduced from Ref. [7], with permission from the Centre National de la Recherche Scientifique (CNRS) and the Royal Society of Chemistry.

The 980 nm band has higher absorption cross-section, so it is generally used when low-noise performances are required; the transition is narrow and therefore wavelength-stabilised laser sources are needed to pump energy into it. The 1,480 nm band features a lower absorption cross-section but is broader, which makes it ideal for high-power amplifiers. Many EDFAs use a combination of both pump wavelengths.

There are two types of networks. In the first one, the signal is conveyed to the end-user by conventional copper-based coaxial cables in the last section of the system; although not entirely satisfying, this avoids costly replacement of existing networks in houses and offices by fibre optics. The second type is entirely based on fibre optics and is implemented in new constructions or low-density population areas. The signals from the broadcast unit are amplified by an EDFA, fed to a wave division multiplexer and distributed to the end users. Return signals are carried by 1,310 or 1,610 nm light generated by InGaAsP diode lasers and are separated from the downstream signals before being converted into radio-frequency packages (Figure 17.8(b)).

One of the challenges faced by these systems is the competition with wireless networks, meaning that future optical fibres will have to deliver more bandwidth at lower cost. This translates into the search for more efficient lasers and optical amplifiers. Indeed, the low absorption cross-section of Er^{III} f-f transitions and concentration quenching when there are more than 10^{20} Er^{III} ions per cm^{-3} limit the performances of EDFAs, so that doped fibres with length of several tens of cm have to be used. There are several solutions to this problem. The first one is co-doping sensitisers to compensate for the low absorbance of Er^{III} at 980 and 1,480 nm; Yb^{III} is particularly adequate for 980 nm pumping since its absorption cross-section at that wavelength is about ten times larger. Moreover, the $Yb(^2F_{5/2})$ excited level efficiently transfers energy onto the resonant $Er(^4I_{11/2})$ level that decays faster to the $^4I_{13/2}$ level than back transfer to Yb^{III}. A second solution consists in using hosts different from silica and co-doping high-refractive index components. A third way out is to dope Er^{III} into polymer fibres, for instance polymethylmethacrylate (PMMA), with the help of organic ligands. The resulting amplifiers have definite technical advantages over silica: better flexibility and larger diameter, allowing optimum coupling to local-scale devices, as well as a much lower pump threshold, up to 660-fold smaller.

Presently, much emphasis is put on developing integrated optical devices combining both EDWAs (erbium-doped waveguide amplifiers) and EDWLs (erbium-doped waveguide lasers) on a small chip (typically 1×1 mm^2). Optimisation of all parameters has led to amplifiers in which the fibre length can be reduced to a few cm thanks to gains up to 935 dB $\cdot$ cm^{-1} and to high-performance lasers, either continuous or pulsed, with ultra-narrow bandwidth. Future possibilities include increasing the functionality of the Er^{III}-based devices by integrating multiple photonic components into microelectronic chips.

17.7.3 *Lighting*

Despite a sharp increase in the number of lighting devices worldwide, the share of electricity devoted to it is rather decreasing. It is estimated (2010–2014) that lighting represents 12–20% of all electricity used, depending on countries. What is really striking is the prominent role played by rare earths in improving the efficiency of lighting devices, as shown in Figure 17.9. The historic electric bulb was more a heating device than a lighting one, with a very poor efficiency. In the 1970s, lanthanide-containing phosphors revolutionised lighting when they were coated on the inner surface of fluorescent tubes. The coating is a carefully designed mixture of three phosphors, typically Y_2O_3:Eu^{III} (red emitter), $LaPO_4$:Ce^{III},Tb^{III} (green emitter,

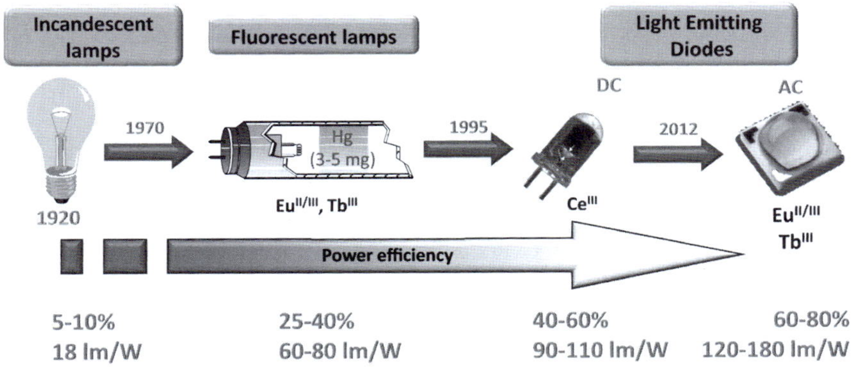

Figure 17.9. Impact of rare earths on lighting. Efficiencies are "typical" values, i.e. average of several sources (technology is evolving and numbers keep improving!); luminous efficiency (lm/W) refers to an input electric power of 100 W.

Ce^{III} acts as sensitiser) and $BaMgAl_{10}O_{17}$:Eu^{II} (BAM, blue emitter). Power efficiency of the luminaires raised by a factor 4 and the same technology was later introduced in compact fluorescent lamps (CFLs). Unfortunately these devices contain a small amount of mercury that is responsible for the electric discharge giving off the UV excitation (254 nm); when not properly disposed of (i.e. broken or not recycled) these lamps cause environmental problems, a reason why their sale is being increasingly banned.

The situation changed in the mid-1990s with the advent of commercial white light-emitting diodes (WLEDs), following the development of GaN blue LEDs, because these devices are much smaller, do not require mercury, have better lifetime and better efficiency. In the initial design white light is generated by combining the blue emission from (In,Ga)N with the yellow d-f luminescence from $Y_{3-x}Ce_xAl_5O_{12}$ ($0.001 < x < 0.03$, YAG:Ce^{III}, quantum yield >90%), a phosphor that was proposed in 1967 for cathode ray tubes (Figure 17.10).[8] Some shortcomings, such as poor colour rendering index and high correlated temperature have since been cured for instance by increasing Ce^{III} concentration, co-doping other Ln^{III} ions, or introducing a red-emitting Eu^{II} phosphor ($Sr_2Si_5N_8$:Eu^{II}, $CaSiAlN_3$:Eu^{II}). Following this success, numerous other designs have been investigated, including three-component (RGB) phosphors and careful nano-patterning of the phosphor structure, with the imperative goal of improving brightness. Presently, laboratory luminaires incorporating WLEDs reach brightness in the range 200–250 lm/W.

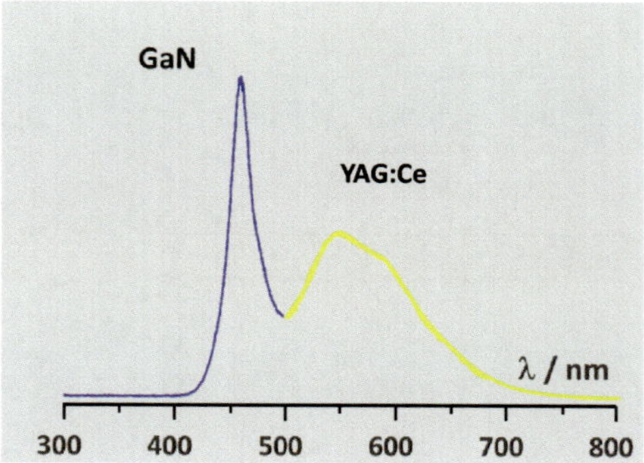

Figure 17.10. Typical emission from a first generation white light-emitting diode.

This is a necessity if the general objective of 30% reduction in lighting electricity consumption looked-for by most states and international institutions is to be met by 2030. Help towards this goal may come from AC-LEDs introduced on the market in 2012. They are devoid of transformer, which improves their efficiency (claims up to 85% power efficiency have been filed). To compensate the flickering due to alternative electro-excitation, phosphors with long lifetimes (long persistence phosphors) are used, that contain $Eu^{II/III}$ and Tb^{III}. It is presently too early to predict if AC-LEDs are really going to supplant conventional ones in view of the large improvements made recently in the latter.

The LED market, worth about \$50–60 billion in 2019 represents 40–45% of the total lighting market and is growing steadily; its largest production share comes from Asia, particularly from Japan (25%), Taiwan (25%) and South Korea (20%), while China, Europe and USA share about equally the remaining 30%. The Chinese market has the fastest growth though. Most scenarios elaborated by governments or international institutions call for the replacement of CFLs and high intensity discharge lamp (high- and low-pressure mercury or sodium lamps for street illumination) by LEDs. For instance the European Union scenario for street, office and industrial lighting calls for a 70% share of LEDs by 2030, with new-generation linear fluorescent lamps (LFLs) keeping a 25% share.

17.7.4 *Displays*

The phosphors of cathode-ray tubes (CRT) displays were very similar to those of fluorescent tubes. CRTs were then replaced with flat-panel displays based on several different technologies, plasma and electro-luminescent panels, LED-backlit liquid-crystal displays (LB-LCDs), or, lately organic light-emitting diodes. Except for the latter, most of these panels still work on the RGB principle and a combination of older and newer RE-containing phosphors make them shine brightly, particularly LB-LCDs which are presently the mainstream displays.

17.7.5 *Security, signage and tagging applications*

The peculiar properties of lanthanide light, i.e. sharp and well-defined emission bands have prompted its use in security inks, counterfeiting tags and safety signage, not to mention pressure sensors or luminescent toys.

Part of these applications are based on "regular" downshifting luminescence, that is excitation in the UV/visible and emission in the visible/NIR. This is for instance the case for some security features implemented in the euro banknotes. The orange-red luminescence clearly arises from a Eu^{III} compound (Figure 17.11(a)) while the bluish-greenish emission could be due to a Eu^{II} pigment. Combining several lanthanide ions into one phosphor material leads to designing bar codes or recognition/counterfeiting tags (Figure 17.11(b)). Several companies are selling rare earth containing tags for marking batches of various goods, among them ammunition. Indeed explosion transforms the tags into a mixture of rare-earth oxides that retain the same composition as the initial tag, allowing tracing the origin of the ammunition or explosive powder. Counterfeiting tags take advantage not only of the spectral signature of luminescent lanthanide mixtures but, also, of their different lifetimes.

In life sciences, lanthanide-based optical encoding materials enable multiplexing of large numbers of samples including assays involving nucleic acids or protein-antibody pairs or the multiplexed immunohistochemical staining of cells and tissues (Parallume® technology).

Growing attention is presently turned to upconversion tags. Upconversion is an optical process in which two low-energy photons (typically NIR photons) are absorbed and a higher energy photon emitted; in this respect, Er^{III} is an ion of choice emitting around 540 (green) and 650 nm (red), as is Tm^{III} with emission at 290, 350, 460, 660 (weak) and 800 nm (Figure 17.11(d)). The most common sensitiser is Yb^{III} (18–20%; see Figure 17.8) and the most common matrix $NaYF_4$. A vast number of papers describe the synthesis and tuning of upconversion phosphors tailored for specific applications by varying the matrix and the co-dopants in order to adjust, for instance, the green-to red ratio in Er^{III} emission. The phosphors are typically used under the form of nanoparticles (UCNPs) and they are ideal for security inks and tags as well as for biological applications, despite their low quantum yield (a few percent at most) and brightness.[9] For instance, yuan banknotes from China include a rectangular spot that emits yellow light under UV excitation; however when 980 nm

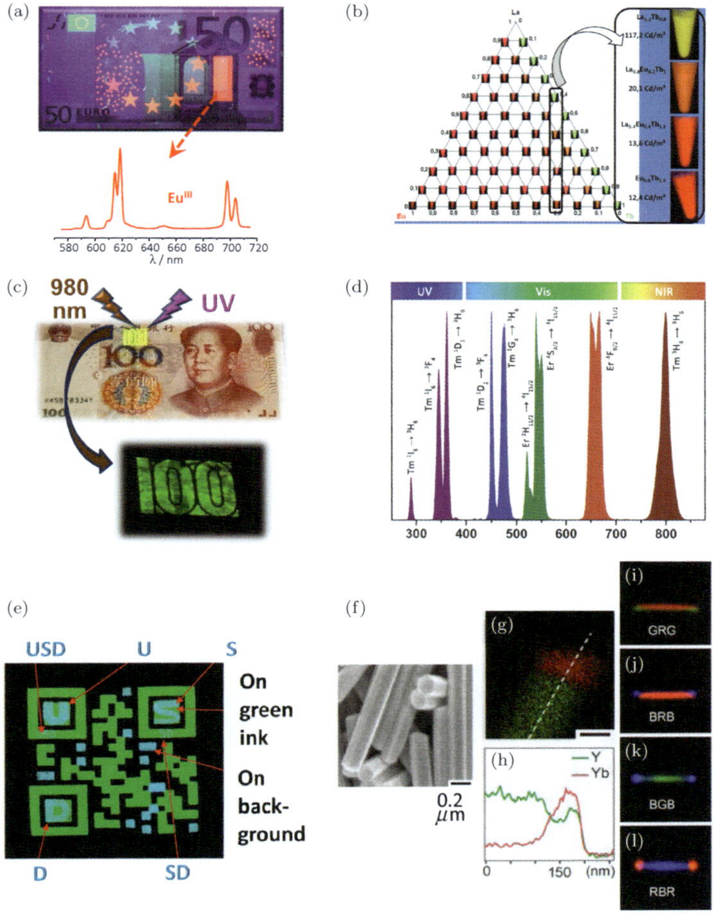

Figure 17.11. (a) 50-euro banknote under UV light; (b) Example of combining three Ln^{III} in a benzene-dicarboxylate metal-organic framework to generate specific luminescent tags. Reproduced with permission from Ref. [22]. Copyright (2013) Wiley & Sons; (c) 100-yuan banknote under UV and NIR illumination (Courtesy of Prof. Liu Xiaogang, NUS, Singapore); (d) Upconverted emission from $NaYF_4$:Yb,Er,Tm co-doped UCNPs. Reproduced with permission from Ref. [23]. Copyright (2015) the Royal Society of Chemistry; (e) QR code under 980 nm illumination: Emission from $NaYF_4$:Yb(20%),Ln(2%), Ln = Er (green, 540 nm), Tm (blue, 450–470 nm). Redrawn from Ref. [24]. Copyright (2012) IOP Publishing Ltd. (f) Epitaxially grown hexagonal $NaYF_4$:Yb,Ln microrods encompassing different combinations of UC phosphors (R: Yb/Er 50/0.05 mol%; G: Yb/Er 5/0.05 mol%; B: Yb/Tm 20/0.2 mol%). Adapted with permission from Ref. [25]. Copyright (2014) American Chemical Society.

illumination is provided, the resulting upconverted green luminescence reveals a number corresponding to the value of the banknote (Figure 17.11(c)). Quite spectacular is the "secret" information that can be embedded into bar codes or QR codes thanks to UCNPs (Figure 17.11(e)). Finally, more sophisticated hexagonal nanorods encompassing different $NaYF_4$:Yb,Ln phosphors allow designing different RGB combinations (Figure 17.11(f)) for specific nanomarking.

In the section on lighting, long persistent phosphors were mentioned. Persistent luminescence is the property of some luminescent materials to continue to emit light long after the excitation source is switched off. Zinc sulphide doped with transition metals, for instance copper and cobalt, has long been used in conjunction with ^{226}Ra as excitation source in luminous paints for watch and clock dials. In 1968, the optical properties of $SrAl_2O_4:Eu^{II}$ were described, in particular a residual luminescence persisting for several seconds, but this emission was too weak to generate practical applications. An important step was achieved in 1994 when it was found that cooping Dy^{III} into this phosphor results in a much brighter material that increasingly replaced the historical phosphor. Applications as diverse as safety signage (e.g. luminous strips indicating emergency issues in airplanes) or luminous toys are now standard. Less convincing are road marking (a test has been made in the Netherlands in 2014 with limited success) or energy storage (for solar cell, but the capacity is rather small). On the other hand, *in vivo* imaging with NIR persistent phosphors is gaining momentum.[10]

17.7.6 *Other applications*

Radiation detection is important to several fields, including defense, medicine (e.g. PET, CT scans), material testing, environmental hazard monitoring or security checks at airports and borders (e.g. doorways for screening trucks for radioactive materials). It turns out that there is no direct sensitive detector for photons with energy larger than a few keV. A wavelength converting material is therefore needed (Figure 17.12). Earlier scintillator crystals were often made of simple lanthanide salts, such as lanthanum or cerium chloride or bromide, or of calcium fluoride doped with divalent europium. More efficient ones are now at hand, featuring various inorganic matrices such as lutetium silicate, aluminate or halides doped with Ce^{III}.[11]

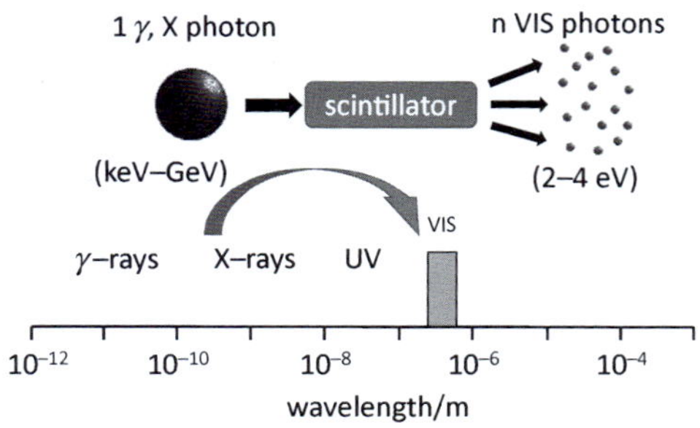

Figure 17.12. Wavelength conversion by a scintillator.

Present record is held by $Lu_3Al_5O_{12}$:Ce with a yield of 25,000 photons per MeV.[12]

Optical limiters are materials that are transparent at normal light intensities and opaque to very bright light, henceforth they help avoiding damages caused to human eyes or optical components by sudden and intense laser pulses. Their characteristics are low ground state absorption and strong excited state absorption, which is the case for porphyrinates and their complexes with lanthanides. Lanthanide fullerenes have also been proposed for this purpose.

About 10,000 ^{171}Yb atoms cooled to 10 μK, confined into an optical lattice and excited into their ultra-narrow transition at 578 nm work as the most precise atomic clock assembled to date with precision around $2{:}10^{-18}$, that is 10 times better than the other known atomic clocks (e.g. caesium clock).

Temperature is an important physical variable affecting the dynamics and viability of practically all natural and engineered systems, from atomic to macroscopic levels Temperature is therefore an crucial parameter to monitor in a broad range of applications, and corresponding sensors are deserving a lot of attention, particularly in nanomedicine. The ideal sensor should be non-invasive (contrary to contact thermometers) and possess a large spatial resolution. Thermal dependence of luminescence (intensity, band shape, lifetimes) is developing as one of the most accurate techniques for sensing temperature because it can achieve large sensitivity ($>1\%$/K), good spatial resolution ($<10\,\mu$m) and short acquisition time (<1 ms). Ratiometric methods using lanthanide luminescence have thus been proposed that work at the nanoscale level and that are becoming more and more popular.[13]

Lanthanide ions, particularly europium and terbium are prone to mechanoluminescence, a phenomenon in which photons are emitted when a material is cracked, crushed or mechanically strained. This gives rise to luminescent pressure sensors that find applications in wind tunnels or for the detection of cracks in astronaut clothes.[28]

17.8 Applications in Biosciences

The development of diagnostic imaging techniques and their associated contrast/enhancing agents has had an impressive impact on the advancement of medicine during the past 30 years, particularly in the case of tumour detection and cure. Lanthanides have greatly contributed to three major fields: X-ray radiography, magnetic resonance imaging (MRI)[14] and optical imaging/analysis.[15] A lanthanide bioprobe is a lanthanide inorganic or coordination compound (or nanoparticles thereof) that exhibit characteristic magnetic or luminescent properties allowing its easy detection by spectroscopy or imaging techniques. Bioconjugates allow more specific detection to be carried out both *in vivo* and *in vitro* and they are usually preferred to simpler probes. Moreover, several lanthanide radioactive isotopes enter into the composition of radiopharmaceuticals for cancer radiotherapy.

17.8.1 *X-ray intensifying screens*

Lanthanide phosphors have long supplanted calcium tungstate in X-ray intensifying screens, allowing to considerably reducing X-ray doses and opening the way to colour contrast medical radiography. Typical phosphors are: Ln_2O_2S:Tb (Ln = La, Ce), BaFCl:Eu^{II}, LaOBr:Ln (Ln = Tb, Tm). Their advantages lie in large absorption cross-section for X-rays leading in noise reduction, adequate conversion efficiencies, as well as adequate refractive index, minimizing refraction and therefore enhancing safety.

17.8.2 *Contrast agents for MRI*

Magnetic resonance imaging is based on the fact that both spin-lattice and spin-spin proton relaxation times take different values depending on the tissue (40–400 ms). Contrast agents increase the proton relaxation rates of water in tissues in which they are distributed and therefore help achieving a better differentiation. Presently 40% of all MRI scans use a Gd^{III}-based contrast agent under the form of a thermodynamically stable, kinetically inert and non-cytotoxic chelates.[16] The high-spin $4f^7$ electronic configuration of Gd^{III} has a profound influence on the spin-lattice relaxation time, up to 10^6-fold. When injected into the patient (0.1 mmol/kg — about 5 g per MRI scan), gadolinium chelates distribute throughout the extracellular space and do not cross the blood-brain barrier; they are thus efficient for detecting pathologic abnormalities. The first experiments using a Gd^{III} chelate to enhance the contrast of brain tumour images collected by nuclear magnetic resonance were performed in 1984 and the first commercial contrast agent (CA), $[Gd(dtpa)(H_2O)]^{2-}$, was authorised in 1988, with a second one, $[Gd(dota)(H_2O)]^-$ following one year later. Since then these contrast agents and others developed subsequently have been administered to millions of patients.

The effect of a CA is defined through its relaxivity r_i that depends on three main factors (Figure 17.13): (i) the exchange rate $1/\tau_M$ of the bound water molecules and of the water molecules hydrogen-bonded to those in the inner coordination sphere; diffusing molecules in the outer sphere also contribute to this parameter; (ii) the rotational correlation rate of the molecule, $1/\tau_R$ and (iii) the longitudinal electron spin relaxation time τ_S:

$$r_i = \frac{1}{\Delta T_i[\mathrm{CA}]} \quad (i = 1, 2) \quad [\mathrm{mM^{-1}s^{-1}}]$$

For $[Gd(dota)]^-$ these values are equal to $4.1 \times 10^7\,\mathrm{s^{-1}}$, $1.25 \times 10^{10}\,\mathrm{s^{-1}}$ and 1 ns, respectively. Basic imaging simply requires the contrast agent to be intravenously injected. However, more sophisticated experiments necessitate a more specific targeting and, moreover, changes in relaxivity upon association with analytes confer to Gd CAs a role as analytical probes.

17.8.3 *Lanthanide luminescent bioprobes: The tools*

Lanthanide luminescent bioprobes (LLBs) have greatly contributed to biosciences during the past 35 years because they present definite advantages over organic probes, such as sharp and easily recognisable

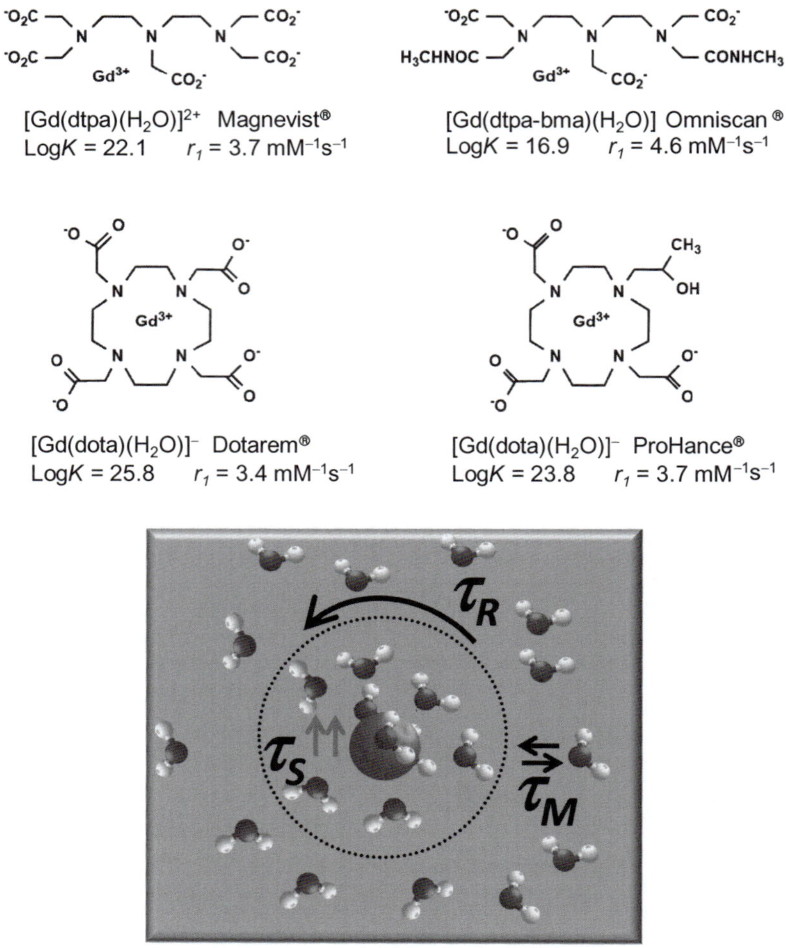

Figure 17.13. Top: Gd^{III} chelates in first-generation contrast agents (trademarks may change depending on the country) with thermodynamic stability constants and relaxivity values. Bottom: Factors influencing relaxivity.

emission bands, little (complexes with organic ligands) or no (inorganic probes) photobleaching and, thanks to time-gated detection, high sensitivity. In the following are listed the main experimental techniques specific to LLBs, as well as the major forms of these probes.[17]

- **Time-resolved detection (TRD).** Ln^{III} excited state have long lifetimes (μs-ms) rendering TRD easy and cheap to implement.

This technique avoids interference from autofluorescence of the sample and scattered light. The sample is illuminated with a pulsed flash lamp or LED and the detector is gated. After the initial excitation flash, the detector is turned off during a time long enough for the unwanted autofluorescence to disappear. It is then turned on for measuring exclusively the luminescence given off by the probe. The experiment lasts about 1–5 ms and can therefore be repeated 200–1,000 times per second, which considerably enhances the signal-to-noise ratio and consequently the sensitivity.

- **Förster resonance energy transfer (FRET or LRET).** Although not specific to LLBs, this phenomenon is a major tool in immunoassays and in the deciphering of molecular interactions. In a typical experiment, the excited Ln^{III} ion transfer energy onto an organic dye, or quantum dot, that, in turn, emits light, but with a lifetime corresponding to that of the Ln^{III} ion, therefore easy to measure by TRD experiments.
- **Complexes with organic ligands** represent a good share of LLBs. They are interesting because of the antenna effect, in which light is absorbed by the ligand and then transferred onto the lanthanide ion, circumvents the low molar absorption coefficients of the latter. Progresses in ligand design reported during the past 20 years have led to quite bright species, with quantum yields in the range 50%–70% for Eu^{III} and 60%–80% for Tb^{III}, for instance. NIR emitting complexes have much lower quantum yields (1%–10%), but are still bright enough for practical applications.
- **Downshifting nanoparticles** (Stokes' emission) draw interest in view of their size-dependent chemical, electronic and photophysical properties. They possess high brightness and surface modification is well mastered, e.g. for bioconjugation or theranostic applications. The core can be silica, latex or metals such as gold, silver or platinum. In some experiments, nanoparticles are built from simple rare-earth salts or oxides. As for other nanoparticles long-persistent luminescent ones (LPNPs) can also be functionalised and bioconjugated. Their advantage is that they are irradiated before recording luminescence microscopy images or before being delivered to the cells, tissues or animals to be imaged. Alternatively, *in situ* illumination can be repeated, making long-term *in vivo* studies feasible.

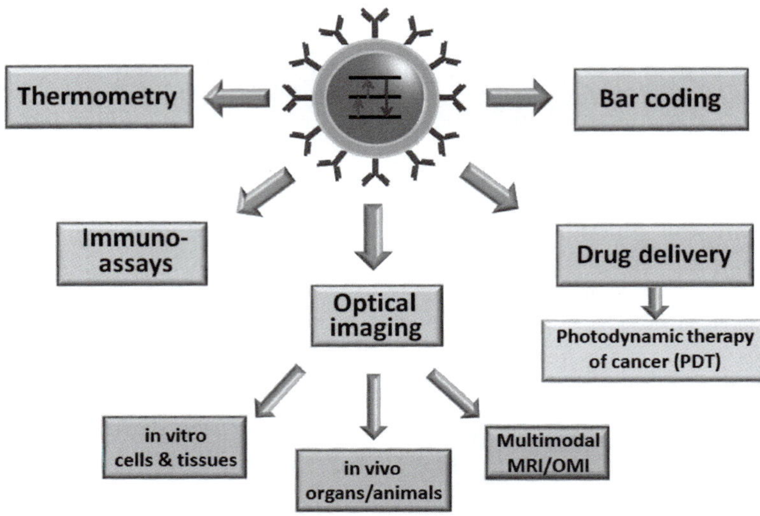

Figure 17.14. Major applications of UCNPs in biosciences. Reproduced with permission from Ref. [17]. Copyright (2016) Elsevier Science B.V. Amsterdam.[17]

- **Upconverting nanoparticles** (anti-Stokes' emission) meet most of the requirements for practical bioprobes, including low toxicity. One disadvantage of UCNPs, common to all nanoparticles, is enhanced non-radiative de-activation due to surface traps and defects that increases with decreasing size, henceforth very low quantum yields. This can be circumvented by designing core-shell UCNPs in which the active core material is coated with a non-luminescent layer of the matrix leading to passivation of the surface; quantum yields up to 5%–7% are reported. Functionalised UCNPs have recently turned into ubiquitous bio nanoprobes, as shown in Figure 17.14.
- **NIR-NIR probes: multi-photon excitation versus UCNPs**. When irradiated with UV and/or visible light, biological tissues display intense light scattering and auto-fluorescence that may obscure the probe signal. In addition UV/blue light is rather phototoxic so that it is desirable to excite bioprobes in the NIR. Another benefit of such excitation is a much deeper penetration depth, a crucial aspect for *in vivo* imaging. The absorption spectrum of human skin along with scattering contribution to the loss coefficient is shown in Figure 17.15. Several "biological windows"

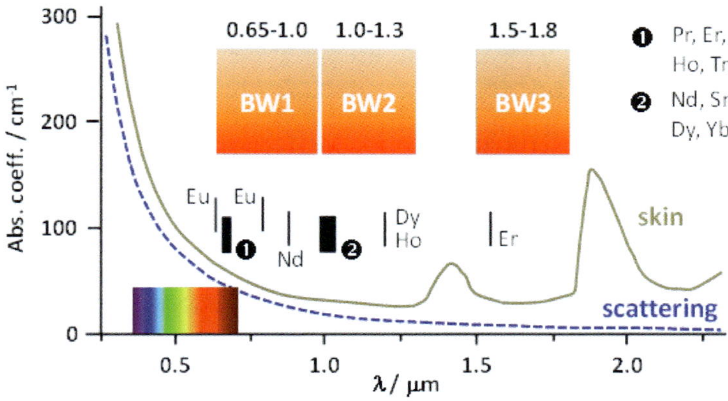

Figure 17.15. Absorption spectrum of human skin (plain green line) with light scattering contribution (dotted blue line), definition of biological windows, and emission wavelengths of some Ln^{III} ions. Reproduced from Ref. [17]. Copyright (2016) with permission from Elsevier Science B.V. Amsterdam.[17]

are referred to in the literature. The first one ($\sim 650 - 1,000$ nm) is presently in use but is less ideal than the second ($1 - 1.3$ μm) and third ($1.5 - 1.8$ μm) ones in view of still sizeable light scattering. For *in vivo* bioimaging, it is essential that both excitation and emission wavelengths fall within the biological windows. NIR excitation can be achieved by (i) multiphoton processes, (ii) upconversion and (iii) NIR excitation of NIR emitting Ln^{III} ions (e.g. Nd^{III}). The first type of excitation requires high laser power and is mainly suitable for *in vitro* experiments. Upconversion is more convenient for *in vivo* experiments because much lower excitation power is needed that can be provided conveniently by inexpensive laser diodes. Practical systems for NIR-NIR imaging of small animals have already been designed.

- **Lanthanide bioconjugates.** Simple lanthanide probes do not display selectivity while this feature is indispensable in bioanalysis. Therefore the luminescent tags are commonly conjugated to a biological molecule, e.g. an antibody or an antigen. Conjugation techniques are numerous and well described in the literature, but delicate to perform in that the resulting products have to be carefully purified and then tested for biological activity. There are mainly two ways of conjugating a LLB to a biological molecule. The first one relies on the very strong streptavidin-biotin interaction

(Log$K\sim$ 15) while the second is a direct — often covalent — link to the targeting molecule of interest.

- **Lanthanide binding peptides (lbp).** Also referred to as lanthanide-binding tags (lbt), these probes have a decisive advantage over other lanthanide reporters in that they can be genetically encoded. They also avoid nonspecific labelling and laborious synthetic and purification procedures. Lanthanide-binding peptides are small peptide sequences (~20 amino acids) that bind Ln^{III} with high affinity. They can be encoded at the DNA level and/or incorporated into recombinant proteins.
- **Multimodal probes.** Probes featuring two (or more) different modalities are prone to be more reliable than conventional unimodal ones. In addition, combining diagnosis with therapy (theranostics) seems to be ideal for better and accelerated cure of the patients. The past decade has witnessed important developments toward the design of multimodal probes, for both biosensing or bioimaging and theranostics. Examples are luminescent probes integrating two different emitting ions or with a contrast agent functionality enabling simultaneous imaging by MRI which allows one to combine high sensitivity of optical sensing with high spatial resolution of MRI. Attention is also driven by the possibility of combining drug delivery (e.g. in photodynamic therapy) with simultaneous monitoring of the drug effect.

17.8.4 *Immunoassays*

Immunoassays are the workhorse of bioanalysis and two main approaches are commonly used. The first one is the two-step dissociation-enhanced procedure DELFIA® that is a more sensitive alternative to conventional enzyme-linked immunosorbent assays (ELISA). Typically in a non-competitive assay, the analyte (often an antigen) is captured on an antibody-coated plate and then reacted with a second specific antibody labelled with a Ln^{III} chelate (step 1, Figure 17.16 top panel). After acid dissociation of the Ln^{III} ions (2a), they are complexed with an "enhancement solution" (step 2b) and the resulting chelates are inserted into micelles resulting in highly luminescent species. Although detection limits for simple Eu^{III} solutions are in the 10^{-12}-10^{-15}M range, the sensitivity of practical assays is commonly in the pM range. It can be boosted by

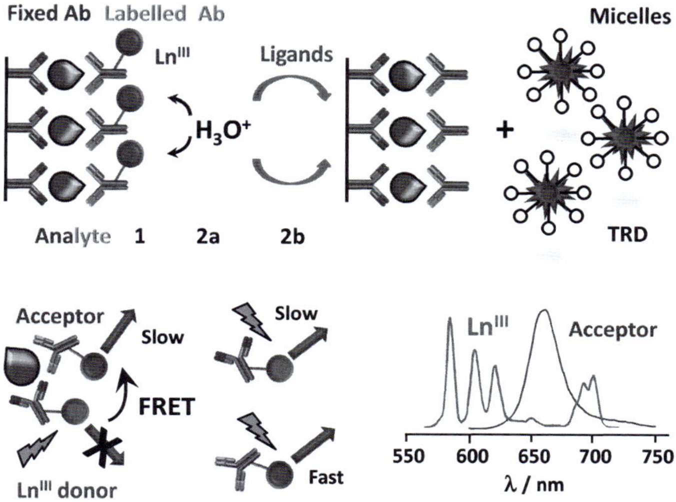

Figure 17.16. Top panel: Principle of heterogeneous immunoassays; 1: specific recognition of the analyte by a fixed and a labelled monoclonal antibodies (Mab); 2a: dissociation of the probe; 2b: re-complexation of the Ln ion and introduction into a micelle; TRD: time-resolved detection. Bottom panel: Principle of homogeneous immunoassays; the analyte is recognised by two labelled monoclonal antibodies; after excitation, the Ln ion transfer energy onto to acceptor dye; labelled antibodies not bound to the analytes are time- and wavelength-discriminated; emission spectra are shown on the right. Reproduced from Ref. [17]. Copyright (2016) with permission from Elsevier Science B.V. Amsterdam.[17]

multi-labelling, co-luminescence or, best, by replacing the lanthanide chelate by nanoparticles; the latter lead to a 500-fold improvement in sensitivity.

The experimental procedure is much simplified in homogeneous assays (Kryptor® technology) based on luminescence energy transfer (LRET or FRET, Figure 17.16, bottom panel). The analyte is reacted with two differently labelled specific antibodies, one bearing a luminescent lanthanide macrocyclic chelate, the donor, and the other fitted with an organic chromophore, the acceptor. Upon light excitation, four types of emission develop: fast autofluorescence from the medium, fast fluorescence from unreacted acceptor molecules, slow phosphorescence from unreacted Ln^{III} probes as well as slow fluorescence from the organic chromophore excited through FRET. Time-resolved detection distinguishes between slow and fast emission, while spectral discrimination takes care of differentiating between the two slow emitters. The technology is far reaching and much brighter macrocyclic complexes of both Eu^{III} and Tb^{III} are now at hand, opening the way to *in cellulo* analysis.

Both types of immunoassays are easily extended to a wealth of bioanalyses, including the detection of punctual mutations in DNA, multiplex analyses, visualisation of membrane protein interactions or ligand binding on the surface of living cells, determination of enzyme activity. In addition, lanthanide bioprobes can be tailored for the specific detection of a plethora of bio-relevant parameters and analytes. Among them are pH, $p(O_2)$, halides, nitrate, phosphates (including HPO_4^{2-}, $H_2PO_4^-$, adenosine mono-, di- and tri-phosphates), acetate, oxalate, malonate, succinate, lactate, citrate, urate and amino-acids.

17.8.5 *Point of care analysis, food and drug control*

The previously described analyses require specific instrumentation and/or data treatment and are not easily transferable to point of care technology. Therefore, lateral-flow paper biosensors represent an appealing and often cheap platform for detecting human and veterinary disease pathogens. The principle is illustrated in Figure 17.17. Multi-channel disks featuring up to 10 strips have been designed for fast analysis of several samples. Optical detection is performed by a small portable device incorporating a diode laser and a small photomultiplier for measuring the emitted visible light, linked to a data acquisition and calculator module (nowadays, simply a smart phone).

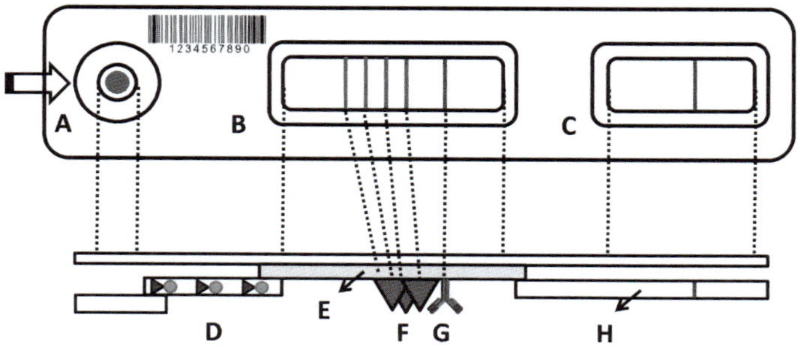

Figure 17.17. Principle of lateral-flow analysis; A = sample window, B = result window, C = ending index window, D = conjugate pad, E = analytical membrane, F = test lines, G = control line, H = absorbent pad. Redrawn from Ref. [26]. Copyright (2009) with permission from Elsevier.

Not only medical analyses benefit from the technology but also food and feed quality control, as well as drug abuse detection.

Novel developments include coupling UCNP-based assays with complementary metal-oxide-semiconductor CMOS technology for signal detection and processing. CMOS microsystems integrate a large array of detectors with signal processing circuits; they have low power consumption and are cheap and easy to manufacture in large quantities. Limit of detection are down to a few tens of nM. Systems biology requires monitoring very large numbers of samples through high-throughput and multiplex protocols. Miniaturisation linked to robotics and efficient data treatment provides the necessary technology for meeting this requirement; it can be achieved thanks to microarrays and microfluidics. The full range of LLB-based analyses is available for these technologies and relevant protocols start to be proposed, both for well-type analysis featuring solid-state arrays and for sophisticated microarrays.

17.8.6 *Bioimaging*

In view of their time-resolved detection ability, lanthanide luminescent bioprobes have far-reaching usefulness in time-resolved luminescence microscopy, immunocytochemistry and immunohistochemistry. Corresponding applications have developed at a slower pace than bioanalyses, mainly in view of the lack of commercially available low-cost instrumentation. The improvement achieved over steady-state measurements depends on the characteristics of the instrument components (light source, detector, optics) and operating parameters (excitation power, repetition rate, width of time-delay and measurement windows, exposure time); enhancements in signal-to-noise ratios (SNR) up to 100-fold are cited but when it comes to low-cost modification of epifluorescence microscopes however, it seems that a 4- to 5-fold improvement is more realistic. There are basically three types of imaging experiments.

The first one simply relies on visible-emitting probes and bioconjugates and is therefore more adequate for *in vitro* experiment rather than for *in vivo* imaging in view of the potentially cytotoxic UV or blue excitation light needed (commonly 320–400 nm). Figure 17.18 illustrates simultaneous detection of two markers expressed by breast cancer cells in a human tissue pressed onto a microfluidic device:

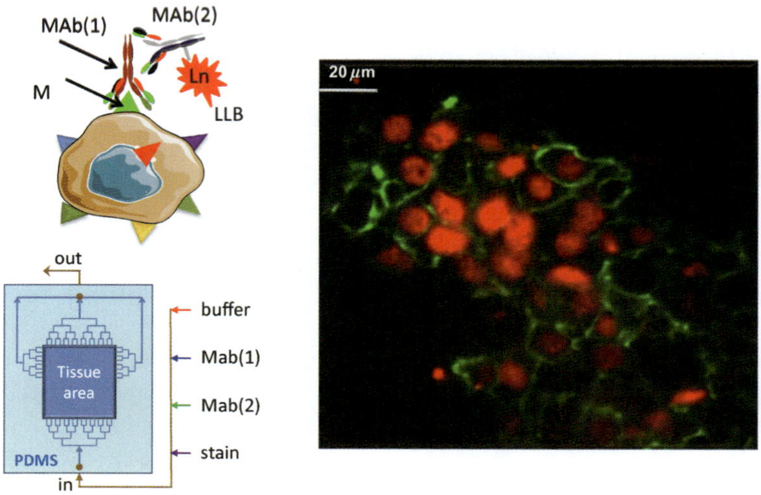

Figure 17.18. Top left: Principle of an indirect immunohistochemical assay: the marker (M) is targeted by a primary antibody Mab(1) itself recognised by a secondary antibody, Mab(2) fitted with the LLB. Bottom left: Sketch of the microfluidic cell used in which networks of distribution channels allow fast flow time (3 s). Right: Time-resolved luminescence microscopy image of oestrogen (nucleus, stained in red) and HER2/neu (membrane, stained in green) receptors expressed by a Human cancerous breast tissue illuminated by Eu^{III} and Tb^{III} LLBs, respectively. Reproduced from Ref. [27]. Copyright (2010) with permission from the Royal Society of Chemistry.

the estrogen receptor, expressed by the nucleus and the HER2/*neu* receptor expressed by the cell membrane. They are detected simultaneously in a very short time, approximately 5 minutes whereas classical analyses at the hospital take several hours and require five times more of the expensive reactants.

The second one relies on so-called NIR-NIR protocols in which the probe is excited in the NIR either by 1- or 2-photon absorption or by multiphoton upconversion and emits in one of the biological windows (BW, see Figure 17.15). This results in deeper tissue penetration, theoretically several cm but in practice resolution is lost after a few mm. NIR-excitable lanthanide ions are usually inserted into inorganic nanoparticles to overcome the poor cross-section of the f-f transitions. Complexes with chromophoric ligands would in principle be better, but their C–H vibrations severely quench the emission of ions such as Nd^{III}, Er^{III}, and, to a lesser extent, Yb^{III}. As for the lanthanide ions, Nd^{III} is ideal in that it can be conveniently excited either at 740 or 800 nm (BW1) and it emits at 880 (BW1), 1,060 and 1,300 nm (BW2); excitation of Er^{III} is at 980 nm and its emission at 1,530 nm (BW3), while Yb^{III} emits at 980–1,000 nm and absorbs in the red (porphyrin complexes). It is noteworthy that 980 nm is not an ideal wavelength since water has sizeable absorption coefficient at that wavelength, leading to heating of the tissue if a laser is used; tissue-friendly irradiation should not exceed 300 $mW \cdot cm^{-2}$.

It is to be stressed that nanophotonic materials are presently allowing cutting-edge technologies in nanoscopy and clinical research.[29] Finally, as mentioned earlier, multimodal imaging is offering complementary performances and is much under scrutiny despite sometimes very different experimental conditions needed for each mode of action.

17.8.7 *Drug delivery*

In therapy, lanthanide probes may function as light transducer for drug release, particularly in photodynamic treatment of cancer. The first step in this therapy is the selective uptake of a photosensitiser in the cancerous cells/tissue; subsequently, irradiation with predetermined doses of light activates the photosensitiser that generates reactive oxygen species themselves killing the tumour cells. One problem is penetration depth of the excitation light. Therefore, UCNPs have been modified with a shell impregnated with the photosensitiser. In this way, deep-penetrating infrared light can be used to initiate up-conversion and the visible light emitted *in situ* by UCNPs then activates the photosensitisers. This technology seems to present a good prospect for the future.

17.8.8 *Nuclear medicine*

Several Ln^{III} ions are β-emitters with a relatively long penetration depth, between 2 and 12 mm, which makes them adequate for treating tumours with high heterogeneity. All of them, baring ^{90}Y, have also a small γ-emission that is useful for measuring the biodistribution of the radiopharmaceutical. The therapeutic agent is a conjugate formed by a targeting monoclonal antibody labelled with a cytotoxic radioisotope embedded into a coordination compound. The latter needs to be kinetically inert on the time-scale of the half-live of the conjugate, about one day, hence the present use of macrocyclic ligands, e.g. cyclen derivatives. Regarding isotopes, ^{90}Y, produced from the decay of ^{90}Sr, has high-energy pure β emission ($t_{1/2}$ = 2.64 days) and is therefore well suited for systemic cancer therapy; it started to be used in the early 1980s. ^{153}Sm is useful when cancer has spread to bones; the most common cancers treated are lung, prostate and breast cancers; radioactive samarium is taken up in the bone cancer area and gives off radiation that helps provide relief of pain. ^{166}Ho emits β rays having a penetration range of 9 mm that are useful for treating skin and hepatic cancers. ^{177}Lu has properties similar to ^{90}Y but its β-ray penetration is smaller so that it is provided for treating smaller tumours; it is presently being tested for curing prostate cancer and generally speaking, its applications are growing. Similarly to ^{166}Ho and ^{177}Lu, ^{149}Pm can be prepared with no added carrier and has similar bio-distributions and properties.

In addition, radio-complexes that also emit γ rays are relevant as tracers in positron emission tomography (PET) imaging.

17.9 Agriculture and Feed for Livestock

In the early 1970s Chinese farmers started to use mixtures of lighter lanthanide salts or complexes as fertilisers. Three main products were developed: *Changle-Yishizu* based on nitrates and containing about 32 wt% equivalent La-Gd REO, *Nongle* made from lanthanide chlorides (38 wt% equiv. REO), and *MAR* comprising a mixture of Ln^{III} complexes (La–Nd) with 17 different amino-acids. These mixtures further contain a number of other elements essential to plant growth and were sometimes applied along with herbicides/pesticides or vitamin C. Initially reported increases in crop or fruit production reached up to 20–30% (50% for tobacco), but these numbers are presently down to about 10–15%, still an appreciable improvement. However due to concerns about the effect of lanthanides on living animals and human beings, the practice seems to be more or less abandoned nowadays.

Another usage relies on wavelength-converting materials, mainly simple europium salts and complexes, as additives to agriculture plastics. Detrimental UV light is absorbed and transformed into red light which has good conversion efficiency in photosynthesis. Here also, increases in crop yields of about 10% are reported.

In 2006, the European Union has banned the use of antibiotics in animal food. This has stirred interest for the development of alternative methods and substances to promote growth. Rare-earth compounds have been tested to this end, particularly in Germany and Switzerland, following Chinese experiments. The main rare earth containing compounds used in these diets are chlorides, nitrates or complexes with ascorbate or citrate and the dosage is usually 100–200 mg/kg feed. In most studies, dietary application of rare earths did not influence the health of animals, meat quality or the safety of the products while improving growth performances of pigs, poultry, sheep, goats, calves, horses, fish and prawns. Reported increase in body weight is around 20% (40% for sheep and goats) and increases in egg or milk production range between 15% and 20%, but there are some contradictory reports. For the time being it seems unlikely that these procedures will get approval from governmental offices though.

17.10 Energy-Related and Futuristic Applications

Several energy production technologies need rare earths: nuclear reactors, wind turbines, hydroelectric production, solar energy conversion, for instance. Here we focus on the latter only.

17.10.1 *Solar energy conversion*

Photovoltaics is a natural target in the quest for alternative sources of energy in view of its availability in large quantity, the sun irradiance at sea level amounting to about $1{,}000\,\mathrm{W}\cdot\mathrm{m}^{-2}$. Commercial silica-based solar panels routinely have conversion efficiency between 18% and 22% while multi-junction semiconductor (e.g. InGaP, GaAs, InGaAs) cells reach 38% (>43% under concentrated illumination). Thin-film, flexible CIGS ($CuIn_{1-x}Ga_xSe_2$) cells have recently attained the 20% efficiency mark. On the other hand, the more convenient dye-sensitised solar cells (DSSCs) experience difficulty in attaining efficiencies >15% although recent introduction of perovskite in them is boosting their efficiency in the range 20%–25%, but presently only for small-sized cells. Why is it so difficult to overtake the 20% efficiency mark? Spectral mismatch between the absorption spectrum of the semiconductor or dye and the solar emission spectrum is a major reason. Indeed, semiconductor materials perform best at a wavelength that depends on their bandgap E_g (in eV) according to:

$$\lambda_{opt} = \frac{1240}{E_g}[nm]$$

For monocrystalline silicon ($E_g \approx 1.1$ eV) this translates to ≈ 1.1 μm. One solution for enhancing photovoltaic conversion is consequently to fit solar cells with wavelength converting layers (WLCs) that absorb light in spectral ranges in which the photosensitive material does not absorb well, and emit in the best-conversion spectral range of the semiconductor. UV/blue light can be converted into visible emission by downconversion (DC), a process in which an energetic photon is "cut" into two longer-wavelength ones, or by downshifting (DS) in which one energetic photon is converted into a visible one, which implies energy loss. Similarly, NIR light can be upconverted by absorption of two (or more) low-energy photons and emission of a visible (or UV) one (Figure 17.19). Multi-electronic level lanthanide ions are among the best candidates as photoactive cores in wavelength converting materials, particularly in view of their ability

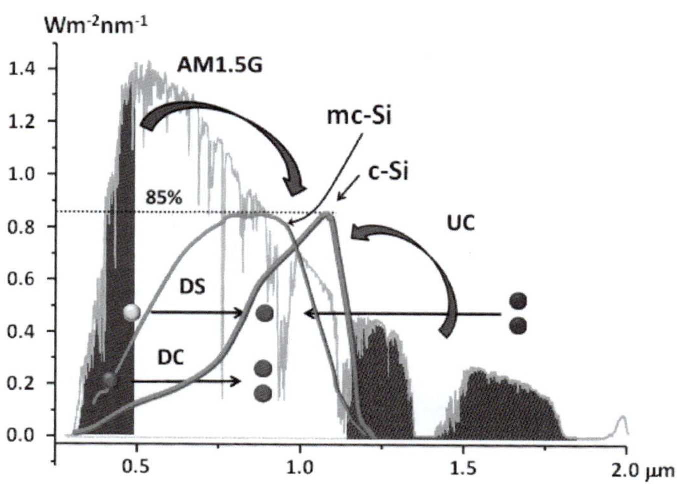

Figure 17.19. Partial AM1.5G solar irradiance spectrum along with conversion efficiencies of crystalline (c-Si) and multi-crystalline (mc-Si) silicon. Portions available for downconversion/shifting (about 16% up to 500 nm) and upconversion (about 16% in the range 1.2–2.5 μm) are highlighted.

to lend themselves to upconversion.[18] Downconversion is difficult to implement because it requires UV excitation and not much UV light is available in the solar spectrum. Downshifting has mainly been tested with Eu^{III} salts and complexes (sometimes under the form of nanoparticles), while many reports on upconversion have relied on $NaYF_4$:Yb,Er or $NaYF_4$:Er.

A brief summary of the best performing materials is given in Table 17.1. Assessing the real role of the WCL is often not easy because adding an extra layer to a solar cell modifies its optical properties (reflection, refraction) in either a detrimental or beneficial way. In addition, reports often lack the necessary control experiments. It is anticipated that with the exception of downshifting, these numbers will be doubled in the forthcoming years (particularly if plasmonics is used to increase upconversion efficiency). As far as DSSCs are concerned, their efficiency can also be substantially boosted by doping the TiO_2 electrode with non-luminescent Ln^{III} ions that modify its bandgap. Improvements in the range +25 to +40 (relative) per cent are reported. Finally, adding a scattering layer containing Ce^{IV}, for better illumination of the photosensitive material in DSSCs, leads to a relative increase of +18% in the conversion yield.

In summary, downshifting has been demonstrated to add 1–2 absolute % to the conversion yield and upconversion 0.5–1%.

Table 17.1. Improvements in absolute yields achieved by implementing a wavelength-converting layer in common silicon and dye-sensitised cells, along with the portion of the solar spectrum available for the chosen mechanism (DS or UC) and an estimate of what could be achieved with current technologies.

Process	% Sun available	Claimed abs %	Feasible abs %[a]
Silicon cells			
DS	19	1–2	3–5
UC, c-Si	21	0.4	1–2
UC, m-Si	7	0.6	1–1.5
Dye-sensitised cells			
DS	4.6	1.5	1.5
UC	12	0.6–1	1.5–2
Scatter	n.r.	1.3	1.5
TiO_2 doping	n.r.	1.5	1.5

Note: [a]Author's estimate.

The latter is suffering from the weak power density of illumination but nanostructures concentrating the solar irradiance onto nanospots up to 70- to 100-fold, boosting UC yield, are presently available. Although the reported improvements may seem to be small, if multi-conversion layers can be technically designed, one may expect an overall improvement on the order of 4–6 absolute %, far from being negligible provided it is cost effective.

17.10.2 *Photocatalysis*

Cerium oxide itself or Ce^{III}-doped Anatase (TiO_2) are efficient photocatalysts for water splitting. Photocatalytic Tm-based UCNPs can produce UV and blue light under IR solar irradiation enhancing the photocatalytic properties of TiO_2 for air and water purification. Thermal reduction of ceria in an aerosol reactor illuminated by concentrated solar light promotes the splitting of both water and carbon dioxide, producing syn gas that is ultimately transformed into kerosene (solar-to-kerosene yield: 1.7%). These applications are still confined to laboratory experiments but may develop into economical processes.

17.10.3 *Optical refrigeration*

An optical detector based on semi-conductors benefit from refrigeration because lowering its temperature considerably reduces the dark current and increases their sensitivity. Cryogenic fluids or thermoelectric coolers are common devices for this purpose but they are rather inconvenient. Similarly, electronic circuits and computer processors generate heat that has to be evacuated, often by noisy fans. The search for smarter refrigeration processes succeeded in 1995 when cooling of a glass sample by a laser was observed. The principle relies on anti-Stokes fluorescence, an all optical mechanism therefore freeing refrigeration from mechanical vibrations. About 15 years later, refrigeration into the cryogenic regime (114 K) was achieved with $YLiF_4{:}Yb^{III}$ and in 2014, a temperature of 93 K (from room temperature) has been reached with the same material (10% Yb^{III}). This is stirring hefty interest for RE-containing optical refrigerants. Until now, only trivalent Er, Tm and Yb have shown cooling effects, but Dy and Ho may be interesting candidates too.[19] Lanthanide ions are interesting for optical refrigeration in view of their weak crystal

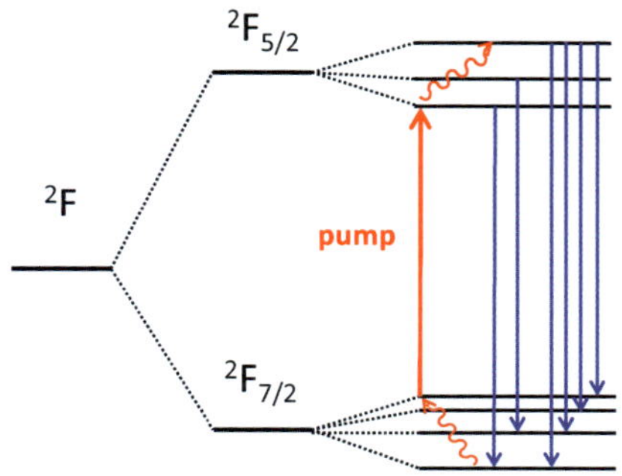

Figure 17.20. Principle of optical cooling with Yb^{III}. Blue arrows represent emitted photons with higher energy than pump photons (not all possible transitions shown). Undulated arrows show energy transfer from the lattice.

field (CF) splitting. Figure 17.20 illustrates the principle for Yb^{III} ($4f^{13}$). The electronic configuration is split into two spin–orbit levels, themselves split by crystal field effects. A pump laser (red arrow) is tuned to a transition starting from a higher CF sub-level of the ground state and terminating onto a lower CF sublevel of the excited state. Higher CF levels are thermally populated and therefore emitted photons (blue arrows) have larger energy than excitation photons; the energy difference is removed from the lattice, inducing cooling.

17.10.4 *Quantum information processing and optical computing*

The discovery in 1984 of a quantum algorithm for encrypting data exchange was the start of quantum information processing, a major research topic in computing sciences. In this concept, classical bits that take only discrete values are replaced with quantum bits (qubits) that can adopt any superposition state. Objects prone to adopt superposition states are atoms, photons, electrons, spins, for instance. Superposition states are combination of two states (e.g. for an atom, ground and excited states) and they persist as long as there is no interaction with other objects, otherwise there is de-coherence. Therefore they are only observed in highly pure isolated systems. The concept is illustrated in Figure 17.21 for an atom "oscillating" between a ground and excited state upon interaction with light. Such a state is made possible because linear combinations of the ground and excited states are also solutions of the Schrödinger equation.

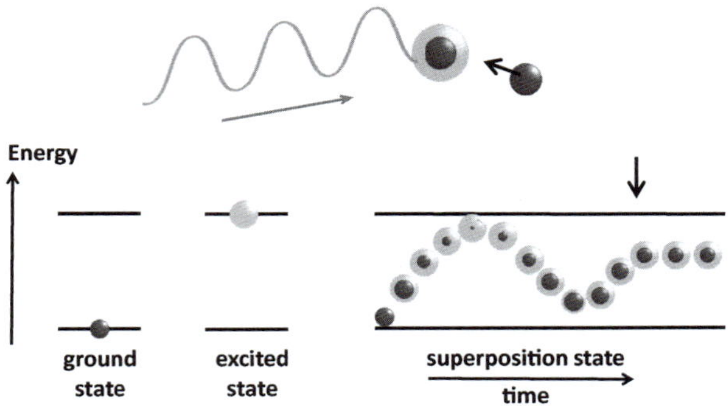

Figure 17.21. Concept of superposition state for an atom interacting with light. The arrow points to de-coherence due to collision with another atom.

Rare-earth-doped crystals have very narrow optical transitions and are well suited as materials for quantum information. Appropriate ions include trivalent Pr, Nd, Eu, Er and Tm. Quantum memories based on these materials have been designed and the realm of quantum computing might soon come.[20]

17.11 Summary of Applications per Rare-Earth Element

Element	Main applications
Scandium Sc, 21	Strengthener in light alloys for aircrafts; anode of solid oxide fuel cells
Yttrium Y, 39	Garnets for photonics (lasers, UCNPs...); stabilisation of zirconia (refractory materials, artificial jewels); superconducting ceramics; ^{90}Y is a β-emitter for cancer therapy
Lanthanum La, 57	Optical glasses; garnets for X-ray phosphors; FCC catalysts; water treatment; rechargeable batteries; ceramics for electronic components; mischmetal
Cerium Ce, 58	FCC catalysts; automotive catalysts; soot filters in diesel engines; glass polishing powder; visible-emitting phosphors (as activator of terbium); scintillators; garnets for X-ray phosphors; ceramics for electronic components; pigments for plastics; mischmetal

(Continued)

Element	Main applications
Praseodymium Pr, 59	Pigments for ceramics (yellow-orange); magnets; mischmetal
Neodymium Nd, 60	Magnets (Fe-Nd-B, audio equipment, computer disks, vehicles, hand tools, wind turbines); lasers (industry, defense, medicine)
Samarium Sm, 62	Magnets (Sm-Co, transportation, defense); radiopharmaceuticals (cancer treatment — lung, prostate, breast, bone); control rods in nuclear reactors
Europium Eu, 63	Red phosphor: CFLs, LEDs, flat panel TV, displays, X-ray screens, security inks, counterfeiting tags, bioanalysis and bioimaging
Gadolinium Gd, 64	Control rods and shields in nuclear reactors; contrast agent in MRI; garnets for X-ray phosphors
Terbium Tb, 65	Green phosphor: CFLs, LEDs, flat panel TV, displays, X-ray screens, security ink, counterfeiting tags, bioanalysis and bioimaging; Tb–Dy alloy for sonars (Terfenol-D)
Dysprosium Dy, 66	Additive in permanent Nd-Fe-B magnets (see neodymium) for high temperature operation
Holmium Ho, 67	Magnetic flux concentrators; pigment for ceramics and glasses; medical lasers; microwave equipment; nuclear control rods (neutron absorption)
Erbium Er, 68	Telecommunications (EDFA); glass and ceramic pink pigment; lasers (medicine, dentistry); UCNPs (bioanalysis and bioimaging, counterfeiting tags); nuclear control rods (neutron absorption)
Thulium Tm, 69	Phosphors (high-intensity discharge lamps); lasers (medicine, defense); UCNPs (bioanalysis and bioimaging, counterfeiting); superconducting ceramics; portable X-ray sources
Ytterbium Yb, 70	Phosphors (activator in UCNPs used in biosciences and counterfeiting); lasers; stress gauges; atomic clock; stainless steel additive
Lutetium Lu, 57	Catalyst (petroleum industry); positron detectors (tomography); scintillators (X-ray detection); ^{177}Lu used in cancer therapy

Further Reading

The series *Handbook on the Physics and Chemistry of Rare Earths* (60 published volumes up to 2021) covers all aspects of rare-earth (and partly actinide) sciences, including applications, through authoritative and comprehensive reviews: http://www.sciencedirect.com/science/handbooks/01681273/.

References

1. J.-C. G. Bünzli, Lanthanides, in *Kirk-Othmer Encyclopedia of Chemical Technology*, Wiley Online Library, 2013, pp. 1–43.
2. P. Vernoux, L. Lizarraga, M. N. Tsampas, F. M. Sapountzi, A. De Lucas-Consuegra, J. L. Valverde, S. Souentie, C. G. Vayenas, D. Tsiplakides, S. Balomenou and E. A. Baranova, *Chem. Rev.*, 2013, *113*, 8192–8260.
3. T. Montini, M. Melchionna, M. Monai and P. Fornasiero, *Chem. Rev.*, 2016, *116*, 5987–6041.
4. H. Zhu, P. Chen, C. F. Yang and Y. X. Wu, *Macromol. React. Eng.*, 2015, *9*, 453–461.
5. K. A. Gschneidner Jr., A. O. Pecharsky and V. K. Pecharsky, *Cryocoolers*, 2005, *13*, 363–371.
6. J.-C. G. Bünzli, *Coord. Chem. Rev.*, 2015, *293–294*, 19–47.
7. S. V. Eliseeva and J.-C. G. Bünzli, *New J. Chem.*, 2011, *35*, 1165–1176.
8. V. Tucureanu, A. Matei and A. M. Avram, *Opto-Electronics Rev.*, 2015, *23*, 239–251.
9. B. Zhou, B. Shi, D. Jin and X. Liu. *Nature Nanotech.*, 2015, *10*, 924–936.
10. P. F. Smet, F. Philippe, K. Van den Eeckhout, O. Q. De Clercq and D. Poelman, Persistent Phosphors, in *Handbook on the Physics and Chemistry of Rare Earths* (Eds. J.-C. G. Bünzli, V. K. Pecharsky), Vol. 48, Ch. 274, Elsevier Science B.V., Amsterdam, 2015, pp. 1–108.
11. M. Nikl and A. Yoshikawa. *Adv. Opt. Mater.*, 2015, *3*, 463–481.
12. S. Liu, J. A. Mares, X. Feng, A. Vedda, M. Fasoli, Y. Shi, H. Kou, A. Beitlerova, L. Wu, C. D'Ambrosio, Y. Pan and M. Nikl, *Adv. Opt. Mater.*, 2016, *4*, 731–739.
13. C. D. S. Brites, A. Millan and L. D. Carlos, Lanthanides in luminescent thermometry, in *Handbook on the Physics and Chemistry of Rare Earths* (Eds. J.-C. G. Bünzli and V. K. Pecharsky), Vol. 49, Ch. 281, Elsevier Science B.V., Amsterdam, 2016, pp. 339–427.

14. E. Terreno, D. D. Castelli, A. Viale and S. Aime, *Chem. Rev.*, 2010, *110*, 3019–3042.
15. S. V. Eliseeva and J.-C. G. Bünzli, *Chem. Soc. Rev.*, 2010, *39*, 189–227.
16. A. E. Merbach, L. Helm and E. Toth, *The Chemistry of Contrast Agents in Medical Magnetic Resonance Imaging*, Wiley & Sons, Chichester, 2013.
17. J.-C. G. Bünzli, *J. Lumin.*, 2016, *170*, 866–878.
18. J.-C. G. Bünzli and A.-S. Chauvin, Lanthanides in Solar Energy Conversion, in *Handbook on the Physics and Chemistry of Rare Earths* (Eds. J.-C. G. Bünzli and V. K. Pecharsky), Vol. 44, Ch. 261, Elsevier Science, B.V., Amsterdam, 2014, pp. 169–281.
19. M. P. Hehlen, M. Sheik-Bahae and R. L. Epstein, Solid-State Optical Refrigeration, in *Handbook on the Physics and Chemistry of Rare Earths* (Eds. J.-C. G. Bünzli and V. K. Pecharsky), Vol. 45, Ch. 265, Elsevier Science, B.V., Amsterdam, 2014, pp. 179–260.
20. P. Goldner, A. Ferrier and O. Guillot-Noël, Rare-Earth doped Crystal for Quantum Information Processing, in *Handbook on the Physics and Chemistry of Rare Earths*, (Eds. J.-C. G. Bünzli and V. K. Pecharsky), Vol. 46, Ch. 267, Elsevier Science B.V., Amsterdam, 2015, pp. 1–78.
21. V. Nicoletopoulos, *Rare Earths Worldwide: An Industry and Policy Analysis*, Natural Resources PC, Athens, Greece, June 2011.
22. V. Haquin, M. Etienne, C. Daiguebonne, S. Freslon, G. Calvez, K. Bernot, L. Le Pollès, S. E. Ashbrook, M. R. Mitchell, J.-C. Bünzli, S. V. Eliseeva and O. Guillou, *Eur. J. Inorg. Chem.*, 2013, 3464.
23. H. Dong, L.-D. Sun and C.-H. Yan, *Chem. Soc. Rev.*, 2015, 1608.
24. J. M. Meruga, W. M. Cross, P. S. May, Q. Lu, G. A. Crawford and J. J. Kellar, *Nanotechnol.*, 2012, *23*, 395201.
25. Y. Zhang, L. Zhang, R. Deng, J. Tian, Y. Zong, D. Jin and X. Liu, *J. Am. Chem. Soc.*, 2014, *136*, 4893.
26. L. Li, L. Y. Yu, Zhou, Z. Zhu, C. Lin, C. Lu and R. Yang, *Diagn. Microbiol. Infect. Diseases*, 2009, *63*, 165.
27. V. Fernández-Moreira, B. Song, V. Sivagnanam, A.-S. Chauvin, C. D. B. Vandevyver, M. Gijs, I. Hemmilä, H.-A. Lehr, J.-C. G. Bünzli, *The Analyst*, 2010, *135*, 42.
28. J.-C. G. Bünzli and K.-L. Wong, *J. Rare Earths*, 2018, *36*, 1.
29. J. Zhou, J. L. Leano Jr., Z. Liu, D. Jin, K.-L. Wong, R. S. Liu and J.-C. G. Bünzli, *Small*, 2018, *14*, Art. Nr. 1801882.

https://doi.org/10.1142/9781800610163_0018

Chapter 18

Applications of Actinides

Jean-Claude G. Bünzli[*,†,¶], *Louise S. Natrajan*[‡,**], and *Mark J. Sarsfield*[§,††]

[*]*Swiss Federal Institute of Technology, Lausanne (EPFL), Institute of Chemical Sciences and Engineering, Lausanne, Switzerland*
[†]*Hong Kong Baptist University, Department of Chemistry, Kowloon Tong, Hong Kong SAR, P. R. China*
[‡]*Centre for Radiochemistry Research, Department of Chemistry, School of Natural Sciences, The University of Manchester, Oxford Road, Manchester, M13 9PL, UK*
[§]*National Nuclear Laboratory, Central Laboratory, B170 Sellafield, Seascale, Cumbria, CA20 1PG, UK*
[¶]*jean-claude.bunzli@epfl.ch*
[**]*louise.natrajan@manchester.ac.uk*
[††]*mark.sarsfield@uknnl.com*

18.1 Basic Properties

Actinides encompass the series of elements actinium (Ac, 89) to lawrencium (Lw, 103).[a] Only a limited number of these elements are found in nature and the first that has been discovered is uranium (U, 92) in 1789 by Martin Klaproth, under the form of an oxide; isolation of the metal had to wait 1841 when E.-M. Péligot treated uranium tetrafluoride with potassium. The main uranium mineral is pitchblende which contains a mixture of UO_2 and U_3O_8 as well as

[a]Actinoids comprise elements 89–103, but is rarely referred to. "Actinides" and 5f-elements" are often used as synonyms. See definition in the IUPAC "red book" of inorganic nomenclature (2005), Section IR 3.5, p. 51.

other elements such as rare earths, thorium, and lead (from uranium radioactive decomposition). Uranium is the actinide element having the greatest number of applications. The discovery of thorium is an interesting story. In 1815, Jöns Berzelius thought he had isolated a new element from a mineral found in Falun (Sweden) and named it thorium; however it turned out that the mineral actually contained yttrium with formula $(Y,Ln)PO_4$, known today as Xenotime. Later, Berzelius was given another mineral to analyse and determined in 1828 that it indeed contained a new actinide element; he kept the name thorium. The mineral was given the name thorite, $(Th,U)SiO_4$, but the most abundant thorium mineral is monazite, a rare earth phosphate mineral. Most of the other members of the series have been man-made in the 20th century, although very small amounts of actinium, protactinium and microscopic amounts of plutonium can be found in nature as decay products of uranium. The existence of transuranium elements was suggested by Enrico Fermi in 1934 while Glen T. Seaborg formulated the "actinide hypothesis", i.e. the existence of a series of elements analogous to the 4f elements, 10 years later.

All of the elements in the actinide series are radioactive resulting in unstable nuclei decaying into daughter nuclides by α- and β,-decays often accompanied by γ-ray emission, henceforth limiting the number of practical applications. Another difficulty in dealing with actinides arises from radioactive decompositions liberating large energy, henceforth facilitating the occurrence of mixtures of both isotopes and oxidation states. In view of their military and energy applications, chemistry of actinides has been thoroughly studied during the 1950s through the 1980s.[1] It is presently undergoing a renewed interest in view of solving the problem of nuclear waste safe disposal.

18.2 Applications of Actinides: Summary

In contrast to rare earth elements, actinide elements have a wider range of available oxidation states, particularly for the earlier members of the series, up to +7 for Np and Pu. While oxidation number +3 exists for all 5f-elements its stability increases along the series and represents the most common oxidation state of the heavier elements. Tetravalent states are particularly stable for Th–Np. The richness of oxidation states is an advantage for catalytic applications which were among the first commercial end uses of actinide compounds, e.g. thorium oxide in the Fischer–Tropsch process.

Most of the other applications of actinides take advantage of their radioactive properties, the major ones being in commercial nuclear reactors and military weaponry, namely for the fabrication of atomic fission bombs. The main elements concerned are thorium, uranium, and plutonium. Production of heat in thermoelectric generators and radioactive sources also profit from radioactive emissions (Table 18.1). The other actinide elements are only available in very small quantities (mg to fractions of μg), so that corresponding applications are rather scarce and even nil for elements 98-103, barring the use of some of them as target nuclides for the synthesis of other transuranium elements.

18.3 Metallurgy and Metals

Early studies on actinide metals were carried out on metals of typical purity ranging between 90 and 99%. Following interest for the unusual properties of lighter actinide metals (Ac–Pu) in the 1970s, purification methods were designed and 99.9 at % purity was reached for Th, U, Np and Pu on the kg scale, while Pa, Am and Cm could be obtained with 99.5 at % purity on the g to kg scale. These are difficult operations in view of the radioactivity and toxicity of these metals. The major procedures used are (i) metallothermal reduction of halides or oxides with lithium, calcium or lanthanum, sometimes followed by distillation, (ii) metallothermal reduction of actinide carbide with transition metals (Ti, Zr, Nb or Ta), (iii) iodide transport (van Arkel–De Boer process) yielding first a volatile iodide which is separated and re-decomposed at higher temperature, or (iv) molten salt electrolysis, especially for U, Np and Pu. Refining methods include vacuum melting with evaporation of the volatile impurities, selective vapourisation, and zone melting, among others.[2]

Thorium is added into multicomponent alloys of magnesium. The resulting "Mag–Thor" alloys have been proposed in the 1980s and they are essentially of three types: Mg–Th-Zr, Mg–Th–Zn–Zr and Mg–Ag–Th–RE–Zr (RE stands for rare earth). These alloys are light but have high strength, large deformation resistance at

Table 18.1. Principal applications of actinide elements.

Element	Form	Main applications
Actinium Ac, 89	^{227}Ac	• Neutron source
	^{228}Ac	• Radioactivity indicator
		• γ-ray source in medicine (with ^{228}Ra)
	^{225}Ac	• Targeted alpha therapy
Thorium Th, 90	Th	• Additive in magnesium alloys for aeronautics and military applications
	ThO_2	• In incandescent (Auer) gas mantles (see Section 17.1.1)
		• Catalyst in Fischer–Tropsch synthesis (pure or mixed with other oxides)
	^{230}Th	• Tracer for geological dating (with ^{231}Pa or ^{234}U)
	^{232}Th	• Nuclear fuel (fertile fuel)
Protactinium Pa, 91	^{231}Pa	• Tracer for geological dating (with ^{230}Th)
Uranium U, 92	UO_x	• Yellow, orange, black pigments for ceramics, porcelain, and glasses
	UO_2[a]	• Catalyst (degradation of volatile organics; oxidation of methane)
		• Semiconductor (potentially for solar cells, integrated circuits)
	^{234}U	• Tracer for geological dating (with ^{230}Th)
	^{235}U	• Nuclear fuel (2%–5% in UO_2)
		• Nuclear fission weapons (>80% in UO_2)
Neptunium Np, 93	^{237}Np	• Precursor in ^{238}Pu synthesis
		• Detectors for high-energy (MeV) neutrons
		• Potential nuclear weapon
Plutonium Pu, 94	PuO_2	• Nuclear fuel (mixture of isotopes) in MO_x reprocessed fuels
	^{238}Pu	• Heat source in thermoelectric generators (satellites, space probes)
	^{239}Pu	• Nuclear weapons (as PuO_2)
Americium Am, 95	^{241}Am	• Ionising source in smoke detectors
		• Heat source in thermoelectric generators (satellites, space probes)
		• Neutron source with ^{9}Be (water determination, radiography, tomography)
		• Precursor in the synthesis of transuranium elements

Table 18.1. (*Continued*)

Element	Form	Main applications
Curium Cm, 96	^{242}Cm	• Heat source in thermoelectric generators (satellites, pacemakers) • Precursor in the synthesis of medical grade ^{238}Pu for pacemakers
	^{244}Cm	• α-particle source for α-particle X-ray spectrometers (in space probes for elemental analysis of planets)
Berkelium Bk, 97	^{249}Bk	• Target nuclide for the synthesis of heavier transuranium elements
Californium Cf, 98	^{252}Cf	• Neutron source (in nuclear reactors, activation analysis, coal analysis, Cervical and brain cancer therapy)

Note: [a]Depleted uranium oxide.

high temperature, typically up to 350 °C, high melting points, and considerable ductility. Their primary application lies in the manufacture of aircraft and spacecraft parts. In addition military applications involve missile and armoured vehicle design. However, the inherent radioactivity of these alloys is seriously hampering their spread.

Depleted uranium (DU) metal (i.e. obtained from nuclear fuel deprived of most of its 235 isotope) is at the centre of several military applications. Due to its large density (1.6 times larger than lead) and hardness, DU is used as armour for tanks in order to increase their defence capability or as counterweight for aircraft control surfaces and in the gyroscopes of inertial guidance systems. Furthermore, it is also attached to the tip of bullets and bombshells because it confers them a larger penetration power with respect to iron plates and concrete. Since metallic uranium is pyrophoric, it burns away at the point of impact, turning the inside of the attacked vehicle or house into a nightmare of hot gases, sparks and high temperature. The toxic nature of DU and its oxidised products is however generating worldwide concerns about these ammunitions. On the other hand, depleted uranium dioxide is a useful material for radiation shielding, e.g. for fabricating casks for nuclear wastes.

18.4 Photonic Applications

About half of the produced thorium, under the form of dioxide, is used as light-emitting material in gas mantles, still in used today (see Section 17.1.1). This is the second oldest industrial application of an actinide element.

Indeed, yellow uranium glass has been identified in a Roman mosaic dated 79 AD, found near Naples in Italy and containing about 1% uranium oxide. In the late Middle Age, pitchblende was extracted from silver mines in Bohemia and used as glass colouring agent. Truly, when incorporated into glass, uranium oxide(s) confer(s) it an orange, yellow or even black colour, depending on the oxidation state and concentration of uranium. In addition, uranium glasses display bright green fluorescence upon UV illumination. Uranium glass was extremely popular in the second half of the 19th century and until the 1920s. Beautiful vases, plates and other ornaments were created (Figure 18.1). Uranium supplies disappeared during the Second World War, until the 1960s. Present production is essentially concentrated on beads. It is noteworthy that uranium glass is not considered as being a health hazard in view of its very low radioactivity and negligible inhalation risk (being embedded in glass), the uranium content being generally small (a few percent). Similarly to glass, uranium pigments were also used in ceramics and porcelain.

The other principle use of the fluorescent properties of uranium, which is increasing in popularity, has been to determine the presence of or the speciation of uranyl(VI) in environmental type conditions, in both solution[3] and sorped onto minerals.[4] Given that uranyl in the +VI oxidation state has an electronic configuation of $5f^0$, the green emission seen in Figure 18.1 arises solely due to formally forbidden ligand to metal charge transfer from largely oxygen based 2p orbitals to the lowest unoccupied molecular orbitals of 5f character on uranium (see Chapter 13, Section 13.3) following UV-visible excitation (typically 250–420 nm). This electronic transition is particularly special as it is unique for a given complex, and much information on the electronic structure of a given compound can be extrapolated from the spectrum. This includes the emission maximum, which is a

Figure 18.1. Fluorescent uranium glass and cake plate under UV illumination. Reproduced from Ref. [20].

direct measurement of the HOMO-LUMO energy gap, the ground state Raman active total symmetric stretch (ν_1), estimated from the energy difference (in cm^{-1}) between the first two highest energy vibrational fingers (if sufficiently vibrationally resolved), which is a measurement of the force constant and hence the uranyl U=O bond strength. Additionally, the emission maximum is often influenced by the donor strength of ligands in the equatorial plane, with strong sigma donors often producing a pronounced red-shift (to lower energy), see Figure 18.2. This indicates that there is some degree of influence on the uranyl bonding scheme itself and may be a consequence of competition of orbital overlap with uranium with the oxo ligands and those of ligands bound in the equatorial plane of the uranyl moeity (although this is not yet proven).

Perhaps a more indicative way to ascertain the presence of a given species is by time resolved methods using lifetime decay kinetics, since each species will give rise to a unique decay lifetime. Both spectral features and lifetimes can be mathematically deconvoluted to indicate the number and type of species present. Note here, that although often referred to as fluorescence, the emission from uranyl is highly variable and temperature dependant and therefore is often classed as phosphorescence; lifetimes can range from nanoseconds to seconds. Indeed, lifetime image mapping techniques have now been used in biological systems to identify uranyl(VI) location and to examine speciation.[5]

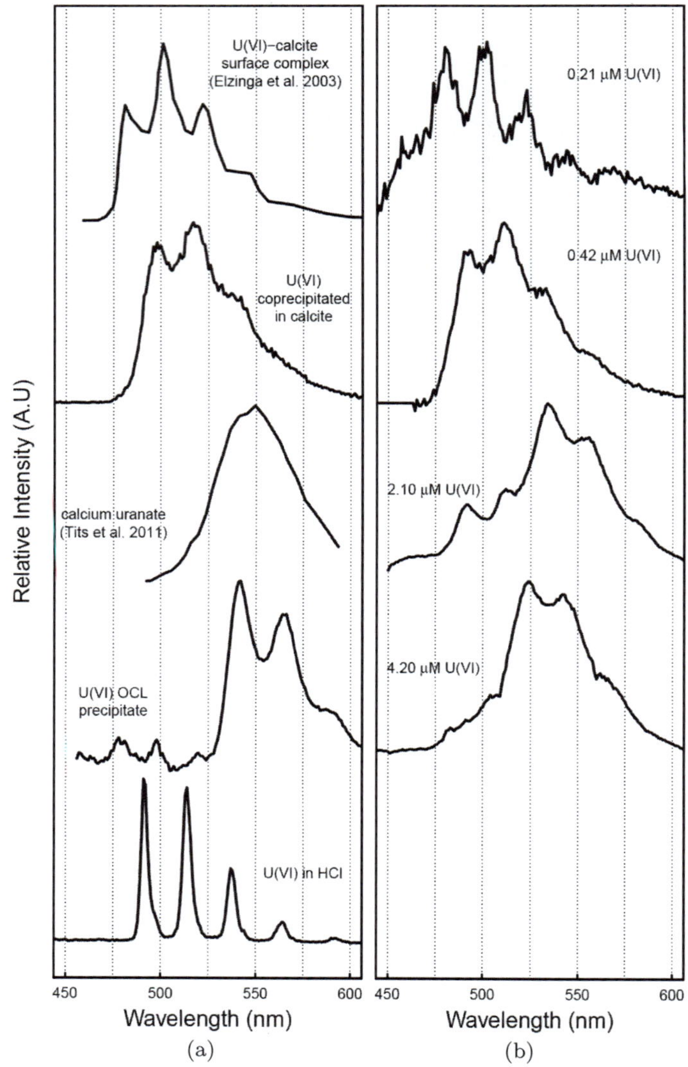

Figure 18.2. (a) Emission spectra of a range of different known uranyl(VI) species in calcite minerals, old cement leachate (OCL-a model for cementitious uranyl(VI) in a geological disposal facility) and uranyl in HCl. Note the changes in emission maxima and vibrational fine structure. (b) Fingerprinting uranyl species by spectral matching with known samples in hyperalkaline calcite phases.[4] Spectra recorded by Adam Swinburne at 77 K on solid samples following 250 nm excitation. Figure taken from the author's own collection.

18.5 Tracers for Dating

Radiometric dating is a technique allowing determining the age of various geological materials such a rocks or carbon deposits. The method is based on comparing abundances of naturally occurring isotopes that were incorporated into the material upon its formation with the concentrations of their decay products. Accurate dating requires that the tracked parent isotopes have long enough half-life so that they are present in significant quantities at the time of measurement, that half-lives are accurately known, and that enough of the daughter isotopes are produced. In addition, possible alteration of the materials (under temperature, pressure) resulting in losses by diffusion of the investigated isotopes must be considered. Analysis of the isotopes is conducted by isotope-ratio mass spectrometry. The most accurate dating methods rely on two naturally occurring parent isotopes and some of the commonly used pairs involve actinides[6] Examples are:

- $^{206}Pb/^{238}U$ and $^{207}Pb/^{235}U$, with a working range between 1 and 4,500 million years and an accuracy of 1%–2%.
- $^{230}Th/^{234}U$ for calcium carbonate materials with an upper age limit of 0.5 million years.
- $^{230}Th/^{231}Pa$ for dating oceanic sediments up to 175,000 years old. This method allowed reconstructing the movements of the North Atlantic Ocean during the melting of the last ice age, the quaternary glaciation (110,000–12,000 years ago).

18.6 Semi-Conductor Properties

The energy band gap for uranium dioxide (≈1.3 eV) lies between those of Si (1.11 eV) and GaAs (1.43 eV) so that development of semiconductors based on uranium dioxide (possibly other actinide compounds) could offer significant improvements in performance as compared to conventional ones. Therefore depleted uranium oxide(s) could be made into efficient solar cells, semiconductors, or other electronic devices. For instance, a Schottky diode based on U_3O_8 and a p-n-p transistor made of UO_2 has been successfully demonstrated in laboratory.

The electrical conductivity of UO_2 is approximately the same as GaAs but its dielectric constant (~22) is larger compared to GaAs (14.1) so that the material has been tested as insulating layer of complementary metal oxide semiconductor (CMOS) devices. Since uranium oxide withstands high operating temperatures (up to 2600 K), less cooling is required. In addition, uranium oxide resists to radiation damages, an ideal property for space and military applications. It is noteworthy that electrical properties of uranium oxides heavily depend on the exact stoichiometry, e.g. the electrical conductivity of $UO_{1.994}$ is largely different from that of $UO_{2.001}$.

18.7 Actinides in Catalysis

One of the main uses of thorium dioxide (thoria) has been as catalyst in the Fischer–Tropsch process. The latter was developed around 1925 and encompasses a bundle of chemical reactions converting a mixture of carbon monoxide and hydrogen (syn gas) into liquid hydrocarbons. In fact, syn gas is produced from gasification of coal, so that the process eventually converts coal into synthetic fuels. The initial catalyst was iron oxide but it did deactivate rapidly. Thoria was then introduced as a catalyst. However, pure thorium dioxide leads to some 10% of unwanted alcohols (mostly water soluble) being obtained during the conversion. It was then found that addition of aluminium oxide (or other dehydrating compounds) in the catalyst yields a product almost free of oxygenated compounds. The process grew in importance during the Second World War and a complex catalyst containing cobalt, thorium and magnesium oxides along with Kieselguhr (silicium oxide) was developed. The long life of this catalyst is remarkable. It may be used several months without sign of ageing. Even in the case of carbon deposits and the resulting increase in pressure drop through the bed, the original activity can be restored by passing air over the catalyst at synthesis temperature. Thoria-based catalysts are also insensitive to sulphur poisoning.

Depleted uranium dioxide can also be used as catalyst, e.g. for the degradation of volatile organic compounds in gaseous phase, oxidation of methane to methanol, and removal of sulphur from petroleum. It has high efficiency and long-term stability.

Organometallic compounds of uranium have expanded the practical role of this element beyond nuclear industry. Indeed the large ionic radius of uranium as well as the availability of the 5f orbitals for bonding interactions, allied with the range of stable oxidation states it can attain, result in unusual chemical reactivity. The latter has been exploited in catalysis, namely in the hydrogenation of alkenes, oligomerisation, hydrosilation and hydroamination of terminal alkynes. In addition, small molecules such as dinitrogen, carbon monoxide or carbon dioxide interact with U(III) rendering their activation and subsequently reduction possible.[7] Similarly, uranium hydrides or hydrido complexes are active catalysts in hydrogenation of alkenes. Some of the reactions are stoichiometric, but can be turned into catalytic ones. The main difficulty though is extracting the functionalised products and regenerating the catalyst, in view of the tendency of uranium ions to form strong bonds with anionic ligands.[8] The field may not be yet ready for commercial applications but, on the other hand it has not revealed all its potential so far.

18.8 Radioactive Sources

A major application of radioactive sources, with respect to the amount of actinide compounds consumed, are electric generators, particularly those based on the thermoelectric effect. This effect appears when two metals or two n-p semiconductors are connected together and submitted to a temperature gradient. There are several applications requiring reliable electric generators working flawless for a very long time, namely space probes, satellites and remote, unmanned facilities such as lighthouses in hostile environment. These devices have their electricity generated by radioisotope thermoelectric generators (RTGs) in which the heat source is a radioactive isotope. Important requirements are a long half-life, so that the generator will release energy at a constant pace for a long time and, needless to say, a favourable power-to-mass ratio. Since alpha particles release about 10 times more energy than beta or gamma decays, α-emitting isotopes are preferred, especially α-particles that are easily absorbed and transformed into heat. Several isotopes qualify, but the best one seems to be ^{238}Pu with a lifetime of about 88 year and a 0.5 kW/kg power density; its β and γ emissions are low so that minimum shielding is required. Since the 1950s USA and Russia have developed processes for generating ^{238}Pu used in civil and military applications.[9] To date, NASA has flown 35 RTGs on 18 NASA missions none of which have failed due to an RTG failure. Much of the detailed information on the outer planets is a direct result of RTG power technology to collect data and send it back to Earth.

The process involves production of neptunium (^{237}Np) targets that are irradiated in a reactor, cooled and then processed to separate the unreacted ^{237}Np and ^{238}Pu product from the fission products. The ^{237}Np is recycled and the purified ^{238}Pu is made into ceramic pellets of the dioxide and encapsulated in a protective clad such as an iridium or platinum alloy. These clad pellets are placed in a protective aeroshell that can withstand rocket launch and re-entry accidents. The aeroshells are stackable and are surrounded by thermocouples that convert the alpha decay heat to electricity. Much of today's ^{238}Pu stocks in the US were made in the 1980s and supplies are running low. Recently, a programme was initiated to restart production at Oak Ridge National Lab at 0.5–1.5 kg/year.[10]

Among the other useful isotopes for RTG technology, ^{241}Am is presently being considered by space agencies, because of a shortage and the expense of ^{238}Pu production. Although the power density of ^{241}Am is four times smaller compared to ^{238}Pu, the mass of the fuel is only a fraction of the mass of the whole system and small electrical devices and heater units are under development.[16] These devices can potentially operate for centuries without noticeable power loss from the isotope ($t_{1/2}$ ^{241}Am $\sim$ 432 years). Sources of very pure ^{241}Am can be found in stored plutonium generated from reprocessing spent nuclear fuel. Separated plutonium contains a range of isotopes (^{238}Pu to ^{242}Pu) one of which (^{241}Pu) beta decays with a half-life of around 14.3 years. In the UK, the 140t stockpile of plutonium contains more than 3,000 kg of ^{241}Am and this is expected to continue growing for at least the next 50 years. It is this large abundance and a relatively simple production process that makes ^{241}Am an attractive alternative.

Alpha particles emitted by ^{241}Am play the role of ionising particles in some smoke detectors. The latter features two ionising chambers, one open to air and a closed, reference one. Alpha particles generated in both chambers ionise air particles and create an electric current. If fume particles enter the measuring chamber, some of the produced ions will be adsorbed by these particles, hence reducing the electric current compared with the reference chamber; the difference in current is detected by an electronic circuit, which sounds the alarm.

Larger amounts of americium are used as neutron sources, often combined with beryllium to exploit the alpha, n reaction and generate neutrons that can be used for well-logging (in the oil industry), neutron activation analysis, moisture levels in soils or as thickness gauges (e.g. glass and paper industry).[11]

In the early times of pacemakers, radioisotope-based batteries were considered (e.g. with ^{238}Pu, made from ^{242}Cm decay, or other non-actinide isotopes), but have been since largely supplanted by lithium batteries.

Medical diagnosis and cancer radiotherapy use several radioactive isotopes but most of them are non-actinide such as ^{226}Ra and ^{222}Rn. One promising avenue being developed is alpha-immunotherapy where short-lived alpha emitters are attached to molecules that can aggregate around, and selectively destroy, cancer cells.[12] An example is the use of actinium-225 and its daughters (e.g. bismuth-213) that can be extracted from thorium-229, which in turn can be found within aged uranium-233 supplies.[13]

18.9 Nuclear Power Generation

Uranium is mined in its ore form from the ground in many continents. The initial ore for the Manhattan project came from Congo but presently the top-producing countries are Kazakhstan (41%), Canada (16%) and Australia (9%) (see Figure 18.3). The ore is then milled and leached with acid or alkali solutions depending on its composition. An important point for successful operations is that uranium must be in its highest oxidation state so that oxidation is sometimes required and performed with chlorate or manganese dioxide. The resulting leachates contain uranyl carbonate or sulfate and are sorbed by ion-exchange resins or extracted by solvent extraction methods. Uranium is then precipitated as diuranate, which is ignited to yield the "yellowcake", a mixture of tri-uranium octoxide (U_3O_8, 70%–90%) with various other uranium oxides. The remaining ore

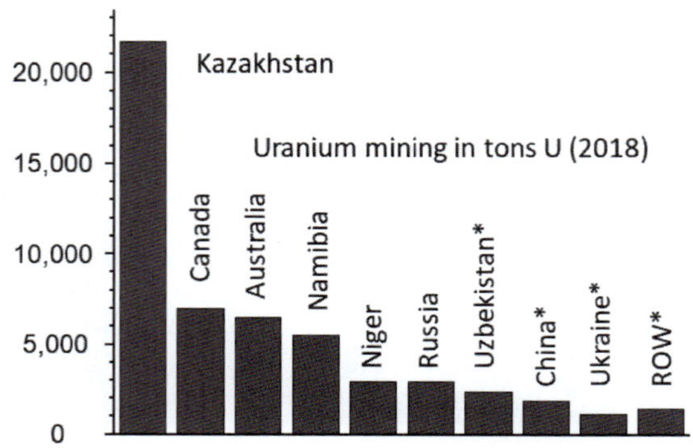

Figure 18.3. Production from uranium mines in 2018 expressed in terms of metric tons of uranium; total mined: 53,500 tons U (1 ton U corresponds to 1.18 tons U_3O_8). Stars indicate estimates. Data from World Nuclear Association, drawn by the author from cited literature data.

"tailings" are disposed of as radioactive waste. Radioactive contamination at these mining and processing sites poses an environmental safety issue, which has been addressed in a similar manner to spent nuclear fuel.

The yellowcake is then further processed to yield uranyl nitrate and then enriched, as only 0.7% of natural uranium is fissile under the form of ^{235}U while the dominant isotope is ^{238}U. The ^{235}U content is increased to approximately 3.5–5%. This is done by converting all the mixed oxides into uranium dioxide and subsequently to uranium hexafluoride gas and separating it into two streams; one of which is enriched in ^{235}U and the other depleted. The enriched UF_6 is then converted back to UO_2 which can be pressed and heated to 1400 °C to form fuel pellets.

The fuel pellets are subsequently encased in metal rods that can then be used in a fuel assembly in a reactor. After 18–36 months of operation, the build-up of fission products is such that the efficiency of the fuel decreases so the fuel rods are removed and replaced. The used fuel is then stored for months or years in water, which absorbs the heat, until the radiation levels decrease sufficiently for it to be disposed of or reprocessed. As there are no disposal facilities at present for nuclear fuel waste, it is simply isolated from the environment and left in storage until facilities become available.

The once-through or "open" fuel cycle, whereby waste is stored for disposal, is presently common in a number of countries including

Canada, parts of Europe including Germany and the USA, although some research on reprocessing techniques is being carried out in these areas as reprocessing is becoming increasingly important for the future of nuclear power. A more sustainable "closed" fuel cycle, where the waste is recycled and reused, is becoming favourable again[14] as a result of this and has been done in some parts of the world for many years, including the UK, France and other parts of Europe, Russia and Japan. Despite the Fukushima incident, many countries in the world are embarking on expanding their nuclear programmes. In April 2020, there were 441 nuclear power plants operating in the world, producing about 10% of total electricity consumed, and 55 under construction, while a further 110 were either on order or planned.[18]

Thorium-232 may be an alternative fertile fuel to uranium. In the reactor, ^{232}Th absorbs a neutron and is transformed into fissile ^{233}U which is then the nuclear fuel. There are several advantages to thorium. In addition to its greater abundance (8.1 ppm in earth crust) than uranium (2.3 ppm) natural thorium contains insufficient amounts of fissile isotopes to start a chain reaction and the cycle produces less toxic radioactive wastes as fewer plutonium and other transuranium isotopes are produced. There are however a number of challenges remaining to be solved before thorium can compete with uranium, especially in light water reactors which represent the majority of operating nuclear power reactors.

18.9.1 *Reprocessing spent nuclear fuel*

A current generation II reactor uses typically 500 kg of uranium in a fuel rod bundle before irradiation, which at the end of its life produces approximately 475–480 kg uranium, 5 kg plutonium and 15–20 kg of corrosion and fission products (Fe, Zn, Mg, Pd, Pt, lanthanides and minor actinides). If nuclear power is to continue in the long term, as a low carbon source of energy production, the reprocessing of spent nuclear fuel (SNF) is essential to prevent the exhaustion of uranium supplies and reduce the volume and radiotoxicity of the waste produced.[15] Based on the known global uranium reserves and the world usage in 2018 of about 65,000 tonnes per year (in terms of U_3O_8), estimates suggest that this will lead to exhaustion of supplies in 60–80 years' time if no recovery is implemented or new resources found. The use of Fast Neutron Reactors will enable more of the uranium and plutonium to undergo fission and can be configured to breed new fuel as energy is produced. A fast reactor fleet could utilise closer to 100% of the original fuel with recycling, compared to the current 3–4% used in modern reactors.

There are at present two major commercial reprocessing plants, Sellafield (UK, capacity ~1,000 tons U/year) and La Hague (France, ~1,700 t/y). Smaller plants are built or are in operation in Japan (Rokkasho-Mura, ~800 t/y); Russia (Chelyabinsk RT1, 400 t/y, India (Trombay, 60 t/y; Tarapor, 100 t/y, Kalpakkham, 100 t/y) while several military plants have been closed.

Existing reprocessing techniques involve the removal of reusable uranium and plutonium present in the waste which can be recycled and reused together in mixed oxide (MOX) reactors to produce more nuclear power. The amount of waste remaining in storage at present worldwide that is suitable for reprocessing is approximately 200,000 tonnes, with a global reprocessing capacity of around 4,000 tonnes per year. Over the last 50 years approximately 90,000 tonnes of spent fuel have been reprocessed.

In addition to the reusable U and Pu in spent fuel, there are also a variety of other fission products present, such as minor actinides: Np, Am and Cm, Ln, and transition metals, neutron absorber additives such gadolinium, in addition to corrosion products as a consequence of high radiation fields, erosion and equipment ageing of steel containers and pipes. The composition of spent fuel in a light water reactor is typically 95.6% U, 0.9% Pu, 0.4% short-lived fission products (e.g. ^{137}Cs, ^{90}Sr), 0.1 % minor actinides (Am, Cm), 0.1% long-lived fission products (e.g. ^{129}I, ^{99}Tc), and 2.9% stable fission products. Recently, research into the removal of the minor actinides has become important in order to transmutate them into shorter-lived radionuclides so that their radioactivity will not persist for as long, keeping the disposal facility design lifetime down to hundreds rather than thousands of years.

Although minor actinides only make up 0.1% of fission products, they are the main contributor to radiotoxicity after a few hundred years and so it is beneficial to separate them from the remaining fission products so that they can be transmutated into shorter lived radionuclides by neutron bombardment. This is known collectively as "partitioning and transmutation". The necessity of the separation arises from the presence of the lanthanides which are known neutron scavengers, meaning that they possess high neutron capture cross-sections, preventing transmutation of other species present. Neptunium can be removed from the mixture of fission products by careful control of its oxidation states.[17] However, the predominantly trivalent minor actinides Am and Cm are much more difficult to separate from the lanthanides owing to the similarities in the chemistries of the elements. Efficient separation strategies for the selective removal of Am and Cm from Ln are now well established and are being developed ready for industrialisation.[19]

18.10 Nuclear Weapons

Plutonium-239 is a key fissile component in nuclear weapons, due to its ease of fission and availability. Special bomb devices decrease the amount of plutonium needed to reach critical mass by reflecting escaping neutrons back into the plutonium core. This reduces the amount of plutonium needed to reach criticality to about 5–6 kg, three times less the critical mass needed for uranium-235. Further sophistication in assembly design may decrease this mass to as little as 4 kg.

The "Fat Man" plutonium bomb which exploded above Nagasaki in 1945 used explosive compression of plutonium to increase the fuel density, combined with a neutron source (^{9}Be-^{210}Po) to start the reaction and increase efficiency. The 6.2 kg of plutonium it contained generated an explosive power equivalent to 20 kilotons of TNT.

Early 2016, it was estimated that the 9 nuclear countries (Russia, USA, France, China, UK, Pakistan, India, Israel and North Korea) possess a total of 15,350 nuclear warheads, with USA and Russia accounting for 93% of them. About one third are "retired", 40% stockpiled while 28% are deployed.

References

1. G. R. Choppin and J. Rydberg, *Nuclear Chemistry: Theory and Applications*, Pergamon Press, Oxford, 1980.

2. J. C. Spirlet, J. R. Peterson and L. B. Asprey. *Adv. Inorg. Chem.*, 1987, *31*, 1–41.
3. L. S. Natrajan, *Coord. Chem. Rev.*, 2012, *256*, 1583–1603.
4. K. F. Smith, N. D. Bryan, A. N. Swinburne, P. Bots, S. Shaw, L. S. Natrajan, J. F. W. Mosselmans, F. R. Livens and K. Morris, *Geochim Cosmochim Ac.*, 2014, *148*, 343–359.
5. D. L. Jones, M. B. Andrews, A. N. Swinburne, S. W. Botchway, A. D. Ward, J. R. Llyod and L. S. Natrajan, *Chemical Science*, 2015, *6*, 5133–5138.
6. D. J. Peppe and A. L. Deino, *Nat. Educ. Knowl.*, 2013, *4*, 1–11.
7. B. M. Gardner and S. T. Liddle, *Eur. J. Inorg. Chem.*, 2013, 3753–3770.
8. A. R. Fox, S. C. Bart, K. Meyer and C. C. Cummins, *Nature*, 2008, *455*, 341–349.
9. R. G. Lange and W. P. Carroll, *Energy Convers. Manage.*, 2008, *49*, 393.
10. A. Witze, *Nature*, 2014, *515*, 484–486.
11. IAEA-TECDOC-1459, Technical Data on Nucleonic Gauges, IAEA, Vienna, 2005.
12. Y. Dekempeneer, M. Keyaerts, A. Krasniqi, J. Puttemans, S. Muyldermans, T. Lahoutte, M. D'huyvetter and N. Devoogdt, *Expert Opini. Biol. Ther.*, 2016, *16*, 1035–1047.
13. Medical Actinium for Therapeutic Treatment (MATT), Idaho National Laboratory, http://www4vip.inl.gov/factsheets/docs/matt.pdf.
14. A. Poullikkas, *Int. J. Energy Enviro.*, 2013, *4*, 743–776.
15. L. S. Natrajan and M. H. Langford Paden, F-block Elements Recovery, in *Element Recovery and Sustainability, RSC Green Chemistry Series* (Ed. A. J. Hunt), Ch. 6, Royal Society of Chemistry, 2013, 140–184.
16. R. M. Ambrosi, H. Williams, B. Foxcroft *et al.*, *Space Sci. Rev.*, 2019, *215*, 55.
17. R. J. Taylor, C. R. Gregson, M. J. Carrott, C. Mason and M. J. Sarsfield, *Solv. Extract. Ion Exch.*, 2013, *31*, 442–462.
18. https://world-nuclear.org/information-library/facts-and-figures/world-nuclear-power-reactors-and-uranium-requireme.aspx.
19. K. L. Nash and G. J. Lumetta, *Advanced Separation Techniques for Nuclear Fuel Reprocessing and Radioactive Wastes*, Woodhead Publishing Series in Energy: Number 2, 2011.
20. J. J. Harrison and Z. Vesoulis, CC BY-SA 2.5, https://commons.wikimedia.org/w/index.php?curid=5720587; https://commons.wikimedia.org/w/index.php?curid=227400.